Weston M. Stacey
Nuclear Reactor Physics

1807–2007 Knowledge for Generations

Each generation has its unique needs and aspirations. When Charles Wiley first opened his small printing shop in lower Manhattan in 1807, it was a generation of boundless potential searching for an identity. And we were there, helping to define a new American literary tradition. Over half a century later, in the midst of the Second Industrial Revolution, it was a generation focused on building the future. Once again, we were there, supplying the critical scientific, technical, and engineering knowledge that helped frame the world. Throughout the 20th Century, and into the new millennium, nations began to reach out beyond their own borders and a new international community was born. Wiley was there, expanding its operations around the world to enable a global exchange of ideas, opinions, and know-how.

For 200 years, Wiley has been an integral part of each generation's journey, enabling the flow of information and understanding necessary to meet their needs and fulfill their aspirations. Today, bold new technologies are changing the way we live and learn. Wiley will be there, providing you the must-have knowledge you need to imagine new worlds, new possibilities, and new opportunities.

Generations come and go, but you can always count on Wiley to provide you the knowledge you need, when and where you need it!

William J. Pesce
President and Chief Executive Officer

Peter Booth Wiley
Chairman of the Board

Weston M. Stacey

Nuclear Reactor Physics

Second Edition, Completely Revised and Enlarged

WILEY-VCH Verlag GmbH & Co. KGaA

The Author

Prof. Weston M. Stacey
Georgia Institute of Technology
Nuclear & Radiological Engineering
900 Atlantic Drive, NW
Atlanta, GA 30332-0425
USA

Cover
Four-assembly fuel module for a boiling water
reactor (Courtesy of General Electric
Company).

All books published by Wiley-VCH are
carefully produced. Nevertheless, authors,
editors, and publisher do not warrant the
information contained in these books,
including this book, to be free of errors.
Readers are advised to keep in mind that
statements, data, illustrations, procedural
details or other items may inadvertently be
inaccurate.

Library of Congress Card No.:
applied for

**British Library Cataloguing-in-Publication
Data**
A catalogue record for this book is available
from the British Library.

**Bibliographic information published by
the Deutsche Nationalbibliothek**
The Deutsche Nationalbibliothek lists this
publication in the Deutsche
Nationalbibliografie; detailed bibliographic
data is available in the Internet at
<http://dnb.ddb.de>.

© 2007 WILEY-VCH Verlag GmbH & Co.
KGaA, Weinheim

Typesetting VTEX, Vilnius, Lithuania
Printing betz-druck GmbH, Darmstadt
Binding Litges & Dopf Buchbinderei
GmbH, Heppenheim
Wiley Bicentennial Logo Richard J. Pacifico

Printed in the Federal Republic of Germany
Printed on acid-free paper

ISBN 978-3-527-40679-1

To Penny, Helen, Billy, and Lucia

Contents

Preface   *xxiii*
Preface to 2nd Edition   *xxvii*

PART 1   BASIC REACTOR PHYSICS

1   Neutron Nuclear Reactions   3
1.1   Neutron-Induced Nuclear Fission   *3*
         Stable Nuclides   *3*
         Binding Energy   *3*
         Threshold External Energy for Fission   *4*
         Neutron-Induced Fission   *5*
         Neutron Fission Cross Sections   *5*
         Products of the Fission Reaction   *8*
         Energy Release   *10*
1.2   Neutron Capture   *13*
         Radiative Capture   *13*
         Neutron Emission   *19*
1.3   Neutron Elastic Scattering   *20*
1.4   Summary of Cross-Section Data   *24*
         Low-Energy Cross Sections   *24*
         Spectrum-Averaged Cross Sections   *24*
1.5   Evaluated Nuclear Data Files   *24*
1.6   Elastic Scattering Kinematics   *27*
         Correlation of Scattering Angle and Energy Loss   *28*
         Average Energy Loss   *29*

2   Neutron Chain Fission Reactors   33
2.1   Neutron Chain Fission Reactions   *33*
         Capture-to-Fission Ratio   *33*
         Number of Fission Neutrons per Neutron Absorbed in Fuel   *33*

Nuclear Reactor Physics. Weston M. Stacey
Copyright © 2007 WILEY-VCH Verlag GmbH & Co. KGaA, Weinheim
ISBN: 978-3-527-40679-1

Neutron Utilization   *34*
Fast Fission   *34*
Resonance Escape   *36*
2.2   Criticality   *37*
Effective Multiplication Constant   *37*
Effect of Fuel Lumping   *37*
Leakage Reduction   *38*
2.3   Time Dependence of a Neutron Fission Chain Assembly   *38*
Prompt Fission Neutron Time Dependence   *38*
Source Multiplication   *39*
Effect of Delayed Neutrons   *39*
2.4   Classification of Nuclear Reactors   *40*
Physics Classification by Neutron Spectrum   *40*
Engineering Classification by Coolant   *41*

3   Neutron Diffusion Theory   43
3.1   Derivation of One-Speed Diffusion Theory   *43*
Partial and Net Currents   *43*
Diffusion Theory   *45*
Interface Conditions   *46*
Boundary Conditions   *46*
Applicability of Diffusion Theory   *47*
3.2   Solutions of the Neutron Diffusion Equation in Nonmultiplying Media   *48*
Plane Isotropic Source in an Infinite Homogeneous Medium   *48*
Plane Isotropic Source in a Finite Homogeneous Medium   *48*
Line Source in an Infinite Homogeneous Medium   *49*
Homogeneous Cylinder of Infinite Axial Extent with Axial Line Source   *49*
Point Source in an Infinite Homogeneous Medium   *49*
Point Source at the Center of a Finite Homogeneous Sphere   *50*
3.3   Diffusion Kernels and Distributed Sources in a Homogeneous Medium   *50*
Infinite-Medium Diffusion Kernels   *50*
Finite-Slab Diffusion Kernel   *51*
Finite Slab with Incident Neutron Beam   *52*
3.4   Albedo Boundary Condition   *52*
3.5   Neutron Diffusion and Migration Lengths   *53*
Thermal Diffusion-Length Experiment   *53*
Migration Length   *55*
3.6   Bare Homogeneous Reactor   *57*
Slab Reactor   *57*
Right Circular Cylinder Reactor   *59*

Interpretation of Criticality Condition   *60*
Optimum Geometries   *61*
3.7   Reflected Reactor   *62*
Reflected Slab Reactor   *62*
Reflector Savings   *64*
Reflected Spherical, Cylindrical, and Rectangular Parallelepiped
Cores   *65*
3.8   Homogenization of a Heterogeneous Fuel–Moderator Assembly   *65*
Spatial Self-Shielding and Thermal Disadvantage Factor   *65*
Effective Homogeneous Cross Sections   *69*
Thermal Utilization   *71*
Measurement of Thermal Utilization   *72*
Local Power Peaking Factor   *73*
3.9   Control Rods   *73*
Effective Diffusion Theory Cross Sections for Control Rods   *73*
Windowshade Treatment of Control Rods   *76*
3.10   Numerical Solution of Diffusion Equation   *77*
Finite Difference Equations in One Dimension   *78*
Forward Elimination/Backward Substitution Spatial Solution
Procedure   *79*
Power Iteration on Fission Source   *79*
Finite-Difference Equations in Two Dimensions   *80*
Successive Relaxation Solution of Two-Dimensional
Finite-Difference Equations   *82*
Power Outer Iteration on Fission Source   *82*
Limitations on Mesh Spacing   *83*
3.11   Nodal Approximation   *83*
3.12   Transport Methods   *85*
Transmission and Absorption in a Purely Absorbing Slab Control
Plate   *87*
Escape Probability in a Slab   *87*
Integral Transport Formulation   *87*
Collision Probability Method   *88*
Differential Transport Formulation   *89*
Spherical Harmonics Methods   *90*
Discrete Ordinates Method   *94*

4   Neutron Energy Distribution   101
4.1   Analytical Solutions in an Infinite Medium   *101*
Fission Source Energy Range   *102*
Slowing-Down Energy Range   *102*
Moderation by Hydrogen Only   *103*
Energy Self-Shielding   *103*
Slowing Down by Nonhydrogenic Moderators with No Absorption   *104*

Slowing-Down Density   *105*
Slowing Down with Weak Absorption   *106*
Fermi Age Neutron Slowing Down   *107*
Neutron Energy Distribution in the Thermal Range   *108*
Summary   *111*
4.2   Multigroup Calculation of Neutron Energy Distribution in an Infinite Medium   *111*
Derivation of Multigroup Equations   *111*
Mathematical Properties of the Multigroup Equations   *113*
Solution of Multigroup Equations   *114*
Preparation of Multigroup Cross-Section Sets   *115*
4.3   Resonance Absorption   *117*
Resonance Cross Sections   *117*
Doppler Broadening   *119*
Resonance Integral   *122*
Resonance Escape Probability   *122*
Multigroup Resonance Cross Section   *122*
Practical Width   *122*
Neutron Flux in Resonance   *123*
Narrow Resonance Approximation   *123*
Wide Resonance Approximation   *124*
Resonance Absorption Calculations   *124*
Temperature Dependence of Resonance Absorption   *127*
4.4   Multigroup Diffusion Theory   *127*
Multigroup Diffusion Equations   *127*
Two-Group Theory   *128*
Two-Group Bare Reactor   *129*
One-and-One-Half-Group Theory   *129*
Two-Group Theory of Two-Region Reactors   *130*
Two-Group Theory of Reflected Reactors   *133*
Numerical Solutions for Multigroup Diffusion Theory   *137*

5   Nuclear Reactor Dynamics   143
5.1   Delayed Fission Neutrons   *143*
Neutrons Emitted in Fission Product Decay   *143*
Effective Delayed Neutron Parameters for Composite Mixtures   *145*
Photoneutrons   *146*
5.2   Point Kinetics Equations   *147*
5.3   Period–Reactivity Relations   *148*
5.4   Approximate Solutions of the Point Neutron Kinetics Equations   *150*
One Delayed Neutron Group Approximation   *150*
Prompt-Jump Approximation   *153*
Reactor Shutdown   *154*

5.5   Delayed Neutron Kernel and Zero-Power Transfer Function   *155*
        Delayed Neutron Kernel   *155*
        Zero-Power Transfer Function   *155*
5.6   Experimental Determination of Neutron Kinetics Parameters   *156*
        Asymptotic Period Measurement   *156*
        Rod Drop Method   *157*
        Source Jerk Method   *157*
        Pulsed Neutron Methods   *157*
        Rod Oscillator Measurements   *158*
        Zero-Power Transfer Function Measurements   *159*
        Rossi-α Measurement   *159*
5.7   Reactivity Feedback   *161*
        Temperature Coefficients of Reactivity   *162*
        Doppler Effect   *162*
        Fuel and Moderator Expansion Effect on Resonance Escape
        Probability   *164*
        Thermal Utilization   *165*
        Nonleakage Probability   *166*
        Representative Thermal Reactor Reactivity Coefficients   *166*
        Startup Temperature Defect   *167*
5.8   Perturbation Theory Evaluation of Reactivity Temperature
        Coefficients   *168*
        Perturbation Theory   *168*
        Sodium Void Effect in Fast Reactors   *169*
        Doppler Effect in Fast Reactors   *169*
        Fuel and Structure Motion in Fast Reactors   *170*
        Fuel Bowing   *171*
        Representative Fast Reactor Reactivity Coefficients   *171*
5.9   Reactor Stability   *171*
        Reactor Transfer Function with Reactivity Feedback   *171*
        Stability Analysis for a Simple Feedback Model   *172*
        Threshold Power Level for Reactor Stability   *174*
        More General Stability Conditions   *175*
        Power Coefficients and Feedback Delay Time Constants   *178*
5.10   Measurement of Reactor Transfer Functions   *179*
        Rod Oscillator Method   *179*
        Correlation Methods   *179*
        Reactor Noise Method   *181*
5.11   Reactor Transients with Feedback   *183*
        Step Reactivity Insertion ($\rho_{ex} < \beta$): Prompt Jump   *184*
        Step Reactivity Insertion ($\rho_{ex} < \beta$): Post-Prompt-Jump Transient   *185*
5.12   Reactor Fast Excursions   *186*
        Step Reactivity Input: Feedback Proportional to Fission Energy   *186*
        Ramp Reactivity Input: Feedback Proportional to Fission Energy   *187*

Step Reactivity Input: Nonlinear Feedback Proportional to
Cumulative Energy Release   *187*
Bethe–Tait Model   *188*
5.13   Numerical Methods   *190*

6   Fuel Burnup   197
6.1   Changes in Fuel Composition   *197*
Fuel Transmutation–Decay Chains   *198*
Fuel Depletion–Transmutation–Decay Equations   *199*
Fission Products   *203*
Solution of the Depletion Equations   *204*
Measure of Fuel Burnup   *205*
Fuel Composition Changes with Burnup   *205*
Reactivity Effects of Fuel Composition Changes   *206*
Compensating for Fuel-Depletion Reactivity Effects   *208*
Reactivity Penalty   *208*
Effects of Fuel Depletion on the Power Distribution   *209*
In-Core Fuel Management   *210*
6.2   Samarium and Xenon   *211*
Samarium Poisoning   *211*
Xenon Poisoning   *213*
Peak Xenon   *215*
Effect of Power-Level Changes   *216*
6.3   Fertile-to-Fissile Conversion and Breeding   *217*
Availability of Neutrons   *217*
Conversion and Breeding Ratios   *219*
6.4   Simple Model of Fuel Depletion   *219*
6.5   Fuel Reprocessing and Recycling   *221*
Composition of Recycled LWR Fuel   *221*
Physics Differences of MOX Cores   *222*
Physics Considerations with Uranium Recycle   *224*
Physics Considerations with Plutonium Recycle   *225*
Reactor Fueling Characteristics   *225*
6.6   Radioactive Waste   *226*
Radioactivity   *226*
Hazard Potential   *226*
Risk Factor   *226*
6.7   Burning Surplus Weapons-Grade Uranium and Plutonium   *233*
Composition of Weapons-Grade Uranium and Plutonium   *233*
Physics Differences Between Weapons- and Reactor-Grade
Plutonium-Fueled Reactors   *234*
6.8   Utilization of Uranium Energy Content   *235*
6.9   Transmutation of Spent Nuclear Fuel   *237*
6.10   Closing the Nuclear Fuel Cycle   *244*

7   Nuclear Power Reactors   249
7.1   Pressurized Water Reactors   249
7.2   Boiling Water Reactors   250
7.3   Pressure Tube Heavy Water–Moderated Reactors   255
7.4   Pressure Tube Graphite-Moderated Reactors   258
7.5   Graphite-Moderated Gas-Cooled Reactors   260
7.6   Liquid-Metal Fast Breeder Reactors   261
7.7   Other Power Reactors   265
7.8   Characteristics of Power Reactors   265
7.9   Advanced Generation-III Reactors   265
       Advanced Boiling Water Reactors (ABWR)   266
       Advanced Pressurized Water Reactors (APWR)   267
       Advanced Pressure Tube Reactor   268
       Modular High-Temperature Gas-Cooled Reactors (GT-MHR)   268
7.10   Advanced Generation-IV Reactors   269
       Gas-Cooled Fast Reactors (GFR)   270
       Lead-Cooled Fast Reactors (LFR)   271
       Molten Salt Reactors (MSR)   271
       Super-Critical Water Reactors (SCWR)   272
       Sodium-Cooled Fast Reactors (SFR)   272
       Very High Temperature Reactors (VHTR)   272
7.11   Advanced Sub-critical Reactors   273
7.12   Nuclear Reactor Analysis   275
       Construction of Homogenized Multigroup Cross Sections   275
       Criticality and Flux Distribution Calculations   276
       Fuel Cycle Analyses   277
       Transient Analyses   278
       Core Operating Data   279
       Criticality Safety Analysis   279
7.13   Interaction of Reactor Physics and Reactor Thermal Hydraulics   280
       Power Distribution   280
       Temperature Reactivity Effects   281
       Coupled Reactor Physics and Thermal-Hydraulics Calculations   281

8   Reactor Safety   283
8.1   Elements of Reactor Safety   283
       Radionuclides of Greatest Concern   283
       Multiple Barriers to Radionuclide Release   283
       Defense in Depth   285
       Energy Sources   285
8.2   Reactor Safety Analysis   285
       Loss of Flow or Loss of Coolant   287
       Loss of Heat Sink   287

Reactivity Insertion   *287*
Anticipated Transients without Scram   *288*
8.3   Quantitative Risk Assessment   *288*
Probabilistic Risk Assessment   *288*
Radiological Assessment   *291*
Reactor Risks   *291*
8.4   Reactor Accidents   *293*
Three Mile Island   *294*
Chernobyl   *297*
8.5   Passive Safety   *299*
Pressurized Water Reactors   *299*
Boiling Water Reactors   *299*
Integral Fast Reactors   *300*
Passive Safety Demonstration   *300*

PART 2   ADVANCED REACTOR PHYSICS

9   Neutron Transport Theory   305
9.1   Neutron Transport Equation   *305*
Boundary Conditions   *310*
Scalar Flux and Current   *310*
Partial Currents   *310*
9.2   Integral Transport Theory   *310*
Isotropic Point Source   *311*
Isotropic Plane Source   *311*
Anisotropic Plane Source   *312*
Transmission and Absorption Probabilities   *314*
Escape Probability   *314*
First-Collision Source for Diffusion Theory   *315*
Inclusion of Isotropic Scattering and Fission   *315*
Distributed Volumetric Sources in Arbitrary Geometry   *316*
Flux from a Line Isotropic Source of Neutrons   *317*
Bickley Functions   *318*
Probability of Reaching a Distance *t* from a Line Isotropic Source
without a Collision   *318*
9.3   Collision Probability Methods   *319*
Reciprocity Among Transmission and Collision Probabilities   *320*
Collision Probabilities for Slab Geometry   *320*
Collision Probabilities in Two-Dimensional Geometry   *321*
Collision Probabilities for Annular Geometry   *322*
9.4   Interface Current Methods in Slab Geometry   *323*
Emergent Currents and Reaction Rates Due to Incident Currents   *323*
Emergent Currents and Reaction Rates Due to Internal Sources   *326*

	Total Reaction Rates and Emergent Currents	*327*
	Boundary Conditions *329*
	Response Matrix *329*
9.5	Multidimensional Interface Current Methods *330*
	Extension to Multidimension *330*
	Evaluation of Transmission and Escape Probabilities *332*
	Transmission Probabilities in Two-Dimensional Geometries *333*
	Escape Probabilities in Two-Dimensional Geometries *335*
	Simple Approximations for the Escape Probability *337*
9.6	Spherical Harmonics (P_L) Methods in One-Dimensional Geometries *338*
	Legendre Polynomials *338*
	Neutron Transport Equation in Slab Geometry *339*
	P_L Equations *339*
	Boundary and Interface Conditions *340*
	P_1 Equations and Diffusion Theory *342*
	Simplified P_L or Extended Diffusion Theory *343*
	P_L Equations in Spherical and Cylindrical Geometries *344*
	Diffusion Equations in One-Dimensional Geometry *347*
	Half-Angle Legendre Polynomials *347*
	Double-P_L Theory *348*
	D-P_0 Equations *349*
9.7	Multidimensional Spherical Harmonics (P_L) Transport Theory *350*
	Spherical Harmonics *350*
	Spherical Harmonics Transport Equations in Cartesian Coordinates *351*
	P_1 Equations in Cartesian Geometry *352*
	Diffusion Theory *353*
9.8	Discrete Ordinates Methods in One-Dimensional Slab Geometry *354*
	P_L and D-P_L Ordinates *355*
	Spatial Differencing and Iterative Solution *357*
	Limitations on Spatial Mesh Size *358*
9.9	Discrete Ordinates Methods in One-Dimensional Spherical Geometry *359*
	Representation of Angular Derivative *360*
	Iterative Solution Procedure *360*
	Acceleration of Convergence *362*
	Calculation of Criticality *362*
9.10	Multidimensional Discrete Ordinates Methods *363*
	Ordinates and Quadrature Sets *363*
	S_N Method in Two-Dimensional x–y Geometry *366*
	Further Discussion *369*
9.11	Even-Parity Transport Formulation *369*
9.12	Monte Carlo Methods *371*
	Probability Distribution Functions *371*

Analog Simulation of Neutron Transport   *372*
Statistical Estimation   *373*
Variance Reduction   *375*
Tallying   *377*
Criticality Problems   *378*
Source Problems   *379*
Random Numbers   *380*

10   Neutron Slowing Down   385
10.1   Elastic Scattering Transfer Function   *385*
Lethargy   *385*
Elastic Scattering Kinematics   *385*
Elastic Scattering Kernel   *386*
Isotropic Scattering in Center-of-Mass System   *388*
Linearly Anisotropic Scattering in Center-of-Mass System   *389*
10.2 P_1 and B_1 Slowing-Down Equations   *390*
Derivation   *390*
Solution in Finite Uniform Medium   *393*
B_1 Equations   *394*
Few-Group Constants   *395*
10.3   Diffusion Theory   *396*
Lethargy-Dependent Diffusion Theory   *396*
Directional Diffusion Theory   *397*
Multigroup Diffusion Theory   *398*
Boundary and Interface Conditions   *399*
10.4   Continuous Slowing-Down Theory   *400*
P_1 Equations in Slowing-Down Density Formulation   *400*
Slowing-Down Density in Hydrogen   *403*
Heavy Mass Scatterers   *404*
Age Approximation   *404*
Selengut–Goertzel Approximation   *405*
Consistent P_1 Approximation   *405*
Extended Age Approximation   *405*
Grueling–Goertzel Approximation   *406*
Summary of P_l Continuous Slowing-Down Theory   *407*
Inclusion of Anisotropic Scattering   *407*
Inclusion of Scattering Resonances   *409*
P_l Continuous Slowing-Down Equations   *410*
10.5   Multigroup Discrete Ordinates Transport Theory   *411*

11   Resonance Absorption   415
11.1   Resonance Cross Sections   *415*

11.2   Widely Spaced Single-Level Resonances in a Heterogeneous
       Fuel–Moderator Lattice   *415*
       Neutron Balance in Heterogeneous Fuel–Moderator Cell   *415*
       Reciprocity Relation   *418*
       Narrow Resonance Approximation   *419*
       Wide Resonance Approximation   *420*
       Evaluation of Resonance Integrals   *420*
       Infinite Dilution Resonance Integral   *422*
       Equivalence Relations   *422*
       Heterogeneous Resonance Escape Probability   *423*
       Homogenized Multigroup Resonance Cross Section   *423*
       Improved and Intermediate Resonance Approximations   *424*
11.3   Calculation of First-Flight Escape Probabilities   *424*
       Escape Probability for an Isolated Fuel Rod   *425*
       Closely Packed Lattices   *427*
11.4   Unresolved Resonances   *428*
       Multigroup Cross Sections for Isolated Resonances   *430*
       Self-Overlap Effects   *431*
       Overlap Effects for Different Sequences   *432*
11.5   Multiband Treatment of Spatially Dependent Self-Shielding   *433*
       Spatially Dependent Self-Shielding   *433*
       Multiband Theory   *434*
       Evaluation of Multiband Parameters   *436*
       Calculation of Multiband Parameters   *437*
       Interface Conditions   *439*
11.6   Resonance Cross-Section Representations   *439*
       R-Matrix Representation   *439*
       Practical Formulations   *441*
       Generalization of the Pole Representation   *445*
       Doppler Broadening of the Generalized Pole Representation   *448*

12   **Neutron Thermalization**   **453**
12.1   Double Differential Scattering Cross Section for Thermal
       Neutrons   *453*
12.2   Neutron Scattering from a Monatomic Maxwellian Gas   *454*
       Differential Scattering Cross Section   *454*
       Cold Target Limit   *455*
       Free-Hydrogen (Proton) Gas Model   *455*
       Radkowsky Model for Scattering from H_2O *455*
       Heavy Gas Model   *456*
12.3   Thermal Neutron Scattering from Bound Nuclei   *457*
       Pair Distribution Functions and Scattering Functions   *457*
       Intermediate Scattering Functions   *458*
       Incoherent Approximation   *459*

Gaussian Representation of Scattering   *459*

Measurement of the Scattering Function   *460*

Applications to Neutron Moderating Media   *460*

12.4   Calculation of the Thermal Neutron Spectra in Homogeneous Media   *462*

Wigner–Wilkins Proton Gas Model   *463*

Heavy Gas Model   *466*

Numerical Solution   *468*

Moments Expansion Solution   *470*

Multigroup Calculation   *473*

Applications to Moderators   *474*

12.5   Calculation of Thermal Neutron Energy Spectra in Heterogeneous Lattices   *474*

12.6   Pulsed Neutron Thermalization   *477*

Spatial Eigenfunction Expansion   *477*

Energy Eigenfunctions of the Scattering Operator   *477*

Expansion in Energy Eigenfunctions of the Scattering Operator   *479*

13   Perturbation and Variational Methods   483

13.1   Perturbation Theory Reactivity Estimate   *483*

Multigroup Diffusion Perturbation Theory   *483*

13.2   Adjoint Operators and Importance Function   *486*

Adjoint Operators   *486*

Importance Interpretation of the Adjoint Function   *487*

Eigenvalues of the Adjoint Equation   *489*

13.3   Variational/Generalized Perturbation Reactivity Estimate   *489*

One-Speed Diffusion Theory   *490*

Other Transport Models   *493*

Reactivity Worth of Localized Perturbations in a Large PWR Core Model   *494*

Higher-Order Variational Estimates   *495*

13.4   Variational/Generalized Perturbation Theory Estimates of Reaction Rate Ratios in Critical Reactors   *495*

13.5   Variational/Generalized Perturbation Theory Estimates of Reaction Rates   *497*

13.6   Variational Theory   *498*

Stationarity   *498*

Roussopolos Variational Functional   *498*

Schwinger Variational Functional   *499*

Rayleigh Quotient   *499*

Construction of Variational Functionals   *500*

13.7   Variational Estimate of Intermediate Resonance Integral   *500*

13.8   Heterogeneity Reactivity Effects   *502*

13.9   Variational Derivation of Approximate Equations   *503*

13.10   Variational Even-Parity Transport Approximations   *505*
         Variational Principle for the Even-Parity Transport Equation   *505*
         Ritz Procedure   *506*
         Diffusion Approximation   *507*
         One-Dimensional Slab Transport Equation   *508*
13.11   Boundary Perturbation Theory   *508*

14   Homogenization   515
14.1   Equivalent Homogenized Cross Sections   *516*
14.2   ABH Collision Probability Method   *517*
14.3   Blackness Theory   *520*
14.4   Fuel Assembly Transport Calculations   *522*
         Pin Cells   *522*
         Wigner–Seitz Approximation   *523*
         Collision Probability Pin-Cell Model   *524*
         Interface Current Formulation   *527*
         Multigroup Pin-Cell Collision Probabilities Model   *528*
         Resonance Cross Sections   *529*
         Full Assembly Transport Calculation   *529*
14.5   Homogenization Theory   *529*
         Homogenization Considerations   *530*
         Conventional Homogenization Theory   *531*
14.6   Equivalence Homogenization Theory   *531*
14.7   Multiscale Expansion Homogenization Theory   *535*
14.8   Flux Detail Reconstruction   *538*

15   Nodal and Synthesis Methods   541
15.1   General Nodal Formalism   *542*
15.2   Conventional Nodal Methods   *544*
15.3   Transverse Integrated Nodal Diffusion Theory Methods   *547*
         Transverse Integrated Equations   *547*
         Polynomial Expansion Methods   *549*
         Analytical Methods   *553*
         Heterogeneous Flux Reconstruction   *554*
15.4   Transverse Integrated Nodal Integral Transport Theory Models   *554*
         Transverse Integrated Integral Transport Equations   *554*
         Polynomial Expansion of Scalar Flux   *557*
         Isotropic Component of Transverse Leakage   *558*
         Double-P_n Expansion of Surface Fluxes   *558*
         Angular Moments of Outgoing Surface Fluxes   *559*
         Nodal Transport Equations   *561*
15.5   Transverse Integrated Nodal Discrete Ordinates Method   *561*
15.6   Finite-Element Coarse Mesh Methods   *563*

Variational Functional for the P_1 Equations   *563*

One-Dimensional Finite-Difference Approximation   *564*

Diffusion Theory Variational Functional   *566*

Linear Finite-Element Diffusion Approximation in One Dimension   *567*

Higher-Order Cubic Hermite Coarse-Mesh Diffusion Approximation   *569*

Multidimensional Finite-Element Coarse-Mesh Methods   *570*

15.7   Variational Discrete Ordinates Nodal Method   *571*

Variational Principle   *571*

Application of the Method   *579*

15.8   Variational Principle for Multigroup Diffusion Theory   *580*

15.9   Single-Channel Spatial Synthesis   *583*

15.10   Multichannel Spatial Synthesis   *589*

15.11   Spectral Synthesis   *591*

16   Space–Time Neutron Kinetics   599

16.1   Flux Tilts and Delayed Neutron Holdback   *599*

Modal Eigenfunction Expansion   *600*

Flux Tilts   *601*

Delayed Neutron Holdback   *602*

16.2   Spatially Dependent Point Kinetics   *602*

Derivation of Point Kinetics Equations   *604*

Adiabatic and Quasistatic Methods   *605*

Variational Principle for Static Reactivity   *606*

Variational Principle for Dynamic Reactivity   *607*

16.3   Time Integration of the Spatial Neutron Flux Distribution   *609*

Explicit Integration: Forward-Difference Method   *610*

Implicit Integration: Backward-Difference Method   *611*

Implicit Integration: θ Method   *612*

Implicit Integration: Time-Integrated Method   *615*

Implicit Integration: GAKIN Method   *616*

Alternating Direction Implicit Method   *619*

Stiffness Confinement Method   *622*

Symmetric Successive Overrelaxation Method   *623*

Generalized Runge–Kutta Methods   *624*

16.4   Stability   *625*

Classical Linear Stability Analysis   *625*

Lyapunov's Method   *627*

Lyapunov's Method for Distributed Parameter Systems   *629*

Control   *631*

Variational Methods of Control Theory   *631*

Dynamic Programming   *633*

Pontryagin's Maximum Principle   *634*

Variational Methods for Spatially Dependent Control Problems   *636*
Dynamic Programming for Spatially Continuous Systems   *638*
Pontryagin's Maximum Principle for a Spatially Continuous
System   *639*
16.5 Xenon Spatial Oscillations   *641*
Linear Stability Analysis   *642*
μ-Mode Approximation   *644*
λ-Mode Approximation   *645*
Nonlinear Stability Criterion   *649*
Control of Xenon Spatial Power Oscillations   *650*
Variational Control Theory of Xenon Spatial Oscillations   *650*
16.6 Stochastic Kinetics   *652*
Forward Stochastic Model   *653*
Means, Variances, and Covariances   *656*
Correlation Functions   *658*
Physical Interpretation, Applications, and Initial and Boundary
Conditions   *659*
Numerical Studies   *660*
Startup Analysis   *663*

APPENDICES

A   Physical Constants and Nuclear Data   669

B   Some Useful Mathematical Formulas   675

C   Step Functions, Delta Functions, and Other Functions   677
C.1 Introduction   *677*
C.2 Properties of the Dirac δ-Function   *678*
A. Alternative Representations   *678*
B. Properties   *678*
C. Derivatives   *679*

D   Some Properties of Special Functions   681

E   Introduction to Matrices and Matrix Algebra   687
E.1 Some Definitions   *687*
E.2 Matrix Algebra   *689*

F   **Introduction to Laplace Transforms**   **691**
F.1   Motivation   *691*
F.2   "Cookbook" Laplace transforms   *694*

Index   *697*

Preface

Nuclear reactor physics is the physics of neutron fission chain reacting systems. It encompasses those applications of nuclear physics and radiation transport and interaction with matter that determine the behavior of nuclear reactors. As such, it is both an applied physics discipline and the core discipline of the field of nuclear engineering.

As a distinct applied physics discipline, nuclear reactor physics originated in the middle of the twentieth century in the wartime convergence of international physics efforts in the Manhattan Project. It developed vigorously for roughly the next third of the century in various government, industrial, and university R&D and design efforts worldwide. Nuclear reactor physics is now a relatively mature discipline, in that the basic physical principles governing the behavior of nuclear reactors are well understood, most of the basic nuclear data needed for nuclear reactor analysis have been measured and evaluated, and the computational methodology is highly developed and validated. It is now possible to accurately predict the physics behavior of existing nuclear reactor types under normal operating conditions. Moreover, the basic physical concepts, nuclear data, and computational methodology needed to develop an understanding of new variants of existing reactor types or of new reactor types exist for the most part.

As the core discipline of nuclear engineering, nuclear reactor physics is fundamental to the major international nuclear power undertaking. As of 2000, there are 434 central station nuclear power reactors operating worldwide to produce 350,442 MWe of electrical power. This is a substantial fraction of the world's electrical power (e.g., more than 80% of the electricity produced in France and more than 20% of the electricity produced in the United States). The world's electrical power requirements will continue to increase, particularly as the less developed countries strive to modernize, and nuclear power is the only proven technology for meeting these growing electricity requirements without dramatically increasing the already unacceptable levels of greenhouse gas emission into the atmosphere.

Nuclear reactors have additional uses other than central station electricity production. There are more than 100 naval propulsion reactors in the U.S. fleet (plus others in foreign fleets). Nuclear reactors are also employed for basic neutron physics research, for materials testing, for radiation therapy, for the production of radio-isotopes for medical, industrial, and national security applications, and

Nuclear Reactor Physics. Weston M. Stacey
Copyright © 2007 WILEY-VCH Verlag GmbH & Co. KGaA, Weinheim
ISBN: 978-3-527-40679-1

as mobile power sources for remote stations. In the future, nuclear reactors may power deep space missions. Thus nuclear reactor physics is a discipline important to the present and future well-being of the world.

This book is intended as both a textbook and a comprehensive reference on nuclear reactor physics. The basic physical principles, nuclear data, and computational methodology needed to understand the physics of nuclear reactors are developed and applied to explain the static and dynamic behavior of nuclear reactors in Part 1. This development is at a level that should be accessible to seniors in physics or engineering (i.e., requiring a mathematical knowledge only through ordinary and partial differential equations and Laplace transforms and an undergraduate-level knowledge of atomic and nuclear physics). Mastery of the material presented in Part 1 provides an understanding of the physics of nuclear reactors sufficient for nuclear engineering graduates at the B.S. and M.S. levels, for most practicing nuclear engineers and for others interested in acquiring a broad working knowledge of nuclear reactor physics.

The material in Part 1 was developed in the process of teaching undergraduate and first-year graduate courses in nuclear reactor physics at Georgia Tech for a number of years. The emphasis in the presentation is on conveying the basic physical concepts and their application to explain nuclear reactor behavior, using the simplest mathematical description that will suffice to illustrate the physics. Numerous examples are included to illustrate the step-by-step procedures for carrying out the calculations discussed in the text. Problems at the end of each chapter have been chosen to provide physical insight and to extend the material discussed in the text, while providing practice in making calculations; they are intended as an integral part of the textbook. Part 1 is suitable for an undergraduate semester-length course in nuclear reactor physics; the material in Part 1 is also suitable for a semester-length first-year graduate course, perhaps with selective augmentation from Part 2.

The purpose of Part 2 is to augment Part 1 to provide a comprehensive, detailed, and advanced development of the principal topics of nuclear reactor physics. There is an emphasis in Part 2 on the theoretical bases for the advanced computational methods of reactor physics. This material provides a comprehensive, though necessarily abridged, reference work on advanced nuclear reactor physics and the theoretical bases for its computational methods. Although the material stops short of descriptions of specific reactor physics codes, it provides the basis for understanding the code manuals. There is more than enough material in Part 2 for a semester-length advanced graduate course in nuclear reactor physics. The treatment is necessarily somewhat more mathematically intense than in Part 1.

Part 2 is intended primarily for those who are or would become specialists in nuclear reactor physics and reactor physics computations. Mastery of this material provides the background for creating the new physics concepts necessary for developing new reactor types and for understanding and extending the computational methods in existing reactor physics codes (i.e., the stock-in-trade for the professional reactor physicist). Moreover, the extensive treatment of neutron transport computational methods also provides an important component of the background necessary for specialists in radiation shielding, for specialists in the applications of neutrons and photons in medicine and industry, and for specialists in neutron, photon, and neutral atom transport in industrial, astrophysical, and thermonuclear plasmas.

Any book of this scope owes much to many people besides the author, and this one is no exception. The elements of the subject of reactor physics were developed by many talented people over the past half-century, and the references can only begin to recognize their contributions. In this regard, I note the special contribution of R.N. Hwang, who helped prepare certain sections on resonance theory. The selection and organization of material has benefited from the example of previous authors of textbooks on reactor physics. The feedback from a generation of students has assisted in shaping the organization and presentation. Several people (C. Nickens, B. Crumbly, S. Bennett-Boyd) supported the evolution of the manuscript through at least three drafts, and several other people at Wiley transformed the manuscript into a book. I am grateful to all of these people, for without them there would be no book.

Atlanta, Georgia   WESTON M. STACEY
October 2000

Preface to 2nd Edition

This second edition differs from the original in two important ways. First, a section on neutron transport methods has been added in Chapter 3 to provide an introduction to that subject in the first section of the book on basic reactor physics that is intended as the text for an advanced undergraduate course. My original intention was to use diffusion theory to introduce reactor physics, without getting into the mathematical complexities of transport theory. I think this works reasonably well from a pedagogical point of view, but it has the disadvantage of sending BS graduates into the workplace without an exposure to transport theory. So, a short section on transport methods in slab geometry was added at the end of the diffusion theory chapter to provide an introduction.

Second, there has been a resurgence in interest and activity in the improvement of reactor designs and in the development of new reactor concepts that are more inherently safe, better utilize the uranium resources, discharge less long-lived waste and are more resistant to the diversion of fuels to other uses. A section has been added in Chapter 7 on the improved Generation-III designs that will be coming online over the next decade or so, and a few sections have been added in Chapters 6 and 7 on the new reactor concepts being developed under the Generation-IV and Advanced Fuel Cycle Initiatives with the objective of closing the nuclear fuel cycle.

The text was amplified for the sake of explication in a few places, some additional homework problems were included, and numerous typos, omissions and other errors that slipped through the final proof-reading of the first edition were corrected. I am grateful to colleagues, students and particularly the translators preparing a Russian edition of the book for calling several such mistakes to my attention.

Otherwise, the structure and context of the book remains unchanged. The first eight chapters on basic reactor physics provide the text for a first course in reactor physics at the advanced undergraduate or graduate level. The second eight chapters on advanced reactor physics provide a text suitable for graduate courses on neutron transport theory and reactor physics.

I hope that this second edition will serve to introduce to the field the new generation of scientists and engineers who will carry forward the emerging resurgence of

Nuclear Reactor Physics. Weston M. Stacey
Copyright © 2007 WILEY-VCH Verlag GmbH & Co. KGaA, Weinheim
ISBN: 978-3-527-40679-1

nuclear power to meet the growing energy needs of mankind in a safe, economical, environmentally sustainable and proliferation-resistant way.

Weston M. Stacey
Atlanta, Georgia
May 2006

Part 1: Basic Reactor Physics

1
Neutron Nuclear Reactions

The physics of nuclear reactors is determined by the transport of neutrons and their interaction with matter within a reactor. The basic neutron nucleus reactions of importance in nuclear reactors and the nuclear data used in reactor physics calculations are described in this chapter.

1.1
Neutron-Induced Nuclear Fission

Stable Nuclides

Short-range attractive nuclear forces acting among nucleons (neutrons and protons) are stronger than the Coulomb repulsive forces acting among protons at distances on the order of the nuclear radius ($R \approx 1.25 \times 10^{-13} A^{1/3}$ cm) in a stable nucleus. These forces are such that the ratio of the atomic mass A (the number of neutrons plus protons) to the atomic number Z (the number of protons) increases with Z; in other words, the stable nuclides become increasingly *neutron-rich* with increasing Z, as illustrated in Fig. 1.1. The various nuclear species are referred to as *nuclides*, and nuclides with the same atomic number are referred to as *isotopes* of the *element* corresponding to Z. We use the notation $^A X_Z$ (e.g., $^{235}U_{92}$) to identify nuclides.

Binding Energy

The actual mass of an atomic nucleus is not the sum of the masses (m_p) of the Z protons and the masses (m_n) of $A - Z$ neutrons of which it is composed. The stable nuclides have a mass defect

$$\Delta = [Zm_p + (A - Z)m_n] - {}^A m_z \tag{1.1}$$

This mass defect is conceptually thought of as having been converted to energy ($E = \Delta c^2$) at the time that the nucleus was formed, putting the nucleus into a negative energy state. The amount of externally supplied energy that would have

Nuclear Reactor Physics. Weston M. Stacey
Copyright © 2007 WILEY-VCH Verlag GmbH & Co. KGaA, Weinheim
ISBN: 978-3-527-40679-1

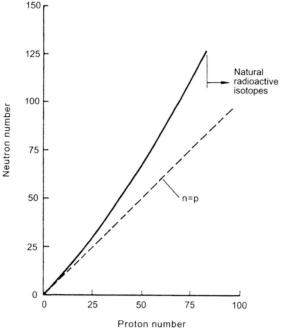

Fig. 1.1 Nuclear stability curve. (From Ref. 1; used with permission of McGraw-Hill.)

to be converted to mass in disassembling a nucleus into its separate nucleons is known as the *binding energy* of the nucleus, $BE = \Delta c^2$. The binding energy per nucleon (BE/A) is shown in Fig. 1.2.

Any process that results in nuclides being converted to other nuclides with more binding energy per nucleon will result in the conversion of mass into energy. The combination of low A nuclides to form higher A nuclides with a higher BE/A value is the basis for the *fusion* process for the release of nuclear energy. The splitting of very high A nuclides to form intermediate-A nuclides with a higher BE/A value is the basis of the *fission* process for the release of nuclear energy.

Threshold External Energy for Fission

The probability of any nuclide undergoing fission (reconfiguring its A nucleons into two nuclides of lower A) can become quite large if a sufficient amount of external energy is supplied to excite the nucleus. The minimum, or *threshold*, amount of such *excitation energy* required to cause fission with high probability depends on the nuclear structure and is quite large for nuclides with $Z < 90$. For nuclides with $Z > 90$, the threshold energy is about 4 to 6 MeV for even-A nuclides, and generally is much lower for odd-A nuclides. Certain of the heavier nuclides (e.g., $^{240}Pu_{94}$ and $^{252}Cf_{98}$) exhibit significant spontaneous fission even in the absence of any externally supplied excitation energy.

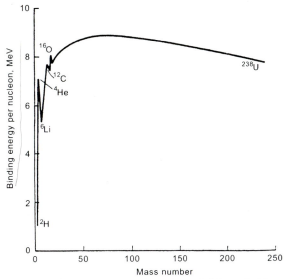

Fig. 1.2 Binding energy per nucleon. (From Ref. 1; used with permission of McGraw-Hill.)

Neutron-Induced Fission

When a neutron is absorbed into a heavy nucleus (A, Z) to form a *compound nucleus* $(A + 1, Z)$, the BE/A value is lower for the compound nucleus than for the original nucleus. For some nuclides (e.g., $^{233}U_{92}$, $^{235}U_{92}$, $^{239}Pu_{94}$, $^{241}Pu_{94}$), this reduction in BE/A value is sufficient that the compound nucleus will undergo fission, with high probability, even if the neutron has very low energy. Such nuclides are referred to as *fissile*; that is, they can be caused to undergo fission by the absorption of a low-energy neutron. If the neutron had kinetic energy prior to being absorbed into a nucleus, this energy is transformed into additional excitation energy of the compound nucleus. All nuclides with $Z > 90$ will undergo fission with high probability when a neutron with kinetic energy in excess of about 1 MeV is absorbed. Nuclides such as $^{232}Th_{90}$, $^{238}U_{92}$, and $^{240}Pu_{94}$ will undergo fission with neutrons with energy of about 1 MeV or higher, with high probability.

Neutron Fission Cross Sections

The probability of a nuclear reaction, in this case fission, taking place can be expressed in terms of a quantity σ which expresses the probable reaction rate for n neutrons traveling with speed v a distance dx in a material with N nuclides per unit volume:

$$\sigma \equiv \frac{\text{reaction rate}}{nvN\,dx} \tag{1.2}$$

The units of σ are area, which gives rise to the concept of σ as a cross-sectional area presented to the neutron by the nucleus, for a particular reaction process, and

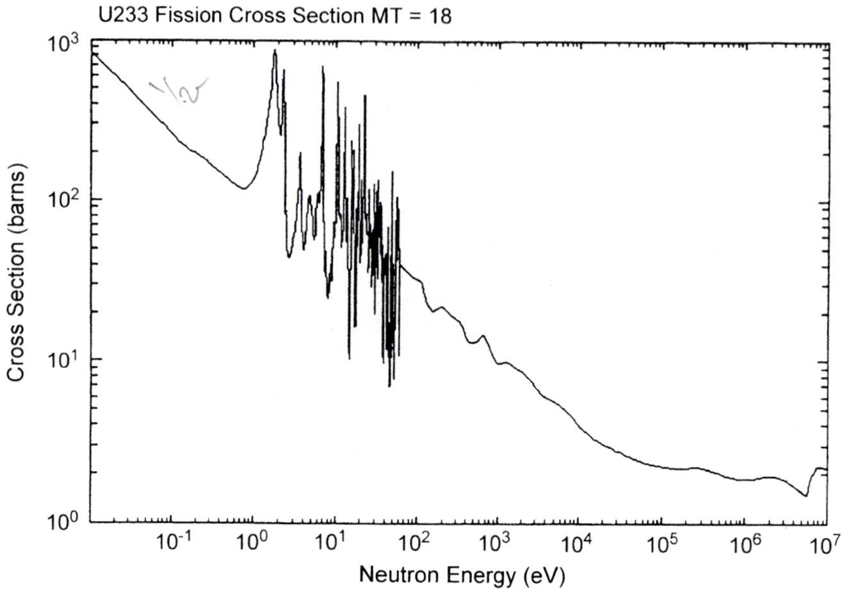

Fig. 1.3 Fission cross sections for $^{233}U_{92}$. (From *http://www.nndc.bnl.gov/*.)

to the designation of σ as a *cross section*. Cross sections are usually on the order of 10^{-24} cm^2, and this unit is referred to as a *barn*, for historical reasons.

The fission cross section, σ_f, is a measure of the probability that a neutron and a nucleus interact to form a compound nucleus which then undergoes fission. The probability that a compound nucleus will be formed is greatly enhanced if the relative energy of the neutron and the original nucleus, plus the reduction in the nuclear binding energy, corresponds to the difference in energy of the ground state and an excited state of the compound nucleus, so that the energetics are just right for formation of a compound nucleus in an excited state. The first excited states of the compound nuclei resulting from neutron absorption by odd-A fissile nuclides are generally lower lying (nearer to the ground state) than are the first excited states of the compound nuclei resulting from neutron absorption by the heavy even-A nuclides, which accounts for the odd-A nuclides having much larger absorption and fission cross sections for low-energy neutrons than do the even-A nuclides.

Fission cross sections for some of the principal fissile nuclides of interest for nuclear reactors are shown in Figs. 1.3 to 1.5. The resonance structure corresponds to the formation of excited states of the compound nuclei, the lowest lying of which are at less than 1 eV. The nature of the resonance cross section can be shown to give rise to a $1/E^{1/2}$ or $1/v$ dependence of the cross section at off-resonance neutron energies below and above the resonance range, as is evident in these figures. The fission cross sections are largest in the thermal energy region $E < \sim 1$ eV. The thermal fission cross section for $^{239}Pu_{94}$ is larger than that of $^{235}U_{92}$ or $^{233}U_{92}$.

U235 Fission Cross Section MT = 18

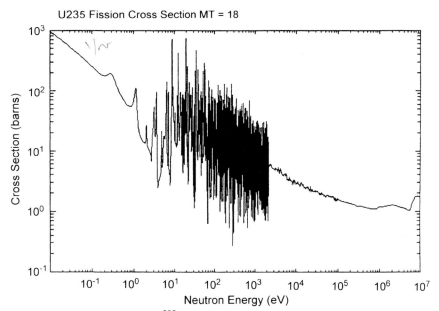

Fig. 1.4 Fission cross sections for $^{235}U_{92}$. (From *http://www.nndc.bnl.gov/*.)

Pu239 Fission Cross Setion MT = 18

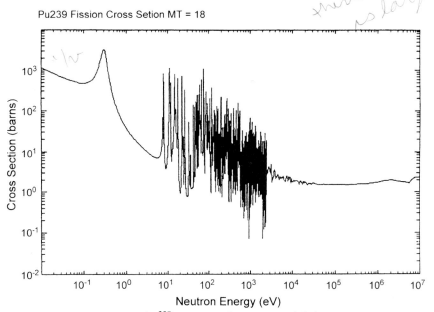

Fig. 1.5 Fission cross sections for $^{239}Pu_{94}$. (From *http://www.nndc.bnl.gov/*.)

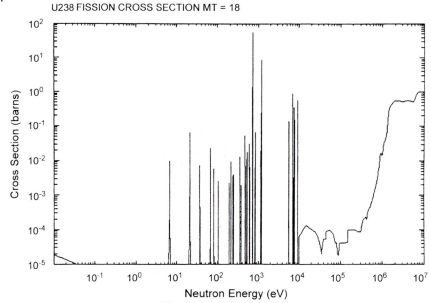

Fig. 1.6 Fission cross sections for $^{238}U_{92}$. (From *http://www.nndc.bnl.gov/*.)

Fission cross sections for $^{238}U_{92}$ and $^{240}Pu_{94}$ are shown in Figs. 1.6 and 1.7. Except for resonances, the fission cross section is insignificant below about 1 MeV, above which it is about 1 barn. The fission cross sections for these and other even-A heavy mass nuclides are compared in Fig. 1.8, without the resonance structure.

Products of the Fission Reaction

A wide range of nuclides are formed by the fission of heavy mass nuclides, but the distribution of these fission fragments is sharply peaked in the mass ranges $90 < A < 100$ and $135 < A < 145$, as shown in Fig. 1.9. With reference to the curvature of the trajectory of the stable isotopes on the n versus p plot of Fig. 1.1, most of these fission fragments are above the stable isotopes (i.e., are neutron rich) and will decay, usually by β-decay (electron emission), which transmutes the fission fragment nuclide (A, Z) to $(A, Z + 1)$, or sometimes by neutron emission, which transmutes the fission fragment nuclide (A, Z) to $(A - 1, Z)$, in both instances toward the range of stable isotopes. Sometimes several decay steps are necessary to reach a stable isotope.

Usually either two or three neutrons will be emitted promptly in the fission event, and there is a probability of one or more neutrons being emitted subsequently upon the decay of neutron-rich fission fragments over the next second or so. The number of neutrons, on average, which are emitted in the fission process, ν, depends on the fissioning nuclide and on the energy of the neutron inducing fission, as shown in Fig. 1.10.

Pu240 FISSION CROSS MT = 18

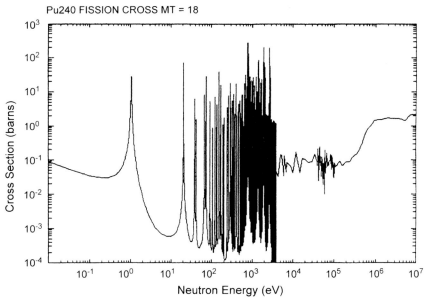

Fig. 1.7 Fission cross sections for $^{240}\text{Pu}_{94}$. (From *http://www.nndc.bnl.gov/*.)

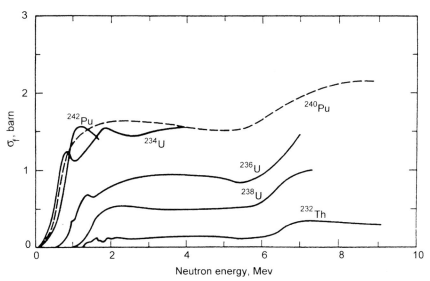

Fig. 1.8 Fission cross sections for principal nonfissile heavy mass nuclides. (From Ref. 15; used with permission of Argonne National Laboratory.)

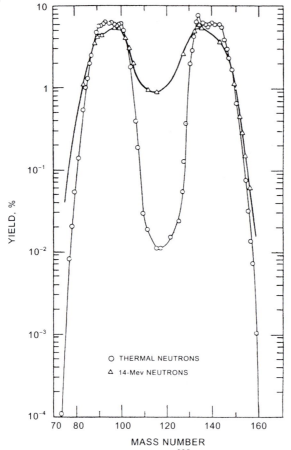

Fig. 1.9 Yield versus mass number for $^{235}U_{92}$ fission. (From Ref. 15.)

Energy Release

The majority of the nuclear energy created by the conversion of mass to energy in the fission event (207 MeV for $^{233}U_{92}$) is in the form of the kinetic energy (168 MeV) of the recoiling fission fragments. The range of these massive, highly charged particles in the fuel element is a fraction of a millimeter, so that the recoil energy is effectively deposited as heat at the point of fission. Another 5 MeV is in the form of kinetic energy of prompt neutrons released in the fission event, distributed in energy as shown in Fig. 1.11, with a most likely energy of 0.7 MeV (for $^{235}U_{92}$). This energy is deposited in the surrounding material within 10 to 100 cm as the neutron diffuses, slows down by scattering collisions with nuclei, and is finally absorbed. A fraction of these neutron absorption events result in neutron capture followed by gamma emission, producing on average about 7 MeV in the form of energetic capture gammas per fission. This secondary capture gamma en-

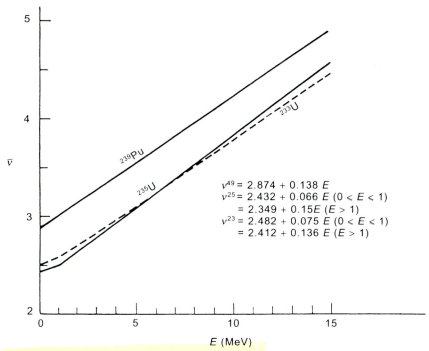

$$v^{49} = 2.874 + 0.138\,E$$
$$v^{25} = 2.432 + 0.066\,E \quad (0 < E < 1)$$
$$= 2.349 + 0.15E \quad (E > 1)$$
$$v^{23} = 2.482 + 0.075\,E \quad (0 < E < 1)$$
$$= 2.412 + 0.136\,E \quad (E > 1)$$

Fig. 1.10 Average number of neutrons emitted per fission. (From Ref. 12; used with permission of Wiley.)

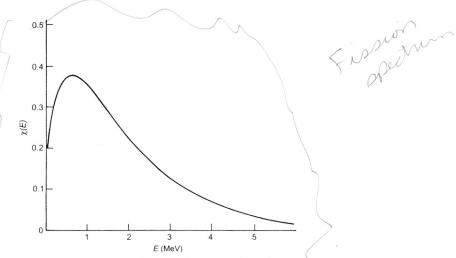

Fig. 1.11 Fission spectrum for thermal neutron-induced fission in $^{235}U_{92}$. (From Ref. 12; used with permission of Wiley.)

Table 1.1 $^{235}U_{92}$ Fission Energy Release

Form	Energy (MeV)	Range
Kinetic energy fission products	168	< mm
Kinetic energy prompt gammas	7	10–100 cm
Kinetic energy prompt neutrons	5	10–100 cm
Kinetic energy capture gammas	7	10–100 cm
Decay of fission products		
Kinetic energy electrons	8	~mm
Kinetic energy neutrinos	12	∞

ergy is transferred as heat to the surrounding material over a range of 10 to 100 cm by gamma interactions.

There is also on average about 7 MeV of fission energy directly released as gamma rays in the fission event, which is deposited as heat within the surrounding 10 to 100 cm. The remaining 20 MeV of fission energy is in the form of kinetic energy of electrons (8 MeV) and neutrinos (12 MeV) from the decay of the fission fragments. The electron energy is deposited, essentially in the fuel element, within about 1 mm of the fission fragment, but since neutrinos rarely interact with matter, the neutrino energy is lost. Although the kinetic energy of the neutrons emitted by the decay of fission products is almost as great as that of the prompt fission neutrons, there are so few delayed neutrons from fission product decay that their contribution to the fission energy distribution is negligible. This fission energy distribution for $^{235}U_{92}$ is summarized in Table 1.1. The recoverable energy released from fission by thermal and fission spectrum neutrons is given in Table 1.2.

Table 1.2 Recoverable Energy from Fission

Isotope	Thermal Neutron	Fission Neutron
^{233}U	190.0	–
^{235}U	192.9	–
^{239}Pu	198.5	–
^{241}Pu	200.3	–
^{232}Th	–	184.2
^{234}U	–	188.9
^{236}U	–	191.4
^{238}U	–	193.9
^{237}Np	–	193.6
^{238}Pu	–	196.9
^{240}Pu	–	196.9
^{242}Pu	–	200.0

Source: Data from Ref. 12; used with permission of Wiley.

Thus, in total, about 200 MeV per fission of heat energy is produced. One Watt of heat energy then corresponds to the fission of 3.1×10^{10} nuclei per second. Since 1 g of any fissile nuclide contains about 2.5×10^{21} nuclei, the fissioning of 1 g of fissile material produces about 1 megawatt-day (MWd) of heat energy. Because some fissile nuclei will also be transmuted by neutron capture, the amount of fissile material destroyed is greater than the amount fissioned.

$1W \approx 3.1 \times 10^{10} \text{ fis}/s$

1.2
Neutron Capture

Radiative Capture

When a neutron is absorbed by a nucleus to form a compound nucleus, a number of reactions are possible, in addition to fission, in the heavy nuclides. We have already mentioned *radiative capture*, in which the compound nucleus decays by the emission of a gamma ray, and we now consider this process in more detail. An energy-level diagram for the compound nucleus formation and decay associated with the prominent $^{238}U_{92}$ resonance for incident neutron energies of about 6.67 eV is shown in Fig. 1.12. The energy in the center-of-mass (CM) system of an incident neutron with energy E_L in the lab system is $E_c = [A/(1 + A)]E_L$. The reduction in binding energy due to the absorbed neutron is ΔE_B. If $E_c + \Delta E_B$ is close to an excited energy level of the compound nucleus, the probability for com-

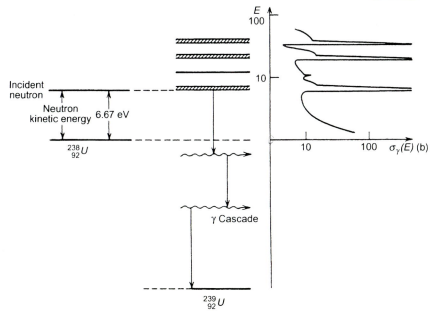

Fig. 1.12 Energy-level diagram for compound nucleus formation. (From Ref. 12; used with permission of Wiley.)

Th232 Capture Cross Section MT = 27

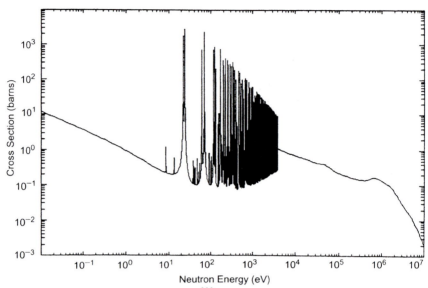

Fig. 1.13 Radiative capture cross section for ^{232}Th$_{90}$. (From *http://www.nndc.bnl.gov/*.)

pound nucleus formation is greatly enhanced. The excited compound nucleus will generally decay by emission of one or more gamma rays, the combined energy of which is equal to the difference in the excited- and ground-state energy levels of the compound nucleus.

Radiative capture cross sections, denoted σ_γ, for some nuclei of interest for nuclear reactors are shown in Figs. 1.13 to 1.21. The resonance nature of the cross sections over certain ranges correspond to the discrete excited states of the compound nucleus that is formed upon neutron capture. These excited states correspond to neutron energies in the range of a fraction of an eV to 10^3 eV for the fissile nuclides, generally correspond to neutron energies of 10 to 10^4 eV for even-*A* heavy mass nuclides (with the notable exception of thermal ^{240}Pu$_{94}$ resonance), and correspond to much higher neutron energies for the lower mass nuclides. The $1/v$ "off-resonance" cross-section dependence is apparent.

The Breit–Wigner single-level resonance formula for the neutron capture cross section is

$$\sigma_\gamma(E_c) = \sigma_0 \frac{\Gamma_\gamma}{\Gamma} \left(\frac{E_0}{E_c}\right)^{1/2} \frac{1}{1+y^2}, \qquad y = \frac{2}{\Gamma}(E_c - E_0) \qquad (1.3)$$

where E_0 is the energy (in the CM) system at which the resonance peak occurs (i.e., $E_c + \Delta E_B$ matches the energy of an excited state of the compound nucleus), Γ the full width at half-maximum of the resonance, σ_0 the maximum value of the total cross section (at E_0), and Γ_γ the radiative capture width (Γ_γ/Γ is the probability that the compound nucleus, once formed, will decay by gamma emission). The

U233 Capture Cross Section MT = 27

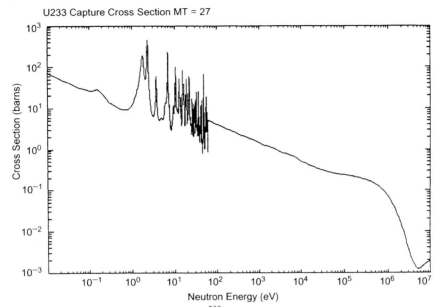

Fig. 1.14 Radiative capture cross section for $^{233}U_{92}$. (From *http://www.nndc.bnl.gov/.*)

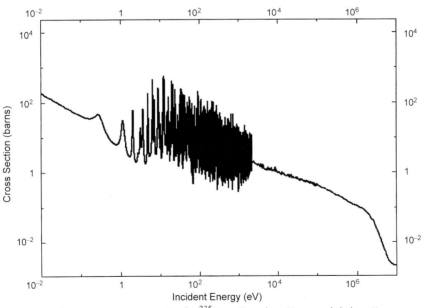

Fig. 1.15 Radiative capture cross section for $^{235}U_{92}$. (From *http://www.nndc.bnl.gov/.*)

Pu239 Capture Cross Section MT = 27

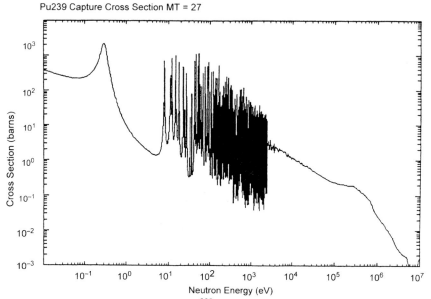

Fig. 1.16 Radiative capture cross section for $^{239}Pu_{94}$. (From *http://www.nndc.bnl.gov/.*)

U238 Capture Cross Section MT = 27

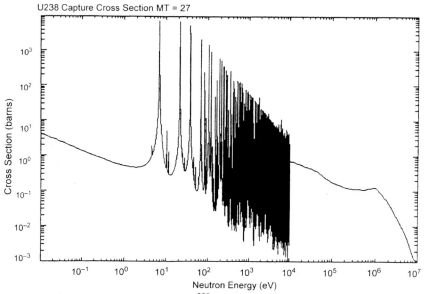

Fig. 1.17 Radiative capture cross section for $^{238}U_{92}$. (From *http://www.nndc.bnl.gov/.*)

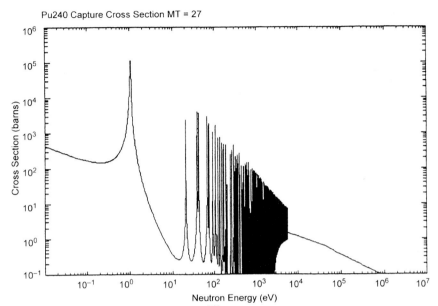

Fig. 1.18 Radiative capture cross section for ^{240}Pu$_{94}$. (From *http://www.nndc.bnl.gov/.*)

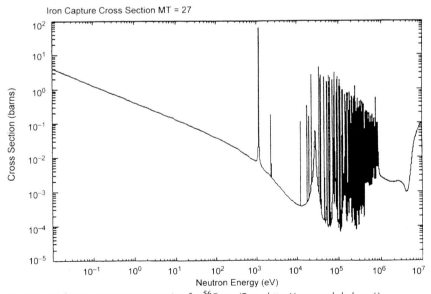

Fig. 1.19 Radiative capture cross section for ^{56}Fe$_{26}$. (From *http://www.nndc.bnl.gov/.*)

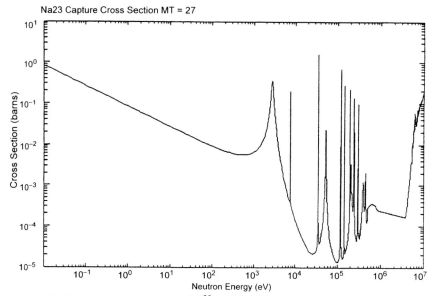

Fig. 1.20 Radiative capture cross section for ^{23}Na$_{11}$. (From *http://www.nndc.bnl.gov/*.)

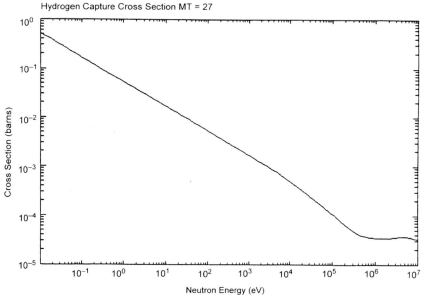

Fig. 1.21 Radiative capture cross section for ^{1}H$_{1}$. (From *http://www.nndc.bnl.gov/*.)

fission resonance cross section can be represented by a similar expression with the fission width Γ_f, defined such that Γ_f/Γ is the probability that the compound nucleus, once formed, will decay by fission.

Equation (1.3) represents the cross section describing the interaction of a neutron and nucleus with relative (CM) energy E_c. However, the nuclei in a material are distributed in energy (approximately a Maxwellian distribution characterized by the temperature of the material). What is needed is a cross section averaged over the motion of the nuclei:

$$\bar{\sigma}(E,T) = \frac{1}{\upsilon(E)} \int dE' \, |\mathbf{v}(E) - \mathbf{v}(E')| \sigma(E_c) f_{max}(E',T) \tag{1.4}$$

where E and E' are the neutron and nuclei energies, respectively, in the lab system, and $f_{max}(E')$ is the Maxwellian energy distribution:

$$f_{max}(E') = \frac{2\pi}{(\pi kT)^{3/2}} \sqrt{E'} e^{-E'/kT} \tag{1.5}$$

Using Eqs. (1.3) and (1.5), Eq. (1.4) becomes

$$\bar{\sigma}_\gamma(E,T) = \frac{\sigma_0 \Gamma_\gamma}{\Gamma} \left(\frac{E_0}{E}\right)^{1/2} \Psi(\xi, x) \tag{1.6}$$

where

$$x = \frac{2}{\Gamma}(E - E_0), \qquad \xi = \frac{\Gamma}{(4E_0 kT/A)^{1/2}} \tag{1.7}$$

A is the atomic mass (amu) of the nuclei, and

$$\Psi(\xi, x) = \frac{\xi}{2\sqrt{\pi}} \int_{-\infty}^{\infty} e^{-(1/4)(x-y)^2 \xi^2} \frac{dy}{1+y^2} \tag{1.8}$$

Neutron Emission

When the compound nucleus formed by neutron capture decays by the emission of one neutron, leaving the nucleus in an excited state which subsequently undergoes further decays, the event is referred to as *inelastic scattering* and the cross section is denoted σ_{in}. Since the nucleus is left in an excited state, the energy of the emitted neutron can be considerably less than the energy of the incident neutron. If the compound nucleus decays by the emission of two or more neutrons, the events are referred to as $n - 2n$, $n - 3n$, and so on, events, and the cross sections are denoted $\sigma_{n,2n}$, $\sigma_{n,3n}$, on so on. Increasingly higher incident neutron energies are required to provide enough excitation energy for single, double, triple, and so on, neutron emission. Inelastic scattering is the most important of these events in nuclear reactors, but it is most important for neutrons 1 MeV and higher in energy.

1.3
Neutron Elastic Scattering

Elastic scattering may take place via compound nucleus formation followed by the emission of a neutron that returns the compound nucleus to the ground state of the original nucleus. In such a resonance elastic scattering event the kinetic energy of the original neutron–nuclear system is conserved. The neutron and the nucleus may also interact without neutron absorption and the formation of a compound nucleus, which is referred to as *potential scattering*. Although quantum mechanical (*s*-wave) in nature, the latter event may be visualized and treated as a classical hard-sphere scattering event, away from resonance energies. Near resonance energies, there is quantum mechanical interference between the potential and resonance scattering, which is constructive just above and destructive just below the resonance energy.

The single-level Breit–Wigner form of the scattering cross section, modified to include potential and interference scattering, is

$$\sigma_s(E_c) = \sigma_0 \frac{\Gamma_n}{\Gamma} \left(\frac{E_0}{E_c} \right)^{1/2} \frac{1}{1+y^2} + \frac{\sigma_0 2R}{\lambda_0} \frac{y}{1+y^2} + 4\pi R^2 \qquad (1.9)$$

where (Γ_n/Γ) is the probability that, once formed, the compound nucleus decays to the ground state of the original nucleus by neutron emission, $R \simeq 1.25 \times 10^{-13} A^{1/3}$ centimeters is the nuclear radius, and λ_0 is the reduced neutron wavelength.

Averaging over a Maxwellian distribution of nuclear motion yields the scattering cross section for neutron lab energy E and material temperature T:

$$\bar{\sigma}_s(E, T) = \sigma_0 \frac{\Gamma_n}{\Gamma} \psi(\xi, x) + \frac{\sigma_0 R}{\lambda_0} \chi(\xi, x) + 4\pi R^2 \qquad (1.10)$$

where

$$\chi(\xi, x) = \frac{\xi}{\sqrt{\pi}} \int_{-\infty}^{\infty} \frac{y e^{-(1/4)(x-y)^2 \xi^2}}{1+y^2} dy \qquad (1.11)$$

The elastic scattering cross sections for a number of nuclides of interest in nuclear reactors are shown in Figs. 1.22 to 1.26. In general, the elastic scattering cross section is almost constant in energy below the neutron energies corresponding to the excited states of the compound nucleus. The destructive interference effects just below the resonance energy are very evident in Fig. 1.26.

The energy dependence of the carbon scattering cross section is extended to very low neutron energies in Fig. 1.27 to illustrate another phenomenon. At sufficiently small neutron energy, the neutron wavelength

$$\lambda_0 = \frac{h}{p} = \frac{h}{\sqrt{2mE}} = \frac{2.86 \times 10^{-9}}{\sqrt{E(\text{eV})}} \text{ cm} \qquad (1.12)$$

Hydrogen Elastic Scattering Cross Section MT = 2

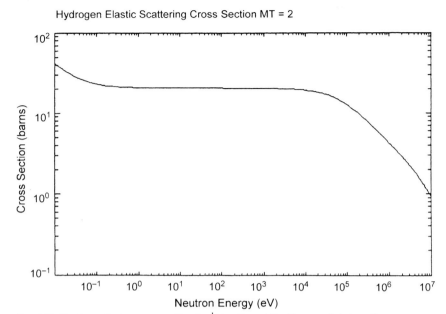

Fig. 1.22 Elastic scattering cross section for $^1H_1$. (From *http://www.nndc.bnl.gov/.*)

Oxygen Elastic Scattering Cross Section MT = 2

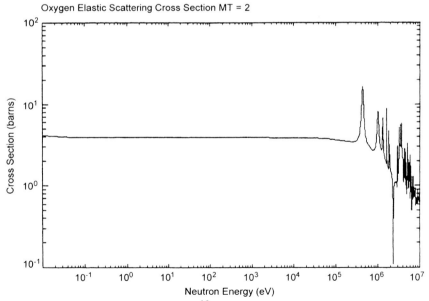

Fig. 1.23 Elastic scattering cross section for $^{16}O_8$. (From *http://www.nndc.bnl.gov/.*)

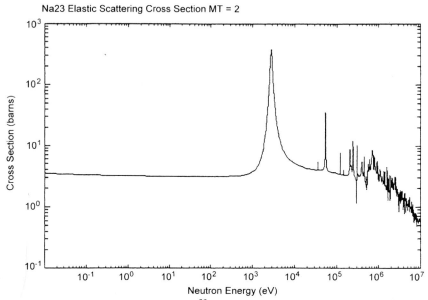

Fig. 1.24 Elastic scattering cross section for ^{23}Na$_{11}$. (From *http://www.nndc.bnl.gov/*.)

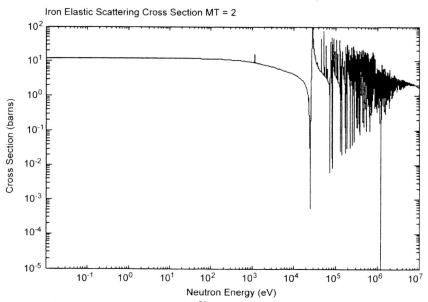

Fig. 1.25 Elastic scattering cross section for ^{56}Fe$_{26}$. (From *http://www.nndc.bnl.gov/*.)

U238 Elastic Scattering Cross Section MT = 2

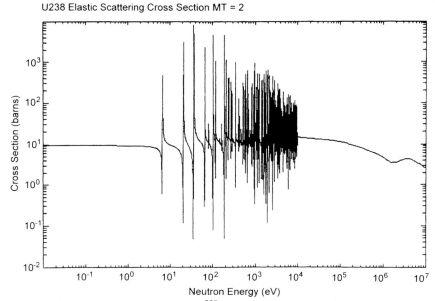

Fig. 1.26 Elastic scattering cross section for $^{238}U_{92}$. (From *http://www.nndc.bnl.gov/*.)

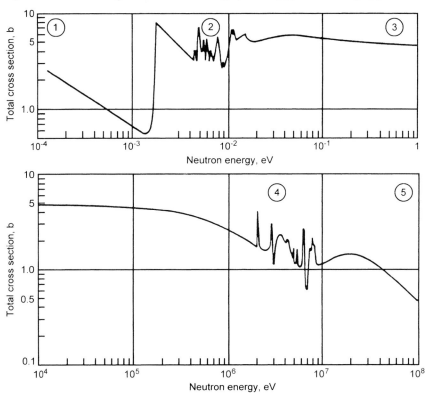

Fig. 1.27 Total scattering cross section of $^{12}C_6$. (From Ref. 12; used with permission of Wiley.)

becomes comparable to the interatomic spacing, and the neutron interacts not with a single nucleus but with an aggregate of bound nuclei. If the material has a regular structure, as graphite does, the neutron will be diffracted and the energy dependence of the cross section will reflect the neutron energies corresponding to multiples of interatomic spacing. For sufficiently small energies, diffraction becomes impossible and the cross section is once again insensitive to neutron energy.

1.4
Summary of Cross-Section Data

Low-Energy Cross Sections

The low-energy total cross sections for several nuclides of interest in nuclear reactors are plotted in Fig. 1.28. Gadolinium is sometimes used as a "burnable poison," and xenon and samarium are fission products with large thermal cross sections.

Spectrum-Averaged Cross Sections

Table 1.3 summarizes the cross-section data for a number of important nuclides in nuclear reactors. The first three columns give fission, radiative capture, and elastic scattering cross sections averaged over a Maxwellian distribution with $T = 0.0253$ eV, corresponding to a representative thermal energy spectrum. The next two columns give the infinite dilution fission and radiative capture resonance integrals, which are averages of the respective resonance cross sections over a $1/E$ spectrum typical of the resonance energy region in the limit of an infinitely dilute concentration of the resonance absorber. The final five columns give cross sections averaged over the fission spectrum.

Example 1.1: Calculation of Macroscopic Cross Section. The macroscopic cross section $\Sigma = N\sigma$, where N is the number density. The number density is related to the density ρ and atomic number A by $N = (\rho/A)N_0$, where $N_0 = 6.022 \times 10^{23}$ is Avogadro's number, the number of atoms in a mole. For a mixture of isotopes with volume fractions v_i, the macroscopic cross section is $\Sigma = \Sigma_i v_i (\rho/A)_i N_0 \sigma_i$; for example, for a 50:50 vol % mixture of carbon and ^{238}U, the macroscopic thermal absorption cross section is $\Sigma_a = 0.5(\rho_C/A_C)N_0\sigma_{aC} + 0.5(\rho_U/A_U)N_0\sigma_{aU} = 0.5(1.60 \text{ g/cm}^3 \text{ per 12 g/mol})(6.022 \times 10^{23} \text{ atom/mol})(0.003 \times 10^{-24} \text{ cm}^2) + 0.5(18.9 \text{ g/cm}^3 \text{ per 238 g/mol})(6.022 \times 10^{23} \text{ atom/mol})(2.4 \times 10^{-24} \text{ cm}^2) = 0.0575 \text{ cm}^{-1}$.

1.5
Evaluated Nuclear Data Files

Published experimental and theoretical results on neutron–nuclear reactions are collected by several collaborating nuclear data agencies worldwide. Perhaps the

Table 1.3 Spectrum-Averaged Thermal, Resonance, and Fast Neutron Cross Sections (barns)

Nuclide	Thermal Cross Section			Resonance Cross Section		Fission Spectrum Cross Section				
	σ_f	σ_γ	σ_{el}	σ_f	σ_γ	σ_f	σ_γ	σ_{el}	σ_{in}	$\sigma_{n,2n}$
$^{233}U_{92}$	469	41	11.9	774	138	1.9	0.07	4.4	1.2	4×10^{-3}
$^{235}U_{92}$	507	87	15.0	278	133	1.2	0.09	4.6	1.8	12×10^{-3}
$^{239}Pu_{94}$	698	274	7.8	303	182	1.8	0.05	4.4	1.5	4×10^{-3}
$^{241}Pu_{94}$	938	326	11.1	573	180	1.6	0.12	5.2	0.9	21×10^{-3}
$^{232}Th_{90}$	–	6.5	13.7	–	84	0.08	0.09	4.6	2.9	14×10^{-3}
$^{238}U_{92}$	0.05	2.4	9.4	2	278	0.31	0.07	4.8	2.6	12×10^{-3}
$^{240}Pu_{94}$	–	264	1.5	8.9	8103	1.4	0.09	4.3	2.0	4×10^{-3}
$^{242}Pu_{94}$	–	16.8	8.3	5.6	1130	1.1	0.09	4.8	1.9	7×10^{-3}
$^{1}H_{1}$	–	0.29	20.5	–	0.15	–	4×10^{-5}	3.9	–	–
$^{2}H_{1}$	–	5×10^{-4}	3.4	–	3×10^{-4}	–	7×10^{-6}	2.5	–	–
$^{10}B_{5}$	–	443	2.1	–	0.22	–	8×10^{-5}	2.1	0.07	–
$^{12}C_{6}$	–	0.003	4.7	–	0.002	–	2×10^{-5}	2.3	0.01	–
$^{16}O_{8}$	–	2×10^{-4}	3.8	–	6×10^{-4}	–	9×10^{-5}	2.7	–	–
$^{23}Na_{11}$	–	0.47	3.0	–	0.31	–	2×10^{-4}	2.7	0.5	–
$^{56}Fe_{26}$	–	2.5	12.5	–	1.4	–	3×10^{-3}	3.0	0.7	–
$^{91}Zr_{40}$	–	1.1	10.6	–	6.9	–	0.01	5.0	0.7	–
$^{135}Xe_{54}$	–	2.7×10^{6}	3.8×10^{5}	–	7.6×10^{3}	–	0.01	4.9	1.0	–
$^{149}Sm_{62}$	–	6.0×10^{4}	373	–	3.5×10^{3}	–	0.22	4.6	2.2	–
$^{157}Gd_{64}$	–	1.9×10^{3}	819	–	761	–	0.11	4.7	2.2	11×10^{-3}

Source: Data from *http://www.nndc.bnl.gov/*.

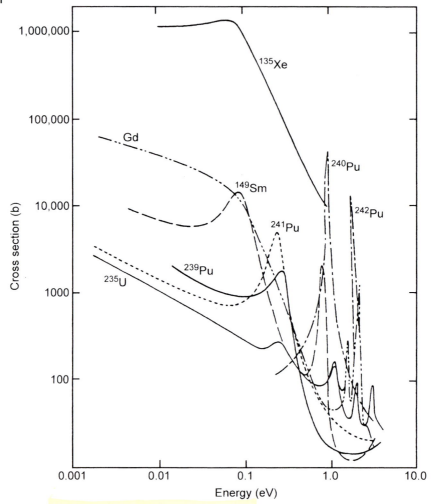

Fig. 1.28 Low-energy absorption (fission + capture) cross sections for several important nuclides. (From Ref. 12; used with permission of Wiley.)

most comprehensive computerized compilation of experimental data is the EXFOR computer library (Ref. 11). The computerized card index file CINDA (Ref. 8), which contains comprehensive information on measurements, calculations, and evaluations of neutron–nuclear data, is updated annually. The plethora of sometimes contradictory nuclear data must be evaluated before it can be used confidently in reactor physics calculations. Such evaluation consists of intercomparison of data, use of data to calculate benchmark experiments, critical assessment of statistical and systematic errors, checks for internal consistency and consistency with standard neutron cross sections, and the derivation of consistent preferred values by appropriate averaging procedures. Several large evaluated nuclear data files are maintained:

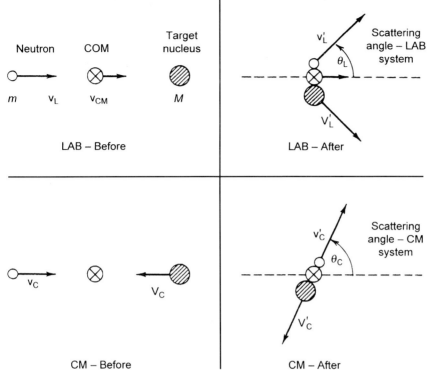

Fig. 1.29 Scattering event in lab and CM systems. (From Ref. 12; used with permission of Wiley.)

(1) United States Evaluated Nuclear Data File (ENDF/B), (2) Evaluated Nuclear Data Library of the Lawrence Livermore National Laboratory (ENDL), (3) United Kingdom Nuclear Data Library (UKNDL), (4) Japanese Evaluated Nuclear Data Library (JENDL), (5) Karlsruhe Nuclear Data File (KEDAK), (6) Russian (formerly Soviet) Evaluated Nuclear Data File (BROND), and (7) Joint Evaluated File of NEA Countries (JEF). Processing codes are used to convert these data to a form that can be used in reactor physic calculations, as discussed in subsequent chapters.

1.6
Elastic Scattering Kinematics

Consider a neutron with energy $E_L = \frac{1}{2}mv_L^2$ in the laboratory (L) system incident upon a stationary nucleus of mass M. Since only the relative masses are important in the kinematics, we set $m = 1$ and $M = A$. It is convenient to convert to the center-of-mass (CM) system, as indicated in Fig. 1.29, because the elastic scattering event is usually isotropic in the CM system.

The velocity of the CM system in the L system is

$$\mathbf{v}_{cm} = \frac{1}{1+A}(\mathbf{v}_L + A\mathbf{V}_L) = \frac{\mathbf{v}_L}{1+A} \tag{1.13}$$

and the velocities of the neutron and the nucleus in the CM system are

$$\mathbf{v}_c = \mathbf{v}_L - \mathbf{v}_{cm} = \frac{A}{A+1}\mathbf{v}_L$$
$$\mathbf{V}_c = -\mathbf{v}_{cm} = \frac{-1}{A+1}\mathbf{v}_L \tag{1.14}$$

The energy of the neutron in the CM system, E_c, is related to the energy of the neutron in the lab, E_L, by

$$E_c = \frac{1}{2}v_c^2 + \frac{1}{2}AV_c^2 = \frac{A}{A+1}\frac{1}{2}v_L^2 = \frac{A}{A+1}E_L \tag{1.15}$$

Correlation of Scattering Angle and Energy Loss

From consideration of conservation of momentum and kinetic energy, it can be shown that the speeds of the neutron and the nucleus in the center-of-mass system do not change during the scattering event:

$$v_c' = v_c = \frac{A}{A+1}v_L$$
$$V_c' = V_c = \frac{-1}{A+1}v_L \tag{1.16}$$

With reference to Fig. 1.30, the scattering angles in the lab and CM systems are related by

$$\tan\theta_L = \frac{v_c' \sin\theta_c}{v_{cm} + v_c' \cos\theta_c} = \frac{\sin\theta_c}{(1/A) + \cos\theta_c} \tag{1.17}$$

The law of cosines yields

$$\cos(\pi - \theta_c) = \frac{(v_c')^2 + (v_{cm})^2 - (v_L')^2}{2v_{cm}v_c'} \tag{1.18}$$

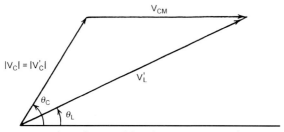

Fig. 1.30 Relation between lab and CM scattering angles.
(From Ref. 12; used with permission of Wiley.)

which may be combined with Eqs. (1.13) and (1.16) to obtain a relationship between the incident and final energies of the neutron in the lab system and the scattering angle in the CM system:

$$\frac{\frac{1}{2}m(v'_L)^2}{\frac{1}{2}m(v_L)^2} \equiv \frac{E'_L}{E_L} = \frac{A^2 + 1 + 2A\cos\theta_c}{(A+1)^2} = \frac{(1+\alpha) + (1-\alpha)\cos\theta_c}{2}, \qquad (1.19)$$

where $\alpha \equiv (A-1)^2/(A+1)^2$.

Average Energy Loss

Equation (1.19) states that the ratio of final to incident energies in an elastic scattering event is correlated to the scattering angle in the CM system, which in turn is correlated via Eq. (1.17) to the scattering angle in the lab system. The maximum energy loss (minimum value of E'_L/E_L) occurs for $\theta_c = \pi$ (i.e., backward scattering in the CM system), in which case $E'_L = \alpha E_L$. For hydrogen $(A = 1)$, $\alpha = 0$ and all of the neutron energy can be lost in a single collision. For other nuclides, only a fraction $(1 - \alpha)$ of the neutron energy can be lost in a single collision, and for heavy nuclides $(\alpha \to 1)$ this fraction becomes very small.

The probability that a neutron scatters from energy E_L to within a differential band of energies dE'_L about energy E'_L is equivalent to the probability that a neutron scatters into a cone $2\pi \sin\theta_c d\theta_c$ about θ_c:

$$\sigma_s(E_L)P(E_L \to E'_L)dE'_L = -\sigma_{cm}(E_L, \theta_c)2\pi \sin\theta_c d\theta_c \qquad (1.20)$$

where the negative sign takes into account that an increase in angle corresponds to a decrease in energy, σ_s is the elastic scattering cross section, and $\sigma_{cm}(\theta_c)$ is the cross section for scattering through angle θ_c. Using Eq. (1.19) to evaluate $dE'_L/d\theta_c$, this becomes

$$P(E_L \to E'_L) = \begin{cases} \dfrac{4\pi\sigma_{cm}(E_L, \theta_c)}{(1-\alpha)E_L\sigma_s(E_L)}, & \alpha E_L \le E'_L \le E_L \\ 0, & \text{otherwise} \end{cases} \qquad (1.21)$$

Except for very high energy neutrons scattering from heavy mass nuclides, elastic scattering in the CM is isotropic, $\sigma_{cm}(\theta_c) = \sigma_s/4\pi$. In this case, Eq. (1.21) may be written

$$\sigma_s(E_L \to E'_L) \equiv \sigma_s(E_L)P(E_L \to E'_L) = \frac{\sigma_s(E_L)}{(1-\alpha)E_L}, \quad \alpha E_L \le E'_L \le E_L \qquad (1.22)$$

$$= 0, \qquad\qquad\qquad \text{otherwise}$$

The average energy loss in an elastic scattering event may be calculated from

$$\langle \Delta E_L \rangle \equiv E_L - \int_{\alpha E_L}^{E_L} dE'_L E'_L P(E_L \to E'_L) = \frac{1}{2}(1-\alpha)E_L \qquad (1.23)$$

Table 1.4 Number of Collisions, on Average, to Moderate a Neutron from 2 MeV to 1 eV

Moderator	ξ	Number of Collisions	$\xi \Sigma_s / \Sigma_a$
H	1.0	14	–
D	0.725	20	–
H_2O	0.920	16	71
D_2O	0.509	29	5670
He	0.425	43	83
Be	0.209	69	143
C	0.158	91	192
Na	0.084	171	1134
Fe	0.035	411	35
^{238}U	0.008	1730	0.0092

and the average logarithmic energy loss may be calculated from

$$\xi \equiv \int_{\alpha E_L}^{E_L} dE_L' \ln\left(\frac{E_L}{E_L'}\right) P(E_L \to E_L')$$

$$= 1 + \frac{\alpha}{1-\alpha} \ln \alpha = 1 - \frac{(A-1)^2}{2A} \ln\left(\frac{A+1}{A-1}\right) \tag{1.24}$$

The number of collisions, on average, required for a neutron of energy E_0 to be moderated to thermal energies, say 1 eV, can be estimated from

$$\langle \text{no. collisions} \rangle \simeq \frac{\ln[E_0(eV)/1.0]}{\xi} \tag{1.25}$$

The results are shown in Table 1.4 for $E_0 = 2$ MeV.

The parameter ξ, which is a measure of the moderating ability, decreases with nuclide mass, with the result that the number of collisions that are needed to moderate a fast neutron increases with nuclide mass. However, the effectiveness of a nuclide (or molecule) in moderating a neutron also depends on the relative probability that a collision will result in a scattering reaction, not a capture reaction, which would remove the neutron. Thus the parameter $\xi \Sigma_s / \Sigma_a$, referred to as the *moderating ratio*, is a measure of the effectiveness of a moderating material. Even though H_2O is the better moderator in terms of the number of collisions required to thermalize a fast neutron, D_2O is the more effective moderator because the absorption cross section for D is much less than that for H.

Example 1.2: Moderation by a Mixture. The moderating parameters for a mixture of isotopes is constructed by weighting the moderating parameters of the individual isotopes by their concentrations in the mixture. For example, in a mixture of ^{12}C and ^{238}U the average value of $\xi \Sigma_s = N_C \xi_C \sigma_{sC} + N_U \xi_U \sigma_{sU} = N_C(0.158)$ $(2.3 \times 10^{-24} \text{ cm}^2) + N_U(0.008)(4.8 \times 10^{-24} \text{ cm}^2)$, where the fission spectrum range elastic scattering cross sections of Table 1.3 have been assumed to hold also in the slowing-down range. The total absorption cross section is $\Sigma_a = N_C \sigma_{aC} + N_U \sigma_{aU} = N_C(0.002 \times 10^{-24}\ \text{cm}^2) + N_U(280 \times 10^{-24}\ \text{cm}^2)$ in the slowing-down range, where the resonance range cross sections from Table 1.3 have been used.

References

1 H. CEMBER, *Introduction to Health Physics*, 3rd ed., McGraw-Hill, New York (**1996**).

2 C. NORDBORG and M. SALVATORES, "Status of the JEF Evaluated Nuclear Data Library," *Proc. Int. Conf. Nuclear Data for Science and Technology*, Gatlinburg, TN, Vol. 2 (**1994**), p. 680.

3 R. W. ROUSSIN, P. G. YOUNG, and R. MCKNIGHT, "Current Status of ENDF/B-VI, *Proc. Int. Conf. Nuclear Data for Science and Technology*, Gatlinburg, TN, Vol. 2 (**1994**), p. 692.

4 Y. KIKUCHI, "JENDL-3 Revision 2: JENDL 3-2," *Proc. Int. Conf. Nuclear Data for Science and Technology*, Gatlinburg, TN, Vol. 2 (**1994**), p. 685.

5 R. A. KNIEF, *Nuclear Engineering*, Taylor & Francis, Washington, DC (**1992**).

6 J. J. SCHMIDT, "Nuclear Data: Their Importance and Application in Fission Reactor Physics Calculations," in D. E. Cullen, R. Muranaka, and J. Schmidt, eds., *Reactor Physics Calculations for Applications in Nuclear Technology*, World Scientific, Singapore (**1990**).

7 A. TRKOV, "Evaluated Nuclear Data Processing and Nuclear Reactor Calculations," in D. E. Cullen, R. Muranaka, and J. Schmidt, eds., *Reactor Physics Calculations for Applications in Nuclear Technology*, World Scientific, Singapore (**1990**).

8 *CINDA: An Index to the Literature on Microscopic Neutron Data*, International Atomic Energy Agency, Vienna; CINDA-A, 1935–1976 (**1979**); CINDA-B, 1977–1981 (**1984**); CINDA-89 (**1989**).

9 D. E. CULLEN, "Nuclear Cross Section Preparation," in Y. Ronen, ed., *CRC Handbook of Nuclear Reactor Calculations I*, CRC Press, Boca Raton, FL (**1986**).

10 J. L. ROWLANDS and N. TUBBS, "The Joint Evaluated File: A New Nuclear Data Library for Reactor Calculations," *Proc. Int. Conf. Nuclear Data for Basic and Applied Science*, Santa Fe, NM, Vol. 2 (**1985**), p. 1493.

11 A. CALAMAND and H. D. LEMMEL, *Short Guide to EXFOR*, IAEA-NDS-1, Rev. 3, International Atomic Energy Agency, Vienna (**1981**).

12 J. J. DUDERSTADT and L. G. HAMILTON, *Nuclear Reactor Analysis*, Wiley, New York (**1976**), Chap. 2.

13 H. C. HONECK, *ENDF/B: Specifications for an Evaluated Data File for Reactor Applications*, USAEC report BNL-50066, Brookhaven National Laboratory, Upton, NY (**1966**).

14 I. KAPLAN, *Nuclear Physics*, 2nd ed., Addison-Wesley, Reading, MA (**1963**).

15 L. J. TEMPLIN, ed., *Reactor Physics Constants*, 2nd ed., ANL-5800, Argonne National Laboratory, Argonne, IL (**1963**).

Problems

1.1. Demonstrate that the speeds of the neutron and nucleus in the CM system do not change in an elastic scattering event by using conservation of momentum and kinetic energy.

1.2. Estimate the probability that a 1-MeV neutron will be moderated to thermal without being captured in a mixture of uranium and water with $N_H/N_U = 1:1$. Repeat for a 1:1 mixture of uranium and carbon.

1.3. Neutrons are slowed down to thermal energies in a 1:1 mixture of H_2O and 4% enriched uranium (4% ^{235}U, 96% ^{238}U). Estimate the thermal value of $\eta = \nu\sigma_f/(\sigma_c + \sigma_f)$. Repeat the calculation for a mixture of (2% ^{235}U, 2% ^{239}Pu, 96% ^{238}U).

1.4. Estimate the probability that a fission neutron will have a scattering collision with H_2O in the mixtures of Problem 1.3.

1.5. Calculate the average energy loss for neutrons at 1-MeV, 100-keV, 10-keV, and 1-keV scattering from carbon. Repeat the calculation for scattering from iron and from uranium.

1.6. Repeat Problem 1.5 for scattering from hydrogen and sodium.

1.7. Calculate the moderating ratio and the average number of collisions required to moderate a fission neutron to thermal for a 1:1 mixture of $^{12}C : ^{238}U$. Repeat for a 10:1 mixture.

1.8. Calculate the thermal absorption cross section for a 1:1 wt% mixture of carbon and 4% enriched uranium (e.g., 4% ^{235}U, 96% ^{238}U).

1.9. Derive Eq. (1.21) from Eqs. (1.20) and (1.19).

1.10. Calculate the average number of scattering events required to moderate a neutron's energy from above the resonance range to below the resonance range of ^{238}U for carbon, H_2O and D_2O moderators.

2
Neutron Chain Fission Reactors

2.1
Neutron Chain Fission Reactions

Since two or three neutrons are released in every neutron-induced fission reaction, the possibility of a sustained neutron chain reaction is obvious, as illustrated in Fig. 2.1. To sustain a fission chain reaction, one or more of the neutrons produced in the fission event must, on average, survive to produce another fission event. There is competition for the fission neutrons in any assembly—some will be absorbed in fuel nuclides as radiative capture events rather than fission events, some will be absorbed by nonfuel nuclides, and some will leak out of the assembly. A scattering event does not compete for a neutron because the scattered neutron remains in the assembly and available for causing a fission event, but a scattering event does change a neutron's energy and thus, because the various cross sections are energy dependent, does change the relative likelihood of the next collision being a fission event.

Capture-to-Fission Ratio

The fission cross sections for the fissile nuclides increase approximately as $1/\upsilon$ with decreasing neutron energy, but then so do the capture cross sections of the fissile nuclides. The probability that a neutron that is captured in a fissile nuclide causes a fission is just $\sigma_f/(\sigma_f + \sigma_\gamma) = 1/(1 + \sigma_\gamma/\sigma_f) = 1/(1 + \alpha)$, where $\alpha \equiv \sigma_\gamma/\sigma_f$ is referred to as the _capture-to-fission ratio_. The capture-to-fission ratio for the principal fissile nuclides decreases as the neutron energy increases. For high neutron energies, the fission probability, which varies as $(1 + \alpha)^{-1}$, is larger for ^{239}Pu than for ^{235}U or ^{233}U, but the situation is reversed for low-energy thermal neutrons.

Number of Fission Neutrons per Neutron Absorbed in Fuel

The product of the fission probability for a neutron absorbed in the fuel and the average number of neutrons released per fission, $\eta \equiv \nu\sigma_f/(\sigma_f + \sigma_\gamma) = \nu/(1 + \alpha)$, provides a somewhat better characterization of the relative capabilities of the var-

Nuclear Reactor Physics. Weston M. Stacey
Copyright © 2007 WILEY-VCH Verlag GmbH & Co. KGaA, Weinheim
ISBN: 978-3-527-40679-1

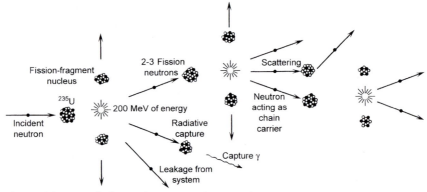

Fig. 2.1 Schematic of a fission chain reaction. (From Ref. 3; used with permission of Wiley.)

ious fissile nuclides to sustain a fission chain reaction. This quantity is plotted in Fig. 2.2 for the principal fissile nuclides. For high neutron energies, η is larger for ^{239}Pu than for ^{235}U or ^{233}U, but the situation is reversed for low-energy thermal neutrons.

Neutron Utilization

The probability that a neutron is absorbed in a fissile nuclide instead of being absorbed in another nuclide or leaking from the assembly is

$$\frac{\text{absorb fissile}}{\text{absorb fissile} + \text{absorb nonfissile} + \text{leak}}$$

$$= \frac{\text{absorb fissile}}{\text{absorb total}} \frac{1}{(1 + \text{leak/absorb total})} \equiv f\, P_{\text{NL}} \qquad (2.1)$$

where f is the fraction of the absorbed neutrons which are absorbed in the fissile nuclides, or the *utilization*:

$$f = \frac{N_{\text{fis}} \sigma_a^{\text{fis}}}{N_{\text{fis}} \sigma_a^{\text{fis}} + N_{\text{other}} \sigma_a^{\text{other}}} \qquad (2.2)$$

and P_{NL} refers to the *nonleakage probability*. Since the absorption cross section, $\sigma_a = \sigma_f + \sigma_\gamma$, is much greater for thermal neutrons than for fast neutrons for the fissile nuclides, but comparable for fast and thermal neutrons for the nonfissile fuel nuclides and for structural nuclides, the utilization for a given composition is much greater for thermal neutrons than for fast neutrons (and, in fact, is usually referred to as the *thermal utilization*).

Fast Fission

The product ηf is the number of neutrons produced, on average, from the fission of fissile nuclides for each neutron absorbed in the assembly. There will also

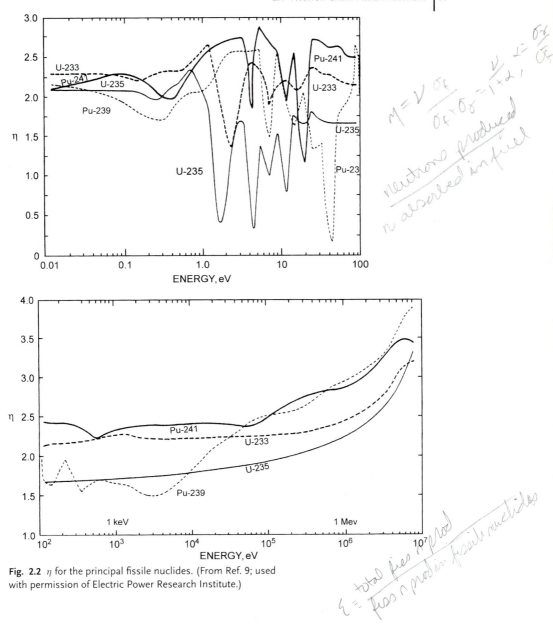

Fig. 2.2 η for the principal fissile nuclides. (From Ref. 9; used with permission of Electric Power Research Institute.)

be neutrons produced by the fission of the nonfissile fuel nuclides, mostly by fast neutrons. Defining the *fast fission factor* $\varepsilon \equiv$ total fission neutron production rate/fission neutron production rate in fissile nuclides, $\eta f \varepsilon$ is the total number of fission neutrons produced for each neutron absorbed in the assembly, and $\eta f \varepsilon P_{NL}$ is the total number of fission neutrons produced, on average, for each neutron introduced into the assembly by a previous fission event.

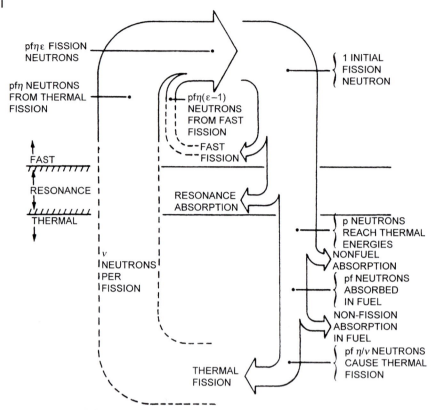

Fig. 2.3 Neutron balance in a thermal neutron fission assembly.
(From Ref. 1; used with permission of Taylor & Francis.)

Resonance Escape

The parameters $\eta f \varepsilon$ must be evaluated by averaging over the energy of the neutrons in the assembly, of course. When the neutron population consists predominantly of thermal neutrons, the thermal spectrum-averaged cross sections given in Table 1.3 may be used to estimate η and f, and the cross sections averaged over the fission spectrum may be used in estimating ε, which should now also include fast fission in the fissile nuclides. In this case, it is necessary to take into account separately the capture of fission neutrons while they are slowing down to the thermal energy range, predominantly by the capture resonances of the fuel nuclides. The probability that a neutron is not captured during the slowing-down process is referred to as the *resonance escape probability* and denoted p. This competition for neutrons is illustrated schematically in Fig. 2.3 (leakage is neglected).

2.2
Criticality

Effective Multiplication Constant

The product $\eta f \varepsilon p P_{\mathrm{NL}}$ is the total number of fission neutrons produced, on average, by one fast neutron from a previous fission event. This quantity is referred to as the *effective multiplication constant* of the assembly:

$$k = \eta f \varepsilon p P_{\mathrm{NL}} \equiv k_\infty P_{\mathrm{NL}} \tag{2.3}$$

where k_∞ refers to the multiplication constant of an infinite assembly with no leakage.

If exactly one neutron, on average, survives to cause another fission, a condition referred to as *criticality* ($k = 1$), the neutron population in the assembly will remain constant. If less than one neutron, on average, survives to produce another fission event, a condition referred to as *subcriticality* ($k < 1$), the neutron population in the assembly will decrease. If more than one fission neutron, on average, survives to cause another fission, a condition referred to as *supercriticality* ($k > 1$), the neutron population in the assembly will increase. The effective multiplication constant depends on the composition (k_∞) and size (P_{NL}) of an assembly and on the arrangement of the materials within the assembly (f and p). The composition affects k both by the relative number of nuclides of different species that are present and by the determination of the neutron energy distribution, which determines the average cross sections for each nuclide. The arrangement of materials determines the spatial neutron distribution and hence the relative number of neutrons at the locations of the various nuclides.

The fissile nuclide ^{235}U is only 0.72% of natural uranium. Fuel enrichment to achieve a higher fissile content, hence larger value of f, is a major means of increasing the multiplication constant. The number of fission neutrons produced for each neutron absorbed in fissile material, η, is significantly larger for fast neutrons than for thermal neutrons, because the capture-to-fission ratio is smaller and the number of neutrons per fission is larger. On the other hand, for a given fuel enrichment, the utilization, f, is greater for thermal neutrons than for fast neutrons because the absorption cross section is much greater for thermal neutrons than for fast neutrons for the fissile nuclides, but comparable for fast and thermal neutrons for the nonfissile fuel nuclides and for structural nuclides. On the whole, the amount of fissile material necessary to achieve a given value of the multiplication constant is substantially less in a fast neutron spectrum than in a thermal neutron spectrum.

Effect of Fuel Lumping

Lumping the fuel rather than distributing it uniformly can have a significant effect on the multiplication constant. For example, if natural uranium is distributed uniformly in a graphite lattice, the values of the various parameters are $\eta \approx 1.33$, $f \approx 0.9$, $\varepsilon \approx 1.05$, and $p \approx 0.7$, yielding $k_\infty \approx 0.88$ (i.e., the assembly is subcritical). If the fuel is lumped, the strong resonance absorption at the exterior of the fuel elements reduces the number of neutrons that reach the interior of the fuel elements, increasing the resonance escape probability to $p \approx 0.9$. Lumping the fuel also reduces the thermal utilization f, for the same reason, but the effect is not so significant. Lumping the fuel was the key to achieving criticality ($k = 1$) in the first graphite-moderated natural uranium reactors and is crucial in achieving criticality in present-day D_2O-moderated natural uranium reactors.

Leakage Reduction

The multiplication constant can be increased by reducing the leakage, most of which is due to fast neutrons. This can be done simply by increasing the size. The leakage can also be reduced by choosing a composition that moderates the neutrons quickly before they can travel far or by surrounding the assembly with a material with a large scattering cross section (e.g., graphite), which will reflect leaking neutrons back into the assembly.

Example 2.1: Effective Multiplication Factor for a PWR. For a typical pressurized water reactor (PWR), the various parameters are $\eta \approx 1.65$, $f \approx 0.71$, $\varepsilon \approx 1.02$, and $p \approx 0.87$, yielding $k_\infty \approx 1.04$. The nonleakage factors for fast and thermal neutrons are typically 0.97 and 0.99, yielding $k \approx 1.00$.

2.3
Time Dependence of a Neutron Fission Chain Assembly

Prompt Fission Neutron Time Dependence

If there are N_0 fission neutrons introduced into an assembly at $t = 0$, and if l is the average time required for a fission neutron to slow down and be absorbed or leak out, $l = \frac{1}{v\Sigma_f v}$, the number of neutrons, on average, in the assembly at time $t = l$ is $(k)N_0$. Continuing in this fashion, the number of neutrons in the assembly at time $t = ml$ (m integer) is $(k)^m N_0$. The quantity l is typically $\approx 10^{-4}$ s for assemblies in which the neutrons slow down to thermal before causing another fission, and is typically $\approx 10^{-6}$ s for assemblies in which the fission is produced by fast neutrons. For example, a $\frac{1}{2}$% change in absorption cross section, which could be produced by control rod motion, causes an approximately 0.005 change in k. The neutron population after 0.1 s in a thermal assembly (0.1 s $= 10^3 l$) in which $k = 1.005$, would be $N(0.1) = (1.005)^{1000} N_0 \approx 150 N_0$. In a thermal assembly with $k = 0.995$, the neutron population after 0.1 s would be $N(0.1) = (0.995)^{1000} N_0 \approx 0.0066 N_0$.

An equation governing the neutron kinetics described above is

$$\frac{dN(t)}{dt} = \frac{k-1}{l} N(t) + S(t) \tag{2.4}$$

$$\frac{\delta N}{\delta t} = P(t) - L(t) + S(t)$$

which simply states that the time rate of change of the neutron population is equal to the excess of neutron production (by fission) minus neutron loss by absorption or leakage in a neutron lifetime plus any external source that is present. For a constant source, Eq. (2.4) has the solution

$$N(t) = N(0)e^{(k-1)t/l} + \frac{Sl}{k-1}[e^{(k-1)t/l} - 1] \tag{2.5}$$

which displays an exponential time behavior. Using the same example as above, with the source set to zero, leads to $N(0.1) = N(0)\exp(5.0) = 148N(0)$ for $k = 1.005$ and $N(0.1) = N(0)\exp(-5.0) = 0.00677N(0)$ for $k = 0.995$.

Source Multiplication

Equation (2.4) does not have a steady-state solution for $k > 1$ and does not have a unique steady-state solution for $k = 1$. However, for $k < 1$, the asymptotic solution is

Source
$k < 0$

$$N_{\text{asymptotic}} = \frac{lS_0}{1-k} \tag{2.6}$$

This equation provides a method to measure the effective multiplication factor k when $k < 1$ by measuring the asymptotic neutron population which results from placing a source S_0 in a multiplying medium.

Effect of Delayed Neutrons

It would be very difficult, if not impossible, to control a neutron fission chain assembly which responded so dramatically to a $\frac{1}{2}\%$ change in absorption cross section. Fortunately, a small fraction ($\beta \approx 0.0075$ for ^{235}U fueled reactors) of the fission neutrons are delayed until the decay ($\lambda \approx 0.08$ s^{-1}) of the fission fragments. For an assembly that was critical prior to $t = 0$, the equilibrium concentration of such delayed neutron precursor fission fragments is found from the balance equation:

delayed
neutron precursor fragments

$$\frac{dC_0}{dt} = 0 = \beta v N_f \sigma_f v N_0 - \lambda C_0 = \frac{\beta}{l}N_0 - \lambda C_0 \tag{2.7}$$

$$\lambda = \frac{1}{v N_f \sigma v}$$

where N_f is the density of fuel nuclei, N_0 the neutron population, and C_0 the population of delayed neutron precursor fission fragments.

When the $\frac{1}{2}\%$ change in cross section occurs at $t = 0$, the multiplication of the prompt neutrons after 0.1 s (1000l) is $[(1 - \beta)k]^{1000}$. During each multiplication interval l there is a source λlC of delayed neutrons from the decay of fission fragments. This source results in $(1 - \beta)k\lambda lC$ neutrons in the following multiplication interval, $[(1 - \beta)k]^2\lambda lC$ neutrons in the second following multiplication interval, and so on. There is such a delayed neutron source in each of the 1000 multiplication intervals in our example. To simplify the problem, we assume that the fission fragment concentration does not change (i.e., $C = C_0$). Thus the number of neutrons after 0.1 s ($1000l$) is

$$N(1000l) = [(1-\beta)k]^m N_0 + \lambda l C_0 [(1-\beta)k]^{m-1} + \lambda l C_0 [(1-\beta)k]^{m-2}$$
$$+ \cdots + \lambda l C_0 (1-\beta)k + \lambda l C_0$$
$$= [(1-\beta)k]^m N_0 + \lambda l C_0 \frac{(1-k[1-\beta])^{m-1}}{1-k(1-\beta)}$$
$$= \left\{ [(1-\beta)k]^m \left[1 - \frac{\beta}{k(1-\beta)[1-k(1-\beta)]} \right] + \frac{\beta}{1-k(1-\beta)} \right\} N_0$$

$$(2.8)$$

where Eq. (2.7) has been used in the last step. Evaluating this expression for $k = 1.005$ yields $N(t = 0.1 \text{ s}) = 3.03 N_0$, instead of the $150 N_0$ found without taking the delayed neutrons into account. If we had taken into account the changing fission fragment population, we would have found a slightly larger number. Nevertheless, the fact that some of the neutrons emitted in fission are delayed results in a rather slow and hence controllable response of a neutron chain fission reacting assembly, provided that $(1-\beta)k < 1$.

2.4
Classification of Nuclear Reactors

Physics Classification by Neutron Spectrum

From the physics viewpoint, the main differences among reactor types arise from differences in the neutron energy distribution, or spectrum, which causes differences in the neutron–nuclear reaction rates and the competition for neutrons. The first level of physics classification categories are then thermal reactors and fast reactors, corresponding to the majority of the neutron–nuclear reactions involving neutrons in the thermal energy range ($E < 1$ eV) and to the majority of the neutron–nuclear reactions involving neutrons in the fast energy range ($E > 1$ keV), respectively. Representative neutron spectra for thermal (LWR) and fast (LMFBR) reactor cores are shown in Fig. 2.4.

There are important physics differences among the different thermal reactors and among the different fast reactors, but these differences are not so great as the physics differences between a thermal reactor and a fast reactor. The capture-to-fission ratio, α, is lower and the number of neutrons produced per fission, ν, is larger in fast reactors than in thermal reactors. This generally results in a larger value of k for a given amount of fuel in a fast reactor than in a thermal reactor, or, more to the point, a smaller *critical mass* of fuel in a fast reactor than in a thermal reactor. Because of the larger neutron–nuclear reaction rates for thermal neutrons than for fast neutrons, the mean distance that a neutron travels before absorption is greater in a fast reactor than in a thermal reactor. This implies that the detailed

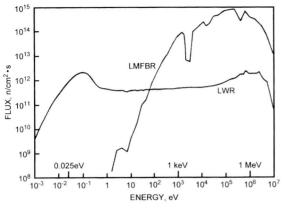

Fig. 2.4 Representative fast (LMFBR) and thermal (LWR) reactor neutron energy distributions. Flux $\equiv nv$ (From Ref. 1; used with permission of Taylor & Francis.)

distribution of fuel, coolant, and control elements has a much greater effect on the local competition for neutrons in a thermal reactor than in a fast reactor and that the neutron populations in the different regions of the core are more tightly coupled in a fast reactor than in a thermal reactor.

Engineering Classification by Coolant

The neutron spectrum is determined primarily by the principal neutron moderating material present, and in many cases this material is the coolant. Because the heat transport system is such a major aspect of a nuclear reactor, it is also common to classify reactors according to coolant. Water-cooled reactors, such as the pressurized water (PWR) and boiling water (BWR) reactors, which use H_2O coolant, and the pressurized heavy water reactor (PHWR), which uses D_2O coolant, have thermal neutron spectra because of the excellent moderating properties of hydrogen. Since gas is too diffuse to serve as an effective moderator, gas-cooled reactors can be either thermal or fast, depending on whether or not a moderator, commonly graphite, is included. The early Magnox and subsequent advanced gas reactors (AGR) are cooled with CO_2, and the advanced high-temperature gas-cooled reactor (HTGR) is cooled with helium; all are moderated with graphite to achieve a thermal spectrum. Designs have been developed for a helium-cooled reactor without graphite, which is known as the gas-cooled fast reactor (GCFR). The pressure tube graphite-moderated reactor (PTGR) is cooled with pressurized or boiling water in pressure tubes, but it is necessary to include graphite to achieve a thermal spectrum. The molten salt breeder reactor (MSBR) employs a molten salt fluid which acts as both the fuel and the primary coolant loop, and is moderated by graphite to achieve a thermal spectrum. The advanced liquid-metal reactor (ALMR) and the liquid-metal fast breeder reactor (LMFBR) are cooled with sodium, which is not a particularly effective moderator, and the neutron spectrum is fast.

References

1 R. A. KNIEF, *Nuclear Engineering*, 2nd ed., Taylor & Francis, Washington, DC (**1992**).

2 J. R. LAMARSH, *Introduction to Nuclear Reactor Theory*, 2nd ed., Addison-Wesley, Reading, MA (**1983**).

3 J. J. DUDERSTADT and L. J. HAMILTON, *Nuclear Reactor Analysis*, Wiley, New York (**1976**).

4 A. F. HENRY, *Nuclear-Reactor Analysis*, MIT Press, Cambridge, MA (**1975**).

5 G. I. BELL and S. GLASSTONE, *Nuclear Reactor Theory*, Van Nostrand Reinhold, New York (**1970**).

6 R. V. MEGHREBLIAN and D. K. HOLMES, *Reactor Analysis*, McGraw-Hill, New York (**1960**), pp. 160–267 and 626–747.

7 A. M. WEINBERG and E. P. WIGNER, *The Physical Theory of Neutron Chain Reactors*, University of Chicago Press, Chicago (**1958**).

8 S. GLASSTONE and M. C. EDLUND, *Nuclear Reactor Theory*, D. Van Nostrand, Princeton, NJ (**1952**).

9 N. L. SHAPIRO et al., *Electric Power Research Institute Report*, EPRI-NP-359, Electric Power Research Institute, Palo Alto, CA (**1977**).

Problems

2.1. Calculate and plot the thermal value of η for a uranium-fueled reactor as a function of enrichment (e.g., percentage ^{235}U in uranium) over the range 0.07 to 5.0%.

2.2. Calculate the thermal utilization in a homogeneous 50:50 volume % mixture of carbon and natural uranium. Repeat the calculation for 4% enriched uranium.

3
Neutron Diffusion Theory

In this chapter we develop a one-speed diffusion theory mathematical description of nuclear reactors. Such a relatively simple description has the great advantage of illustrating many of the important features of nuclear reactors without the complexity that is introduced by the treatment of important effects associated with the neutron energy spectrum and with highly directional neutron transport, which are the subjects of subsequent chapters. Moreover, diffusion theory is sufficiently accurate to provide a quantitative understanding of many physics features of nuclear reactors and is, in fact, the workhorse computational method of nuclear reactor physics.

3.1
Derivation of One-Speed Diffusion Theory

Calculation of the rates of the different reactions of neutrons with the materials in the various parts of a nuclear reactor is the fundamental task of nuclear reactor physics. This calculation requires a knowledge of nuclear cross sections and their energy dependence (Chapter 1) and of the distribution of neutrons in space and energy throughout the reactor. The neutron distribution depends on the neutron source distribution, which in the case of the fission source depends on the neutron distribution itself, and on the interactions with atomic nuclei experienced by the neutrons as they move away from the source. The simplest and most widely used mathematical description of the neutron distribution in nuclear reactors is provided by neutron diffusion theory. For simplicity of explication, the neutrons are treated as if they are all of one effective speed, and effects associated with changes in neutron energy are suppressed. Such a simplification would be justified in practice if the cross sections were averaged over the appropriate neutron energy distribution. As a further simplification, the medium is initially assumed to be uniform.

Partial and Net Currents

With respect to Fig. 3.1, the rate at which neutrons are scattering in the differential volume element $d\mathbf{r} = r^2 dr\, d\mu\, d\psi$ is $\Sigma_s \phi\, d\mathbf{r}$, where $\mu \equiv \cos\theta$, the macroscopic

Nuclear Reactor Physics. Weston M. Stacey
Copyright © 2007 WILEY-VCH Verlag GmbH & Co. KGaA, Weinheim
ISBN: 978-3-527-40679-1

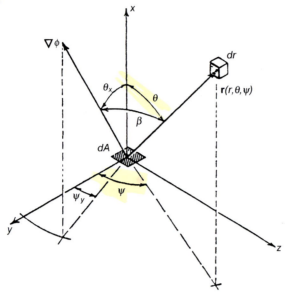

Fig. 3.1 Definition of coordinate system. (From Ref. 10; used with permission of McGraw-Hill.)

cross section $\Sigma \equiv N\sigma$ is the product of the number density of atomic nuclei and the microscopic cross section discussed previously, and the neutron scalar flux $\phi \equiv n\upsilon$ is the product of the neutron density and the neutron speed. The fraction of the isotropically scattered neutrons leaving $d\mathbf{r}$ headed toward the differential area dA at the origin is $-(\mathbf{r}/r) \cdot d\mathbf{A}/4\pi r^2 = \mu dA/4\pi r^2$. Not all of these neutrons reach dA, however; some are absorbed and others are scattered again so that they do not cross dA. The probability that a neutron leaving $d\mathbf{r}$ in the direction of dA actually reaches dA is $e^{-\Sigma r}$. The differential current $j_-(0 : r, \mu, \psi)d\mathbf{r}\, dA$ of neutrons passing downward through dA which had their last scattering collision in $d\mathbf{r}$ is thus

$$j_-(0 : r, \mu, \psi)dr\, dA = \frac{\mu e^{-\Sigma r}\Sigma_s \phi(r, \mu, \psi)d\mathbf{r}\, dA}{4\pi r^2} \tag{3.1}$$

The total current passing downward through dA is found by integrating this expression over the entire upper half-plane ($x > 0$):

$$j_-(0)dA = \Sigma_s \int_0^\infty dr \int_0^{2\pi} d\psi \int_0^1 d\mu\, \mu e^{-\Sigma_s r}\phi(r, \mu, \psi)\frac{dA}{4\pi} \tag{3.2}$$

Now, the first major approximation leading to diffusion theory is made—for the purpose of evaluating the integral in Eq. (3.2), the flux is assumed to be sufficiently slowly varying in space that it can be approximated by expansion in a Taylor series about the origin:

$$\phi(\mathbf{r}) = \phi(0) + \mathbf{r} \cdot \nabla\phi(0) + \frac{1}{2}[r^2\nabla^2\phi(0)] + \cdots \tag{3.3}$$

in which only the first two terms are retained. Using this approximation and the trigonometric identity $\cos\beta = \cos\theta_x \cos\theta + \sin\theta_x \sin\theta\cos(\psi_y - \psi)$, and making the second major approximation—that absorption is small relative to scattering (e.g., $\Sigma \sim \Sigma_s$)—Eq. (3.2) can be integrated to obtain the diffusion theory expression for the partial downward current density:

$$j_-(0) = \frac{1}{4}\phi(0) + \frac{1}{6\Sigma_s}|\nabla\phi(0)|\cos\theta_x = \frac{1}{4}\phi(0) + \frac{1}{6\Sigma_s}\frac{d\phi(0)}{dx}$$

$$\equiv \frac{1}{4}\phi(0) + \frac{1}{2}D\frac{d\phi(0)}{dx} \tag{3.4}$$

A similar derivation leads to an expression for the partial upward current density,

$$j_+(0) = \frac{1}{4}\phi(0) - \frac{1}{2}D\frac{d\phi(0)}{dx} \tag{3.5}$$

where D is known as the *diffusion coefficient*.

The diffusion theory expression for the net current at the origin (positive sign up) is

$$J_x(0) = j_+(0) - j_-(0) = \frac{1}{3\Sigma_s}\frac{d\phi(0)}{dx} = -D\frac{d\phi(0)}{dx} \tag{3.6}$$

Carrying out a similar derivation for dA in the x–y and x–z planes leads immediately to the three-dimensional generalization

$$\mathbf{J}(0) = -\frac{1}{3\Sigma_s}\nabla\phi(0) \equiv -D\nabla\phi(0) \tag{3.7}$$

A third assumption—that the neutrons are scattered isotropically—was used in the derivation above. The last form of Eq. (3.7) is known as *Fick's law*, which governs the diffusion of many other quantities as well as neutrons. A more accurate derivation of diffusion theory from transport theory (Section 3.12) reveals that a better approximation for the diffusion coefficient which takes into account anisotropy in scattering is given by

$$D = \frac{1}{3(\Sigma_t - \bar{\mu}_0\Sigma_s)} \equiv \frac{1}{3\Sigma_{tr}} \tag{3.8}$$

where Σ_t and Σ_s are the total and scattering cross sections and $\bar{\mu}_0 \approx \frac{2}{3A}$ is the average cosine of the scattering angle (A is the atomic mass number of the scattering nuclei).

Diffusion Theory

The mathematical formulation of neutron diffusion theory is then obtained by using the diffusion theory expression for the neutron current in the neutron balance equation on a differential volume element:

$$\frac{\partial n}{\partial t} = S + v\Sigma_f\phi - \Sigma_a\phi - \nabla\cdot\mathbf{J}$$

$$= S + v\Sigma_f\phi - \Sigma_a\phi + \nabla\cdot D\nabla\phi \tag{3.9}$$

which states that the time rate of change of the neutron density within a differential volume is equal to the external rate at which neutrons are produced in the volume by an external source (S) and by fission ($v\Sigma_f\phi$) minus the rate at which neutrons are lost by absorption ($\Sigma_a\phi$) and minus the net leakage of neutrons out of the volume ($\nabla\cdot\mathbf{J}$). Proof that the net leakage out of a differential volume element is $\nabla\cdot\mathbf{J}$ follows from considering the difference of outward and inward currents in a cube of dimensions $\Delta_x\Delta_y\Delta_z$. The net transport of particles out of the cube is

$$[J_x(x+\Delta x) - J_x(x)]\Delta_y\Delta_z + [J_y(y+\Delta_y) - J_y(y)]\Delta_x\Delta_z + [J_z(z+\Delta_z)$$

$$- J_z(z)]\Delta_x\Delta_y \simeq \left(\frac{\partial J_x}{\partial x}\Delta_x\right)\Delta_y\Delta_z + \left(\frac{\partial J_y}{\partial y}\Delta_y\right)\Delta_x\Delta_z + \left(\frac{\partial J_z}{\partial z}\Delta_y\right)\Delta_x\Delta_y$$

$$\equiv \nabla\cdot\vec{J}\Delta_x\Delta_y\Delta_z$$

where a Taylor's series expansion of the current has been made.

Interface Conditions

At an interface between regions 1 and 2 at which there is an isotropic source S_0, the partial currents on both side of the interface must be related by

$$j_+^{(2)}(0) = \tfrac{1}{2}S_0 + j_+^{(1)}(0)$$

$$j_-^{(1)}(0) = \tfrac{1}{2}S_0 + j_-^{(2)}(0) \tag{3.10}$$

Subtracting these two equations and using Eqs. (3.4) and (3.5) yields an interface condition of continuity of neutron flux:

$$\phi_2(0) = \phi_1(0) \tag{3.11}$$

Adding the two equations yields

$$J_2(0) = J_1(0) + S_0 \tag{3.12}$$

which, in the absence of an interface source, is a continuity of neutron net current condition.

Boundary Conditions

At an external boundary, the appropriate boundary condition is found by equating the expression for the inward partial current to the known incident current, j^{in}, for example, from the right at x_b,

$$j_-^{in} = \frac{1}{4}\phi(x_b) + \frac{1}{2}D\frac{d\phi(x_b)}{dx} \tag{3.13}$$

When the diffusing medium is surrounded by a vacuum or nonreflecting region, $j^{in} = 0$ and Eq. (3.13) may be written

$$\frac{1}{\phi}\frac{d\phi}{dx}\bigg|_{x_b} = -\frac{1}{2D} = -\frac{3\Sigma_{tr}}{2} \tag{3.14}$$

A widely used but more approximate vacuum boundary condition is obtained by noting that this expression relates the flux and the flux slope at the boundary. If the slope of the flux versus x at the boundary (x_b) is used to extrapolate the flux outside the boundary, the extrapolated flux will vanish at a distance $\lambda_{extrap} = \frac{2}{3}\lambda_{tr} = \frac{2}{3}\Sigma_{tr}^{-1}$ outside the external boundary. A more accurate result from neutron transport theory is $\lambda_{extrap} = 0.7104\lambda_{tr}$. This result gives rise to the approximate vacuum boundary condition of zero neutron flux at a distance λ_{extrap} outside the physical boundary at $x = a$, or $\phi(a + \lambda_{extrap}) \equiv \phi(a_{ex}) = 0$, where we have defined the extrapolated boundary

$$a_{ex} \equiv a + \lambda_{extrap} \tag{3.15}$$

Since λ_{extrap} is usually very small compared to the typical dimensions of a diffusing medium encountered in reactor physics, it is common to use the even more approximate vacuum boundary condition of zero flux at the physical external boundary.

Example 3.1: Typical Values of Thermal Extrapolation Distance. The thermal neutron extrapolation distance $\lambda_{extrap} = 0.7104/\Sigma_{tr} = 0.7104/[\Sigma_a + (1 - \mu_0)\Sigma_s]$ for some typical diffusing media are 0.30 cm for H_2O, 1.79 cm for D_2O, 1.95 cm for C, and 6.34 cm for Na. The approximation that the neutron flux vanishes at the boundary of the diffusing medium is valid when the dimension L of the diffusing medium is much larger than the extrapolation distance, $L \gg \lambda_{extrap}$.

Applicability of Diffusion Theory

Diffusion theory provides a strictly valid mathematical description of the neutron flux when the assumptions made in its derivation—absorption much less likely than scattering, linear spatial variation of the neutron distribution, isotropic scattering—are satisfied. The first condition is satisfied for most of the moderating (e.g., water, graphite) and structural materials found in a nuclear reactor, but not for the fuel and control elements. The second condition is satisfied a few mean free paths away from the boundary of large (relative to the mean free path) homogeneous media with relatively uniform source distributions. The third condition is satisfied for scattering from heavy atomic mass nuclei. One might well ask at this point how diffusion theory can be used in reactor physics when a modern nuclear reactor consists of thousands of small elements, many of them highly absorbing, with dimensions on the order of a few mean free paths or less. Yet diffusion theory is widely used in nuclear reactor analysis and makes accurate predictions. The secret is that a more accurate transport theory is used to "make diffusion theory work" where it would be expected to fail. The many small elements in a large region are replaced by a homogenized mixture with effective averaged cross sections and diffusion coefficients, thus creating a computational model for which diffusion theory is valid. Highly absorbing control elements are represented by effective diffusion theory cross sections which reproduce transport theory absorption rates.

3.2
Solutions of the Neutron Diffusion Equation in Nonmultiplying Media

Plane Isotropic Source in an Infinite Homogeneous Medium

Consider an infinite homogeneous nonmultiplying ($\Sigma_f = 0$) medium in which a plane isotropic source (infinite in the y–z plane) with strength S_0 is located at $x = 0$. Everywhere except at $x = 0$ the time-independent diffusion equation can be written

$$\frac{d^2\phi(x)}{dx^2} - \frac{1}{L^2}\phi(x) = 0 \tag{3.16}$$

where $L^2 \equiv D/\Sigma_a$ is the *neutron diffusion length* squared. This equation has a general solution $\phi = A\exp(x/L) + B\exp(-x/L)$. For $x > 0$, the physical requirement for a finite solution at large x requires that $A = 0$, and the physical requirement that the net current must approach $\frac{1}{2}S_0$ as x approaches 0 requires that $B = LS_0/2D$. Following a similar procedure for $x < 0$ leads to similar results, so that the solution may be written

$$\phi(x) = \frac{S_0 L e^{-|x|/L}}{2D} \tag{3.17}$$

Plane Isotropic Source in a Finite Homogeneous Medium

Consider next a finite slab medium extending from $x = 0$ to $x = +a$ with an isotropic plane source at $x = 0$. In this case, the general solution of Eq. (3.16) is more conveniently written as $\phi = A\sinh(x/L) + B\cosh(x/L)$. The appropriate boundary conditions are that the inward partial current vanishes at $x = a$ [i.e., $j^-(a) = 0$] and that the outward partial current equals $\frac{1}{2}$ the isotropic source strength as $x \to 0$ [i.e., $j^+(0) = \frac{1}{2}S_0$]. The resulting solution is

$$\phi(x) = 4S_0 \frac{\sinh[(a-x)/L] + (2D/L)\cosh[(a-x)/L]}{[2(D/L)+1]^2 e^{a/L} - [2(D/L)-1]^2 e^{-a/L}} \tag{3.18}$$

If instead of $j^-(a) = 0$, the extrapolated boundary condition $\phi(a_{\text{ex}}) = 0$ is used, the resulting solution is

$$\phi(x) = 4S_0 \frac{\sinh[(a_{\text{ex}}-x)/L]}{\sinh(a_{\text{ex}}/L) + (2D/L)\cosh(a_{\text{ex}}/L)} \tag{3.19}$$

When $0.71\lambda_{tr}/a \ll 1$ and $2(\Sigma_a/3\Sigma_{tr})^{1/2} \ll 1$ (i.e., when the transport mean free path is small compared to the dimension of the medium and the absorption cross section is small relative to the scattering cross section), these two solutions agree. These conditions must also be satisfied in order for diffusion theory to be valid, so we conclude that use of the extrapolated zero flux boundary condition instead of the zero inward current boundary condition is acceptable.

Line Source in an Infinite Homogeneous Medium

Consider an isotropic line source (e.g., infinite along the z-axis) of strength S_0 (per centimeter per second) located at $r = 0$. The general solution of

$$\frac{1}{r}\frac{d}{dr}\left[r\frac{d\phi(r)}{dr}\right] - \frac{1}{L^2}\phi(r) = 0 \tag{3.20}$$

is $\phi = AI_0(r/L) + BK_0(r/L)$, where I_0 and K_0 are the modified Bessel functions of order zero of the first and second kind, respectively. The physical requirement for a finite solution at large r requires that $A = 0$. The source condition is $\lim_{r\to 0} 2\pi r J = S_0$. The resulting solution for an isotropic line source in an infinite homogeneous nonmultiplying medium is

$$\phi(r) = \frac{S_0 K_0(r/L)}{2\pi D} \tag{3.21}$$

Homogeneous Cylinder of Infinite Axial Extent with Axial Line Source

Consider an infinitely long cylinder of radius a with an isotropic source on axis. The source condition $\lim_{r\to 0} 2\pi r J = S_0$ still obtains, but now $A = 0$ no longer holds and the other boundary condition is a zero incident current condition at $r = a$ or a zero flux condition at $r = a + \lambda_{extrap}$. The latter vacuum boundary condition leads to the solution for the neutron flux distribution in an infinite homogeneous nonmultiplying cylinder with an isotropic axial line source:

$$\phi(r) = \frac{S_0[I_0(a_{ex}/L)K_0(r/L) - K_0(a_{ex}/L)I_0(r/L)]}{2\pi D I_0(a_{ex}/L)} \tag{3.22}$$

Point Source in an Infinite Homogeneous Medium

The neutron diffusion equation in spherical coordinates is

$$\frac{1}{r^2}\frac{d}{dr}\left[r^2\frac{d\phi(r)}{dr}\right] - \frac{1}{L^2}\phi(r) = 0 \tag{3.23}$$

This equation has the general solution $\phi = (Ae^{r/L} + Be^{-r/L})/r$. The source condition is $\lim_{r\to 0} 4\pi r^2 J = S_0$, and the physical requirement for a finite solution at large r requires that $A = 0$, yielding

$$\phi(r) = \frac{S_0 e^{-r/L}}{4\pi r D} \tag{3.24}$$

Point Source at the Center of a Finite Homogeneous Sphere

Consider a finite sphere of radius a with a point source at the center. The same general solution $\phi = (Ae^{r/L} + Be^{-r/L})/r$ of Eq. (3.23) is applicable, but the $A = 0$ condition must be replaced by a vacuum boundary condition at $r = a$. Using an extrapolated zero flux condition yields

$$\phi(r) = \frac{S_0 \sinh[(a_{ex} - r)/L]}{4\pi r D \sinh(a_{ex}/L)} \tag{3.25}$$

for the neutron distribution in a finite sphere of homogeneous nonmultiplying material with a point source at the center.

3.3
Diffusion Kernels and Distributed Sources in a Homogeneous Medium

Infinite-Medium Diffusion Kernels

The previous solutions for plane, line, and point sources at the origin of slab, cylindrical, and spherical coordinate systems in an infinite medium can be generalized immediately to slab, line, and point sources located away from the origin (i.e., the location of the coordinate axis in an infinite medium can be offset without changing the functional form of the result). The resulting solutions for the neutron flux at location x or r due to a unit isotropic source at x' and r' may be thought of as kernels. The infinite-medium kernels for a plane isotropic source of one neutron per unit area per second, a point isotropic source of one neutron per second, a line isotropic source of one neutron per unit length, a cylindrical shell source of one neutron per shell per unit length per second, and a spherical shell source of one neutron per shell per second are

Plane: $\phi_{pl}(x : x') = \dfrac{L}{2D} e^{-|x-x'|/L}$

Line: $\phi_l(r : r') = \dfrac{K_0(|r - r'|/L)}{2\pi D}$

Point: $\phi_{pt}(\mathbf{r} : \mathbf{r}') = \dfrac{e^{-|\mathbf{r}-\mathbf{r}'|/L}}{4\pi |\mathbf{r} - \mathbf{r}'| D}$ (3.26)

Cylindrical shell: $\phi_{cyl}(r : r') = \dfrac{1}{2\pi D} \times \begin{cases} K_0(r/L) I_0(r'/L), & r > r', \\ K_0(r'/L) I_0(r/L), & r < r' \end{cases}$

Spherical shell: $\phi_{sph}(r : r') = \dfrac{L}{8\pi r r' D} (e^{-|r-r'|/L} - e^{-|r+r'|/L})$

These kernels may be used to construct the neutron flux in an infinite homogeneous nonmultiplying medium due to an arbitrary source distribution S_0:

$$\phi(\mathbf{r}) = \int \phi(\mathbf{r} : \mathbf{r}') S_0(\mathbf{r}') d\mathbf{r}' \tag{3.27}$$

For a planar source distribution this takes the form

$$\phi(x) = \int_{-\infty}^{\infty} \frac{S_0(x')L}{2D} e^{-|x-x'|/L} dx' \tag{3.28}$$

and for the more general point source,

$$\phi(\mathbf{r}) = \int_0^{\infty} S_0(\mathbf{r}') \frac{e^{-|\mathbf{r}-\mathbf{r}'|/L}}{4\pi |\mathbf{r} - \mathbf{r}'| D} d\mathbf{r}' \tag{3.29}$$

Finite-Slab Diffusion Kernel

Consider a slab infinite in the y- and z-directions extending from $x = -a$ to $x = +a$ with a unit isotropic source at x'. The neutron diffusion equation

$$\frac{d^2\phi(x)}{dx^2} - \frac{1}{L^2}\phi(x) = 0 \tag{3.30}$$

holds everywhere in $-a < x < +a$ except at $x = x'$, the source plane. The interface conditions at the source plane, $x = x'$, are, from Eqs. (3.11) and (3.12),

$$\phi(x' + \varepsilon) = \phi(x' - \varepsilon)$$

$$J(x' + \varepsilon) = J(x' - \varepsilon) + 1 \tag{3.31}$$

where $x' + \varepsilon$ indicates an infinitesimal distance to the right of x', and so on. For the vacuum boundary conditions at $x = -a$ and $x = a$ we use the approximate zero flux conditions

$$\phi(-a) = \phi(a) = 0 \tag{3.32}$$

Solving Eq. (3.30) as before and using these source and boundary conditions yields the following expressions for the flux at x due to a unit isotropic source at x', or the finite-slab diffusion kernel:

$$\phi_+(x:x') = \frac{\sinh[(a+x')/L]\sinh[(a-x)/L]}{(D/L)\sinh(2a/L)}, \quad x > x'$$

$$\phi_-(x:x') = \frac{\sinh[(a-x')/L]\sinh[(a+x)/L]}{(D/L)\sinh(2a/L)}, \quad x < x' \tag{3.33}$$

These kernels may be used to calculate the neutron flux distribution in the slab due to a distributed source, $S_0(x')$:

$$\phi(x) = \int_{-a}^{x} S_0(x')\phi_+(x:x')dx' + \int_{x}^{a} S_0(x')\phi_-(x:x')dx' \tag{3.34}$$

Finite Slab with Incident Neutron Beam

As a further relevant example, consider the first-collision source distribution in a slab due to a beam incident from the left at $x = -a$:

$$S_0(x) = q_0 \Sigma_s e^{-\Sigma_t(x+a)} \tag{3.35}$$

Using this source in Eq. (3.34) yields the neutron flux distribution within the slab:

$$\phi(x) = \frac{q_0 \Sigma_s e^{-\Sigma_t a}}{D(\Sigma_t^2 - (1/L^2)) \sinh(2a/L)}$$

$$\times \left[e^{\Sigma_t a} \sinh\left(\frac{a-x}{L}\right) + e^{-\Sigma_t a} \sinh\left(\frac{a+x}{L}\right) - e^{-\Sigma_t x} \sinh\left(\frac{2a}{L}\right) \right] \tag{3.36}$$

By using a first-collision source, the highly anisotropic incident beam neutrons are treated by first-flight transport theory until they have had a scattering collision which (at least partially) converts the beam to a nearly isotropic neutron distribution which is amenable to treatment by diffusion theory. The solution for the nearly isotropic neutron distribution given by Eq. (3.36) has a maximum some distance into the slab at $0 > x > -a$.

3.4
Albedo Boundary Condition

reflection coefficient

for n entering slab from the left at $x=0$

Consider a slab that is infinite in the y- and z-directions located between $x = 0$ and $x = a$ with a known inward partial current $j^+(0) = j_{in}^+$. Upon solving for the neutron flux distribution for an extrapolated zero flux vacuum boundary condition $\phi(a + \lambda_{extrap}) \equiv \phi(a_{ex}) = 0$, it is possible to evaluate the reflection coefficient, or albedo, for neutrons entering the slab from the left at $x = 0$.

$$\alpha \equiv \frac{j_-(0)}{j_+(0)} = \frac{1 - (2D/L)\coth(a_{ex}/L)}{1 + (2D/L)\coth(a_{ex}/L)} \tag{3.37}$$

As a/L becomes large, $\coth[(a+\lambda_{extrap})/L] \to 1$, and $\alpha \to (1-2D/L)/(1+2D/L)$, the infinite-medium value.

Now consider two adjacent slabs, one denoted B and located in the range $-b \leq x \leq 0$ and the other denoted A and located in the range $0 \leq x \leq a$. If we are not interested in the neutron flux distribution in slab A but only in the effect of slab A on the neutron flux distribution in slab B, the albedo of slab A can be used as an albedo boundary condition for the neutron flux solution in slab B. From Eqs. (3.4) and (3.5),

$$\frac{1}{\phi_B} D_B \frac{d\phi_B}{dx}\bigg|_{x=0} = -\frac{j_+(0) - j_-(0)}{2[j_+(0) + j_-(0)]} = -\frac{1}{2}\left(\frac{1 - \alpha_A}{1 + \alpha_A}\right) \tag{3.38}$$

This albedo boundary condition can also be simplified by a geometric interpretation. If the flux in slab B at the interface between slabs A and B ($x = 0$) is extrapolated (into slab A) to zero using the slope at the interface given by Eq. (3.38), an approximate albedo boundary condition for the flux solution in slab B ($-b < x < 0$) becomes $\phi_B(\lambda_{\text{albedo}}) = 0$, where

$$\lambda_{\text{albedo}} \equiv 0.71 \lambda_{\text{tr}}^B \left(\frac{1 + \alpha_A}{1 - \alpha_A} \right) \tag{3.39}$$

3.5
Neutron Diffusion and Migration Lengths

The distribution of neutrons within a finite or infinite medium is determined by the source distribution, the geometry (in a finite medium), and the neutron diffusion length, $L = (D/\Sigma_a)^{1/2}$. The (thermal) diffusion length is related to the mean-squared distance that a thermal neutron travels from the source point to the point at which it is absorbed, as may be seen by computing the mean-squared distance to capture for (thermal) neutrons emitted by a point source in an infinite medium:

$$\bar{r}^2 \equiv \frac{\int_0^\infty r^2 (4\pi r^2 \Sigma_a \phi) dr}{\int_0^\infty (4\pi r^2 \Sigma_a \phi) dr} = \frac{\int_0^\infty r^3 e^{-r/L} dr}{\int_0^\infty r e^{-r/L} dr} = 6L^2 \tag{3.40}$$

where Eq. (3.24) has been used for the neutron flux due to a point source at $r = 0$. It is also apparent from the $\exp(\pm x/L)$ nature of many of the solutions above that L is the physical distance over which the neutron flux can change by a significant amount (i.e., e^{-1}).

Thermal Diffusion-Length Experiment

The thermal neutron diffusion length can be determined experimentally by measuring the axial neutron flux distribution in a long (with respect to mean free path) block of material with an isotropic thermal neutron flux incident on one end (e.g., from the thermal column of a reactor). With reference to Fig. 3.2, consider a rectangular parallelepiped of length c and cross section $2a \times 2b$ with an incident isotropic thermal neutron source $S_0(x, y)$ at $z = 0$ which is symmetric in x and y about $x = 0$ and $y = 0$. The neutron flux in the block satisfies

$$\frac{\partial^2 \phi}{\partial x^2} + \frac{\partial^2 \phi}{\partial y^2} + \frac{\partial^2 \phi}{\partial z^2} - \frac{1}{L^2} = 0 \tag{3.41}$$

and the boundary conditions

$$j_+(x, y, 0) = \frac{1}{2} S_0(x, y) \tag{3.42a}$$

$$\phi(\pm a_{\text{ex}}, y, z) = 0 \tag{3.42b}$$

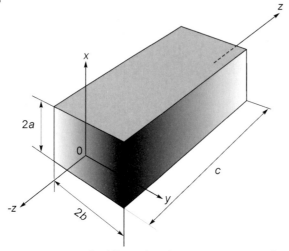

Fig. 3.2 Geometry for diffusion-length experiment. (From Ref. 10.)

$$\phi(x, \pm b_{ex}, z) = 0 \tag{3.42c}$$

$$\phi(x, y, c_{ex}) = 0 \tag{3.42d}$$

We seek a separable solution to Eq. (3.41) of the form $\phi(x, y, z) = X(x)Y(y)Z(z)$. Substitution of this form into Eq. (3.41) and division by XYZ yields

$$\frac{X''(x)}{X(x)} + \frac{Y''(y)}{Y(y)} + \frac{Z''(z)}{Z(z)} = \frac{1}{L^2} \tag{3.43}$$

where the double prime indicates a second derivative with respect to the respective spatial variables. In general, this equation can only be satisfied if each of the terms on the left is separately equal to a constant:

$$\frac{X''(x)}{X(x)} = -k_1^2, \qquad \frac{Y'(y)}{Y(y)} = -k_2^2, \qquad \frac{Z''(z)}{Z(z)} = k_3^2 \tag{3.44}$$

in which case Eq. (3.43) becomes

$$k_3^2 = \frac{1}{L^2} + k_1^2 + k_2^2 \tag{3.45}$$

The general solutions to Eqs. (3.44) are

$$X(x) = A_1 \sin k_1 x + C_1 \cos k_1 x$$

$$Y(y) = A_2 \sin k_2 y + C_2 \cos k_2 y \tag{3.46}$$

$$Z(z) = A_3 e^{-k_3 z} + C_3 e^{k_3 z}$$

The x–y symmetry requirement determines that $A_1 = A_2 = 0$. The end condition of Eq. (3.42d) may be used to eliminate C_3 to obtain

$$Z(z) = A_3 e^{-k_3 z} [1 - e^{-2k_3(c_{ex} - z)}] \tag{3.47}$$

The extrapolated boundary conditions of Eqs. (3.42b) and (3.42c) require that $\cos k_1 a_{\mathrm{ex}} = \cos k_2 b_{\mathrm{ex}} = 0$, which can only be satisfied if k_1 and k_2 have the discrete values

$$k_{1n} = \frac{\pi}{2a_{\mathrm{ex}}}(2n+1), \quad n = 0, 1, \ldots$$

$$k_{2m} = \frac{\pi}{2b_{\mathrm{ex}}}(2m+1), \quad m = 0, 1, \ldots \tag{3.48}$$

This result, together with Eq. (3.45), requires that k_3 can only take on discrete values

$$k_3^2 \rightarrow k_{3nm}^2 = k_{1n}^2 + k_{2m}^2 + \frac{1}{L^2} \tag{3.49}$$

Thus the most general solution of the neutron diffusion equation that satisfies the extrapolated boundary conditions of Eqs. (3.42b) to (3.42d) is

$$\phi(x, y, z) = \sum_{n,m=0}^{\infty} A_{mn} \cos k_{1n}x \cos k_{2m}y \, e^{-k_{3nm}z}[1 - e^{-2k_{3nm}(c_{\mathrm{ex}}-z)}] \tag{3.50}$$

where A_{mn} is a constant that can be determined from Eq. (3.42a), but that is not necessary for our purposes.

Noting that k_{3nm} increases with m and n, the asymptotic form of the neutron flux distribution along the z-axis that persists at large distances from $z = 0$ is

$$\phi(0, 0, z) \simeq A_{00} e^{-k_{300}z}[1 - e^{-2k_{300}(c_{\mathrm{ex}}-z)}] \tag{3.51}$$

For very long blocks (large c_{ex}), the term in brackets is unimportant except near the end, and the flux decreases exponentially, so that a measurement of the axial flux distribution far away from both the source at $z = 0$ and the end at $z = c_{\mathrm{ex}}$ should provide for experimental determination of k_{300}. The diffusion length then is determined from

$$\frac{l}{L^2} = k_{300}^2 - \left(\frac{\pi}{2a_{\mathrm{ex}}}\right)^2 - \left(\frac{\pi}{2b_{\mathrm{ex}}}\right)^2 \tag{3.52}$$

The measured diffusion lengths L for thermal neutrons in H_2O, D_2O, and graphite are about 2.9, 170, and 60 cm, respectively. The implication of these measurements is that thermal neutrons would diffuse a root-mean-square distance from the point at which they appear (are thermalized) to the point at which they are absorbed of 7.1, 416, and 147 cm, respectively, in these three moderators.

Migration Length

In a water- or graphite-moderated reactor, the fission neutrons are born fast (average energy about 1.0 MeV) and diffuse as fast neutrons while they are in the process of slowing down to become thermal neutrons. In fast reactors, the neutrons are absorbed before thermalizing. In a later chapter we return to calculation

Table 3.1 Diffusion Parameters for Common Moderators

Moderator	Density (g/cm^3)	D (cm)	Σ_a (cm^{-1})	L (cm)	$\tau_{th}^{1/2}$ (cm)	M (cm)
H_2O	1.00	0.16	2.0×10^{-2}	2.9	5.1	5.8
D_2O	1.10	0.87	2.9×10^{-5}	170	11.4	170
Graphite	1.60	0.84	2.4×10^{-4}	59	19	62

Source: Data from Ref. 4; used with permission of Wiley.

of the diffusion of these fast neutrons, but for now we simply indicate that there is an equivalent for fast neutrons of the thermal diffusion length, which for historical reasons is identified as the square root of the "age to thermal," τ_{th}. For intermediate to heavy mass moderators, this quantity can be shown to be equal to one-sixth the mean-squared distance a fast neutron diffuses before it thermalizes (for hydrogenous moderators, this is the definition of the quantity).

The mean-squared distance that a neutron travels from birth as a fast fission neutron until capture as a thermal neutron is given by

$$\bar{r}^2 = 6(\tau_{th} + L^2) \equiv 6M^2 \tag{3.53}$$

where $M = (\tau_{th} + L^2)^{1/2}$ is known as the *migration length*.

Example 3.2: Characteristic Diffusion Parameters. Diffusion characteristics for some common moderators are given in Table 3.1. The values of D, Σ_a, and L are for thermal neutrons. Diffusion characteristics for compositions representative of pressurized water (H_2O) reactors (PWRs), boiling water (H_2O) reactors (BWRs), high-temperature graphite thermal reactors (HTGRs), sodium-cooled fast reactors (LMFRs), and gas-cooled fast reactors (GCFRs) are given in Table 3.2. Typical core diameters, measured in thermal diffusion lengths and in migration lengths for the thermal reactors and measured in fast diffusion lengths for the fast reactors, are also given. It is clear from these numbers that most of the diffusion displacement undergone by a fission neutron occurs during the slowing-down process.

Table 3.2 Diffusion Parameters for Representative Reactor Core Types

Reactor	L (cm)	$\tau_{th}^{1/2}$ (cm)	M (cm)	Diameter (L)	Diameter (M)
PWR	1.8	6.3	6.6	190	56
BWR	2.2	7.1	7.3	180	50
HTGR	12	17	21	63	40
LMFR	5.0*		5.0	35	35
GCFR	6.6*		6.6	35	35

Source: Data from Ref. 4; used with permission of Wiley.
* Fast neutron diffusion length.

3.6
Bare Homogeneous Reactor

In a fission chain reacting medium (i.e., a medium in which neutron absorption can lead to fission and the production of more neutrons), the diffusion equation may or may not have an equilibrium steady-state solution, depending on the precise amount of multiplication. Thus we must consider the time-dependent diffusion equation

$$\frac{1}{v}\frac{\partial\phi(\mathbf{r},t)}{\partial t} - D\nabla^2\phi(\mathbf{r},t) + \Sigma_a\phi(\mathbf{r},t) = v\Sigma_f\phi(\mathbf{r},t) \tag{3.54}$$

In a finite homogeneous medium (i.e., a bare reactor) the appropriate boundary condition is the extrapolated zero flux condition

$$\phi(\mathbf{a}_{ex},t) = 0 \tag{3.55}$$

where $\mathbf{a}_{ex}$ denotes the external boundaries. We further specify an initial condition

$$\phi(\mathbf{r},0) = \phi_0(\mathbf{r}) \tag{3.56}$$

where ϕ_0 denotes the initial spatial flux distribution at $t = 0$.

We use the separation-of-variables technique and look for a solution of the form

$$\phi(\mathbf{r},t) = \psi(r)T(t) \tag{3.57}$$

Substituting Eq. (3.57) into Eq. (3.54) and dividing by $\phi = \psi T$ yields

$$\frac{v}{\psi}[D\nabla^2\psi + (v\Sigma_f - \Sigma_a)\psi] = \frac{1}{T}\frac{\partial T}{\partial t} = -\lambda \tag{3.58}$$

where we have indicated that an expression which depends only on the spatial variable and an expression which depends only on the time variable can be equal at all spatial locations and times only if both expressions are equal to the same constant, $-\lambda$. The second form of Eq. (3.58) has the solution

$$T(t) = T(0)e^{-\lambda t} \tag{3.59}$$

We look for spatial solutions ψ that satisfy

$$\nabla^2\psi(\mathbf{r}) = -B_g^2\psi(\mathbf{r}) \tag{3.60}$$

and the extrapolated spatial boundary conditions of Eq. (3.55). The constant B_g, known as *geometric buckling*, depends only on the geometry.

Slab Reactor

For example, in a slab reactor extending from $x = -a/2$ to $x = +a/2$ and infinite in the y- and z-directions, Eqs. (3.60) and (3.55) become

$$\frac{d^2\psi}{dx^2}(x) = -B_g^2\psi(x), \qquad \psi\left(\frac{a_{ex}}{2}\right) = \psi\left(\frac{-a_{ex}}{2}\right) = 0 \tag{3.61}$$

which have solutions $\psi = \psi_n$ only for the (infinite) set of discrete spatial eigenvalues of $B_g = B_n$:

$$\psi_n(x) = \cos B_n x, \quad B_n^2 = \left(\frac{n\pi}{a_{\text{ex}}}\right)^2, \quad n = 1, 3, 5, \ldots \tag{3.62}$$

Using this result in Eq. (3.58) implies that solutions of that equation exist only for discrete-time eigenvalues λ_n given by

$$\lambda_n = v\left(\Sigma_a + DB_n^2 - v\Sigma_f\right) \tag{3.63}$$

Thus the solution of Eq. (3.54) for a slab reactor is

$$\phi(x, t) = \sum_{n=\text{odd}} A_n T_n(t) \cos \frac{n\pi x}{a_{\text{ex}}} \tag{3.64}$$

where T_n is given by Eq. (3.59) with $\lambda = \lambda_n$ and A_n is a constant which may be determined from the initial condition of Eq. (3.56) and orthogonality:

$$A_n(x) = \frac{2}{a_{\text{ex}}} \int_{-a_{\text{ex}}/2}^{a_{\text{ex}}/2} dx \, \phi_0(x) \cos \frac{n\pi x}{a_{\text{ex}}} \tag{3.65}$$

Since $B_1^2 < B_3^2 < \cdots < B_n^2 = (n\pi/a_{\text{ex}})^2$, the time eigenvalues are ordered $\lambda_1 < \lambda_3 < \cdots < \lambda_n = v(\Sigma_a + DB_n^2 - v\Sigma_f)$. Thus, after a sufficiently long time ($t \gg 1/\lambda_3$), the solution becomes

$$\phi(x, t) \to A_1 e^{-\lambda_1 t} \cos B_1 x = A_1 e^{-\lambda_1 t} \cos \frac{\pi x}{a_{\text{ex}}} \tag{3.66}$$

This result implies that, independent of the initial distribution (as long as $A_1 \neq 0$), the asymptotic shape will be the fundamental mode solution corresponding to the smallest spatial and time eigenvalues. The asymptotic solution is steady-state only if $\lambda_1 = 0$. If $\lambda_1 > 0$, the asymptotic solution is decaying in time, and if $\lambda_1 < 0$, it is increasing in time. When the neutron population is sustained precisely in steady-state by the fission chain reaction, the reactor is said to be *critical*; when the neutron population is increasing in time, the reactor is said to be *supercritical*; and when the neutron population is dying away in time, the reactor is said to be *subcritical*. Defining the *material* buckling, B_m,

$$B_m^2 \equiv \frac{v\Sigma_f - \Sigma_a}{D} = \frac{v\Sigma_f/\Sigma_a - 1}{L^2} \tag{3.67}$$

the *criticality condition* for a bare homogeneous reactor may be written:

Supercritical: $\quad \lambda_1 < 0, \quad B_m^2 > B_1^2$

Critical: $\qquad\; \lambda_1 = 0, \quad B_m^2 = B_1^2 \tag{3.68}$

Subcritical: $\quad\; \lambda_1 > 0, \quad B_m^2 < B_1^2$

Right Circular Cylinder Reactor

The slab reactor results can be extended immediately to more general geometries by replacing Eqs. (3.61) and (3.62) with the corresponding equations for the other geometries. For example, for the more realistic core geometry of a right circular cylinder of radius a and height H, the equation corresponding to Eq. (3.60) is

$$\frac{1}{r}\frac{\partial}{\partial r}\left[r\frac{\partial \psi(r,z)}{\partial r}\right] + \frac{\partial^2 \psi(r,z)}{\partial z^2} = -B_g^2 \psi(r,z) \tag{3.69}$$

and the extrapolated boundary conditions are

$$\psi(a_{ex},z) = \psi\left(r,\pm\frac{H_{ex}}{2}\right) = 0 \tag{3.70}$$

We make further use of the separation-of-variables technique to write

$$\psi(r,z) = R(r)Z(z) \tag{3.71}$$

Substituting Eq. (3.71) into Eq. (3.69) and dividing by RZ yields

$$\frac{1}{R(r)}\frac{1}{r}\frac{\partial}{\partial r}\left[r\frac{\partial R(r)}{\partial r}\right] + \frac{1}{Z(z)}\frac{\partial^2 Z(z)}{\partial z^2} = -\nu^2 - \kappa^2 = -B_g^2 \tag{3.72}$$

where the second form of the equation indicates that the only way in which the sum of an expression which depends only on the r-variable plus an expression which depends only on the z-variable can everywhere equal a constant is if the two expressions separately are equal to constants. Solutions of these two equations— the first expression equal to the first constant and the second expression equal to the second constant—which satisfy the corresponding boundary condition of Eqs. (3.70), exist only for discrete values of the constants ν_m (the roots of $J_0(\nu_m) = 0$, $m = 1, 2, \ldots$) and κ_n ($\kappa_n = n\pi/H_{ex}$, $n = 1, 3, \ldots$). Since the roots of J_0 are ordered, $\nu_1 < \nu_2 < \cdots < \nu_n$, the smallest of the corresponding discrete eigenvalues $B_{mn}^2 = (\frac{\nu_m}{R_{ex}})^2 + (n\pi/H_{ex})^2$ is $B_{11}^2 = (\frac{\nu_1}{R_{ex}})^2 + (\pi/H_{ex})^2$, and the smallest time eigenvalue is

$$\lambda_1 = v\left\{\Sigma_a + D\left[\left(\frac{\nu_1}{R_{ex}}\right)^2 + \left(\frac{\pi}{H_{ex}}\right)^2\right] - \nu\Sigma_f\right\} = v\left(B_{11}^2 - B_m^2\right)$$

$$= v\left(B_g^2 - B_m^2\right) \tag{3.73}$$

The corresponding asymptotic solution is

$$\phi(r,z,t) \rightarrow A_{11}J_0\left(\frac{\nu_1 r}{a_{ex}}\right)\cos\left(\frac{\pi z}{H_{ex}}\right)e^{-\lambda_1 t} \tag{3.74}$$

The criticality condition, $\lambda_1 = 0$, corresponds to $B_m^2 = B_g^2 = B_{11}^2$. The first zero-crossing for the Bessel function $J_0(\nu)$ occurs at $\nu = \nu_1 = 2.405$.

The geometric bucklings and asymptotic flux solutions are given for the common geometries in Table 3.3.

Table 3.3 Geometric Bucklings and Critical Flux Profiles Characterizing Some Common Core Geometries

Geometry		Geometric Buckling B_g^2	Flux Profile
Slab		$\left(\frac{\pi}{a_{ex}}\right)^2$	$\cos\left(\frac{\pi x}{a_{ex}}\right)$
Infinite cylinder		$\left(\frac{\nu_\perp}{R_{ex}}\right)^2$	$J_0\left(\frac{\nu_\perp r}{R_{ex}}\right)$
Sphere		$\left(\frac{\pi}{R_{ex}}\right)^2$	$r^{-1}\sin\left(\frac{\pi r}{R_{ex}}\right)$
Rectangular parallelepiped		$\left(\frac{\pi}{a_{ex}}\right)^2+\left(\frac{\pi}{b_{ex}}\right)^2+\left(\frac{\pi}{c_{ex}}\right)^2$	$\cos\left(\frac{\pi x}{a_{ex}}\right)\cos\left(\frac{\pi y}{b_{ex}}\right)\cos\left(\frac{\pi z}{c_{ex}}\right)$
Finite cylinder		$\left(\frac{\nu_\perp}{R_{ex}}\right)^2+\left(\frac{\pi}{H_{ex}}\right)^2$	$J_0\left(\frac{\nu_\perp r}{R_{ex}}\right)\cos\left(\frac{\pi z}{H_{ex}}\right)$

Source: Adapted from Ref. 4; used with permission of Wiley.

Interpretation of Criticality Condition

The criticality condition $\lambda_1 = 0$, or $B_m^2 = B_g^2$, can be rearranged to yield

$$1 = \frac{\nu\Sigma_f/\Sigma_a}{1+L^2 B_g^2} \equiv \frac{k_\infty}{1+L^2 B_g^2} = k_\infty P_{NL} \tag{3.75}$$

where k_∞ is the infinite-medium multiplication constant and $P_{NL} = (1+L^2 B_g^2)^{-1}$ is interpreted as the nonleakage probability.

If $\lambda_1 \neq 0$, the reactor is not critical and the asymptotic solution will either grow indefinitely or decay away in time, because the multiplication of neutrons (the ratio of the neutron population in successive generations) is greater or less than, respectively, unity. Since Eq. (3.75) applies only when $k = 1$, we can more generally write

$$k = \frac{\nu\Sigma_f/\Sigma_a}{1+L^2 B_g^2} \equiv k_\infty P_{NL} \tag{3.76}$$

The situation $\lambda_1 < 0$, in which the asymptotic solution increases in time, corresponds to $k > 1$, and the situation $\lambda_1 > 0$, in which the asymptotic solution decays in time, corresponds to $k < 1$. From Eqs. (3.63) and (3.76),

$$\lambda_1 = v\left(B_1^2 - B_m^2\right)D = v\Sigma_a\left(1 + L^2 B_1^2\right)\left(1 - \frac{v\Sigma_f/\Sigma_a}{1 + L^2 B_1^2}\right) \tag{3.77}$$

Since the mean free path to absorption is $1/\Sigma_a$, the lifetime of a neutron that remains in the reactor until absorption is $1/v\Sigma_a$. Defining an effective lifetime of a neutron in the reactor which takes into account the possibility of leakage before absorption,

$$l = \frac{1}{v\Sigma_a(1 + L^2 B_1^2)} = \frac{P_{NL}}{v\Sigma_a} \tag{3.78}$$

enables Eq. (3.77) to be written

$$\lambda_1 = \frac{1 - k}{l} \qquad \text{lifetime of the neutron} \qquad \text{for } \lambda_1 = 0 \Rightarrow k = 1 \tag{3.79}$$

Thus the asymptotic solution of Eq. (3.54) that satisfies the extrapolated boundary conditions of Eq. (3.55) can be written

$$\phi_{asy}(\mathbf{r}, t) \to A_1 \psi_1(\mathbf{r}) e^{[(k-1)/l]t} \tag{3.80}$$

where ψ is the fundamental mode spatial distribution for the specific geometry given in Table 3.3.

Optimum Geometries

The minimum size for a bare reactor of a given composition that will be critical depends on the leakage, hence on the surface-to-volume ratio. The minimum critical volume for a rectangular parallelepiped bare reactor occurs for a cube and is $V \approx 161.11/B_m^3$. For a right circular cylinder, the minimum critical volume bare reactor occurs for a radius $a = 2^{1/2} \times 2.405 H/\pi \approx 1.08 H$ and is $V \approx 148.31/B_m^3$. The minimum critical volume for a spherical bare reactor is $129.88/B_m^3$.

It is generally desirable for the neutron flux to be distributed as uniformly as possible over the reactor core. A measure of non-uniformity is the peak-to-volume average value. For a homogeneous bare core, the peak value occurs at the center, and the peak-to-volume average is $(\pi/2)^2 = 3.88$ for a rectangular parallelepiped, $-2.405\pi v_1/4 J_1(v_1) = 3.65$ for a right circular cylinder, and $\pi^2/3 = 3.29$ for a sphere.

Example 3.3: Critical Size of a Bare Cylindrical Reactor. Although the above formalism has been developed for a one-speed description of neutron diffusion, it can be generalized to energy-dependent diffusion by using cross sections that are averaged over the neutron energy distribution. A typical composition and set of

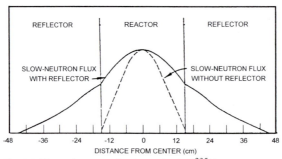

Fig. 3.3 Thermal neutron flux in a spherical ^{235}U water-moderated reactor with and without a beryllium oxide reflector. (From Ref. 11; used with permission of University of Chicago Press.)

spectrum-averaged cross sections for a PWR are given in Table 3.4. From the table a number of important materials parameters can be determined: $D = \frac{1}{3\Sigma_{tr}} =$ 9.21 cm, $L^2 = D/\Sigma_a = 60.1$ cm^2, $B_m^2 = (\nu\Sigma_f - \Sigma_a)/D = 4.13 \times 10^{-4}$ cm^{-2}, $k_\infty = \nu\Sigma_f/\Sigma_a = 1.025$, and $\lambda_{extrap} = 19.6$ cm. The criticality condition is $B_m^2 = B_g^2 = (\pi/H_{ex})^2 + (2.405/R_{ex})^2$. Fixing the height at 370 cm, the criticality condition requires that $R_{ex} = 127.6$ cm or $R = 108$ cm.

3.7
Reflected Reactor

Since the dimensions of a critical core of a given composition depend on the fraction of the neutrons that leak out, these dimensions can be reduced if some of the leaking neutrons are reflected back into the core. A reflector has the added benefit of making the neutron flux distribution in the core more uniform by increasing the neutron population in the outer region due to reflected neutrons which otherwise would have escaped. Figure 3.3 illustrates the neutron flux distributions in bare and reflected cores of the same composition and dimension.

Reflected Slab Reactor

The mathematical treatment of a reflected reactor can be illustrated most simply by considering a slab core of thickness a extending from $x = -a/2$ to $x = +a/2$ reflected on both sides by a nonmultiplying slab of thickness b. If we were to solve the time-dependent equations in both the core and reflector as we did for the bare core, but now also requiring that the solutions satisfied continuity of flux and current conditions at $x = +a/2$, we would find a similar but more complicated result as before—that the solution consists of a sum of spatial eigenfunctions corresponding to discrete geometrical eigenvalues, and at long times the dominant component is the fundamental mode. Rather than carry through the entire calculation, we examine the fundamental mode that obtains at long times.

3.7 Reflected Reactor | 63

Table 3.4 Typical PWR Core Composition and Spectrum-Averaged Cross Sections

Isotope	n (10^{24} cm^{-3})	σ_{tr} (10^{-24} cm^2)	σ_a (10^{-24} cm^2)	σ_f (10^{-24} cm^2)	ν	Σ_{tr} (cm^{-1})	Σ_a (cm^{-1})	$\nu\Sigma_f$ (cm^{-1})
H	2.748×10^{-2}	0.650	0.294	0	0	1.79×10^{-2}	8.08×10^{-3}	0
O	2.757×10^{-2}	0.260	1.78×10^{-4}	0	0	7.16×10^{-3}	4.90×10^{-6}	0
Zr	3.694×10^{-3}	0.787	0.190	0	0	2.91×10^{-3}	7.01×10^{-4}	0
Fe	1.710×10^{-3}	0.554	2.33	0	0	9.46×10^{-4}	3.99×10^{-3}	0
235U	1.909×10^{-4}	1.62	484.0	312.0	2.43	3.08×10^{-4}	9.24×10^{-2}	0.145
238U	6.592×10^{-3}	1.06	2.11	0.638	2.84	6.93×10^{-3}	1.39×10^{-2}	1.20×10^{-2}
10B	1.001×10^{-5}	0.877	3.41×10^3	0	0	8.77×10^{-6}	3.41×10^{-2}	0
Sum						3.62×10^{-2}	0.1532	0.1570

Source: Data from Ref. 4; used with permission of Wiley.

The neutron diffusion equations in the core and reflector are

$$\text{Core:} \qquad -D_C \frac{d^2\phi_C}{dx^2} + (\Sigma_{aC} - \nu\Sigma_{fC})\phi_C = 0$$

$$\text{Reflector:} \qquad -D_R \frac{d^2\phi_R}{dx^2} + \Sigma_{aR}\phi_R = 0$$

(3.81)

The appropriate interface and boundary conditions are symmetry at $x = 0$, continuity of flux and current at $x = a/2$, and zero flux at the extrapolated boundary $a/2 + b_{ex}$:

$$\left.\frac{d\phi_C}{dx}\right|_{x=0} = 0 \tag{3.82a}$$

$$\phi_C\left(\frac{a}{2}\right) = \phi_R\left(\frac{a}{2}\right) \tag{3.82b}$$

$$J_C\left(\frac{a}{2}\right) = J_R\left(\frac{a}{2}\right) \tag{3.82c}$$

$$\phi_R\left(\frac{a}{2} + b_{ex}\right) = 0 \tag{3.82d}$$

The solution in the core satisfying the symmetry boundary condition Eq. (3.82a) is

$$\phi_C(x) = A_C \cos B_{mC}x \tag{3.83}$$

and the solution in the reflector satisfying the extrapolated boundary condition Eq. (3.82d) is

$$\phi_R = A_R \sinh \frac{(a/2) + b_{ex} - x}{L_R} \tag{3.84}$$

where $B_{mC}^2 = (\nu\Sigma_{fC} - \Sigma_{aC})/D_C$ and $L_R^2 = D_R/\Sigma_{aC}$. Using these general solutions in the interface conditions of Eqs. (3.82b) and (3.82c), dividing the two equations, and rearranging leads to the criticality condition which must be satisfied in order for a steady-state solution to exist:

$$\frac{B_{mC}a}{2} \tan \frac{B_{mC}a}{2} = \frac{D_R a}{2D_C L_R} \coth \frac{b_{ex}}{L_R} \tag{3.85}$$

The smallest value of a for which a solution of this equation exists is less than π/B_{mC}, as can be seen by plotting both sides of Eq. (3.85), in Fig. 3.4. Since the criticality condition for the bare slab was $B_{mC} = \pi/a_{ex}$, this result confirms that the addition of a reflector reduces the dimension necessary for criticality.

Reflector Savings

The difference in the reflected and unreflected critical dimensions is known as the reflector savings, δ:

$$\delta \equiv a(\text{bare}) - a(\text{reflected}) = \frac{1}{B_{mC}} \tan^{-1}\left(\frac{D_C B_{mC}}{D_R} L_R \tanh \frac{b_{ex}}{L_R}\right) \tag{3.86}$$

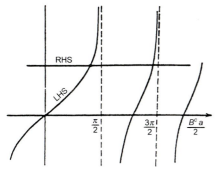

Fig. 3.4 Plot of criticality equation for reflected reactor. (From Ref. 10; used with permission of McGraw-Hill.)

In the limit of a reflector that is thick in comparison to the neutron diffusion length $(b \gg L_R)$, this reduces to $\delta \approx D_C L_R / D_R$.

Reflected Spherical, Cylindrical, and Rectangular Parallelepiped Cores

A similar calculation can be performed for other core geometries, but with reflection in only one direction. The resulting criticality conditions are given in Table 3.5.

3.8
Homogenization of a Heterogeneous Fuel–Moderator Assembly

In our previous treatment of a homogeneous core, we have implicitly assumed that the actual core—consisting of thousands of fuel and control elements, coolant, and structure (Fig. 3.5)—can be represented by some effective homogeneous mixture.

Spatial Self-Shielding and Thermal Disadvantage Factor

We might be tempted to construct this homogeneous mixture by simply volume-weighting the number densities of the various fuel, control, moderator, coolant, and structural materials, but this procedure would fail to take into account the reduction of the neutron population in the region of strong absorbers, a phenomenon known as *spatial self-shielding*. We illustrate this phenomenon by considering the thermal neutron flux distribution in a large fuel–moderator assembly consisting of a repeating array of slab fuel elements of width $2a$ interspersed with moderating regions of thickness $2(b - a)$. Since the moderator is much more effective than the fuel at slowing down neutrons, we specify a uniform source S_M of thermal neutrons in the moderator and no thermal neutron source in the fuel. We take as a calculational model one-half of the slab fuel element, extending from $x = 0$ to

Table 3.5 Criticality Condition for Reflected Reactors

Geometry	
Sphere	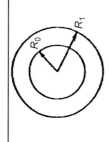 $$D_c(BR_0 \cot BR_0 - 1) = -D_R\left(\frac{R_0}{L_R}\coth R_{1ex}\frac{R_{1ex}-R_0}{L_R} + 1\right)$$ $$B^2 = \frac{(v\Sigma_f - \Sigma_a)}{D}$$
Cylinder: side-reflected	$$\frac{D_c J_0'(\kappa_c \rho_0)}{J_0(\kappa_c \rho_0)} = \frac{D_R L_0'(\rho_0)}{L_0(\rho_0)}$$ $$L_0(\rho) = I_0(\kappa_R \rho_{1ex})K_0(\kappa_R \rho) - I_0(\kappa_R \rho)K_0(\kappa_R \rho_{1ex})$$ $$B_{mC}^2 = \kappa_c^2 + \left(\frac{\pi}{2h_{ex}}\right)^2$$ $$\kappa_R^2 = \frac{1}{L_R^2} + \left(\frac{\pi}{2h_{ex}}\right)^2 , \quad \kappa_c^2 = \frac{v\Sigma_f - \Sigma_a}{D} - \left(\frac{\pi}{2h_{ex}}\right)^2$$

(Continued)

Table 3.5 (Continued)

	Geometry
Cylinder: end-reflected	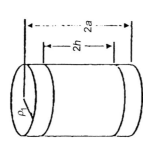 $D_C\mu_C\tan\mu_C h = D_R\coth\mu_R(a_{ex}-h)$ $\mu_R^2 = \dfrac{1}{L_R^2} + \left(\dfrac{\nu_1}{\rho_{1ex}}\right)^2$, $\quad B_{mC}^2 = \mu_C^2 + \left(\dfrac{\nu_1}{\rho_{1ex}}\right)^2$ $\nu_1 = 2.405$ $\mu_C^2 = \left(\dfrac{\nu\Sigma_f - \Sigma_n}{D}\right) - \left(\dfrac{\nu_1}{\rho_{1ex}}\right)^2$
Block: end-reflected	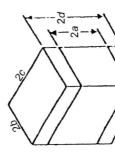 $D_C\kappa_1\tan\kappa_1 a = D_R\mu_1\coth\mu_1(d_{ex}-a)$ $\mu_1^2 = \dfrac{1}{L_R^2} + \left(\dfrac{\pi}{2b_{ex}}\right)^2 + \left(\dfrac{\pi}{2c_{ex}}\right)^2$ $B_{mC}^2 = \kappa_1^2 + \left(\dfrac{\pi}{2b_{ex}}\right)^2 + \left(\dfrac{\pi}{2c_{ex}}\right)^2$ $\kappa_1^2 = \dfrac{\nu\Sigma_f - \Sigma_a}{D} - \left(\dfrac{\pi}{2b_{ex}}\right)^2 - \left(\dfrac{\pi}{2c_{ex}}\right)^2$

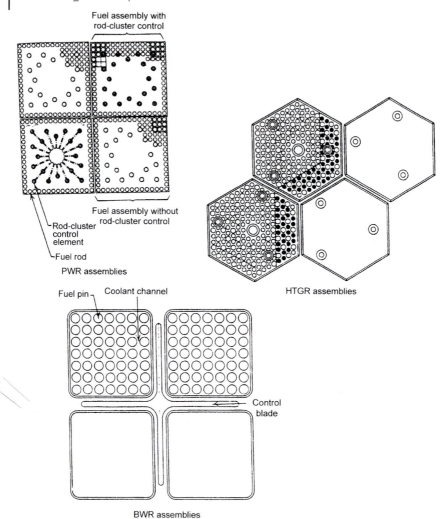

Fig. 3.5 Heterogeneous nuclear reactor fuel assemblies. (From Ref. 4; used with permission of Wiley.)

$x = a$, and one-half of the moderating region, extending from $x = a$ to $x = b$. The neutron diffusion equations in the fuel and moderator are

$$\text{Fuel:} \qquad -D_F \frac{d^2 \phi_F(x)}{dx^2} + \Sigma_{aF} \phi_F(x) = 0$$

$$\text{Moderator:} \quad -D_M \frac{d^2 \phi_M(x)}{dx^2} + \Sigma_{aM} \phi_M(x) = S_M$$

(3.87)

The appropriate boundary conditions are symmetry at the fuel and moderator midplanes at $x = 0$ and $x = b$, respectively. The other two conditions that must be satisfied are continuity of flux and current at the fuel–moderator interface at $x = a$

$$\frac{d\phi_F(0)}{dx} = 0 \tag{3.88a}$$

$$\frac{d\phi_M(b)}{dx} = 0 \tag{3.88b}$$

$$\phi_F(a) = \phi_M(a) \tag{3.88c}$$

$$D_F \frac{d\phi_F(a)}{dx} = D_M \frac{d\phi_M(a)}{dx} \tag{3.88d}$$

The solutions to Eqs. (3.87) that satisfy the conditions of Eqs. (3.88) are

$$\phi_F(x) = \frac{S_M \cosh(x/L_F)}{\{(L_F/D_F)\coth(a/L_F) + (L_M/D_M)\coth[(b-a)/L_M]\}(D_F/L_F)\Sigma_{aM}\sinh(a/L_F)}$$

$$\phi_M(x) = \frac{S_M}{\Sigma_{aM}}\left[1 - \frac{\cosh[(b-x)/L_M]}{\{(L_F/D_F)\coth(a/L_F) + (L_M/D_M)\coth[(b-a)/L_M]\}(D_M/L_M)\sinh[(b-a)/L_M]}\right]$$

$$\tag{3.89}$$

The *thermal flux disadvantage factor* is defined as the ratio of the average flux in the moderator to the average flux in the fuel:

$$\xi \equiv \frac{\bar{\phi}_M}{\bar{\phi}_F} = \frac{a\int_a^b \phi_M(x)dx}{(b-a)\int_0^a \phi_F(x)dx} = \frac{V_F\Sigma_{aF}}{V_M\Sigma_{aM}}\left(\frac{V_M\Sigma_{aM}}{V_F\Sigma_{aF}}F + E - 1\right) \tag{3.90}$$

where $V_F = a$, $V_M = b - a$, and

$$F = \frac{a}{L_F}\coth\frac{a}{L_F}, \qquad E = \frac{b-a}{L_M}\coth\left(\frac{b-a}{L_M}\right) \tag{3.91}$$

for slab geometry.

Thermal flux disadvantage factors for repeating arrays formed by other simple geometries can be calculated in the same manner and represented by the second form of Eq. (3.90). The results for the *lattice functions E and F* in other geometries are given in Table 3.6. The volumes are $V_F = \pi\rho_F^2$ and $\frac{4}{3}\pi r_f^3$ and $V_M = \pi\rho_M^2 - \pi\rho_F^2$ and $\frac{4}{3}\pi(r_M^3 - r_F^3)$, for the cylinder and sphere, respectively.

Effective Homogeneous Cross Sections

An effective homogeneous fuel cross section averaged over the fuel–moderator lattice can be constructed by using the thermal disadvantage factor of Eq. (3.90) in the definition

$$\Sigma_{aF}^{\text{eff}} \equiv \frac{\Sigma_{aF}\bar{\phi}_F V_F}{\bar{\phi}_F V_F + \bar{\phi}_M V_M} = \frac{\Sigma_{aF}V_F}{V_F + V_M}\frac{V_F + V_M}{V_F + \xi V_M}$$

$$= \Sigma_{aF}^{\text{hom}}\left(\frac{1 + V_M/V_F}{1 + \xi V_M/V_F}\right) \tag{3.92}$$

Table 3.6 Functions E and F for Various Cell Geometries

Geometry		E and F Functions
Slab		$F = \dfrac{a}{L_F} \coth \dfrac{a}{L_F}$ $E = \dfrac{b-a}{L_M} \coth \dfrac{b-a}{L_M}$
Cylindrical		$F = \dfrac{(\rho_F/L_F) I_0(\rho_F/L_F)}{2 I_1(\rho_F/L_F)}$ $E = \dfrac{(1/L_M)(\rho_M^2 - \rho_F^2)}{2\rho_F} \left[\dfrac{I_0(\rho_F/L_M) K_1(\rho_M/L_M) + K_0(\rho_F/L_M) I_1(\rho_M/L_M)}{I_1(\rho_M/L_M) K_1(\rho_F/L_M) - K_1(\rho_M/L_M) I_1(\rho_F/L_M)} \right]$
Spherical		$F = \dfrac{(r_F/L_F)^2 \tanh(r_F/L_F)}{3[(r_F/L_F) - \tanh(r_F/L_F)]}$ $E = \dfrac{r_M^3 - r_F^3}{3 r_F L_M^2} \dfrac{1 - (r_M/L_M) \coth[(r_M - r_F)/L_M]}{1 - r_M r_F/L_M^2 - [(r_M - r_F)/L_M] \coth[(r_M - r_F)/L_M]}$

Source: Adapted from Ref. 10; used with permission of McGraw-Hill.

An effective homogeneous absorption cross section for the moderator can obviously be constructed by exchanging the F and M subscripts and replacing ξ by ξ^{-1}. These fuel and moderator effective cross sections can then be combined $(\Sigma_a^{\text{eff}} = \Sigma_{aF}^{\text{eff}} + \Sigma_{aM}^{\text{eff}})$ to obtain an effective homogeneous cross section for the fuel–moderator assembly to be used in one of the previous homogeneous core calculations. Effective homogeneous scattering and transport cross sections can be constructed in a similar manner.

Example 3.4: Flux Disadvantage Factor and Effective Homogenized Cross Section in a Slab Lattice. Consider a lattice consisting of a large number of 1-cm-thick slab fuel plates separated by 1 cm of water at room temperature. The fuel is 10% enriched uranium. The fuel and water number densities are $n_{235} = 0.00478 \times 10^{24}$ cm^{-3}, $n_{238} = 0.0430 \times 10^{24}$ cm^{-3}, and $n_{H_2O} = 0.0334 \times 10^{24}$ cm^{-3}. Using the spectrum-averaged cross sections of Table 3.4 (and constructing effective H_2O σ's as two times the H σ's plus the O σ) yields the following material properties for the uranium fuel: $\Sigma_{\text{tr}} = 0.0534$ cm^{-1}, $\Sigma_a = 3.220$ cm^{-1}, $D = 6.17$ cm, and $L = 1.38$ cm, and for water: $\Sigma_{\text{tr}} = 0.0521$ cm^{-1}, $\Sigma_a = 0.0196$ cm^{-1}, $D = 6.40$ cm, and $L = 18.06$ cm. The geometric parameters in Eqs. (3.90) and (3.91) are $V_F = V_M = a = b - a = 0.5$ cm.

Evaluating Eq. (3.90) yields $\xi = 1.04$ for the thermal disadvantage factor. The effective homogenized fuel absorption and transport cross sections calculated from Eq. (3.92) are $\Sigma_{aF}^{\text{eff}} = 1.575$ cm^{-1} and $\Sigma_{\text{tr}F}^{\text{eff}} = 0.0264$ cm^{-1}. A simple homogenization (implicitly assuming that $\xi = 1$) yields $\Sigma_{aF}^{\text{hom}} = 1.610$ cm^{-1} and $\Sigma_{\text{tr}F}^{\text{hom}} = 0.0267$ cm^{-1}, so the effect of the spatial self-shielding (ξ) is significant.

The effective homogenized cross section for the water (moderator) is derived by a procedure similar to that in Eq. (3.92) and results in an expression similar to Eq. (3.92) but with the M and F subscripts interchanged and ξ replaced by ξ^{-1}. The effective homogenized water absorption and transport cross sections are $\Sigma_{aM}^{\text{eff}} = 0.010$ cm^{-1} and $\Sigma_{\text{tr}M}^{\text{eff}} = 0.0266$ cm^{-1}, so that the total effective absorption and transport cross sections for the lattice are $\Sigma_a^{\text{eff}} = \Sigma_{aF}^{\text{eff}} + \Sigma_{aM}^{\text{eff}} = 1.575 + 0.010 = 1.585$ cm^{-1} and $\Sigma_{\text{tr}}^{\text{eff}} = \Sigma_{\text{tr}F}^{\text{eff}} + \Sigma_{\text{tr}M}^{\text{eff}} = 0.0264 + 0.0266 = 0.053$ cm^{-1}.

Note that diffusion theory is not really suitable for calculating the diffusion of neutrons in such a lattice because $\lambda_{\text{tr}} = 1/\Sigma_{\text{tr}} \gg 0.5$ cm, the dimension of the diffusing medium, in both the fuel and the water; and that this example serves more to illustrate the application of the methodology than to provide accurate quantitative results. The transport methods introduced in Section 3.12 and described more fully in Chapter 9 must be used to calculate flux disadvantage factors, in general.

Thermal Utilization

Another use of the thermal disadvantage factor is to calculate the thermal utilization for the fuel–moderator lattice:

$$f_{\text{het}} \equiv \frac{\Sigma_{aF} \bar{\phi}_F V_F}{\Sigma_{aF} \bar{\phi}_F V_F + \Sigma_{aM} \bar{\phi}_M V_M} = \frac{\Sigma_{aF} V_F}{\Sigma_{aF} V_F + \Sigma_{aM} V_M} \frac{\Sigma_{aF} V_F + \Sigma_{aM} V_M}{\Sigma_{aF} V_F + \xi \Sigma_{aM} V_M}$$

$$= f_{\text{hom}} \frac{\Sigma_{aF} V_F + \Sigma_{aM} V_M}{\Sigma_{aF} V_F + \xi \Sigma_{aM} V_M} \tag{3.93}$$

In both Eqs. (3.92) and (3.93), the first term is the result that would be obtained with simple volume-weighted homogenization of the fuel and moderator number densities, and the second term is a correction that accounts for the flux self-shielding in the fuel.

Measurement of Thermal Utilization

In a finite fuel–moderator assembly with geometry characterized by the geometric buckling B_g and neutrons becoming thermal at a rate q_M (per second per cubic centimeter) in the moderator, the thermal neutron balance is

$$q_M V_M = (\Sigma_{aM} \bar{\phi}_M V_M + \Sigma_{aF} \bar{\phi}_F V_F)(1 + L^2 B_g^2) \tag{3.94}$$

and the thermal utilization is just the fraction of those thermal neutrons which are absorbed that are absorbed by the fuel:

$$f = \frac{\Sigma_{aF} \bar{\phi}_F V_F}{\Sigma_{aF} \bar{\phi}_F V_F + \Sigma_{aM} \bar{\phi}_M V_M} \tag{3.95}$$

The ratio of the slowing-down source to the thermal flux at some point in the moderator, $q_M/\phi_M(x)$ can be determined by irradiating an indium foil (indium has an absorption resonance just above thermal) at that point and then measuring the total foil activation A_{tot}. Then another indium foil clad in a cadmium jacket, which will absorb all the thermal neutrons before they can reach the foil but will pass the epithermal neutrons, is irradiated at the same location to determine the epithermal activation A_{epi}. The thermal component of the total activation, $A_{\text{th}} = A_{\text{tot}} - A_{\text{epi}}$, is proportional to the thermal flux at the location of the foil, $A_{\text{th}} = c_{\text{th}} \phi_M(x)$. The epithermal activation is proportional to the slowing-down source, $A_{\text{epi}} = c_{\text{epi}} q_M$. Thus $q_M/\phi_M(x) = (c_{\text{epi}}/c_{\text{th}})(A_{\text{epi}}/A_{\text{th}})$. The quantity $CR = A_{\text{epi}}/A_{\text{th}}$ is determined by the foil measurements and is known as the *cadmium ratio*.

The ratio of constants $(c_{\text{epi}}/c_{\text{th}})$ can be determined by irradiating many clad and unclad indium foils in a large block of pure moderator that has a source emitting Q neutrons per second. The neutron balance is

$$\Sigma_{aM} \int \phi(x)dx \simeq \int q(x)dx = Q \tag{3.96}$$

and the ratio of integrated thermal and epithermal activities is

$$\rho \equiv \frac{\int A_{\text{th}}(x)dx}{\int A_{\text{epi}}(x)dx} = \frac{c_{\text{epi}} \int \phi(x)dx}{c_{\text{th}} \int q(x)dx} = \frac{c_{\text{epi}}}{c_{\text{th}}} \frac{1}{\Sigma_{aM}} \tag{3.97}$$

These results can be combined to write an expression for the thermal utilization,

$$f = 1 - \frac{\Sigma_{aM} \bar{\phi}_M (1 + L^2 B^2)}{q_M} = 1 - \frac{1 + L^2 B^2}{\rho CR} \frac{\bar{\phi}_M}{\phi_M(x)} \tag{3.98}$$

in terms of the experimentally determined quantities CR and ρ and the local-to-average moderator flux ratio, which can be calculated using the foregoing formalism.

Local Power Peaking Factor

Once effective homogenized cross sections are constructed, the fuel–moderator assembly may be treated as a homogeneous region, and the average flux distribution in the assembly may be calculated using one of the other techniques discussed in this chapter. The average power density in the fuel–moderator assembly is then $\Sigma_{fF}^{\text{eff}}\phi_{\text{av}}$, where Σ_{fF}^{eff} is given by an expression such as Eq. (3.92) and ϕ_{av} is the average flux in the fuel–moderator assembly:

$$\phi_{\text{av}} = \frac{\bar{\phi}_F V_F + \bar{\phi}_M V_M}{V_F + V_M} = \bar{\phi}_F\frac{1+\xi(V_M/V_F)}{1+(V_M/V_F)} \tag{3.99}$$

The peak power density will occur at the location of the maximum neutron flux in the fuel element, which is at $x = a$, as may be seen from Eq. (3.89). The power peaking factor—the ratio of the peak to average power densities in the assembly—is given by

$$F_{pp} \equiv \frac{\Sigma_{fF}\phi_F(a)}{\Sigma_{fF}^{\text{eff}}\phi_{\text{av}}} = \left(1+\frac{V_M}{V_F}\right)\frac{\phi_F(a)}{\bar{\phi}_F} = \left(1+\frac{V_M}{V_F}\right)\frac{a}{L_F}\coth\frac{a}{L_F}$$

$$= \left(1+\frac{V_M}{V_F}\right)\left[1+\frac{1}{3}\left(\frac{a}{L_F}\right)^2 - \frac{1}{45}\left(\frac{a}{L_F}\right)^4 + \cdots\right], \quad \frac{a}{L_F} < \pi \tag{3.100}$$

where the form of $\phi_F(a)/\phi_F$ for a slab fuel–moderator lattice has been used to arrive at the second form of the equation. The power peaking is minimized by minimizing a/L_F and V_M/V_F.

3.9
Control Rods

Effective Diffusion Theory Cross Sections for Control Rods

Localized highly absorbing control elements such as control rods cannot be calculated directly using diffusion theory. However, transport theory can be used to determine effective diffusion theory cross sections for use with diffusion theory. We illustrate this by considering the BWR example shown in Fig. 3.5 of a core consisting of a repeating array of four fuel–moderator assemblies surrounding a cruciform control rod. First, the fuel–moderator assemblies must be homogenized, using the procedure of Section 3.8 or some more sophisticated procedure based on transport theory, yielding a model of a cruciform control rod embedded in a square cell of homogeneous fuel–moderator, as shown in Fig. 3.6. If the span, l, of the control blade is large compared to the neutron diffusion length in the fuel–moderator

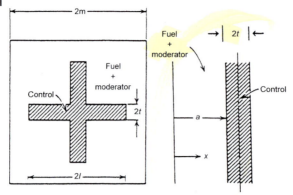

Fig. 3.6 One-dimensional model of a cruciform control blade cell. (From Ref. 4; used with permission of Wiley.)

region, the diffusion of neutrons into the rod is essentially one-dimensional. We take advantage of this fact to replace the two-dimensional problem by an equivalent one-dimensional problem that preserves both the ratio of the control rod surface to the fuel–moderator volume and the thickness of the control blade. We construct an equivalent model consisting of a repeating array of fuel–moderator slabs of thickness $2a$ and control rod slabs of thickness $2t$, where $a = (m^2 - 2tl + t^2)/2l$, as shown in Fig. 3.6. Our calculational model then is a fuel–moderator (half) slab from $x = 0$ to $x = a$ and a control (half) slab from $x = a$ to $x = a + t$, with symmetry boundary conditions at $x = 0$ and $x = a + t$. The neutron diffusion equation

$$-D\frac{d^2\phi(x)}{dx^2} + \Sigma_a\phi(x) = S_0 \tag{3.101}$$

is valid in the fuel–moderator slab, where S_0 is a uniform source of neutrons slowing down in the fuel–moderator region. The symmetry boundary condition for the diffusion theory calculation is

$$\frac{d\phi(0)}{dx} = 0 \tag{3.102}$$

and a transport boundary condition

$$\frac{J(a)}{\phi(a)} = \alpha \tag{3.103}$$

is used at the fuel–moderator interface with the control rod. The parameter α must be determined from a transport theory calculation of the control rod region (Section 3.12). For a slab of width $2t$, such a calculation yields

$$\alpha = \frac{1 - 2E_3(2\Sigma_{ac}t)}{2[1 + 3E_4(2\Sigma_{ac}t)]} \tag{3.104}$$

where Σ_{ac} is the control rod absorption cross section and E_n is the exponential integral function:

$$E_n(g) = \int_1^\infty e^{-gu} u^{-n} du \qquad (3.105)$$

The solution to Eq. (3.101) which satisfies Eqs. (3.102) and (3.103) is

$$\phi(x) = \frac{S_0}{\Sigma_a}\left[1 - \frac{\alpha \cosh(x/L)}{\alpha \cosh(a/L) + (D/L)\sinh(a/L)}\right] \qquad (3.106)$$

We now define an effective diffusion theory cross section for the control rod by requiring that the diffusion theory and transport theory calculations of the neutron absorption rate in the control rod agree:

$$\Sigma_c^{\text{eff}} \phi_{\text{av}} A_{\text{cell}} = P_c J_c \qquad (3.107)$$

where ϕ_{av} is the average diffusion theory flux in the fuel–moderator region, $A_{\text{cell}} = (a+t)b$ is the area of the fuel–moderator plus control rod cell of arbitrary transverse direction b, $P_c = b$ is the perimeter of the control rod interface with the fuel–moderator region, and J_c is the neutron current from the fuel–moderator region at the surface of the control rod. It is assumed that all neutrons which enter the control rod are absorbed. Combining the results above yields

$$\Sigma_c^{\text{eff}} = \frac{P_c}{A_{\text{cell}}}\frac{J_c}{\bar\phi} = \frac{1}{a}\alpha\frac{\phi(a)}{\bar\phi} = \frac{\Sigma_a}{a[\Sigma_a/\alpha + (1/L)\coth(a/L)] - 1} \qquad (3.108)$$

for the effective homogeneous control rod cross section to be used in a diffusion theory calculation. Note that the Σ_a in this equation is the effective fuel–moderator homogenized cross section, and the control rod cross section is hidden in the parameter α.

Example 3.5: Slab Control Plate Effective Cross Section. Consider again the lattice of alternating 10% enriched uranium fuel and water slabs, each 1 cm thick, discussed in Section 3.8. The effective homogenized lattice cross sections are $\Sigma_a^{\text{eff}} = 0.4144$ cm^{-1} and $\Sigma_{\text{tr}}^{\text{eff}} = 0.0525$ cm^{-1}, leading to $D^{\text{eff}} = 6.35$ cm and $L^{\text{eff}} = 3.91$ cm in the fuel–water lattice. Now consider the placement of 1-cm-thick slab natural boron plates (19.9% ^{10}B) every 10.5 cm in the lattice. With respect to Fig. 3.6, $t = 0.5$ cm and $a = 5$ cm. The ^{10}B density in the control slab is $n_{\text{B}10} = 0.199(2.45/10.8)(0.6022 \times 10^{24}) = 0.0271 \times 10^{24}$ cm^{-3}, the absorption cross section from Table 3.4 is $\sigma_{\text{B}10} = 3.41 \times 10^{-21}$ cm^2, and the macroscopic control slab absorption cross section is $\Sigma_{ac} = 92.411$ cm^{-1}. For such large values of $2t\Sigma_{ac}$, the exponential integrals approach zero and the transport boundary condition parameter $\alpha \to 0.5$. Evaluation of Eq. (3.108) with these parameters yields for the effective homogenized control cross section $\Sigma_c^{\text{eff}} = 0.0894$ cm^{-1}. Thus, in the homogenized representation of the lattice, the effective macroscopic absorption cross section is 0.414 cm^{-1} with the control plates removed and 0.493 cm^{-1} with the control plates inserted. The effective transport cross section is 0.0525 cm^{-1} and is assumed to be the same with or without the control plates.

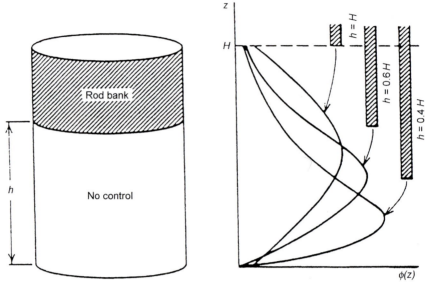

Fig. 3.7 Insertion of a control rod bank into a bare cylindrical core. (From Ref. 4; used with permission of Wiley.)

Windowshade Treatment of Control Rods

Now that we know how to obtain effective homogenized cross sections for the fuel–moderator assemblies and for control rods, we can represent the partial insertion of a bank of control rods (from the top) into a bare cylindrical core as a two-region core diffusion problem, as indicated in Fig. 3.7. The lower, unrodded region is represented by the homogenized fuel–moderator cross sections, and the upper rodded region is represented by the homogenized fuel–moderator cross sections plus the effective control rod cross section.

The neutron diffusion equation in both the rodded and unrodded regions is of the form of Eq. (3.69), and we can anticipate from the development of Section 3.6 that a separation of variables solution that satisfies a zero flux boundary condition at $r = R$ (we assume that the reactor is sufficiently large that the zero flux condition at the external boundary is equivalent to the zero flux condition at the extrapolated boundary) will be of the form

$$\psi(r, z) = Z(z) J_0\left(\frac{v_1 r}{R}\right) \tag{3.109}$$

and the function $Z(z)$ will satisfy

$$\frac{d^2 Z}{dz^2} + B_z^2(Z)(z) = 0 \tag{3.110}$$

where

$$B_z^2 \equiv \frac{v \Sigma_f / \Sigma_a - 1}{L^2} - \left(\frac{v_1}{R_{ex}}\right)^2 \tag{3.111}$$

and $v_1 = 2.405$ is the smallest root of $J_0(v) = 0$.

Solving the diffusion equation separately in the rodded and unrodded regions and requiring that the solutions vanish at $z = 0$ and $z = H$ yields

$$\begin{aligned} Z_{un}(z) &= A_{un} \sin\left(B_z^{un} z\right), & 0 \leq z \leq h \\ Z_{rod}(z) &= A_{rod} \sinh\left[B_z^{rod}(H - z)\right], & h \leq z \leq H \end{aligned}$$

(3.112)

We require continuity of flux and current at $z = h$, the interface between the rodded and unrodded regions,

$$Z_{un}(h) = Z_{rod}(h)$$

$$D_{un} \frac{dZ_{un}(h)}{dz} = D_{rod} \frac{dZ_{rod}(h)}{dz}$$

(3.113)

The first condition leads to the relationship

$$\frac{A_{rod}}{A_{un}} = \frac{\sin(B_z^{un} h)}{\sinh[B_z^{rod}(H - h)]}$$

(3.114)

and dividing the two conditions leads to the criticality condition,

$$\frac{1}{D_{un} B_z^{un}} \tan B_z^{un} h = -\frac{1}{D_{rod} B_z^{rod}} \tanh\left[B_z^{rod}(H - h)\right]$$

(3.115)

which may be solved for the rod insertion distance $(H - h)$, for which the reactor is just critical.

The axial neutron flux solution is sketched in Fig. 3.7 for several rod insertions. As might be expected, the axial flux distribution is symmetric when the rod bank is fully withdrawn and becomes progressively more peaked toward the bottom of the core as the rod bank is inserted farther downward. Note that in case of rod insertion from the bottom, the situation is just reversed.

3.10
Numerical Solution of Diffusion Equation

Although the semianalytical techniques for solving the neutron diffusion equation that we have developed can be extended to treat reactor models consisting of a larger number of different homogeneous regions than we have considered, realistic reactor models may consist of hundreds or thousands of different homogenized regions, even after the local fuel–moderator homogenization has taken place. The fuel concentration may vary from assembly to assembly and within an assembly in order to make the power distribution more uniform, and even within initially uniform assemblies the composition will change differently from location to location with fuel burnup. The standard practice today is to use numerical techniques to solve the neutron diffusion equation.

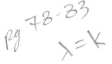

Pg 78-33   λ=k

Finite Difference Equations in One Dimension

The neutron diffusion equation in a one-dimensional slab reactor model is

$$-\frac{d}{dx}D(x)\frac{d\phi(x)}{dx} + \Sigma_a(x)\phi(x) = \frac{1}{\lambda}\nu\Sigma_f(x)\phi(x) \tag{3.116}$$

The first step in developing a numerical solution procedure is to replace the continuous spatial dependence of the flux, $\phi(x)$, with the values of the flux at a number of discrete spatial locations, $\phi_i \equiv \phi(x_i)$, the solution for which will be the objective of the numerical technique. There are many ways to do this, and we will use a simple finite-difference approximation. We subdivide the interval $0 \leq x \leq a$ of interest into I subintervals of length $\Delta = a/I$. (A more general development would use nonuniform subintervals.) A general rule of thumb is that $\Delta < L$ (the neutron diffusion length) sets an upper limit on the subinterval length (or mesh spacing).

| | | $\leftarrow \Delta \rightarrow$ | | | | | | |
| x_0 | x_1 | | x_2 | $\ldots$ | x_{i-1} | x_i | x_{i+1} | $\ldots$ | x_{I-1} | x_I |

$$x_{i-1/2} \quad x_{i+1/2}$$

Next, the terms in Eq. (3.116) are each integrated from $x_{i-1/2}$ to $x_{i+1/2}$, using the following approximations:

$$\int_{x_i-(1/2)\Delta}^{x_i+(1/2)\Delta} dx\,\Sigma_a(x)\phi(x) \simeq \Sigma_{ai}\phi_i\,\Delta$$

$$\int_{x_i-(1/2)\Delta}^{x_i+(1/2)\Delta} dx\left(\frac{d}{dx}D(x)\frac{d\phi}{dx}\right) \simeq D\frac{d\phi}{dx}\bigg|_{x_{i+1/2}\Delta} - D\frac{d\phi}{dx}\bigg|_{x_{i+1/2}\Delta} \tag{3.117}$$

$$\simeq \frac{1}{2}(D_i + D_{i+1})\frac{\phi_{i+1} - \phi_i}{\Delta}$$

$$-\frac{1}{2}(D_{i-1} + D_i)\frac{\phi_i - \phi_{i-1}}{\Delta}$$

where we have associated Σ_{ai}, D_i, and so on, with the subinterval $x_{i-1/2} \leq x \leq x_{i+1/2}$. The discrete equation associated with x_i may be written

$$a_{i,i-1}\phi_{i-1} + a_{i,i}\phi_i + a_{i,i+1}\phi_{i+1} = \frac{1}{\lambda}f_i\phi_i \equiv S_i, \quad i = 1,\ldots,I-1 \tag{3.118}$$

where

$$a_{i,i-1} = -\frac{1}{2}\left(\frac{D_i + D_{i-1}}{\Delta^2}\right)\left(1 - \frac{c}{2i-1}\right)$$

$$a_{i,i} = \Sigma_{ai} + \frac{1}{2}\left(\frac{D_{i-1} + 2D_i + D_{i+1}}{\Delta^2}\right)$$

$$a_{i,i+1} = -\frac{1}{2}\left(\frac{D_{i+1} + D_i}{\Delta^2}\right)\left(1 + \frac{c}{2i-1}\right) \tag{3.119}$$

$$f_i = \nu\Sigma_{fi}$$

We have generalized to other one-dimensional geometries, where $c = 0, 1$, and 2 for slab, cylindrical, and spherical, respectively. The significant feature of the set of Eqs. (3.118) is nearest-neighbor coupling—the flux at any x_i is only directly coupled to the flux at the adjacent points x_{i-1} and x_{i+1}, which greatly facilitates their solution.

Note that the difference equations are formulated only for the $I - 1$ interior mesh points at $x_1, x_2, \ldots, x_{I-1}$. The boundary conditions determine the exterior mesh points. A zero flux boundary condition at the left boundary corresponds to $\phi_0 = 0$, for example. A zero current or symmetry boundary condition at the left boundary corresponds to $\phi_0 = \phi_1$ and would be implemented by setting $a_{1,0} = 0$ and $a_{1,1} = \Sigma_{a1} + (D_1 + D_2)/\Delta^2$.

Forward Elimination/Backward Substitution Spatial Solution Procedure

The set of $I - 1$ Eqs. (3.118) can readily be solved by *Gaussian elimination*, or *forward elimination backward substitution*, for a known fission source S_i. The Gaussian elimination solution is implemented by subtracting $a_{i,i-1}/a_{i-1,i-1}$ times the $(i - 1)$th equation from the ith equation to eliminate the $a_{i,i-1}$ element in the ith equation. The modified ith equation is then divided by $a_{i,i}$. This process is repeated successively for $i = 1$ through $i = I - 1$. Then the manipulated equations can be solved successively from $i = I - 1$ to $i = 1$ using the algorithms

$$\phi_{I-1} = \alpha_{I-1}$$

$$\phi_{I-2} = -A_{I-2}\phi_{I-1} + \alpha_{I-2}$$

$$\vdots \tag{3.120}$$

$$\phi_i = -A_i\phi_{i+1} + \alpha_i$$

for the backward substitution, where

$$A_1 = \frac{a_{1,2}}{a_{1,1}}, \qquad A_i = \frac{a_{i,i+1}}{a_{i,i} - a_{i,i-1}A_{i-1}}$$

$$\tag{3.121}$$

$$\alpha_1 = \frac{S_1}{a_{1,1}}, \qquad \alpha_i = \frac{S_i - a_{i,i-1}\alpha_{i-1}}{a_{i,i} - a_{i,i-1}A_{i-1}}$$

had previously been constructed on the forward elimination.

Power Iteration on Fission Source

The fission source is not known a priori, of course, so the Gaussian elimination must be embedded in an iteration on the fission source term, as follows. An initial guess of the flux $\phi_i^{(0)}$ at each point and of the eigenvalue $\lambda^{(0)}$ is made and an initial fission source at each point is constructed $S_i^{(0)} = \nu\Sigma_{fi}\phi_i^{(0)}/\lambda^{(0)}$. The Gaussian elimination is performed to determine $\phi_i^{(1)}$. A new estimate of the eigenvalue is made from

$$\lambda^{(1)} = \frac{\sum_{i=1}^{I-1} \nu \Sigma_{fi} \phi_i^{(1)} \Delta}{\sum_{i=1}^{I-1} \Delta[a_{i,i-1}\phi_{i-1}^{(0)} + a_{i,i}\phi_i^{(0)} + a_{i,i+1}\phi_{i+1}^{(0)}]}$$

$$\simeq \frac{\sum_{i=1}^{I-1} \nu \Sigma_{fi} \phi_i^{(1)} \Delta}{\sum_{i=1}^{I-1} \nu \Sigma_{fi} \phi_i^{(0)} \Delta / \lambda^{(0)}} \qquad (3.122)$$

and a new fission source is constructed from

$$S_i^{(1)} = \frac{1}{\lambda^{(1)}} \nu \Sigma_{fi} \phi_i^{(1)} \qquad (3.123)$$

This iteration process is continued [using Eqs. (3.122) and (3.123) with $0 \to n-1$ and $1 \to n$] until the eigenvalues obtained on two successive iterates differ by less than some convergence criterion, say $\varepsilon = 10^{-5}$:

$$\left| \frac{\lambda^{(n)} - \lambda^{(n-1)}}{\lambda^{(n-1)}} \right| < \varepsilon \qquad (3.124)$$

Finite-Difference Equations in Two Dimensions

In rectangular geometry, the neutron diffusion equation is

$$-\frac{\partial}{\partial x} D(x,y) \frac{\partial \phi(x,y)}{\partial x} - \frac{\partial}{\partial y} D(x,y) \frac{\partial \phi(x,y)}{\partial y} + \Sigma_a(x,y)\phi(x,y)$$

$$= \frac{1}{\lambda} \nu \Sigma_f(x,y)\phi(x,y) \qquad (3.125)$$

To extend the procedure for developing finite-difference equations which was discussed for the one-dimensional case, we consider a rectangle with x-dimension a and y-dimension b. We subdivide a into I intervals of length $\Delta_x = a/I$ and subdivide b into J intervals of length $\Delta_y = b/J$.

The diffusion equation is integrated over the mesh box $(x_{i-1/2} \leq x \leq x_{i+1/2},$ $y_{j-1/2}/ \leq y \leq y_{j+1/2})$ and the approximations of Eqs. (3.117) are extended to two dimensions to obtain the finite-difference equations:

$$-\frac{1}{2}\left(\frac{D_{i-1,j} + D_{i,j}}{\Delta_x^2}\right)\phi_{i-1,j} - \frac{1}{2}\left(\frac{D_{i,j} + D_{i+1,j}}{\Delta_x^2}\right)\phi_{i+1,j}$$

$$-\frac{1}{2}\left(\frac{D_{i,j-1} + D_{i,j}}{\Delta_y^2}\right)\phi_{i,j-1} - \frac{1}{2}\left(\frac{D_{i,j} + D_{i,j+1}}{\Delta_y^2}\right)\phi_{i,j+1}$$

$$+\left(\Sigma_{aij} + \frac{\frac{1}{2}D_{i-1,j} + D_{i,j} + \frac{1}{2}D_{i+1,j}}{\Delta_x^2} + \frac{\frac{1}{2}D_{i,j-1} + D_{i,j} + \frac{1}{2}D_{i,j+1}}{\Delta_y^2}\right)\phi_{i,j}$$

$$= \frac{1}{\lambda}\nu\Sigma_{fi,j}\phi_{i,j}, \quad i = 1, \ldots, I-1; \; j = 1, \ldots, J-1 \qquad (3.126)$$

The significant feature of these equations is, once again, nearest-neighbor coupling —the flux at (i, j) is only directly coupled to the fluxes at $(i, j+1)$, $(i, j-1)$, $(i+1, j)$, and $(i-1, j)$.

The boundary conditions are used to specify $\phi_{0,j}$, $\phi_{I,j}$, $\phi_{i,0}$, and $\phi_{i,J}$, as discussed for the one-dimensional case.

In order to simplify the notation somewhat, we replace the (i, j) identification of a spatial location with a (p) identification. The total number of spatial locations is $P = (I-1) \times (J-1)$. We will choose $p = 1$ for $(i = 1, j = 1)$, $p = 2$ for $(i = 2, j = 1)$, $\ldots$, $p = I-1$ for $(i = I-1, j = 1)$, $p = I$ for $(i = 1, j = 2)$, $\ldots$, $p = 2(I-1)$ for $(i = I-1, j = 2)$, and so on. Then the set of finite difference equations may be written

$$a_{1,1}\phi_1 + a_{1,2}\phi_2 + a_{1,3}\phi_3 + \cdots + a_{1,p}\phi_p + \cdots + a_{1,P}\phi_P = S_{f1}$$

$$a_{2,1}\phi_1 + a_{2,2}\phi_2 + a_{2,3}\phi_3 + \cdots + a_{2,p}\phi_p + \cdots + a_{2,P}\phi_P = S_{f2}$$

$$a_{3,1}\phi_1 + a_{3,2}\phi_2 + a_{3,3}\phi_3 + \cdots + a_{3,p}\phi_p + \cdots + a_{3,P}\phi_P = S_{f3} \qquad (3.127)$$

$$\vdots$$

$$a_{P,1}\phi_1 + a_{P,2}\phi_2 + a_{P,3}\phi_3 + \cdots + a_{P,p}\phi_p + \cdots + a_{P,P}\phi_P = S_{fP}$$

where

$$a_{p,p} = \Sigma_{ap} + \frac{\frac{1}{2}D_{p-1} + D_p + \frac{1}{2}D_{p+1}}{\Delta_x^2} + \frac{\frac{1}{2}D_{p-I} + D_p + \frac{1}{2}D_{p+I}}{\Delta_y^2}$$

$$a_{p,p-1} = -\frac{1}{2}\left(\frac{D_{p-1} + D_p}{\Delta_x^2}\right), \qquad a_{p,p+1} = -\frac{1}{2}\left(\frac{D_p + D_{p+1}}{\Delta_x^2}\right)$$

$$a_{p,p-I} = -\frac{1}{2}\left(\frac{D_p + D_{p-I}}{\Delta_y^2}\right), \qquad a_{p,p+I} = -\frac{1}{2}\left(\frac{D_{p+I} + D_p}{\Delta_y^2}\right) \qquad (3.128)$$

$$a_{p,q} = 0, \quad q \neq p-1, p+1, p-I, p+I$$

$$S_{fp} = \frac{1}{\lambda}\nu\Sigma_{fp}\phi_p$$

Successive Relaxation Solution of Two-Dimensional Finite-Difference Equations

There are a number of possible ways to solve the set of Eqs. (3.127). We describe here the widely used *Gauss–Seidel* or *successive relaxation method*. This is an iterative method that proceeds by solving the first equation for ϕ_1, assuming S_1 is known and guessing a value for $\phi_2, \ldots, \phi_P$; then solving the second equation for ϕ_2, assuming that S_2 is known, using the value just calculated for ϕ_1, and using the same guessed values for $\phi_3, \ldots, \phi_P$; then solving the third equation for ϕ_3, assuming that S_3 is known, using the just calculated values for ϕ_1 and ϕ_2, and using the same guessed values for $\phi_4, \ldots, \phi_P$; and continuing thusly until the last equation is solved for ϕ_P, assuming that S_P is known, and using the just calculated values for $\phi_1, \ldots, \phi_{P-1}$. The set of new values of $\phi_1, \ldots, \phi_P$ thus calculated provides a new guess to be used in a repeated iteration. The general algorithm for the solution at each step is

$$\phi_p^{(m+1)} = \frac{1}{a_{p,p}} \left[S_{fp} - \sum_{q=1}^{p-1} a_{p,q} \phi_q^{(m+1)} - \sum_{q=p+1}^{P} a_{p,q} \phi_q^{(m)} \right] \tag{3.129}$$

where m is the iteration index. This inner iteration is continued until the flux solution at each location has converged to within a specified tolerance, $\varepsilon \approx 10^{-2}$, which may be chosen smaller in regions where exact knowledge of the neutron flux is important than in, for example, reflector regions:

$$\left| \frac{\phi_p^{(m+1)} - \phi_p^{(m)}}{\phi_p^{(m)}} \right| < \varepsilon_p \tag{3.130}$$

It is possible to accelerate the convergence of the relaxation iteration by using as a new flux guess a mixture of the previous flux and the relaxation result of Eq. (3.129):

$$\phi_p^{(m+1)} = (1-\omega)\phi_p^{(m)} + \frac{\omega}{a_{p,p}} \left[S_{fp} - \sum_{q=1}^{p-1} a_{p,q} \phi_q^{(m+1)} - \sum_{q=p+1}^{P} a_{p,q} \phi_q^{(m)} \right] \tag{3.131}$$

The acceleration parameter ω may be chosen in a number of ways (see Ref. 8), but generally varies between 1 and 2. The algorithm of Eq. (3.131) is known as *successive overrelaxation* (SOR).

Another widely used method for solving the two-dimensional diffusion equations is the *alternating direction implicit* iteration scheme described in Section 16.3.

Power Outer Iteration on Fission Source

The power iteration on the fission source proceeds as described above [i.e., in Eqs. (3.122) to (3.124)] but with i replaced by p in the notation of this section, and with Δ replaced by Δ_x, Δ_y.

Thus the solution of the finite-difference equations has a two-level iteration hierarchy. There is an outer power iteration on the fission source and the eigenvalue, described by Eqs. (3.122) to (3.124). Then for each of the outer iterations, there is a series of inner relaxation iterations—described by Eq. (3.129) or (3.131) and (3.130)—to converge the flux solution for that outer iterate of the fission source.

Limitations on Mesh Spacing

We can obtain some insight as to limitations on mesh spacing by considering the source-free diffusion equation in one dimension:

$$\frac{d^2\phi}{dx^2} - \frac{\phi}{L^2} = 0 \tag{3.132}$$

which can be solved exactly over the mesh interval $\Delta = x_{i+1/2} - x_{i-1/2}$ centered on x_i:

$$\phi(x_{i+1/2}) = e^{-\Delta/2L}\phi(x_i) = e^{-\Delta/L}\phi(x_{i-1/2}) \tag{3.133}$$

The central difference finite-difference approximation (which we have been using) of Eq. (3.132) on this interval can be written

$$\phi(x_{i+1/2}) + \phi(x_{i-1/2}) = \left[2 + \left(\frac{\Delta}{L}\right)^2\right]\phi(x_i) \tag{3.134}$$

Comparing the right side of Eq. (3.134) to the exact expression for the left side constructed from Eq. (3.133) allows us to define the difference as a measure of the error in the finite-difference approximation:

$$\text{error} = \frac{3}{4}\left(\frac{\Delta}{L}\right)^2 + \cdots \tag{3.135}$$

Clearly, the mesh spacing should be less than the diffusion length.

3.11
Nodal Approximation

In principle, once the local fuel cell heterogeneity in each fuel assembly is replaced by effective homogenized cross sections and effective cross sections are constructed for the control rods, the three-dimensional finite-difference diffusion equations can be solved for the effective multiplication constant and the neutron flux distribution everywhere in a reactor. In practice, it is seldom practical to do so because of the large number of simultaneous equations that must be solved. As we have seen, accuracy in the finite-difference solution requires that the mesh spacing be smaller than the diffusion length, and a typical LWR core is about 200 thermal diffusion lengths in each of the three dimensions, which results in several million mesh points, hence several million simultaneous equations.

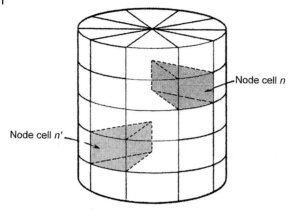

Fig. 3.8 Division of a reactor into nodes. (From Ref. 4; used with permission of Wiley.)

One means to deal with this situation is to divide the flux solution into two parts. The reactor core (and reflector, etc.) is divided into a relatively small number (on the order of 100 or less) large regions, or nodes, as depicted in Fig. 3.8. The detailed flux distribution within each node is determined from a finite-difference calculation just within the node (or set of contiguous nodes); such calculations need be performed only for every different type of node, since the solution for different nodes that have the same internal material distribution and the same boundary conditions will be identical. The global flux distribution (i.e., the average value of the flux in the different nodes) and the effective multiplication factor are then determined from a nodal calculation.

The general derivation of nodal diffusion theory methods may be illustrated by integrating the diffusion equation

$$-\nabla \cdot D\nabla\phi + \Sigma_a\phi = \frac{v}{k}\Sigma_f\phi \tag{3.136}$$

over the spatial domain of each node n to obtain

$$-\int_{S_n} d\vec{s} \cdot D\nabla\phi + \int_{V_n} \Sigma_a\phi\, dr = \frac{1}{k}\int_{V_n} v\Sigma_f\phi\, dr, \quad n = 1, \ldots, N \tag{3.137}$$

where Gauss's law has been used to replace the volume integral over node n, V_n, of the divergence with the surface integral over the surface S_n bounding node n of the normal component of the current. In general, the surface S_n bounding node n consists of the several interfaces $S_{nn'}$ between node n and the contiguous nodes n'.

Defining the average nodal flux as

$$\phi_n \equiv \frac{1}{V_n}\int_{V_n} \phi(r)\, dr \tag{3.138}$$

the definition of average nodal cross section follows immediately:

$$\Sigma_{a_n} \equiv \frac{1}{\phi_n V_n}\int_{V_n} \Sigma_a(r)\phi(r)\, dr, \qquad v\Sigma_{f_n} \equiv \frac{1}{\phi_n V_n}\int_{V_n} v\Sigma_f(r)\phi(r)\, dr \tag{3.139}$$

The treatment of the surface integral term, which represents node-to-node leakage, is not so obvious. However, it is plausible that the gradient of the flux across the surface between two adjacent nodes is proportional to the difference in the two average nodal fluxes:

$$-\alpha_{nn'}(\phi_n - \phi_{n'}) \equiv \int_{S_{nn'}} d\vec{s} \cdot D\nabla\phi \tag{3.140}$$

The accuracy of the nodal methods depends to a large extent on the actual evaluation of the nodal coupling coefficients $\alpha_{nn'}$, which is discussed in some detail in Chapter 15. A simple approximation results from using an average value $\frac{1}{2}(D_n + D_{n'})$ for the diffusion coefficient on the interface between nodes n and n', and assuming the average diffusion coefficient and the flux gradient are both constant over the interface, which yields

$$\alpha_{nn'} \simeq \frac{S_{nn'} \cdot \frac{1}{2}(D_n + Dn')}{l_{nn'}} \tag{3.141}$$

where $l_{nn'}$ is the distance between the centers of contiguous nodes n and n'.

Collecting these results leads to the set of N nodal equations for the nodal average fluxes and the effective multiplication constant:

$$-\sum_{n' \in n} \alpha_{nn'}\phi_{n'} \left(\sum_{n' \in n} \alpha_{nn'} + \Sigma_{an} V_n \right) \phi_n = \frac{1}{k} \nu \Sigma_{fn} V_n \phi_n, \quad n = 1, \ldots, N \tag{3.142}$$

plus sign

where $n' \in n$ indicates that the sum is over nodes n' which are contiguous to node n.

For those nodes n located adjacent to the exterior boundary of the reactor, the nodal equations contain the flux $\phi_{n'}$ for a nonexistent node on the other side of the boundary. For vacuum boundary conditions, this flux $\phi_{n'}$, would be set to zero in the equation for node n. For symmetry boundary conditions, $\phi_{n'} = \phi_n$ would be used in the equation for node n.

3.12
Transport Methods

For many situations of interest (e.g. a strongly absorbing control rod), the conditions for the validity of neutron diffusion theory are not satisfied, and a more rigorous approximation for the transport of neutrons than the diffusion approximation developed in Section 3.1 is required. Limiting consideration to a "one-dimensional" medium, i.e. one which is uniform in two dimensions (y and z in Fig. 3.9), the number of uncollided neutrons arising from an isotropic source, S_0, in the differential area $dA = \rho d\theta d\rho$ in the y–z plane at $x = 0$ that passes through unit area at a point on the x-axis per unit time is $(S_0 e^{-\Sigma_t R} dA)/4\pi R^2$. Defining the cosine of the angle between R and x, $\mu \equiv \mathbf{n}_x \cdot \mathbf{n}_R$, we can write $R = x/\mu$ and make use of $\rho^2 + x^2 = R^2$ to write the total flux of uncollided neutrons arising from a ring of a

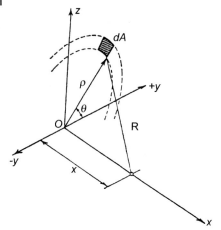

Fig. 3.9 Coordinate system for plane isotropic source calculation. (From Ref. 10; used with permission of McGraw-Hill.)

uniform plane isotropic source that passes through a point on the x-axis as

$$\psi(x, \mu : 0) = \frac{1}{2} S_0 e^{-\Sigma_t R} \frac{dR}{R} = \frac{1}{2} S_0 e^{-\Sigma_t(x/\mu)} \frac{d(\frac{x}{\mu})}{(\frac{x}{\mu})} = -\frac{1}{2} S_0 e^{-\Sigma_t(x/\mu)} \frac{d\mu}{\mu}$$

(3.143)

This expression can be used to construct the total flux of uncollided neutrons passing through unit area at a point on the x-axis by integrating over all such rings to obtain

$$\phi(x : 0) = \frac{1}{2} S_0 \int_x^\infty e^{-\Sigma_t R} \frac{dR}{R} = \frac{1}{2} S_0 \int_0^1 e^{-\Sigma_t x/\mu} \frac{d\mu}{\mu} \equiv \frac{1}{2} S_0 E_1(\Sigma_t x) \quad (3.144)$$

and a similar expression for the total current of uncollided neutrons

$$J(x : 0) = \frac{1}{2} S_0 \int_x^\infty \mu e^{-\Sigma_t R} \frac{dR}{R} = \frac{1}{2} S_0 \int_0^1 e^{-\Sigma_t x/\mu} d\mu = \frac{1}{2} S_0 E_2(\Sigma_t x)$$

(3.145)

where

$$E_n(y) \equiv \int_0^1 \mu^{n-2} e^{-y/\mu} d\mu$$

(3.146)

is the exponential integral function, which has the useful differentiation property

$$\frac{dE_n(y)}{dy} = -E_{n-1}(y), \quad n = 1, 2, 3, \dots$$

(3.147)

Transmission and Absorption in a Purely Absorbing Slab Control Plate

Consider a purely absorbing slab of thickness a inserted in a diffusing medium; i.e. a medium in which diffusion theory is appropriate, such as a homogenized fuel–moderator lattice. Assuming that the scattering source external to the slab is isotropic, the probability that a neutron incident on the slab is transmitted across the absorbing slab is $T \equiv J(a:0)/J(0:0) = E_2(\Sigma_t a)$, and the probability that an incident neutron is absorbed in the slab is $A \equiv 1 - T = 1 - E_2(\Sigma_t a)$. Such an approximation was used in deriving the effective control cross section in Section 3.9.

Escape Probability in a Slab

The probability that a neutron that enters a slab region of thickness a from a diffusing medium and suffers an isotropic scattering event at a location $0 \le x' \le a$ and then escapes from the slab across the surface at $x = a$ without another collision can be calculated by treating the scattering event as an isotropic source; i.e. the escaping current at $x = a$ of 'uncollided' neutrons arising from the isotropic scattering at $x = x'$ is $J(a:x') = (1/2)S_0 E_2(\Sigma_t(a - x'))$. If the scattering source of incident neutrons is uniform over the slab, then the total current of scattered neutrons which escapes across the surface at $x = a$ without another collision is

$$J_{out}(a) = \int_0^a J(a:x')dx'' = \frac{1}{2}S_0 \int_0^a E_2(\Sigma_t(a-x'))dx'$$

$$= \frac{1}{2}\frac{S_0}{\Sigma_t}[E_3(0) - E_3(\Sigma_t a)] \tag{3.148}$$

Noting that $E_3(0) = 1/2$ and that the current of neutrons which escapes across the surface at $x = 0$ without another collision must be the same as escape across the surface at $x = a$, the 'first-flight' escape probability is the sum of these two escaping currents divided by the total scattering source, aS_0

$$P_0 \equiv \frac{J_{out}(a) + J_{out}(0)}{a S_0} = \frac{1}{a\Sigma_t}\left[\frac{1}{2} - E_3(\Sigma_t a)\right] \tag{3.149}$$

1st flight escape prob

Integral Transport Formulation

Consider again the slab with a distributed isotropic source of neutrons, but now with isotropic elastic scattering and fission, as well as absorption, represented explicitly. The flux of uncollided source neutrons is

$$\phi_0(x) = \int_0^a \frac{1}{2}S_0(x')E_1(\Sigma_t(|x - x'|))dx' \tag{3.150}$$

If the first-collision rate at $x = x'$ is considered as a plane isotropic source of once-collided neutrons at x', then the flux of once-collided neutrons at x due to the once-collided source at x' is

$$\phi_1(x:x') = \frac{1}{2}(\Sigma_s(x') + \nu\Sigma_f(x'))\phi_0(x')E_1(\Sigma_t(|x - x'|)) \tag{3.151}$$

and the total flux of once-collided neutrons at x is found by integrating over the distribution of first-collision sources

$$\phi_1(x) = \int_0^a \phi_1(x:x')dx'$$

$$= \int_0^a \left(\frac{1}{2}(\Sigma_s(x') + \nu\Sigma_f(x'))\phi_0(x') \right) E_1\left(\Sigma_t(|x-x'|)\right)dx' \qquad (3.152)$$

Continuing in this vein, the flux of n-collided neutrons is given by

$$\phi_n(x) = \int_0^a \left(\frac{1}{2}(\Sigma_s(x') + \nu\Sigma_f(x'))\phi_{n-1}(x') \right) E_1\left(\Sigma_t(|x-x'|)\right)dx',$$

$$n = 1, 2, 3, \ldots \qquad (3.153)$$

The total neutron flux is the sum of the uncollided, once-collided, twice-collided, etc. fluxes

$$\phi(x) \equiv \phi_0(x) + \sum_{n=1}^{\infty} \phi_n(x)$$

$$= \int_0^a \frac{1}{2}(\Sigma_s(x') + \nu\Sigma_f(x')) \sum_{n=0}^{\infty} \phi_{n-1}(x') E_1\left(\Sigma_t(|x-x'|)\right)dx'$$

$$+ \int_0^a (S_0(x')E_1\left(\Sigma_t(|x-x'|)\right))dx'$$

$$= \int_0^a \frac{1}{2}(\Sigma_s(x') + \nu\Sigma_f(x'))\phi(x')E_1\left(\Sigma_t(|x-x'|)\right)dx'$$

$$+ \int_0^a S_0(x')E_1\left(\Sigma_t(|x-x'|)\right)dx' \qquad (3.154)$$

Thus, we have found an integral equation for the neutron flux in a slab with isotropic scattering and fission, with a kernel $(1/2)(\Sigma_s(x') + \nu\Sigma_f(x'))E_1(\Sigma_t(|x-x'|))$ and a first collision or external source $S_0 E_1(\Sigma_t x')$.

Collision Probability Method

If the volume of the slab is partitioned into discrete slab regions centered at x_i with widths $\Delta_i \equiv x_{i+1/2} - x_{i-1/2}$ within each of which uniform average cross sections and a flat flux is assumed, Eq. (3.154) can be integrated over the volume of each region, $V_i = \text{const.} \times \Delta_i$ and the resulting equation can be divided by V_i to obtain

$$\phi_i = \sum_j T^{j \to i}\left[(\Sigma_{sj} + \nu\Sigma_{fj})\phi_j + S_{0j}\right] \qquad (3.155)$$

which relates the fluxes in the various volumes by the 'first-flight transmission probabilities' $T^{j \to i}$

$$T^{j \to i} \equiv \frac{1}{\Delta_j} \int_{\Delta_i} dx' \int_{\Delta_j} dx \frac{1}{2} E_1\left(\alpha(x', x)\right) \qquad (3.156)$$

where the "optical thickness" is defined

$$\alpha(x', x) \equiv \left| \int_{x'}^{x} \Sigma_t(x'')dx'' \right| \tag{3.157}$$

Since $\alpha(x_j, x_i) = \alpha(x_i, x_j)$—i.e. the optical path is the same no matter which way the neutron traverses the straight-line distance between x_i and x_j—there is a reciprocity relation between the transmission probabilities

$$\Delta_i T^{j \to i} = \Delta_j T^{i \to j} \tag{3.158}$$

Upon multiplication by $\Sigma_{ti}\Delta_i$, Eq. (3.155) can be written

$$\Sigma_{ti}\Delta_i\phi_i = \sum_j P^{ji} \frac{[(\Sigma_{sj} + \nu\Sigma_{fj})\phi_j + S_{0j}]}{\Sigma_{tj}} \tag{3.159}$$

where the collision rate in cell i is related to the neutrons introduced by scattering, fission and an external source in all cells j by the 'collision probabilities'

$$P^{ji} = \Sigma_{ti}\Sigma_{tj} \int_{\Delta_i} dx' \int_{\Delta_j} dx \frac{1}{2} E_1(\alpha(x', x)) \tag{3.160}$$

Because $\alpha(x_i, x_j) = \alpha(x_j, x_i)$, there is also reciprocity between the collision probabilities; i.e.

$$P^{ij} = P^{ji} \tag{3.161}$$

For $j \neq i$, the probability that a neutron introduced in cell j has its next collision in cell i is

$$P^{ji} = \frac{1}{2}\left[E_3(\alpha_{i+1/2, j+1/2}) - E_3(\alpha_{i-1/2, j+1/2}) - E_3(\alpha_{i+1/2, j-1/2}) \right.$$
$$\left. + E_3(\alpha_{i-1/2, j-1/2}) \right] \tag{3.162}$$

where $\alpha_{i,j} \equiv \alpha(x_i, x_j)$.

For $j = i$, a similar development leads to an expression for the probability that a neutron introduced in cell i has its next collision in cell i

$$P^{ii} = \Sigma_{ti}\Delta_i\left[1 - \frac{1}{2\Sigma_{ti}\Delta_i}\left(1 - 2E_3(\Sigma_{ti}\Delta_i) \right) \right] \tag{3.163}$$

The set of Eqs. (3.159) can be solved for the neutron flux in all the cells.

Differential Transport Formulation

Another formulation of neutron transport theory follows an approach similar to that used in Section 3.1 to derive diffusion theory, but without some of the limiting assumptions. Referring to Fig. 3.9, the change in the flux of particles moving along the cone of radius vectors R making the same angle ($\mu = \cos\theta$) with respect to a point on the x-axis within the differential distance ΔR can be written

$$\psi(R + \Delta R, \mu) = \psi(R, \mu) + \Delta R \int_{-1}^{1} \Sigma_s(R, \mu' \to \mu)\psi(R, \mu')d\mu'$$

$$+ \frac{1}{2}\Delta R \int_{-1}^{1} \nu\Sigma_f(R)\psi(R, \mu')d\mu'$$

$$- \Delta R[\Sigma_s(R) + \Sigma_a(R)]\psi(R, \mu) \tag{3.164}$$

The second term on the right is the source of neutrons within ΔR with direction μ due to scattering within ΔR by neutrons with other directions μ' (including μ). The third term on the right is the number of source neutrons within ΔR with direction μ due to isotropic fission produced by neutrons with other directions μ' (including μ). The last term on the right represents the rate at which neutrons within ΔR with direction μ are being lost by scattering to some other direction μ' (including μ) and by absorption. The first assumption made is that the directional flux can be represented by the first two terms of a Taylor's series, $\psi(R + \Delta R, \mu) \simeq \psi(R, \mu) + \Delta R[d\psi(R, \mu)/dR]$. Noting that the spatial non-uniformity depends on the variable x and that $\mu = x/R$, Eq. (3.164) becomes in the differential limit $\Delta R \to dR$ the one-dimensional slab version of the Boltzmann transport equation

$$\mu\frac{d\psi(R, \mu)}{dx} + [\Sigma_s(R) + \Sigma_a(R)]\psi(R, \mu)$$

$$= \int_{-1}^{1} \Sigma_s(R, \mu' \to \mu)\psi(R, \mu')d\mu' + \frac{1}{2}\int_{-1}^{1} \nu\Sigma_f(R)\psi(R, \mu')d\mu' \tag{3.165}$$

Spherical Harmonics Methods

The spherical harmonics, or P_L, approximation is developed by expansion of the angular dependence of the angular flux and of the differential scattering cross section in *Legendre polynomials*. The first few Legendre polynomials are

$$P_0(\mu) = 1, \qquad P_2(\mu) = \frac{1}{2}\left(3\mu^2 - 1\right)$$

$$P_1(\mu) = \mu, \qquad P_3(\mu) = \frac{1}{2}\left(5\mu^3 - 3\mu\right) \tag{3.166}$$

and higher order polynomials can be generated from the recursion relation

$$(2n + 1)\mu P_n(\mu) = (n + 1)P_{n+1}(\mu) + nP_{n-1}(\mu) \tag{3.167}$$

The Legendre polynomials satisfy the orthogonality relation

$$\int_{-1}^{1} d\mu\, P_m(\mu)P_n(\mu) = \frac{2\delta_{mn}}{2n + 1} \tag{3.168}$$

The cosine of the scattering angle between μ' and μ, can be expressed in terms of the Legendre polynomials of μ' and μ by the addition theorem in one-dimension

$$P_n(\mu_0) = P_n(\mu)P_n(\mu') \tag{3.169}$$

The P_L equations are based on the approximation that the angular dependence of the neutron flux can be expanded in the first $L+1$ Legendre polynomials

$$\psi(x,\mu) = \sum_{l=0}^{L} \frac{2l+1}{2}\phi_l(x)P_l(\mu) \tag{3.170}$$

The angular dependence of the differential scattering cross section is also expanded in Legendre polynomials

$$\Sigma_s(x,\mu_0) = \sum_{m=0}^{M} \frac{2m+1}{2}\Sigma_{sm}(x)P_m(\mu_0) \tag{3.171}$$

When these expansions are used in Eq. (3.165), the addition theorem of Eq. (3.169) is used to replace $P_m(\mu_0)$ with $P_m(\mu)P_m(\mu')$, the recursion relation of Eq. (3.167) is used to replace $\mu P_n(\mu)$ terms with $P_{n\pm1}(\mu)$ terms, the resulting equation is multiplied in turn by $P_k(\mu)$ ($k=0$ to L) and integrated over $-1 \le \mu \le 1$, and the orthogonality relation of Eq. (3.168) is used, the $L+1$ P_L equations

$$\frac{d\phi_1(x)}{dx} + (\Sigma_t - \Sigma_{so})\phi_0(x) = S_0(x), \quad n=0$$
$$\frac{(n+1)}{(2n+1)}\frac{d\phi_{n+1}(x)}{dx} + \frac{n}{(2n+1)}\frac{d\phi_{n-1}}{dx} + (\Sigma_t - \Sigma_{sn})\phi_n(x) = S_n(x), \tag{3.172}$$
$$n=1,\ldots,L$$

are obtained. The n subscript indicates the nth Legendre moment of the direction flux, source and scattering cross section

$$\phi_n(x) \equiv \int_{-1}^{1} d\mu\, P_n(\mu)\psi(x,\mu)$$

$$S_n(x) \equiv \int_{-1}^{1} d\mu\, P_n(\mu)S(x,\mu) \tag{3.173}$$

$$\Sigma_{sn}(x) \equiv \int_{-1}^{1} d\mu_0\, P_n(\mu_0)\Sigma_s(x,\mu_0)$$

This set of $L+1$ equations has a closure problem—they involve $L+2$ unknowns. This problem is usually resolved by ignoring the term $d\phi_{L+1}/dx$ which appears in the $n=L$ equation.

Boundary and interface conditions
The true boundary condition at the left boundary x_L

$$\psi(x_L,\mu) = \psi_{in}(x_L,\mu), \quad \mu > 0 \tag{3.174}$$

where $\psi_{in}(x_L, \mu > 0)$ is a known incident flux ($\psi_{in}(x_L, \mu > 0) = 0$ is the vacuum boundary condition), cannot be satisfied exactly by the angular flux approximation of Eq. (3.170), for finite L.

The most obvious way to develop approximate boundary conditions which are consistent with the flux approximation is to substitute Eq. (3.170) into the exact boundary condition of (3.174), multiply by $P_m(\mu)$, and integrate over $0 \leq \mu \leq 1$. Since it is the odd Legendre polynomials which represent directionality (i.e. are different for μ and $-\mu$), this procedure is repeated for all the odd Legendre polynomials $m = 1, 3, \ldots, L$ (or $L-1$) as weighting functions to obtain, with the use of the orthogonality relation of Eq. (3.168), the *Marshak boundary conditions*

$$\int_0^1 d\mu \, P_m(\mu) \sum_{n=0}^{N} \left(\frac{2n+1}{2} \right) \phi_n(x_L) P_n(\mu) \equiv \phi_m(x_L)$$

$$= \int_0^1 d\mu \, P_m(\mu) \psi_{in}(x_L, \mu), \quad m = 1, 3, \ldots, L \text{ (or } L-1) \tag{3.175}$$

Equations (3.175) constitute a set of $(L+1)/2$ boundary conditions. An additional $(L+1)/2$ boundary conditions are similarly obtain for the right boundary. The Marshak boundary conditions insure that the exact inward partial current at the boundary is incorporated into the solution; i.e.

$$J^+(x_L) \equiv \int_0^1 d\mu \, P_1(\mu) \sum_{n=0}^{N} \left(\frac{2n+1}{2} \right) \phi_n(x_L) P_n(\mu)$$

$$= \int_0^1 d\mu \, P_1(\mu) \psi_{in}(x_L, \mu) \equiv J_{in}^+(x_L) \tag{3.176}$$

A symmetry, or reflective, boundary condition $\psi(x_L, \mu) = \psi(x_L, -\mu)$ obviously requires that all odd moments of the flux vanish; i.e. $\phi_n(x_L) = 0$ for $n = 1, 3, \ldots$ odd.

The exact *interface condition* of continuity of angular flux

$$\psi(x_s - \varepsilon, \mu) = \psi(x_s + \varepsilon, \mu) \tag{3.177}$$

where ε is a vanishingly small distance, cannot, of course, be satisfied exactly by the flux approximation of Eq. (3.170), for finite L. Following the same procedure as for Marshak boundary conditions, we replace the exact flux with the expansion of Eq. (3.170) and require that the first $L+1$ Legendre moments of this relation be satisfied (i.e. multiply by P_m and integrate over $-1 \leq \mu \leq 1$, for $m = 0, \ldots, L$). Using the orthogonality relation of Eq. (3.168) then leads to the interface conditions of continuity of the moments

$$\phi_n(x - \varepsilon) = \phi_n(x_s + \varepsilon), \quad n = 0, 1, 2, \ldots, L \tag{3.178}$$

P_1 equations and diffusion theory

Neglecting the $d\phi_2/dx$ term, the first two of Eqs. (3.172) constitute the P_1 equations

$$\frac{d\phi_1}{dx} + (\Sigma_t - \Sigma_{so})\phi_0 = S_0$$

$$\frac{1}{3}\frac{d\phi_0}{dx} + (\Sigma_t - \Sigma_{s1})\phi_1 = S_1 \tag{3.179}$$

Noting that $\Sigma_{s0} = \Sigma_s$, the total scattering cross section, and that $\Sigma_{s1} = \bar{\mu}_0\Sigma_s$, where $\bar{\mu}_0$ is the average cosine of the scattering angle, and assuming that the source is isotropic (i.e. $S_1 = 0$), the second of these P_1 equations yields a *Fick's law* for neutron diffusion

$$\phi_1(x) = \int_{-1}^{1} \mu\psi(x,\mu)\,d\mu \equiv J(x) = -\frac{1}{3(\Sigma_t - \bar{\mu}_0\Sigma_s)}\frac{d\phi_0}{dx} \tag{3.180}$$

which, when used in the first of the P_1 equations yields the neutron diffusion equation

$$-\frac{d}{dx}\left(D_0(x)\frac{d\phi_0(x)}{dx}\right) + (\Sigma_t - \Sigma_s)\phi_0(x) = S_0(x) \tag{3.181}$$

where the diffusion coefficient and the transport cross section are defined by

$$D_0 \equiv \frac{1}{3(\Sigma_t - \bar{\mu}_0\Sigma_s)} \equiv \frac{1}{3\Sigma_{\mathrm{tr}}} \tag{3.182}$$

The basic assumptions made in this derivation of diffusion theory are that the angular dependence of the neutron flux is linearly anisotropic

$$\psi(x,\mu) \simeq \frac{1}{2}\phi_0(x) + \frac{3}{2}\mu\phi_1(x) \tag{3.183}$$

and that the neutron source is isotropic, or at least has no linearly anisotropic component ($S_1 = 0$). Diffusion theory should be a good approximation when these assumptions are valid; i.e. in media for which the distribution is almost isotropic because of the preponderance of randomizing scattering collisions, away from interfaces with dissimilar media, and in the absence of anisotropic sources.

The boundary conditions for diffusion theory follow directly from the Marshak condition (3.175)

$$J_{\mathrm{in}}^+(x_L) = \int_0^1 d\mu\, P_1(\mu)\left[\frac{1}{2}\phi_0(x_L) + \frac{3}{2}\mu\phi_1(x_L)\right]$$

$$= \frac{1}{4}\phi_0(x_L) - \frac{1}{2}D\frac{d\phi_0(x_L)}{dx} \tag{3.184}$$

When the prescribed incident current, $J_{\mathrm{in}}^+ = 0$, the vacuum boundary condition for diffusion theory can be constructed from a geometrical interpretation of the ratio of the flux gradient to the flux in this equation to obtain the condition that the extrapolated flux vanishes a distance $\lambda_{tr} = 1/\Sigma_{tr}$ outside the boundary

$$\phi(x_L - \lambda_{ex}) = 0, \qquad \lambda_{ex} = \frac{2}{3\Sigma_{tr}} \tag{3.185}$$

The interface conditions of Eq. (3.178) become in the diffusion approximation

$$\phi_0(x_s + \varepsilon) = \phi(x_s + \varepsilon)$$

$$-D(x_s + \varepsilon)\frac{d\phi_0(x_s + \varepsilon)}{dx} = -D(x_s - \varepsilon)\frac{d\phi_0(x_s - \varepsilon)}{dx} \tag{3.186}$$

Discrete Ordinates Method

The discrete ordinate method is based on a conceptually straightforward evaluation of the transport equation at a few discrete angular directions, or ordinates, and the use of quadrature relationships to replace scattering and fission neutron source integrals over angle with summations over ordinates. The essence of the method is the choice of ordinates, quadrature weights, differencing schemes and iterative solution procedures. In one dimension, the ordinates can be chosen such that the discrete ordinates methods are completely equivalent to the P_L method discussed above, and in fact the use of discrete ordinates is probably the most effective way to solve the P_L and $D - P_L$ equations in one dimension.

Making use of the spherical harmonics expansion of the differential scattering cross section of Eq. (3.171) and the addition theorem for Legendre polynomials of Eq. (3.169), the one dimensional neutron transport equation (3.165) in slab geometry becomes

$$\mu\frac{d\psi}{dx}(x,\mu) + \Sigma_t(x)\psi(x,\mu)$$

$$= \sum_{l'=0}\left(\frac{2l'+1}{2}\right)\Sigma_{sl'}(x)P_{l'}(\mu)\int_{-1}^{1}d\mu' P_{l'}(\mu')\psi(x,\mu') + S(x,\mu) \tag{3.187}$$

where the source term includes an external source and, in the case of a multiplying medium such as a reactor core, a fission source. We will first discuss the solution of the fixed external source problem (which implicitly assumes a subcritical reactor) and then return to the solution of the critical reactor problem, in which the solution of the fixed source problem constitutes part of the iteration strategy.

Defining N ordinate directions, μ_n, and corresponding quadrature weights, w_n, the integral over angle in Eq. (3.187) can be replaced by

$$\phi_l(x) \equiv \int_{-1}^{1}d\mu P_l(\mu)\psi(x,\mu)\sum_n w_n P_l(\mu_n)\psi_n(x) \tag{3.188}$$

where $\psi_n \equiv \psi(\mu_n)$. The quadrature weights are normalized by

$$\sum_{n=1}^{N} w_n = 2 \tag{3.189}$$

It is convenient to choose ordinates and quadrature weights that are symmetric about $\mu = 0$, hence providing equal detail in the description of forward and backward neutron fluxes. This can be accomplished by choosing

$$\mu_{N+1-n} = -\mu_n, \quad \mu_n > 0, \quad n = 1, 2, \ldots, N/2$$

$$w_{N+1-n} = w_n, \quad w_n > 0, \quad n = 1, 2, \ldots, N/2$$

(3.190)

With such even ordinates, reflective boundary conditions are simply prescribed

$$\psi_n = \psi_{N+1-n}, \quad n = 1, 2, \ldots, N/2 \tag{3.191}$$

Known incident flux, $\psi_{in}(\mu)$, boundary conditions, including vacuum conditions when $\psi_{in}(\mu) = 0$, are

$$\psi_n = \psi_{in}(\mu_n), \quad n = 1, 2, \ldots, N/2 \tag{3.192}$$

Normally, an even number of ordinates is used ($N =$ even), because this results in the correct number of boundary conditions and avoids certain other problems encountered with $N =$ odd. Even with these restrictions, there remains considerable freedom in the choice of ordinates and weights.

If the ordinates are chosen to be the L roots of the Legendre polynomial of order N

$$P_N(\mu_i) = 0 \tag{3.193}$$

and the weights are chosen to correctly integrate all Legendre polynomials up to P_{N-1}

$$\int_{-1}^{1} P_l(\mu) d\mu = \sum_{n=1}^{N} w_n P_l(\mu_n) = 2\delta_{l0}, \quad l = 0, 1, \ldots, N-1 \tag{3.194}$$

then the discrete ordinates equations with N ordinates are entirely equivalent to the P_{N-1} equations. To establish this, we multiply Eq. (3.187) by $w_n P_l(\mu_n)$ for $0 \leq l \leq N-1$, in turn, and use the recursion relation of Eq. (3.167) to obtain

$$w_n \left[\left(\frac{l+1}{2l+1} \right) P_{l+1}(\mu_n) + \left(\frac{l}{2l+1} \right) P_{l-1}(\mu_n) \right] \frac{d\psi_n}{dx} + w_n \Sigma_t \psi_n$$

$$= \sum_{l'=0}^{N-1} \left(\frac{2l'+1}{2} \right) \Sigma_{sl'} w_n P_{l'}(\mu_n) P_l(\mu_n) \phi_{l'} + w_n P_l(\mu_n) S(\mu_n),$$

$$l = 0, \ldots, N-1, \; n = 1, \ldots, N \tag{3.195}$$

Summing these equations over $1 \leq n \leq N$ yields

$$\left(\frac{l+1}{2l+1} \right) \frac{d\phi_{l+1}}{dx} + \left(\frac{l}{2l+1} \right) \frac{d\phi_{l-1}}{dx} + \Sigma_t \phi_l$$

$$
= \sum_{l'=0}^{N-1} \left(\frac{2l'+1}{2} \right) \Sigma_{sl'} \phi_{l'} \left[\sum_{n=0}^{N} w_n P_{l'}(\mu_n) P_l(\mu_n) \right] + \sum_{n=1}^{N} w_n P_l(\mu_n) S(\mu_n),
$$

$$
l = 0, \ldots, N-1 \tag{3.196}
$$

Weights chosen to satisfy Eqs. (3.194) obviously correctly integrate all polynomials through order N (any polynomial of order n can be written as a sum of Legendre polynomials through order n), but fortuitously they also integrate correctly all polynomials through order less than $2N$. Thus, the term in the scattering integral becomes

$$
\sum_{n=1}^{N} w_n P_{l'}(\mu_n) P_l(\mu_n) = \int_{-1}^{1} P_{l'}(\mu) P_l(\mu) d\mu = \frac{2\delta_{ll'}}{2l+1} \tag{3.197}
$$

and assuming that the angular dependence of the source term can be represented by a polynomial of order $< 2N$

$$
\sum_{n=1}^{N} w_n P_l(\mu_n) S(\mu_n) = \int_{-1}^{1} P_l(\mu) S(\mu) d\mu = \frac{2S_l}{2l+1} \tag{3.198}
$$

where S_l is the Legendre moment of the source given by Eq. (3.173).

Using Eqs. (3.197) and (3.198), Eqs. (3.196) become

$$
\left(\frac{l+1}{2l+1} \right) \frac{d\phi_{l+1}}{dx} + \left(\frac{l}{2l+1} \right) \frac{d\phi_{l-1}}{dx} + (\Sigma_t - \Sigma_{sl}) \phi_l = S_l
$$

$$
l = 0, \ldots, N-2
$$

$$
\frac{N-1}{2(N-1)+1} \frac{d\phi_{(N-1)-1}}{dx} + (\Sigma_t - \Sigma_{s,N-1}) \phi_{N-1} = S_{N-1}
$$

$$
l = N-1
$$

$$(3.199)$$

which, when ϕ_{-1} is set to zero, are identically the P_L equations (3.172) for $L = N-1$.

References

1 D. R. Vondy, "Diffusion Theory," in Y. Ronen, ed., *CRC Handbook of Nuclear Reactor Calculations I*, CRC Press, Boca Raton, FL (1986).

2 R. J. J. Stamm'ler and M. J. Abbate, *Methods of Steady-State Reactor Physics in Nuclear Design*, Academic Press, London (1983), Chap. 5.

3 J. R. Lamarsh, *Introduction to Nuclear Reactor Theory*, 2nd ed., Addison-

Wesley, Reading, MA (1983), Chaps. 5, 6, 8, 9, and 10.

4 J. J. Duderstadt and L. J. Hamilton, *Nuclear Reactor Analysis*, Wiley, New York (1976), pp. 149–232 and 537–556.

5 A. F. Henry, *Nuclear-Reactor Analysis*, MIT Press, Cambridge, MA (1975), pp. 115–199.

6 G. I. Bell and S. Glasstone, *Nuclear Reactor Theory*, Van Nostrand Reinhold, New York (1970), pp. 89–91, 104–105, and 151–157.

7 M. K. BUTLER and J. M. COOK, "One Dimensional Diffusion Theory," and A. HASSITT, "Diffusion Theory in Two and Three Dimensions," in H. Greenspan, C. N. Kelber, and D. Okrent, eds., *Computing Methods in Reactor Physics*, Gordon and Breach, New York (1968).

8 E. L. WACHSPRESS, *Iterative Solution of Elliptic Systems*, Prentice Hall, Englewood Cliffs, NJ (1966). Another widely used method for solving the two-dimensional diffusion equations is the *alternating direction implicit* iteration scheme described in Section 14.3.

9 M. CLARKE and K. F. HANSEN, *Numerical Methods of Reactor Analysis*, Academic Press, New York (1964).

10 R. V. MEGHREBLIAN and D. K. HOLMES, *Reactor Analysis*, McGraw-Hill, New York (1960), pp. 160–267 and 626–747.

11 A. M. WEINBERG and E. P. WIGNER, *The Physical Theory of Neutron Chain Reactors*, University of Chicago Press, Chicago (1958), pp. 181–218, 495–500, and 615–655.

12 S. GLASSTONE and M. C. EDLUND, *Nuclear Reactor Theory*, D. Van Nostrand, Princeton, NJ (1952), pp. 90–136, 236–272, and 279–289.

Problems

3.1. Plot the neutron flux distribution given by Eq. (3.24) from $r = 0$ to $r = 25$ cm away from a point thermal neutron source in an infinite medium of (a) H_2O ($L = 2.9$ cm, $D = 0.16$ cm); (b) D_2O ($L = 170$ cm, $D = 0.9$ cm); and (c) graphite ($L = 60$ cm, $D = 0.8$ cm).

3.2. Plot the neutron flux distribution in a finite slab of width $2a = 10$ cm with an incident thermal neutron beam from the left, as given by Eq. (3.36), for an iron slab ($\Sigma_t = 1.15$ cm^{-1}, $D = 0.36$ cm, $L = 1.3$ cm).

3.3. Derive the albedo boundary condition of Eq. (3.38) from the definition of the albedo, $\alpha \equiv j_-/j_+$, and the diffusion theory expressions for partial currents, Eqs. (3.4) and (3.5).

2 lines work

3.4. A thermal diffusion-length experiment is performed by placing a block of diffusing medium with $a_{ex} = b_{ex} = 175.7$ cm adjacent to a reactor thermal column port and irradiating a series of indium foils placed along the z-axis of the block. The saturation activity (disintegrations/min) of foils at various locations is (40,000 at $z = 28$ cm), (29,000 at $z = 40$ cm), (20,000 at $z = 45$ cm), (17,000 at z = 56 cm), (10,000 at $z = 70$ cm), (8500 at $z = 76$ cm), (5800 at 90 cm), and (3500 at 100 cm). The experimental error is $\pm 10\%$. Determine the thermal neutron diffusion length.

Section 3.5

3.5. Derive the criticality condition for a bare rectangular parallelepiped core of x-dimension a, y-dimension b, and z-dimension c.

3.6. A typical composition for a PWR core is: H, 2.75×10^{22} cm^{-3}; O, 2.76×10^{22} cm^{-3}; Zr, 3.69×10^{21} cm^{-3}; Fe, 1.71×10^{21} cm^{-3}; ^{235}U 1.91×10^{20} cm^{-3}; ^{238}U, 6.59×10^{21}cm^{-3}; and ^{10}B, 1×10^{19} cm^{-3}. Appropriate spectrum-averaged microscopic cross sections (barns) for these isotopes are $\sigma_{\mathrm{tr}}/\sigma_a/\nu\sigma_f = 0.65/0.29/0.0$ for H, $0.26/0.0002/0.0$ for O, $0.79/0.19/0.0$ for Zr, $0.55/2.33/0.0$ for Fe, $1.62/484.0/758.0$ for ^{235}U, $1.06/2.11/1.82$ for ^{238}U, and $0.89/3410.0/0.0$ for ^{10}B. Calculate the critical radius for a right circular cylindrical bare core of fixed height $H = 375$ cm.

3.7. Calculate the critical radius for the right circular cylindrical core of Problem 3.6 with a 20-cm-thick side reflector with $D_R = 1$ cm and $\Sigma_{aR} = 0.01$ cm^{-1}.

3.8. Calculate the thermal flux disadvantage factor for UO_2 rods varying from 0.5 to 2.0 cm in diameter in an H_2O moderator for V_M/V_F varying from 1.0 to 4.0. Calculate the corresponding effective homogeneous absorption cross sections and thermal utilization. Plot the results.

3.9. Derive an expression analogous to Eq. (3.100) for the power peaking factor in a fuel–moderator assembly with cylindrical fuel elements.

3.10. Derive an expression for the effective diffusion theory absorption cross section for a cylindrical control rod of radius a surrounded by an annular region of fuel–moderator extending from $r = a$ to $r = R$. The transport parameter for this geometry is given by $(1/3\alpha) = 0.7104 + 0.2524/a\Sigma_{aC} + 0.0949/(a\Sigma_{aC})^2 + \cdots$.

3.11.* Jezebel is a bare, critical, spherical fast reactor assembly with radius 6.3 cm constructed of ^{239}Pu metal (density 15.4 g/cm^3). Using the one-group constants $\nu = 2.98$, $\sigma_f = 1.85$ barns (1 barn $= 10^{-24}$ cm^2) $\sigma_a = 2.11$ barns, and $\sigma_{\mathrm{tr}} = 6.8$ barns and the finite-difference numerical method, calculate the effective multiplication constant, $\lambda = k_{\mathrm{eff}}$, predicted by diffusion theory. $\lambda - 1$ is a measure of the accuracy of diffusion theory for this assembly. Should diffusion theory be valid for this assembly?

---

* Problems 11 to 13 are longer problems suitable for take-home projects.

3.12.* Solve numerically for the eigenvalue and neutron flux distribution in a slab reactor consisting of two adjacent core regions each of thickness 50 cm, with a 25-cm-thick reflector on each side. The nuclear parameters of the two core regions are ($D = 0.65$ cm, $\Sigma_a = 0.12$ cm^{-1}, and $\nu\Sigma_f = 0.125$ cm$^-1$) and ($D = 0.75$ cm, $\Sigma_a = 0.10$ cm^{-1}, and $\nu\Sigma_f = 0.12$ cm^{-1}), and the parameters of the reflector are ($D = 1.15$ cm, $\Sigma_a = 0.01$ cm^{-1}, and $\nu\Sigma_f = 0.0$ cm^{-1}). Solve this problem analytically and compare the answers.

3.13.* Calculate numerically the effective multiplication constant and the flux distribution in a reactor with rectangular (x, y) cross section which is sufficiently tall that axial (z) leakage can be neglected. The core cross section in the x–y plane consists of four symmetric quadrants. The upper right quadrant consists of core region 1, rectangular ($0 < x < 50$ cm, $60 < y < 100$ cm); core region 2, rectangular ($0 < x < 50$ cm, $0 < y < 60$ cm); and reflector region 3, also rectangular ($50 < x < 100$ cm, $0 < y < 100$ cm). The nuclear parameters are: core region 1 ($D = 0.65$ cm, $\Sigma_a = 0.12$ cm^{-1}, $\nu\Sigma_f = 0.185$); core region 2 ($D = 0.75$ cm, $\Sigma_a = 0.10$ cm^{-1}, $\nu\Sigma_f = 0.15$); and reflector region 3 ($D = 1.15$ cm, $\Sigma_a = 0.01$ cm^{-1}, $\nu\Sigma_f = 0.0$ cm^{-1}). Use vacuum boundary conditions except on the boundaries ($x = 0$, $0 < y < 100$ cm) and ($0 < x < 100$ cm, $y = 0$), where a symmetry condition should be used. (This is a model for one-half of the symmetric reactor cross section.) Plot the x-direction flux distribution at $y = 30$ and 80 cm.

3.14. Calculate the thermal extrapolation distance λ_{extrap} for H_2O and for a 1:1 wt% homogeneous mixture of H_2O and 4% enriched uranium.

3.15. Estimate the maximum size of the mesh spacing that can be used in a finite-difference solution for the thermal neutron flux distribution in an H_2O medium and in a 1:1 wt% homogeneous mixture of H_2O and 4% enriched uranium.

3.16. Calculate and plot the thermal neutron flux distribution arising from a plane neutron source in an H_2O medium and in a 1:1 wt% homogeneous mixture of H_2O and 4% enriched uranium.

3.17. Repeat the calculation of Problem 3.16 in a carbon medium and in a 1:1 wt% homogeneous mixture of carbon and 4% enriched uranium.

3.18. Calculate the albedo boundary condition for the thermal neutron flux in a 1-m-thick slab medium with a 1:1 wt%

homogeneous mixture of H_2O and 4% enriched uranium which is bounded on both sides by very thick graphite slabs.

3.19. Using the microscopic cross sections and number densities (except for ^{235}U) of Table 3.4, determine the critical ^{235}U enrichment for a bare cylindrical core of height $H = 350$ cm and radius $R = 110$ cm. Repeat the calculation for $R = 100$ and 120 cm.

3.20. Repeat the calculation in Section 3.8 (Example 3.4) of flux disadvantage factor and effective homogenized fuel absorption cross section for a water thickness of 2 and 5 cm between fuel plates.

3.21. Calculate the power peaking factor in the slab lattices of Problem 3.20.

3.22. Repeat the calculation of the effective control slab cross section given in Section 3.9 (Example 3.5) for a control blade that contains only 2% natural boron.

3.23. Solve Problem 3.12 using a four-node model, one node for each reflector and core region. Compare the result with the results of Problem 3.12.

3.24. Discuss the approximations made in the derivation of neutron diffusion theory and how these approximations limit the validity of diffusion theory in reactor calculations. Explain why diffusion theory is not directly valid for the calculation of the flux distribution within the highly heterogeneous assembly of fuel, moderator, structure and control elements that make up a typical LWR fuel region. Describe how an effective homogeneous model of this heterogeneous assembly can be constructed for which diffusion theory is valid.

3.25. Calculate the critical ($k = 1$) dimension of a bare cube of multiplying medium with average macroscopic cross sections $\nu\Sigma_f = 0.115$ cm^{-1}, $\Sigma_a = 0.113$ cm^{-1}, $\Sigma_{\text{tr}} = 0.3$ cm^{-1}.

3.26. A single rod containing a spontaneous fission source is placed in a large volume of moderating material and surrounded at different radial distances with foils containing a material with a large $1/\upsilon$ capture cross section. The foil activation is measured. Explain how the radial distribution of the foil activation rate could be used to determine the migration length of neutrons in the moderating material.

4
Neutron Energy Distribution

Because the cross sections for neutron–nucleus reactions depend on energy, it is necessary to determine the energy distribution of neutrons in order to determine the rate of interactions of neutrons with matter, which in turn determines the transport of neutrons. We first address this problem by considering the neutron energy distribution in an infinite homogeneous medium, for which some analytical results can be obtained to provide physical insight. Then the important multigroup method for calculating an approximate neutron energy distribution is described. Methods for dealing with the rapidly varying neutron energy distribution in the energy range of cross-section resonances are described. Then the multigroup calculation of the neutron energy distribution is combined with the diffusion theory calculation of the spatial neutron distribution to obtain a powerful method for calculating the space- and energy-dependent neutron flux distribution in a nuclear reactor.

4.1
Analytical Solutions in an Infinite Medium

We start our investigation of the neutron energy distribution in a nuclear reactor by considering an infinite homogeneous medium in which spatial effects may be ignored. The neutron flux within a differential energy interval dE is determined by a balance between the source of fission neutrons being created within dE plus neutrons being scattered into dE from some other energy interval dE' and the loss of neutrons from within dE due to absorption and to scattering from dE into some other energy interval dE':

$$\left[\Sigma_a(E) + \Sigma_s(E)\right]\phi(E)dE$$
$$= \left[\int_0^\infty dE' \Sigma_s(E' \to E)\phi(E') + \frac{\chi(E)}{k_\infty}\int_0^\infty dE' \nu\Sigma_k(E')\phi(E')\right]dE \quad (4.1)$$

where we have included the infinite medium multiplication constant which may be adjusted to ensure that a steady-state solution exists.

Nuclear Reactor Physics. Weston M. Stacey
Copyright © 2007 WILEY-VCH Verlag GmbH & Co. KGaA, Weinheim
ISBN: 978-3-527-40679-1

Fission Source Energy Range

At very high energies, the direct source of fission neutrons into dE is much larger than the source of fission neutrons which have been created at higher energies and are slowing down into dE, in which case the first term on the right in Eq. (4.1) can be neglected in comparison to the second term, leading to

$$\phi^{(0)}(E) \simeq \frac{\chi(E)}{K_\infty \Sigma_t(E)} \int_0^\infty dE' \nu \Sigma_f(E')\phi(E') = \frac{\chi(E)}{\Sigma_t(E)} \times \text{const.} \qquad (4.2)$$

where $\Sigma_t = \Sigma_a + \Sigma_s$. Thus the neutron flux distribution at the higher energies, where direct fission neutrons are the principal source, is proportional to the fission spectrum divided by the total cross section.

This solution can be improved by using Eq. (4.2) as a first iterate on the right side of Eq. (4.1) to evaluate

$$\phi^{(1)}(E) = \frac{1}{\Sigma_t(E)} \left[\int_E^\infty dE' \Sigma_s(E' \to E)\phi^{(0)}(E') + \chi(E) \times \text{const.} \right]$$

$$= \frac{1}{\Sigma_t(E)} \left[\int_E^\infty dE' \Sigma_s(E' \to E)\frac{\chi(E')}{\Sigma_t(E')} + \chi(E) \right] \times \text{const.}$$

$$= \frac{\chi(E)}{\Sigma_t(E)} \times \text{const.} \left[1 + \frac{1}{\chi(E)} \int_E^\infty dE' \Sigma_s(E' \to E)\frac{\chi(E')}{\Sigma_t(E')} \right] \qquad (4.3)$$

where we have taken advantage of the fact that scattering of very energetic neutrons with much less energetic nuclei will result in an energy loss for the neutron to place a lower limit of E on the energies from which a neutron can scatter into dE. At the higher energies, where the fission source is important, inelastic scattering is also important and must be included in calculation of the correction factor. The improved energy distribution is also of the form of the fission spectrum divided by the total cross section times 1 plus a correction factor that obviously becomes large at lower energies where $\chi(E)$ becomes small. Numerical evaluation of the correction factor for typical compositions indicates that $\phi(E) = \chi(E)/\Sigma_t(E)$ represents the energy distribution rather well for energies $E > 0.5$ MeV.

Slowing-Down Energy Range

Very few fission neutrons are produced with energy less than about 50 keV. There is very little inelastic scattering in this energy range, so the elastic scattering transfer function

$$\Sigma_s(E' \to E) = \begin{cases} \dfrac{\Sigma_s(E')}{E'(1-\alpha)}, & E \le E' \le \dfrac{E}{\alpha} \\ 0, & \text{otherwise} \end{cases} \qquad (4.4)$$

can be used, where $\alpha \equiv [(A-1)/(A+1)]^2$ and A is the mass of the scattering nucleus in amu. If we limit our attention further to neutron energies greater than only about 1 eV, the neutrons will lose energy in a scattering collision, and we can write the slowing-down equation for the neutron energy distribution

$$\Sigma_t(E)\phi(E) = \sum_j \int_E^{E/\alpha_j} dE' \frac{\Sigma_s^j(E')\phi(E')}{E'(1-\alpha_j)} \tag{4.5}$$

where the sum is over the various nuclear species present.

Moderation by Hydrogen Only

Consider a mixture of hydrogen $\alpha_H \equiv [(A_H - 1)/(A_H + 1)]^2 = 0$ and very heavy mass nuclei $\alpha \equiv [(A - 1)/(A + 1)]^2 \approx 1$, for which Eq. (4.5) becomes

$$\Sigma_t(E)\phi(E) = \int_E^{\infty} \Sigma_s^H \phi(E') \frac{dE'}{E'} + \sum_{j \neq H} \int_E^{E/\alpha_j} dE' \frac{\Sigma_s^j(E')\phi(E')}{E'(1-\alpha_j)}$$

$$\simeq \int_E^{\infty} \Sigma_s^H \phi(E') \frac{dE'}{E'} + \sum_{j \neq H} \frac{1}{\alpha_j} \Sigma_s^j(E)\phi(E) \tag{4.6}$$

where the range of integration $E < E' < E/\alpha_j$ is so small for the heavy mass nuclei that the approximation $\Sigma_s^j(E')\phi(E')/E' \approx \Sigma_s^j(E)\phi(E)/E$ can be made. This equation can be rearranged:

$$[\Sigma_a(E) + \Sigma_s^H]\phi(E) = \int_E^{\infty} \Sigma_s^H \frac{\phi(E')}{E'} dE' \tag{4.7}$$

Differentiating Eq. (4.7), dividing both sides by $(\Sigma_a + \Sigma_s^H)\phi$ and integrating from E to some arbitrary upper energy E_1 leads to

$$\phi(E) = \frac{[\Sigma_a(E_1) + \Sigma_s^H]E_1\phi(E_1)}{[\Sigma_a(E) + \Sigma_s^H]E} \exp\left[-\int_E^{E_1} \frac{\Sigma_a(E')dE'}{[\Sigma_a(E') + \Sigma_s^H]E'}\right] \tag{4.8}$$

The neutron energy distribution varies with energy as $\phi(E) \sim 1/(\Sigma_a(E) + \Sigma_s^H)E$ and is exponentially attenuated in magnitude relative to the value at E_1 by any absorption that occurs over the interval $E_1 > E' > E$. The overall $1/E$ energy dependence of the flux is modified by the energy dependence of $\Sigma_a(E)$.

Energy Self-Shielding

In particular, if a large narrow resonance is present, $\Sigma_a(E)$ will increase sharply over the range of the resonance, causing $\phi(E) \sim 1/(\Sigma_a(E) + \Sigma_s^H)$ to dip sharply over this range where the resonance cross section is large, as indicated in Fig. 4.1. At energies just below the resonance, where $\Sigma_a(E)$ becomes small again, the flux recovers almost to the value just above the resonance, the difference being due to the absorption in the resonance. Physically, only those neutrons that are scattered into the energy range of the resonance will be absorbed, but those neutrons that

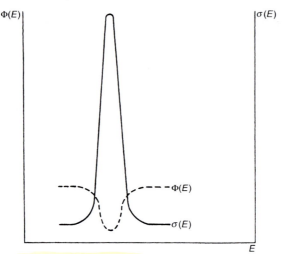

Fig. 4.1 Energy self-shielding of the neutron flux in a large absorption resonance. (From Ref. 6; used with permission of MIT Press.)

are scattered from energies above the resonance to energies below the resonance will not be affected by the presence of the resonance. This reduction in the neutron flux in the energy range of the resonance reduces the resonance absorption relative to what it would be if the effect of the resonance on the flux was not present, a phenomenon known as *energy self-shielding*.

We can obtain a rough estimate of the importance of energy self-shielding by calculating the exponential attenuation due to the resonance under the simplifying assumption that the resonance is very large over an energy width ΔE. Then the attenuation factor can be approximated:

$$\exp\left[-\int_E^{E_1} \frac{\Sigma_a(E')dE'}{[\Sigma_a(E') + \Sigma_s^H]E'} \right] \simeq \exp\left(-\int_E^{E+\Delta E} \frac{dE'}{E'} \right) = \frac{E}{E+\Delta E} \quad (4.9)$$

For the first large absorption resonance in ^{238}U at $E = 6.67$ eV, the width of the resonance is about $\Delta E = 0.027$ eV, which would absorb only about 4% of the neutrons slowing down past the resonance energy according to Eq. (4.9).

Slowing Down by Nonhydrogenic Moderators with No Absorption

The case of slowing down by only hydrogen provides valuable physical insight into features of the neutron energy distribution in the slowing-down energy range, most notably $\phi \sim 1/E$ and energy self-shielding of resonances, which remains valid under other moderating conditions. To gain some insight into the effect of various moderators on the neutron energy distribution, we now consider the case of moderation by nonhydrogen isotopes, first in the absence of absorption. The slowing-down balance equation is

$$\Sigma_s(E)\phi(E) \equiv \sum_j \Sigma_s^j(E)\phi(E) = \sum_j \int_E^{E/\alpha_j} dE' \frac{\Sigma_s^j(E')\phi(E')}{(1-\alpha_j)E'} \qquad (4.10)$$

Guided by the result for slowing down from hydrogen, we look for a solution of the form

$$\phi(E) = \frac{C}{E\Sigma_s(E)} = \frac{C}{E\sum_j \Sigma_s^j(E)} \qquad (4.11)$$

Substituting this into Eq. (4.10) leads to the identity

$$\frac{C}{E} = \sum_j C \int_E^{E/\alpha_j} \frac{\Sigma_s^j(E')}{\Sigma_s(E')} \frac{dE'}{(1-\alpha_j)(E')^2} \simeq C \sum_j \frac{\Sigma_s^j}{\Sigma_s} \frac{1}{E} = \frac{C}{E} \qquad (4.12)$$

when it is assumed that the energy dependence of the scattering cross section is the same for all isotopes present, establishing that a solution of the form of Eq. (4.11) satisfies Eq. (4.10) under this assumption.

Slowing-Down Density

The slowing-down density at energy E, $q(E)$, is defined as the rate at which neutrons are scattered from energies E' above E to energies E'' below E. With reference to Fig. 4.2, this may be written

$$q(E) \equiv \sum_j \int_E^{E/\alpha_j} dE' \int_{\alpha_j E'}^E dE'' \frac{\Sigma_s^j(E')\phi(E')}{E'(1-\alpha_j)} \qquad (4.13)$$

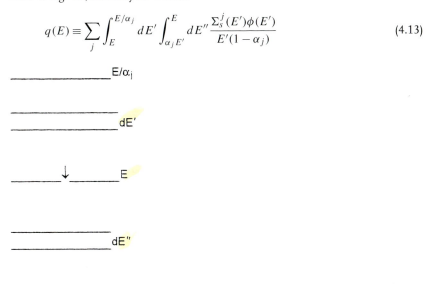

Fig. 4.2 Energy intervals for neutron slowing-down density.

The maximum energy E' from which a neutron may scatter elastically below E is E/α, and the minimum energy E'' of a neutron that scatters from an energy $E' > E$ to an energy $E'' < E$ is $E'' = E'\alpha$. Without absorption, the slowing-down density is obviously constant in energy.

Substituting Eq. (4.11) into Eq. (4.13) leads to

$$q(E) = \sum_j \int_E^{E/\alpha_j} dE' \int_{\alpha_j E'}^{E} dE'' \frac{C\Sigma_s^j(E')}{\Sigma_s(E')(1-\alpha_j)(E')^2}$$

$$= C\sum_j \left(1 + \frac{\alpha_j \ln \alpha_j}{1-\alpha_j}\right)\frac{\Sigma_s^j}{\Sigma_s} \equiv C\sum_j \xi_j \frac{\Sigma_s^j(E)}{\Sigma_s(E)} \equiv C\bar{\xi}(E) \qquad (4.14)$$

where the assumption of a common energy dependence of all scattering cross sections has been used again. The quantity ξ_j is the average logarithmic energy loss in a collision with a nucleus of species j that was discussed in Chapter 1, and $\bar{\xi}$ is the effective logarithmic energy loss for the mixture of moderators. Using this result with Eq. (4.11) leads to the very important relationship between neutral slowing-down density and neutron flux:

$$q(E) = \bar{\xi}\Sigma_s(E)E\phi(E) \qquad (4.15)$$

Slowing Down with Weak Absorption

Absorption removes neutrons from the slowing-down process and thereby reduces the slowing-down density as the energy decreases. Noting that decreasing energy corresponds to $-dE$, the change in slowing-down density due to absorption is described by

$$\frac{dq(E)}{dE} = \Sigma_a(E)\phi(E) \qquad (4.16)$$

Assuming weak absorption or localized (resonance) absorption near energy E, so that the flux given by Eq. (4.15) can be used to evaluate the scattering-in integral, the neutron balance equation yields a generalization of Eq. (4.15) for the case of weak or resonance absorption

$$[\Sigma_a(E) + \Sigma_s(E)]\phi(E) = \sum_j \int_E^{E/\alpha_j} dE' \frac{\Sigma_s^j(E')\phi(E')}{E'(1-\alpha_j)}$$

$$\simeq q\sum_j \int_E^{E/\alpha_j} \frac{1}{\bar{\xi}(1-\alpha_j)}\frac{\Sigma_s^j(E')}{\Sigma_s(E')}\frac{dE'}{(E')^2} = \frac{q}{\bar{\xi}E} \qquad (4.17)$$

where again the assumption of similar energy dependence of the scattering cross sections for all species present has been used. Combining Eqs. (4.16) and (4.17) yields

$$\frac{dq(E)}{dE} = \frac{\Sigma_a(E)q(E)}{E\bar{\xi}[\Sigma_a(E) + \Sigma_s(E)]} \qquad (4.18)$$

which may be integrated from energy E up to some energy E_1 to obtain

$$q(E) = q(E_1)\exp\left\{-\int_E^{E_1} \frac{\Sigma_a(E')dE'}{\bar{\xi}[\Sigma_a(E') + \Sigma_s(E')]E'}\right\} \tag{4.19}$$

which describes the attenuation of the neutron slowing-down density due to absorption. Making use of Eq. (4.17) yields an expression for the energy dependence of the neutron flux

$$\phi(E) = \frac{[\Sigma_a(E_1) + \Sigma_s(E_1)]\bar{\xi}(E_1)E_1\phi(E_1)}{[\Sigma_a(E) + \Sigma_s(E)]\bar{\xi}(E)E}$$

$$\times \exp\left\{-\int_E^{E_1} \frac{\Sigma_a(E')dE'}{\bar{\xi}(E')[\Sigma_a(E') + \Sigma_s(E')]E'}\right\} \tag{4.20}$$

The neutron flux with nonhydrogenic moderators and weak or resonant absorption has an energy dependence $\phi \sim 1/\bar{\xi}\Sigma_t(E)E$ and is exponentially attenuated, a result very similar to that obtained for moderation by hydrogen only [Eq. (4.8)—note that $\xi = 1$ for hydrogen]. In particular, the energy self-shielding of resonances discussed previously is contained in the $1/\Sigma_t(E)$ dependence of the neutron flux.

Fermi Age Neutron Slowing Down

The assumption that the scattering cross sections of all moderating isotopes had the same energy dependence, which was made to obtain a relatively simple solution for slowing down by nonhydrogenic moderators, can be avoided in the case of heavy moderators. The neutron balance equation for slowing down by a mixture of moderators is

$$\Sigma_t(E)\phi(E) = \sum_j \int_E^{E/\alpha_j} dE' \frac{\Sigma_s^j(E')\phi(E')}{E'(1 - \alpha_j)} \tag{4.21}$$

Based on the previous results, we expect that $\bar{\xi}\Sigma_t(E)E\phi$ is a slowly varying function of E. Thus we make a Taylor's series expansion of $\Sigma_s^j(E')E'\phi(E')$ about $\Sigma_s^j(E)E\phi(E)$:

$$E'\Sigma_s^j(E')\phi(E') = E\Sigma_s^j(E)\phi(E) + \frac{d}{d\ln E}\left[E\Sigma_s^j(E)\phi(E)\right]\ln\frac{E'}{E} + \cdots \tag{4.22}$$

in the scattering-in integrals on the right of Eq. (4.21). If the scattering-in interval E/α_j to E is small (i.e., if $\alpha_j \equiv [(A_j - 1)/(A_j + 1)]^2 \approx 1$), it should be sufficient to retain only the first two terms in the Taylor's series expansion, leading to

$$\Sigma_t(E)\phi(E) = \sum_j \int_E^{E/\alpha_j} \frac{dE'}{(E')^2(1 - \alpha_j)}$$

$$\times \left\{E\Sigma_s^j(E)\phi(E) + \ln\frac{E'}{E}\frac{d}{d\ln E}\left[E\Sigma_s^j(E)\phi(E)\right] + \cdots\right\}$$

$$= \sum_j \left\{ \Sigma_s^j(E)\phi(E) \right.$$

$$\left. + \frac{1}{E}\left(1 + \frac{\alpha_j \ln \alpha_j}{1 - \alpha_j}\right)\frac{d}{d\ln E}\left[E\Sigma_s^j(E)\phi(E)\right] + \cdots \right\}$$

$$= \left\{ \Sigma_s(E)\phi(E) + \frac{d}{dE}\left[E\bar{\xi}(E)\Sigma_s(E)\phi(E)\right] + \cdots \right\} \qquad (4.23)$$

which can be integrated to obtain

$$\phi(E) = \frac{E_1\bar{\xi}(E_1)\Sigma_s(E_1)\phi(E_1)}{E\bar{\xi}(E)\Sigma_s(E)}\exp\left[-\int_E^{E_1}\frac{\Sigma_a(E')dE'}{E'\bar{\xi}(E')\Sigma_s(E')}\right] \qquad (4.24)$$

This result for the energy distribution of the neutron flux is identical to the result obtained in Eq. (4.20) when $\Sigma_a \ll \Sigma_s$, but obtained under quite different assumptions. The assumptions in deriving Eq. (4.20) were that the absorption was weak, so that the no-absorption relationship between the slowing-down density and the flux could be used and that the energy dependence of the scattering cross sections was the same for all moderators in the mixture, in order to evaluate the scattering-in integrals. The only assumption in deriving Eq. (4.24) was that $\Sigma_s^j(E')E'\phi(E')$ varied slowly over the scattering-in interval E to E/α_j.

The important results we have obtained about the neutron energy distribution in the slowing-down region are $\phi(E) \sim 1/\bar{\xi}(E)\Sigma_t(E)E$, $q \approx \bar{\xi}(E)\Sigma_t(E)E\phi(E)$ and that both the neutron slowing-down density, q, and the neutron flux, ϕ, are attenuated exponentially by absorption during the slowing-down process. The expressions that we have developed for this exponential attenuation are qualitatively correct, but need to be refined to explicitly treat the resonance absorption which dominates in the slowing-down region. We return to this in Section 4.3.

Neutron Energy Distribution in the Thermal Range

Determination of the neutron energy distribution in the "thermal" range (E less than 1 eV or so) is complicated by a number of factors. The thermal motion of the nuclei is comparable to the neutron motion, with the consequences that the cross sections must be averaged over the nuclear motion and that a scattering event can increase, as well as decrease, the energy of the neutron. Since the thermal neutron energy is comparable to the binding energy of nuclei in material lattices, the recoil of the nucleus will be affected by the binding of the nucleus in the lattice, and the neutron scattering kinematics is more complex. Inelastic scattering in which the molecular rotational or vibrational states or the crystal lattice vibration state is changed also affects the scattering kinematics. At very low energies the neutron wavelength is comparable to the interatomic spacing of the scattering nuclei, and diffraction effects become important. Accurate calculation of thermal reaction rates requires both the calculation of appropriate cross sections characterizing thermal neutron scattering and calculation of the energy distribution of neutrons in the thermal range. Fortunately, most of the complex details of thermal neutron cross sections are not of great importance in nuclear reactor calculations, and reasonable accuracy can be obtained with relatively simple models. In this section we characterize the thermal neutron distribution and reaction rates from relatively simple physical considerations. We return to a more detailed discussion of thermal neutron cross sections and distributions in Chapter 12.

The neutron balance equation in the thermal energy range is

$$[\Sigma_a(E) + \Sigma_s(E)]\phi(E) = \int_0^{E_{th}} dE' \Sigma_s(E' \to E)\phi(E') + S(E) \tag{4.25}$$

where the scattering-in integral is from all energies in the thermal range $E < E_{th}$, and $S(E)$ is the source of neutrons scattered into the thermal energy range from $E > E_{th}$. An equilibrium solution requires that the total number of neutrons down-scattered into the thermal energy range be absorbed, assuming for the moment no leakage and no upscatter above E_{th}:

$$\int_0^{E_{th}} dE \, \Sigma_a(E)\phi(E) = q(E_{th}) \tag{4.26}$$

where $q(E_{th})$ is the neutron slowing-down density past E_{th}.

Consider the situation that would obtain if there were no absorption and slowing-down source; that is, the neutron flux balance is

$$\Sigma_s(E)\phi(E) = \int_0^{\infty} dE \, \Sigma_s(E' \to E)\phi(E') \tag{4.27}$$

where we have extended the upper limit on the integral to infinity under the assumption that the scattering to energies greater than E_{th} is zero. The principle of detailed balance places the following constraint on the scattering transfer cross section for a neutron distribution that is in equilibrium, regardless of the physical details of the scattering event:

$$v' \Sigma_s(E' \to E) M(E', T) = v \Sigma_s(E \to E') M(E, T) \tag{4.28}$$

where $M(E, T)$ is the Maxwellian neutron distribution

$$M(E, T) = \frac{2\pi}{(\pi kT)^{3/2}} \sqrt{E} \exp\left(-\frac{E}{kT}\right) \tag{4.29}$$

It can be shown that the Maxwellian neutron flux distribution,

$$\phi_M(E, T) = nv(E)M(E) = \frac{2\pi n}{(\pi kT)^{3/2}} \left(\frac{2}{m}\right)^{1/2} E \exp\left(-\frac{E}{kT}\right)$$

$$\equiv \phi_T \frac{E}{(kT)^2} \exp\left(-\frac{E}{kT}\right) \tag{4.30}$$

satisfies Eq. (4.27). Thus the principal of detailed balance is sufficient to ensure that the equilibrium neutron distribution, in the absence of absorption, leakage, or sources, is a Maxwellian distribution characterized by the temperature T of the medium (i.e., the neutrons will come into thermal equilibrium with the moderator nuclei). The most probable energy of neutrons in a Maxwellian distribution is kT, and the corresponding neutron speed is $v_T = (2kT/m)^{1/2}$.

However, absorption, leakage, and a slowing-down source will distort the actual neutron distribution from a Maxwellian. Since most absorption cross sections vary as $1/v = 1/(E)^{1/2}$, absorption preferentially removes lower-energy neutrons, effectively shifting the neutron distribution to higher energies than a Maxwellian at the moderator temperature T. A shift to higher energies can be represented approximately by a Maxwellian distribution with an effective "neutron temperature"

$$T_n = T\left(1 + \frac{C\Sigma_a}{\xi\Sigma_s}\right) \tag{4.31}$$

where C must be determined experimentally. Leakage can be represented by modifying the absorption cross section to $\Sigma_a \Rightarrow \Sigma_a + DB^2$. Since $D = 1/3\Sigma_{tr}$, leakage will preferentially remove higher-energy neutrons, offsetting the effect of absorption to some extent.

In the slowing-down region $E > E_{th}$, the neutron flux distribution is $1/E$, and the slowing-down source into the upper part of the thermal energy range will tend to make the flux $1/E$. Thus the hardened Maxwellian must be corrected by the addition of a joining term Δ which is about unity for values of $E/kT_n > 10$ and vanishes for values of $E/kT_n < 5$:

$$\phi(E) = \phi_M(E, T_n) + \lambda\frac{\Delta(E/kT_n)}{E} \tag{4.32}$$

where λ is a normalization factor

$$\lambda = \phi_T\frac{(\sqrt{\pi}/2)}{1 - \xi\Sigma_s/\Sigma_a} \tag{4.33}$$

The Maxwellian distribution has some useful properties insofar as calculation of the neutron absorption rate in the thermal energy range is concerned. Most absorption cross sections are $1/v$; that is,

$$\Sigma_a(E, T) = \frac{\Sigma_a^0}{v} = \frac{\Sigma_a(E_0)v_0}{v} \tag{4.34}$$

where $E_0 = kT = 0.025$ eV and $v_0 = (2kT/m_n)^{1/2} = 2200$ m/s. The total absorption rate integrated over the thermal energy range is

$$R_a = \int_0^{E_{th}} dE\,\Sigma_a(E, T)vn\,M(E, T_n) = \Sigma_a(E_0)v_0n_0 \equiv \Sigma_a(E_0)\phi_0 \tag{4.35}$$

The quantity $\phi_0 = nv_0$ is the 2200 m/s flux, which when multiplied by the cross section evaluated at $E = 0.025$ eV yields the total absorption rate integrated over the thermal energy range. Most thermal data compilations include the 2200 m/s value of the cross section (see Appendix A). From the definitions of $\phi_T = (2/\pi^{1/2})nv$ [Eq. (4.30)] and of $\phi_0 = nv_0$, the appropriate thermal group absorption cross section (the quantity that is multiplied by the integral of the neutron flux over the thermal energy range to recover R_a) for a 1/v absorber in a Maxwellian neutron distribution at neutron temperature T_n is

$$\Sigma_a^{th} = \frac{\sqrt{\pi}}{2}\left(\frac{T_0}{T_n}\right)^{1/2}\Sigma_a(E_0) \tag{4.36}$$

[handwritten: $T_0 = 293\,K$]

[handwritten: $k = .861735 \times 10^{-4}\ \frac{eV}{K}$]

[handwritten: $M_n = 939.5731\ \frac{MeV}{c^2}$]

Non-1/v correction factors have been developed to correct this expression for absorbers that are not 1/v.

Summary

The fission spectrum divided by the total cross section, $\phi(E) = \chi(E)/\Sigma_t(E)$, represents the energy distribution rather well for energies $E > 0.5$ MeV. In the slowing-down range below the fission spectrum, $E < 50$ keV, and above the thermal range, $E > 1$ eV, $\phi(E) \sim 1/\bar{\xi}(E)\Sigma_t(E)E$ represents the neutron energy distribution. In the thermal range, $E < 1$ eV, a hardened Maxwellian plus a $1/E$ correction at higher energies, $\phi(E) = \phi_M(E, T_n) + \lambda\Delta(E/kT_n)/E$, represents the neutron energy distribution.

[handwritten: $\chi(E) \cdot$ fission spectrum]

4.2
Multigroup Calculation of Neutron Energy Distribution in an Infinite Medium

Derivation of Multigroup Equations

While the neutron energy dependences derived in Section 4.1 provide a qualitative, even semiquantitative description of the neutron energy distribution in nuclear reactors, the multigroup method is widely used for the quantitative calculation of the neutron energy distribution. As we will see, the qualitative results of Section 4.1 will provide valuable insight as to the choice of weighting functions to be used in the preparation of multigroup constants.

To develop a multigroup calculational method for the energy distribution, we divide the energy interval of interest, say 10 MeV down to zero, into G intervals, or groups, as indicated in Fig. 4.3. The equation describing the neutron energy distribution in a very large homogeneous region of a nuclear reactor (where spatial and leakage effects may be neglected) is

$$\left[\Sigma_a(E) + \Sigma_s(E)\right]\phi(E)$$
$$= \int_0^\infty dE' \Sigma_s(E' \to E)\phi(E') + \frac{\chi(E)}{k_\infty}\int_0^\infty dE' v\Sigma_f(E')\phi(E') \tag{4.37}$$

This equation can be integrated over the energy interval $E_g < E < E_{g-1}$ of group g to obtain

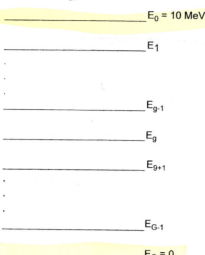

$E_0 = 10$ MeV

E_1

E_{g-1}

E_g

E_{g+1}

E_{G-1}

$E_G = 0$

Fig. 4.3 Multigroup energy structure.

$$\int_{E_g}^{E_{g-1}} dE \left[\Sigma_a(E) + \Sigma_s(E) \right] \phi(E)$$

$$= \int_{E_g}^{E_{g-1}} dE \sum_{g'=1}^{G} \int_{E_{g'}}^{E_{g'-1}} dE' \Sigma_s \left(E' \to E \right) \phi(E')$$

$$+ \frac{1}{k_\infty} \int_{E_g}^{E_{g-1}} dE \chi(E) \sum_{g'=1}^{G} \int_{E_{g'}}^{E_{g'-1}} dE' \nu \Sigma_f (E') \phi(E') \qquad (4.38)$$

where we have made use of the fact that the sum of integrals over the groups is equal to the integral over $0 < E < \infty$. Defining the integral terms in Eq. (4.38) in a natural way,

$$\phi_g \equiv \int_{E_g}^{E_{g-1}} dE \, \phi(E), \qquad\qquad \chi_g \equiv \int_{E_g}^{E_{g-1}} dE \, \chi(E)$$

$$\Sigma_a^g \equiv \frac{\int_{E_g}^{E_{g-1}} dE \, \Sigma_a(E)\phi(E)}{\phi_g}, \qquad\qquad \nu\Sigma_f^g \equiv \frac{\int_{E_g}^{E_{g-1}} dE \, \nu\Sigma_f(E)}{\phi_g}$$

$$\Sigma_s^{g' \to g} \equiv \frac{\int_{E_g}^{E_{g-1}} dE \int_{E_{g'}}^{E_{g'-1}} dE' \Sigma_s (E' \to E)\phi(E')}{\phi_{g'}}, \qquad \Sigma_s^g \equiv \sum_{g'=1}^{G} \Sigma_s^{g \to g'}$$

$$(4.39)$$

Eq. (4.38) can be written as

$$\left(\Sigma_a^g + \Sigma_s^g \right) \phi_g = \sum_{g'=1}^{G} \Sigma_s^{g' \to g} \phi_{g'} + \frac{\chi_g}{k_\infty} \sum_{g'=1}^{G} \nu\Sigma_f^{g'} \phi_{g'}, \quad g = 1, \dots, G \qquad (4.40)$$

Equations (4.40) are the multigroup neutron spectrum equations for an *infinite medium*, one in which spatial and leakage effects are unimportant. There are G equations and G unknowns, the group fluxes ϕ_g, so the problem is well posed. This overlooks the fact that the group constants Σ^g depend on the neutron flux and hence are also unknown. Actually, the group constants depend only on the energy dependence of the neutron flux within the group, not on the magnitude of the neutron flux, which appears in both the numerator and denominator of the definition of the group constants. In practice, some assumption is made about this energy dependence, so that the group constants are known. From the results of the preceding section, we have a pretty good idea about the energy dependence of the neutron flux in the fission, slowing-down, and thermal energy ranges, which can be used to evaluate group constants.

Summing Eqs. (4.40) over groups yields

$$k_\infty = \frac{\sum_{g=1}^{G} \nu \Sigma_f^g \phi_g}{\sum_{g=1}^{G} \Sigma_a^g \phi_g} \tag{4.41}$$

which identifies k_∞ as the ratio of the total neutron production rate by fission to the total neutron absorption rate, in accord with our previous discussion of the multiplication constant.

Mathematical Properties of the Multigroup Equations

The set of equations (4.40) may be written in matrix notation as

$$A\phi - \frac{1}{k_\infty} F\phi = \left(A - \frac{1}{k_\infty} F \right) \phi = 0 \tag{4.42}$$

where A and F are $G \times G$ matrices and ϕ is a G-element column vector:

$$A = \begin{bmatrix} \Sigma_a^1 + \Sigma_s^1 - \Sigma_s^{1\to1} & -\Sigma_s^{2\to1} & -\Sigma_s^{3\to1} & \cdots & -\Sigma_s^{G\to1} \\ -\Sigma_s^{1\to2} & \Sigma_a^2 + \Sigma_s^2 - \Sigma_s^{2\to2} & -\Sigma_s^{3\to2} & \cdots & -\Sigma_s^{G\to2} \\ \vdots & \vdots & \vdots & \vdots & \vdots \\ -\Sigma_s^{1\to G} & -\Sigma_s^{2\to G} & -\Sigma_s^{3\to G} & \cdots & \Sigma_a^G + \Sigma_s^G - \Sigma_s^{G\to G} \end{bmatrix} \tag{4.43}$$

$$F = \begin{bmatrix} \chi_1 \nu \Sigma_f^1 & \chi_1 \nu \Sigma_f^2 & \chi_1 \nu \Sigma_f^3 & \cdots & \chi_1 \nu \Sigma_f^G \\ \chi_2 \nu \Sigma_f^1 & \chi_2 \nu \Sigma_f^2 & \chi_2 \nu \Sigma_f^3 & \cdots & \chi_2 \nu \Sigma_f^G \\ \vdots & \vdots & \vdots & \vdots & \vdots \\ \chi_G \nu \Sigma_f^1 & \chi_G \nu \Sigma_f^2 & \chi_G \nu \Sigma_f^3 & \cdots & \chi_g \nu \Sigma_f^G \end{bmatrix}, \qquad \phi = \begin{bmatrix} \phi_1 \\ \phi_2 \\ \vdots \\ \phi_G \end{bmatrix}$$

Note that the scattering terms on the diagonal are of the form $\Sigma_s^g - \Sigma_s^{g\to g}$, leading to the concept of a *removal cross section* $\Sigma_r^g \equiv \Sigma_a^g + \Sigma_s^g - \Sigma_{ss}^{g\to g}$ to represent the net loss of neutrons from group g by absorption plus scattering.

Equations (4.40) or (4.42) are homogeneous equations and thus, by Cramer's rule, have nontrivial solutions only if the determinant of the coefficient matrix vanishes:

$$\det\left(A - \frac{1}{k_\infty}F\right) = 0 \tag{4.44}$$

This condition defines an eigenvalue problem for the determination of k_∞—there are only a certain set of G discrete values of k_∞ for which a nontrivial solution exists. [Note that we have included k_∞ in the formulation for just this reason. If we had not included k_∞, Eq. (4.44) would be a requirement on the composition of the reactor for criticality, and we would be faced with the cumbersome requirement to adjust the composition by trial and error until Eq. (4.44) was satisfied.]

It is possible to prove that the inverse of the matrix A exists for any physically real set of cross sections and number densities. Multiplying Eq. (4.42) by $k_\infty A^{-1}$ yields

$$k_\infty\boldsymbol{\phi} = A^{-1}F\boldsymbol{\phi} \tag{4.45}$$

which is the standard form for a matrix eigenvalue problem. It is possible to prove (e.g., Refs. 8, 11, and 12) for this equation that (1) there is a unique real, positive eigenvalue greater in magnitude than any other eigenvalue; (2) all of the elements—the group fluxes—of the eigenvector corresponding to this largest eigenvalue are real and positive; and (3) the eigenvectors corresponding to all other eigenvalues have zero or negative elements. Thus the largest value of k_∞ for which Eq. (4.44) is satisfied is real and positive and the associated group fluxes given by Eq. (4.45) are real and positive (i.e., physical).

Solution of Multigroup Equations

The multigroup equations have been written in their full generality, allowing upscatter (the terms above the diagonal in A) as well as downscatter (the terms below the diagonal in A) and a fission spectrum contribution in every group. In fact, upscatter takes place only for those groups that are in the thermal energy range $E \lesssim 1$ eV, and the fission spectrum contributes only to the higher-energy groups $E \gtrsim 50$ keV. Taking these physical considerations into account greatly simplifies solution of the multigroup equations.

Consider, as the simplest example of a multigroup description, the representation of the neutrons in a nuclear reactor as being either in a thermal group ($E \lesssim 1$ eV) or in a fast group ($E \gtrsim 1$ eV). All of the fission neutrons are deposited in the fast group, and there is no upscatter from the thermal to the fast group. The two-group equations are

$$\left(\Sigma_a^1 + \Sigma_s^{1\to2}\right)\phi_1 = \frac{1}{k_\infty}\left(\nu\Sigma_f^1\phi_1 + \nu\Sigma_f^2\phi_2\right)$$

$$\Sigma_a^2\phi_2 = \Sigma_s^{1\to2}\phi_1 \tag{4.46}$$

which may readily be solved for

$$\phi_1 = \frac{\Sigma_a^2}{\Sigma_s^{1\to2}}\phi_2, \qquad k_\infty = \frac{\nu\Sigma_f^1 + (\Sigma_s^{1\to2}/\Sigma_a^2)\nu\Sigma_f^2}{\Sigma_a^1 + \Sigma_s^{1\to2}} \tag{4.47}$$

Note that a critical reactor may operate at many power levels, so the absolute magnitude of the group fluxes quite properly cannot be determined by the set of homogeneous multigroup equations, but the relative magnitudes of the different group fluxes can be determined.

A somewhat better multigroup description results from representing the fission interval ($E > 50$ keV) as a fast group into which all fission neutrons are introduced, the slowing-down interval (50 keV $> E > 1$ eV) as an intermediate group, and the thermal region ($E < 1$ eV) as a thermal group. There would be no upscattering in such a group structure, allowing the three-group equations to be written

$$\left(\Sigma_a^1 + \Sigma_s^{1\to2} + \Sigma_s^{1\to3}\right)\phi_1 = \frac{1}{k_\infty}\left(\nu\Sigma_f^1\phi_1 + \nu\Sigma_f^2\phi_2 + \nu\Sigma_f^3\phi_3\right)$$

$$\left(\Sigma_a^2 + \Sigma_s^{2\to3}\right)\phi_2 = \Sigma_s^{1\to2}\phi_1 \tag{4.48}$$

$$\Sigma_a^3\phi_3 = \Sigma_s^{1\to3}\phi_1 + \Sigma_s^{2\to3}\phi_2$$

with solutions

$$\phi_2 = \left[\Sigma_s^{1\to2}/\left(\Sigma_a^2 + \Sigma_s^{2\to3}\right)\right]\phi_1,$$

$$\phi_3 = \left[\left(\Sigma_s^{1\to3} + \frac{\Sigma_s^{2\to3}}{\Sigma_a^2 + \Sigma_s^{2\to3}}\Sigma_s^{1\to2}\right)\Big/\Sigma_a^3\right]\phi_1$$

$$k_\infty = \left[\nu\Sigma_f^1 + \nu\Sigma_f^2\frac{\Sigma_s^{1\to2}}{\Sigma_a^2 + \Sigma_s^{2\to3}}\right. \tag{4.49}$$

$$\left. + \nu\Sigma_f^3\left(\Sigma_s^{1\to3} + \frac{\Sigma_s^{2\to3}}{\Sigma_a^2 + \Sigma_s^{2\to3}}\Sigma_s^{1\to2}\right)\right]\Big/\left(\Sigma_a^1 + \Sigma_s^{1\to2} + \Sigma_s^{1\to3}\right)$$

Example 4.1: Two-Group Fluxes and k_∞. A representative set of two-group cross sections for a PWR fuel assembly are ($\Sigma_s^{1\to2} = 0.0241$ cm^{-1}, $\Sigma_a^1 = 0.0121$ cm^{-1}, $\nu\Sigma_f^1 = 0.0085$) and ($\Sigma_a^2 = 0.121$ cm^{-1}, $\nu\Sigma_f^2 = 0.185$). From Eq. (4.47) the fast/thermal flux ratio is $\phi_1/\phi_2 = 0.121/0.241 = 5.02$, and $k_\infty = (0.0085 + 0.185/5.02)/(0.0121 + 0.0241) = 1.253$. The spectrum-averaged one-group absorption cross section is $\Sigma_a = (\Sigma_a^1\phi_1 + \Sigma_a^2\phi_2)/(\phi_1 + \phi_2) = 0.0302$ cm^{-1}.

Preparation of Multigroup Cross-Section Sets

There exist in the world several sets of evaluated nuclear data (e.g., Refs. 7 and 9), which have been both checked for consistency and benchmarked extensively in the calculation of experiments designed for data testing. Representation of the cross-section data in such data files is generally as follows:

1. $\sigma(E_i)$ are tabulated pointwise in energy at low energies below the resonance region.
2. Resolved resonance parameters and background cross sections in the resolved resonance region.
3. Unresolved resonance statistical parameters and background cross section in the unresolved resonance region.
4. $\sigma(E_i)$ are tabulated pointwise in energy at energies above the resonance region.
5. Scattering transfer functions $p(E_i, \mu_s)$ are tabulated pointwise in energy and either pointwise in angle (μ_{sj}) or as Legendre coefficients.

The resonance parameters and the construction of multigroup cross sections from them are discussed in Section 4.3 and in Chapter 11.

The scattering transfer function—the probability that a neutron will undergo a scattering event which changes its direction from direction Ω to direction Ω' ($\mu_s = \Omega \cdot \Omega'$) and its energy from E to E'—is represented as

$$\sigma_s\left(\mu_s, E \to E'\right) = m(E)\sigma_s(E)p(E, \mu_s)g\left(\mu_s, E \to E'\right) \tag{4.50}$$

where $m(E) = 1$ for elastic and inelastic scattering, 2 for $(n, 2n)$, ν for fission; $p(E, \mu_s)$ is the angular distribution for scattering of a neutron of energy E; and $g(\mu_s, E \to E')$ is the final energy distribution of a neutron at energy E which has scattered through μ_s. When the scattering angle and energy loss are correlated, as they are for elastic scattering, $E'/E = [(1 + \alpha) + (1 - \alpha)\cos\theta]/2$ and $g(\mu_s, E \to E') = \delta(\mu_s - \mu(E, E'))$. Otherwise, $g(\mu_{si}, E_j \to E'_k)$ is tabulated. The angular distribution may be tabulated as $p(E_i, \mu_{sj})$, or the Legendre components may be tabulated pointwise in energy

$$p_n(E_i) = \int_{-1}^{1} d\mu_s \, P_n(\mu_s)p(E_i, \mu_s) \tag{4.51}$$

where P_n is the Legendre polynomial.

There are a number of codes (e.g., Refs. 2, 4, and 5) which directly process the evaluated nuclear data files to prepare multigroup cross sections. These codes numerically calculate integrals of the type

$$\sigma^g = \frac{\int_{E_g}^{E_{g-1}} dE\sigma(E)W(E)}{\int_{E_g}^{E_{g-1}} dE\,W(E)}$$

$$\sigma_n^{g \to g'} = \int_{E_g}^{E_{g-1}} dE\sigma_s(E)W(E) \int_{E_{g'}}^{E_{g'-1}} dE'\, p_n\left(E'\right) \Big/ \int_{E_g}^{E_{g-1}} dE\,W(E) \tag{4.52}$$

for a specified weighting function, $W(E)$, which may be a constant, $1/E$, $\chi(E)$, and so on. These codes are used to calculate *fine-group* cross sections in a few hundred groups for thermal reactors or *ultrafine-group* cross sections in a few thousand groups for fast reactors. These fine- or ultrafine-group structures are chosen such that the results of calculations using the fine- or ultrafine-group cross sections are essentially independent of the choice of weighting function, $W(E)$, used in the cross-section preparation.

Once the fine- or ultrafine-group cross sections are prepared, a fine- or ultrafine-group spectrum (ϕ_g) is calculated for a representative homogenized medium. The unit cell heterogeneous structure of the region must be taken into account in homogenizing the medium. Resonances must be treated specially, as discussed in Chapter 11. This fine- or ultrafine-group spectrum can then be used to weight the fine- or ultrafine-group cross sections to obtain *few-group* (2 to 10) cross sections for thermal reactors or *many-group* (20 to 30) cross sections for fast reactors:

$$\sigma^k = \frac{\sum_{g \in k} \sigma^g \phi_g}{\sum_{g \in k} \phi_g}$$

$$\sigma_n^{k \to k'} = \frac{\sum_{g \in k} \sum_{g' \in k'} \sigma_n^{g \to g'} \phi_g}{\sum_{g \in k} \phi_g}$$

(4.53)

The notation $g \in k$ indicates that the sum is over all fine or ultrafine groups g within few or many group k.

The few- or many-group cross sections may be calculated for several different large regions in a reactor. They are then used in a few- or many-group diffusion or transport theory calculation of the entire reactor to determine the effective multiplication constant, power distribution, and so on. Because many such calculations must be made, a number of parameterizations of few- and many-group cross sections have been developed (e.g., Ref. 10) to avoid the necessity of making the fine- or ultrafine-group spectrum calculation numerous times.

4.3
Resonance Absorption

Resonance Cross Sections

When the relative (center-of-mass) energy of an incident neutron and a nucleus plus the neutron binding energy match an energy level of the compound nucleus that would be formed upon neutron capture, the probability of capture is quite large. The lowest-energy excited states are only a fraction of 1 eV above the ground state and extend up to about 100 keV for heavy mass fuel nuclei (fissile and fertile), but start at about 10 eV for intermediate mass nuclei and at about 10 keV for lighter mass nuclei. The heavier mass isotopes have many relatively low energy excited states, which give rise to resonances in the neutron absorption and scattering cross sections (Fig. 4.4).

The neutron resonance absorption phenomena constitute one of the most fundamental subjects in nuclear reactor physics. One of the most effective means of treating these phenomena is in terms of the resonance integral concept, which has a fundamental premise that the resonance cross sections are representable

U238 CAPTURE CROSS SECTION MT = 27

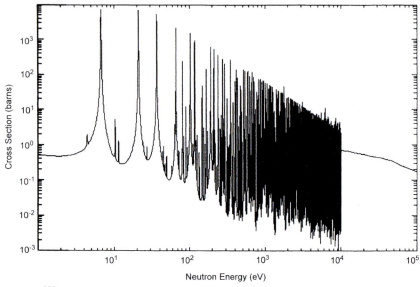

Fig. 4.4 ^{238}U capture cross section. (From http://www.nndc.bnl.gov/.)

by superposition of many Breit–Wigner resonances with known parameters. This premise allows the complex resonance structure to be characterized in a reasonably simple way by calculating the contributions of each individual resonance. The discussion in this section concentrates on *s*-wave neutron cross sections in the low-energy range.

As shown in Chapter 1, the (n, γ) capture cross section averaged over the motion of the nucleus is given by

$$\sigma_\gamma(E, T) = \sigma_0 \frac{\Gamma_\gamma}{\Gamma} \left(\frac{E_0}{E} \right)^{1/2} \psi(\xi, x) \qquad (4.54)$$

and the total scattering cross section, including resonance and potential scattering and interference between the two, can be written

$$\sigma_s(E, T) = \sigma_0 \frac{\Gamma_n}{\Gamma} \psi(\xi, x) + \frac{\sigma_0 R}{\lambda_0} \chi(\xi, x) + 4\pi R^2 \qquad (4.55)$$

where R is the nuclear radius, λ_0 the neutron DeBroglie wavelength, the functions

$$\psi(\xi, x) = \frac{\xi}{2\sqrt{\pi}} \int_{-\infty}^{\infty} e^{-1/4(x-y)^2 \xi^2} \frac{dy}{1 + y^2} \qquad \textit{Table 4.1} \qquad (4.56)$$

$$\chi(\xi, x) = \frac{\xi}{\sqrt{\pi}} \int_{-\infty}^{\infty} e^{-1/4(x-y)^2 \xi^2} \frac{y\,dy}{1 + y^2} \qquad \textit{Table 4.2} \qquad (4.57)$$

are integrals over the relative motion of the neutron and nucleus, $x = 2(E - E_0)/\Gamma$, it has been assumed that the nuclear motion can be characterized by a Maxwellian

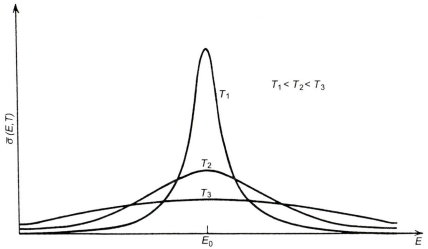

Fig. 4.5 Temperature broadening of the ψ-function. (From Ref. 3; used with permission of Wiley.)

distribution with temperature T, and E is the energy of the neutron in the lab system. The parameters characterizing the resonance are σ_0, the peak value of the cross section; E_0, the neutron energy in the center-of-mass system at which it occurs; Γ, the resonance width; Γ_γ, the partial width for neutron capture; Γ_f, the partial width for fission; and Γ_n, the partial width for scattering. The resonance occurs when the center of mass energy $E_{cm} = (A/(A + 1))E$ plus the change in binding energy upon neutron capture equals the energy above ground state of an excited level of the compound nucleus.

Doppler Broadening

The temperature characterizing the nuclear motion is contained in the parameter

$$\xi = \frac{\Gamma}{(4E_0kT/A)^{1/2}} \qquad (4.58)$$

where A is the atomic mass (amu) and k is the Boltzmann constant. The general dependence of the ψ-function on temperature is indicated in Fig. 4.5. As the temperature increases, the peak magnitude of ψ at E_0 decreases and the magnitude away from peak increases. This broadening of the cross section is known as *Doppler broadening*. It can be shown that the area under the curve of the ψ-function remains constant as the temperature changes. Similar behavior results for the χ-function. The ψ- and χ-functions are tabulated in Tables 4.1 and 4.2.

The assumption that the nuclear motion can be characterized by a Maxwellian is only approximately correct for atoms bound in a crystalline state. Investigation of this point indicates that a Maxwellian is a good approximation, but with a slightly higher temperature which corresponds to the average energy per vibrational degree

Table 4.1 ψ-Function

ξ	x									
	0	0.5	1	2	4	6	8	10	20	40
0.05	0.04309	0.04308	0.04306	0.04298	0.04267	0.04216	0.04145	0.04055	0.03380	0.01639
0.10	0.08384	0.08379	0.08364	0.08305	0.08073	0.07700	0.07208	0.06623	0.03291	0.00262
0.15	0.12239	0.12223	0.12176	0.11989	0.11268	0.10165	0.08805	0.07328	0.01695	0.00080
0.20	0.15889	0.15854	0.15748	0.15331	0.13777	0.11540	0.09027	0.06614	0.00713	0.00070
0.25	0.19347	0.19281	0.19086	0.18324	0.15584	0.11934	0.08277	0.05253	0.00394	0.00067
0.30	0.22624	0.22516	0.22197	0.20968	0.16729	0.11571	0.07042	0.03880	0.00314	0.00065
0.35	0.25731	0.25569	0.25091	0.23271	0.17288	0.10713	0.05724	0.02815	0.00289	0.00064
0.40	0.28679	0.28450	0.27776	0.25245	0.17359	0.09604	0.04566	0.02109	0.00277	0.00064
0.45	0.31477	0.31168	0.30261	0.26909	0.17052	0.08439	0.03670	0.01687	0.00270	0.00064
0.50	0.34135	0.33733	0.32557	0.28286	0.16469	0.07346	0.03025	0.01446	0.00266	0.00063

Source: Data from Ref. 3; used with permission of Wiley.

Table 4.2 χ-Function

ξ	0	0.5	1	2	4	6	8	10	20	40
0.05	0	0.00120	0.00239	0.00478	0.00951	0.01415	0.01865	0.02297	0.04076	0.05221
0.10	0	0.00458	0.00915	0.01821	0.03573	0.05192	0.06626	0.07833	0.10132	0.05957
0.15	0	0.00986	0.01968	0.03894	0.07470	0.10460	0.12690	0.14096	0.12219	0.05341
0.20	0	0.01680	0.03344	0.06567	0.12219	0.16295	0.18538	0.19091	0.11754	0.05170
0.25	0	0.02515	0.04994	0.09714	0.17413	0.21909	0.23168	0.22043	0.11052	0.05103
0.30	0	0.03470	0.06873	0.13219	0.22694	0.26757	0.26227	0.23199	0.10650	0.05069
0.35	0	0.04529	0.08940	0.16976	0.27773	0.30564	0.27850	0.23236	0.10437	0.05049
0.40	0	0.05674	0.11160	0.20890	0.32442	0.33286	0.28419	0.22782	0.10316	0.05037
0.45	0	0.06890	0.13498	0.24880	0.36563	0.35033	0.28351	0.22223	0.10238	0.05028
0.50	0	0.08165	0.15927	0.28875	0.40075	0.35998	0.27979	0.21729	0.10185	0.05022

Source: Data from Ref. 3; used with permission of Wiley.

of freedom of the lattice, including the zero-point energy. In practice, the actual material temperature is widely used.

Resonance Integral

The total absorption rate per nuclei by a resonance absorber is known as the *resonance integral*,

$$I_\gamma = \int \sigma_\gamma(E)\phi(E)\,dE \tag{4.59}$$

Resonance Escape Probability

The absorption probability for a single resonance depends on the balance between absorption and moderation and is given by $R_{abs} = N_{res}I/q_0$, where $q_0 = \xi \Sigma_s E\phi_{asy}$ is the asymptotic slowing-down density above the resonance and N_{res} is the number density of the resonance absorber. If we use $\phi_{asy} = 1/E$ to evaluate the resonance integral, then $R_{abs} = I/\xi\sigma_s$, where the denominator is the moderating power per absorber nucleus. The resonance escape probability is $p = 1 - R_{abs} = 1 - I/\xi\sigma_s \approx \exp(-I/\xi\sigma_s)$, where R_{abs} is assumed small for any one resonance. The quantity $\sigma_s = (N_{res}\sigma_s^{res} + N_{non}\sigma_s^{non})/N_{res}$ is the effective scattering cross section per resonance absorbing atom, including both the resonance and non-resonance species present.

The total resonance integral for all resonances is a sum over the individual resonance integrals, and the total resonance escape probability is

$$p = \prod_i p_i = \exp\left(-\frac{1}{\xi\sigma_s}\sum_i I_i\right) \tag{4.60}$$

Multigroup Resonance Cross Section

The resonances within a given energy group in a multigroup treatment can be treated as a group capture cross section given by

$$\sigma_\gamma^g = \frac{\int_{E_g}^{E_{g-1}} dE\,\sigma_\gamma(E)\phi(E)}{\int_{E_g}^{E_{g-1}} dE\,\phi(E)} = \frac{\sum_{i \in g} I_i}{\ln(E_{g-1}/E_g)} \tag{4.61}$$

where $\phi(E) \sim 1/E$ has been used.

Practical Width

The practical width of a resonance is defined as the energy range over which the resonance cross section is larger than the nonresonance part of the cross section of the resonance nuclide, which from the Breit–Wigner formula is

$$\Gamma_p \simeq \sqrt{\frac{\sigma_0}{4\pi R^2}}\,\Gamma = \sqrt{\frac{\sigma_0}{\sigma_p}}\,\Gamma \tag{4.62}$$

Typically, for low-energy resonances $\sigma_0/4\pi R^2 \equiv \sigma_0/\sigma_p \sim 10^3$, so the practical width is much larger than the total width. The practical width provides a measure of the range of influence of the resonance, which we will see is important in evaluating the neutron flux in the resonance.

Neutron Flux in Resonance

The resonance region is well below the fission spectrum, so the neutron balance in the vicinity of the resonance can be written

$$\left[\Sigma_t^{\mathrm{res}}(E) + \Sigma_s^M\right]\phi(E) = \int_E^{E/\alpha_M} \frac{dE}{E'} \frac{\Sigma_s^M \phi(E')}{1 - \alpha_m}$$
$$+ \int_E^{E/\alpha_{\mathrm{res}}} \frac{dE'}{E'} \frac{\Sigma_s^{\mathrm{res}}(E')\phi(E')}{1 - \alpha_{\mathrm{res}}} \tag{4.63}$$

where the moderator scattering cross section is assumed to be much larger than its absorption cross section and to be effectively constant. The practical width of the resonance will generally be much less than the scattering-in interval of the moderator, $\Gamma_p \ll E_0(1 - \alpha_M)$. For widely spaced resonances, this allows the approximate evaluation of the moderator scattering source term with the asymptotic form of the neutron flux in the absence of resonances, $\phi_{\mathrm{asy}} \sim 1/\xi\Sigma_s^M E$. We choose the normalization $\phi_{\mathrm{asy}} = 1/E$ above the resonance energy to obtain

$$\left[\Sigma_t^{\mathrm{res}}(E) + \Sigma_s^M\right]\phi(E) = \frac{\Sigma_s^M}{E} + \int_E^{E/\alpha_{\mathrm{res}}} \frac{dE'}{E'} \frac{\Sigma_s^{\mathrm{res}}(E')\phi(E')}{1 - \alpha_{\mathrm{res}}} \tag{4.64}$$

Narrow Resonance Approximation

If the practical width of the resonance is also small compared to the scattering-in interval of the resonance absorber, $\Gamma_p \ll E_0(1 - \alpha_{\mathrm{res}})$, then the second scattering source term can be approximated in the same fashion to obtain

$$\phi_{\mathrm{NR}}(E) = \frac{\Sigma_s^M + \Sigma_p^{\mathrm{res}}}{[\Sigma_t^{\mathrm{res}}(E) + \Sigma_s^M]E} \tag{4.65}$$

which can be used in Eq. (4.59) to evaluate the resonance integral:

$$I_{\mathrm{NR}}^\gamma = \int \frac{dE}{E} \sigma_\gamma(E) \frac{\Sigma_s^M + \Sigma_p^{\mathrm{res}}}{\Sigma_t^{\mathrm{res}}(E) + \Sigma_s^M}$$
$$= \frac{\Gamma_\gamma}{2E_0}(\sigma_p^{\mathrm{res}} + \sigma_s^M) \int_{-\infty}^\infty \frac{\psi(\xi, x)\,dx}{\psi(\xi, x) + \theta\chi(\xi, x) + \beta} \tag{4.66}$$

where

$$\beta = \frac{\sigma_s^M + \sigma_p^{\mathrm{res}}}{\sigma_0}, \qquad \theta = \left(\frac{\Gamma_n}{\Gamma}\frac{\sigma_p^{\mathrm{res}}}{\sigma_0}\right)^{1/2} \tag{4.67}$$

σ_s^M is the moderator scattering cross section per absorber nucleus and $\sigma_p^{res} = 4\pi R^2$ is the potential scattering cross section of the resonance absorber. If interference between resonance and potential scattering is neglected, the resonance integral can be written

$$I_{NR}^\gamma = \frac{\Gamma_\gamma}{E_0}(\sigma_p^{res} + \sigma_s^M)J(\xi, \beta) \tag{4.68}$$

where the function

$$J(\xi, \beta) \equiv \int_0^\infty \frac{\psi(\xi, x)\,dx}{\psi(\xi, x) + \beta} \tag{4.69}$$

is tabulated in Table 4.3. A generalization of the J-function which includes the interference scattering term has been devised, but the form given above is more commonly used.

Wide Resonance Approximation

If the practical width of the resonance is large compared to the scattering-in interval of the resonance absorber, $\Gamma_p \gg E_0(1 - \alpha_{res})$, the second scattering source term in Eq. (4.64) can be approximated by assuming that $\Sigma_s^{res}(E')\phi(E')/E' \approx \Sigma_s^{res}(E)\phi(E)/E$, which leads to

$$\phi_{WR}(E) = \frac{\Sigma_s^M}{[\Sigma_t^{res}(E) - \Sigma_s^{res}(E) + \Sigma_s^M]E} \tag{4.70}$$

Using this result to evaluate the resonance integral defined by Eq. (4.59) yields

$$I_{WR} = \int \frac{dE}{E}\sigma_\gamma(E)\frac{\Sigma_s^M}{\Sigma_t^{res}(E) - \Sigma_s^{res}(E) + \Sigma_s^M} = \frac{\Gamma}{E_0}\sigma_s^M J(\xi, \beta') \tag{4.71}$$

where

$$\beta' = \frac{\sigma_s^M}{\sigma_0}\frac{\Gamma}{\Gamma_\gamma} \tag{4.72}$$

Resonance Absorption Calculations

Data for several of the low-energy resonances in ^{238}U are given in Table 4.4. Also shown is a comparison of the absorption probabilities calculated with the narrow and wide resonance approximations with an "exact" solution obtained numerically, for a representative fuel-to-moderator ratio for a thermal reactor. The WR approximation is more suitable for the lowest-energy resonances, but the narrow resonance approximation generally is preferable for all but the lowest-energy resonances.

Table 4.3 J-Function $(\beta = 2^j \times 10^{-5})^*$

j					$J(\xi; \beta)$					
	$\xi = 0.1$	$\xi = 0.2$	$\xi = 0.3$	$\xi = 0.4$	$\xi = 0.5$	$\xi = 0.6$	$\xi = 0.7$	$\xi = 0.8$	$\xi = 0.9$	$\xi = 1.0$
0	4.979(2)	4.970(2)	4.969(2)	4.968(2)	4.968(2)	4.968(2)	4.967(2)	4.967(2)	4.967(2)	4.967(2)
1	3.532	3.517	3.514	3.513	3.513	3.513	3.513	3.513	3.513	3.513
2	2.514	2.491	2.487	2.485	2.485	2.484	2.484	2.484	2.484	2.484
3	1.801	1.767	1.761	1.759	1.758	1.757	1.757	1.757	1.757	1.757
4	1.307	1.257	1.248	1.245	1.244	1.243	1.243	1.243	1.242	1.242
5	9.667(1)	8.993(1)	8.872(1)	8.831(1)	8.812(1)	8.802(1)	8.796(1)	8.792(1)	8.790(1)	8.788(1)
6	7.355	6.501	6.335	6.278	6.252	6.238	6.230	6.225	6.221	6.218
7	5.773	4.777	4.562	4.485	4.450	4.430	4.419	4.412	4.407	4.403
8	4.647	3.589	3.328	3.230	3.183	3.158	3.143	3.133	3.126	3.121
9	3.781	2.759	2.471	2.354	2.297	2.265	2.245	2.232	2.223	2.217
10	3.045	2.153	1.867	1.741	1.675	1.638	1.614	1.598	1.587	1.579
11	2.367	1.676	1.423	1.301	1.235	1.194	1.168	1.151	1.138	1.129
12	1.730	1.268	1.074	9.718(0)	9.119(0)	8.739(0)	8.484(0)	8.304(0)	8.174(0)	8.077(0)
13	1.164	9.081(0)	7.815(0)	7.087	6.629	6.322	6.107	5.950	5.833	5.744
14	7.172(0)	6.014	5.342	4.914	4.624	4.419	4.268	4.154	4.066	3.997
15	4.088	3.658	3.371	3.169	3.022	2.911	2.826	2.759	2.706	2.663
16	2.204	2.067	1.966	1.889	1.829	1.781	1.743	1.712	1.687	1.666
17	1.148	1.109	1.078	1.053	1.033	1.016	1.002	9.904(-1)	9.805(-1)	9.722(-1)
18	5.862(-1)	5.757(-1)	5.671(-1)	5.599(-1)	5.539(-1)	5.488(-1)	5.445(-1)	5.408	5.376	5.348
19	2.963	2.936	2.913	2.894	2.877	2.863	2.851	2.840	2.831	2.823
20	1.490	1.483	1.477	1.472	1.468	1.464	1.461	1.458	1.455	1.453

Table 4.3 (Continued)

j	$J(\xi, \beta)$									
	$\xi = 0.1$	$\xi = 0.2$	$\xi = 0.3$	$\xi = 0.4$	$\xi = 0.5$	$\xi = 0.6$	$\xi = 0.7$	$\xi = 0.8$	$\xi = 0.9$	$\xi = 1.0$
21	7.468(−2)	7.452(−2)	7.437(−2)	7.424(−2)	7.413(−2)	7.403(−2)	7.395(−2)	7.388(−2)	7.381(−2)	7.375(−2)
22	3.739	3.735	3.732	3.728	3.726	3.723	3.721	3.719	3.718	3.716
23	1.871	1.870	1.869	1.868	1.868	1.867	1.867	1.866	1.866	1.865
24	9.358(−3)	9.356(−3)	9.355(−3)	9.352(−3)	9.350(−3)	9.349(−3)	9.348(−3)	9.346(−3)	9.345(−3)	9.344(−3)
25	4.680	4.680	4.679	4.679	4.678	4.678	4.678	4.677	4.677	4.677
26	2.340	2.340	2.340	2.340	2.340	2.340	2.340	2.340	2.340	2.340
27	1.170	1.170	1.170	1.170	1.170	1.170	1.170	1.170	1.170	1.170
28	5.851(−4)	5.851(−4)	5.851(−4)	5.851(−4)	5.851(−4)	5.851(−4)	5.851(−4)	5.851(−4)	5.851(−4)	5.851(−4)
29	2.925	2.926	2.926	2.926	2.926	2.926	2.926	2.926	2.926	2.926
30	1.463	1.463	1.463	1.463	1.463	1.463	1.463	1.463	1.463	1.463
31	7.314(−5)	7.314(−5)	7.315(−5)	7.315(−5)	7.315(−5)	7.315(−5)	7.314(−5)	7.314(−5)	7.314(−5)	7.314(−5)

Source: Data from Ref. 3; used with permission of Wiley.

* Numbers in parentheses are powers of 10, which multiply the entry next to which they stand and all unmarked entries below it.

Table 4.4 Low-Energy ^{238}U Resonances

E_0 (eV)	Γ_n (eV)	Γ_γ (eV)	σ_0 (barns)	Γ_p (eV)	$(1-\alpha_{res})E_0)$ (eV)	NR	$1-p$ WR	Exact
6.67	0.00152	0.026	2.15×10^5	1.26	0.110	0.2376	0.1998	0.1963
20.90	0.0087	0.025	3.19×10^4	1.95	0.348	0.07455	0.07059	0.06755
36.80	0.032	0.025	3.98×10^4	3.65	0.612	0.04739	0.06110	0.05820
116.85	0.030	0.022	1.30×10^4	1.32	0.966	0.00904	0.00950	0.00917
208.46	0.053	0.022	8.86×10^3	2.63	1.73	0.00444	0.00769	0.00502

Source: Data from Ref. 3; used with permission of Wiley.

Example 4.2: Group Capture Cross Section for 6.67-eV ^{238}U Resonance. The contribution of the 6.67-eV ^{238}U resonance to the capture cross section of an energy group extending from 1 to 10 eV is calculated in the narrow resonance approximation from $\sigma_\gamma^g = I_{NR}^\gamma/\ln(10/1)$, where $I_{NR}^\gamma = (\Gamma_\gamma/E_0)(\sigma_p^{res} + \sigma_s^M)J(\xi,\beta)$. For uranium, $\sigma_p^{res} = 8.3$ barns. For a moderator cross section per fuel atom of $\sigma_s^M = \Sigma_s^M/N_F = 60$ barns and a temperature $T = 330°C$, $\xi = (\Gamma/2)/(E_0kT/A)^{1/2} = (0.0275 \text{ eV}/2)/[(6.67 \text{ eV} \times 603 \text{ K} \times 0.86 \times 10^{-4} \text{ eV/K})/238]^{1/2} = 0.361$ and $\beta = (\sigma_p^{res} + \sigma_s^M)/\sigma_0 = (60 + 8.3)/2.15 \times 10^5 = 31.8 \times 10^{-5} = 2^j \times 10^5$, or $j = 4.98$. Interpolating on j and ξ in Table 4.3 yields $J \approx 88$. With these values, $I_{NR}^\gamma \approx 23$ and as $\sigma_\gamma^g \approx 10$ barns.

Temperature Dependence of Resonance Absorption

Examination of the function $J(\xi, \beta)$ of Eq. (4.69) reveals that for any value of β, the value of J increases or remains constant as ξ decreases. Since $\xi \sim 1/T^{1/2}$, the resonance absorption must increase or remain unchanged when the temperature increases. The physical reason for this is that as the temperature increases, the cross section (averaged over nuclear motion) decreases in peak value and broadens in energy in such a manner as to preserve the area under the cross-section curve, as indicated in Fig. 4.5, but the decreasing value of the cross section results in a decreasing depression in the neutron flux in the resonance region. This increase in absorption cross section with increasing fuel temperature introduces an important negative-feedback Doppler temperature coefficient of reactivity, which is important for reactor safety, as discussed in Chapter 5.

4.4
Multigroup Diffusion Theory

Multigroup Diffusion Equations

We consider cohorts of neutrons of different energies diffusing within a nuclear reactor. The basic diffusion equation for each cohort, or group, of neutrons is the same as derived in Chapter 3, but with absorption generalized to all processes that remove the neutron from the cohort or group (i.e., absorption plus scattering to another group) and with the source of neutrons for each group specialized to include the in-scatter of neutrons from other groups, which are also diffusing within the reactor:

$$-\nabla \cdot D^g(r)\nabla\phi_g(r) + \Sigma_r^g(r)\phi_g(r)$$

$$= \sum_{g' \neq g}^{G} \Sigma_s^{g' \to g}(r)\phi(r)_{g'} + \frac{1}{k}\chi^g \sum_{g'=1}^{G} \nu\Sigma_f^{g'}(r)\phi_{g'}(r), \quad g = 1, \ldots, G \quad (4.73)$$

The definition of group constants given by Eqs. (4.39) is applicable. For the group diffusion coefficient there are two plausible definitions:

$$D^g(r) = \int_{E_g}^{E_{g-1}} dE\, D(r, E)\phi(r, E)/\phi_g(r)$$

$$= \frac{1}{3}\int_{E_g}^{E_{g-1}} dE\, \frac{\phi(r, E)}{\Sigma_{tr}(r, E)}\Big/\phi_g(r) \quad (4.74)$$

or

$$D^g(r) = \frac{1}{3\Sigma_{tr}^g(r)} = \frac{1}{3\int_g^{E_{g-1}} dE\, \Sigma_{tr}(r, E)\phi(r, E)/\phi_g(r)} \quad (4.75)$$

We return to this issue in Chapter 10, where the multigroup diffusion equations are formally derived from energy-dependent transport theory.

Equations (4.73) constitute a set of homogeneous equations, the solutions of which are nontrivial only for certain discrete values of the effective multiplication constant, k. It has been shown (Refs. 8 and 12) that the mathematical properties of the multigroup diffusion equations are such that the largest such discrete eigenvalue is real and positive. The corresponding eigenfunction is unique and nonnegative everywhere within the reactor. In other words, mathematically, these equations have a physically correct solution corresponding to the largest value of the eigenvalue.

Two-Group Theory

The simplest example of multigroup diffusion theory is two-group theory in which the fast group contains all neutrons with $E \gtrsim 1$ eV and the thermal group contains the neutrons that have slowed down into the thermal interval $E \lesssim 1$ eV. This model is described by

$$-\nabla \cdot D^1\nabla\phi_1 + (\Sigma_a^1 + \Sigma_s^{1 \to 2})\phi_1 = \frac{1}{k}(\nu\Sigma_f^1\phi_1 + \nu\Sigma_f^2\phi_2)$$

$$-\nabla \cdot D^2\nabla\phi_2 + \Sigma_a^2\phi_2 = \Sigma_s^{1 \to 2}\phi_1 \quad (4.76)$$

and the boundary conditions of the neutron fluxes in both groups vanishing on the boundary of the reactor.

Two-Group Bare Reactor

For a uniform reactor, the vanishing of the neutron flux on the boundary requires that the neutron flux in both groups satisfies

$$\nabla^2 \psi(r) + B_g^2 \psi(r) = 0 \tag{4.77}$$

where B_g is the geometric buckling of Table 3.3. Using this form for the group fluxes in Eqs. (4.76) leads to a pair of homogeneous algebraic equations that can be solved for the effective multiplication constant

$$k = \frac{\nu \Sigma_f^1}{\Sigma_a^1 + \Sigma_s^{1 \to 2} + D^1 B_g^2} + \frac{\Sigma_s^{1 \to 2}}{\Sigma_a^1 + \Sigma_s^{1 \to 2} + D^1 B_g^2} \cdot \frac{\nu \Sigma_f^2}{\Sigma_a^2 + D^2 B_g^2} \tag{4.78}$$

and the flux ratio

$$\phi_1 = \frac{\Sigma_a^2 + D^2 B_g^2}{\Sigma_s^{1 \to 2}} \phi_2 \tag{4.79}$$

Extending the definition of the diffusion length for the fast group to include removal by scattering to the thermal group

$$L_1^2 = \frac{D^1}{\Sigma_a^1 + \Sigma_s^{1 \to 2}} = \frac{D^1}{\Sigma_r^1} \tag{4.80}$$

Eq. (4.78) for the effective multiplication constant can be rearranged into a form from which the definition of terms in the six-factor formula are apparent:

$$k = \left(\frac{\nu \Sigma_f^2}{\Sigma_a^2} \right) \left(1 + \frac{\nu \Sigma_f^1 \phi_1}{\nu \Sigma_f^2 \phi_2} \right) \left(\frac{\Sigma_s^{1 \to 2}}{\Sigma_a^1 + \Sigma_s^{1 \to 2}} \right) \left(\frac{1}{1 + L_1^2 B_g^2} \right) \left(\frac{1}{1 + L_2^2 B_g^2} \right)$$

$$= (\eta f)(\varepsilon)(p)(P_{NL}^1)(P_{NL}^2) \tag{4.81}$$

where the fast (P_{NL}^1) and thermal (P_{NL}^2) non-leakage probabilities are identified separately.

One-and-One-Half-Group Theory

Because the thermal group absorption cross section is generally much larger than the fast-group cross section, $D^2 \ll D^1$. This suggests approximating the two-group equations by neglecting D^2 and using the resulting solution of the thermal group equation $\phi^2 = (\Sigma_s^{1 \to 2} / \Sigma_a^2) \phi^1$ in the fast-group equation to obtain

$$-\nabla \cdot D^1 \nabla \phi_1 + \Sigma_r^1 \phi_1 = \frac{1}{k} \left(\nu \Sigma_f^1 + \nu \Sigma_f^2 \frac{\Sigma_s^{1 \to 2}}{\Sigma_a^2} \right) \phi_1 \tag{4.82}$$

which has the form of a one-group diffusion equation for the fast neutrons. This method may be extended to account for the diffusion of thermal neutrons by using an effective value of the fast diffusion coefficient,

$$D_{eff}^1 = D^1 + \frac{\Sigma_a^1 + \Sigma_s^{1 \to 2}}{\Sigma_a^2} D^2 \tag{4.83}$$

which has the effect of replacing the fast diffusion length L_1 by the migration length; that is,

$$L_1^2 \to M^2 = \frac{D^1}{\Sigma_a^1 + \Sigma_s^{1\to2}} + \frac{D^2}{\Sigma_a^2} \tag{4.84}$$

The solutions discussed in Chapter 3 for the one-speed neutron diffusion equation can be applied immediately to $1\frac{1}{2}$-group theory merely by replacing $\nu\Sigma_f \to \nu\Sigma_f^1 + \nu\Sigma_f^2(\Sigma_s^{1\to2}/\Sigma_a^2)$ and $D \to D_{\text{eff}}^1$.

Two-Group Theory of Two-Region Reactors

Consider a rectangular parallelepiped core consisting of a uniform central region (material 1) bounded on both ends by regions of the same composition (material 2), as depicted in Fig. 4.6. The two-group equations in each material (subscript k) are

$$-D_k^1\nabla^2\phi_{1k}(x,y,z) + \Sigma_{rk}^1\phi_{1k}(x,y,z)$$
$$= \frac{1}{k}\left[\nu\Sigma_{fk}^1\phi_{1k}(x,y,z) + \nu\Sigma_f^2\phi_{2k}(x,y,z)\right] \tag{4.85}$$
$$-D_k^2\nabla^2\phi_{2k}(x,y,z) + \Sigma_{ak}^2\phi_{2k}(x,y,z) = \Sigma_{sk}^{1\to2}\phi_1(x,y,z)$$

where group 2 is assumed to be below the fission spectrum. We seek a solution by separation of variables, and recalling the results of Chapter 3 look for a solution of the form

$$\phi_{gk}(x,y,z) = X_{gk}(x)\cos\frac{\pi y}{2Y_1}\cos\frac{\pi y}{2Z_1} \tag{4.86}$$

The y- and z-components of the gradient operators acting on the trial solutions of Eq. (4.86) give rise to a *transverse buckling* term,

$$B_{yz}^2 = \left(\frac{\pi}{2Y_1}\right)^2 + \left(\frac{\pi}{2Z_1}\right)^2 \tag{4.87}$$

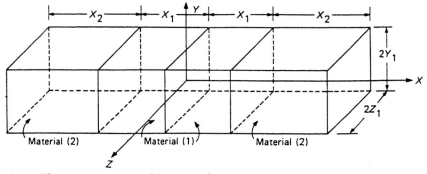

Fig. 4.6 Three-region reactor model. (From Ref. 6; used with permission of MIT Press.)

These trial solutions are substituted into Eqs. (4.85) to obtain equations for the X_{gk}:

$$-D_k^1 \frac{\partial^2}{\partial x^2} X_{1k}(x) + (\Sigma_{rk}^1 + D_k^1 B_{yz}^2) X_{1k}(x) = \frac{1}{k} \left[\nu \Sigma_{fk}^1 X_{1k}(x) + \nu \Sigma_{fk}^2 X_{2k}(x) \right]$$
(4.88)

$$-D_k^2 \frac{\partial^2 X_{2k}(x)}{\partial x^2} + (\Sigma_{ak}^2 + D_k^2 B_{yz}^2) X_{2k}(x) = \Sigma_{sk}^{1 \to 2} X_{1k}(x)$$

These equations must satisfy symmetry boundary conditions at $x = 0$, continuity of flux and current interface conditions at $x = x_1$, and zero flux at $x = x_1 + x_2$:

$$\frac{dX_{g1}(0)}{dx} = 0, \qquad X_{g2}(x_1 + x_2) = 0$$
(4.89)

$$X_{g1}(x_1) = X_{g2}(x_1), \qquad -D_1^g \frac{dX_{g1}(x_1)}{dx} = -D_2^g \frac{dX_{g2}(x_1)}{dx}$$

The procedure for solving Eqs. (4.88) is to look for solutions of a particular form with arbitrary constants and then to establish conditions on the arbitrary constants by requiring the form to satisfy Eqs. (4.88). In particular, we look for solutions that satisfy

$$\frac{d^2 X_{gk}(x)}{dx^2} + B_k^2 X_{gk}(x) = 0$$
(4.90)

in each region k. Note that we require that Eq. (4.90) be satisfied with the same value of B_k^2 by both the fast (X_{1k}) and thermal (X_{2k}) fluxes in each region k. Substituting the solution of the form that satisfies Eqs. (4.90) into Eqs. (4.88) leads to a set of equations for each region k:

$$\left(D_k^1 B_k^2 + D_k^1 B_{yz}^2 + \Sigma_{rk}^1 - \frac{1}{k} \nu \Sigma_{fk}^1 \right) X_{1k}(x) - \frac{1}{k} \left(\nu \Sigma_{fk}^2 \right) X_{2k}(x) = 0$$
(4.91)

$$\left(-\Sigma_{sk}^{1 \to 2} \right) X_{1k}(x) + \left(D_k^2 B_k^2 + D_k^2 B_{yz}^2 + \Sigma_{ak}^2 \right) X_{2k}(x) = 0$$

which must be satisfied if the solution of Eqs. (4.88) within each material is to have the form that satisfies Eqs. (4.90). These are homogeneous equations, which have a nontrivial solution only if the determinant of the coefficient matrix vanishes, which defines two values $B_k^2 = \mu_k^2$ and $B_k^2 = -\nu_k^2$ for which Eqs. (4.88) have solutions of the form that satisfies Eqs. (4.90):

$$\mu_k^2 = -B_{yz}^2 - \frac{1}{2} \left(\frac{\Sigma_{ak}^2}{D_k^2} + \frac{\Sigma_{rk}^1 - k^{-1} \nu \Sigma_{fk}^1}{D_k^1} \right)$$

$$+ \left[\left(\frac{\Sigma_{ak}^2}{2D_k^2} - \frac{\Sigma_{rk}^1 + k^{-1} \nu \Sigma_{fk}^1}{2D_k^1} \right)^2 + \frac{k^{-1} \nu \Sigma_{fk}^2 \Sigma_{sk}^{1 \to 2}}{D_k^1 D_k^2} \right]^{1/2}$$
(4.92)

$$\nu_k^2 = B_{yz}^2 + \frac{1}{2} \left(\frac{\Sigma_{ak}^2}{D_k^2} + \frac{\Sigma_{rk}^1 - k^{-1} \nu \Sigma_{fk}^1}{D_k^1} \right)$$

$$+ \left[\left(\frac{\Sigma_{ak}^2}{2D_k^2} - \frac{\Sigma_{rk}^1 - k^{-1} \nu \Sigma_{fk}^1}{D_k^1} \right)^2 + \frac{k^{-1} \nu \Sigma_{fk}^2 \Sigma_{sk}^{1 \to 2}}{D_k^1 D_k^2} \right]^{1/2}$$

The quantity $-v_k^2$ is always negative, but μ_k^2 can be positive or negative, depending on the value of the two-group constants. Thus there are solutions of Eqs. (4.88) that satisfy Eqs. (4.90), which are of the form

$$X_{2k}(x) = A_{2k}^1 \sin \mu_k x + A_{2k}^2 \cos \mu_k x + A_{2k}^3 \sinh v_k x + A_{2k}^4 \cosh v_k x$$

$$X_{1k}(x) = s_k A_{2k}^1 \sin \mu_k x + s_k A_{2k}^2 \cos \mu_k x + t_k A_{2k}^3 \sinh v_k x + t_k A_{2k}^4 \cosh v_k x \tag{4.93}$$

where the second of Eqs. (4.91) has been used to determine the ratio of fast-to-thermal group components:

$$s_k = \frac{D_k^2(\mu_k^2 + B_{yz}^2) + \Sigma_{ak}^2}{\Sigma_{sk}^{1 \to 2}}$$

$$t_k = \frac{D_k^2(-v_k^2 + B_{yz}^2) + \Sigma_{ak}^2}{\Sigma_{sk}^{1 \to 2}} \tag{4.94}$$

The symmetry conditions $x = 0$ require that $A_{21}^1 = A_{21}^3 = 0$, and the zero flux conditions at $x = x_1 + x_2$ require that the solution in region 2 be of the form

$$X_{22}(x) = C_{22}^1 \sin \mu_2(x_1 + x_2 - x) + C_{22}^2 \sinh v_2(x_1 + x_2 - x)$$

$$X_{12}(x) = s_2 C_{22}^1 \sin \mu_2(x_1 + x_2 - x) + t_2 C_{22}^2 \sinh v_2(x_1 + x_2 - x) \tag{4.95}$$

Requiring the solution in region 1 given by Eqs. (4.93) and the solution in region 2 given by Eqs. (4.95) to satisfy the continuity of flux and current interface conditions results in a set of four homogeneous equations for the constants A_{21}^2, A_{21}^4, C_{22}^1, and C_{22}^2. The requirement for a nontrivial solution, the vanishing of the determinant of the coefficients, then, is the criticality condition

$$\det \begin{bmatrix} s_1 \cos \mu_1 x_1 & t_1 \cosh v_1 x_1 & -s_2 \sin \mu_2 x_2 & -t_2 \sinh v_2 x_2 \\ s_1 D_1^1 \mu_1 \sin \mu_1 x_1 & -t_1 D_1^1 v_1 \sinh v_1 x_1 & -s_2 D_2^1 \mu_2 \cos \mu_2 x_2 & -t_2 D_2^1 \cosh v_2 x_2 \\ \cos \mu_1 x_1 & \cosh v_1 x_1 & -\sin \mu_2 x_2 & -\sinh v_2 x_2 \\ D_1^2 \mu_1 \sin \mu_1 x_1 & -D_1^2 v_1 \sinh v_1 x_1 & -D_2^2 \mu_2 \cos \mu_2 x_2 & -D_2^2 v_2 \cosh v_2 x_2 \end{bmatrix}$$

$$= 0 \tag{4.96}$$

which may be solved for the effective multiplication constant, k. The four equations can be solved for three of the constants in terms of the one remaining constant, which must be determined from the total reactor power level.

This procedure could be extended to multiregion reactors, but it becomes extremely cumbersome, and direct numerical solution of Eqs. (4.85) becomes preferable.

Two-Group Theory of Reflected Reactors

The results above can be specialized to the situation of a reflected reactor by setting $\Sigma_f = 0$ in region 2, in which case Eq. (4.92) reduces to

$$\mu_2^2 = -B_{yz}^2 - \frac{\Sigma_{r2}^1}{D_2^1}, \qquad -\nu_2^2 = -B_{yz}^2 - \frac{\Sigma_{a2}^1}{D_2^2} \tag{4.97}$$

A solution of the type just described can be carried out in spherical and cylindrical geometry (reflected axially or radially, but not both), as well as in the block geometry. The results are summarized in Table 4.5, where $Z(\mathbf{R} = R, \{R, z\}$ or $[x, y, z])$ and W are spatial flux shapes in the core and U and V are spatial flux shapes in the reflector.

The thermal flux in the core of a spherical reflected reactor is given by

$$\phi_c^2(r) = \frac{\phi_{0c}}{r}(\sin \mu_c r + a \sinh \nu_c r) \tag{4.98}$$

and the thermal flux in the spherical shell reflector is given by

$$\phi_R^2(r) = \frac{\phi_{0R}}{r}\left[\sinh \mu_R (R' - r) + b \sinh \nu_R (R' - r)\right] \tag{4.99}$$

The corresponding fast fluxes are related to the thermal fluxes by the factors s_k and t_k given by Eqs. (4.94). These fluxes are plotted for a representative set of two-group constants in Fig. 4.7. The much larger ratio $\Sigma_s^{1\rightarrow 2}/\Sigma_a^2$ in the reflector than in the core causes a peaking of the thermal flux in the reflector at the core–reflector interface. Physically, fast neutrons are diffusing out of the core and being slowed down into the thermal group in the reflector, where the thermal absorption is greatly reduced relative to the core. This same type of peaking of the thermal flux would occur in a water gap next to a fuel assembly within the core.

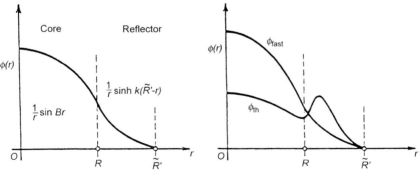

Fig. 4.7 Fast and thermal fluxes in a reflected spherical reactor with properties (core: $D_1 = D_2 = 1$ cm, $\Sigma_s^{1\rightarrow 2} = 0.009$ cm^{-1}, $\Sigma_a^1 = 0.001$, $\Sigma_a^2 = 0.05$ cm^{-1}, $\nu\Sigma_f^2 = 0.057$; reflector: $D_1 = D_2 = 1$ cm, $\Sigma_s^{1\rightarrow 2} = 0.009$ cm^{-1}, $\Sigma_a^1 = 0.001$, $\Sigma_a^2 = 0.0049$ cm^{-1}, $\nu\Sigma_f^2 = 0.0$. (From Ref. 13; used with permission of McGraw-Hill.)

Table 4.5 Flux Shapes in Reflected Reactors in Two-Group Theory

Geometry

Spherical

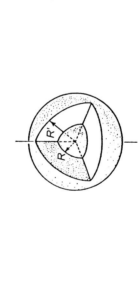

$$Z(\mathbf{R}) = \frac{\sin \mu R}{R}$$

$$Z'(\mathbf{R}) = \frac{\mu \cos \mu R}{R} - \frac{\sin \mu R}{R^2}$$

$$W(\mathbf{R}) = \frac{\sinh \lambda R}{R}$$

$$W'(\mathbf{R}) = \frac{\lambda \cosh \lambda R}{R} - \frac{\sinh \lambda R}{R^2}$$

$$U(\mathbf{R}) = \frac{\sinh \kappa_3 (\tilde{R}' - R)}{R}$$

$$U'(\mathbf{R}) = -\frac{\kappa_3 \cosh \kappa_3 (\tilde{R}' - R)}{R} - \frac{\sinh \kappa_3 (\tilde{R}' - R)}{R^2}$$

$$V(\mathbf{R}) = \frac{\sinh \kappa_4 (\tilde{R}' - R)}{R^2}$$

$$V'(\mathbf{R}) = -\frac{\kappa_4 \cosh \kappa_4 (\tilde{R}' - R)}{R} - \frac{\sinh \kappa_4 (\tilde{R}' - R)}{R^2}$$

Table 4.5 (Continued)

Geometry

Cylinder: Side Reflectors

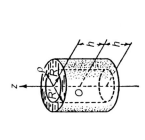

$$Z(\mathbf{R}) = J_0(l_1 R)\cos l_2 z$$
$$Z'(\mathbf{R}) = -l_1 J_1(l_1 R)\cos l_2 z$$
$$W(\mathbf{R}) = I_0(l_3 R)\cos l_2 z$$
$$W'(\mathbf{R}) = l_3 I_1(l_3 R)\cos l_2 z$$

$$l_1^2 = \mu^2 - l_2^2 \qquad l_2 = \frac{\pi}{2h} \qquad l_3^2 = \lambda^2 + l_2^2$$

$$U(\mathbf{R}) = [I_0(l_4 R)K_0(l_4 \tilde{R}') - I_0(l_4 \tilde{R}')K_0(l_4 R)]\cos l_2 z$$
$$U'(\mathbf{R}) = l_4[I_1(l_4 R)K_0(l_4 \tilde{R}') + I_0(l_4 \tilde{R}')K_1(l_4 R)]\cos l_2 z$$
$$V(\mathbf{R}) = [I_0(l_5 R)K_0(l_5 \tilde{R}') - I_0(l_5 \tilde{R}')K_0(l_5 R)]\cos l_2 z$$
$$V'(\mathbf{R}) = l_5[I_1(l_5 R)K_0(l_5 \tilde{R}') + I_0(l_5 \tilde{R}')K_1(l_5 R)]\cos l_2 z$$

$$l_4^2 = \kappa_3^2 + l_2^2 \qquad l_5^2 = \kappa_4^2 + l_2^2$$

Cylinder: End Reflectors

$$Z(\mathbf{R}) = J_0(m_1 \rho)\cos m_2 h$$
$$Z'(\mathbf{R}) = -m_2 J_0(m_1 \rho)\sin m_2 h$$
$$W(\mathbf{R}) = J_0(m_1 \rho)\cosh m_3 h$$
$$W'(\mathbf{R}) = m_3 J_0(m_1 \rho)\sinh m_3 h$$

$$m_1 = \frac{2.405}{\tilde{R}'} \qquad m_2^2 = \mu^2 - m_1^2 \qquad m_3^2 = \lambda^2 + m_1^2$$

$$U(\mathbf{R}) = J_0(m_1 \rho)\sinh m_4(\tilde{d} - h)$$
$$U'(\mathbf{R}) = -m_4 J_0(m_1 \rho)\cosh m_4(\tilde{d} - h) \qquad m_4^2 = \kappa_3^2 + m_1^2$$
$$V(\mathbf{R}) = J_0(m_1 \rho)\sinh m_5(\tilde{d} - h)$$
$$V'(\mathbf{R}) = -m_5 J_0(m_1 \rho)\cosh m_5(\tilde{d} - h) \qquad m_5^2 = \kappa_4^2 + m_1^2$$

Table 4.5 (Continued)

Geometry	
Block	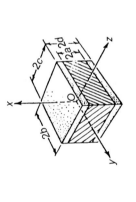

$$Z(\mathbf{R}) = \cos n_1 a \cos n_2 y \cos n_3 z$$

$$Z'(\mathbf{R}) = -n_1 \sin n_1 a \cos n_2 y \cos n_3 z$$

$$n_1^2 \equiv \mu^2 - n_2^2 - n_3^2$$

$$W(\mathbf{R}) = \cosh n_4 a \cos n_2 y \cos n_3 z$$

$$W'(\mathbf{R}) = n_4 \sinh n_4 a \cos n_2 y \cos n_3 z$$

$$n_4^2 \equiv \lambda^2 + n_2^2 + n_3^2 \qquad n_3 \equiv \frac{\pi}{2\tilde{c}} \qquad n_2 \equiv \frac{\pi}{2\tilde{b}}$$

$$U(\mathbf{R}) = \sinh n_5 (\tilde{d} - a) \cos n_2 y \cos n_3 z$$

$$U'(\mathbf{R}) = -n_5 \cosh n_5 (\tilde{d} - a) \cos n_2 y \cos n_3 z$$

$$V(\mathbf{R}) = \sinh n_6 (\tilde{d} - a) \cos n_2 y \cos n_3 z$$

$$V'(\mathbf{R}) = -n_6 \cosh n_6 (\tilde{d} - a) \cos n_2 y \cos n_3 z$$

$$n_5^2 \equiv \kappa_3^2 + n_2^2 + n_3^2 \qquad n_6^2 \equiv \kappa_4^2 + n_2^2 + n_3^2$$

Source: Adapted from Ref. 13; used with permission of McGraw-Hill.

Prime on Z', W', U', V' indicates spatial derivative.

Tilde on symbol indicates extrapolated boundary.

Numerical Solutions for Multigroup Diffusion Theory

The numerical solution procedures discussed for the one-speed diffusion equation in Section 3.10 are readily extended to the solution of the multigroup diffusion equations. The G multigroup equations for the case of $G - 1$ fast groups and a thermal group G are

$$-\nabla \cdot D^1 \nabla \phi_1 + \Sigma_r^1 \phi_1 = \frac{1}{k} \chi_1 S_f \quad \text{\textit{fast}}$$

$$-\nabla \cdot D^2 \nabla \phi_2 + \Sigma_r^2 \phi_2 = \frac{1}{k} \chi_2 S_f + \Sigma_s^{1\to2} \phi_1$$

$$\tag{4.100}$$

$$-\nabla \cdot D^3 \nabla \phi_3 + \Sigma_r^3 \phi_3 = \frac{1}{k} \chi_3 S_f + \Sigma_s^{1\to3} \phi_1 + \Sigma_s^{2\to3} \phi_2$$

$$-\nabla \cdot D^G \nabla \phi_G + \Sigma_a^G \phi_G = \Sigma_s^{1\to G} \phi_1 + \Sigma_s^{2\to G} \phi_2 + \cdots + \Sigma_s^{G-1\to G} \phi_{G-1} \quad \text{\textit{thermal}}$$

where the fission source is

$$S_f(r) = \sum_{g=1}^{G} \nu \Sigma_f^g(r) \phi_g(r) \tag{4.101}$$

The solution procedure is initiated by guessing a fission source distribution, $S_f^{(0)}$, and an effective multiplication constant, $k^{(0)}$, and solving the group 1 equation for the first iterate flux, $\phi_1^{(1)}$:

$$-\nabla \cdot D^1 \nabla \phi_1^{(1)} + \Sigma_r^1 \phi_1^{(1)} = \frac{1}{k^{(0)}} \chi_1 S_f^{(0)} \tag{4.102}$$

Equation (4.102) is solved iteratively (e.g., by the successive relaxation method described in Section 3.10). Next, the group 2 equation is solved for the first iterate flux, $\phi_2^{(1)}$:

$$-\nabla \cdot D^2 \nabla \phi_2^{(1)} + \Sigma_r^2 \phi_2^{(1)} = \frac{1}{k^{(0)}} \chi_2 S_f^{(0)} + \Sigma_s^{1\to2} \phi_1^{(1)} \tag{4.103}$$

using the just calculated $\phi_1^{(1)}$ and an iteration procedure of the type described in Section 3.10. This procedure is continued successively to all the lower groups, using the just calculated values of the fluxes for higher-energy groups to calculate the scattering-in source, to determine the first iterate of all G group fluxes $[\phi_1^{(1)}, \phi_2^{(1)}, \ldots, \phi_G^{(1)}]$, which are then used to compute a first iterate fission source:

$$S_f^{(1)}(r) = \sum_{g=1}^{G} \nu \Sigma_f^g(e) \phi_g^{(1)}(r) \tag{4.104}$$

and a first iterate effective multiplication constant:

$$k^{(1)} = \frac{k^{(0)} \int dr \, S_f^{(1)}(r)}{\int dr \, S_f^{(0)}(r)} \tag{4.105}$$

The iterations are continued until the effective multiplication constant converges, as described in Section 3.10.

If a multigroup structure is chosen in which there is more than one group in the thermal energy interval $E < 1$ eV, there is upscattering among the thermal groups and the successive-group solution procedure above must be modified by solving simultaneously for the fluxes in the thermal groups or by an iterative solution for the thermal group fluxes.

References

1 D. E. CULLEN, "Nuclear Cross Section Preparation," in Y. Ronen, ed., *CRC Handbook of Nuclear Reactor Calculations I*, CRC Press, Boca Raton, FL (**1986**).

2 R. E. MACFARLANE, D. W. MUIR, and R. M. BOICOURT, *The NJOY Nuclear Data Processing System*, Vols. I and II, LA-9303-M, Los Alamos National Laboratory, Los Alamos, NM (**1982**).

3 J. J. DUDERSTADT and L. J. HAMILTON, *Nuclear Reactor Analysis*, Wiley, New York (**1976**).

4 B. J. TOPPEL, "The New Multigroup Cross Section Code, MC^2-II," in *Proc. Conf. New Developments of Reactor Mathematics and Applications*, CONF-710302, Idaho Falls, ID (**1971**); H. HENRYSON et al., MC^2-*II: A Code to Calculate Fast Neutron Spectra and Multigroup Cross Sections*, ANL-8144, Argonne National Laboratory, Argonne, IL (**1976**).

5 C. R. WEISBIN et al., *MINX: A Multigroup Interpretation of Nuclear Cross Sections from ENDF/B*, LA-6486-MS-(ENDF-237), Los Alamos National Laboratory, Los Alamos, NM (**1976**).

6 A. F. HENRY, *Nuclear-Reactor Analysis*, MIT Press, Cambridge, MA (**1975**).

7 R. KINSEY, *Data Formats and Procedures for the Evaluated Nuclear Data File, ENDF*, BNL-NCS-50496, ENDG-1021, 2nd ed., ENDF/B-V, Brookhaven National Laboratory, Upton, NY (**1970**); C. BREWSTER, *ENDF/B Cross Sections*, BNL-17100 (ENDF-200), 2nd ed., Brookhaven National Laboratory, Upton, NY (**1975**).

8 E. L. WACHSPRESS, *Iterative Solutions of Elliptic Systems and Applications to the Neutron Diffusion Equations of Reactor Physics*, Prentice Hall, Englewood Cliffs, NJ (**1973**).

9 R. H. HOWERTON et al., *Evaluation Techniques and Documentation of Specific Evaluations of the LLL Evaluated Nuclear Data Library (ENDL)*, UCRL-50400, Vol. 15, Lawrence Livermore Laboratory, Livermore, CA (**1970**).

10 I. I. BONDARENKO et al., *Group Constants for Nuclear Reactor Calculations*, Consultants Bureau, New York (**1964**).

11 R. S. VARGA, *Matrix Iterative Analysis*, Prentice Hall, Englewood Cliffs, NJ (**1962**).

12 G. J. HABETLER and M. A. MARTINO, *Proc. Symp. Appl. Math. IX*, 127 (**1961**).

13 R. V. MEGHREBLIAN and D. K. HOLMES, *Reactor Analysis*, McGraw-Hill, New York (**1960**).

Problems

4.1. Solve the neutron balance equation in the slowing-down range for the neutron flux, and determine the neutron slowing-down density, for a mixture of nonhydrogenic moderators and no absorption. Compare the result with Eq. (4.20) in the no-absorption limit.

4.2. Consider a very large block of material with composition $(^{235}\text{U} = 0.002 \times 10^{24}$ at/cm^3, $^{238}\text{U} = 0.040 \times 10^{24}$ at/cm^3, $\text{H}_2\text{O} = 0.022 \times 10^{24}$ at/cm^3, Fe $= 0.009 \times 10^{24}$ at/cm^3) and temperature $T = 400°\text{C}$. Calculate and plot the neutron flux energy distribution in the fission, slowing-down, and thermal regions.

4.3. Carry out the steps to demonstrate that the Maxwellian distribution of Eq. (4.29) satisfies the equilibrium neutron balance equation of Eq. (4.27).

4.4. Calculate the thermal group absorption cross section for ^{235}U at $T_n = 300, 400$, and $500°\text{C}$.

4.5. Calculate the infinite multiplication constant and the relative group fluxes in a very large fuel assembly with the four-group constants given in Table P4.5.

4.6. The radiative capture cross section for a certain isotope is measured at the following energies: 50 eV, 200 barns; 100 eV, 245 barns; 150 eV, 275 barns; 300 eV, 200 barns; 350 eV, 180 barns; 400 eV, 210 barns. Calculate a multigroup capture cross section for the group $E_g = 75$ eV, $E_{g-1} = 425$ eV.

4.7. Calculate the resonance escape probability for the 6.67 eV ^{238}U resonance at $T = 300°\text{C}$ when the moderator scattering cross section per uranium nucleus is $\Sigma_s^M / N_{\text{res}} = 50$ barns. Calculate the resonance integral using either the narrow or wide resonance approximation; explain your choice.

4.8. Calculate the contribution of each of the resonances in Table 4.4 to the multigroup capture cross section for a group extending from 1 to 300 eV when the moderator scattering cross section per uranium nucleus is $\Sigma_s^M / N_{\text{res}} = 75$ barns and the temperature is $300°\text{C}$.

$$\bar{\sigma}_\gamma^g = \frac{\sum\limits_{i \in g} I_i}{\ell n \left(\frac{E_{g-1}}{E_g} \right)}$$

Table P4.5

Group Constant	Group 1: 1.35–10 MeV	Group 2: 9.1 keV–1.35 MeV	Group 3: 0.4 eV–9.1 keV	Group 4: 0.0–0.4 eV
χ	0.575	0.425	0	0
$\nu\Sigma_f$ (cm^{-1})	0.0096	0.0012	0.0177	0.1851
Σ_a (cm^{-1})	0.0049	0.0028	0.0305	0.1210
$\Sigma_s^{g \to g+1}$ (cm^{-1})	0.0831	0.0585	0.0651	–
D (cm)	2.162	1.087	0.632	0.354

Table P4.11

Group Constant	Core		Water/Structure	
	Group 1	Group 2	Group 1	Group 2
χ	1.0	0.0	0.0	0.0
$\nu\Sigma_f$ (cm^{-1})	0.0085	0.1851	0.0	0.0
Σ_a (cm^{-1})	0.0121	0.121	0.0004	0.020
$\Sigma_s^{1\to2}$ (cm^{-1})	0.0241	–	0.0493	–
D (cm)	1.267	0.354	1.130	0.166

4.9. Repeat the calculation of Problem 4.8 for $\Sigma_s^M/N_{res} = 25$ barns. Repeat the calculation for 500°C.

4.10. Calculate the total resonance escape probability for the resonances in Table 4.4 when the moderator scattering cross section per uranium nucleus is $\Sigma_s^M/N_{res} = 75$ barns and the temperature is 300°C.

4.11. Consider a large repeating array of slab fuel assemblies of width 50 cm separated by 10 cm water–structure slabs. Calculate the thermal and fast flux distributions and the infinite multiplication factor for the fuel–water–structure array using the two-group cross sections given in Table P4.11.

4.12.* Write a computer code to solve numerically for the fast and thermal flux distributions and the effective multiplication constant in a two-dimensional cut through a very tall reactor core. The reactor core extends from -50 cm $< x < +50$ cm. Region 1 of the core extends from 15 cm $< y < 55$ cm, and region 2 of the core extends from 55 cm $< y < 105$ cm. The core is entirely surrounded by a 15-cm-thick reflector. The two-group constants for the core and reflector are given in Table P4.12.

Table P4.12

Group Constant	Core 1		Core 2		Reflector	
	Group 1	Group 2	Group 1	Group 2	Group 1	Group 2
χ	1.0	0.0	1.0	0.0	0.0	0.0
$\nu\Sigma_f$ (cm^{-1})	0.0085	0.1851	0.006	0.150	0.0	0.0
Σ_a (cm^{-1})	0.0121	0.121	0.010	0.100	0.0004	0.020
$\Sigma_s^{1\to2}$ (cm^{-1})	0.0241	–	0.016	0.000	0.0493	–
D (cm)	1.267	0.354	1.280	0.400	1.130	0.166

* Problem 4.12 is a longer problem suitable for a take-home project.

4.13. Calculate the reduction of the slowing-down density as a function of energy below 50 keV in a 1:1 homogeneous mixture of H_2O and 3% enriched uranium. (Use the resonance range cross sections of Table 1.3, assuming them to be constant in energy.)

4.14. Calculate and plot the hardened Maxwellian component of the thermal spectrum for in a 1:1 homogeneous mixture of H_2O and uranium for natural uranium and 4% enriched uranium. (Use the thermal range cross sections of Table 1.3, assuming them to be constant in energy, and use $C = 1.5$.)

4.15. Calculate the spectrum averaged one group cross sections for Problem 4.5.

4.16. Extend the development of Section 3.11 to derive the equations for a multigroup nodal model.

4.17. Calculate the node average fluxes and the effective multiplication constant of Problem 4.12 using a two-group nodal model. Compare with the results of Problem 4.12.

4.18. Calculate in two-group theory the critical radius of a 3.5-m-high bare cylindrical core with the cross sections given for core 1 in Problem 4.12.

4.19. Repeat the calculation of Problem 4.18 for the situation in which the core is surrounded by a 15-cm-thick annular reflector with the properties given in Problem 4.12. Compare the result with the result that would be obtained by subtracting the reflector savings from the critical radius for the bare core calculated in Problem 4.18.

4.20. Solve Problem 4.18 in $1\frac{1}{2}$ group theory.

4.21. Calculate the multiplication constant for a bare, cylindrical reactor of height $H = 3.5$ m and radius $R = 1.1$ m using 2-group theory and the group constants given for the core in Table P4.11 on p. 140 of the text.

4.22. You have access to an evaluated nuclear data library in which the cross-sections are tabulated pointwise in energy, except in the resonance region where single-level resonance parameters are given. You need to construct a 3-group cross section set for analyzing a PWR, but you do not have access to a many-group cross section processing code, nor do you have time to write one. Describe how you might construct a 3-group cross section set from the tabulated cross-section data and the resonance parameters. In particular, describe how you would weight the pointwise data.

4.23. Describe how you would construct a set of 2-group cross-sections for a fast reactor in which there were essentially no neutrons moderated below 1 keV. What group structure would you use? What weighting functions? Define the group cross sections that could be calculated from a library of cross sections given at discrete energies.

4.24. Calculate the critical radius of a bare, homogeneous cylindrical reactor of height $H = 3$ m. Use 2-group theory and the core cross sections ($\nu \Sigma_f = 0.0085$, $\Sigma_s^{1 \to 2} = 0.0241$, $\Sigma_a = 0.0121$ cm^{-1}, $D = 1.267$ cm, $\chi = 1$ for fast group 1) and ($\nu \Sigma_f = 0.1851$, $\Sigma_a = 0.121$ cm^{-1}, $D = 0.354$ cm, $\chi = 0$ for thermal group 2).

5
Nuclear Reactor Dynamics

An understanding of the time-dependent behavior of the neutron population in a
nuclear reactor in response to either a planned change in the reactor conditions
or to unplanned and abnormal conditions is of the utmost importance to the safe
and reliable operation of nuclear reactors. We saw in Chapter 2 that the response
of the prompt neutrons is very rapid indeed. However, unless the reactor is supercritical on prompt neutrons alone, the delayed emission of a small fraction of the
fission neutrons can slow the increase in neutron population to the delayed neutron precursor decay time scale of seconds, providing time for corrective control
measures to be taken. If a change in conditions makes a reactor supercritical on
prompt neutrons alone, only intrinsic negative feedback mechanisms that automatically provide a compensating change in reactor conditions in response to an
increase in the neutron population can prevent a runaway increase in neutron population (and fission power level). However, some of the intrinsic changes in reactor
conditions in response to a change in power level may enhance the power excursion (positive feedback), and others may be negative but delayed sufficiently long
that the compensatory reactivity feedback is out of phase with the actual condition
of the neutron population in the reactor, leading to power-level instabilities. These
reactor dynamics phenomena, the methods used for their analysis, and the experimental techniques for determining the basic kinetics parameters that govern them
are discussed in this chapter.

5.1
Delayed Fission Neutrons

Neutrons Emitted in Fission Product Decay

$E_n \uparrow, \nu_d \uparrow$

The dynamics of a nuclear reactor or any other fission chain-reacting system under normal operation is determined primarily by the characteristics of the delayed
emission of neutrons from the decay of fission products. The total yield of delayed
neutrons per fission, ν_d, depends on the fissioning isotope and generally increases
with the energy of the neutron causing fission. Although there are a relatively large
number of fission products which subsequently decay via neutron emission, the

Nuclear Reactor Physics. Weston M. Stacey
Copyright © 2007 WILEY-VCH Verlag GmbH & Co. KGaA, Weinheim
ISBN: 978-3-527-40679-1

$\beta = v_d/v$

pcm/$

6 groups of delayed neutron precursor fission products

Table 5.1 Delayed Neutron Parameters

	Fast Neutrons		Thermal Neutrons	
	Decay Constant	Relative Yield	Decay Constant	Relative Yield
Group	λ_i (s^{-1})	β_i/β	λ_i (s^{-1})	β_i/β
^{233}U		$v_d = 0.00731$		$v_d = 0.00667$
		$\beta = 0.0026$		$\beta = 0.0026$
1	0.0125	0.096	0.0126	0.086
2	0.0360	0.208	0.0337	0.299
3	0.138	0.242	0.139	0.252
4	0.318	0.327	0.325	0.278
5	1.22	0.087	1.13	0.051
6	3.15	0.041	2.50	0.034
^{235}U		$v_d = 0.01673$		$v_d = 0.01668$
		$\beta = 0.0064$		$\beta = 0.0067$
1	0.0127	0.038	0.0124	0.033
2	0.0317	0.213	0.0305	0.219
3	0.115	0.188	0.111	0.196
4	0.311	0.407	0.301	0.395
5	1.40	0.128	1.14	0.115
6	3.87	0.026	3.01	0.042
^{239}Pu		$v_d = 0.0063$		$v_d = 0.00645$
		$\beta = 0.0020$		$\beta = 0.0022$
1	0.0129	0.038	0.0128	0.035
2	0.0311	0.280	0.0301	0.298
3	0.134	0.216	0.124	0.211
4	0.331	0.328	0.325	0.326
5	1.26	0.103	1.12	0.086
6	3.21	0.035	2.69	0.044
^{241}Pu		$v_d = 0.0152$		$v_d = 0.0157$
				$\beta = 0.0054$
1	—	—	0.0128	0.010
2	—	—	0.0297	0.229
3	—	—	0.124	0.173
4	—	—	0.352	0.390
5	—	—	1.61	0.182
6	—	—	3.47	0.016
^{232}Th		$v_d = 0.0531$		
		$\beta = 0.0203$		
1	0.0124	0.034		
2	0.0334	0.150		
3	0.121	0.155		
4	0.321	0.446		
5	1.21	0.172		
6	3.29	0.043		

$$S_d(t) = \sum_{i=1}^{n_{dC}} \nu_{di} \lambda_i \exp\left[-\lambda_i t\right]$$

Table 5.1 (Continued)

	Fast Neutrons		Thermal Neutrons	
	Decay Constant	Relative Yield	Decay Constant	Relative Yield
Group	λ_i (s^{-1})	β_i/β	λ_i (s^{-1})	β_i/β
^{238}U		$\nu_d = 0.0460$		
		$\beta = 0.0164$		
1	0.0132	0.013		
2	0.0321	0.137		
3	0.139	0.162		
4	0.358	0.388		
5	1.41	0.225		
6	4.02	0.075		
^{240}Pu		$\nu_d = 0.0090$		
		$\beta = 0.0029$		
1	0.0129	0.028		
2	0.0313	0.273		
3	0.135	0.192		
4	0.333	0.350		
5	1.36	0.128		
6	4.04	0.029		

observed composite emission characteristics can be well represented by defining six effective groups of delayed neutron precursor fission products. Each group can be characterized by a decay constant, λ_i, and a relative yield fraction, β_i/β. The fraction of the total fission neutrons that are delayed is $\beta = \nu_d/\nu$. The parameters of delayed neutrons emitted by fission product decay of several relevant isotopes are given in Table 5.1.

Effective Delayed Neutron Parameters for Composite Mixtures

The delayed neutrons emitted by the decay of fission products are generally less energetic (average energy about 0.5 MeV) than the prompt neutrons (average energy about 1 MeV) released directly in the fission event. Thus these delayed neutrons will slow down quicker than the prompt neutrons and experience less probability for absorption and leakage in the process (i.e., the delayed and prompt neutrons have a difference in their effectiveness in producing a subsequent fission event). Since the energy distribution of the delayed neutrons differs from group to group, the different groups of delayed neutrons will also have a different effectiveness. Furthermore, a nuclear reactor will, of course, contain a mixture of fissionable isotopes (e.g., a uranium-fueled reactor will initially contain ^{235}U and ^{238}U, and after operation for some time will also contain some admixture of ^{239}Pu, ^{240}Pu, and so on; see Chapter 6).

To deal with this situation, it is necessary to define an *importance function*, $\phi^+(r, E)$, which is the probability that a neutron introduced at position r and energy E will ultimately result in a fission (Chapter 13). Then the relative importance (to the production of a subsequent fission) of delayed neutrons in group i emitted with energy distribution $\chi_{di}^q(E)$ and prompt neutrons from the fission of isotope q emitted with energy distribution $\chi_p^q(E)$ are

$$I_{di}^q = \int dV \int_0^\infty dE\, \chi_{di}^q(E)\phi^+(r,E) \int_0^\infty dE'\, v\sigma_f^q(E')N_q(r)\phi(r,E') \quad (5.1)$$

$$I_p^q = \int dV \int_0^\infty dE\, \chi_p^q(E)\phi^+(r,E) \int_0^\infty dE'\, v\sigma_f^q(E')N_q(r)\phi(r,E') \quad (5.2)$$

The relative effective delayed neutron yield of group i delayed neutrons for fissionable isotope q is $I_{di}^q\beta_i^q$, where β_i^q is the group i delayed neutron yield of fissionable isotope q given in Table 5.1. The effective group i delayed neutron fraction for isotope q in a mixture of fissionable isotopes is then

$$\overline{\gamma_i\beta_i^q} = I_{di}^q\beta_i^q \bigg/ \sum_q \left[I_p^q\left(1 - \sum_{i=1}^6 \beta_i^q\right) + \sum_{i=1}^6 I_{di}^q\beta_i^q \right] \quad (5.3)$$

The effectiveness of delayed neutron group i of fissionable isotope q in a specific admixture of fissionable isotopes and reactor geometry is then $\gamma_i^q = \overline{\gamma_i^q\beta_i^q}/\beta_i^q$. In the remainder of the book, except when specifically stated otherwise, it is assumed that the delayed neutron effectiveness is included in the evaluation of β_i and β, and the effectiveness parameter will be suppressed.

Photoneutrons

Fission products also emit gamma rays when they undergo β-decay. A photon can eject a neutron from a nucleus when its energy exceeds the neutron binding energy. Although most nuclei have neutron binding energies in excess of 6 MeV, which is above the energy of most gamma rays from fission, there are four nuclei that have sufficiently low neutron binding energy, E_n, to be of practical interest: ^{2}D ($E_n = 2.2$ MeV), ^{9}Be ($E_n = 1.7$ MeV), ^{6}Li ($E_n = 5.4$ MeV), and ^{13}C ($E_n = 4.9$ MeV). The photoneutrons can be considered as additional groups of delayed neutrons. Since the β-decay of fission products is generally much slower than the direct neutron decay, the photoneutron precursor decay constants are much smaller than the delayed neutron precursor decay constants shown in Table 5.1. The only reactors in which photoneutrons are of practical importance are D_2O-moderated reactors. As we shall see, the dynamic response time of a reactor under normal operation is largely determined by the inverse decay constants, and consequently, D_2O reactors are quite sluggish compared to other reactor types.

5.2
Point Kinetics Equations

The delayed neutron precursors satisfy an equation of the form

$$\frac{\partial \hat{C}_i}{\partial t}(r,t) = \beta_i \nu \Sigma_f(r,t)\phi(r,t) - \lambda_i \hat{C}_i(r,t), \quad i = 1,\ldots,6 \tag{5.4}$$

The one-speed neutron diffusion equation is now written

$$\frac{1}{\mathrm{v}}\frac{\partial \phi(r,t)}{\partial t} - D(r,t)\nabla^2\phi(r,t) + \Sigma_a(r,t)\phi(r,t)$$

$$= (1-\beta)\nu\Sigma_F(r,t)\phi(r,t) + \sum_{i=1}^{6}\lambda_i\hat{C}_i(r,t) \tag{5.5}$$

where we have taken into account that a fraction β of the fission neutrons is delayed and that there is a source of neutrons due to the decay of the delayed neutron precursors.

Based on the results of Chapter 3, we assume a separation-of-variables solution

$$\phi(r,t) = \mathrm{v}n(t)\psi_1(r), \qquad \hat{C}_i(r,t) = C_i(t)\psi_1(r) \tag{5.6}$$

where ψ_1 is the fundamental mode solution of

$$\nabla^2\psi_n + B_g^2\psi_n = 0 \tag{5.7}$$

and B_g is the geometric buckling appropriate for the reactor geometry, as discussed in Chapter 3. Using this in Eqs. (5.4) and (5.5) leads to the point kinetics equations

$$\frac{dn(t)}{dt} = \frac{\rho(t)-\beta}{\Lambda}n(t) + \sum_{i=1}^{6}\lambda_i C_i(t)$$

$$\frac{dC_i(t)}{dt} = \frac{\beta_i}{\Lambda}n(t) - \lambda_i C_i(t), \quad i = 1,\ldots,6 \tag{5.8}$$

where

$$\Lambda \equiv (\nu\nu\Sigma_F)^{-1} \tag{5.9}$$

mean generation time — *fission n born* ⟶ *absorption leads to another fiss*

is the mean generation time between the birth of a fission neutron and the subsequent absorption leading to another fission, and

$$\rho(t) \equiv \frac{\nu\Sigma_F - \Sigma_a(1+L^2B_g^2)}{\nu\Sigma_F} \equiv \frac{k(t)-1}{k(t)} \tag{5.10}$$

is the reactivity. The quantity k is the effective multiplication constant, given by

$$k \equiv \frac{\nu\Sigma_F/\Sigma_a}{1+L^2B_g^2} \tag{5.11}$$

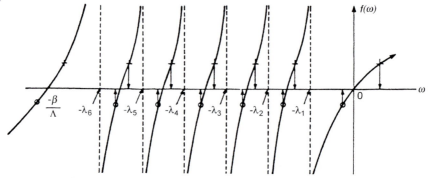

Fig. 5.1 Plot of the function $R(\omega) = \omega[\Lambda + \Sigma\beta_i/(\omega + \lambda_i)]$, which appears in the in-hour equation. (From Ref. 4; used with permission of MIT Press.)

For predominantly thermal reactors, $\nu\Sigma_f$ and Σ_a are thermal cross sections, and L^2 should be replaced by $M^2 = L^2 + \tau_{\text{th}}$ to include the fast diffusion while the neutron is slowing down, τ_{th}, as well as the thermal diffusion length, L^2. For fast reactors, all cross sections are averaged over the appropriate fast spectrum.

The limiting assumption for the validity of the point kinetics equations is the assumption of a constant spatial shape. As we will see, this assumption is reasonable for transients caused by uniform changes in reactor properties or for reactors with dimensions that are only a few migration lengths, M (or diffusion lengths, L), but is poor for reactors with dimensions that are very large compared to M in which the transient is caused by localized changes in reactor properties (e.g., a nonsymmetric control rod withdrawal). However, as we will see in Chapter 16, such spatial shape changes can be taken into account in computation of the reactivity and the mean generation time, and the point kinetics equations can be extended to have a much wider range of validity.

5.3
Period–Reactivity Relations

Equations (5.8) may be solved for the case of an initially critical reactor in which the properties are changed at $t = 0$ in such a way as to introduce a reactivity ρ_0 which is then constant over time, by Laplace transforming, or equivalently assuming an exponential time dependence e^{-st}. The equations for the time-dependent parts of n and C_i are

$$sn(s) = \frac{\rho_0 - \beta}{\Lambda}n(s) + \sum_{i=1}^{6}\lambda_i C_i(s) + n_0$$

$$sC_i(s) = \frac{\beta_i}{\Lambda}n(s) - \lambda_i C_i(s) + C_{i0}, \quad i = 1, \ldots, 6$$

(5.12)

which can be reduced to

$$n(s) = \frac{f(s, n_0, C_{i0})}{Y(s)} \tag{5.13}$$

where

$$Y(s) \equiv \rho_0 - s\left(\Lambda + \sum_{i=1}^{6} \frac{\beta_i}{s + \lambda_i}\right) \tag{5.14}$$

The poles of the right side—the roots of $Y(s) = 0$—determine the time dependence of the neutron and precursor populations. $Y(s) = 0$ is a seventh-order equation, known as the *inverse hour*, or more succinctly, the *inhour*, equation, the solutions of which are best visualized graphically, as indicated in Fig. 5.1, where the right-hand side of

$$\rho_0 = s\left(\Lambda + \sum_{i=1}^{6} \frac{\beta_i}{s + \lambda_i}\right) \tag{5.15}$$

$R(\omega) = \omega\left[\Lambda + \sum \beta_i / (\omega + \lambda_i)\right]$

is plotted. The left-hand side, ρ_0, would plot as a straight horizontal line, of course, and the points at which it crosses the right-hand side are the solutions (roots of the equation). For $\rho_0 < 0$, indicated by the circles in Fig. 5.1, all the solutions $s_j < 0$. For $\rho_0 > 0$, indicated by the crosses, there are one positive and six negative solutions.

The solution for the time-dependent neutron flux is of the form

$$n(t) = \sum_{j=0}^{6} A_j e^{s_j t} \tag{5.16}$$

where the s_j are the roots of $Y(s) = 0$ and the A_j are given by

$$A_j = \left(\Lambda + \sum_{i=1}^{6} \frac{\beta_i}{s_j + \lambda_i}\right) \Big/ \left[1 + k \sum_{i=1}^{6} \frac{\beta_i \lambda_i}{(s_j + \lambda_i)^2}\right] \quad \text{wrong} \tag{5.17}$$

After a sufficient time, the solution will be dominated by the largest root s_0 ($s_0 > 0$ when $\rho_0 > 0$, s_0 is the least negative root when $\rho_0 < 0$):

$$n(t) \simeq A_0 e^{s_0 t} \equiv A_0 e^{t/T} \tag{5.18}$$

where $T \equiv s_0^{-1}$ is referred to as the *asymptotic period*. Measurement of the asymptotic period then provides a means for the experimental determination of the reactivity

$$\rho_0 = \frac{1}{T}\left[\Lambda + \sum_{i=1}^{6} \frac{\beta_i}{(1/T) + \lambda_i}\right] \tag{5.19}$$

5.4
Approximate Solutions of the Point Neutron Kinetics Equations

One Delayed Neutron Group Approximation

To simplify the problem so that we can gain insight into the nature of the solution of the point kinetics equations, we assume that the six groups of delayed neutrons can be replaced by one delayed neutron group with an effective yield fraction $\beta = \Sigma_i \gamma_i \beta_i$ and an effective decay constant $\lambda = \Sigma_i \beta_i \lambda_i / \beta$, so that the point kinetics equations become

$$\frac{dn}{dt} = \frac{\rho - \beta}{\Lambda} n + \lambda C, \qquad \frac{dC}{dt} = \frac{\beta}{\Lambda} n - \lambda C \qquad (5.20)$$

Proceeding as in Section 5.3 by Laplace transforming or assuming an e^{st} form of the solution, the equivalent of Eq. (5.13) for the determination of the roots of the reduced in-hour equation is

$$s^2 - \left(\frac{\rho - \beta}{\Lambda} - \lambda \right) s - \frac{\lambda \rho}{\Lambda} = 0 \qquad (5.21)$$

which has the solution

$$s_{1,2} = \frac{1}{2} \left(\frac{\rho - \beta}{\Lambda} - \lambda \right) \pm \sqrt{\frac{1}{4} \left(\frac{\rho - \beta}{\Lambda} + \lambda \right)^2 + \frac{\beta \lambda}{\Lambda}}$$

$$= \frac{1}{2} \left(\frac{\rho - \beta}{\Lambda} - \lambda \right) \pm \sqrt{\frac{1}{4} \left(\frac{\rho - \beta}{\Lambda} - \lambda \right)^2 + \frac{\lambda \rho}{\Lambda}} \qquad (5.22)$$

For $\rho > 0$, one root is positive and the other negative; for $\rho = 0$, one root is zero and the other is negative; and for $\rho < 0$, both roots are negative.

The assumed e^{st} time dependence, when used in Eqs. (5.20), requires that for each of the two roots, s_1 and s_2, there is a fixed relation between the precursor and the neutron populations:

$$\frac{C(t)}{n(t)} = \frac{\beta}{\Lambda(s_{1,2} + \lambda)} = - \left(\frac{\rho - \beta}{\Lambda} - s_{1,2} \right) \bigg/ \lambda \qquad (5.23)$$

which means that the solution of Eqs. (5.20) is of the form

$$n(t) = A_1 e^{s_1 t} + A_2 e^{s_2 t}, \qquad C(t) = A_1 \frac{\beta}{\Lambda(s_1 + \lambda)} e^{s_1 t} + A_2 \frac{\beta}{\Lambda(s_2 + \lambda)} e^{s_2 t} \qquad (5.24)$$

Now, let us take some parameters typical of a light water reactor: $\beta = 0.0075$, $\lambda = 0.08 \text{ s}^{-1}$, $\Lambda = 6 \times 10^{-5} \text{ s}$. Except for $|\rho - \beta| \approx 0$, one root of Eq. (5.21) will be of very large magnitude, and the other will be of very small magnitude. For the larger root, $s_1^2 \gg \lambda\rho/\Lambda$ and $\lambda\rho/\Lambda$ can be neglected in Eq. (5.21); and for the smaller root, $s_2^2 \ll \lambda\rho/\Lambda$ and s_2^2 can be neglected. Assuming that $|\rho - \beta|/\Lambda \gg \lambda$, the solutions of Eq. (5.21) are

$$s_1 = \frac{\rho - \beta}{\Lambda}, \qquad s_2 = -\frac{\lambda\rho}{\rho - \beta} \tag{5.25}$$

The constants A_1 and A_2 can be evaluated by requiring that the solution satisfy the initial condition at $t = 0$ that is determined by setting $\rho = 0$ in Eqs. (5.24), which identifies $A_1 \approx n_0\rho/(\rho - \beta)$ and $A_2 \approx -n_0\beta/(\rho - \beta)$, where n_0 is the initial neutron population before the reactivity insertion, so that the solutions of Eqs. (5.24) become

$$n(t) = n_0\left[\frac{\rho}{\rho - \beta}\exp\left(\frac{\rho - \beta}{\Lambda}t\right) - \frac{\beta}{\rho - \beta}\exp\left(-\frac{\lambda\rho}{\rho - \beta}t\right)\right]$$

$$c(t) = n_0\left[\frac{\rho\beta}{(\rho - \beta)^2}\exp\left(\frac{\rho - \beta}{\Lambda}t\right) + \frac{\beta}{\Lambda\lambda}\exp\left(-\frac{\lambda\rho}{\rho - \beta}t\right)\right] \tag{5.26}$$

At $t = 0$, before the reactivity insertion, $C_0 = \beta n_0/\Lambda\lambda \approx 1600n_0$. Thus the population of delayed neutron precursors, hence the latent source of neutrons, is about 1600 times greater than the neutron population in a critical reactor. It is not surprising that this large latent neutron source controls the dynamics of the neutron population under normal conditions, as we shall now see.

Example 5.1: Step Negative Reactivity Insertion, $\rho < 0$. Equations (5.26) enable us to investigate the neutron kinetics of a nuclear reactor. We first consider the case of a large negative reactivity insertion $\rho = -0.05$ into a critical reactor at $t = 0$, such as might be produced by scramming (rapid insertion) of a control rod bank. With the representative light water reactor parameters ($\beta = 0.0075$, $\lambda = 0.08$ s^{-1}, $\Lambda = 6 \times 10^{-5}$ s), Eqs. (5.26) become

$$n(t) = n_0(0.87e^{-958t} + 0.13e^{-0.068t})$$

$$C(t) = n_0(0.0113e^{-958t} + 1563e^{-0.068t}) \tag{5.27}$$

which is plotted in Fig. 5.2, with $T \equiv n$. The first term goes promptly to zero on a time scale $\Delta t \approx \Lambda$, corresponding physically to readjustment of the prompt neutron population to the subcritical condition of the reactor on the neutron generation time scale. The second term decays slowly, corresponding to the slow decay of the delayed neutron precursor source of neutrons. The neutron population drops promptly from n_0 to $n_0/(1 - \rho/\beta)$—the *prompt jump*—then slowly decays as $e^{-[\lambda/(1-\beta/\rho)]t}$. Thus, scramming a control rod bank cannot immediately shut down (reduce the neutron population or the fission rate to near zero) a nuclear reactor or other fission chain reacting medium. The delayed neutron precursors decay as $e^{-[\lambda/(1-\beta/\rho)]t}$.

Example 5.2: Subprompt-Critical (Delayed Critical) Step Positive Reactivity Insertion, $0 < \rho < \beta$. Next, consider a positive reactivity insertion $\rho = 0.0015 < \beta$, such as

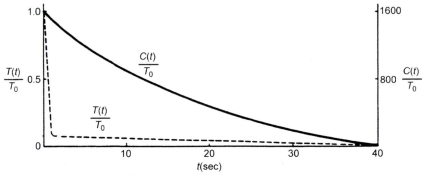

Fig. 5.2 Neutron and delayed neutron precursor decay following negative reactivity insertion $\rho = -0.05$ into a critical nuclear reactor. (From Ref. 4; used with permission of MIT Press.)

might occur as a result of control rod withdrawal. Equations (5.26) now become

$$n(t) = n_0(-0.25e^{-100t} + 1.25e^{0.02t})$$

$$C(t) = n_0(0.3125e^{-100t} + 1562.5e^{0.02t})$$

(5.28)

which is plotted in Fig. 5.3. The neutron population increases promptly, on the neutron generation time scale—the prompt jump—from n_0 to $n_0/(1 - \rho/\beta)$, as the prompt neutron population adjusts to the supercritical condition of the reactor, then increases as $e^{-[\lambda/(1-\beta/\rho)]t}$, governed by the rate of increase in the delayed neutron source. The relatively slow rate of increase of the neutron population, following the prompt jump, allows time for corrective control action to be taken before the fission rate becomes excessive.

Example 5.3: Superprompt-Critical Step Positive Reactivity Insertion, $\rho > \beta$. Now consider a step increase of reactivity $\rho = 0.0115 > \beta$, such as might occur as the result

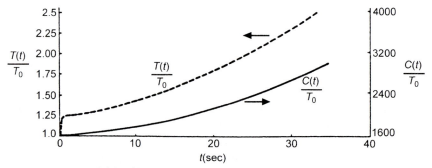

Fig. 5.3 Neutron and delayed neutron precursor increase following subprompt-critical positive reactivity insertion $\rho = 0.0015 < \beta$ into a critical nuclear reactor. (From Ref. 4; used with permission of MIT Press.)

of the ejection of a bank of control rods from a reactor. Equations (5.26) now become

$$n(t) = n_0(2.9e^{66.7t} - 1.9e^{-0.23t})$$

$$C(t) = n_0(5.4e^{66.7t} + 1563e^{-0.23t})$$

(5.29)

The neutron population in the reactor grows exponentially on the neutron generation time scale, $n \sim e^{[(\rho-\beta)/\Lambda]t}$, because the reactor is supercritical on prompt neutrons alone [i.e., $k(1-\beta) > 1$]. In this example, the neutron population would increase by almost a factor of 800 in a tenth of a second, and it would be impossible to take corrective action quickly enough to prevent excessive fission heating and destruction of the reactor. Fortunately, there are inherent feedback mechanisms that introduce negative reactivity instantaneously in response to an increase in the fission heating (e.g., the Doppler effect discussed in Sections 5.7 and 5.8), and the neutron population will first increase rapidly, then decrease. However, conditions that would lead to superprompt-critical reactivity insertion are to be avoided for reasons of safety. Since $\beta = 0.0026$ for ^{233}U, 0.0067 for ^{235}U, and 0.0022 ^{239}Pu, the safe operating range for positive reactivity insertions, $0 < \rho < \beta$, is much larger for reactors fueled with ^{235}U than for reactors fueled with ^{233}U or ^{239}Pu.

Prompt-Jump Approximation

We found that with a reactivity insertion for which the reactor condition is less than prompt critical ($\rho < \beta$) the neutron population changed sharply on the neutron generation time scale, then changed slowly on the delayed neutron inverse decay constant time scale. If we are not interested in the details of the prompt neutron kinetics during the prompt jump, we can simplify the equations by assuming that the prompt jump takes place instantaneously in response to any reactivity change, and afterward, the neutron population changes instantaneously in response to changes in the delayed neutron source (i.e., we set the time derivative to zero in the neutron equation).

$$0 = [\rho(t) - \beta]n(t) + \Lambda \sum_{i=1}^{6} \lambda_i C_i(t)$$

(5.30)

Since the delayed neutron precursor population does not respond instantaneously to a change in reactivity, Eq. (5.30) is valid with the same delayed precursor population both before and just after a change in reactivity from ρ_0 to $\rho_1 < \beta$, from which we conclude that the ratio of the neutron populations just after and before the reactivity change is

$$\frac{n_1}{n_0} = \frac{\beta - \rho_0}{\beta - \rho_1}$$

(5.31)

Use of Eq. (5.30) to eliminate $n(t)$ in the second of Eqs. (5.8) yields a coupled set of equations for the time dependence of the precursor density:

$$\frac{dC_i}{dt} = \frac{\beta_i}{\rho(t) - \beta_i} \sum_{j=1}^{6} \lambda_j C_j(t) - \lambda_i C_i(t) \tag{5.32}$$

which in the one delayed precursor group approximation takes on the simple form

$$\frac{dC(t)}{dt} = \frac{-\lambda C(t)}{1 - \beta/\rho(t)} \tag{5.33}$$

The prompt-jump approximation is convenient for numerical solutions because it eliminates the fast time scale due to Λ, which introduces difficulties in time differencing methods. Numerical solutions of the point kinetics equations with and without the prompt-jump approximation for a variety of reactivity insertions indicate that the prompt-jump approximation is accurate to within about 1% for reactivities $\rho < 0.5\beta$.

Using the one delayed precursor group approximation, the equivalent of Eq. (5.30) can be solved for $C(t)$ and used in the second of Eqs. (5.20) to obtain

$$[\rho(t) - \beta]\frac{dn(t)}{dt} + \left[\frac{d\rho(t)}{dt} + \lambda\rho(t)\right]n(t) = 0 \tag{5.34}$$

which for a given reactivity variation $\rho(t)$ can be solved for the neutron population

$$n(t) = n_0 \exp\left[\int_0^t \frac{\dot{\rho}(t') + \lambda\rho(t')}{\beta - \rho(t')} dt'\right] \tag{5.35}$$

Example 5.4: Reactivity Worth of Rod Insertion. The neutron flux measured by a detector is observed to drop instantaneously from n_0 to $0.5\, n_0$ when a control rod is dropped into a cold highly enriched critical nuclear reactor, in which $\rho_0 = 0$. Using the one-delayed group model with $\beta = 0.0065$, Eq. (5.31) yields $\rho_1 = \beta(1 - n_0/n_1) = 0.0065(1 - 2) = -0.0065\Delta k/k$.

Reactor Shutdown

We mentioned that the large step negative reactivity insertion considered previously might be representative of the situation encountered in a reactor shutdown, or scram. However, the time required to fully insert control rods is very long compared to the prompt neutron generation time that governs the time scale of the prompt jump. We can improve on the representation of the control rod insertion by considering a ramp reactivity insertion $\rho(t) = -\varepsilon t$. If we are only interested in calculating the initial rapid decrease in the neutron population, we can make the assumption that the initial precursor concentration remains constant; hence the precursor source of delayed neutrons remains constant at its pre-insertion value

$$\sum_{i=1}^{6} \lambda_i C_i(0) = \frac{\beta}{\Lambda} n_0 \tag{5.36}$$

$$n(t) = n_0 (1 - \gamma t)^{-(1 + \gamma/\delta)} e^{-\lambda t} \quad for \quad \rho(t) = \gamma \beta t \quad ramp \ reactivity$$

Using this approximation, the equation governing the prompt neutron response to the reactivity insertion—the first of Eqs. (5.8)—can be integrated to obtain

$$n(t) = n_0 \left[\exp\left[-\frac{1}{\Lambda}\left(\frac{1}{2}\varepsilon t^2 + \beta t \right) \right] \right.$$
$$\left. + \frac{\beta}{\Lambda} \int_0^t \exp\left\{ -\frac{1}{\Lambda}\left[\frac{\varepsilon}{2}(t^2 - (t')^2) + \beta(t - t') \right] \right\} dt' \right] \quad (5.37)$$

This provides a somewhat better description of the initial reduction in the neutron population than do Eqs. (5.27), which, however, would still govern the long-time decay after completion of the rod insertion.

5.5
Delayed Neutron Kernel and Zero-Power Transfer Function

Delayed Neutron Kernel

The delayed neutron precursor equations, the second of Eqs. (5.8), can be formally integrated to obtain (assuming that $C_i = 0$ at $-\infty$)

$$C_i(t) = \int_{-\infty}^t \frac{\beta_i}{\Lambda} n(t') e^{-\lambda_i(t-t')} dt' = \int_0^\infty \frac{\beta_i}{\Lambda} e^{-\lambda_i \tau} n(t - \tau) d\tau. \quad (5.38)$$

Using this result in the neutron kinetics equation, the first of Eqs. (5.8), yields

$$\frac{dn(t)}{dt} = \left(\frac{\rho(t) - \beta}{\Lambda} \right) n(t) + \int_0^\infty \frac{\beta}{\Lambda} D(\tau) n(t - \tau) d\tau \quad (5.39)$$

where we have defined the delayed neutron kernel

$$D(\tau) \equiv \sum_{i=1}^6 \frac{\lambda_i \beta_i}{\beta} e^{-\lambda_i \tau} \quad (5.40)$$

Zero-Power Transfer Function

If the neutron population is expanded about the initial neutron population in the critical reactor at $t = 0$,

$$n(t) = n_0 + n_1(t) \quad (5.41)$$

Eq. (5.39) may be rewritten

$$\frac{dn_1(t)}{dt} = \frac{\rho(t)n_0}{\Lambda} + \frac{\rho(t)n_1(t)}{\Lambda} + \int_0^\infty \frac{\beta}{\Lambda} D(\tau)[n_1(t - \tau) - n_1(t)] d\tau \quad (5.42)$$

The Laplace transform of a function of time $A(t)$ is defined as

$$A(s) = \int_0^\infty A(t) e^{-st} dt \quad (5.43)$$

Laplace transforming Eq. (5.42) and using the convolution theorem

$$\mathcal{L}\left[\int_0^\infty A(t)B(\tau - t)\,dt\right] = A(s)B(s) \tag{5.44}$$

yields, upon assuming that the term $\rho(t)n_1(t)$ is a product of two small terms and can be neglected relative to $\rho(t)n_0$,

$$n_1(s) = n_0 Z(s)\rho(s) \tag{5.45}$$

where

$$Z(s) \equiv \frac{1}{s}\left(\Lambda + \sum_{i=1}^6 \frac{\beta_i}{s + \lambda_i}\right)^{-1} \tag{5.46}$$

is the *zero-power transfer function*, which defines the response of the density n_1 to the reactivity.

The inverse Laplace transformation of Eq. (5.45) and the convolution theorem yield the solution for the time dependence of the neutron population as a function of the time dependence of the reactivity:

$$n_1(t) = n_0 \int_0^t d\tau\, Z(t - \tau)\rho(\tau) \tag{5.47}$$

where the inverse Laplace transform of the zero-power transfer function is

$$Z(t - \tau) = \frac{1}{\Lambda} + \sum_{j=2}^7 \frac{e^{s_j(t-\tau)}}{s_j\{\Lambda + \sum_{i=1}^6 [\beta_i\lambda_i/(s_j + \lambda_i)^2]\}} \tag{5.48}$$

and the s_j are the roots of the inhour equation, $Y(s) = 0$, with $Y(s)$ given by Eq. (5.14).

5.6
Experimental Determination of Neutron Kinetics Parameters

Asymptotic Period Measurement

When a critical reactor is perturbed by a step change in properties, the asymptotic period may be determined from the response $R(t)$ of neutron detectors by $T^{-1} = d(\ln R)/dt$; then the period–reactivity relation of Eq. (5.19) can be used to infer the reactivity. For negative reactivities, the asymptotic period, the largest root of the inhour equation, is dominated by the largest delayed neutron period and is relatively insensitive to the value of the reactivity, so this method is limited practically to supercritical reactivity $(0 < \rho)$ measurements, for which Eq. (5.19) may be written

$$\frac{\rho}{\beta} = \frac{\Lambda}{\beta T} + \sum_{i=1}^6 \frac{\beta_i/\beta}{1 + \lambda_i T} \simeq \sum_{i=1}^6 \frac{\beta_i/\beta}{1 + \lambda_i T} \tag{5.49}$$

where the fact that safety considerations further limit the practical applicability of this method to the delayed critical regime ($0 < \rho < \beta$) has been taken into account in writing the second form of the equation.

Rod Drop Method

The responses of a neutron detector immediately before ($R_0 \sim n_0$) and after ($R_1 \sim n_1$) a control rod is dropped into a critical reactor ($\rho_0 = 0$) are related by Eq. (5.31), which allows determination of the reactivity worth of the rod

$$\frac{\rho_1}{\beta} = 1 - \frac{R_0}{R_1} \qquad (5.50)$$

Source Jerk Method

Consider a subcritical system that is maintained at equilibrium neutron, n_0, and precursor, C_{i0}, populations by an extraneous neutron source rate, S. The neutron balance equation is

$$\left(\frac{\rho - \beta}{\Lambda}\right)n_0 + \sum_{i=1}^{6} \lambda_i C_{i0} + S = 0 \qquad (5.51)$$

If the source is jerked, the prompt-jump approximation for the neutron density immediately after the source jerk is

$$\left(\frac{\rho - \beta}{\Lambda}\right)n_1 + \sum_{i=1}^{6} \lambda_i C_{i0} = 0 \qquad (5.52)$$

because the delayed neutron precursor population will not change immediately. These equations and the equilibrium precursor concentrations $C_{i0} = \beta_i n_0 / \lambda_i \Lambda$ may be used to relate the responses of a neutron detector immediately before ($R_0 \sim n_0$) and after ($R_1 \sim n_1$) the source jerk to the reactivity of the system:

$$\frac{\rho}{\beta} = 1 - \frac{R_0}{R_1} \qquad (5.53)$$

Pulsed Neutron Methods

The time dependence of the prompt neutron population in a subcritical fission chain reacting medium following the introduction of a burst of neutrons is described by

$$\frac{1}{v}\frac{\partial \phi(r,t)}{\partial t} = D\nabla^2 \phi(r,t) - [\Sigma_a - (1-\beta)v\Sigma_f]\phi(r,t) \qquad (5.54)$$

since the delayed neutrons will not contribute until later. As discussed in Section 3.6, the asymptotic solution that remains after higher-order spatial transients decay is the fundamental mode, which decays exponentially:

$$n(r,t) \simeq A_1 \psi_1(r) e^{-v[\Sigma_a - (1-\beta)v\Sigma_f + DB_g^2]t} \qquad (5.55)$$

where B_g is the fundamental mode geometric buckling for the geometry of the system.

If the neutron detector response, $R(\mathbf{r}, t) \sim n(\mathbf{r}, t)$, is measured as a function of time, then

$$\alpha_0 = \frac{1}{R}\frac{dR}{dt} = v\left[\nu\Sigma_f(1-\beta) - \Sigma_a - DB_g^2\right] = \frac{\rho - \beta}{\Lambda} \tag{5.56}$$

Thus the pulsed neutron method can be used to determine $\sim \rho/\Lambda$, assuming that β/Λ is known. If the experiment is performed in a critical system ($\rho = 0$), the measurement yields a value for β/Λ. In practice, a correction must be made to account for transport- and energy-dependent effects which have been neglected in this analysis, so that

$$\alpha_0 = v\left[\nu\Sigma_f(1-\beta) - \Sigma_a - DB_g^2 - CB_g^4 + \cdots\right] \tag{5.57}$$

Rod Oscillator Measurements

The response of the neutron population, as measured by a neutron detector $R(t) \sim n(t)$, to a sinusoidal oscillation of a control rod that produces a sinusoidal reactivity perturbation

$$\rho(t) = \rho_0 \sin\omega t \tag{5.58}$$

can be used to determine a number of neutron kinetics parameters. The response of the neutron population to a sinusoidal reactivity perturbation can be calculated from Eq. (5.45) by first computing the Laplace transform of Eq. (5.58):

$$\rho(s) = \frac{\rho_0\omega}{s^2 + \omega^2} = \frac{\rho_0\omega}{(s + i\omega)(s - i\omega)} \tag{5.59}$$

and then Laplace inverting Eq. (5.45), or equivalently, by using Eq. (5.58) in Eq. (5.47), to obtain

$$n_1(t) = n_0\rho_0[|Z(i\omega)|\sin(\omega t + \phi)] + \omega\sum_{j=0}^{6} \frac{e^{s_j t}}{(\omega^2 + s_j^2)(dY/ds)_{s_j}} \tag{5.60}$$

where ϕ is the *phase angle*, defined by

$$\tan\phi \equiv \frac{\text{Im}\{Z(i\omega)\}}{\text{Re}\{Z(i\omega)\}} \tag{5.61}$$

The first term in Eq. (5.60) arises from the poles of the reactivity [Eq. (5.59)] at $s = \pm i\omega$, and the remaining terms arise from the poles of the zero-power transfer function $Z(s)$ [i.e., the roots of the inhour equation $Y(s) = 0$ given by Eq. (5.13)]. For a critical system, the largest root of the inhour equation is $s = 0$, so that after sufficient time the solution given by Eq. (5.60) approaches

$$n_1(t) \simeq n_0\rho_0\left[|Z(i\omega)|\sin(\omega t + \phi) + \frac{1}{\omega\Lambda}\right] \tag{5.62}$$

The average neutron detector response will be $(\rho_0/\omega\Lambda)R_0$, where R_0 is the average detector response before the oscillation began. At high oscillation frequency, the contribution of the first term in Eq. (5.62) to the detector response will average to zero and the detector response will reflect the second term. In both cases, this provides a means for the experimental determination of ρ_0/Λ in terms of the average detector response $\langle R \rangle$:

$$\frac{\rho_0}{\Lambda} = \omega \frac{\langle R \rangle - R_0}{R_0} \tag{5.63}$$

Zero-Power Transfer Function Measurements

By varying the frequency of rod oscillation, ω, the zero-power transfer function, $Z(i\omega)$, can be measured for a reactor or other critical fission chain reacting system by interpreting the detector reading $R(t)$ as

$$R(t) - R_0 = R_0\rho_0\left[|Z(i\omega)|\sin(\omega t + \phi) + \frac{1}{\omega\Lambda}\right] \tag{5.64}$$

Such measurements, when compared with calculation of the transfer function, provide an indirect means of determining or confirming the parameters Λ, β_i, and λ_i. At low frequencies the amplitude of the transfer function approaches

$$|Z(i\omega)| \rightarrow \left|\omega\sum_{i=1}^{6}\frac{\beta_i/\beta}{\lambda_i^2}\right| \tag{5.65}$$

for $\omega \ll \lambda_i$, and the phase angle ϕ approaches

$$\tan\phi \rightarrow -\sum_{i=1}^{6}\frac{\beta_i/\beta}{\lambda_i}\bigg/\omega\sum_{i=1}^{6}\frac{\beta_i/\beta}{\lambda_i^2} \tag{5.66}$$

Rossi-α Measurement

The prompt neutron decay constant

$$\alpha \equiv \frac{1}{n}\frac{dn}{dt} = \frac{k(1-\beta)-1}{l} \tag{5.67}$$

can be measured by observing the decay of individual fission reaction chains in succession if the process is continued long enough to observe a statistically significant number of decay chains. Assume that a neutron count from a decay chain is observed at $t = 0$. The probability of another neutron count being observed at a later time t is the sum of the probability of a count from a chain-related neutron, $Q\exp(\alpha t)\Delta t$, plus the probability of a neutron from another chain, $C\Delta t$, where C is the average counting rate:

$$P(t)\,dt = C\,dt + Qe^{\alpha t}\,dt \tag{5.68}$$

We use a statistical argument to determine Q. The probability of a count occurring at t_0 is $F\,dt_0$, where F is just the average fission rate in the system. The probability of another detector count at $t_1 > t_0$ that is chain related to the count at t_0 is

$$P(t_1)dt_1 = \varepsilon v_p v \Sigma_f e^{\alpha(t_1 - t_0)} dt_1 \tag{5.69}$$

where v_p is the number of prompt neutrons per fission and ε is the detector efficiency. The probability of a second chain-related count at $t_2 > t_1$ is

$$P(t_2)dt_2 = \varepsilon(v_p - 1)v\Sigma_f e^{\alpha(t_2 - t_0)} dt_2 \tag{5.70}$$

where $(v_p - 1)$ takes account of the chain-related fission required to produce the count at t_1. The three probabilities $F\,dt_0$, $P(t_1)dt_1$ and $P(t_2)dt_2$ are treated as independent probabilities. Hence the probability for a count in dt_1 followed by a count in dt_2, both in the chain that produced the count in dt_0, is obtained by multiplying the three probabilities and integrating over $-\infty < t < t_1$:

$$P(t_1, t_2)dt_1 dt_2 = \int_{-\infty}^{t_1} F\varepsilon^2 \left(\overline{v_p^2} - \bar{v}_p\right)(v\Sigma_f)^2 e^{\alpha(t_1 + t_2 - 2t_0)} dt_0 dt_1 dt_2$$

$$= F\varepsilon^2 \left(\overline{v_p^2} - \bar{v}_p\right) \frac{(v\Sigma_f)^2}{-2\alpha} e^{\alpha(t_2 - t_1)} dt_1 dt_2 \tag{5.71}$$

where an overbar indicates an average over the prompt neutron emission distribution function.

Noting that $v_p = k_p \Sigma_a / \Sigma_f = k_p / (v l \Sigma_f)$ and including the probability $F^2 \varepsilon^2 dt_1 dt_2$ of a random pair of counts, this becomes

$$P(t_1, t_2)dt_1 dt_2 = F^2 \varepsilon^2 dt_1 dt_2 + F\varepsilon^2 \frac{\left(\overline{v_p^2} - \bar{v}_p\right)k_p^2 e^{\alpha(t_2 - t_1)} dt_1 dt_2}{2 \overline{v_p^2}(1 - k_p)l} \tag{5.72}$$

Since the overall probability of a count in dt_1 is $F\varepsilon dt_1$, we need to normalize this conditional probability by division by $F\varepsilon dt_1$, which yields, upon rescaling time from $t_1 = 0$,

$$P(t_1, t_2)dt_1 dt_2 = \frac{\varepsilon\left(\overline{v_p^2} - \bar{v}_p\right)}{\overline{v_p^2}} \frac{k_p^2}{2(1 - k_p)l} e^{\alpha t} dt \tag{5.73}$$

This is the $Q \exp(\alpha t)dt$ term in Eq. (5.68), so

$$Q = \frac{\varepsilon\left(\overline{v_p^2} - \bar{v}_p\right)}{\overline{v_p^2}} \frac{k_p^2}{2(1 - k_p)l} \tag{5.74}$$

In a Rossi-α experiment, the function $P(t)$ of Eq. (5.68) is measured by a time analyzer and the random count rate $C\,dt$ is subtracted. The parameter α is then determined from the remaining $Q \exp(\alpha t)dt$ term.

5.7
Reactivity Feedback

Up to this point, we have discussed neutron kinetics—the response of the neutron population in a nuclear reactor or other fission chain reacting system to an external reactivity input—under the implicit assumption that the level of the neutron population does not affect the properties of the system that determine the neutron kinetics, most notably the reactivity. This is the situation when the neutron population is sufficiently small that the fission heat does not affect the temperature of the system (i.e., at zero power). However, in an operating nuclear reactor the neutron population is large enough that any change in fission heating resulting from a change in neutron population will produce changes in temperature, which in turn will produce changes in reactivity, or reactivity feedback. The combined and coupled response of the neutron population and of the temperatures, densities, and displacements of the various materials in a nuclear reactor is properly the subject of reactor dynamics, but the term is commonly used to also include neutron kinetics.

When the neutron population increases, the fission heating increases. Since this heating is deposited in the fuel element, the fuel temperature will increase immediately. An increase in fuel temperature will broaden the effective resonance absorption (and fission) cross section, generally resulting in an increase in neutron absorption and a corresponding reduction in reactivity—the *Doppler effect*. The fuel element will also expand and, depending on the constraints, bend or bow slightly, thus changing the local fuel–moderator geometry and *flux disadvantage factor* (the ratio of the flux in the fuel to the flux in the moderator), thereby producing a change in reactivity. If the increase in fission heating is large enough to raise the fuel temperature above the melting point, fuel slumping will occur, resulting in a large change in the local fuel–moderator geometry and a corresponding change in flux disadvantage factor and fuel absorption, producing a further change in reactivity.

Some of the increased fission heat will be transported out of the fuel element (time constant of tenths of seconds to seconds) into the surrounding moderator/coolant and structure, causing a delayed increase in moderator/coolant and structure temperature. An increase in moderator/coolant temperature will produce a decrease in moderator/coolant density, which causes a change in the local fuel–moderator properties and a corresponding change in both the moderator absorption and the flux disadvantage factor. In addition, a decrease in moderator density will reduce the moderating effectiveness and produce a hardening (shift to higher energies) in the neutron energy distribution, which will change the effective energy-averaged absorption cross sections for the fuel, control elements, and so on. An increase in structure temperature will cause expansion and deformation, producing a change in the local geometry that will further affect the flux disadvantage factor. These various moderator/coolant changes all produce changes in reactivity.

The reduction in moderator/coolant density increases the diffusion of neutrons, and the increase in temperature causes an expansion of the reactor. The effect of increased diffusion is to increase the leakage, and the effect of increased size is to reduce the leakage, producing offsetting negative and positive reactivity effects. In addition to these internal (to the core) reactivity feedback effects, there are external feedback effects caused by changes in the coolant outlet temperature that will produce changes in the coolant inlet temperature.

Temperature Coefficients of Reactivity

The temperature coefficient of reactivity is defined as

$$\alpha_T \equiv \frac{\partial \rho}{\partial T} = \frac{\partial}{\partial T}\left(\frac{k-1}{k}\right) = \frac{1}{k^2}\frac{\partial k}{\partial T} \simeq \frac{1}{k}\frac{\partial k}{\partial T} \tag{5.75}$$

To gain physical insight into the various physical phenomena that contribute to the reactivity feedback, we first use the one-speed diffusion theory expression for the effective multiplication constant for a bare reactor, but extend it to account for fast fission by including the ratio $\varepsilon = $ total fission/thermal fission, to account for the resonance absorption of neutrons during the slowing down to thermal energies by including the resonance escape probability p, and to account for the leakage of fast as well as thermal neutrons by replacing the diffusion length with the migration length M:

$$k = k_\infty P_{\mathrm{NL}} = \frac{\nu \Sigma_f}{\Sigma_a}\frac{1}{1+L^2 B^2} = \frac{\nu \Sigma_f}{\Sigma_a^F}\frac{\Sigma_a^F}{\Sigma_a}\frac{1}{(1+L^2 B^2)}$$

$$= \eta f P_{\mathrm{NL}} \rightarrow \eta f \varepsilon p P_{\mathrm{NL}} \tag{5.76}$$

which allows us to write

$$\alpha_T = \frac{1}{\eta}\frac{\partial \eta}{\partial T} + \frac{1}{\varepsilon}\frac{\partial \varepsilon}{\partial T} + \frac{1}{f}\frac{\partial f}{\partial T} + \frac{1}{p}\frac{\partial p}{\partial T} + \frac{1}{P_{\mathrm{NL}}}\frac{\partial P_{\mathrm{NL}}}{\partial T} \tag{5.77}$$

This formalism lends itself to physical interpretation and can provide quantitative estimates of reactivity coefficients for thermal reactors, but it is not directly applicable to fast reactors. We discuss fast reactor reactivity coefficients in the next section, where a perturbation theory formalism that is more appropriate for the quantitative evaluation of reactivity coefficients in both fast and thermal reactors is introduced. We now discuss reactivity feedback effects on p, f, and P_{NL}; there are also smaller reactivity effects associated with η due to shifts in the thermal neutron energy distribution and associated with ε, which latter are similar to the effects associated with the thermal utilization factor.

Doppler Effect

The resonance capture cross section (one-level Breit–Wigner) is

$$\sigma_\gamma = \sigma_0 \sqrt{\frac{E_0}{E}}\frac{\Gamma_\gamma}{\Gamma}\psi(x,\xi) \tag{5.78}$$

where ψ is the Doppler broadening shape function, which takes into account the averaging of the neutron–nucleus interaction cross section over the thermal motion of the nucleus,

$$\psi(x,\xi) = \frac{\xi}{\sqrt{4\pi}} \int_{-\infty}^{\infty} e^{-[(x-y)^2\xi^2/4]} \frac{dy}{1+y^2} \qquad (5.79)$$

σ_0 is the peak resonance cross section, Γ_γ and Γ are the capture and total widths of the resonance, $x = (E - E_0)/\Gamma$, $\xi = \Gamma/(4E_0kT/A)^{1/2}$, E and E_0 are the energies of the neutron and of the resonance peak, and A is the mass of the nucleus in amu. The total capture in the resonance is given by the resonance integral

$$I_\gamma \equiv \int \sigma_\gamma(E)\phi(E)dE \qquad (5.80)$$

The function ψ broadens with increasing temperature, T, characterizing the motion of the nucleus. A broadening of the ψ function reduces the energy self-shielding in the resonance and increases the resonance integral. Thus an increase in fuel temperature due to an increase in fission heating will cause an increase in the effective capture cross section $\langle\sigma_\gamma\rangle \sim I_\gamma$. A similar result is found for the fission resonances.

In thermal reactors, the Doppler effect is due primarily to epithermal capture resonances in the nonfissionable fuel isotopes (^{232}Th, ^{238}U, ^{240}Pu) and can be estimated by considering the change in resonance escape probability

$$p = e^{-(N_F I_\gamma/\xi\Sigma_p)} \qquad (5.81)$$

where $\xi\Sigma_p/N_F$ is the average moderating power per fuel atom, with a sum over resonance integrals for all fuel resonances implied, the function

$$J(\xi,\beta') \equiv \int_0^\infty \frac{\psi(x,\xi)}{\psi(x,\xi)+\beta'} dx \qquad (5.82)$$

is tabulated in Table 4.3, and $\beta' = (\Sigma_p/N_F)(\Gamma/\sigma_0\Gamma_\gamma)$. The Doppler temperature coefficient of reactivity for a thermal reactor can then be calculated as

$$\alpha_{T_F}^D = \frac{\partial\rho}{\partial T_F} \simeq \frac{1}{k}\frac{\partial k}{\partial T_F} = \frac{1}{p}\frac{\partial p}{\partial T_F} = \ln p \left(\frac{1}{I}\frac{\partial I}{\partial T_F}\right) \qquad (5.83)$$

Since the additional fission heating is deposited in the fuel, the fuel temperature, T_F, increases immediately, and the Doppler effect immediately reduces the reactivity. The Doppler effect is a very strong contributor to the safety and operational stability of thermal reactors.

There are useful fits to the total resonance integrals for ^{238}UO$_2$ and ^{232}ThO$_2$:

$$I(300\ \text{K}) = 11.6 + 22.8\left(\frac{S_F}{M_F}\right)$$

$$I(T_F) = I(300\ \text{K})\left[1 + \beta''(\sqrt{T(\text{K})} - \sqrt{300})\right]$$

$$^{238}UO_2 : \beta'' = 61 \times 10^{-4} + 47 \times 10^{-4} \left(\frac{S_F}{M_F} \right) \tag{5.84}$$

$$^{232}ThO_2 : \beta'' = 97 \times 10^{-4} + 120 \times 10^{-4} \left(\frac{S_F}{M_F} \right)$$

where S_F and M_F are surface area and mass of the fuel element. Using this fit, Eq. (5.83) becomes

$$\alpha_{T_F}^D = -\ln \left[\frac{1}{p(300\ K)} \right] \frac{\beta''}{2\sqrt{T_F(K)}} \tag{5.85}$$

Fuel and Moderator Expansion Effect on Resonance Escape Probability

When the fuel temperature increases, the fuel will expand, causing among other things a decrease in the fuel density, which affects the resonance escape probability and contributes an immediate temperature coefficient of reactivity:

$$\alpha_{T_F}^p = \frac{1}{p} \frac{\partial p}{\partial N_F} \frac{\partial N_f}{\partial T_f} = \ln p \left(\frac{1}{N_F} \frac{\partial N_F}{\partial T_F} \right) = -3\theta_F \ln p \tag{5.86}$$

where $(dN/dT)/T = -3(dl/dT)/l = -3\theta$, with θ being the linear coefficient of expansion of the material. Since the fuel density decreases upon expansion, the resonance absorption decreases, and this reactivity coefficient contribution is positive (note that since $p < 1$, $\ln p < 0$).

After the increase in fission heating has been transported out of the fuel element into the coolant/moderator, the moderator temperature, T_M, will increase, which causes the moderator to expand and contributes a delayed temperature coefficient of reactivity:

$$\alpha_{T_M}^p = \frac{1}{p} \frac{\partial p}{\partial N_M} \frac{\partial N_M}{\partial T_M} = -\ln p \left(\frac{1}{N_M} \frac{\partial N_M}{\partial T_M} \right) = 3\theta_M \ln p \tag{5.87}$$

The decreased moderator density reduces the moderating power, reducing the probability that the neutrons will be scattered to energies beneath the resonance, hence increasing the resonance absorption and contributing a negative reactivity coefficient.

Example 5.5: Resonance Escape Probability Fuel Temperature Coefficient for UO_2. The prompt feedback resulting immediately from an increase in power is associated with the increase in fuel temperature, the most significant part of which is due to the change in the resonance escape probability due to the Doppler broadening of resonances, as given by Eq. (5.85), and due to the fuel expansion, as given by Eq. (5.86). For a UO_2 reactor consisting of assemblies of 1-cm-diameter fuel pins of height H in a water lattice with $\Sigma_p/N_F = 100$ and fuel density $\rho = 10\ g/cm^3$, $S_F/M_F = \pi dH/\pi(d/2)^2 H\rho = 0.4$, $I(300\ K) = 11.6 + 22.8 \times 0.4 = 20.72$, and $\beta'' = 61 + 47(S_F/M_F) \times 10^{-4} = 79.8 \times 10^{-4}$. The resonance escape probability at

300 K is $p = \exp(-N_F I / \xi \Sigma_p) = \exp[-20.72/(100 \times 0.948)] = 0.8036$, and $\ln(p) = -0.2186$. The Doppler temperature coefficient of reactivity at 300 K is $\alpha_{TF}^D = \ln(p)\beta''/2T^{1/2} = (-0.2186)(79.8 \times 10^{-4})/(2)(17.32) = -5.036 \times 10^{-5} \Delta k / k$. The linear thermal expansion coefficient for UO_2 is $\theta_F = 1.75 \times 10^{-5} \text{ K}^{-1}$, and the fuel expansion contribution to the resonance escape probability temperature coefficient of reactivity is $\alpha_{TF}^p = -3\theta_F \ln(p) = -3(1.75 \times 10^{-5}) \cdot (-0.2186) = 1.148 \times 10^{-5} \Delta k / k$. Thus the total prompt fuel temperature coefficient of reactivity due to the resonance escape probability is $\alpha_{TF}^D + \alpha_{TF}^p = -3.888 \times 10^{-5} \Delta k / k$.

Thermal Utilization

The thermal utilization can be written simply in terms of the effective cell-averaged fuel and moderator absorption cross sections discussed in Section 3.8:

$$f = \frac{\Sigma_{aF}^{\text{eff}}}{\Sigma_{aF}^{\text{eff}} + \Sigma_{aM}^{\text{eff}}} \rightarrow \frac{\Sigma_a^F}{\Sigma_a^F + \Sigma_a^M} \tag{5.88}$$

Recalling that $\Sigma \equiv N\sigma$, the reactivity coefficient associated with the thermal utilization has an immediate negative component associated with the fuel temperature increase and a delayed positive contribution associated with the moderator density decrease:

$$
\begin{aligned}
\frac{1}{f} \frac{\partial f}{\partial T} &= (1-f)\left[\left(\frac{1}{\sigma_a^F}\frac{\partial \sigma_a^F}{\partial T_F} + \frac{1}{\Sigma_a^F}\frac{\partial \Sigma_a^F}{\partial \xi}\frac{\partial \xi}{\partial T_F} + \frac{1}{N_F}\frac{\partial N_F}{\partial T_F}\right)\right. \\
&\quad \left. - \left(\frac{1}{\sigma_a^M}\frac{\partial \sigma_a^M}{\partial T_M} + \frac{1}{\Sigma_a^M}\frac{\partial \Sigma_a^M}{\partial \xi}\frac{\partial \xi}{\partial T_M} + \frac{1}{N_M}\frac{\partial N_M}{\partial T_M}\right)\right] \\
&\simeq (1-f)\left[\left(\frac{1}{2T_F} + \frac{1}{\Sigma_a^F}\frac{\partial \Sigma_a^F}{\partial \xi}\frac{\partial \xi}{\partial T_F} + 3\theta_F\right)\right. \\
&\quad \left. - \left(+\frac{1}{\Sigma_a^M}\frac{\partial \Sigma_a^M}{\partial \xi}\frac{\partial \xi}{\partial T_M} + 3\theta_M\right)\right] \\
&\equiv \alpha_{T_F}^f + \alpha_{T_M}^f
\end{aligned}
\tag{5.89}
$$

Account has been taken in writing Eq. (5.89) of the fact that the thermal disadvantage factor, ξ, which is used in the definition of effective homogenized fuel and moderator cross sections, will also be affected by a change in temperature. An increase in fuel temperature hardens (makes more energetic) the thermal neutron energy distribution, which reduces the spectrum average of the $1/v$ thermal fuel cross section and thus reduces the thermal utilization. An increase in the fuel temperature also reduces the fuel density, further reducing the thermal utilization. An increase in moderator temperature has little effect on the moderator cross section but reduces the moderator density, which increases the thermal utilization.

Nonleakage Probability

The nonleakage probability can be represented by

$$P_{\mathrm{NL}} \simeq \frac{1}{1 + M^2 B_g^2} \tag{5.90}$$

Temperature increases can affect the nonleakage probability by changing the characteristic neutron migration length, or the mean distance that a neutron is displaced before absorption, and by changing the size of the reactor. Assuming that both of these effects are associated primarily with changes in the moderator temperature, we write

$$\frac{1}{P_{\mathrm{NL}}} \frac{\partial P_{\mathrm{NL}}}{\partial T_M} = -\frac{M^2 B_g^2}{1 + M^2 B_g^2} \left(\frac{1}{M^2} \frac{\partial M^2}{\partial T_M} + \frac{1}{B_g^2} \frac{\partial B_g^2}{\partial T_M} \right) \tag{5.91}$$

An increase in moderator temperature causes a decrease in moderator density, which affects the migration area as

$$\frac{1}{M^2} \frac{\partial M^2}{\partial T_M} = \frac{1}{D_M} \frac{\partial D_M}{\partial T_M} - \frac{1}{\Sigma_a^M} \frac{\partial \Sigma_a^M (1-f)}{\partial T_M}$$

$$= 6\theta_M + \frac{1}{2T_F} - \frac{1}{1-f} \frac{\partial f}{\partial T} \tag{5.92}$$

where we have used $\Sigma_a = \Sigma_a^M + \Sigma_a^F = \Sigma_a^M (1-f)$.

The geometric buckling $B_g = G/l_R$, where G is a constant depending on geometry (Table 3.3) and l_R is a characteristic physical dimension of the reactor. Thus

$$\frac{1}{B_g^2} \frac{\partial B_g^2}{\partial T_M} = \left(\frac{l_R}{G} \right)^2 \frac{\partial (G/l_R)^2}{\partial T_M} = -2 \left(\frac{1}{l_R} \frac{\partial l_R}{\partial T_M} \right) \tag{5.93}$$

and Eq. (5.91) becomes

$$\alpha_{T_M}^{P_{\mathrm{NL}}} = \frac{1}{P_{\mathrm{NL}}} \frac{\partial P_{\mathrm{NL}}}{\partial T_M}$$

$$= \frac{M^2 B_g^2}{1 + M^2 B_g^2} \left(\frac{2}{l_R} \frac{\partial l_R}{\partial T_M} - 6\theta_M - \frac{1}{2T_F} + \frac{1}{1-f} \frac{\partial f}{\partial T} \right) \tag{5.94}$$

A decrease in moderator density allows neutrons to travel farther before absorption, which increases the leakage and contributes a negative reactivity coefficient component. Expansion of the reactor means that a neutron must travel farther to escape, which contributes a positive reactivity coefficient component.

Representative Thermal Reactor Reactivity Coefficients

Reactivity coefficients calculated for representative thermal reactors are given in Table 5.2.

Table 5.2 Representative Reactivity Temperature Coefficients in Thermal Reactors

	BWR	PWR	HTGR
Doppler ($\Delta k/k \times 10^{-6}$ K^{-1})	−4 to −1	−4 to −1	−7
Coolant void ($\Delta k/k \times 10^{-6}$/% void)	−200 to −100	—	—
Moderator ($\Delta k/k \times 10^{-6}$ K^{-1})	−50 to −8	−50 to −8	+1
Expansion ($\Delta k/k \times 10^{-6}$ K^{-1})	≈ 0	≈ 0	≈ 0

Source: Data from Ref. 3; used with permission of Wiley.

Example 5.6: UO$_2$ Fuel Heat Removal Time Constant. It is important to emphasize that the temperature reactivity feedback associated with the various mechanisms that have been discussed take place at different times. The feedback associated with changes in the fuel temperature take place essentially instantaneously, since an increase in fission rate produces an immediate increase in fuel temperature. However, the increase in moderator/coolant temperature occurs later, after some of the additional heat is conducted out of the fuel element. The heat balance equation in the fuel element,

$$\rho C\left(\frac{\partial T}{\partial t}\right) = r^{-1}\frac{\partial}{\partial r}\left(\frac{r\kappa\partial T}{\partial r}\right) + q''' \tag{5.95}$$

where ρ is the fuel density, κ the heat conductivity, C the heat capacity, and q''' the volumetric fission heat source, can be used to estimate a time constant characterizing the conduction of heat out of the fuel element to the interface with the coolant/moderator for a fuel pin of radius a, $\tau \approx \rho C a^2/\kappa$.

Typical parameters for a UO$_2$ fuel element in a thermal reactor are $a = 0.5$ cm, $\kappa = 0.024$ W/cm · K, $\rho = 10.0$ g/cm^3, and $C = 220$ J/kg · K. The heat conduction time constant for heat removal from the fuel into the coolant is $\tau = \rho C a^2/\kappa = (10$ g/cm$^3)(220$ J/kg · K)$/(0.024$ J/s · cm · K)$(10^3$ g/kg$) = 22.9$ s. For a smaller fuel pin characteristic of a fast reactor with $a = 0.25$ cm, the UO$_2$ fuel time constant would be about 6 s. With a metal fuel instead of UO$_2$, the heat conductivity is much larger, and the heat removal time constants are on the order of 0.1 to 1.0 s.

Startup Temperature Defect

A reactor is initially started up from a cold condition by withdrawing control rods until the reactor is slightly supercritical, thus producing an exponentially increasing neutron population on a very long period. As the neutron population increases, the fission heating and thus the reactor temperature increase. This increase in temperature produces a decrease in reactivity (almost all reactors are designed to have a negative temperature coefficient) that would cause the neutron population to decrease and the reactor to shut down if the control rods were not withdrawn further to maintain an increasing neutron population. The total amount of feedback reactivity that must be offset by control rod withdrawal during the course of the startup to operating power level is known as the *temperature defect*. The temperature defects for water-moderated reactors, graphite-moderated reactors, and sodium-cooled fast reactors are about $\Delta k/k = 2\text{–}3 \times 10^{-2}, 0.7 \times 10^{-2}$, and 0.5×10^{-2}, respectively.

5.8
Perturbation Theory Evaluation of Reactivity Temperature Coefficients

Perturbation Theory

The multigroup diffusion equations (Chapter 4) are

$$-\nabla \cdot D_g \nabla \phi_g + \Sigma_{rg}\phi_g = \sum_{\substack{g'=1 \\ g'\neq g}}^{G} \Sigma_{g'\to g}\phi_{g'} + \frac{1}{k}\chi_g \sum_{g'=1}^{G} \nu\Sigma_{fg'}\phi_{g'},$$

$$g = 1,\ldots,G \tag{5.96}$$

where $\Sigma_{g'\to g}$ is the cross section for scattering a neutron from group g' to group g, Σ_{rg} is the removal cross section for group g, which is equal to the absorption cross section plus the cross section for scattering to all other groups, χ_g is the fraction of the fission neutrons in group g, D_g and $\nu\Sigma_{fg}$ are the diffusion coefficient and the nu-fission cross section in group g, and ϕ_g is the neutron flux in group g.

We now consider a perturbation in materials properties (e.g., as would be caused by a change in local temperature) such that the reactor is described by an equation like Eq. (5.96), but with $D_g \to D_g + \Delta D_g$, $\Sigma_g \to \Sigma_g + \Delta\Sigma_g$, where the Δ terms include changes in densities, changes in the energy averaging of the cross-section data and energy self-shielding, changes in spatial self-shielding, and changes in geometry. If we assume that the perturbation in materials properties is sufficiently small that it does not significantly alter the group fluxes, we can multiply the unperturbed and perturbed equations by ϕ_g^+, subtract the two, integrate over the volume of the reactor, and sum the resulting equations for all groups to obtain the perturbation theory estimate for the change in reactivity associated with the perturbation in material properties:

$$\frac{\Delta k}{k} \simeq \sum_{g=1}^{G} \int dr \left[\phi_g^+ \nabla \cdot (\Delta D_g \nabla \phi_g) - \phi_g^+ \Delta\Sigma_{rg}\phi_g + \phi_g^+ \sum_{\substack{g'=1 \\ g'\neq g}}^{G} \Delta\Sigma_{g'\to g}\phi_{g'} \right.$$
$$\left. + \phi_g^+ \chi_g \sum_{g'=1}^{G} \Delta(\nu\Sigma_{fg'})\phi_{g'} \right]$$
$$\div \sum_{g=1}^{G} \int dr \left(\phi_g^+ \chi_g \sum_{g'=1}^{G} \nu\Sigma_{fg'}\phi_{g'} \right) \tag{5.97}$$

The quantity ϕ_g^+, the importance of neutrons in group g in producing a subsequent fission event, is discussed in Chapter 13. This expression, together with the subsidiary calculation of the $\Delta \Sigma_g$ and ΔD_g terms, including all the effects mentioned above, provides a practical means for the quantitative evaluation of reactivity coefficients in nuclear reactors.

Example 5.7: Reactivity Worth of Uniform Change in Thermal Absorption Cross Section. With the assumption that all of the fission occurs in the thermal group, the reactivity worth of a uniform change in thermal absorption cross section in a uniform thermal reactor is $\Delta k / k = \Delta \Sigma_a^{th} I_{th} / \nu \Sigma_f^{th} I_{th} \approx \Delta \Sigma_a^{th} / \Sigma_a^{th}$, because I_{th}, the integral over the reactor of the product of the thermal group importance function and flux, appears in both the numerator and denominator, and because in a critical reactor $\Sigma_a^{th} \approx \nu \Sigma_f^{th}$.

We now discuss some fast reactor reactivity coefficients that could not be treated by the more approximate method of the preceding section, although we emphasize that this perturbation theory calculation is also used for thermal reactor reactivity coefficient evaluation.

Sodium Void Effect in Fast Reactors

The reactivity change that occurs when sodium is voided from a fast reactor can be separated into leakage, absorption, and spectral components. The *leakage* and *spectral components* correspond to the first (ΔD_g) and third ($\Delta \Sigma_{g' \to g}$) terms, respectively, in Eq. (5.97). The *absorption component* corresponds to the second ($\Delta \Sigma_{rg}$) and fourth ($\Delta \nu \Sigma_f$) terms in Eq. (5.97), although the change in fission cross section is usually small and therefore neglected, and this component is usually referred to as the *capture component*. The spectral and capture components are normally largest in the center of the core, where the neutron flux and importance function are largest, and the leakage component is normally largest in the outer part of the core, where the flux gradient is largest.

The magnitude of the sodium void coefficient varies directly with the ratio of the number of sodium atoms removed to the number of fuel atoms present. The spectral component of the sodium void coefficient is generally positive, is more positive for ^{239}Pu than for ^{235}U, and becomes increasingly positive as fissile material concentration decreases relative to sodium content. The capture component tends to become more positive with softer neutron spectra because of the 2.85-keV resonance in ^{23}Na, hence to become more positive with increasing sodium concentration relative to fuel concentration. The negative leakage component is generally smaller than the other two components, although the leakage component can be enhanced by the choice of geometrical configuration. As a result, the overall reactivity effect of voiding the central part of the core is positive, and may be positive for voiding of the entire core. This poses a serious safety concern that must be offset by proper design to ensure that other negative reactivity coefficients are dominant.

Doppler Effect in Fast Reactors

In fast reactors, the neutron energy spectrum includes the resonance regions of both the fissionable (^{235}U, ^{233}U, ^{239}Pu, ^{241}Pu) and nonfissionable (^{232}Th, ^{238}U,

^{240}Pu) fuel isotopes. The Doppler effect in fast reactors is due almost entirely to resonances below about 25 keV. An increase in fuel temperature will produce an increase in both the fission and absorption cross sections, and the resulting change in reactivity can be positive or negative, depending on the exact composition. The temperature coefficient of reactivity can be estimated from

$$\frac{\partial k}{\partial T_F} = \int N_F \left[\phi_f^+ \frac{\partial \sigma_f}{\partial T_F} - \phi^+(E) \left(\frac{\partial \sigma_\gamma}{\partial T_F} + \frac{\partial \sigma_f}{\partial T_F} \right) \right] \phi(E) dE$$

$$\simeq N_F \int \frac{1}{\nu} \frac{\partial \sigma_f}{\partial T_F} (\nu - 1 - \alpha) \phi(E) dE \tag{5.98}$$

where N_F is the density of fuel nuclei (sum over species implied), $\phi^+(E)$ and ϕ_f^+ are the importance of a neutron at energy E and of a fission neutron (i.e., the number of fissions the neutron subsequently produces). Since in a critical system each neutron will on average produce $1/\nu$ fissions, $\phi^+ \approx \phi_f^+ \approx 1/\nu$ is used in the second form of the estimate, and $\alpha \equiv \sigma_\gamma/\sigma_f$ has also been used. Since α generally decreases with increasing neutron energy (Chapter 2), the reactivity change will tend to be more positive/less negative for metal-fueled cores with a relatively hard spectrum. The oxygen in UO_2 fuel softens (makes less energetic) the energy spectrum and thereby makes the reactivity change more negative/less positive. Detailed design calculations, using methods benchmarked against critical experiments, indicate that in larger reactors with a high fertile-to-fissile ratio the Doppler coefficient is sufficiently negative to provide a prompt shutdown mechanism in the event of excess fission heating of the fuel.

Fuel and Structure Motion in Fast Reactors

The increased fission heating coincident with an increase in the neutron population causes the fuel to expand radially and axially and to distort (e.g., bow) due to constraints. The expanding fuel first compresses, then ejects, sodium. The additional fission heat is transferred to the structure, producing a delayed expansion and distortion of the structure. The radial expansion, which is cumulative from the core center outward, results in a general outward radial movement of the fuel and in an expansion of the size of the reactor. The reactivity effect of this fuel and structure motion is highly dependent on the details of the design. However, a few simple estimates provide a sense of the magnitude of the effects.

Example 5.8: Reactivity Effects of Fuel and Structure Expansion. Radial motion of the fuel by an amount Δr from an initial radial location r causes a reduction in local fuel density which varies as r^2, leading to a local density change $\Delta N_F/N_F \approx (r^2 - (r + \Delta r)^2/r^2 \approx -2\Delta r/r$. Axial fuel expansion leads to linear fuel density decreases. The overall expansion reactivity coefficient is a combination of the negative effect of reduced fuel density and the positive effect of increased core size, hence reduced leakage. An overall expansion reactivity coefficient is of the form

$$\alpha_{T_M}^{\exp} = \left(a \frac{\Delta R}{R} + b \frac{\Delta N_F}{N_F} \right)_{\text{radial}} + \left(c \frac{\Delta H}{H} + d \frac{\Delta N_F}{N_F} \right)_{\text{axial}} \tag{5.99}$$

Table 5.3 Reactivity Coefficients in a 1000-MWe Oxide-Fueled Fast Reactor

	Temperature $\Delta k/k \times 10^{-6}$ $°C^{-1}$	Power: $\Delta k/k \times 10^{-6}$ MW^{-1}
Sodium expansion core	+3.0	+0.085
Sodium expansion reflector	−1.6	−0.081
Doppler	−3.2	−0.628
Radial fuel pin expansion	+0.4	+0.117
Axial core expansion	−4.1	−0.181
Radial core expansion	−6.8	−0.182

Source: Data from Ref. 9; used with permission of American Nuclear Society.

where, for the example of a 1000-MWe UO_2 reactor with H/D = 0.6, the constants are ($a = 0.143$, $b = 0.282$, $c = 0.131$, $d = 0.281$).

Fuel Bowing

Fuel distortion (e.g., bowing) is very much a function of how the fuel is constrained. The calculated reactivity effect of inward bowing in the metal fueled EBR-II was $\Delta k/k \approx -0.35\Delta V/V \approx -0.7\Delta R/R \approx 0.0013$. This predicted positive reactivity due to bowing exceeded the combined negative reactivity from all other effects at full flow and intermediate power, suggesting the possibility of a positive reactivity coefficient over the intermediate power range, consistent with experimental observation.

Representative Fast Reactor Reactivity Coefficients

Reactivity coefficients calculated for a representative fast reactor design are given in Table 5.3.

5.9
Reactor Stability

Reactor Transfer Function with Reactivity Feedback

Since the reactor power is related directly to the neutron population, we can rewrite the neutron kinetics equations, in particular Eq. (5.39), in terms of the power, $P = E_f n v v \Sigma_f \cdot$ Vol, where E_f is the energy release per fission. If we expand the power about the equilibrium power P_0 as $P(t) = P_0 + P_1(t)$ and limit consideration to the situation $|P_1/P_0| \ll 1$, we find that

$$\frac{dP_1(t)}{dt} = \frac{1}{\Lambda}\left\{\rho(t)P_0 + \int_0^\infty d\tau \beta D(\tau)[P_1(t-\tau) - P_1(t)]\right\} \qquad (5.100)$$

Representing the reactivity as the sum of an external reactivity, ρ_{ex}, such as may be caused by control rod motion, and a feedback reactivity, ρ_f, caused by the inherent reactivity feedback mechanisms discussed in the preceding two sections, the total reactivity may be written

$$\rho(t) = \rho_{ex}(t) + \rho_f(t)$$

$$= \rho_{ex}(t) + \int_{-\infty}^{t} f(t-\tau)P_1(\tau)d\tau$$

$$= \rho_{ex}(t) + \int_{0}^{\infty} f(\tau)P_1(t-\tau)d\tau \tag{5.101}$$

where $f(t-\tau)$ is the feedback kernel that relates the power deviation $P_1 = P - P_0$ at time $t - \tau$ to the resulting reactivity at time t.

Using the last form of Eq. (5.101) in Eq. (5.100), Laplace transforming (equivalently, assuming an e^{st} time dependence), and rearranging yields a transfer function, $H(s)$, relating the external reactivity input to the power deviation from equilibrium:

$$P_1(s) = \frac{Z(s)}{1 - P_0 F(s)Z(s)} P_0\rho_{ex}(s) \equiv H(s)P_0\rho_{ex}(s) \tag{5.102}$$

This new transfer function contains the zero-power transfer function, $Z(s)$, which relates the prompt and delayed neutron response to the external reactivity, and the feedback transfer function, $F(s)$, which relates the feedback reactivity to the power deviation $P_1 = P - P_0$:

$$\rho_f(s) = F(s)P_1(s) \tag{5.103}$$

Note that when $P_0 \to 0$, $H(s) \to Z(s)$.

The linear stability of a nuclear reactor can be determined by locating the poles of $H(s)$ in the complex s-plane. This follows from noting that when Eq. (5.102) is Laplace inverted, the solutions for $P_1(t) \sim \exp(s_j t)$, where the s_j are the poles of $H(s)$. Any poles located in the right half of the complex s-plane (i.e., with a positive real part) indicate a growing value of $P_1(t)$—an instability. Since $Z(s)$ appears in the numerator and denominator of $H(s)$, its poles (the roots of the inhour equation) cancel in $H(s)$, and the poles of $H(s)$ are the roots of

$$1 - P_0 F(s)Z(s) = 0 \tag{5.104}$$

We can anticipate from Eq. (5.104) that the poles of $H(s)$, hence the linear stability of the reactor, will depend on the equilibrium power level, P_0.

Stability Analysis for a Simple Feedback Model

To determine the roots of Eq. (5.104), we must first specify a feedback model in order to determine the feedback transfer function, $F(s)$. We consider a two-temperature model in which the deviation in the fuel temperature from the equilibrium value satisfies

$$\frac{dT_F(t)}{dt} = a P_1(t) - \omega_F T_F(t) \tag{5.105}$$

where a involves the heat capacity and density of the fuel and ω_F is the inverse of the heat transfer time constant of the fuel element (i.e., the time constant for removal of heat from the fuel element into the coolant/moderator). The temperature deviation about the equilibrium value in the coolant/moderator satisfies

$$\frac{dT_M(t)}{dt} = b T_F(t - \Delta t) - \omega_M T_M(t) \tag{5.106}$$

where b involves the mechanism governing the response of the coolant/moderator temperature to a change in the fuel temperature, ω_M is the inverse of the heat removal time constant for the moderator, and for the sake of generality we assume that the coolant mass flow rate is varied in response to the fuel temperature at an earlier time $(t - \Delta t)$. The same model could be applied to any two-temperature representation of a reactor core. For example, we could consider T_F to be the temperature of a simultaneously heated fuel–coolant region and T_M to represent the temperature of the structure in a fast reactor model. Writing

$$\rho(t) = \rho_{ex}(t) + \alpha_F T_F(t) + \alpha_M T_M(t) \equiv \rho_{ex}(t) + \int_0^t f(t - \tau) P_1(\tau) d\tau \tag{5.107}$$

defines the feedback kernel, $f(t - \tau)$, where $T_F(t)$ and $T_M(t)$ are deviations from the equilibrium temperatures.

Laplace transforming these three equations, using the convolution theorem, and combining leads to identification of the feedback transfer function:

$$F(s) = \frac{X_F}{1 + s/\omega_F} + \frac{X_M e^{-s\Delta t}}{(1 + s/\omega_F)(1 + s/\omega_M)} \tag{5.108}$$

where $X_F = a\alpha_F/\omega_F$ and $X_M = (ab\alpha_M/\omega_F\omega_M)$ are the steady-state reactivity power coefficient for the fuel and coolant/moderator, respectively. Using the zero-power transfer function, $Z(s)$, of Eq. (5.46), but in the one-delayed neutron group approximation, and the feedback transfer function, $F(s)$, of Eq. (5.108), Eq. (5.104) for the poles of the reactor transfer function with feedback, $H(s)$, becomes

$$1 - \frac{P_0}{s[\Lambda + \beta/(s + \lambda)]} \left[\frac{X_F}{1 + s/\omega_F} + \frac{X_M e^{-s\Delta t}}{(1 + s/\omega_F)(1 + s/\omega_M)} \right] = 0 \tag{5.109}$$

There are a number of powerful mathematical techniques from the field of linear control theory (Nyquist diagrams, root-locus plots, Routh–Hurwitz criterion, iterative root finding methods, etc.) for finding the roots of Eq. (5.109), or of the more complex equations that would result from more detailed reactivity feedback models. Some simplification results from limiting attention to growth rates that are small compared to the inverse neutron generation time ($s \ll \Lambda^{-1}$), allowing neglect of the Λ term. We now consider two additional approximations which allow us to obtain valuable physical insights.

If we set $X_M \sim \alpha_M = 0$ (i.e., neglect the coolant/moderator feedback), Eq. (5.109) can be solved analytically to obtain

$$s_\pm = \frac{1}{2}\omega_F\left(\frac{P_0 X_F}{\beta} - 1\right)\left[1 \pm \sqrt{1 + \frac{4(P_0 X_F/\beta)(\lambda/\omega_F)}{(P_0 X_F/\beta - 1)^2}}\right] \tag{5.110}$$

If the fuel power coefficient is positive ($X_F \sim \alpha_F > 0$), the term under the radical is positive and greater than unity, both roots are real, and one root is positive, indicating an instability. If the fuel power coefficient is negative ($X_F \sim \alpha_F < 0$), the real parts of both roots are negative, indicating stability.

Threshold Power Level for Reactor Stability

If we retain X_M finite but restrict our consideration to instabilities with growth rates much less than the inverse fuel heat removal time constant, $s \ll \omega_F$, and set the time delay to zero, $\Delta t = 0$, we can again solve Eq. (5.109) analytically for the poles of the reactor transfer function, $H(s)$:

$$s_\pm = -\frac{1}{2}\omega_M \frac{1 - \frac{P_0 X_F}{\beta}\left(\frac{X_M}{X_F} + 1 + \frac{\lambda}{\omega_M}\right)}{1 - \frac{P_0 X_F}{\beta}}$$

$$\times \left[1 \pm \left\{1 + \frac{4\left(\frac{\lambda}{\omega_M}\right)\left(\frac{P_0 X_F}{\beta}\right)\left(1 - \frac{P_0 X_F}{\beta}\right)}{\left[1 - \frac{P_0 X_F}{\beta}\left(\frac{X_M}{X_F} + 1 + \frac{\lambda}{\omega_M}\right)\right]^2}\right\}^{1/2}\right] \tag{5.111}$$

This expression reveals the existence of a threshold equilibrium power level, P_0, above which a reactor becomes unstable. As $P_0 \to 0$, the two roots approach 0 and $-\omega_M$, a marginally stable condition, and do not depend on the reactivity power coefficients X_M and X_F. As P_0 increases, the nature of the solution depends on X_M and X_F. Suppose that the fuel power coefficient is positive, $X_F > 0$, and the moderator power coefficient is negative, $X_M < 0$; this situation might arise, for example, in a fast reactor when X_F represents the combined Doppler, fuel expansion, and sodium void coefficients of the fuel–coolant mixture and X_M represents the structure expansion coefficient. Taking $X_F/X_M = -\frac{1}{2}$ and $\omega_M = \frac{1}{4}$, the roots of Eq. (5.111) are plotted as a function of $|X_M|P_0/\beta$ (denoted at P_0) in Fig. 5.4. As P_0 increases from zero, the marginally stable ($s = 0$) root moves into the left-half complex s-plane and the ($s = \omega_M$) root becomes less negative, indicating that the reactor would be stable. At $|X_M|P_0/\beta = 0.0962$, the roots become complex conjugates with a real part that increases with P_0. At $|X_M|P_0/\beta > \frac{2}{3}$, the real part of the two roots becomes positive, indicating that the reactor would become unstable above a certain threshold operating power level. At $|X_M|P_0/\beta > 1.664$, the roots become real and positive, with one increasing and the other decreasing with increasing P_0, continuing to indicate instability.

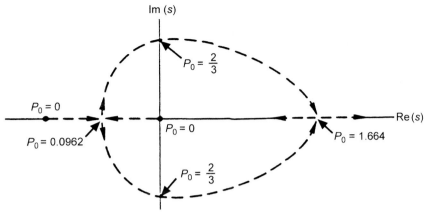

Fig. 5.4 Characteristic roots s_+ and s_- of Eq. (5.111) as a function of critical power level P_0 ($|X_M|P_0/\beta$) ($X_F > 0$, $X_F/X_M = -\frac{1}{2}$, $W_M = \frac{1}{4}S$). (From Ref. 8; used with permission of Van Nostrand.)

The total power coefficient at steady state is negative ($F(0) = X_F + X_M < 0$), but the reactor in this example was unstable above a certain threshold power level. The positive fuel power feedback was instantaneous because the fuel temperature increases instantaneously in response to an increase in fission heating. However, the coolant/moderator temperature does not increase instantaneously because of moderator heat removal, but increases on a time scale governed by the moderator heat removal time constant ω_M^{-1} following a change in fuel temperature, as may be seen by solving Eq. (5.106) for a step increase ΔT_F at $t = 0$:

$$\Delta T_M = \begin{cases} 0, & t < \Delta t \\ \dfrac{b\Delta T_F}{\omega_M}(1 - e^{-\omega_M(t+\Delta t)}), & t \geq \Delta t \end{cases} \qquad (5.112)$$

The delay of the moderator temperature response to an increase in the temperature of the fuel was neglected; its inclusion would contribute further to the possibility of instabilities. It is clear that heat removal time constants play an important role in the stability of a reactor.

The two-temperature feedback model can be generalized to investigate the stability of a variety of different feedback models that can be characterized by a fast (f)- and a slow (s)- responding temperature. For a fast temperature response that was either prompt ($\omega_f = 0$) or zero ($X_f = 0$) plus a slow temperature response with a finite time constant ($\omega_s \neq 0$) determined either by heat conduction or heat convection, the results are given in Table 5.4.

More General Stability Conditions

A necessary condition for stability is

$$F(0) = \int_0^\infty f(t)dt < 0 \qquad (5.113)$$

Table 5.4 Instability Conditions for Some Simple Two-Temperature Feedback Models

Reactivity Coefficients		Heat Removal	$F(s)$	Instability
Fast ($\omega_f = 0$)	Slow ($\omega_s \neq 0$)			
$X_f = 0$	$X_s < 0$	Conduction	$\frac{X_s}{1+s/\omega_s}$	None
$X_f = 0$	$X_s < 0$	Convection	$X_s e^{-s/\omega_s}$	$P_0 > P_{thresh}$
$X_f > 0$	$X_s < 0$	Conduction	$X_f + \frac{X_s}{1+s/\omega_s}$	$P_0 > P_{thresh}$
$X_f < 0$	$X_s < 0$	Conduction	$X_f + \frac{X_s}{1+s/\omega_s}$	None
$X_f > 0$	$X_s < 0$	Convection	$X_f + X_s e^{-s/\omega_s}$	$P_0 > P_{thresh}$
$X_f < 0$	$X_s < 0$	Convection	$X_f + X_s e^{-s/\omega_s}$	$P_0 > P_{thresh}$
$X_f = 0$	$X_{s1} < 0$	Convection	$X_{s1} e^{-s/\omega_{s1}} + \frac{X_{s2}}{1+s/\omega_{s2}}$	$P_0 > P_{thresh}$
	$X_{s2} <$ or > 0	Conduction		

Source: Data from Ref. 9; used with permission of American Nuclear Society.

However, this is not a sufficient condition, as the analysis above, in which $F(0) = X_F + X_M < 0$, demonstrates. The result discussed in the preceding example suggests a useful generalization—a reactor is on the verge of becoming unstable when the transfer function, $H(s)$, has a pole with purely imaginary s [i.e., when Eq. (5.104) has a purely imaginary root $s = i\omega$]. Except for values of ω for which $Z(i\omega) = 0$, Eq. (5.104), which determines the poles of the transfer function, can be rewritten in the case $s = i\omega$:

$$G(i\omega) = \frac{1}{Z(i\omega)} - P_0 F(i\omega) = i\omega\Lambda + \sum_{j=1}^{6} \frac{\beta_j i\omega}{i\omega + \lambda_j} - P_0 F(i\omega) = 0 \quad (5.114)$$

If this equation has a solution, it corresponds to a condition for which the reactor is on the verge of instability. A necessary condition for a solution is that $Z^{-1}(i\omega)$ and $F(i\omega)$ have the same ratio of real to imaginary parts (i.e., the same phase). If $Z^{-1}(i\omega)$ and $F(i\omega)$ do have the same phase at some $\omega = \omega_{res}$, there will be some value of P_0 for which Eq. (5.114) has a solution. If this value of P_0 is physically reasonable ($P_0 \geq 0$), there is instability onset at this (P_0, ω_{res}) condition. The real and imaginary parts of $1/Z(i\omega)$ are

$$\mathrm{Re}\left\{\frac{1}{Z(i\omega)}\right\} = \sum_{j=1}^{6} \frac{\omega^2 \beta_j}{\omega^2 + \lambda_j^2}$$

$$\mathrm{Im}\left\{\frac{1}{Z(i\omega)}\right\} = \omega\Lambda + \sum_{j=1}^{6} \frac{\omega \beta_j \lambda_j}{\omega^2 + \lambda_j^2} \quad (5.115)$$

which are both real and positive, thus are in the upper right quadrant of the complex plane. Therefore, a necessary condition for $G(i\omega) = 0$ to have a solution is that

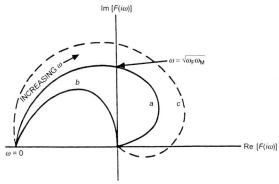

Fig. 5.5 Plot of $R = \text{Re}\{F(i\omega)\} + iI\{F(i\omega)\}$ of Eq. (5.108) with $\Delta t = 0$: case (a) $X_F = 0$, $X_M < 0$; case (b), $X_F < 0$, $X_M < 0$; case (c), $|X_M| > X_F > 0$, $X_M < 0$. (From Ref. 8; used with permission of Van Nostrand.)

the real and imaginary parts of the feedback transfer function, $F(i\omega)$, also lie in the same quadrant (i.e., both be real and positive). Hence a necessary condition for an instability is

$$\text{Re}\{F(i\omega)\} > 0 \quad \text{and} \quad \text{Im}\{F(i\omega)\} > 0 \tag{5.116}$$

We now consider the example above with the simple feedback model of Eqs. (5.105) to (5.108), but with the delay term $\Delta t = 0$. The qualitative behavior of the real and imaginary parts of $F(i\omega)$ of Eq. (5.108) are plotted in Fig. 5.5 for three different cases, all of which have a negative moderator power coefficient, $X_M < 0$. Case (a) corresponds to no reactivity feedback from the fuel ($X_F = 0$); the instability criterion of Eq. (5.116) is satisfied for $\omega > (\omega_M \omega_f)^{1/2}$, even though the steady-state power coefficient $X(0) = X_M < 0$. For case (b), with a sufficiently large negative value of the fuel power coefficient, $X_F < 0$, the criterion of Eq. (5.116) is never satisfied and the reactor is stable. In case (c), the fuel reactivity power coefficient is positive but smaller in magnitude than the negative moderator reactivity power coefficient, $|X_M| > |X_F| > 0$, which is the situation leading to the solution of Eq. (5.111); the reactor can become unstable, as found above from examination of the roots given by Eq. (5.111).

A sufficient condition for unconditional stability (i.e., no power threshold) has been shown to be

$$\text{Re}\{F(i\omega)\} = \int_0^\infty f(t)\cos(\omega t)\,dt \leq 0 \tag{5.117}$$

which is a requirement that the phase angle of the feedback transfer function, $-F(s)$, along the $i\omega$-axis is between $-90° < \phi < +90°$; thus the feedback response is negative and less than 90° out of phase with the power change that produced it. This phase constraint places constraints on the time delays. This sufficient criterion for stability has been found to be over restrictive, however.

Table 5.5 Sufficient Conditions for Unconditional Stability of Two-Temperature Feedback Models

Reactivity Coefficients	$F(i\omega)$	Stability Criterion
Coupled prompt X_f, conduction X_s	$X_f + \dfrac{X_s}{1+i\omega/\omega_s}$	$X_f \leq 0$ and $X_f + X_s < 0$
Uncoupled conduction X_f and X_s	$\dfrac{X_f}{1+i\omega/\omega_f} + \dfrac{X_s}{1+i\omega/\omega_s}$	$X_f + X_s \leq 0$, $X_f\omega_f + X_s\omega_s < 0$, and $X_f\omega_f^2 + X_s\omega_s^2 \leq 0$
Coupled conduction X_f and X_s	$\dfrac{X_f}{1+i\omega/\omega_f} + \dfrac{X_s}{(1+i\omega/\omega_f)(1+i\omega/\omega_s)}$	$X_f < 0$, $X_f + X_s \leq 0$, and $X_f\omega_f - X_s\omega_s \leq 0$
Coupled prompt X_f, convection X_s	$X_f + X_s e^{-i\omega/\omega_s}$	$X_f < 0$ and $-X_f \geq \lvert X_s \rvert$
Coupled conduction X_f, convection X_s	$\dfrac{X_f}{1+i\omega/\omega_f} + \dfrac{X_s e^{-i\omega/\omega_s}}{1+i\omega/\omega_s}$	Never unconditionally stable

The unconditional stability sufficient condition of Eq. (5.117) has been used to determine unconditional stability criteria for a variety of feedback models that can be characterized by a fast (f) and a slow (s) responding temperature. The fast temperature response was either prompt ($\omega_f = 0$) or determined by heat conduction, and the slow temperature response was with a finite time constant ($\omega_s \neq 0$) determined by either heat conduction or heat convection. The results are given in Table 5.5.

Power Coefficients and Feedback Delay Time Constants

It is clear from the previous discussion that the reactivity temperature coefficients actually enter the analysis as reactor power coefficients, associated with which there are time delays related to heat transfer and removal time constants, and that the results of the analysis are dependent on the delay times as well as on the temperature coefficients. We can generalize the two-temperature model to define a general reactor power coefficient:

$$X(t) = \sum_j \left(\frac{\partial \rho}{\partial T_j} \frac{\partial T_j(t)}{\partial P} + \frac{\partial \rho}{\partial T_j'} \frac{\partial T_j'(t)}{\partial P} \right) \tag{5.118}$$

where $\partial \rho/\partial T_j$ are the reactivity temperature coefficients corresponding to a change in local temperature T_j. The quantities $\partial \rho/\partial T_j'$ are reactivity temperature gradient coefficients denoting the change in reactivity due to a change in temperature gradient (e.g., as would produce bowing of a fuel element). These reactivity coefficients can be calculated as discussed in the two preceding sections. The quantities $\partial T_j/\partial P$ and $\partial T_j'/\partial P$ are the time-dependent changes in local temperature and temperature gradients resulting from a change in reactor power and must be calculated from models of the distributed temperature response to a change in reactor power.

The time constants that determine the time delays in the various local temperature responses to a power increase depend on the specific reactor design. Some simple estimates suffice to establish orders of magnitude. The time constant for heat transfer out of a fuel pin of radius r or plate of thickness r, density ρ, heat capacity C, and thermal conductivity κ is $\tau_f = \rho C r^2 / \kappa$, which generally varies from a few tenths to a few tens of seconds. The effect of cladding and the surface film drop is to increase the time constant for the fuel element. The lumped time constant for the coolant temperature is $\tau_c = C_c/h + (Z/2v)(1 + C_f/C_c)$, where C_c and C_f are the heat capacities per unit length of the coolant and fuel, respectively, h is the heat transfer coefficient between fuel and coolant, Z is the core height, and v is the coolant flow speed. Typical values of τ_c vary from a few tenths to a few seconds.

5.10
Measurement of Reactor Transfer Functions

Measurement of the reactor transfer function provides useful information about a reactor. A measurement at low power can identify incipient instabilities which produce peaks in the transfer function. Provided that the feedback mechanisms do not change abruptly with power, the low-power transfer function measurements can identify conditions that would be hazardous at high power, thus allowing for their correction. Information about the feedback mechanisms can be extracted from measurement of the amplitude and phase of the transfer function. Any component malfunction that altered the heat removal characteristics of the reactor would affect the transfer function, so periodic transfer function measurements provide a means to monitor for component malfunction.

Rod Oscillator Method

The sinusoidal oscillation of a control rod over a range of frequencies can be used to measure the transfer function, as described in Section 5.6. The results of Eqs. (5.60) to (5.64) apply to a reactor with feedback when $n_0 Z(i\omega)$ is replaced by $P_0 H(i\omega)$. There are some practical problems in measuring the transfer function with rod oscillation. There will be noise in the detector response, which will require a sufficiently large reactivity oscillation for the detector response to be separable from the noise, and nonlinear effects [i.e., the term ρn_1 which was neglected in Eq. (5.42)] may invalidate the interpretation. Furthermore, the oscillation will not be perfectly sinusoidal, and it will be necessary to Fourier analyze the detector response to isolate the fundamental sinusoidal component.

Correlation Methods

It is possible to measure the reactor transfer function with a nonperiodic rod oscillation. Consider the inverse Laplace transform of Eq. (5.102):

$$P_1(t) = \int_{-\infty}^{t} \rho_{ex}(\tau)h(t-\tau)d\tau = \int_{0}^{\infty} \rho_{ex}(t-\tau)h(\tau)d\tau \qquad (5.119)$$

which relates the relative power variation from equilibrium $[P_1/P_0 = (P-P_0)/P_0]$ to the time history of the external reactivity—the rod oscillation in this case—including the effect of feedback. The kernel $h(t)$ is the inverse Laplace transform of the transfer function, $H(s)$. The cross correlation between the external reactivity and the power variation is defined as

$$\phi_{\rho P} \equiv \frac{1}{2T} \int_{-T}^{T} \rho_{ex}(t)P_1(t+\tau)dt = \frac{1}{2T} \int_{-T}^{T} \rho_{ex}(t-\tau)P_1(t)dt \qquad (5.120)$$

where T is the period if ρ_{ex} and P_1 are periodic and T goes to infinity if not.

Using Eq. (5.119) in Eq. (5.120) yields

$$\phi_{\rho P} = \frac{1}{2T} \int_{-T}^{T} \rho_{ex}(t-\tau)\left[\int_{0}^{\infty} \rho_{ex}(t-t')h(t')dt'\right]dt$$

$$= \int_{0}^{\infty} h(t')\left[\frac{1}{2T} \int_{-T}^{T} \rho_{ex}(t-\tau)\rho_{ex}(t-t')dt\right]dt'$$

$$\equiv \int_{0}^{\infty} h(t')\phi_{\rho\rho}(\tau-t')dt' \qquad (5.121)$$

where $\phi_{\rho\rho}$ is the reactivity autocorrelation function. Taking the Fourier transform of Eq. (5.121) yields an expression for the transfer function

$$H(-i\omega) = \mp \frac{\mathcal{F}\{\phi_{\rho P}(\tau)\}}{\mathcal{F}\{\phi_{\rho\rho}(\tau)\}} \qquad (5.122)$$

where the transforms

$$\mathcal{F}\{\phi_{\rho P}(\tau)\} \equiv \int_{-\infty}^{\infty} e^{i\omega\tau}\phi_{\rho P}(\tau)d\tau$$

$$\mathcal{F}\{\phi_{\rho\rho}(\tau)\} \equiv \int_{-\infty}^{\infty} e^{i\omega\tau}\phi_{\rho\rho}(\tau)d\tau \qquad (5.123)$$

are known as the *cross spectral density* and the *input* or *reactivity spectral density*, respectively.

If the control rod (or other neutron absorber) position is varied randomly over a narrow range and a neutron detector response is recorded, the reactivity autocorrelation function, $\phi_{\rho\rho}$, and the reactivity-power cross-correlation function, $\phi_{\rho P}$, can be constructed by numerically evaluating the defining integrals over a period of about 5 min using a series of delay intervals, τ, increasing in discrete steps of about $\Delta\tau = 0.01$ s. The cross spectral density and reactivity spectral density can then be calculated by numerically evaluating the defining Fourier transform; for example,

$$\mathcal{F}\{\phi_{\rho P}(\tau)\} \simeq \sum_{n} \phi_{\rho P}(n\Delta\tau)(\cos n\omega\Delta\tau + i\sin n\omega\Delta\tau)\Delta\tau \qquad (5.124)$$

where n varies from a large negative integer to a large positive integer. There are sophisticated fast Fourier transform methods which are used in practice for evaluation of the cross and reactivity spectral densities.

Experimentally, it is convenient to use a reactivity variation that changes from positive to negative at definite times, so that the reactivity autocorrelation function is nearly a delta function. For such a pseudorandom binary reactivity variation,

$$\phi_{\rho\rho}(\tau - t') \simeq \text{const}\,\delta(t - t') \tag{5.125}$$

In this case, it follows from Eq. (5.121) that

$$\phi_{\rho P}(\tau) \simeq \text{const}\,h(\tau)$$

$$\mathcal{F}\{\phi_{\rho P}(\tau)\} \simeq \text{const}\,H(-i\omega) \tag{5.126}$$

and that the amplitude and phase of the transfer function can be extracted from the computation of only the cross correlation function. By repeating the Fourier transforms of Eq. (5.123) for different values of ω, the frequency dependence of $H(-i\omega)$ can be determined.

Reactor Noise Method

Minor and essentially random variations in temperature and density within a nuclear reactor, such as bubble formation in boiling water reactors, produce small and essentially random reactivity variations. Autocorrelation of the response of a neutron detector, which is proportional to the reactor neutron population or power, provides a means of determining the amplitude of the reactor transfer function from this noise. Writing the power autocorrelation function

$$\phi_{PP}(\tau) \equiv \frac{1}{2T} \int_{-T}^{T} P_1(t) P_1(t + \tau) dt \tag{5.127}$$

and using Eq. (5.119) yields

$$\phi_{PP}(\tau) = \frac{1}{2T} \int_{-T}^{T} dt \int_{0}^{\infty} h(t') \rho_{\text{ex}}(t - t') dt' \int_{0}^{\infty} h(t'') \rho_{\text{ex}}(t + \tau - t'') dt''$$

$$= \int_{0}^{\infty} h(t') \int_{0}^{\infty} h(t'') \left[\frac{1}{2T} \int_{-T}^{T} \rho_{\text{ex}}(t - t') \rho_{\text{ex}}(t + \tau - t'') dt \right] dt' dt''$$

$$= \int_{0}^{\infty} h(t') dt' \int_{0}^{\infty} h(t'') [\phi_{\rho\rho}(\tau + t' - t'')] dt'' \tag{5.128}$$

Fourier transformation then gives

$$H(-i\omega) H(i\omega) = |H(i\omega)|^2 = \frac{\mathcal{F}\{\phi_{PP}(\tau)\}}{\mathcal{F}\{\phi_{\rho\rho}(\tau)\}} \simeq \frac{\mathcal{F}\{\phi_{PP}(\tau)\}}{\text{const}} \tag{5.129}$$

where the fact that the autocorrelation function of a random reactivity input is a delta function, the Fourier transform of which is a constant, has been used in writing the final form. Thus the amplitude, but not the phase, of the reactor transfer function can be determined from autocorrelation of the reactor noise. Again, the frequency dependence is determined by taking the Fourier transform with respect to various frequencies, ω. This provides a powerful technique for online, nonintrusive monitoring of an operating reactor for component malfunction and incipient problems.

Example 5.9: Reactor Transfer Function Measurement in EBR-I. The reactor transfer function measurement on the early EBR-I sodium-cooled, metal fuel fast reactor provides a good example of the physical insight provided by transfer function measurements. The Mark II core was stable at lower power levels, but at moderate power levels an oscillatory power was observed. The measured transfer function is shown in Fig. 5.6: in part (a) for several values of the coolant flow rate (gallons

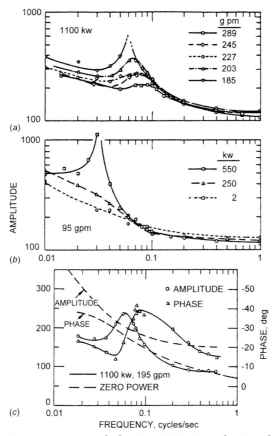

Fig. 5.6 Reactor transfer function EBR-1: (a) as a function of coolant flow rate; (b, c) as a function of reactor power. (From Ref. 9; used with permission of American Nuclear Society.)

per minute), and in parts (*b*) and (*c*) for several values of the reactor power level. At the lower coolant flow rates and the higher power levels there is a pronounced resonance in the transfer function, suggesting an incipient instability, which is not present at the higher flow rates and lower power levels.

The Mark II core was known to have a prompt reactivity feedback which added reactivity with an increase in power or a decrease in coolant flow. However, when steady state was achieved following an increase in power at constant flow, the net change in reactivity was negative, indicating an overall asymptotic power coefficient that was negative. Calculations indicated that the Doppler effect was negligible, that bowing of the fuel rods toward the center of the core contributed significant positive reactivity, and that the outward expansion of the structural plates supporting the fuel rods led to a delayed outward movement of the fuel rods that contributed negative reactivity.

A three-temperature model was used to explain the phenomena observed. The fast positive reactivity was modeled as due to the fuel bowing, and the delayed negative reactivity was modeled as the fuel motion due to the delayed outward motion of the fuel rods upon expansion of the structural plates. Heat conduction plus convection for the two separate structural effects led to a three-term representation of the power feedback. After correcting for the frequency dependence of the oscillatory heat flow, the model achieved very good agreement with the transfer function measurements.

5.11
Reactor Transients with Feedback

The dynamics equations are intrinsically nonlinear when feedback effects are included. The calculation of reactor transients is carried out with very sophisticated computer codes which model in detail the coupled dynamics of the neutrons, temperature, flow, structural motion, change of state, and so on. However, some physical insight as to the effects of feedback can be obtained by considering the simple model of Section 5.4 in the presence of feedback.

The point kinetics equations with feedback may be written in the one delayed neutron group approximation as

$$\frac{dn(t)}{dt} = \left(\frac{\rho_{ex} + \alpha_F T(t) - \beta}{\Lambda} \right) n(t) + \lambda C(t)$$

$$\frac{dC(t)}{dt} = \frac{\beta}{\Lambda} n(t) - \lambda C(t)$$

(5.130)

where a feedback reactivity $\rho_f(t) = \alpha_f T(t)$ has been added to the step reactivity insertion ρ_{ex}. We will treat the temperature, T, as either a fuel temperature or a lumped fuel–moderator temperature which satisfies

$$\rho C_P \frac{dT(t)}{dt} = E_f \Sigma_f \nu n(t) - \theta T(t)$$

(5.131)

where ρ is the density, E_f the deposited energy per fission, and $\theta \approx \kappa/($heat transfer distance) account for conductive heat removal. In Section 5.4 we found that the response to a step subprompt-critical ($\rho_{ex} < \beta$) reactivity insertion into a critical reactor was a prompt jump that changed the neutron density from n_0 to $n_0/(1 - \rho_{ex}/\beta)$ in a time on the order of the neutron generation time, Λ, followed by a slow rise ($\rho_{ex} > 0$) or decay ($\rho_{ex} < 0$) of the neutron density on the delayed neutron decay constant time scale. We examine these two phases of the transient separately in the presence of feedback.

Step Reactivity Insertion ($\rho_{ex} < \beta$): Prompt Jump

During the initial phase of the transient for a few Λ following the reactivity insertion, the delayed neutron precursor decay source is constant at the critical equilibrium value $\lambda C_0 = (\beta/\Lambda)n_0$. In the absence of feedback, the solution of Eq. (5.130) in this case is

$$n(t) = n_0 \exp\left(\frac{\rho_{ex} - \beta}{\Lambda}t\right)\left[1 + \frac{\beta}{\Lambda}\int_0^t \exp\left(-\frac{\rho_{ex} - \beta}{\Lambda}t'\right)dt'\right]$$

$$\simeq \frac{n_0}{1 - \rho_{ex}/\beta} \tag{5.132}$$

Assuming that the feedback is on the fuel temperature, which responds instantaneously to an increase in the fission rate, the corresponding solution with feedback reactivity is

$$n(t) = n_0 \exp\left(\frac{\rho_{ex} + \alpha_f T(t) - \beta}{\Lambda}t\right)$$

$$\times \left[1 + \frac{\beta}{\Lambda}\int_0^t \exp\left(-\frac{\rho_{ex} + \alpha_f T(t') - \beta}{\Lambda}t'\right)dt'\right] \tag{5.133}$$

On this short time scale $t \sim \Lambda \ll \rho C_p/\theta$, the solution of Eq. (5.131) is

$$T(t) \simeq \frac{E_f \nu \Sigma_f}{\rho C_p}\int_0^t n(t')dt' \tag{5.134}$$

If the feedback is negative ($\alpha_f < 0$), the effect of the feedback is to reduce the magnitude of the input reactivity step. If $\rho_{ex} > 0$, n and T increase in time and $\rho_f = \alpha_f T < 0$; if $\rho_{ex} < 0$, n and T decrease in time and $\rho_f = \alpha_f T > 0$; ($T_0 = 0$). If the feedback is positive ($\alpha_f > 0$), the effect of the feedback is to enhance the magnitude of the input reactivity step. If $\rho_{ex} > 0$, n and T increase in time and $\rho_f = \alpha_f T > 0$; if $\rho_{ex} < 0$, n and T decrease in time and $\rho_f = \alpha_f T < 0$. Thus negative feedback reactivity would reduce the magnitude of the prompt jump and perhaps reverse the sign if the feedback reactivity exceeds the input reactivity; positive feedback reactivity would enhance the magnitude of the prompt jump.

Step Reactivity Insertion ($\rho_{ex} < \beta$): Post-Prompt-Jump Transient

We saw in Section 5.4 that in the absence of feedback, after the initial prompt jump in the neutron density on the prompt neutron time scale, the subsequent transient evolves on the slower time scale of the delayed neutron precursor decay:

$$n(t) = \frac{n_0 \exp\{(\lambda \rho_{ex}/\beta)t/(1 - \rho_{ex}/\beta)\}}{1 - \rho_{ex}/\beta} \tag{5.135}$$

For the problem with feedback, we make use of the prompt-jump approximation (set $dn/dt = 0$) and solve Eqs. (5.130) to obtain

$$n(t) \simeq \frac{n_0 \exp\left\{-\lambda\left(t - \int_0^t \frac{dt'}{1 - [\rho_{ex} + \alpha_f T(t')]/\beta}\right)\right\}}{1 - [\rho_{ex} + \alpha_f T(t)]/\beta} \tag{5.136}$$

which reduces to Eq. (5.135) when $\alpha_f = 0$. Note that Eq. (5.136) is valid only for the time after the prompt jump in neutron density between $t = 0$ and $t = t_{pj} \approx \Lambda$. This equation evaluated at t_{pj} implies an effective prompt jump from $n_0 \rightarrow n_0/[1 - (\rho_{ex} + \alpha_f T(t_{pj}))/\beta]$, to be compared with the effective prompt jump from $n_0 \rightarrow n_0/(1 - \rho_{ex}/\beta)$ in the case without feedback implied by Eq. (5.135). Equation (5.131) can be solved formally for the temperature

$$T(t) = \frac{E_f \nu \Sigma_f}{\rho C_p} \int_0^t n(t') \exp -(\theta/\rho C_p)(t - t')dt' \tag{5.137}$$

The presence of feedback can have a dramatic effect on the course of the transient. Consider a positive step reactivity insertion, $0 < \rho_{ex} < \beta$, which without feedback would result in an exponentially increasing neutron density with period $(\beta/\rho_{ex} - 1)/\lambda$. With negative reactivity feedback ($\alpha_f < 0$), the period becomes longer (the rate of increase is slower), or even becomes negative (the neutron density decreases in time) if $|\alpha_f|T(t)$ becomes greater than ρ_{ex}. For a negative step reactivity insertion, $\rho_{ex} < 0$, and negative reactivity feedback, the presence of feedback with the decreasing temperature causes the decay in the neutron density to become slower and even reverse and start increasing if $|\alpha_f T(t)|$ becomes greater than $|\rho_{ex}|$. Thus a reactor with a negative temperature coefficient of reactivity will adjust automatically to a step reactivity insertion by seeking a new critical condition. For example, when a cold reactor is started up by withdrawing the control rods to produce an increasing neutron population and increasing fission heating, the negative reactivity will increase also, until the reactor reaches a new temperature and neutron population at which it is just critical. A negative temperature coefficient of reactivity also allows a reactor to automatically *load follow* (an increase in power output demand will result in a decrease in coolant inlet temperature, which produces a positive reactivity that causes the neutron population and the fission rate to increase until a new critical condition is reached at higher power).

5.12
Reactor Fast Excursions

The examination of hypothetical accidents requires the analysis of fast, supercritical excursions in the neutron population in a reactor. Although this analysis is done with sophisticated computer codes, which solve the coupled neutron–thermodynamics–hydrodynamics equation of state equations, there are several analytical models which provide physical insight into the phenomena of fast supercritical reactor excursions. Delayed neutron precursors respond too slowly to be important in such transients and may be neglected.

Step Reactivity Input: Feedback Proportional to Fission Energy

The prompt neutron kinetics equation for a step reactivity input $\Delta k_0 > k\beta$ and a feedback negative reactivity proportional to the cumulative fission energy release is described by

$$\frac{1}{P}\frac{dP(t)}{dt} = \frac{k-1}{\Lambda} = \frac{\Delta k_0 - \alpha_E E(t)}{\Lambda} = \frac{\Delta k_0}{\Lambda} - \frac{\alpha_E}{\Lambda}\int_0^t P(t')dt' \tag{5.138}$$

where Δk_0 is measured relative to prompt critical and

$$E(t) \equiv \int_0^t P(t')dt', \qquad \alpha_E \equiv \frac{\partial k}{\partial E} \tag{5.139}$$

The solution of Eq. (5.139) is

$$E(t) = \frac{(\Delta k_0/\Lambda) + R}{\alpha E/\Lambda}\frac{1 - e^{-Rt}}{\frac{[(R+\Delta k_0/\Lambda)]}{[(R-\Delta k_0/\Lambda)]}e^{-Rt} + 1} \tag{5.140}$$

where

$$R \equiv \sqrt{\left(\frac{\Delta k_0}{\Lambda}\right)^2 + 2\left(\frac{\alpha_E}{\Lambda}\right)P_0} \tag{5.141}$$

For transients initiated from low initial power level, P_0, $R \approx \Delta k_0/\Lambda$ and

$$E(t) \simeq \left(2(\Delta k_0/\alpha_E)(1 - e^{-(\Delta k_0/\Lambda)t})\right) \bigg/ \left(\frac{2(\Delta k_0/\Lambda)^2}{(\alpha_E/\Lambda)P_0}e^{-(\Delta k_0/\Lambda)t} + 1\right) \tag{5.142}$$

The instantaneous power is

$$P(t) = \dot{E}(t) = \frac{2R^2}{\alpha_E/\Lambda}\left(\frac{R+(\Delta k_0/\Lambda)}{R-(\Delta k_0/\Lambda)}\right)e^{-Rt}\bigg/\left(\left(\frac{R+(\Delta k_0/\Lambda)}{R-(\Delta k_0/\Lambda)}\right)e^{-Rt} + 1\right)^2$$

$$\simeq \frac{4(\Delta k_0/\Lambda)^4}{(\alpha_E/\Lambda)^2 P_0}e^{-(\Delta k_0/\Lambda)t}\bigg/\left[2\frac{(\Delta k_0/\Lambda)^2}{(\alpha_E/\Lambda)P_0}e^{-(\Delta k_0/\Lambda)t} + 1\right]^2 \tag{5.143}$$

where the second form is valid only for low initial power.

Equation (5.143) describes a symmetrical power excursion that increases to a maximum power $P_{\max} = (\Delta k_0/\Lambda)^2/2(\alpha_E/\Lambda)$ at $t \approx 1.3/(\Delta k_0/\Lambda)$ and then decreases to zero. The width of the power burst at half maximum is $\approx 3.52/(\Delta k_0/\Lambda)$, and the total fission energy produced in the burst is $2\Delta k_0/\alpha_E$.

Ramp Reactivity Input: Feedback Proportional to Fission Energy

If, instead of a step reactivity input, the external reactivity input is a ramp (e.g., as might occur in rod withdrawal), Eq. (5.138) becomes

$$\frac{1}{P}\frac{dP(t)}{dt} = \frac{at - \alpha_E E(t)}{\Lambda} = \frac{at}{\Lambda} - \frac{\alpha_E}{\Lambda}\int_0^t P(t')dt' \tag{5.144}$$

which has a solution of the form

$$E(t) = \frac{a}{\alpha_E}t + \text{periodic function} \tag{5.145}$$

The power level has a background (a/α_E) upon which is superimposed a series of oscillations as the net external plus feedback reactivity oscillates about prompt critical $(\rho = \beta)$. We now examine one of the power oscillations. Differentiating Eq. (5.144) yields an equation for the instantaneous period $\theta \equiv (dP/dt)/P$:

$$\frac{d\theta(t)}{dt} = \frac{a}{\Lambda} - \frac{\alpha_E}{\Lambda}P(t) \tag{5.146}$$

which may be combined with Eq. (5.144) to obtain

$$\frac{dP}{d\theta} = \frac{\theta P}{a/\Lambda - (\alpha_E/\Lambda)P} \tag{5.147}$$

This equation has the solution

$$\frac{1}{2}\theta^2(t) = \frac{a}{\Lambda}\ln\frac{P(t)}{P_0} - \frac{\alpha_E}{\Lambda}[P(t) - P_0] \tag{5.148}$$

The maximum power at the peak of the oscillation occurs when $\theta = 0$ and thus satisfies

$$P_{\max} = P_0 + \frac{a}{\alpha_E}\ln\frac{P_{\max}}{P_0} \simeq \frac{a}{\alpha_E}\ln\frac{P_{\max}}{P_0} \tag{5.149}$$

where the second form is only valid for $P_0 \ll P_{\max}$, where $P_0 = a/\alpha_E$ now refers to the background power at the beginning of the oscillation.

Step Reactivity Input: Nonlinear Feedback Proportional to Cumulative Energy Release

The Doppler feedback coefficient in large fast power reactors is not constant but is calculated to vary approximately inversely with fuel temperature, and theoretical considerations suggest that it varies inversely with fuel temperature to the $\frac{3}{2}$ power.

If we assume no heat loss from the fuel and constant specific heat to relate the fuel temperature increase during a transient to the cumulative fission energy release, we can represent a broad class of temperature-dependent feedback reactivities as $\alpha_E E^n$, where α_E now refers to the value of the feedback coefficient at the temperature at which the transient is initiated. In this case, the prompt neutron dynamics equation for a step external reactivity input Δk_0 is

$$\frac{1}{P}\frac{dP(t)}{dt} = \frac{\Delta k_0}{\Lambda} - \frac{\alpha'_E}{\Lambda}[E(t)]^n \tag{5.150}$$

This equation has the solution for the cumulative fission energy release

$$E(t) = \left[(n+1)\frac{\Delta k_0}{\alpha'_E}\right]^{1/n} \Big/ \left[1 + ne^{-(n\Delta k_0/\Lambda)t}\right]^{1/n} \tag{5.151}$$

which can be differentiated to obtain the instantaneous power

$$P(t) = \dot{E}(t)$$

$$= \left[(n+1)\frac{\Delta k_0}{\alpha'_E}\right]^{1/n} \left(n^2\frac{\Delta k_0}{\Lambda}\right)e^{-(n\Delta k_0/\Lambda)t} \Big/ \left[1 + ne^{-(n\Delta k_0/\Lambda)t}\right]^{1/(n+1)}$$

$$\tag{5.152}$$

Once again, the power increases to a maximum value, in this case

$$P_{\max} = \frac{n}{1+n}\left[\frac{(\Delta k_0/\Lambda)^{n+1}}{\alpha'_E/\Lambda}\right]^{1/n}$$

and then decreases to zero. The total energy release in the burst is $E_{\text{tot}} = [(1+n)\Delta k_0/\alpha'_E]^{1/n}$.

Bethe–Tait Model

It is clear that the course of a reactor excursion produced by a given external reactivity insertion is very sensitive to the feedback reactivity, hence to the evolution of the thermodynamic, hydrodynamic, and geometric condition of the reactor. The coupled evolution of these variables is calculated numerically in modern analyses. However, we can gain valuable physical insight by considering an early semianalytical model developed for fast metal fuel reactors. The prompt neutron dynamics are determined by

$$\frac{1}{P}\frac{dP(t)}{dt} = \frac{k-1-\beta}{\Lambda} = \frac{\Delta k}{\Lambda}$$

$$= \Delta k_0 + \Delta k_{\text{input}}(t) + \Delta k_{\text{displ}}(t) + \Delta k_{\text{other}}(t) \tag{5.153}$$

where Δk_0 is the initiating step reactivity (relative to prompt critical), Δk_{input} is any control rod input, Δk_{displ} is the reactivity associated with a displacement of core material due to pressure buildup, and Δk_{other} includes the Doppler effect and other nonhydrodynamic reactivity changes.

The displacement reactivity is given by

$$\Delta k_{\text{displ}}(t) = \int \rho(r, t) \mathbf{u}(r, t) \cdot \nabla w^+(r) dr \tag{5.154}$$

Here ρ is the material density, $\mathbf{u}(r, t)$ represents a material displacement from r to $r + \Delta r$, and $w^+(r, t)$ is the importance of a unit mass of material at location r to producing subsequent fission events. (The importance function is discussed in Chapter 13.)

The displacement is related to the pressure by the hydrodynamic equations

$$\rho \frac{\partial^2 \mathbf{u}(r, t)}{\partial t^2} = -\nabla p(r, t) \tag{5.155}$$

and

$$\frac{\partial \rho(r, t)}{\partial t} + \nabla \cdot \left[\rho(r, t) \frac{\partial \mathbf{u}(r, t)}{\partial t} \right] = 0 \tag{5.156}$$

An equation of state, represented symbolically as

$$p(r, t) = p(e(r, t), \rho(r, t)) \tag{5.157}$$

relates the pressure to the energy density, $e(r, t)$, and to the density. We neglect changes in density and work done in expansion or compression. Differentiating Eq. (5.154) twice and using Eq. (5.155) yields

$$\frac{\partial^2 \Delta k_{\text{displ}}}{\partial t^2} = -\int \nabla p(r, t) \cdot \nabla w^+(r) dr \tag{5.158}$$

The analysis proceeds by postulating that there is no feedback, except the Doppler effect, until the total energy generated in the core reaches a threshold value, E^*, at which point the core material begins to vaporize, thereby building up pressure, which causes the core to expand until the negative reactivity associated with expansion eventually terminates the excursion. Rather than carry through the rather involved derivation (see Ref. 9), we summarize the main results for a spherical core. When the energy, E, exceeds the threshold value, it subsequently increases as

$$E - E^* = E^* \left(e^{(\Delta k / \Lambda)(t - t^*)} - 1 \right) \tag{5.159}$$

The pressure near the center of the core is proportional to $E - E^* \approx E$, so that once it becomes large the pressure varies as

$$p \sim E N e^{(\Delta k / \Lambda) t} \tag{5.160}$$

The pressure gradient that tends to blow the core apart is proportional to p/R. Thus the radial acceleration produced by the pressure gradient goes as

$$\ddot{R} \sim |\nabla p| = \frac{C_1}{R} e^{(\Delta k/\Lambda)t} \tag{5.161}$$

Integrating this expression twice yields an expression for the instantaneous core radius

$$R'(t) \simeq R\left[1 + \frac{C_1 \Lambda^2}{(\Delta k)^2 R^2} e^{(\Delta k/\Lambda)t}\right] \tag{5.162}$$

The excursion terminates when the expansion increases the negative reactivity sufficiently to offset the initiating reactivity less any negative Doppler or rod input reactivity:

$$\Delta k_{\mathrm{displ}}(R' - R) = \Delta k_0 - \Delta k_{\mathrm{other}} - \Delta k_{\mathrm{input}} = \Delta k' \tag{5.163}$$

which occurs at time t given by

$$e^{(\Delta k'/\Lambda)t} = \frac{(\Delta k')^3 R^2}{C_1 \Delta k_{\mathrm{displ}} \Lambda^2} \tag{5.164}$$

The energy generated up to the time of termination is

$$E \sim \frac{(\Delta k')^3 R^2}{\Lambda^2} \tag{5.165}$$

Numerical calculations indicate that the approximate relationships above represent quite well excursions resulting from large initial reactivity insertions. For modest initiating reactivities, the expression

$$\left(\frac{E}{E^*} - 1\right) \sim \left[\frac{(\Delta k')^3 R^2}{\Lambda^2}\right]^{2/9} \tag{5.166}$$

is in better qualitative agreement with numerical results.

5.13
Numerical Methods

In practice, numerical methods are used to solve the neutron dynamics equations. The solution is made difficult by the difference in time scales involved. The prompt neutron time scale is on the order of $\Lambda = 10^{-4}$ to 10^{-5} s for thermal reactors or 10^{-6} to 10^{-7} s for fast reactors, while the delayed neutron time scales vary from tenths of seconds to tens of seconds. When ρ is significantly less than β, making the prompt jump approximation removes the prompt neutron time scale from the problem, and straightforward time-differencing schemes are satisfactory. When it is necessary to retain the prompt neutron dynamics (i.e., for transients near or above prompt critical), the usual numerical methods for solving ordinary differential equations (e.g., Runge–Kutta) are limited by solution stability to extremely small time steps over which there is little change in the neutron population. However, a class of methods for solving stiff sets of ordinary differential equations (sets with very different time constants) have been developed (Refs. 2 and 7) and are now widely used for solution of the neutron dynamics equations.

References

1 D. SAPHIER, "Reactor Dynamics," in Y. Ronen, ed., *CRC Handbook of Nuclear Reactor Calculations II*, CRC Press, Boca Raton, FL (**1986**).

2 G. HALL and J. M. WATTS, *Modern Numerical Methods for Ordinary Differential Equations*, Clarendon Press, Oxford (**1976**).

3 J. L. DUDERSTADT and L. J. HAMILTON, *Nuclear Reactor Analysis*, Wiley, New York (**1976**), Chap. 6 and pp. 556–565.

4 A. F. HENRY, *Nuclear-Reactor Analysis*, MIT Press, Cambridge, MA (**1975**), Chap. 7.

5 D. L. HETRICK, ed., *Dynamics of Nuclear Systems*, University of Arizona Press, Tucson, AZ (**1972**).

6 A. Z. AKCASU, G. S. LELLOUCHE, and M. L. SHOTKIN, *Mathematical Methods in Nuclear Reactor Dynamics*, Academic Press, New York (**1971**).

7 C. W. GEAR, *Numerical Initial Value Problems in Ordinary Differential Equations*, Prentice Hall, Englewood Cliffs, NJ (**1971**).

8 G. I. BELL and S. GLASSTONE, *Nuclear Reactor Theory*, Wiley (Van Nostrand Reinhold), New York (**1970**), Chap. 9.

9 H. H. HUMMEL and D. OKRENT, *Reactivity Coefficients in Large Fast Power Reactors*, American Nuclear Society, La Grange Park, IL (**1970**).

10 L. E. WEAVER, *Reactor Dynamics and Control*, Elsevier, New York (**1968**).

11 H. P. FLATT, "Reactor Kinetics Calculations," in H. Greenspan, C. N. Kelber,

and D. Okrent, eds., *Computational Methods in Reactor Physics*, Gordon and Breach, New York (**1968**).

12 D. L. HETRICK and L. E. WEAVER, eds., *Neutron Dynamics and Control*, USAEC-CONF-650413, U.S. Atomic Energy Commission, Washington, DC (**1966**).

13 M. ASH, *Nuclear Reactor Kinetics*, McGraw-Hill, New York (**1965**).

14 G. R. KEEPIN, *Physics of Nuclear Kinetics*, Addison-Wesley, Reading, MA (**1965**).

15 A. RADKOWSKY, ed., *Naval Reactors Physics Handbook*, U.S. Atomic Energy Commission, Washington, DC (**1964**), Chap. 5.

16 T. J. THOMPSON and J. G. BECKERLY, eds., *The Technology of Nuclear Reactor Safety*, MIT Press, Cambridge, MA (**1964**).

17 L. E. WEAVER, ed., *Reactor Kinetics and Control*, USAEC-TID-7662, U.S. Atomic Energy Commission, Washington, DC (**1964**).

18 J. A. THIE, *Reactor Noise*, Rowman & Littlefield, Totowa, NJ (**1963**).

19 J. LASALLE and S. LEFSCHETZ, *Stability by Liapunov's Direct Methods and Applications*, Academic Press, New York (**1961**).

20 R. V. MEGHREBLIAN and D. K. HOLMES, *Reactor Analysis*, McGraw-Hill, New York (**1960**), Chap. 9.

Problems

5.1. The absorption cross section in a bare, critical thermal reactor is decreased by 0.5% by removing a purely absorbing material. Calculate the associated reactivity.

5.2. A bare metal sphere of essentially pure ^{235}U is assembled, and the output of a neutron detector is observed, after an initial transient, to be increasing exponentially with a period $T = 1$ s. The neutron effectiveness values for the six delayed neutron groups are calculated to be $\gamma_i = 1.10, 1.03, 1.05, 1.03, 1.01$, and 1.01. What is the effective multiplication constant, k, for the assembly?

5.3 Using the one-delayed precursor group approximation, prompt-jump approximation, and the reactor parameters $\beta = 0.0075, \lambda = 0.08$ s^{-1}, $\Lambda = 6 \times 10^{-5}$ s, solve for the time dependence of the neutron population over the interval $0 < t < 10$ s following the introduction of a ramp reactivity $\rho(t) = 0.1\beta t$ into a critical reactor for $0 < t < 5$ s. Such a reactivity insertion might result from partial withdrawal of a control rod bank.

5.4. A pulsed neutron measurement was performed in an assembly with $\beta = 0.0075$ and $\Lambda = 6 \times 10^{-5}$. An exponential prompt neutron decay constant $\alpha_0 = -100$ s^{-1} was measured. What are the reactivity and effective multiplication constant of the assembly?

5.5. A control rod was partially withdrawn from a critical nuclear reactor for 5 s, then reinserted to bring the reactor back to critical. The reactivity worth of the partial rod withdrawal was $\rho = 0.0025$. Use the prompt-jump approximation and a one delayed neutron group approximation to calculate the neutron and precursor populations, relative to the initial critical populations, for times $0 < t < 10$ s. Use the neutron kinetic parameters $\beta = 0.0075, \lambda = 0.08$ s^{-1}, and $\Lambda = 6 \times 10^{-5}$ s.

5.6. A control rod bank is scrammed in an initially critical reactor. The signal of a neutron detector drops instantaneously to one-third of its prescram level, then decays exponentially. Assume one group of delayed neutrons with $\beta = 0.0075$ and $\lambda = 0.08$ s, and use $\Lambda = 10^{-4}$ s for the reactor lifetime. What is the reactivity worth of the control rod bank? How long is needed for the power level to reach 1% of the initial prescram level?

5.7. Plot the real and imaginary parts of the zero-power transfer function versus ω ($s = i\omega$) for a ^{235}U reactor using a one delayed neutron group model with $\beta = 0.0075$, $\lambda = 0.08$ s^{-1}, and $\Lambda = 6 \times 10^{-5}$ s.

5.8 Calculate the Doppler reactivity temperature coefficient for a UO_2-fueled, H_2O-cooled thermal reactor with long fuel rods 1 cm in diameter operating with a fuel temperature of 450 K. The moderator macroscopic scattering cross section per atom of ^{238}U is 100. Take the resonance integral at 300 K as $I = 10$ barns.

5.9. Derive an expression for the calculation of a void temperature coefficient of reactivity for a pressurized water reactor (i.e., the temperature coefficient associated with a small fraction of the moderator being replaced with void). Repeat the calculation for when the water contains 1000 ppm ^{10}B as a "chemical shim."

5.10. Calculate the nonleakage reactivity temperature coefficient for a bare cylindrical graphite reactor with height-to-diameter ratio $H/D = 1.0$, $k_\infty = 1.10$, migration area $M^2 = 400$ cm^2, and moderator linear expansion coefficient $\theta_M = 1 \times 10^{-5}$ °C^{-1}.

5.11. Calculate the reactivity defect in a PWR with fuel and moderator temperature coefficients of $\alpha_F = -1.0 \times 10^{-5}$ $\Delta k/k/$°F and $\alpha_M = -2.0 \times 10^{-4}$ $\Delta k/k/$°F when the reactor goes from hot zero power ($T_F = T_M = 530$°F) to hot full power ($T_F = 1200$°F and $T_M = 572$°F).

5.12. A critical reactor is operating at steady state when there is a step reactivity insertion $\rho = \Delta k/k = 0.0025$. Use one group of delayed neutrons, the parameters $\beta = 0.0075$, $\lambda = 0.08$ s^{-1}, and $\Lambda = 6 \times 10^{-5}$ s, and a temperature coefficient of reactivity $\alpha_T = -2.5 \times 10^{-4}$ °C^{-1}. Assume that the heat removal is proportional to the temperature. Write the coupled set of equations that describe the dynamics of the prompt and delayed neutrons and the temperature. Linearize and solve these equations (e.g., by Laplace transform).

5.13. Calculate the power threshold for linear stability (in units of $P_0 X_F/\beta$) from Eq. (5.111) for $X_F/X_M = -0.25$ and -0.50 and for $\omega_M = 0.1, 0.25$, and 0.5.

5.14. Analyze the linear stability of a one-temperature model for a nuclear reactor in which the heat is removed by conduction with time constant ω_R^{-1} and in which there is an overall negative steady-state power coefficient, $X_R < 0$. Is the reactor stable at all power levels?

5.15. Repeat problem 5.14 for convective heat removal.

5.16. Calculate and plot the power burst described by Eq. (5.143) for a fast reactor with generation time $\Lambda = 1 \times 10^{-6}$ s and negative energy feedback coefficient $\alpha_E = -0.5 \times 10^{-6}$

$\Delta k/k/$MJ into which a step reactivity insertion of $\Delta k_0 = +0.02$ takes place at $t = 0$. Use $P_0 = 100$ MW.

5.17. A control rod is partially withdrawn (assume instantaneously) from a ^{235}U-fueled nuclear reactor that is critical and at low power at room temperature. The signal measured by a neutron detector is observed to increase immediately to 125% of its value prior to rod withdrawal, and then to increase approximately exponentially. What is the reactivity worth of the control rod? What is the value of the exponent that governs the long-time exponential increase of the signal measured by the neutron detector?

5.18. In a cold critical PWR fueled with 4% enriched UO_2, the control rod bank is withdrawn a fraction of a centimeter, introducing a positive reactivity of $\rho = 0.0005$. The neutron flux begins to increase, increasing the fission rate. Discuss the feedback reactivity effects that occur as a result of the increasing fission heating.

5.19. Use the temperature coefficients of reactivity given in Table 5.3 to calculate the change in reactivity when the core temperature in an oxide-fueled fast reactor increases from $300°C$ to $500°C$. Assume uniform temperatures in fuel, coolant, and structure. Repeat the calculation for a fuel temperature increase to $800°C$ and a coolant and structure temperature increase to $350°C$.

5.20.* Solve Eqs. (5.133) and (5.134) to calculate the response of the neutron population in a UO_2-fueled PWR to step rod withdrawal with reactivity worth $\rho = 0.002$, taking into account a negative fuel Doppler feedback coefficient of -2×10^{-6} $\Delta k/k/$K. The reactor has neutronics properties ($\beta = 0.0065$, $\lambda = 0.08$ s^{-1}, $\Lambda = 1.0 \times 10^{-4}$), fission heat deposition in the fuel $vn\Sigma_f E_f = 250$ W/cm^3, and fuel properties $\rho = 10.0$ g/cm^3 and $C_p = 220$ J/kg. (*Hint*: It is probably easiest to do this numerically.)

5.21. Evaluate the resonance escape probability moderator temperature coefficient of reactivity of Eq. (5.87) for a UO_2 reactor consisting of assemblies of 1-cm-diameter fuel pins of height H in a water lattice with $\Sigma_p/N_M = 100$ and fuel density $\rho = 10$ g/cm^3. Use $\theta_M = 1 \times 10^{-4}$/K for the linear coefficient of expansion for water.

5.22. Derive an explicit expression for the thermal utilization temperature coefficient of reactivity of Eq. (5.89) by using Eqs. (3.90) and (3.92) to evaluate the $\partial\Sigma_a^F/\partial\xi$ and $\partial\xi/\partial T_F$

---

* Problem 5.20 is a longer problem suitable for a take-home project.

terms and equivalent relations to evaluate the $d\Sigma_a^M/d\xi$ and $d\xi/dT_M$ terms.

5.23. In a "rod-drop" experiment, a control rod is dropped into a cold, critical reactor. The neutron flux is observed to immediately drop to one-half of its value prior to the rod drop and then decay slowly. Using a one-delayed-group model with delayed neutron fraction $\beta = 0.0065$ and decay constant $\lambda = 0.08/s$, determine the reactivity worth of the control rod.

6
Fuel Burnup

The long-term changes in the properties of a nuclear reactor over its lifetime are determined by the changes in composition due to fuel burnup and the manner in which these are compensated. The economics of nuclear power is strongly affected by the efficiency of fuel utilization to produce power, which in turn is affected by these long-term changes associated with fuel burnup. In this chapter we describe the changes in fuel composition that take place in an operating reactor and their effects on the reactor, the effects of the samarium and xenon fission products with large thermal neutron cross sections, the conversion of fertile material to fissionable material by neutron transmutation, the effects of using plutonium from spent fuel and from weapons surplus as fuel, the production of radioactive waste, the extraction of the residual energy from spent fuel, and the destruction of long-lived actinides.

6.1
Changes in Fuel Composition

The initial composition of a fuel element will depend on the source of the fuel. For reactors operating on the uranium cycle, fuel developed directly from natural uranium will contain a mixture of ^{234}U, ^{235}U, and ^{238}U, with the fissile ^{235}U content varying from 0.72% (for natural uranium) to more than 90%, depending on the enrichment. Recycled fuel from reprocessing plants will also contain the various isotopes produced in the transmutation–decay process of uranium. Reactors operating on the thorium cycle will contain ^{232}Th and ^{233}U or ^{235}U, and if the fuel is from a reprocessing plant, isotopes produced in the transmutation–decay process of thorium.

During the operation of a nuclear reactor a number of changes occur in the composition of the fuel. The various fuel nuclei are transmuted by neutron capture and subsequent decay. For a uranium-fueled reactor, this process produces a variety of transuranic elements in the actinide series of the periodic table. For a thorium-fueled reactor, a number of uranium isotopes are produced. The fission event destroys a fissile nucleus, of course, and in the process produces two intermediate mass *fission products*. The fission products tend to be neutron-rich and

Nuclear Reactor Physics. Weston M. Stacey
Copyright © 2007 WILEY-VCH Verlag GmbH & Co. KGaA, Weinheim
ISBN: 978-3-527-40679-1

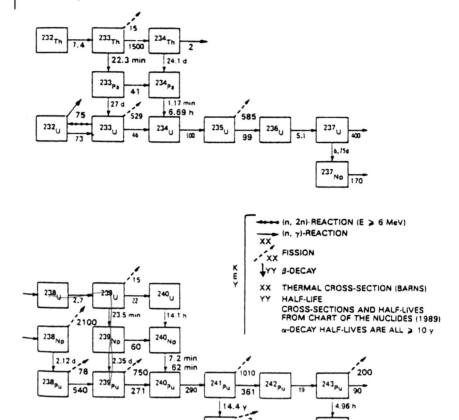

Fig. 6.1 Transmutation–decay chains for ^{238}U and ^{232}Th.
(From Ref. 3; used with permission of Taylor & Francis.)

subsequently decay by beta or neutron emission (usually accompanied by gamma emission) and undergo neutron capture to be transmuted into a heavier isotope, which itself undergoes radioactive decay and neutron transmutation, and so on. The fissile nuclei also undergo neutron transmutation via radiative capture followed by decay or further transmutation.

Fuel Transmutation–Decay Chains

Uranium-235, present 0.72% in natural uranium, is the only naturally occurring isotope that is fissionable by thermal neutrons. However, three other fissile (fissionable by thermal neutrons) isotopes of major interest as nuclear reactor fuel are produced as the result of transmutation–decay chains. Isotopes that can be converted to fissile isotopes by neutron transmutation and decay are known as *fertile isotopes.* ^{239}Pu and ^{241}Pu are products of the transmutation–decay chain beginning with the fertile isotope ^{238}U, and ^{233}U is a product of the transmutation–decay chain beginning with the fertile isotope ^{232}Th. These two transmutation–decay chains are shown in Fig. 6.1. Isotopes are in rows with horizontal arrows representing (n, γ) transmutation reactions, with the value of the cross section (in barns) shown. Downward arrows indicate β-decay, with the half-lives shown. Thermal neutron fission is represented by a dashed diagonal arrow, and the thermal cross section is shown. (Fast fission also occurs but is relatively less important in thermal reactors.) Natural abundances, decay half-lifes, modes of decay, decay energies, spontaneous fission yields, thermal capture and fission cross sections averaged over a Maxwellian distribution with $kT = 0.0253$ eV (σ^{th}), infinite-dilution capture and fission resonance integrals (RIs), and capture and fission cross sections averaged over the fission spectrum (σ^X) are given in Table 6.1.

Fuel Depletion–Transmutation–Decay Equations

Concentrations of the various fuel isotopes in a reactor are described by a coupled set of production–destruction equations. We adopt the two-digit superscript convention for identifying isotopes in which the first digit is the last digit in the atomic number and the second digit is the last digit in the atomic mass. We represent the neutron reaction rate by $\sigma_x^{nm} \varphi n^{nm}$, although the actual calculation may involve a sum over energy groups of such terms.

For reactors operating on the uranium cycle, the isotopic concentrations are described by

$$\frac{\partial n^{24}}{\partial t} = -\sigma_a^{24} \phi n^{24}$$

$$\frac{\partial n^{25}}{\partial t} = \sigma_\gamma^{24} \phi n^{24} - \sigma_a^{25} \phi n^{25}$$

$$\frac{\partial n^{26}}{\partial t} = \sigma_\gamma^{25} \phi n^{25} - \sigma_a^{26} \phi n^{26} + \lambda_{ec}^{36} n^{36}$$

$$\frac{\partial n^{27}}{\partial t} = \sigma_\gamma^{26} \phi n^{26} + \sigma_{n,2n}^{28} \phi n^{28} - \lambda^{27} n^{27}$$

$$\frac{\partial n^{28}}{\partial t} = -\sigma_a^{28} \phi n^{28}$$

$$\frac{\partial n^{29}}{\partial t} = \sigma_\gamma^{28} \phi n^{28} - \left(\lambda^{29} + \sigma_a^{29} \phi\right) n^{29}$$

$$\frac{\partial n^{36}}{\partial t} = \sigma_{n,2n}^{37} \phi n^{37} - \left(\lambda^{36} + \sigma_a^{36} \phi\right) n^{36}$$

$$\frac{\partial n^{37}}{\partial t} = \lambda^{27} n^{27} - \sigma_a^{27} \phi n^{37}$$

$$\frac{\partial n^{38}}{\partial t} = \sigma_\gamma^{37} \phi n^{37} - \left(\lambda^{38} + \sigma_a^{38} \phi\right) n^{38}$$

$$\frac{\partial n^{39}}{\partial t} = \lambda^{29} n^{29} - \left(\lambda^{39} + \sigma_a^{39} \phi\right) n^{39} \tag{6.1}$$

$$\frac{\partial n^{48}}{\partial t} = \lambda^{38} n^{38} - \sigma_a^{48} \phi n^{48}$$

$$\frac{\partial n^{49}}{\partial t} = \lambda^{39} n^{39} - \sigma_a^{49} \phi n^{49} + \sigma_\gamma^{48} \phi n^{48}$$

$$\frac{\partial n^{40}}{\partial t} = \sigma_\gamma^{49} \phi n^{49} - \sigma_a^{40} \phi n^{40} + \sigma_\gamma^{29} \phi n^{29} + \sigma_\gamma^{39} \phi n^{39}$$

$$\frac{\partial n^{41}}{\partial t} = \sigma_\gamma^{40} \phi n^{40} - \left(\lambda^{41} + \sigma_a^{41} \phi \right) n^{41}$$

$$\frac{\partial n^{42}}{\partial t} = \sigma_\gamma^{41} \phi n^{41} - \sigma_a^{42} \phi n^{42}$$

$$\frac{\partial n^{43}}{\partial t} = \sigma_\gamma^{42} \phi n^{42} - \left(\lambda^{43} + \sigma_a^{43} \phi \right) n^{43}$$

$$\frac{\partial n^{51}}{\partial t} = \lambda^{41} n^{41} - \left(\lambda^{51} + \sigma_a^{51} \phi \right) n^{51}$$

$$\frac{\partial n^{52}}{\partial t} = \sigma_\gamma^{51} \phi n^{51} - \sigma_a^{52} \phi n^{52}$$

$$\frac{\partial n^{53}}{\partial t} = \lambda^{43} n^{43} - \sigma_a^{53} \phi n^{53} + \sigma_\gamma^{52} \phi n^{52}$$

With respect to Fig. 6.1, a few approximations have been made in writing Eqs. (6.1). The neutron capture in ^{239}U to produce ^{240}U followed by the decay ($t_{1/2} = 14$ h) into ^{240}Np and the subsequent decay ($t_{1/2} = 7$ min) into ^{240}Pu is treated as the direct production of ^{240}Pu by neutron capture in ^{239}U, and the production of ^{240}Np by neutron capture in ^{239}Np followed by the subsequent decay ($t_{1/2} = 7$ min) of ^{240}Np into ^{240}Pu is treated as the direct production of ^{240}Pu by neutron capture in ^{239}Np. These approximations have the beneficial effect for numerical solution techniques of removing short time scales from the set of equations, without sacrificing information of interest on the longer time scale of fuel burnup.

For reactors operating on the thorium cycle, the isotopic concentrations are described by

$$\frac{\partial n^{02}}{\partial t} = -\sigma_a^{02} \phi n^{02}$$

$$\frac{\partial n^{03}}{\partial t} = \sigma_\gamma^{02} \phi n^{02} - \left(\lambda^{03} + \sigma_a^{03} \phi \right) n^{03}$$

$$\frac{\partial n^{13}}{\partial t} = \lambda^{03} n^{03} - \left(\lambda^{13} + \sigma_a^{13} \phi \right) n^{13}$$

$$\frac{\partial n^{22}}{\partial t} = -\left(\lambda^{22} + \sigma_a^{22} \phi \right) n^{22}$$

$$\frac{\partial n^{23}}{\partial t} = \sigma_\gamma^{22} \phi n^{22} + \lambda^{13} n^{13} - \sigma_a^{25} \phi n^{23}$$

$$\frac{\partial n^{24}}{\partial t} = \sigma_\gamma^{23} \phi n^{23} + \sigma_\gamma^{13} \phi n^{13} - \sigma_a^{24} \phi n^{24}$$

(6.2)

Table 6.1 Cross Section and Decay Data for Fuel Isotopes

Isotope	Abundance (%)	$t_{1/2}$	Decay Mode	Energy (MeV)	Spontaneous Fission Yield (%)	σ_γ^{th} (barns)	σ_f^{th} (barns)	RI_γ (barns)	RI_f (barns)	σ_γ^x (barns)	σ_f^x (barns)
232Th	100	1.41×10^{10} y	α	4.1	$<1 \times 10^{-9}$	7	–	84	–	0.09	0.08
233Th	–	22.3 m	β	1.2	–	1285	13	643	11	0.09	0.11
234Th	–	24.1 d	β	0.27	–	2	–	94	–	0.11	0.04
233Pa	–	27.0 d	β	0.57	–	35	–	864	–	0.28	0.33
234Pa	–	6.7 h	β	2.2	–	–	–	–	–	–	–
232U	–	68.9 y	α	5.4	–	64	66	173	364	0.03	2.01
233U	–	1.59×10^5 y	α	4.9	$<6 \times 10^{-9}$	41	469	138	774	0.07	1.95
234U	0.0057	2.46×10^5 y	α	4.9	1.7×10^{-9}	88	6	631	7	0.22	1.22
235U	0.719	7.04×10^8 y	α	4.7	7.0×10^{-9}	87	507	133	278	0.09	1.24
236U	–	2.34×10^6 y	α	4.6	9.6×10^{-8}	5	54	346	8	0.11	0.59
237U	–	6.75 d	β	0.52	–	392	1	1084	49	0.93	0.74
238U	99.27	4.47×10^9 y	α	4.3	5×10^{-5}	2	10	278	2	0.07	0.31
239U	–	23.5 m	β	1.3	–	–	–	–	–	–	–
240U	–	14.1 h	β	0.39	–	–	–	–	–	–	–
236Np	–	1.54×10^5 y	ec*	0.94	–	621	2453	259	1032	0.19	1.92
			β^*	0.49	–	–	–	–	–	–	–
237Np	–	2.14×10^6 y	α	5.0	$<2 \times 10^{-10}$	144	20	661	7	0.17	1.33
238Np	–	2.12 d	β	1.3	–	399	1835	201	940	0.11	1.42

(Continued)

Table 6.1 (Continued)

Isotope	Abundance (%)	$t_{1/2}$	Decay Mode	Energy (MeV)	Spontaneous Fission Yield (%)	σ_γ^{th} (barns)	σ_f^{th} (barns)	RI_γ (barns)	RI_f (barns)	σ_γ^X (barns)	σ_f^X (barns)
239Np	–	236 d	β	0.72	–	33	–	445	–	0.19	1.46
240Np	–	–	–	–	–	–	–	–	–	–	–
236Pu	–	2.86 y	α	5.9	1.4×10^{-7}	126	146	401	59	0.15	2.08
237Pu	–	45	ec*	0.22	–	–	–	–	–	–	–
238Pu	–	87.7 y	α	5.6	1.9×10^{-7}	458	15	154	33	0.10	1.99
239Pu	–	2.41×10^4 y	α	5.2	3×10^{-10}	274	698	182	303	0.05	1.80
240Pu	–	6.56×10^3 y	α	5.3	5.7×10^{-6}	264	53	8103	9	0.10	1.36
241Pu	–	14.4 y	β	0.02	$<2 \times 10^{-14}$	326	938	180	576	0.12	1.65
242Pu	–	3.73×10^5 y	α	5.0	$>5.5 \times 10^{-4}$	17	–	1130	–	0.09	1.13
241Am	–	432 y	α	5.6	4×10^{-10}	532	3	1305	14	0.23	1.38

Source: Brookhaven National Laboratory Nuclear Data Center,
http://www.dne.bnl.gov/CoN/index.html.
* 87.3% electron capture, 12.5% β.

$$\frac{\partial n^{25}}{\partial t} = \sigma_\gamma^{24} \phi n^{24} - \sigma_a^{25} \phi n^{25}$$

$$\frac{\partial n^{26}}{\partial t} = \sigma_\gamma^{25} \phi n^{25} - \sigma_a^{26} \phi n^{26}$$

$$\frac{\partial n^{27}}{\partial t} = \sigma_\gamma^{26} \phi n^{26} - \left(\lambda^{27} + \sigma_a^{27} \phi\right) n^{27}$$

$$\frac{\partial n^{37}}{\partial t} = \lambda^{27} n^{27} - \sigma_a^{37} \phi n^{37}$$

Another short-time-scale elimination approximation that neutron capture in ^{233}Pa leads directly to ^{234}U has been made.

Example 6.1: Depletion of a Pure ^{235}U-Fueled Reactor. As an example of the nature of the solution of the equations above, consider the hypothetical case of a reactor initially fueled with pure ^{235}U which operates for 1 year with a constant neutron flux of 10^{14} n/cm^2·s. The solution of the second of Eqs. (6.1) is $n^{25}(t) = n^{25}(0) \exp(-\sigma_a^{25} \phi t)$, where at the end of 1 year, $\sigma_a^{25} \phi t = (594 \times 10^{-24}$ cm$^2) \times (1 \times 10^{14}/$cm^2·s$) \times (3.15 \times 10^7$ s$) = 1.87$ and $n^{25}(t) = 0.154 n^{25}(0)$. The number of atoms that have fissioned in this 1 year is $(n(1) - n(0)) \times [\sigma_f/(\sigma_f + \sigma_\gamma)] = [0.846 n^{25}(0)](507/594) = 0.722 n^{25}(0)$. Each fission event releases 192.9 MeV of recoverable energy, so the total recoverable fission energy release is $[0.722 n^{25}(0)$ fissions$] \times (192.9$ MeV/fission$) \times (1.6 \times 10^{-19}$ MJ/MeV$) = 2.23 \times 10^{-17} \times n^{25}(0)$ MJ. If the initial core loading is 100 kg of ^{235}U, this corresponds to $(2.23 \times 10^{-17}) \times (10^5/235) \times (6.02 \times 10^{23}) = 0.95 \times 10^9$ MJ $= 1.1 \times 10^4$ MWd of recoverable fission energy.

Neglecting the production of ^{236}U by electron capture decay of ^{236}Np, the solution for $n^{25}(t)$ can be used to solve the third of Eqs. (6.1) to obtain $n^{26}(t) = [n^{25}(0)\sigma_\gamma^{25}/(\sigma_a^{25} - \sigma_a^{26})][\exp(-\sigma_a^{26} \phi t) - \exp(-\sigma_a^{25} \phi t)]$. This expression for $n^{26}(t)$ can be used in the fourth of Eqs. (6.1) to obtain a similar, but more complicated solution for $n^{27}(t)$, since we have assumed that $n^{28} = 0$; and so on.

Fission Products

The fission event usually produces two intermediate mass nuclei, in addition to releasing two or three neutrons. Interestingly, the fission product masses are not usually equal to about half the mass of the fissioning species, but are distributed in mass with peaks at about 100 and 140 amu, as shown in Fig. 6.2. The isotopes produced by fission tend to be neutron-rich and undergo radioactive decay. They also undergo neutron capture, with cross sections ranging from a few tenths of a barn to millions of barns. The general production–destruction equation satisfied by a fission product species j is

$$\frac{dn_j}{dt} = \gamma_j \Sigma_f \phi + \sum_i \left(\lambda^{i \to j} + \sigma^{i \to j} \phi\right) n_i - \left(\lambda^j + \sigma_a^j \phi\right) n_j \tag{6.3}$$

where γ_j is the fraction of fission events that produces a fission product species j, $\lambda^{i \to j}$ is the decay rate of isotope i to produce isotope j (beta, alpha, neutron, etc.,

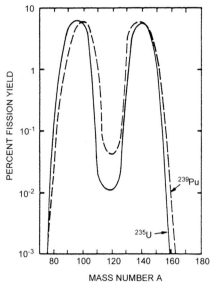

Fig. 6.2 Fission yields for ^{235}U and ^{239}Pu. (From Ref. 15.)

decay) and $\sigma^{i \rightarrow j}$ is the transmutation cross section for the production of isotope j by neutron capture in isotope j. Even though the fission products undergo transmutation and decay, the total inventory of direct fission products plus their progeny increases in time as

$$\frac{dn_{fp}}{dt} = \sum_j \frac{dn_j}{dt} = \sum_j \gamma_j \Sigma_f \phi \qquad (6.4)$$

Solution of the Depletion Equations

The equations above can be integrated to determine composition changes over the lifetime of the reactor core loading if the time dependence of the flux is known. However, the flux distribution depends on the composition. In practice, a neutron flux distribution is calculated for the beginning-of-cycle composition and critical control rod position or soluble boron concentration (PWR), and this flux distribution is used to integrate the composition equations above over a depletion-time step, Δt_{burn}. Then the new critical control rod position or soluble boron concentration is determined (by trial and error) and the flux distribution is recalculated for use in integrating the production–destruction equations over the next depletion time step, and so on, until the end of cycle is reached. The maximum value of Δt_{burn} depends on how fast the composition is changing and the effect of that composition change on the neutron flux distribution and on the accuracy of the numerical integration scheme. Excluding, for the moment, the relatively short time scale phenomena associated with the xenon and samarium fission products, the time scale of significant composition and flux changes is typically several hundred hours or more.

The typical process of advancing the depletion solution from time t_i, at which the composition is known, to time t_{i+1} is: (1) determine the multigroup constants appropriate for the composition at t_i, (2) determine the critical control rod positions or soluble poison concentration by solving the multigroup diffusion equations for the flux at t_i (adjusting the control rod positions or boron concentration until the reactor is critical), and (3) integrate the various fuel and fission product production–destruction equations from t_i to t_{i+1}. (The neutron flux solution could be made with a multigroup transport calculation or with multigroup or continuous-energy Monte Carlo calculation, and the preparation of cross sections could involve infinite media spectra and unit cell homogenization calculations or could be based on fitted, precomputed constants.) The integration of the production–destruction equations can be for a large number of points, using the neutron flux at each point; for each fuel pin, using the average flux in the fuel pin; for each fuel assembly, using the average flux over the fuel assembly; and so on.

Assuming that the flux is constant in the interval $t_i < t < t_{i+1}$, the production–destruction equations can be written in matrix notation as

$$\frac{d\mathbf{N}(t)}{dt} = \mathbf{A}(\phi(t_i))\mathbf{N}(t) + \mathbf{F}(\phi(t_i)), \quad t_i \leq t \leq t_{i+1} \tag{6.5}$$

The general solution to these equations is of the form

$$\mathbf{N}(t_{i+1}) = \exp[\mathbf{A}(t_i)\Delta t]\mathbf{N}(t_i) + \mathbf{A}^{-1}(t_i)\{\exp[\mathbf{A}(t_i)\Delta t] - 1\}\mathbf{F}(t_i) \tag{6.6}$$

In general, the accuracy of the solution depends on Δt_{burn} being chosen so that $(\lambda^i + \sigma_a^i \varphi)\Delta t_{\text{burn}} \ll 1$ for all of the isotopes involved. For this reason, it is economical to reformulate the physical production–destruction equations to eliminate short-time-scale phenomena that do not act the overall result, as discussed previously. There exist a number of computer codes that solve the production–destruction equations for input neutron fluxes (e.g., Ref. 7).

Measure of Fuel Burnup

The most commonly used measure of fuel burnup is the fission energy release per unit mass of fuel. The fission energy release in megawatt-days divided by the total mass (in units of 1000 kg or 1 tonne) of fuel nuclei (fissile plus fertile) in the initial loading is referred to as *megawatt-days per tonne* (MWd/T). An equivalent unit is MWd/kg—10^{-3} MWd/T. For example, a reactor with 100,000 kg of fuel operating at 3000 MW power level for 1000 days would have a burnup of 30,000 MWd/T.

For LWRs the typical fuel burnup is 30,000 to 50,000 MWd/T. Fuel burnup in fast reactors is projected up to be about 100,000 to 150,000 MWd/T.

Fuel Composition Changes with Burnup

The original fissionable isotope (e.g., ^{235}U) naturally decreases as the reactor operates. However, the neutron transmutation of the fertile isotope (e.g., ^{238}U) produces

Mass [g/kg HM initial]

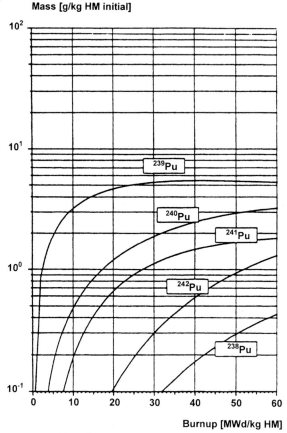

Burnup [MWd/kg HM]

Fig. 6.3 Buildup of Pu isotopes in 4 wt% enriched UO_2 in an LWR. (From Ref. 1; used with permission of Nuclear Energy Agency, Paris.)

the fissionable isotope ^{239}Pu, which in turn is transmuted by neutron capture into ^{240}Pu and higher actinide isotopes. The buildup of the various Pu isotopes as a function of fuel burnup for a typical LWR is shown in Fig. 6.3.

Compositions of spent fuel discharged from representative LWR and LMFBR designs are given in Table 6.2. The units are densities (cgs units) times 10^{-24}, which allows construction of macroscopic cross section upon multiplication by the microscopic cross section in barns. The composition for the average enrichment and burnup of PWR spent fuel is shown in the first column for fuel discharged before 1995 and in the second column for fuel discharged after 1995.

Reactivity Effects of Fuel Composition Changes

There are a variety of reactivity effects associated with the change in fuel composition. The fission of fuel nuclei produces two negative reactivity effects; the number

Table 6.2 Heavy Metal Composition of Spent UO_2 Fuel at Discharge*

Reactor Type	LWR	LWR	LMFBR	LMFBR
Initial enrichment (wt %)	3.13	4.11	20	20
Power (MW/MTU)	21.90	27.99	54.76	54.76
Burnup (GWd/T)	32	46	100	150
Actinides (1×10^{24} cm^{-3})				
^{234}U	3.92×10^{-6}	4.51×10^{-6}	3.37×10^{-5}	2.88×10^{-5}
^{235}U	1.92×10^{-4}	1.72×10^{-4}	2.17×10^{-3}	1.37×10^{-3}
^{236}U	8.73×10^{-5}	1.23×10^{-4}	4.58×10^{-4}	5.62×10^{-4}
^{237}U	†	2.48×10^{-7}	5.71×10^{-7}	7.89×10^{-7}
^{238}U	2.12×10^{-2}	2.08×10^{-2}	1.63×10^{-2}	1.53×10^{-2}
^{237}Np	1.01×10^{-5}	1.64×10^{-5}	5.11×10^{-5}	1.01×10^{-4}
^{239}Np	1.25×10^{-6}	1.55×10^{-6}	2.93×10^{-6}	3.16×10^{-6}
^{238}Pu	3.36×10^{-6}	6.56×10^{-6}	3.84×10^{-6}	1.20×10^{-5}
^{239}Pu	1.23×10^{-4}	1.23×10^{-4}	1.04×10^{-3}	1.36×10^{-3}
^{240}Pu	4.05×10^{-5}	4.28×10^{-5}	7.83×10^{-5}	1.71×10^{-4}
^{241}Pu	3.44×10^{-5}	4.07×10^{-5}	2.60×10^{-6}	8.37×10^{-6}
^{242}Pu	1.05×10^{-5}	1.69×10^{-5}	†	4.70×10^{-7}
^{241}Am	1.45×10^{-6}	1.62×10^{-6}	1.50×10^{-7}	6.87×10^{-7}
^{243}Am	2.12×10^{-6}	4.46×10^{-6}	†	†
^{242}Cm	3.71×10^{-7}	5.66×10^{-7}	†	†
^{244}Cm	4.81×10^{-7}	1.39×10^{-6}	†	†

* Calculated with ORIGEN (Ref. 7).
† < 0.001%.

of fuel nuclei is reduced and fission products are created, many of which have large neutron capture cross sections. The transmutation–decay chain of fertile fuel nuclei of a given species produces a sequence of actinides (uranium-fueled reactor) or uranium isotopes (thorium-fueled reactor), some of which are fissile. The transmutation of one fertile isotope into another nonfissile isotope can have a positive or negative reactivity effect, depending on the cross sections for the isotopes involved, but the transmutation of a fertile isotope into a fissile isotope has a positive reactivity effect. Depending on the initial enrichment, the transmutation–decay process generally produces more fissile nuclei than are destroyed early in the cycle, causing a positive reactivity effect, until the concentration of transmuted fissile nuclei comes into equilibrium.

The buildup of ^{239}Pu early in life of a uranium-fueled reactor produces a large positive reactivity effect which may be greater than the negative reactivity effect of ^{235}U depletion and fission product buildup. For thermal reactors, $\eta^{49} < \eta^{25}$, so the buildup of ^{239}Pu must exceed the burnup of ^{235}U in order for a positive reactivity effect. For fast reactors, $\eta^{49} > \eta^{25}$ for neutron energies in excess of about 10 keV, and there can be an initial positive reactivity effect even if the decrease in ^{235}U is greater than the buildup of ^{239}Pu. However, the ^{239}Pu concentration will saturate at a value determined by the balance between the ^{238}U transmutation rate and the ^{239}Pu depletion rate, at which point the continued depletion of ^{235}U and buildup of fission products produce a negative reactivity effect that accrues over the lifetime of the fuel in the reactor.

Compensating for Fuel-Depletion Reactivity Effects

The reactivity effects of fuel depletion must be compensated to maintain criticality over the fuel burnup cycle. The major compensating elements are the control rods, which can be inserted to compensate positive depletion reactivity effects and withdrawn to compensate negative depletion reactivity effects. Adjustment of the concentration of a neutron absorber (e.g., boron in the form of boric acid) in the water coolant is another means used to compensate for fuel-depletion reactivity effects. *Soluble poisons* are used to compensate fuel-depletion reactivity in PWRs but not in BWRs, because of the possibility that they will plate out on boiling surfaces. Since a soluble poison introduces a positive coolant temperature reactivity coefficient because an increase in temperature decreases the density of the soluble neutron absorber, the maximum concentration (hence the amount of fuel depletion reactivity that can be compensated) is limited.

Burnable poisons (e.g., boron, erbium, or gadolinium elements located in the fuel lattice), which themselves deplete over time, can be used to compensate the negative reactivity effects of fuel depletion. The concentration of burnable poison can be described by

$$\frac{dn^{\mathrm{bp}}}{dt} = -f_{\mathrm{bp}}n^{\mathrm{bp}}\sigma_{\mathrm{bp}}\phi \tag{6.7}$$

where f_{bp} is the self-shielding of the poison element (i.e., the ratio of the neutron flux in the poison element to the neutron flux in the adjacent fuel assembly). The poison concentration is chosen so that the spatial self-shielding of the poison elements is large enough ($f_{\mathrm{bp}} \ll 1$) early in the burnup cycle to shield the poison from neutron capture, and the neutron capture rate remains constant in time. After a certain time the concentration of the poison nuclei is sufficiently reduced that f_{bp} increases and the poison burns out, resulting in an increasing reactivity. If the poison starts to burn out at about the same time that the overall fuel depletion reactivity effect starts to become progressively more negative (i.e., when the ^{239}Pu concentration saturates), the burnout of the poison will at least partially compensate the fuel-depletion reactivity decrease.

Reactivity Penalty

The buildup of actinides in the ^{238}U transmutation–decay process introduces a fuel reactivity penalty because some of actinides act primarily as parasitic absorbers. While ^{239}Pu and ^{241}Pu are fissionable in a thermal reactor, and ^{240}Pu transmutes into ^{241}Pu, ^{242}Pu transmutes into ^{243}Pu with a rather small cross section, and ^{243}Pu

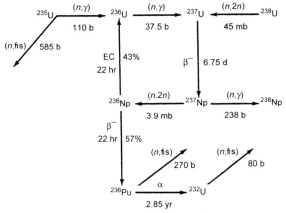

Fig. 6.4 ^{235}U neutron transmutation–decay chain. (From Ref. 4; used with permission of American Nuclear Society.)

has a rather small fission cross section, so that ^{242}Pu is effectively a parasitic absorber that builds up in time. The ^{243}Am also accumulates and acts primarily as a parasitic absorber. Whereas the ^{243}Am, which is produced by the decay of ^{243}Pu, can be separated readily, it is difficult to separate the different plutonium isotopes from each other, so the negative ^{242}Pu reactivity effect is exacerbated if the plutonium is recycled with uranium. A similar problem arises with the ^{236}U produced by radiative capture in ^{235}U, as shown in Fig. 6.4, which is difficult to separate from ^{235}U, and with ^{237}Np, which is produced by transmutation of ^{236}U into ^{237}U followed by beta decay. The ^{237}Np can be separated readily, however, and does not need to accumulate in recycled fuel.

End-of-cycle reactivity penalties calculated for the recycle of BWR fuel are shown in Table 6.3 after one, two, and three cycles. It was assumed that the ^{237}Np and ^{243}Am were removed between cycles, but there was a cycle-to-cycle increase in the ^{237}Np and ^{243}Am reactivity penalties due to the accumulation of ^{236}U and ^{242}Pu, respectively.

Effects of Fuel Depletion on the Power Distribution

Fuel depletion and the compensating control actions affect the reactor power distribution over the lifetime of the fuel in the core. Depletion of fuel will be greatest

Table 6.3 Reactivity Penalties with Recycled BWR Fuel (% $\Delta k/k$)

End of Cycle:	^{236}U	^{237}Np	^{242}Pu	^{243}Am
1	0.62	0.13	0.65	0.36
2	0.90	0.59	1.53	0.57
3	1.12	0.73	2.04	0.89

Source: Data from Ref. 16.

where the power is greatest. The initial positive reactivity effect of depletion will then enhance the power peaking. At later times, the negative reactivity effects will cause the power to shift away to regions with higher k_{infinity}. Any strong tendency of the power distribution to peak as a result of fuel depletion must be compensated by control rod movement. However, the control rod movement to offset fuel depletion reactivity effects itself produces power peaking; the presence of the rods shields the nearby fuel from depletion and when the rods are withdrawn, the higher local k_∞ causes power peaking. Similarly, burnable poisons shield the nearby fuel, producing local regions of higher k_∞ and power peaking when they burn out. Determination of the proper fuel concentration zoning and distribution of burnable poisons and of the proper control rod motion to compensate fuel depletion reactivity effects without unduly large power peaking is a major nuclear analysis task.

In-Core Fuel Management

At any given time, the fuel in a reactor core will consist of several batches that have been in the core for different lengths of time. The choice of the number of batches is made on the basis of a trade-off between maximizing fuel burnup and minimizing the number of shutdowns for refueling, which reduces the plant capacity factor. At each refueling, the batch of fuel with the highest burnup is discharged, the batches with lower burnup may be moved to different locations, and a fresh or partially depleted batch is added to replace the discharged batch. The analysis leading to determination of the distribution of the fuel batches within the core to meet the safety, power distribution and burnup, or cycle length constraints for fuel burn cycle is known as *fuel management analysis*. Although fuel management may be planned in advance, it must be updated online to adjust to higher or lower capacity factors than planned (which result in lower or higher reactivity than planned at the planned refueling time) and unforeseen outages (which result in higher reactivity than planned at the planned refueling time).

Typically, a PWR will have three fuel batches, and a BWR will have four fuel batches in the core at any given time and will refuel every 12 to 18 months. A number of different loading patterns have been considered, with the general conclusion that more energy is extracted from the fuel when the power distribution in the core is as flat as possible. In the *in–out loading pattern*, the reactor is divided into concentric annular regions loaded with different fuel batches. The fresh fuel batch is placed at the periphery, the highest burnup batch is placed at the center, and intermediate burnup batches are placed in between to counter the natural tendency of power to peak in the center of the core. At refueling, the central batch is discharged, the other batches are shifted inward, and a fresh batch is loaded on the periphery. The in–out loading pattern has been found to go too far in the sense that the power distribution is depressed in the center and peaked at the periphery. An additional difficulty is the production of a large number of fast neutrons at the periphery that leak from the core and damage the pressure vessel.

In the *scatter loading pattern* the reactor core is divided into many small regions of four to six assemblies from different batches. At refueling, the assemblies within each region with the highest burnup are discharged and replaced by fresh fuel assemblies. This loading pattern has been found to produce a more uniform power distribution and to result in less fast neutron leakage than the in–out pattern.

Since the pressure vessel damage by fast neutrons became recognized as a significant problem, a number of different loading patterns have been developed with the specific objective of minimizing neutron damage to the pressure vessel. These include placement of only partially depleted assemblies at the core periphery, placement of highly depleted assemblies near welds and other critical locations, using burnable poisons in peripheral assemblies, replacing peripheral fuel assemblies with dummy assemblies, and others.

Better utilization of resources argues for the highest possible fuel burnup consistent with materials damage limitations, and a new higher enrichment fuel has been developed that can achieve burnups of up to 50,000 MWd/T in LWRs. The higher fuel burnup produces more actinides and fission products with large thermal neutron cross sections, which compete more effectively with control rods for thermal neutrons and reduces control rod worth, and which produces larger coolant temperature reactivity coefficients. The higher-enrichment higher-burnup fuel also provides the possibility of longer refueling cycles, which improves plant capacity factor and reduces power costs.

6.2
Samarium and Xenon

The short-term time dependence of two fission product progeny, ^{149}Sm and ^{135}Xe, which have very large absorption cross sections, introduces some interesting reactivity transients when the reactor power level is changed.

Samarium Poisoning

Samarium-149 is produced by the beta decay of the fission product ^{149}Nd, as described in Fig. 6.5. It has a thermal neutron absorption cross section of 4×10^4 barns and a large epithermal absorption resonance. The 1.7-h half-life of ^{149}Nd is sufficiently short that ^{149}Pm can be assumed to be formed directly from fission in writing the production–destruction equations for ^{149}Sm:

$$\frac{dP}{dt} = \gamma^{Nd} \Sigma_f \phi - \lambda^P P$$

$$\frac{dS}{dt} = \lambda^P P - \sigma_a^S \phi S$$

(6.8)

where P and S refer to ^{149}Pm and ^{149}Sm, respectively. These equations have the solution, for constant ϕ,

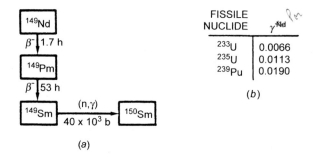

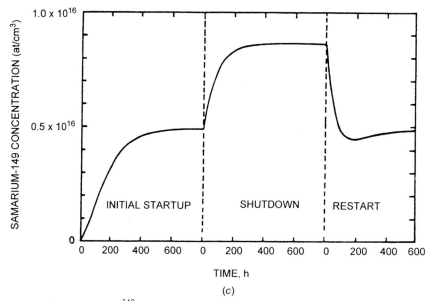

Fig. 6.5 Characteristics of ^{149}Sm under representative LWR conditions: (a) transmutation–decay chain; (b) fission yields; (c) time dependence. (From Ref. 3; used with permission of Taylor & Francis/Hemisphere Publishing.)

$$P(t) = \frac{\gamma^{\text{Nd}} \Sigma_f \phi}{\lambda^P} \left(1 - e^{-\lambda^P t}\right) + P(0)e^{-\lambda^P t}$$

$$S(t) = S(0)e^{-\sigma_a^S \phi t} + \frac{\gamma^{\text{Nd}} \Sigma_f}{\sigma_a^S} \left(1 - e^{-\sigma_a^S \phi t}\right) \tag{6.9}$$

$$- \frac{\gamma^{\text{Nd}} \Sigma_f \phi - \lambda^P P(0)}{\lambda^P - \sigma_a^S \phi} \left(e^{-\sigma_a^S \phi t} - e^{-\lambda^P t}\right)$$

At the beginning of life in a fresh core, $P(0) = S(0) = 0$, and the promethium and samarium concentrations build up to equilibrium values:

$$P_{\text{eq}} = \frac{\gamma^{\text{Nd}} \Sigma_f \phi}{\lambda^{P}}, \qquad S_{\text{eq}} = \frac{\gamma^{\text{Nd}} \Sigma_f}{\sigma_a^{S}} \tag{6.10}$$

The equilibrium value of ^{149}Pm depends on the neutron flux level. However, the equilibrium value of ^{149}Sm is determined by a balance between the fission production rate of ^{149}Pm and the neutron transmutation rate of ^{149}Sm, both of which are proportional to the neutron flux, and consequently, does not depend on the neutron flux level. The time required for the achievement of equilibrium concentrations depends on ϕ, σ_a^S and λ^P. For typical thermal reactor flux levels (e.g., 5×10^{13} n/cm$^2 \cdot$ s), equilibrium levels are achieved in a few hundred hours.

When a reactor is shut down after running sufficiently long to build up equilibrium concentrations, the solutions of Eqs. (6.9) with $P(0) = P_{\text{eq}}$, $S(0) = S_{\text{eq}}$, and $\phi = 0$ are

$$P(t) = P_{\text{eq}} e^{-\lambda^{P} t}$$
$$S(t) = S_{\text{eq}} + P_{\text{eq}}\left(1 - e^{-\lambda^{P} t}\right) \rightarrow S_{\text{eq}} + P_{\text{eq}} \tag{6.11}$$

indicating that the ^{149}Sm concentration will increase to $S_{\text{eq}} + P_{\text{eq}}$ as the ^{149}Pm decays into ^{149}Sm with time constant $1/\lambda^P = 78$ h. If the reactor is restarted, the ^{149}Sm burns out until the ^{149}Pm builds up; then the ^{149}Sm returns to its equilibrium value. This time dependence of the samarium concentration is illustrated in Fig. 6.5.

The perturbation theory estimate for the reactivity worth of ^{149}Sm is

$$\rho_{\text{Sm}}^{(t)} = -\frac{S(t)\sigma_a^{S}}{\Sigma_a} \tag{6.12}$$

which for the equilibrium concentration becomes

$$\rho_{\text{Sm}}^{\text{eq}} = -\frac{\gamma^{\text{Nd}} \Sigma_f \, \sigma_a^{S}}{\sigma_a^{S} \, \Sigma_a} = -\gamma^{\text{Nd}} \frac{\Sigma_f}{\Sigma_a} = -\frac{\gamma^{\text{Nd}}}{\nu} \tag{6.13}$$

where we have used the approximation that $k \approx \nu \Sigma_f / \Sigma_a = 1$. For a ^{235}U-fueled reactor, $\rho_{\text{Sm}}^{\text{eq}} \approx 0.0045$.

Xenon Poisoning

Xenon-135 has a thermal absorption cross section of 2.6×10^6 barns. It is produced directly from fission, with yield γ^{Xe}, and from the decay of ^{135}I, which in turn is produced by the decay of the direct fission product ^{135}Te, with yield γ^{Te}, as indicated in Fig. 6.6. The production–destruction equations may be written, with the assumption that ^{135}I is produced directly from fission with yield γ^{Te},

$$\frac{dI(t)}{dt} = \gamma^{\text{Te}} \Sigma_f \phi - \lambda^{I} I$$
$$\frac{dX(t)}{dt} = \gamma^{\text{Xe}} \Sigma_f \phi + \lambda^{I} I - \left(\lambda^{X} + \sigma_a^{X} \phi\right) X \tag{6.14}$$

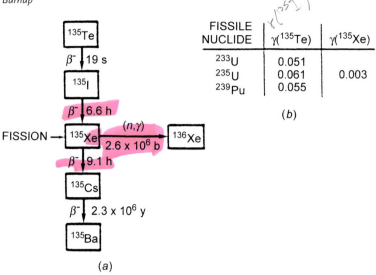

FISSILE NUCLIDE	$\gamma(^{135}\text{Te})$	$\gamma(^{135}\text{Xe})$
^{233}U	0.051	
^{235}U	0.061	0.003
^{239}Pu	0.055	

(b)

(a)

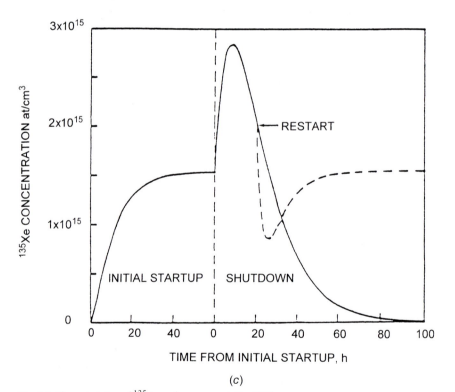

(c)

Fig. 6.6 Characteristics of ^{135}Xe under representative LWR conditions: (*a*) transmutation–decay chain; (*b*) fission yields; (*c*) time dependence. (From Ref. 3; used with permission of Taylor & Francis/Hemisphere Publishing.)

These equations have the solutions

$$I(t) = \frac{\gamma^{\text{Te}} \Sigma_f \phi}{\lambda^I} \left(1 - e^{-\lambda^I t}\right) + I(0)e^{-\lambda^I t}$$

$$X(t) = \frac{(\gamma^{\text{Te}} + \gamma^{\text{Xe}}) \Sigma_f \phi}{\lambda^X + \sigma_a^X \phi} \left[1 - e^{-(\lambda^X + \sigma_a^X \phi)t}\right] \tag{6.15}$$

$$+ \frac{\gamma^{\text{Te}} \Sigma_f \phi - \lambda^I I(0)}{\lambda^X - \lambda^I + \sigma_a^X \phi} \left[e^{-(\lambda^X + \sigma_a^X \phi)t} - e^{-\lambda^I t}\right] + X(0)e^{-(\lambda^X + \sigma_a^X \phi)t}$$

When the reactor is started up from a clean condition in which $X(0) = I(0) = 0$, or the reactor power level is changed, the ^{135}I and ^{135}Xe concentrations approach equilibrium values:

$$I_{\text{eq}} = \frac{\gamma^{\text{Te}} \Sigma_f \phi}{\lambda^I}, \qquad X_{\text{eq}} = \frac{(\gamma^{\text{Te}} + \gamma^{\text{Xe}}) \Sigma_f \phi}{\lambda^X + \sigma_a^X \phi} \tag{6.16}$$

with time constants $1/\lambda^I = 0.1$ h and $1/(\lambda^X + \sigma_a^X \varphi) \approx 30$ h, respectively. The perturbation theory estimate of the reactivity worth of equilibrium xenon is

$$\rho_{\text{Xe}}^{\text{eq}} = -\frac{\sigma_a^X (\gamma^{\text{Te}} + \gamma^{\text{Xe}}) \Sigma_f \phi}{\Sigma_a (\lambda^X + \sigma_a^X \phi)} \simeq -\frac{\gamma^{\text{Te}} + \gamma^{\text{Xe}}}{\nu(1 + \lambda^X/\sigma_a^X \phi)}$$

$$= -\frac{0.026}{1 + (0.756 \times 10^{13})/\phi} \tag{6.17}$$

Peak Xenon

When a reactor is shut down from an equilibrium xenon condition, the iodine and xenon populations satisfy Eqs. (6.15) with $I(0) = I_{\text{eq}}$, $X(0) = X_{\text{eq}}$, and $\phi = 0$:

$$I(t) = I_{\text{eq}} e^{-\lambda^I t}$$

$$X(t) = X_{\text{eq}} e^{-\lambda^X t} + I_{\text{eq}} \frac{\lambda^I}{\lambda^I - \lambda^X} \left(e^{-\lambda^X t} - e^{-\lambda^I t}\right) \tag{6.18}$$

If $\phi > (\gamma^X/\gamma^{\text{Te}})(\lambda^X/\sigma_a^X)$, the xenon will build up after shutdown to a peak value at time

$$t_{\text{Pk}} = \frac{1}{\lambda^I - \lambda^X} \ln \frac{\lambda^I/\lambda^X}{1 + (\lambda^X/\lambda^I)(\lambda^I/\lambda^X - 1)(X_{\text{eq}}/I_{\text{eq}})} \tag{6.19}$$

and then decay to zero unless the reactor is restarted. For ^{235}U- and ^{233}U-fueled reactors $\phi > 4 \times 10^{11}$ and 3×10^{12} n/cm$^2 \cdot$ s, respectively, is sufficient for an increase in the xenon concentration following shutdown. Typical flux values (e.g., 5×10^{13} n/cm$^2 \cdot$ s) in thermal reactors are well above these threshold levels, and for typical flux values, Eq. (6.19) yields a peak xenon time of ≈ 11.6 h. If the reactor is restarted before the xenon has entirely decayed, the xenon concentration will initially decrease because of the burnout of xenon and then gradually build up again because of the decay of a growing iodine concentration, returning to values of I_{eq} and X_{eq} for the new power level. This time-dependence of the xenon concentration is illustrated in Fig. 6.6.

Effect of Power-Level Changes

When the power level changes in a reactor (e.g., in load following) the xenon concentration will change. Consider a reactor operating at equilibrium iodine $I_{eq}(\phi_0)$ and xenon $X_{eq}(\phi_0)$ at flux level ϕ_0. At $t = t_0$ the flux changes from ϕ_0 to ϕ_1. Equations (6.16) can be written

$$I(t) = I_{eq}(\phi_1)\left(1 - \frac{\phi_1 - \phi_0}{\phi_1}e^{-\lambda^I t}\right)$$

$$X(t) = X_{eq}(\phi_1)\left(1 - \frac{\phi_1 - \phi_0}{\phi_1}\left\{\frac{\lambda^X}{\lambda^X + \sigma_a^X \phi_0}e^{-(\lambda^X + \sigma_a^X \phi_1)t}\right.\right.$$

$$\left.\left. + \frac{\gamma^{Te}}{\gamma^{Te} + \gamma^{Xe}}\frac{\lambda^X + \sigma_a^X \phi_1}{\lambda^X - \lambda^I + \sigma_a^X \phi_1}\left[e^{-\lambda^I t} - e^{-(\lambda^X + \sigma_a^X \phi_1)t}\right]\right\}\right) \tag{6.20}$$

The xenon concentration during a transient of this type is shown in Fig. 6.7.

The perturbation theory estimate for the reactivity worth of xenon at any time during the transient discussed above is

$$\rho_{Xe}(t) = -\frac{\sigma_a^X X(t)}{\Sigma_a} \simeq -\frac{\sigma_a^X X(t)}{\nu \Sigma_f} \tag{6.21}$$

Example 6.2: Xenon Reactivity Worth. As an example of xenon buildup, consider a ^{235}U-fueled reactor that has operated at a thermal flux level of $5 \times 10^{13}\,\mathrm{cm^{-2}\,s^{-1}}$ for two months such that equilibrium xenon and iodine have built in to the levels given by Eqs. (6.16). Using $\sigma_a^X = 2.6 \times 10^{-18}\,\mathrm{cm^2}$, $t_{1/2}^I = 6.6\,\mathrm{h}$, $t_{1/2}^X = 9.1\,\mathrm{h}$, $\lambda = \ln 2/t_{1/2}$, $\gamma_{Te} = 0.061$, and $\gamma_{Xe} = 0.003$, the equilibrium values of xenon and iodine are $X^{eq} = 0.0203 \times 10^{18}\Sigma_f\,\mathrm{cm^{-3}}$ and $I^{eq} = 0.1051 \times 10^{18}\Sigma_f\,\mathrm{cm^{-3}}$. The reactivity worth of equilibrium xenon is $\rho_{Xe}^{eq} \approx \sigma_a^X X^{eq}/\Sigma_a \approx 0.022\Delta k/k$, where the approximate criticality condition $\nu\Sigma_f \approx \Sigma_a$ has been used.

If the reactor is shut down for 6 h and then restarted, the xenon reactivity worth that must be compensated is, from Eqs. (6.16) and (6.21), $\rho_{Xe}(t = 6\,\mathrm{h}) \approx \sigma_a^X X(t = 6\,\mathrm{h})/\nu\Sigma_f = (0.634X^{eq} + 0.367I^{eq}) \times \sigma_a^X/\nu\Sigma_f = 0.0171 + 0.04 = 0.0571\Delta k/k$. The largest contribution to the xenon worth at 6 h after shutdown clearly comes from buildup of xenon from the decay of the iodine concentration at shutdown at a faster rate than the resulting xenon decays.

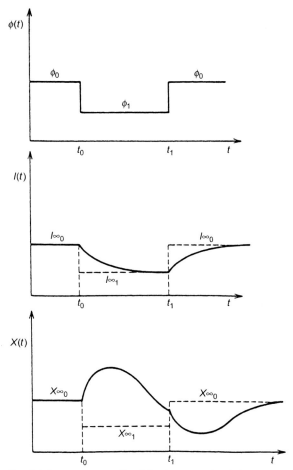

Fig. 6.7 Xenon concentration following power-level changes.
(From Ref. 9; used with permission of Wiley.)

6.3
Fertile-to-Fissile Conversion and Breeding

Availability of Neutrons

The transmutation–decay processes depicted in Fig. 6.1 hold out the potential for increasing the recoverable energy content from the world's uranium and thorium resources by almost two orders of magnitude by converting the fertile isotopes ^{238}U and ^{232}Th, which only fission at very high neutron energies, into fissile isotopes, ^{239}Pu and ^{241}Pu in the case of ^{238}U, and ^{233}U in the case of ^{232}Th, which have large fission cross sections for thermal neutrons and substantial fission cross sections for fast neutrons. The rate of transmutation of fertile-to-fissile isotopes depends on the number of neutrons in excess of those needed to maintain the chain fission reaction that are available. In the absence of neutron absorption by anything other

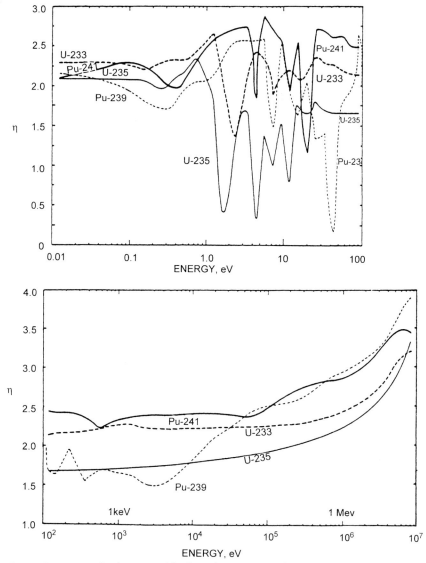

Fig. 6.8 Parameter η for the principal fissile nuclei. (From Ref. 17; used with permission of Electric Power Research Institute.)

than fuel and in the absence of leakage, the number of excess neutrons is $\eta - 1$. The quantity η is plotted in Fig. 6.8 for the principal fissile isotopes.

The fertile-to-fissile conversion characteristics depend on the fuel cycle and the neutron energy spectrum. For a thermal neutron spectrum ($E < 1$ eV), ^{233}U has the largest value of η of the fissile nuclei. Thus the best possibility for fertile-to-fissile conversion in a thermal spectrum is with the ^{232}Th–^{233}U fuel cycle. For a

Table 6.4 Conversion/Breeding Ratios in Different Reactor Systems

Reactor System	Initial Fuel	Conversion Cycle	Conversion Ratio
BWR	2–4 wt% ^{235}U	^{238}U–^{239}Pu	0.6
PWR	2–4 wt% ^{235}U	^{238}U–^{239}Pu	0.6
PHWR	Natural U	^{238}U–^{239}Pu	0.8
HTGR	≈5 wt% ^{235}U	^{232}Th–^{233}U	0.8
LMFBR	10–20 wt% Pu	^{238}U–^{239}Pu	1.0–1.6

Source: Data from Ref. 3; used with permission of Taylor & Francis/Hemisphere Publishing.

fast neutron spectrum ($E > 5 \times 10^4$ eV), ^{239}Pu and ^{241}Pu have the largest values of η of the fissile nuclei. The LMFBR, based on the ^{238}U–^{239}Pu fuel cycle, is intended to take advantage of the increase of η^{49} at high neutron energy.

Conversion and Breeding Ratios

The instantaneous *conversion ratio* is defined as the ratio of the rate of creation of new fissile isotopes to the rate of destruction of fissile isotopes. When this ratio is greater than unity, it is conventional to speak of a *breeding ratio,* because the reactor would then be producing more fissile material than it was consuming. Average conversion or breeding ratios calculated for reference reactor designs of various types are shown in Table 6.4.

The values of the conversion ratios for the PWR and BWR are the same because of design similarities. The HTGR conversion ratio is somewhat higher because of the higher value of η for ^{233}U than for ^{235}U. The improved conversion ratio for the CANDU–PHWR is due to the better neutron economy provided by online refueling and consequent reduced requirements for control poisons to compensate excess reactivity.

The breeding ratio in an LMFBR can vary over a rather wide range, depending on the neutron energy spectrum. Achieving a large value of η and hence a large breeding ratio favors a hard neutron spectrum. However, a softer spectrum is favored for safety reasons—the lower-energy neutrons which are subject to resonance absorption become more likely to be radiatively captured than to cause fission as the neutron energy is reduced, as discussed in Chapter 5.

6.4
Simple Model of Fuel Depletion

The concepts involved in fuel depletion and the compensating control adjustment can be illustrated by a simple model in which the criticality requirement is written as

$$k = \eta f = \frac{\eta \Sigma_a^F(t)}{\Sigma_a^F(t) + \Sigma_a^M + \Sigma_a^{\text{fp}}(t) + \Sigma_c(t)} = 1 \tag{6.22}$$

where Σ_a^F is the fuel macroscopic absorption cross section, Σ_a^M the moderator macroscopic absorption cross section, and Σ_a^C the combined (soluble and burnable poisons plus control rod) control absorption cross section. Assuming that the reactor operates at constant power $\nu\Sigma_f^F(t)\phi(t) = \nu\Sigma_f^F(0)\phi(0)$ and that $\eta = \nu\Sigma_f^F/\Sigma_a^F$ is constant in time, the fuel macroscopic absorption cross section at any time is

$$\Sigma_a^F(t) = N_F(t)\sigma_a^F = \sigma_a^F \left[N_F(0) - \varepsilon\sigma_a^F \int_0^t N_F(t')\phi(t')dt' \right]$$

$$= N_F(0)\sigma_a^F \left[1 - \varepsilon\phi(0)\sigma_a^F t \right] \tag{6.23}$$

The neutron flux is related to the beginning-of-cycle neutron flux by

$$\phi(t) = \frac{\phi(0)}{1 - \varepsilon\sigma_a^F \phi(0)t} \tag{6.24}$$

where $\varepsilon < 1$ is a factor that accounts for the production of new fissionable nuclei via transmutation–decay.

The fission product cross section is the sum of the equilibrium xenon and samarium cross sections constructed using Eqs. (6.16) and (6.10), respectively, and an effective cross section for the other fission products,

$$\Sigma_{\text{fp}'} = \sigma_{\text{fp}'}\gamma_{\text{fp}'}\Sigma_f(t)\phi(t)t = \sigma_{\text{fp}'}\gamma_{\text{fp}'}\Sigma_f(0)\phi(0)t \tag{6.25}$$

which accumulate in time from fission with yield $\gamma_{\text{fp}'}$. The quantity $\gamma_{\text{fp}'}\sigma_{\text{fp}'}$ is about 40 to 50 barns per fission. Using these results, Eq. (6.22) can be solved for the value of the control cross section that is necessary to maintain criticality:

$$\Sigma_c(t) = (\eta - 1)\Sigma_a^F(0)\left[1 - \sigma_a^F \varepsilon\phi(0)t\right] - \Sigma_a^M - \frac{(\gamma^{\text{Te}} + \gamma^{\text{Xe}})\Sigma_f(0)\phi(0)}{\lambda^X/\sigma_a^X + \phi(t)}$$

$$- \gamma^{\text{Nd}}\Sigma_f(0)\left[1 - \varepsilon\sigma_a^F \phi(0)t\right] - \sigma_{\text{fp}'}\gamma_{\text{fp}'}\Sigma_f(0)\phi(0)t \tag{6.26}$$

The soluble poison will be removed by the end of cycle, and the burnable poisons should be fully depleted by that time. Thus the lifetime, or cycle time, is the time at which the reactor can no longer be maintained critical with the control rods withdrawn as fully as allowed by safety considerations. This minimum control cross section is small, and we set it to zero. The end-of-cycle time can be determined from Eq. (6.26) by setting $\Sigma_a^C = 0$ and solving for t_{EOC}:

$$t_{\text{EOC}} = \begin{cases} \dfrac{\eta\rho_{\text{ex}}(1 + \alpha) - (\gamma^{\text{Te}} + \gamma^{\text{Xe}})\phi(0)\sigma_a^X/\lambda^X - \gamma^{\text{Nd}}}{[(\eta - 1)(1 + \alpha)\sigma_a^F - \gamma^{\text{Nd}}\sigma_a^F + \gamma_{\text{fp}'}]\phi(0)}, & \phi(t) \ll \dfrac{\lambda^X}{\sigma_a^X} \\[2em] \dfrac{\eta\rho_{\text{ex}}(1 + \alpha) - (\gamma^{\text{Te}} + \gamma^{\text{Xe}} + \gamma^{\text{Nd}})}{[(\eta - 1)(1 + \alpha)\sigma_a^F - (\gamma^{\text{Te}} + \gamma^{\text{Xe}} + \gamma^{\text{Nd}})\sigma_a^F + \gamma_{\text{fp}'}\sigma_{\text{fp}'}]\phi(0)}, \\[1em] \quad \phi(t) \gg \dfrac{\lambda^X}{\sigma_a^X} \end{cases} \tag{6.27}$$

where α is the capture-to-fission ratio for the fuel, and

$$\rho_{\text{ex}} \equiv \frac{k_\infty(0) - 1}{k_\infty(0)} \tag{6.28}$$

is the excess reactivity at beginning-of-cycle without xenon, samarium, fission products, or control cross section. The initial control cross section (including soluble and burnable poisons) must be able to produce a negative reactivity greater than ρ_{ex}. It is clear from Eq. (6.27) that the cycle lifetime is inversely proportional to the power, or flux, level.

6.5
Fuel Reprocessing and Recycling

A substantial amount of plutonium is produced by neutron transmutation of ^{238}U in LWRs. About 220 kg of fissionable plutonium (mainly ^{239}Pu and ^{241}Pu) is present in the spent fuel discharged from an LWR at a burnup of 45 MWd/T. The spent fuel can be reprocessed to recover the plutonium (and remaining enriched uranium) for recycling as new fuel.

Composition of Recycled LWR Fuel

The potential energy content of the fissile and fertile isotopes remaining in spent reactor fuel (Table 6.2) constitutes a substantial fraction of the potential energy content of the initial fuel loading, providing an incentive to recover the uranium and plutonium isotopes for reuse as reactor fuel. The recycled plutonium concentrations calculated for successive core reloads of a PWR are shown in Table 6.5. The initial core loading and the first reload were slightly enriched UO_2. The plutonium

Table 6.5 Plutonium Concentrations in a PWR Recycling Only Self-Generated Plutonium (wt%)

Loading: Recycle:	1	2	3 1	4 2	5 3	6 4	7 5
^{235}U in UO_2	2.14	3.0	3.0	3.0	3.0	3.0	3.0
Pu in MOX	—	—	4.72	5.83	6.89	7.51	8.05
MOX of fuel	—	—	18.4	23.4	26.5	27.8	28.8
^{235}U discharged	0.83	—	—	—	—	—	—
Discharged Pu							
^{239}Pu	56.8	56.8	49.7	44.6	42.1	40.9	40.0
^{240}Pu	23.8	23.8	27.0	38.7	29.4	29.6	29.8
^{241}Pu	14.3	14.3	16.2	17.2	17.4	17.4	17.3
^{242}Pu	5.1	5.1	7.1	9.5	11.1	12.1	12.9

Source: Data from Ref. 3; used with permission of Taylor & Francis/Hemisphere Publishing.

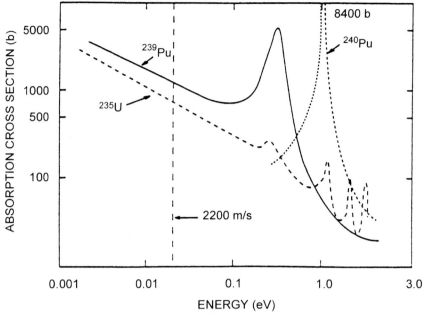

Fig. 6.9 Thermal absorption cross section for ^{239}Pu. (From Ref. 4; used with permission of American Nuclear Society.)

discharged from the first cycle was recycled in the third cycle, that in the second cycle in the fourth cycle, and so on, in separate mixed oxide (MOX) UPuO$_2$ pins. The proportion of MOX increases from about 18% in the second reload to just under 30% in the sixth and subsequent reloads, for which reloads the plutonium recovered from spent MOX and UO$_2$ fuel is about the same as was loaded into this fuel at beginning-of-cycle (i.e., the plutonium concentration reaches equilibrium). The percentage of plutonium in MOX increases from less than 5% on the initial recycle load to about 8% in equilibrium, in order to offset the reactivity penalty.

Physics Differences of MOX Cores

The use of MOX fuels in PWRs changes the physics characteristics in several ways. The variation with energy of the cross sections for the plutonium isotopes is more complex than for the uranium isotopes, as shown in Fig. 6.9. The absorption cross sections for the plutonium isotopes are about twice those of the uranium isotopes in a thermal spectrum and are characterized by large absorption resonances in the epithermal (0.3 to 1.5 eV) range and by overlapping resonances. Representative thermal neutron spectra in UO$_2$ and MOX fuel cells are compared in Fig. 6.10.

Thermal parameters for ^{235}U and ^{239}Pu, averaged over a representative LWR thermal neutron energy distribution, are given in Table 6.6. Because of the larger thermal absorption cross section for ^{239}Pu, the reactivity worth of control rods, burnable poisons, and soluble poisons (PWRs) will be less with MOX fuel than with UO$_2$, unless the MOX rods can be placed well away from control rods and

Table 6.6 Thermal Parameters for ^{235}U and ^{239}Pu in a LWR

Parameter	^{235}U	^{239}Pu
Fission cross section σ_f (barns)	365	610
Absorption cross section σ_a (barns)	430	915
Nu-fission to absorption η	2.07	1.90
Delayed neutron fraction β	0.0065	0.0021
Generation time Λ (s)	4.7×10^{-5}	2.7×10^{-5}

Source: Data from Ref. 4; used with permission of American Nuclear Society.

burnable poisons. The higher ^{239}Pu fission cross section will lead to greater power peaking with MOX than with UO_2, unless the MOX rods are placed well away from water gaps.

There are reactivity differences between MOX and UO_2. The buildup of ^{240}Pu and ^{242}Pu with the recycling MOX fuel accumulates parasitic absorbers that results in a reactivity penalty, as discussed in Section 6.1. The average thermal value of η

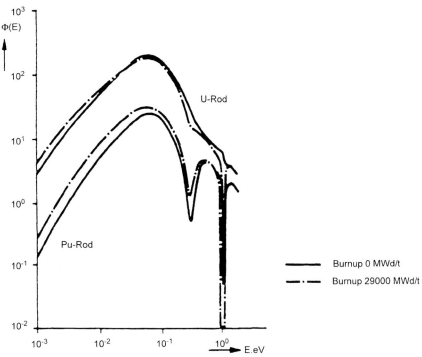

Fig. 6.10 Thermal neutron spectra in UO_2 and MOX PWR fuel cells. (From Ref. 1; used with permission of Nuclear Energy Agency, Paris.)

is less for ^{239}Pu than for ^{235}U, which requires a larger fissile loading to achieve the same initial excess reactivity with MOX as with UO$_2$. Furthermore, the temperature defect is greater for MOX because of the large low-energy resonances in ^{239}Pu and ^{240}Pu shown in Fig. 6.9. However, the reactivity decrease with burnup is less for MOX than for UO$_2$, because of the lower η for ^{239}Pu than for ^{235}U, and because of the transmutation of ^{240}Pu into fissionable ^{241}Pu, so that less excess reactivity is needed.

The delayed neutron fractions for ^{239}Pu, ^{241}Pu, and ^{235}U are in the ratio 0.0020/0.0054/0.0064, which means that the reactivity insertion required to reach prompt critical runaway conditions is less for MOX than for UO$_2$ by a factor that depends on the ^{239}Pu/^{241}Pu/^{235}U ratio. As the ^{241}Pu builds up with repeated recycle, the difference between MOX and UO$_2$ decreases. The neutron generation time is also shorter for MOX than for UO$_2$, so that any prompt supercritical excursion would have a shorter period. The fission spectrum neutrons are more energetic for ^{239}Pu than for ^{235}U. On the other hand, because of the large epithermal absorption resonances in the plutonium isotopes, the moderator and fuel Doppler temperature coefficients of reactivity tend to be more negative for MOX cores than for UO$_2$ cores. Accumulation of actinides, which are strong emitters of energetic alpha particles, leads to higher radioactive decay heat removal requirements with MOX. These considerations would tend to limit the MOX fraction in a reload core.

The yield of ^{135}Xe is about the same for the fission of plutonium as for the fission of uranium. Due to the higher thermal absorption cross section of the plutonium isotopes, the excess reactivity needed to start up at peak xenon conditions and the propensity for spatial flux oscillations driven by xenon oscillations (Chapter 16) are less in a MOX than a UO$_2$ core.

For plutonium recycle in other reactor types, similar types of physics considerations would enter. However, the different relative values of η for ^{235}U and ^{239}Pu in different spectra (e.g., the epithermal spectrum of a HTGR and the fast spectrum of a LMFBR) would lead to different conclusions about reactivity penalties. In fact, LMFBRs have been designed from the outset with the concept of switching from ^{235}U to ^{239}Pu as the latter was bred.

Physics Considerations with Uranium Recycle

Although it is relatively straightforward to separate uranium from other chemically distinct isotopes, it is impractical to separate the various uranium isotopes from each other in the reprocessing step. So recycling uranium means recycling all of the uranium isotopes, some of which are just parasitic absorbers and another of which leads through subsequent decay to the emission of an energetic gamma.

Two isotopes present in relatively small concentrations in fresh fuel (^{234}U and ^{236}U) necessitate adding ^{235}U to enrich reprocessed uranium to a higher enrichment than is required with fresh uranium fuel. Uranium-234 has a large absorption resonance integral and, while only a tiny fraction in natural uranium, will tend to be enriched along with ^{235}U. Uranium-236 is produced by neutron capture in ^{235}U and by electron capture in ^{236}Np, as shown in Fig. 6.4, and is a parasitic

Table 6.7 Representative Fueling Characteristics of 1000-MWt Reactors

Characteristic	Reactor Type		
	LWR	LWR	LMFBR
Fuel cycle	^{232}Th–^{233}U	^{238}U–^{239}Pu	^{238}U–^{239}Pu
Conversion ratio	0.78	0.71	1.32
Initial core load (kg)	1,580	2,150	3,160
Burnup (MWd/T)	35,000	33,000	100,000
Annual reload (kg)	720	1,000	1,480
Annual discharge (kg)	435	650	1,690
Annual makeup (kg)	285	350	(−210)

Source: Data from Ref. 8; used with permission of
International Atomic Energy Agency.

neutron absorber with a significant capture resonance integral. Reprocessed uranium is made difficult to handle by the decay product ^{208}Tl, which emits a 2.6-MeV gamma with $t_{1/2} = 3.1$ min. This radioisotope is produced by a series of alpha decays of ^{232}U, which is produced by the chain shown in Fig. 6.4.

Physics Considerations with Plutonium Recycle

The same type of difficulties exists for plutonium reprocessing as discussed for uranium—all of the plutonium isotopes must be recycled. Plutonium-236 decays into ^{232}U, which leads to the emission of a 2.6-MeV gamma, as described above. Plutonium-238 is produced through neutron transmutation of ^{237}Np; it alpha-decays with $t_{1/2} = 88$ years and constitutes a large shutdown heat source if present in sufficient quantity. Plutonium-240 has an enormous capture resonance integral. Both ^{238}Pu and ^{240}Pu contribute a large spontaneous fission neutron source. Plutonium-241, while having a large fission cross section, also decays into ^{241}Am, which has a large thermal capture cross section and a large capture resonance integral. Americium-241 also decays into daughter products which are energetic gamma emitters. Stored plutonium loses its potency as a fuel over time because of the decay of ^{241}Pu into ^{241}Am. Plutonium from spent LWR fuel at a typical burnup of about 35,000 MWd/T must be utilized within 3 years after discharge or it will be necessary to reprocess it again to remove the ^{241}Am and daughter products.

Reactor Fueling Characteristics

Nuclear fuel cycles with plutonium recycle have been studied extensively (e.g., Ref. 1). Representative equilibrium fueling characteristics for LWRs operating on the ^{238}U–^{239}Pu and ^{232}Th–^{233}U fuel cycles and for a LMFBR operating on the ^{238}U–^{239}Pu fuel cycle are shown in Table 6.7. Fuel is partially discharged and replenished each year (*annual discharge* and *annual reload*), requiring a net amount of new fuel (*annual makeup*) from outside sources. In the absence of reprocessing and recycling, the annual reload would have to be supplied from outside sources. The LMFBR produces more fuel than it uses and could provide the extra fuel needed by the LWRs from the transmutation of ^{238}U if LMFBRs and LWRs were deployed in the ratio of about 7:5.

6.6
Radioactive Waste

Radioactivity

The actinides produced in the transmutation–decay of the fuel isotopes and the fission products are the major contributors to the radioactive waste produced in nuclear reactors, although activated structure and other materials are also present. The activity per ton of fuel for representative LWR and LMFBR discharges are given in Table 6.8. The fission products account for almost the entire radioactivity of spent fuel at reactor shutdown, but because of their short half-lives, this radioactivity level decays relatively quickly. In fact, the radioactivity of the spent fuel decreases substantially within the first 6 months after removal from the reactor, as shown in Table 6.8. The more troublesome fission products from the waste management point of view are those with long half-lives like ^{99}Tc ($t_{1/2} = 2.1 \times 10^5$ years) and ^{129}I ($t_{1/2} = 1.59 \times 10^7$ years) and those that are gamma emitters, such as ^{90}Sr and ^{137}Cs, which produce substantial decay heating. The actinides constitute a relatively small part of the total radioactivity at reactor shutdown but become relatively more important with time because of the longer half-lives of ^{239}Pu and ^{240}Pu and dominate the radioactivity of spent fuel after about 1000 years.

Hazard Potential

A simple, but useful, measure of the hazard potential of radioactive material is the *hazard index*, defined as the quantity of water required to dilute the material to the maximum permissible concentration for human consumption. The hazard index for spent LWR fuel is plotted against time after shutdown in Fig. 6.11. Fission products dominate the hazard index up to about 1000 years after shutdown, beyond which time the transuranics (actinides) become dominant. Including the plutonium in the recycled uranium fuel increases the hazard potential because of the continued buildup of ^{239}Pu and ^{240}Pu. Beyond 1000 to 10,000 years after shutdown, depending on burnup, the hazard potential of spent reactor fuel is less than the hazard potential of uranium ore as it is mined from the earth.

Risk Factor

In an effort to relate the radioactivity of a given radioisotope to a health hazard, the number of curies of radiation from a given radioisotope that would cause cancer (on the average) if swallowed by a person has been estimated and is shown as the cancer dose per curie (CD/Ci) in Table 6.9. The CD/Ci is not an absolute measure of

Table 6.8 Radioactivity of Representative LWR and LMFBR Spent Fuel at Discharge and at 180 Days (LWR) or 30 Days (LMFBR) After Discharge*

| Nuclide | Half-Life $t_{1/2}$ | Radiations† | Activity (Ci/tonne Heavy Metal) | | | |
| | | | LWR Fuel | | LMFBR Fuel | |
			Discharge	180 d	Discharge	30 d
^{3}H	12.3 y	β	5.744×10^2	5.587×10^2	1.648×10^3	1.640×10^3
^{85}Kr	10.73 y	β, γ	1.108×10^4	1.074×10^4	1.473×10^4	1.466×10^4
^{89}Sr	50.5 d	β, γ	1.058×10^6	9.603×10^4	1.333×10^6	8.939×10^5
^{90}Sr	29.0 y	β, γ	8.425×10^4	8.323×10^4	9.591×10^4	9.572×10^4
^{90}Y	64.0 h	β, γ	8.850×10^4	8.325×10^4	1.214×10^5	9.572×10^4
^{91}Y	59.0 d	β, γ	1.263×10^6	1.525×10^5	1.794×10^6	1.269×10^6
^{95}Zr	64.0 d	β, γ	1.637×10^6	2.437×10^5	3.215×10^6	2.340×10^6
^{95}Nb	3.50 d	β, γ	1.557×10^6	4.689×10^5	3.149×10^6	2.954×10^6
^{99}Mo	66.0 h	β, γ	1.875×10^6	3.780×10^{-14}	4.040×10^6	2.108×10^3
^{99m}Tc	6.0 h	γ	1.618×10^6	3.589×10^{-14}	3.487×10^6	2.002×10^3
^{99}Tc	2.1×10^5 y	β, γ	1.435×10^1	1.442×10^1	3.278×10^1	3.293×10^1
^{103}Ru	40.0 d	β, γ	1.560×10^6	6.680×10^4	4.617×10^6	2.730×10^6
^{106}Ru	369.0 d	β, γ	4.935×10^5	3.519×10^5	2.248×10^6	2.125×10^6
^{103m}Rh	56.0 min	γ	1.561×10^6	6.686×10^4	4.619×10^6	2.733×10^6
^{111}Ag	7.47 d	β, γ	5.375×10^4	3.005×10^{-3}	2.294×10^5	1.422×10^4
^{115}Cd	44.6 d	β, γ	1.483×10^3	9.042×10^1	7.041×10^3	4.418×10^3

Table 6.8 (Continued)

Nuclide	Half-Life $t_{1/2}$	Radiations[†]	Activity (Ci/tonne Heavy Metal)			
			LWR Fuel		LMFBR Fuel	
			Discharge	180 d	Discharge	30 d
125Sn	9.65 d	β, γ	1.081×10^4	2.624×10^{-2}	3.404×10^4	3.946×10^3
124Sb	60.2 d	β, γ	4.147×10^2	5.219×10^1	2.329×10^3	1.649×10^3
125Sb	2.73 y	β, γ	9.525×10^3	8.498×10^3	5.251×10^4	5.171×10^4
125mTe	58.0 d	γ	1.976×10^3	2.031×10^3	1.121×10^4	1.144×10^4
127mTe	109.0 d	β, γ	1.384×10^4	4.595×10^3	4.969×10^4	4.265×10^4
127Te	9.4 h	β, γ	9.920×10^4	4.500×10^3	3.247×10^5	4.308×10^4
129mTe	33.4 d	β, γ	8.508×10^4	2.041×10^3	2.316×10^5	1.249×10^5
129Te	70.0 min	β, γ	3.211×10^5	1.296×10^3	8.454×10^5	7.932×10^4
132Te	78.0 h	β, γ	1.486×10^6	3.159×10^{-11}	3.473×10^6	5.783×10^3
129I	1.59×10^7 y	β, γ	3.219×10^{-2}	3.268×10^{-2}	1.033×10^{-1}	1.040×10^{-1}
131I	8.04 d	β, γ	1.028×10^6	1.933×10^{-1}	2.602×10^6	2.020×10^5
132I	2.285 h	β, γ	1.511×10^6	3.254×10^{-11}	3.546×10^6	5.956×10^3
133Xe	5.29 d	β, γ	2.098×10^6	1.612×10^{-4}	4.414×10^6	1.076×10^5
134Cs	2.06 y	β, γ	2.718×10^5	2.303×10^5	8.283×10^4	8.058×10^4
136Cs	13.0 d	β, γ	6.962×10^4	4.719×10^0	2.577×10^5	5.204×10^4
137Cs	30.1 y	β, γ	1.115×10^5	1.102×10^5	2.522×10^5	2.518×10^5
140Ba	12.79 d	β, γ	1.953×10^6	1.133×10^2	3.636×10^6	7.153×10^5
140La	40.23 h	β, γ	2.019×10^6	1.303×10^2	3.698×10^4	8.238×10^5

Table 6.8 (Continued)

Nuclide	Half-Life $t_{1/2}$	Radiations[*]	Activity (Ci/tonne Heavy Metal)			
			LWR Fuel		LMFBR Fuel	
			Discharge	180 d	Discharge	30 d
141Ce	32.53 d	β, γ	1.784×10^6	3.876×10^4	3.730×10^6	1.979×10^6
144Ce	284.0 d	β, γ	1.229×10^6	7.925×10^5	2.148×10^6	1.996×10^6
143Pr	13.58 d	β	1.657×10^6	1.887×10^2	3.044×10^6	7.349×10^5
147Nd	10.99 d	β, γ	7.902×10^5	9.278×10^0	1.513×10^6	2.283×10^5
147Pm	2.62 y	β, γ	1.031×10^5	9.859×10^4	6.344×10^5	6.353×10^5
149Pm	53.1 h	β, γ	3.919×10^5	1.326×10^{-19}	9.842×10^5	8.451×10^1
151Sm	93.0 y	β^+, β^-, γ	8.658×10^2	8.696×10^2	9.693×10^3	9.703×10^3
152Eu	13.4 y	β^+, β^-, γ	7.838×10^0	7.635×10^0	4.759×10^1	4.738×10^1
155Eu	4.8 y	β, γ	2.540×10^3	2.365×10^3	4.305×10^4	4.255×10^4
160Tb	72.3 d	β, γ	1.418×10^3	2.525×10^2	4.880×10^3	3.661×10^3
239Np	2.35 d	β, γ	2.435×10^7	2.050×10^1	5.990×10^7	8.727×10^3
238Pu	87.8 y	α, γ	2.899×10^3	3.021×10^3	2.770×10^4	2.820×10^4
239Pu	2.44×10^4 y	α, γ, SF	3.250×10^2	3.314×10^2	6.247×10^3	6.263×10^3
240Pu	6.54×10^3 y	α, γ, SF	4.842×10^2	4.843×10^2	8.323×10^3	8.323×10^3
241Pu	15.0 y	α, β, γ	1.098×10^5	1.072×10^5	7.280×10^5	7.252×10^5
241Am	433.0 y	α, γ, SF	8.023×10^1	1.657×10^2	9.091×10^3	9.186×10^3
242Cm	163.0 d	α, γ, SF	3.666×10^4	1.717×10^4	8.467×10^5	7.489×10^5
244Cm	17.9 d	α, γ	2.772×10^3	2.720×10^3	8.032×10^3	8.007×10^3

* Calculated with ORIGEN (Ref. 7).

† α, alpha particle; β, electron; γ, gamma ray; SF, spontaneous fission.

Table 6.9 Cancer Dose per Curie for Radioisotopes Present in Spent Fuel[*]

Isotope	Toxicity Factor (CD/Ci)	Half-Life (years)	Toxicity Factor (CD/g)
Actinides and Their Daughters			
^{210}Pb	455.0	22.3	3.48×10^4
^{223}Ra	15.6	0.03	7.99×10^5
^{226}Ra	36.3	1.60×10^3	3.59×10^1
^{227}Ac	1185.0	21.8	8.58×10^4
^{229}Th	127.3	7.3×10^3	2.72×10^1
^{230}Th	19.1	7.54×10^4	3.94×10^{-1}
^{231}Pa	372.0	3.28×10^4	1.76×10^{-1}
^{234}U	7.59	2.46×10^5	4.71×10^{-2}
^{235}U	7.23	7.04×10^8	1.56×10^{-5}
^{236}U	7.50	2.34×10^7	4.85×10^{-4}
^{238}U	6.97	4.47×10^9	2.34×10^{-6}
^{237}Np	197.2	2.14×10^6	1.39×10^{-1}
^{238}Pu	246.1	87.7	4.22×10^3
^{239}Pu	267.5	2.41×10^4	1.66×10^1
^{240}Pu	267.5	6.56×10^3	6.08×10^1
^{242}Pu	267.5	3.75×10^5	1.65×10^0
^{241}Am	272.9	433	9.36×10^2
^{242m}Am	267.5	141	2.80×10^4
^{243}Am	272.9	7.37×10^3	5.45×10^1
^{242}Cm	6.90	0.45	2.29×10^4
^{243}Cm	196.9	29.1	9.96×10^3
^{244}Cm	163.0	18.1	1.32×10^4
^{245}Cm	284.0	8.5×10^3	4.88×10^1
^{246}Cm	284.0	4.8×10^3	8.67×10^1
Short-Lived Fission Products			
^{90}Sr	16.7	29.1	2.28×10^3
^{90}Y	0.60	7.3×10^{-3}	3.26×10^5
^{137}Cs	5.77	30.2	4.99×10^2
Long-Lived Fission Products			
^{99}Tc	0.17	2.13×10^5	2.28×10^{-3}
^{129}I	64.8	1.57×10^7	1.15×10^{-2}
^{93}Zr	0.095	1.5×10^6	2.44×10^{-4}
^{135}Cs	0.84	2.3×10^6	9.68×10^{-4}
^{14}C	0.20	5.73×10^3	8.92×10^{-1}
^{59}Ni	0.08	7.6×10^4	6.38×10^{-3}
^{63}Ni	0.03	100	1.70×10^0
^{126}Sn	1.70	1.0×10^5	4.83×10^{-2}

Source: Data from Ref. 1: used with permission of Nuclear Energy Agency, Paris.

[*] The toxicity factors are constructed using the methodology described by Bernard L. Cohn, "Effects of the ICRP Publication 30 and the 1980 BEIR Report of Hazard

Table 6.9 (Continued)

Assessments of High Level Waste," *Health Phys. 42* (2)
133–143 (1982) with the following data: ICRP Publication 30,
Part 4, 88, 19, and *BEIR III*, 80, 19. The factors stand for the
fatal cancer doses per gram of isotope injected orally. They
denote the hazard of the material rather than the risk because
they do not include any account of pathway attenuation
processes, but simply assume oral ingestion.

the biological hazard of a given amount of radioactive material, because it does not
contain any measure of the probability for a sequence of events that would result in
individual members of a population actually swallowing exactly the amount of radioisotopes that would produce a cancer, and no more. However, the CD/Ci can be
used to construct a *relative* measure of the biological hazard potential. The CD/ton
(of heavy metal in the fuel) due to the radioisotopes in discharged fuel given in
Table 6.8 can be multiplied by the number of tons of heavy metal spent fuel discharged from a reactor to construct a total cancer dose (TCD). A similar total cancer
dose of natural uranium (TCDNU) as it is mined from the earth can be constructed
by multiplying the Ci/ton for the radioisotopes in natural uranium by the mass of
natural uranium that was required to produce the discharged fuel for which the

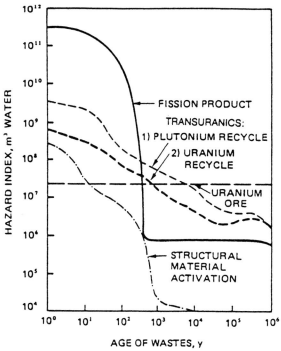

Fig. 6.11 Hazard index for spent LWR fuel as a function of time
since reactor shutdown. (From Ref. 18; used with permission of
Woods Hole Oceanographic Institute.)

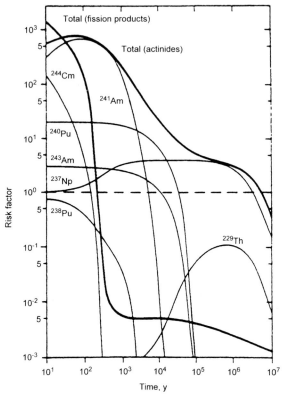

Fig. 6.12 Risk factor for LWR spent fuel without recycle. (From Ref. 5; used with permission of Elsevier Science Publishers.)

TCD is calculated. (Typically, about 5 tons of natural uranium is needed to produce the fuel for a PWR.) The *risk factor* is then defined as $RF \equiv TCD/TCDNU$, which may be interpreted as the ratio of the number of cancers that would be caused by individual members of a population swallowing all of the discharged fuel in portions just sufficient to produce a cancer (on average) to the number of cancers that would be caused by individual members of a population swallowing 5 tons of natural uranium in portions just sufficient to produce a cancer (on average). The advantage of the risk factor is that the highly uncertain probability of ingestion of radioisotopes is normalized out by being treated in the same (highly questionable) way in the numerator (TCD) and denominator (TCDNU), so that RF is a measure of the relative cancer potential of spent fuel and of the natural uranium from which it was produced.

The risk factor is plotted for a typical spent fuel loading from a LWR in Fig. 6.12. The short-lived fission products are dominant in the decades following discharge, but the fission product activity becomes negligible relative to the actinide activity after about 200 to 300 years. The potential α-toxicity of the actinide concentration is dominated by ^{241}Am over the first 5000 years, then by ^{240}Pu up to about 100,000

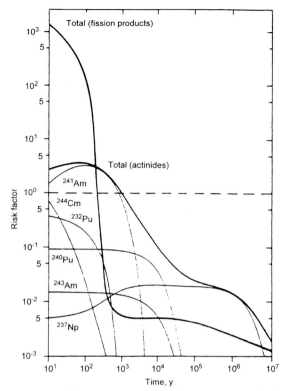

Fig. 6.13 Risk factor for LWR spent fuel with 99.5% recycle of Pu, Am, and Np. (From Ref. 5; used with permission of Elsevier Science Publishers.)

years, and thereafter by ^{237}Np. Note that when the risk factor becomes less than unity, the cancer potential of the spent fuel is less than the cancer potential of the natural uranium ore from which it was originally made.

The long-term potential α-toxicity of spent fuel can be reduced dramatically by recycling the fuel. Figure 6.13 illustrates risk factor for the same LWR fuel as in Fig. 6.12, but now with the Pu, Am, and Np recycled to 99.5% annihilation. After about 200–300 years the potential α-toxicity of the spent fuel is less than that of the natural uranium from which it was originally produced. As discussed in Section 6.8, repeatedly recycling the spent fuel to 99.5% annihilation may be feasible, from neutron balance considerations, in a fast spectrum reactor, but does not appear to be feasible in a thermal reactor.

6.7
Burning Surplus Weapons-Grade Uranium and Plutonium

Composition of Weapons-Grade Uranium and Plutonium

With the reduction in nuclear weapons worldwide, surplus highly enriched, *weapons-grade* uranium and plutonium become available for use as fuel in nuclear

Table 6.10 Composition of Weapons- and Reactor-Grade Uranium and Plutonium (wt%)

	Weapons-Grade Pu	Reactor-Grade Pu		Weapons-Grade U (HEU)	Reactor-Grade U (LEU)	Natural U
^{238}Pu	0.01	1.30	^{234}U	0.12	0.025	0.0057
^{239}Pu	93.80	60.30	^{235}U	94.00	3.500	0.7193
^{240}Pu	5.80	24.30	^{238}U	5.88	96.475	99.2750
^{241}Pu	0.13	5.60				
^{242}Pu	0.02	5.00				
^{241}Am	0.22	3.50				

Source: Data from Ref. 2; used with permission of National Academy Press.

reactors. The composition of typical weapons-grade uranium and plutonium is compared with the composition of reactor-grade uranium and plutonium in Table 6.10. *Reactor-grade* here refers to the typical enriched uranium used in LWRs and the plutonium composition created by transmutation in LWR fuel. Although it is feasible to de-enrich the weapons-grade uranium, the weapons-grade plutonium would be used as is.

Physics Differences Between Weapons- and Reactor-Grade Plutonium-Fueled Reactors

There are some similarities and some important differences between using weapons- and reactor-grade plutonium in an LWR designed for low-enrichment uranium fuel. The delayed neutron fractions for thermal fission of ^{239}Pu, ^{241}Pu, and ^{235}U are in the ratio 0.0021:0.0049:0.0065. Because the delayed neutron fraction is smaller in ^{239}Pu than in ^{235}U, the subprompt-critical reactivity range is much less for plutonium-fueled reactors than for uranium-fueled reactors, as discussed in Section 6.5; and because the delayed neutron fraction is much smaller in ^{239}Pu than in ^{241}Pu, reactors fueled with weapons-grade plutonium will have an even smaller subprompt-critical reactivity range than reactors fueled with reactor-grade plutonium.

The large resonance integral of ^{240}Pu contributes a significant negative Doppler coefficient when reactor-grade plutonium is used, but which is absent when weapons-grade plutonium is used. Similarly, the use of weapons-grade uranium with the low ^{238}U content would substantially reduced the negative ^{238}U Doppler coefficient relative to the use of reactor-grade uranium. A resonance absorber such as tungsten can be added to weapons-grade fuel in order to recover part of the negative Doppler coefficient. Calculated Doppler coefficients for a standard LWR UO_2 lattice with reactor-grade uranium and for various combinations of UO_2ZrO_2 and W with weapons-grade uranium and plutonium are given in Table 6.11. Because of the higher fission cross section and higher value of η for ^{239}Pu than for

Table 6.11 Fuel Doppler Temperature Coefficients of Reactivity with Weapons-Grade Plutonium

Composition	$\Delta k/k$ ($\times 10^{-5}$)
R-G UO$_2$ (3% ^{235}U)	−2.4720
W-G UO$_2$-ZrO$_2$ (0.6% UO$_2$)	−0.0017
W-G UO$_2$-ZrO$_2$ + W (3% UO)	−1.0357
W-G MOX-ZrO$_2$ (2.7% UO$_2$, 0.3% PuO$_2$)	−0.9588
W-G PuO$_2$-ZrO$_2$ (0.34% PuO$_2$)	−0.0009
W-G PuO$_2$-ZrO$_2$ + W (3% PuO$_2$)	−1.2003

^{235}U in a fast neutron spectrum, weapons-grade plutonium fuel projects superior performance to uranium fuel in fast breeder reactors.

6.8
Utilization of Uranium Energy Content

Only about 1% of the energy content of the uranium used to produce the fuel is extracted (via fission) in a typical LWR fuel cycle. About 3% of the energy content of the mined uranium is stored as tails from the original uranium fuel production process, and about 96% remains in the depleted uranium from the enrichment process and in the discharged spent fuel in the form of uranium, plutonium, and higher-actinide isotopes. With continued reprocessing and recycling of spent fuel and depleted uranium, there is the possibility of recovering much of this remaining energy. The projected worldwide consumption of uranium based on the LWR once-through fuel cycle is shown in Fig. 6.14.

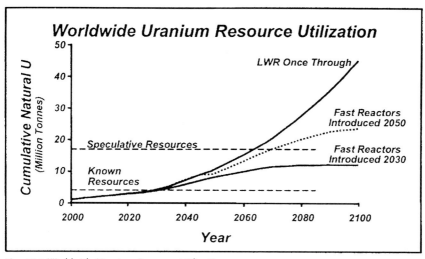

Fig. 6.14 Worldwide Uranium Resource Utilization.

Table 6.12 Equilibrium Distribution of Transuranic Isotopic Masses (%) for Continuously Recycled Fuel in Thermal and Fast Reactor Neutron Spectra

Isotope	Thermal Reactor Spectrum	Fast Reactor Spectrum
^{237}Np	5.51	0.75
^{238}Pu	4.17	0.89
^{239}Pu	23.03	66.75
^{240}Pu	10.49	24.48
^{241}Pu	9.48	2.98
^{242}Pu	3.89	1.86
^{241}Am	0.54	0.97
^{242}Am	0.02	0.07
^{243}Am	8.11	0.44
^{242}Cm	0.18	0.40
^{243}Cm	0.02	0.03
^{244}Cm	17.85	0.28
^{245}Cm	1.27	0.07
^{246}Cm	11.71	0.03
^{247}Cm	0.75	$2. \times 10^{-3}$
^{248}Cm	2.77	$6. \times 10^{-4}$
^{249}Bk	0.05	$1. \times 10^{-5}$
^{249}Cf	0.03	$4. \times 10^{-5}$
^{250}Cf	0.03	$7. \times 10^{-6}$
^{251}Cf	0.02	$9. \times 10^{-7}$
^{252}Cf	0.08	$4. \times 10^{-8}$

Source: Data from Ref. 14.

To fully consume the initial uranium feedstream, for each transuranic atom fissioned there must be one neutron released to sustain the chain reaction and one neutron available for capture in ^{238}U (or ^{240}Pu) to produce ^{239}Pu (or ^{241}Pu) to replace the fissioned atom. There is, of course unavoidable parasitic capture in fission products, structure, and the transuranic elements. The continued recycling of spent fuel would lead, after long exposure, to equilibrium distributions of the transuranic isotopes in the recycled fuel, as shown for thermal and fast neutron spectra in Table 6.12. (Note that these concentrations could be altered by blending spent fuels from different numbers of recycles.)

The number of neutrons per fission lost to parasitic capture in the transuranics can be estimated from their capture and fission probabilities, which are shown in Fig. 6.15 for typical LWR and LMR spectra. For the equilibrium distribution of Table 6.12, the number of neutrons per fission lost to parasitic capture is typically about 0.25 in a fast neutron spectrum and 1.25 in a thermal neutron spectrum. This means that a minimum (not accounting for parasitic capture in fission products, control elements, and structure or leakage) number of neutrons released per fission to maintain the chain reaction and transmute a fertile isotope into a fis-

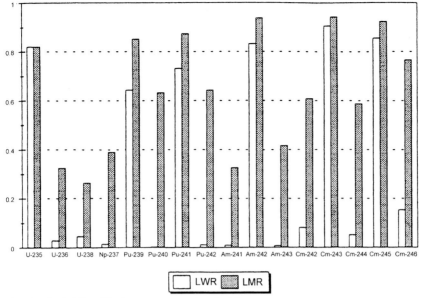

Fig. 6.15 Probability of fission per neutron absorbed for actinide isotopes in thermal and fast neutron spectra. (From Ref. 1; used with permission of Nuclear Energy Agency, Paris.)

sionable isotope to replace each fissioned isotope is 2.25 for the LMR spectrum and 3.25 for the LWR spectrum. Physically, more neutrons are wasted transmuting a transuranic nuclide into another transuranic nuclide in a thermal spectrum than in a fast spectrum. Since $2.5 < \eta \ll 3.25$, total energy extraction by repeated recycling in a thermal reactor is not possible, but it may be in a fast reactor.

The projected uranium utilization of LWRs together with fast breeder reactors that would transmute the non-fissile ^{238}U into fissile ^{239}Pu for burning in the LWRs is compared with the utilization of LWRs operating on the once-through cycle in Fig. 6.14. Clearly the early deployment of fast breeder reactors is essential if nuclear power is to achieve its potential in meeting the world's growing energy demands.

6.9
Transmutation of Spent Nuclear Fuel

The once-through cycle (OTC), in which slightly enriched UO_2 fuel (^{235}U increased from 0.72% in natural U to 3 to 5%) is irradiated to 30 to 50 GWd/T in a commercial reactor and then disposed of *in toto* as high-level waste (HLW), is the reference nuclear fuel cycle in the United States and a few other countries. With the present low uranium prices, this is the cheapest fuel cycle in the short term. Moreover, until recently (2006) U.S. government policy against reprocessing, motivated by proliferation concerns, was consistent only with the OTC. However, the long-term implications of the OTC are rather unfavorable. The potential energy content of the

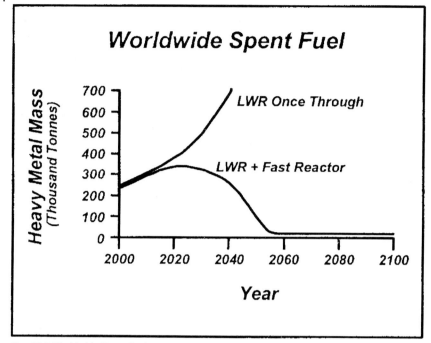

Fig. 6.16 Worldwide Spent Fuel.

residual fissile material (about 1% each Pu and ^{235}U) and of the ^{238}U (>90%) in the spent fuel and in the depleted uranium, which constitutes >90% of the potential energy content of the mined uranium, is lost in the OTC. Moreover, all the nuclides that can contribute to the potential radiotoxicity of the spent fuel are retained, together with the much greater volume of depleted U (mostly ^{238}U), which makes a relatively small contribution to the potential radiotoxicity, resulting in the largest possible volume of HLW, which must be stored in geological repositories for hundreds of thousands to millions of years. The projected worldwide accumulation of spent nuclear fuel in the once-through cycle is shown in Fig. 6.16.

Today, there are large inventories of plutonium and other minor actinides that have accumulated in discharged spent nuclear fuel. Presently, 40,000 tonnes initial uranium of spent nuclear fuel has accumulated in the Unites States. This inventory continues to grow at a rate of 2000 tonnes/yr. At the current level of nuclear energy production in the United States using the OTC, a new repository on the scale of the presently proposed Yucca Mountain site would have to be installed about every 30 years. The objective of transmutation of spent fuel is to reduce both the mass of HLW that must be stored in geological repositories and the time of high radiotoxicity of that HLW, thus reducing the requirements for both the number of repositories and the duration of secured storage. A National Research Council (NRC) study recently concluded that the need for a geological repository could be reduced, but would not be eliminated, by transmutation.

The short-term radiotoxicity of the spent fuel is dominated by fission products, but after 300 to 500 years only the long-lived radionuclides (particularly ^{99}Tc and ^{129}I, but also ^{135}Cs, ^{93}Zr, and others) remain—unfortunately, some of these are relatively mobile and contribute disproportionately to the potential radiological hazard from spent fuel. However, the long-term potential radiotoxicity of spent fuel arises principally from the presence of transuranic actinides (Pu and the so-called minor actinides Np, Am, Cm, etc.) produced by transmutation–decay chains originating with neutron capture in ^{238}U, which constitute a significant radiation source for hundreds of thousands of years. The contributions to the radiotoxicity of typical spent nuclear fuel from actinides, fission products, and activated structure are shown in Fig. 6.17.

Processing of spent UO_2 fuel to recover the residual U and Pu reduces the potential long-term radiotoxicity of the remaining HLW (minor actinides, fission products, activated structure, etc.) by a factor of 10 and reduces the volume by a much larger factor, and processing technology (PUREX) capable of 99.9% efficient recovery of U and Pu is commercially available in a number of countries (United Kingdom, France, Japan, India, Russia, and China). A fuel cycle in which the recovered Pu and U was recycled as a mixed oxide (MOX) UO_2–PuO_2 commercial reactor fuel has been envisioned since the beginning of the nuclear energy era, and at present a number of commercial reactors are operating with recycled Pu in western Europe. (Reprocessed uranium is not being recycled significantly because of the low cost of fresh uranium, which does not contain the neutron-absorbing ^{236}U that decreases the reactivity of recycled U.) Taking into account further production of minor actinides and fission products in the recycled Pu, a single recycle of the Pu in spent fuel reduces the potential radiotoxicity of the HLW associated with the original spent fuel only by a factor of 3 (rather than 10). Repeated recycling of the MOX fuel is technically feasible and would result in better fuel utilization, but the potential radiotoxicity of the HLW associated with the original spent fuel would actually increase relative to the OTC because of the further production of minor actinides and fission products.

It is clear from the discussion above that to reduce the potential radiological hazard associated with spent fuel or the length of time that hazard exists, it is necessary (1) to destroy the actinides (Pu and the minor actinides) and (2) to destroy the potentially hazardous long-lived fission products. The destruction of the minor actinides and long-lived fission products, as well as the Pu, by neutron transmutation implies the requirement for separation of these nuclides from the waste stream of processed spent fuel for recycling with subsequent fuel loadings. Effective separation of Pu with 99.9% efficiency is achieved commercially with the PUREX process. The effective separation of Np is technically feasible with a modified PUREX process, but practical separation methods for Am, Cm, and the long-lived fission products are still in the research stage. The pyrometallurgical (PYRO) separation technology presently under development would, unlike the PUREX process, allow separation of Np, Am, and Cm along with Pu into a code-posited metallic product that could be recycled in a metal-fuel fast reactor, resulting in a waste stream essentially free of actinides.

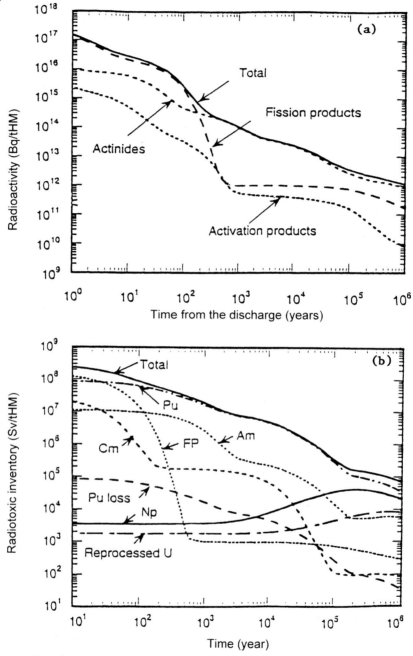

Fig. 6.17 Radiotoxic inventory of UO_2 fuel as a function of time [3.7% ^{235}U, 45 GWD/tonne heavy metal; Becquerel (Bq) = 1 disintegration per second = 2.7×10^{-11} Ci; Sievert (Sv) = 100 rad equivalent]. (From Ref. 12; used with permission of Nuclear Energy Agency, Paris.)

Since all of the actinides are potentially radiotoxic and since neutron capture (n, γ) reactions in the actinides just produce other actinides, the only effective way to destroy actinides is by neutron fission (n, f) reactions. Some of the actinides are effectively not fissionable in a thermal neutron spectrum, such as exists in almost all commercial nuclear reactors, and the probability of fission per neutron absorbed is greater for all the actinides in a fast neutron spectrum (see Fig. 6.15). The neutron absorption cross sections for the troublesome long-lived fission products are small in a thermal neutron spectrum and even smaller in a fast neutron spectrum, implying the advantage of a very high flux of thermal neutrons for their effective destruction (effective destruction of ^{135}Cs may prove impractical because of the presence of other neutron-absorbing Cs isotope fission products).

Several studies of minor actinide transmutation in nuclear reactors have been performed. They indicate that recycling of industrial levels of minor actinides as well as Pu in thermal neutron spectrum commercial reactors does not significantly reduce the overall radiotoxicity and requires an increase in fuel enrichment, with a corresponding increase in the cost of energy. On the other hand, recycling minor actinides as well as Pu in fast reactors is predicted to reduce the overall radiotoxicity of the HLW, but the maximum loading of minor actinides is limited by reactor safety considerations. The possibility of recycling Pu and the minor actinides first in thermal neutron spectrum commercial light water reactors (LWRs) and then in dedicated fast reactors has been calculated to be able to reduce the radio-toxic inventory in the HLW by a factor of about 100 relative to the OTC.

Such studies generally indicate that the transmutation of Pu, minor actinides and fission products in critical nuclear reactors would ultimately be limited by criticality or safety constraints. While fast reactors could, in principle, burn the mix of Pu plus minor actinides and some of the fission products, the available PUREX process does not separate the minor actinides with the plutonium from the waste stream for recycling. Moreover, it would be difficult to fabricate MOX fuel containing the highly radioactive minor actinides in existing facilities. This has led in Europe and Japan to consideration of remote fuel fabrication facilities to supply fuel containing minor actinides for destruction in dedicated subcritical transmuter reactors driven by accelerator spallation neutron sources, while the Pu would be consumed in dedicated fast reactors.

The U.S. ATW concept is to use remote fabrication of fuel containing separated Pu plus minor actinides (but no ^{238}U) for destruction in a subcritical transmuter reactor driven by an external neutron source. A variant of this concept would involve first irradiating this Pu plus minor actinide fuel by repeated recycling in a critical reactor before the final irradiation in a subcritical transmuter reactor.

The small delayed neutron fraction of the minor actinides and the generally positive reactivity coefficient of fast reactors without ^{238}U dictates that these actinide destruction, or transmuter, reactors must remain well subcritical. The reactivity coefficient could be made negative by the addition of ^{238}U, which would allow the possibility of actinide destruction in critical fast reactors, but that would lead to the production of additional Pu and minor actinides by transmutation of ^{238}U, hence to a decreased net actinide destruction rate.

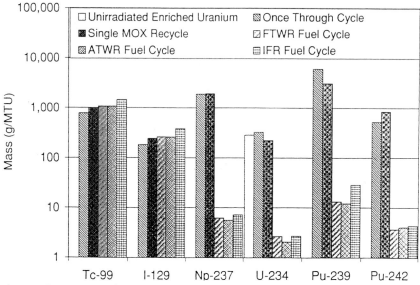

Fig. 6.18 Concentration of repository isotopes in HLW.

Development of the PYRO separation technology would allow separation of Np, Am, and Cm along with Pu, all of which could be recycled in a metal-fuel fast reactor, resulting in a waste stream essentially free of actinides. However, it would be necessary to include ^{238}U in the fuel to avoid the safety problems mentioned in the preceding paragraph, which would reduce the net destruction rate of the actinides. Thus safety or net destruction rate constraints on transmutation of actinides in critical reactors could be relaxed by operating the reactors subcritical with a neutron source. Several studies of subcritical reactors driven by accelerator spallation neutron sources and a few studies of subcritical reactors driven by fusion neutron sources have predicted significantly higher levels of Pu, minor actinide, and/or long-lived fission product destruction than are predicted to be achievable in critical nuclear reactors. The optimum scenario for recycling Pu, minor actinides, and long-lived fission products in commercial thermal neutron spectrum reactors, in dedicated fast neutron spectrum reactors, and in subcritical transmuter reactors driven by neutron sources remains the subject of active investigation.

The neutron spectrum in a subcritical reactor driven by a neutron source will depend more on the moderating and absorption properties, hence the material composition, of the subcritical reactor than on the energy spectrum of the source neutrons. Thus the material composition in the subcritical reactor can be optimized for the transmutation task at hand, without the criticality and safety constraints that would be present in a critical reactor.

Figure 6.18 shows the mass, per metric tonne of uranium used in fabricating the original fuel, of two long-lived fission products and four actinides that would contribute significantly to the dose rates of spent fuel in a HLW repository at long times after discharge. The "Once Through Cycle" indicates LWR fuel sent directly

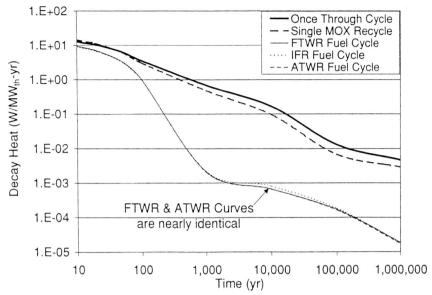

Fig. 6.19 Rate of production of decay heat.

to a HLW repository after discharge from a once-through cycle. The "Single MOX Recycle" indicates the same LWR spent fuel, but with the plutonium separated out and recycled once in a LWR before being sent to the HLW repository, and with the uranium separated out and sent to a low-level-waste (LLW) repository. In the "IFR Fuel Cycle," the original LWR spent fuel is separated into fission products (sent to HLW repository), uranium (sent to LLW repository) and transuranics (TRU) which are repeatedly recycled in a TRU-U metal fuel Integral Fast Reactor (with fission products removed and sent to a HLW repository).

The presence of U in the IFR fuel is necessary to provide a prompt negative Doppler coefficient of reactivity, but also reduces the net TRU burnup because of the production of TRU by the transmutation of ^{238}U. If the IFR is operated sub-critical to provide a large reactivity margin to prompt critical, it could be fueled with pure TRU metal fuel, in which case the net transmutation rate would be greater. The "FTWR Fuel Cycle" and "ATWR Fuel Cycle" indicate variants of the IFR Fuel Cycle in which the IFR is fueled with pure TRU and operated sub-critical with a D–T fusion and accelerator-spallation, respectively, neutron source. The small differences between the FTWR and ATWR are due to the better neutron utilization for actinide fission with Na coolant (IFR and ATWR) than with Li coolant (FTWR).

Clearly, a single MOX recycle does not significantly reduce the mass of long-lived radioactive material sent to the HLW repository. On the other hand, repeatedly recycling the actinides from the LWR spent fuel in an IFR (critical or sub-critical) can significantly reduce the actinide content in the material ultimately sent to the HLW repository.

The HLW repository capacity will be limited by the amount of decay heat that can be removed passively. Thus, a more meaningful comparison among the above fuel cycles is in terms of the decay heat at a given time after discharge of the material placed in the HLW repository, normalized by the total nuclear energy produced from the original LWR fuel in the respective fuel cycle. As shown in Fig. 6.19, the repository capacity required to achieve a given level of nuclear energy production can be reduced by a couple of orders of magnitude by repeatedly recycling the actinides from LWR spent fuel in a fast reactor, relative to depositing the LWR spent fuel directly in the repository.

6.10
Closing the Nuclear Fuel Cycle

The discussions of the previous two sections on uranium utilization and transmutation of long-lived actinides in spent nuclear fuel are obviously related. The common thread is fissioning the transuranics to simultaneously recover their energy content and eliminate them as long-lived "waste." The full utilization of the energy content of uranium can only be recovered by transmutation of ^{238}U into transuranics that can be fissioned. This "closing of the nuclear fuel cycle" has long been the ultimate goal of nuclear power development, but short-term economics, misguided reactions to proliferation and environmental concerns, and an apparently temporary stagnation in the development of nuclear power in the U.S. and some parts of Europe have until recently conspired to make realization of this goal but a dimly perceived possibility for the distant future.

However, with the recent (2006) initiation of the Global Nuclear Energy Partnership to develop an international closed nuclear fuel cycle by development and deployment of technologies that enable recycling and consumption of spent nuclear fuel in a manner that promotes non-proliferation, enhances energy security and is environmentally responsible, this goal once again appears achievable. The Generation-IV reactor design studies to identify reactors and separations processes that could achieve this goal are discussed in Section 7.10.

References

1 *Physics of Plutonium Recycling*, Vols. I–V, Nuclear Energy Agency, Paris (**1995**).

2 *Management and Disposition of Excess Weapons Plutonium*, National Academy Press, Washington, DC (**1995**).

3 R. A. KNIEF, *Nuclear Engineering*, Taylor & Francis, Washington, DC (**1992**), Chaps. 2 and 6.

4 R. G. COCHRAN and N. TSOULFANIDIS, *The Nuclear–Fuel Cycle: Analysis*

and Management, American Nuclear Society, LaGrange Park, IL (**1990**).

5 L. KOCH, "Formation and Recycling of Minor Actinides in Nuclear Power Stations," in A. J. Freeman and C. Keller, eds., *Handbook of the Physics and Chemistry of Actinides*, Vol. 4, Elsevier Science Publishers, Amsterdam (**1986**), Chap. 9.

6 S. H. LEVINE, "In-Core Fuel Management of Four Reactor Types," in Y. Ronen, ed., *CRC Handbook of Nuclear*

Reactor Calculations II, CRC Press, Boca Raton, FL (**1986**).

7 A. G. Croff, *ORIGEN2: A Revised and Updated Version of the Oak Ridge Isotope Generation and Depletion Code*, ORNL-5621, Oak Ridge National Laboratory, Oak Ridge, TN (**1980**).

8 *International Nuclear Fuel Cycle Evaluation*, STI/PUB/534, International Atomic Energy Agency, Vienna (**1980**).

9 J. J. Duderstadt and L. J. Hamilton, *Nuclear Reactor Analysis*, Wiley, New York (**1976**), Chap. 15.

10 A. F. Henry, *Nuclear-Reactor Analysis*, MIT Press, Cambridge, MA (**1975**), Chap. 6.

11 National Research Council, *Nuclear Wastes Technologies for Separations and Transmutation*, National Academy Press, Washington, DC (**1996**).

12 *First Phase P&T Systems Study: Status and Assessment Report on Actinide and*

Fission Product Partitioning and Transmutation, OECD/NEA, Paris (**1999**).

13 *Proc. 1st–5th NEA International Exchange Meetings*, OECD/NEA, Paris (**1990**, **1992**, **1994**, **1996**, **1998**).

14 D. C. Wade and R. N. Hill, "The Design Rationale of the IFR," *Prog. Nucl. Energy* **31**, 13 (**1997**).

15 L. J. Templin, ed., *Reactor Physics Constants*, 2nd ed., ANL-5800, Argonne National Laboratory, Argonne, IL (**1963**).

16 A. Sesonske, *Nuclear Power Plant Design Analysis*, USAEC-TID-26241, U.S. Atomic Energy Commission, Washington, DC (**1973**).

17 N. L. Shapiro et al., *Electric Power Research Institute Report*, EPRI-NP-359, Electric Power Research Institute, Palo Alts, CA (**1977**).

18 *Oceanus, 20*, Woods Hole Oceanographic Institute, Wood Hole, MA (**1977**).

Problems

6.1. A reactor loaded initially with 125 kg of 93% enriched ^{235}U in the form of UO_2 depletes in a constant neutron flux of $\phi = 5 \times 10^{13}$ n/cm$^2 \cdot$ s for one effective full power year. Assuming a thermal absorption cross section of 450 barns for ^{235}U, calculate the average fuel burnup in MWD/T.

6.2. Calculate the maximum enrichment at which a mixture of ^{235}U and ^{238}U will initially breed (i.e., the fissile concentration $n^{25} + n^{49}$ will increase in time). Use $\sigma_a^{25} = 700$ barns, $\sigma_a^{49} = 1050$ barns, $\sigma_\gamma^{28} = 8$ barns, $\eta^5 = 2.08$ and $\eta^{49} = 2.12$ and assume that ^{239}Pu is produced instantaneously by neutron capture in ^{238}U.

6.3.* Consider a thermal reactor with initial fuel composition 93% ^{235}U and 7% ^{238}U and fuel density 18.9 g/cm^3 with an initial thermal neutron flux of 3×10^{14} n/cm$^2 \cdot$ s. Assume a flux disadvantage factor of 2. Write a computer code to calculate the depletion of ^{235}U and the buildup of ^{239}Pu over the first 2000 h of operation, assuming operation at constant power. Estimate from perturbation theory the reactivity decrease

---

* Problems 3 and 4 are longer problems suitable for take-home projects.

associated with the ^{235}U depletion and ^{239}Pu buildup over 2000 h of operation. Plot your results as a function of time.

6.4.* Estimate from perturbation theory the equilibrium xenon and samarium reactivity worth and the reactivity worth of the other fission products as a function of time in the reactor of Problem 6.1. Assume that the fuel occupies 80% of the core and that $\gamma^{fp}\sigma_a^{fp} = 50$ barns per fission. Plot your results as a function of time.

6.5. A ^{235}U-fueled nuclear reactor operates with a thermal flux level of 1×10^{14} n/cm$^2 \cdot$ s. The reactor has been operating at constant power level for 2 weeks when it becomes necessary to scram the control rods to shut down the core. After detailed investigation it is determined that the scram signal was erroneous and it is now necessary to return the reactor to full-power operation: 12 h has passed since shutdown. The control rods are withdrawn to the critical prescram position and the reactor is brought to temperature, but the reactor is not critical. How much further must the control rods be withdrawn to achieve criticality if the control rod bank worth is $\Delta\rho = 0.001$ cm^{-1}.

6.6. Derive the equations that determine the time dependence of the samarium concentration in a reactor that has achieved equilibrium samarium conditions at a flux ϕ_0 when the flux is changed to ϕ_1.

6.7. Calculate the initial excess reactivity needed for the reactor of Problem 6.2 to have a cycle lifetime of 1.0 years operating at a flux level of 5×10^{13} n/cm$^2 \cdot$ s.

6.8. Calculate the nuclear (fission) heating density in a PWR UO$_2$ fuel element (density 10 g/cm^3, 4% enriched) operating in a thermal neutron flux of 5×10^{13} n/cm$^2 \cdot$ s.

6.9. Calculate the α-decay heating due to plutonium at the end-of-cycle for each recycle core loading indicated in Table 6.5.

6.10. A thermal reactor loaded with 100,000 kg of 3% enriched UO$_2$ depletes in a constant thermal neutron flux of 5×10^{13} n/cm$^2 \cdot$ s for 1 year. Using a thermal absorption cross section of 500 barns and a capture-to-fission ratio of $\alpha = 0.2$ for ^{235}U and a density of 10 g/cm^3 for UO$_2$, calculate the average fuel burnup in MWd/T.

6.11. A ^{235}U-fueled reactor has been operated at a thermal flux level of 5×10^{12} n/cm$^2 \cdot$ s for 2 months, when the power level is reduced by one-half for 10 h, then returned to full power. Calculate the reactivity worth of xenon just before the reactor is shut down; 10 h later, before it is returned to full power;

and then again after it has been operating at full power for 10 h.

6.12. Repeat the calculation of Problem 6.11 for full power thermal neutron flux levels of 1×10^{13}, 5×10^{13}, and $1 \times 10^{14} \ n/cm^2 \cdot s$.

6.13. The equilibrium concentration of ^{149}Sm is independent of power level, and when the reactor is shut down, the ^{149}Sm concentration increases. Can the ^{149}Sm concentration ever be lower than the equilibrium concentration, once that concentration has been attained?

6.14. Calculate the equilibrium and peak xenon concentrations in cores fueled with ^{233}U, ^{235}U, and ^{239}Pu, all operating at a thermal flux level of 1×10^{14}.

6.15. A uniform bare cylindrical reactor, containing an initial loading of 125 kg of ^{235}U, operates until the maximum local ^{235}U depletion reaches 50%. Estimate the total fission energy release from the core.

6.16. Calculate and plot the activity (Ci/tonne fuel) and the toxicity (cancer dose/tonne fuel) of ^{99}Tc, ^{129}I, ^{90}Sr, and ^{137}Cs in spent fuel from a LWR from the time of discharge to 10^4 years later.

6.17. Calculate the toxicity (cancer dose/tonne fuel) of the equilibrium concentrations of the transuranic isotopes given in Table 6.12 for continuously recycled spent fuel in fast and thermal reactor spectra.

6.18. Calculate the change in isotopic composition of weapons-grade plutonium that is irradiated in a thermal neutron flux of $10^{14} \ n/cm^2 \cdot s$ for 1 year.

6.19. A thermal reactor fueled with ^{235}U and ^{232}Th in the ratio 1:20 is operated for 1 year with a neutron flux of $8 \times 10^{13} \ n/cm^2 \cdot s$. Calculate the concentrations of ^{233}U and ^{235}U, in terms of the initial ^{235}U concentration, at the end of the year. What is the annual conversion ratio?

6.20. Calculate the energy content per unit mass of the original fuel loadings for the reactors in Table 6.7. Calculate the fraction of this energy content that is released by fission in a single cycle.

7
Nuclear Power Reactors

As of 2000, there are 434 central station nuclear power reactors operating worldwide to produce 350,442 MWe of electrical power. Of this number, 252 are pressurized water reactors (PWRs), 92 are boiling water reactors (BWRs), 34 are gas-cooled
reactors (GCRs) of all types, 39 are heavy water-cooled reactors of all types (mostly
CANDUs), 15 are graphite-moderated light-water pressure tube reactors (RBMKs),
and 2 are liquid-metal fast breeder reactors (LMFBRs). The general physics-related
characteristics of such reactors are described in the following sections. To be quantitative, specific reactors that produce 900 to 1300 MWe (650 MWe in the case of
CANDUs) were chosen, but it should be noted that reactors of each type can vary
greatly in size and power output, so the numbers should be understood to be only
representative. In addition to the central station power reactors mentioned above,
there are more than 100 pressurized water naval propulsion reactors in the U.S.
fleet (plus others in foreign fleets) and numerous research and special purpose
reactors of various types worldwide.

7.1
Pressurized Water Reactors

Pressurized water reactors (PWRs) were first developed in the United States based
on experience from the naval reactor program. The first commercial electric power–
producing unit started operation at Shippingport, Pennsylvania in 1957. The PWR
is now widely distributed worldwide. The basic structure of the PWR core is the
approximately 20 cm × 20 cm × 4 m high fuel assembly shown in Fig. 7.1, consisting of an array of zircaloy-clad UO_2 fuel pins, or rods, of about 1 cm diameter. The
enrichment varies from about 2 to 4% or more, depending on the burnup objective. A typical fuel assembly may consist of a 17 × 17 array of fuel pins of about
1 cm diameter. The coolant flows in an open lattice structure which permits some
flow mixing and is under sufficient pressure that no boiling occurs under normal
operation.

Long-term reactivity control is provided by adjustment of the boric acid content in the coolant. The soluble poison concentration decreases with fuel burnup
to compensate fuel reactivity loss and must be reduced to compensate ^{135}Xe and

Nuclear Reactor Physics. Weston M. Stacey
Copyright © 2007 WILEY-VCH Verlag GmbH & Co. KGaA, Weinheim
ISBN: 978-3-527-40679-1

^{149}Sm buildup following reactor startup. Boron addition and dilution may be used to minimize control rod motion for startup and shutdown. Soluble poisons make a positive contribution to the moderator temperature coefficient of reactivity (an increase in temperature reduces the absorption cross section), so their maximum concentration is limited, and fixed burnable poisons are used to reduce the control requirements that must be met by adjustment of the boric acid concentration.

Burnable poisons consists of separate shim rods substituted for a fuel rod in the fuel assembly. These rods may consist of borosilicate glass rods with stainless steel cladding or B_4C pellets in an Al_2O_3 matrix with zircaloy cladding. The shim rods burn out as the fuel depletes, which constitutes a positive reactivity contribution to compensate the negative reactivity contribution of fuel depletion, thus reducing the requirement for adjustment of the boric acid concentration.

Because of the relatively short migration length (about 6 cm) of thermal neutrons in a PWR, the active control rods must be distributed. Short-term and rapid insertion (scram) reactivity control is provided by an assembly of full-length control rods driven down into the fuel assembly. For example, the control rod assembly for a 17×17 pin lattice consists of 24 control fingers connected by a spider, as shown in Fig. 7.1. The control rod material is either B_4C or a Ag–In–Cd mixture of somewhat weaker absorbers that produces less flux peaking upon rod withdrawal. "Part-length" rods in which only the lower 25% or so contains poison are used for controlling the axial flux distribution, which is necessary to control axial xenon oscillations (Chapter 16) as well as to minimize axial power peaking.

Full-length control rods are normally designated as *regulating rods*, used for the normal short-term reactivity adjustments that cannot be handled by adjustment of the boric acid concentration, and *shutdown* or *scram rods*, which are held out of the core to be available for a rapid negative reactivity insertion if required for safety or a more gradual negative reactivity insertion required for normal shutdown. A typical distribution of control rods among the assemblies in a PWR core is shown in Fig. 7.2.

About 190 to 240 fuel assemblies containing 90,000 to 125,000 kg of UO_2 constitute a typical PWR core, which is about 3.5 m in diameter and 3.5 to 4.0 m high and is located inside a pressure vessel, as shown in Fig. 7.3. Coolant typically enters the pressure vessel near the top, flows downward between the vessel and the core, is distributed at the lower core plate, flows upward through the core, and exits the vessel at the top. The coolant, which is pressurized to about 15.5 MPa (2250 psi), typically enters the vessel with a temperature of about 290°C and exits at about 325°C.

7.2
Boiling Water Reactors

Boiling water reactors (BWRs) were first developed in the United States and are now found worldwide. The physics of BWRs is similar in many respects to that of PWRs. The basic structure of the BWR core is an approximately

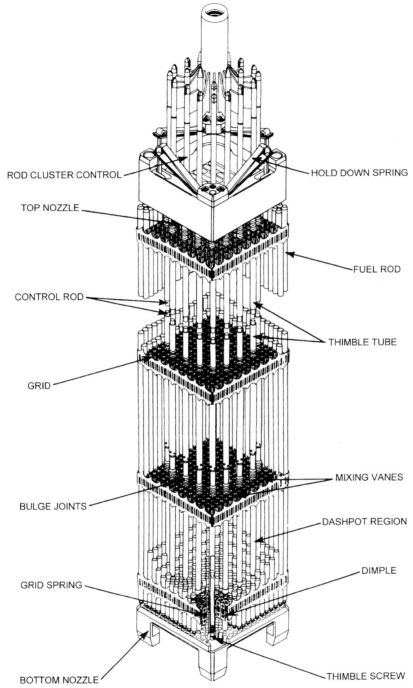

ROD CLUSTER CONTROL

HOLD DOWN SPRING

TOP NOZZLE

FUEL ROD

CONTROL ROD

THIMBLE TUBE

GRID

MIXING VANES

BULGE JOINTS

DASHPOT REGION

DIMPLE

GRID SPRING

BOTTOM NOZZLE

THIMBLE SCREW

Fig. 7.1 Fuel assembly for a pressurized water reactor. (Courtesy of Westinghouse Electric Corporation.)

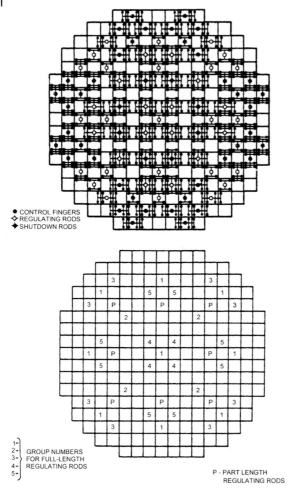

CONTROL FINGERS
REGULATING RODS
SHUTDOWN RODS

1-
2-) GROUP NUMBERS
3-) FOR FULL-LENGTH
4-) REGULATING RODS
5-

P - PART LENGTH
REGULATING RODS

Fig. 7.2 Representative control element pattern in a pressurized water reactor. (Courtesy of ABB Combustion Engineering, Inc.)

14 cm × 14 cm × 4 m high fuel assembly (Fig. 7.4) consisting of an 8 × 8 array of zircaloy-clad UO_2 fuel pins, or rods, of about 1.3 cm diameter. The enrichment varies from 2 to 4% ^{235}U. The 8 × 8 fuel pin array is surrounded by a zircaloy fuel channel to prevent cross-flow between assemblies. A group of four fuel assemblies plus an included cruciform control rod constitutes a fuel module, out of which a typical BWR core is built up, as indicated in Fig. 7.5.

Fuel pins of different enrichment are loaded into each assembly. Fuel pins of lower enrichment are located next to the control rod to suppress the flux peaking that would otherwise occur when the control rod was withdrawn, leaving a substantial water gap. The other pins are arranged to flatten the power distribution within the assembly. Long-term reactivity changes to compensate fuel depletion

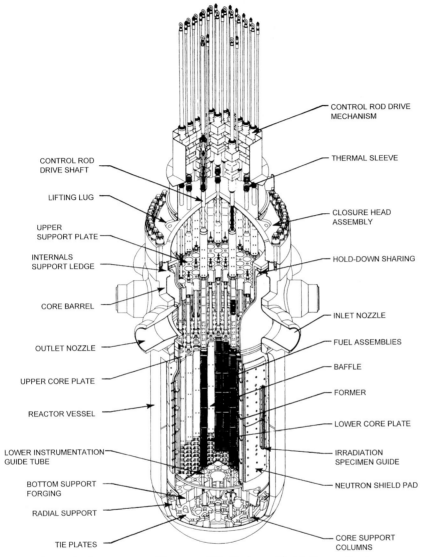

CONTROL ROD DRIVE
MECHANISM

THERMAL SLEEVE

CONTROL ROD
DRIVE SHAFT

LIFTING LUG

CLOSURE HEAD
ASSEMBLY

UPPER
SUPPORT PLATE

INTERNALS
SUPPORT LEDGE

HOLD-DOWN SHARING

CORE BARREL

INLET NOZZLE

OUTLET NOZZLE

FUEL ASSEMBLIES

BAFFLE

UPPER CORE PLATE

FORMER

REACTOR VESSEL

LOWER CORE PLATE

LOWER INSTRUMENTATION
GUIDE TUBE

IRRADIATION
SPECIMEN GUIDE

BOTTOM SUPPORT
FORGING

NEUTRON SHIELD PAD

RADIAL SUPPORT

TIE PLATES

CORE SUPPORT
COLUMNS

Fig. 7.3 Pressurized water reactor. (Courtesy of Westinghouse Electric Corporation.)

and reactivity changes needed for large power level changes are provided by the B_4C cruciform control rods, which are driven up from the bottom of the core because the reactivity worth is greater with the single-phase coolant in the lower part of the core than with the two-phase coolant in the upper part. Long-term compensation of the negative reactivity associated with fuel depletion is provided by mixing Gd_2O_3 uniformly with the UO_2 in several fuel pins in each assembly to provide a positive reactivity contribution as it burns out.

Fig. 7.4 Fuel assembly for a boiling water reactor. 1, top fuel guide; 2, channel fastener; 3, upper tie plate; 4, expansion spring; 5, locking tab; 6, channel; 7, control rod; 8, fuel rod; 9, spacer; 10, core plate assembly; 11, lower tie plate; 12, fuel support piece; 13, fuel pellets; 14, end plug; 15, channel spacer; and 16, plenum spring. (Courtesy of General Electric Company.)

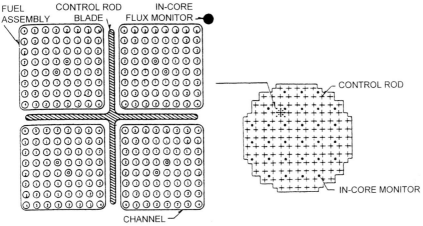

Fig. 7.5 Four-assembly fuel module for a boiling water reactor.
(Courtesy of General Electric Company.)

Short-term reactivity control is provided by recirculation flow and by control rods. Because of the negative coolant/moderator temperature coefficient of reactivity, coolant flow rate can be increased to decrease coolant temperature and the amount of boiling, making neutron moderation more effective and thus increasing reactivity. This causes the power level and the coolant temperature to increase, which in turn decreases the reactivity, until the reactor is again critical at a higher power level. Decreasing the coolant flow rate reduces the power level by a similar mechanism. Typically, about 750 fuel assemblies containing about 140,000 to 160,000 kg of UO_2 constitute a BWR core, which is similar in size to a PWR core and is located inside a pressure vessel, as shown in Fig. 7.6. Coolant enters the vessel at about 7.2 MPa (1000 psi), flows downward between the vessel wall and the shroud, is distributed by the core plate, flows upward through the core and upper structure, and exits the core as steam at about 290°C. About 30% of the coolant flow is recirculated, which has the net effect of increasing the total coolant flow rate in the core.

7.3
Pressure Tube Heavy Water–Moderated Reactors

The use of heavy water and online refueling to maintain criticality with natural uranium fuel is fundamental to CANDU reactors, which are pressure tube heavy water–moderated reactors developed in Canada but now are located in several other countries. The basic structure of the CANDU core is the fuel bundle shown in Fig. 7.7, which contains natural UO_2 in 37 zircaloy-clad fuel pins about 1.3 cm in diameter and 49 cm long which are separated with spacers. Twelve fuel bundles are placed end to end in a pressure tube through which flows pressurized (10 MPa, 1450 psi) D_2O. The reactor core consists of 380 fixed calandria tubes in a vessel

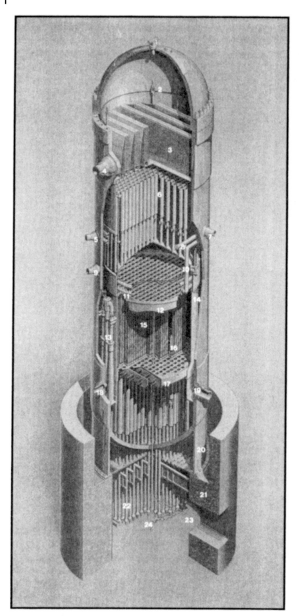

Fig. 7.6 Boiling water reactor. 1, vent and head spray; 2, steam dryer lifting lug; 3, steam dryer assembly; 4, steam outlet; 5, core spray inlet; 6, steam separator assembly; 7, feedwater inlet; 8, feedwater sparger; 9, low-pressure coolant injection inlet; 10, core spray line; 11, core spray sparger; 12, top guide; 13, jet pump assembly; 14, core shroud; 15, fuel assemblies; 16, core blade; 17, core plate; 18, jet pump/recirculation water inlet; 19, recirculation water outlet; 20, vessel support skirt; 21, shield wall; 22, control rod drives; 23, control rod drive hydraulic lines; and 24, in-core flux monitor. (Courtesy of General Electric Company.)

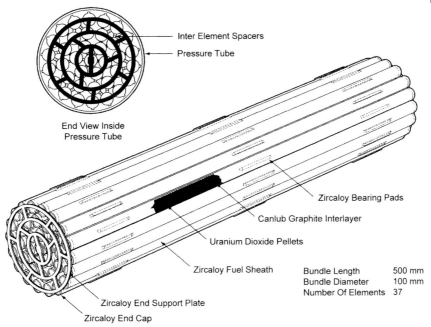

Fig. 7.7 Fuel assembly for a CANDU pressure tube heavy water reactor. (Courtesy of Atomic Energy of Canada, Ltd.)

filled with D_2O moderator, as shown in Fig. 7.8. A pressure tube containing the 12 fuel bundles is loaded into each calandria tube, resulting in a core loading of about 100,000 kg of natural UO_2. The coolant enters each pressure tube at about 265°C and exits at about 310°C. A typical CANDU core is about 7 m in diameter and about 4 m high.

On-line refueling is the primary means of long-term reactivity control in CANDU reactors. This is augmented by addition of soluble poison to the moderator D_2O and by the use of boron and gadolinium as burnable poisons admixed with the fuel. Because the D_2O in the reactor vessel, not the D_2O coolant in the pressure tubes, is the primary moderator, the usual negative coolant temperature coefficient of reactivity present in PWRs and BWRs is not present in the CANDU, and in fact the temperature coefficient of reactivity tends to be positive. This requires a much more precise active reactivity control system than for PWRs and BWRs. Reactivity control in each of 14 chambers is achieved by controlling the amount of H_2O (which is a poison in a D_2O system) in response to local neutron flux detector measurements.

Control rods are also employed. Adjuster rods are used for flattening the power distribution and for short-term reactivity adjustments. Four cadmium rods clad in stainless steel are located above the core, which may be used to supplement the adjuster rods in achieving reactivity control or dropped to effect a rapid shutdown, or scram. A backup shutdown system consists of injection of a gadolinium nitrate solution into the moderator.

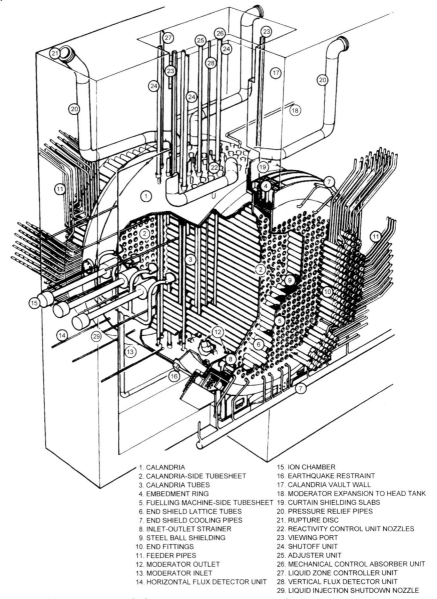

1. CALANDRIA
2. CALANDRIA-SIDE TUBESHEET
3. CALANDRIA TUBES
4. EMBEDMENT RING
5. FUELLING MACHINE-SIDE TUBESHEET
6. END SHIELD LATTICE TUBES
7. END SHIELD COOLING PIPES
8. INLET-OUTLET STRAINER
9. STEEL BALL SHIELDING
10. END FITTINGS
11. FEEDER PIPES
12. MODERATOR OUTLET
13. MODERATOR INLET
14. HORIZONTAL FLUX DETECTOR UNIT

15. ION CHAMBER
16. EARTHQUAKE RESTRAINT
17. CALANDRIA VAULT WALL
18. MODERATOR EXPANSION TO HEAD TANK
19. CURTAIN SHIELDING SLABS
20. PRESSURE RELIEF PIPES
21. RUPTURE DISC
22. REACTIVITY CONTROL UNIT NOZZLES
23. VIEWING PORT
24. SHUTOFF UNIT
25. ADJUSTER UNIT
26. MECHANICAL CONTROL ABSORBER UNIT
27. LIQUID ZONE CONTROLLER UNIT
28. VERTICAL FLUX DETECTOR UNIT
29. LIQUID INJECTION SHUTDOWN NOZZLE

Fig. 7.8 CANDU pressure tube heavy water reactor. (Courtesy of Atomic Energy of Canada, Ltd.)

7.4
Pressure Tube Graphite-Moderated Reactors

The world's first commercial nuclear electricity was generated near Moscow in 1954 by a graphite-moderated pressure tube light water reactor which evolved to

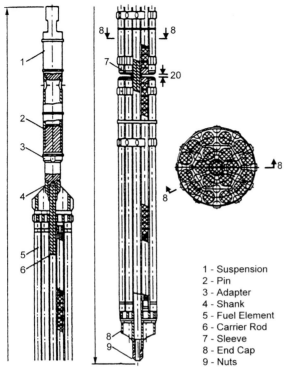

1 - Suspension
2 - Pin
3 - Adapter
4 - Shank
5 - Fuel Element
6 - Carrier Rod
7 - Sleeve
8 - End Cap
9 - Nuts

Fig. 7.9 Fuel assembly for a RMBK pressure tube graphite reactor. (From Ref. 8.)

the reactor generally known by the acronym RBMK from the Russian for "high-power pressure tube reactor." Reactors of this type are located in the countries of the former Soviet Union. The basic structure of the RBMK core is the fuel channel tube, made of zirconium alloyed with 2.5% niobium, shown in Fig. 7.9. Each channel tube consists of two fuel strings, which are separately cooled with H_2O at 7.2 MPa, which enters at 270°C and exits at 284°C, placed end to end. Each fuel string contains 1.8 to 2.0% enriched UO_2 in 18 fuel pins about 1.3 cm in diameter and 3.6 m long, separated with spacers. Each channel tube is placed vertically into a square graphite block 0.25 m on a side and 7 m high. The graphite blocks, 1661 containing a fuel channel tube and 222 containing control rod channels, are set side by side to form an upright cylinder 12.2 m in diameter containing about 200,000 kg of UO_2.

Since the migration length in graphite is about 60 cm, the core is very loosely coupled and subject to flux tilting. Furthermore, since the neutron moderation is provided by the graphite, the coolant temperature coefficient of reactivity is positive because the effect of increased coolant temperature and reduced coolant density is to reduce the coolant absorption cross section. As a consequence, the RMBK reactor is inherently unstable to power oscillations and it is necessary to control the power distribution region by region. Two hundred and eleven cylindrical B_4C control rods, with graphite extenders to enhance rod effectiveness by displacing

H_2O that would otherwise fill the rod channel when the rod was withdrawn, are dispersed in the core. Of these, 24 are normally withdrawn from the core to be available to produce rapid shutdown. An additional 24 short absorbing rods that enter from below are used to control axial xenon oscillations (Chapter 16) and to reduce axial power peaking. With a fresh fuel loading, up to 240 additional control rods must displace fuel in the tubes in order to hold down reactivity. These control channel tubes are replaced with fuel channel tubes as fuel burnup decreases reactivity.

7.5
Graphite-Moderated Gas-Cooled Reactors

The first man-made sustained fission chain reaction took place in a pile of graphite in Chicago—with air cooling—which was the prototype for the first experimental and production reactors. The original gas-cooled power reactors developed in France and England used CO_2 as a coolant and graphite moderator. The original MAGNOX reactors consisted of natural uranium bars clad in a low- neutron-absorbing magnesium alloy known as magnox which were placed in holes in graphite blocks through which the CO_2 coolant flows at 300 psi, leaving the core at about 400°C. A typical MAGNOX core is about 14 m in diameter and 8 m high. To achieve higher coolant outlet temperatures (650°C), the subsequent advanced gas-cooled reactors (AGRs) operate at 600 psi, which requires that the cladding consist of a material that can operate at higher temperature, which in turn requires that the fuel be enriched. The AGR fuel element consists of 36 tubes made up of pellets of 2.3% enriched UO_2 in stainless steel cladding, which are ribbed to improve heat transfer, as shown in Fig. 7.10, encased in a graphite sleeve, and inserted in holes in graphite blocks. Excessive corrosion in piping and steam generators have led to the abandonment of CO_2 as a coolant; most advanced gas-cooled reactors use helium as a coolant.

As an example of a modern gas-cooled reactor, we consider the high-temperature gas-cooled reactor (HTGR). The basic structure of the HTGR core is a hexagonal graphite block containing small channels for stacks of fuel pins and for coolant flow, as shown in Fig. 7.11. The fuel consists of coated microspheres of 93% enriched UC/ThO_2 contained in fuel pins about 1.6 cm in diameter and 6 cm long. About 490 fuel assemblies, each with 6.3 m active fuel height, are placed upright side by side to form a core that has a diameter of 8.4 m and contains 1,720 kg of U and 37,500 kg of Th, as shown in Fig. 7.12. The fuel assemblies are arranged in rings of six about a control assembly. Long-term reactivity control is provided by B_4C loaded into carbon rods which may be loaded into the corner locations of each fuel assembly to serve as burnable poison. Short-term reactivity control is provided by pairs of control rods that can be inserted into the two larger channels in special control assemblies. HTGRs have been deployed only on a limited scale.

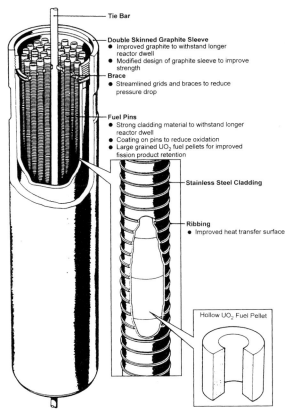

Tie Bar

Double Skinned Graphite Sleeve
● Improved graphite to withstand longer
  reactor dwell
● Modified design of graphite sleeve to improve
  strength
Brace
● Streamlined grids and braces to reduce
  pressure drop

Fuel Pins
● Strong cladding material to withstand longer
  reactor dwell
● Coating on pins to reduce oxidation
● Large grained UO₂ fuel pellets for improved
  fission product retention

Stainless Steel Cladding

Ribbing
● Improved heat transfer surface

Hollow UO₂ Fuel Pellet

Fig. 7.10 Fuel assembly for advanced gas reactor. (From Ref. 4; used with permission of Taylor & Francis/Hemisphere Publishing.)

7.6
Liquid-Metal Fast Breeder Reactors

The first generation of electricity from nuclear fission took place at the light bulb level in 1952 in a liquid-metal fast breeder reactor an (LMFBR), the EBR-1 in Idaho. Several LMFBRs have been operated since then, but this reactor type has not yet been deployed on a substantial scale. The physics of the LMFBR, which has a fast neutron spectrum, differs significantly from the physics of the previously discussed reactors, all of which have a thermal neutron spectrum.

The basic structure of a modern LMFBR core is the fuel assembly, as indicated in Fig. 7.13. The primary fissile nuclide for fast breeders is ^{239}Pu, and the primary fertile nuclide is ^{238}U. The fuel assembly consists of about 270 fuel pins containing 10 to 30% Pu in PuO₂–UO₂ in small pellet form encased in stainless steel cladding. The pins, which are about 0.9 cm in diameter and 2.7 m long, are wrapped in wire to maintain interpin spacing and placed within a stainless steel tube. The flow of liquid sodium is directed by the channel around the array. About 350 such

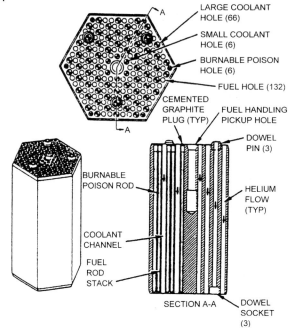

LARGE COOLANT
HOLE (66)

SMALL COOLANT
HOLE (6)

BURNABLE POISON
HOLE (6)

FUEL HOLE (132)

CEMENTED
GRAPHITE
PLUG (TYP)

FUEL HANDLING
PICKUP HOLE

DOWEL
PIN (3)

BURNABLE
POISON ROD

HELIUM
FLOW
(TYP)

COOLANT
CHANNEL

FUEL
ROD
STACK

SECTION A-A

DOWEL
SOCKET
(3)

Fig. 7.11 Fuel assembly for a high-temperature gas-cooled reactor. (Courtesy of General Atomics Company.)

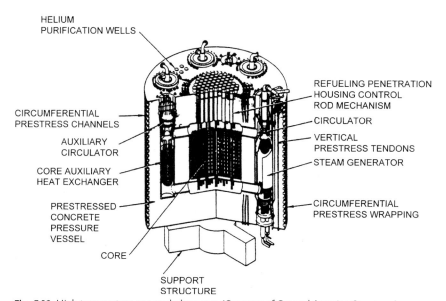

HELIUM
PURIFICATION WELLS

REFUELING PENETRATION
HOUSING CONTROL
ROD MECHANISM

CIRCUMFERENTIAL
PRESTRESS CHANNELS

CIRCULATOR

AUXILIARY
CIRCULATOR

VERTICAL
PRESTRESS TENDONS

CORE AUXILIARY
HEAT EXCHANGER

STEAM GENERATOR

PRESTRESSED
CONCRETE
PRESSURE
VESSEL

CIRCUMFERENTIAL
PRESTRESS WRAPPING

CORE

SUPPORT
STRUCTURE

Fig. 7.12 High-temperature gas-cooled reactor. (Courtesy of General Atomics Company.)

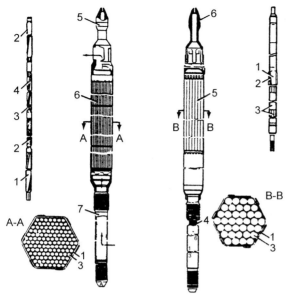

Fig. 7.13 Fuel assemblies for a liquid-metal fast breeder reactor. (Courtesy of Nuclear Engineering International.)

assemblies makes up the core of an LMFBR. Another 230 similar assemblies, but with only UO_2 or with a lower Pu content, are placed in a blanket around the core, as shown in Fig. 7.14. The total mass of PuO_2/UO_2 is about 32,000 kg. A typical LMFBR core is about 1 m high and 2 m in diameter.

Reactivity control is achieved by control bundles of B_4C rods which replace fuel assemblies, located in roughly inner and outer (radially) concentric circles. Typically, the bundles are separated into two groups, each of which is capable of shutting down the core. The fuel depletion reactivity effect of thermal reactors is reversed in LMFBRs, which produce more fissile nuclei than they consume. In addition, the negative reactivity effects of fission products, which are primarily thermal neutron absorbers such as samarium and xenon, are much less in an LMFBR than in a thermal reactor.

Like the RBMK and CANDU pressure tube reactors, in which the moderator is separate from the coolant, the LMFBR tends to have a positive coolant temperature coefficient of reactivity, but for a different reason. Reduction of sodium density hardens the neutron spectrum, which results in a lower capture-to-fission ratio in the fuel and reduces the number of neutrons absorbed in the large ^{23}Na resonance in the keV energy range. The fast neutron spectrum means a shorter neutron lifetime (the mean time from fission to absorption or leakage of the neutron) than in a thermal reactor because the neutron is absorbed or leaks before it slows down in an LMFBR ($\Lambda \approx 10^{-6}$ s for LMFBRs as contrasted to 10^{-4} to 10^{-5} s for thermal reactors). This implies a more rapid response to superprompt-critical ($\rho > \beta$) reactivity insertions. Furthermore, the prompt-critical reactivity level ($\rho = \beta$) with plutonium in a fast spectrum ($\beta = 0.0020$ for ^{239}Pu and $\beta = 0.0054$ for ^{241}Pu) is

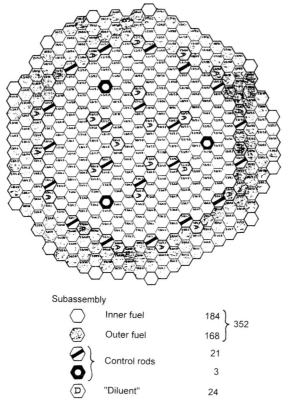

Subassembly		
⬡ Inner fuel	184	} 352
⬡ Outer fuel	168	
⬗ } Control rods	21	
⬢	3	
Ⓓ "Diluent"	24	

Fig. 7.14 Super Phenix liquid-metal fast breeder reactor. (From Ref. 6; used with permission of CRC Press.)

smaller than with ^{235}U in a thermal spectrum ($\beta = 0.0067$). On the other hand, the reactivity worth of perturbations such as inadvertent control rod withdrawal is generally smaller in a fast spectrum because of the smaller value of the absorption cross section for fast than for thermal neutrons.

A variant of the LMFBR is the advanced liquid metal reactor (ALMR), which employs a Pu/U metal alloy fuel. The configuration consists of pairs of core modules constituting a 606 MWe power block; a core can be built up of 1, 2 or 3 power blocks. The ALMR design is based on the integral fast reactor (IFR) actinide recycle concept (see Section 7.11). An IFR would generate less actinide "waste" than do light water reactors and can recycle its own actinide waste and the actinide waste of light water reactors to recover energy which would otherwise be lost, at the same time reducing the waste disposal burden. The passive safety features of the ALMR allow extreme off-normal transients—loss of primary coolant flow without scram and loss of heat removal by the intermediate system without scram—with benign consequences to the reactor core. As will be discussed in Section 8.5, tests have shown that the ALMR can undergo these extreme events without damage.

7.7
Other Power Reactors

There are also a number of other reactors, most of which have been designed to achieve enhanced production of fissile nuclei by neutron transmutation, which have been developed through the demonstration stage but not yet implemented on a significant scale as power reactors. Two of these are basically modifications of conventional thermal light water reactors. The light water breeder reactor (LWBR) operates on a ^{232}Th–^{233}U cycle, which is more favorable than the ^{238}U–^{239}Pu cycle for the production of fissile nuclei by thermal neutron transmutation (Chapter 6). The spectral shift light water reactor operates with a mixed D_2O–H_2O coolant to achieve a slightly harder neutron spectrum to enhance the transmutation of ^{238}U into fissile plutonium early in the cycle and reduces the D_2O/H_2O ratio with burn-up to soften the spectrum and increase the reactivity to offset reactivity loss due to fuel depletion.

There are also two graphite-moderated thermal reactors designed to achieve enhanced production of fissile nuclei. The thermal molten salt breeder reactor (MSBR), which operates on the ^{232}Th–^{233}U cycle with the fuel contained in a circulating molten salt (typically, LiF–BeF$_2$–ThF$_4$–UF$_4$), which also serves as the heat removal system, achieves additional enhancement of neutron utilization for fissile production by continuous removal of fission products from the recirculating fuel. The pebble bed reactor, a variant of the helium-cooled HTGR, contains the ^{232}Th–^{233}U fuel in 6-cm-diameter graphite spheres that can be poured into and drained from a core hopper.

Designs have been developed for gas-cooled fast reactors (GCFRs) which are similar to LMFBR designs, with PuO_2/UO_2 fuel pins clad with stainless steel. The fuel pins are ribbed to enhance heat transfer and their spacing is about twice that of an LMFBR assembly.

7.8
Characteristics of Power Reactors

Typical parameters relevant to power production are summarized for a number of reactor types in Table 7.1.

7.9
Advanced Generation-III Reactors

The reactors discussed in the previous sections of this chapter may be considered to be of the first and second generations of nuclear reactors, the designs of which matured in the 1960–1990s period. Designs of a third generation of power reactors have evolved starting in the 1990s, and the following Generation III reactors

Table 7.1 Representative Parameters Relevant to Power Production for the Major Reactor Types

Type	Thermal Power (MWt)	Core Diameter (m)	Core Height (m)	Average Power Density (MW/m^3)	Linear Fuel Rating (kW/m)	Average Fuel Burnup (MWd/T)
MAGNOX	1875	17.37	9.14	0.87	33.0	3,150
AGR	1500	9.1	8.3	2.78	16.9	11,000
CANDU	3425	7.74	5.94	12.2	27.9	26,400
PWR	3800	3.6	3.81	95.0	17.5	38,800
BWR	3800	5.0	3.81	51.0	19.0	24,600
RBMK	3140	11.8	7.0	4.1	14.3	15,400
LMFBR	612	1.47	0.91	380	27.0	153,000

Source: Date from Ref. 4; used with permission of Taylor & Francis/Hemisphere Publishing.

are now being deployed over roughly the 1995–2015 period. These designs have, of course, benefited from the extensive operating experience with the previous generation of power reactors. In the U.S., Europe and Japan the designs also have been driven by the major objective to incorporate passive safety features to insure safety without reliance on active control actions. In Europe, there has also been a separate emphasis on accommodating mixed oxide (MOX) fuels to take a first step towards closing the nuclear fuel cycle by recycling plutonium from spent nuclear fuel.

Advanced Boiling Water Reactors (ABWR)

The advanced BWR designs (and the advanced PWR designs discussed in the next section) are based on conventional UO_2 fuel assemblies with negative temperature coefficients. Passive safety is enhanced by designing so that, in the event of a loss of coolant accident, the core would be flooded with enough water to provide cooling for three days, without operator action (present designs require operator response in about 20 minutes).

The first two ABWRs were built in Japan and began operation in 1996 and 1997. These reactors had active core heights of 3.71 m and diameters of 5.16 m and were fueled with 3.2% enriched UO_2 designed to achieve 32,000 MWd/t discharge burnup. Operating at a power density of 50.6 MW/m^3 these reactors produce about 1350 MWe. The ABWRs have three independent and redundant safety systems which are physically and electronically isolated. These systems have the capability to keep the core covered at all times. This capability plus the substantial thermal margin in the fuel design is predicted to significantly reduce the frequency of transients which will lead to a scram (to less than one per year). Plant response to a LOCA has been completely automated, and operator action is not required for 72 hours (the same capability specified for passively safe plants). A US evolution of the ABWR concept, the Economically Simplified BWR (ESBWR) relies on natural circulation and passive safety features to simplify the design and enhance

Typical initial core loading

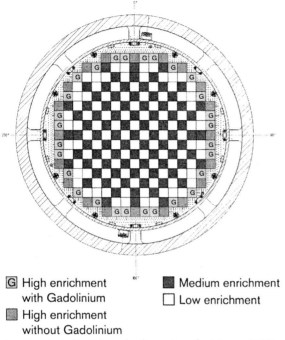

| G High enrichment with Gadolinium | ■ Medium enrichment |
| ■ High enrichment without Gadolinium | □ Low enrichment |

Fig. 7.15 Typical initial core loading pattern for Advanced PWR.

performance in a 1560 MWe plant with 4.2% enriched UO_2 fuel that is designed to achieve 50,000 MWd/t burnup.

Advanced Pressurized Water Reactors (APWR)

Advanced PWRs are being developed in Europe (EPR [1600 MWe class]), the US (AP-600 [600 MWe class] and AP-1000 [1000 MWe class]), Japan (APWR [1500 MWe class]) and Korea (APR-1400 [1400 MWe class]). An EPR is presently (2006) under construction in Europe, and there are plans to construct other APWRs worldwide during the coming decade.

The EPR will produce 4500 MWt in 241 standard 17×17 array PWR fuel assemblies each containing 265 fuel rods. The fuel rods are composed of a stack of sintered pellets of UO_2 or mixed U–Pu-oxide (MOX) enriched up to 5%, with or without Gd_2O_3 burnable poison, within a hermetically sealed zirconium alloy cladding. The fuel rods are 0.95 cm outside diameter and 4.2 m long, and the cladding thickness is 0.57 mm. The reactor core has an active height of 4.2 m and an equivalent diameter of 3.77 m. Fuel cycle lengths up to 24 months, with in-out or out-in fuel management, are possible. A typical initial core loading pattern is indicated in Fig. 7.15.

Soluble boron concentration in the coolant is varied to control relatively slow reactivity changes, including the effects of fuel burnup. The fast shutdown control system comprises 89 control assemblies, each located within a fuel assembly and consisting of 24 absorber rods fastened to a common driver assembly. The absorber rods consist of Ag, In, Cd-alloy and sintered B_4C pellets.

Advanced Pressure Tube Reactor

The Canadian CANDU line of reactors described in Section 7.3 has been improved in several respects, most notably from the reactor physics point of view by replacement of the D_2O coolant with H_2O to achieve a negative void reactivity coefficient. The new ACR-700 reactor will produce 700 MWe.

Modular High-Temperature Gas-Cooled Reactors (GT-MHR)

Two helium-cooled, modular, thermal reactor concepts are under development. These two concepts, while quite different in configuration, both take advantage of the heat capacity of a large amount of graphite to achieve passive safety, take advantage of triple coated TRISO fuel particles to confine fission products, and would build up full scale power plants from modular units.

The Pebble Bed Modular Reactor (PBMR) illustrated in Fig. 7.16 consists of a vertical steel pressure vessel lined with a layer of graphite bricks which reflect neutrons and provides a large heat capacity. Control rods are inserted downward into vertical holes in the graphite reflector. The reactor core is 3.7 m in diameter and 9.0 m high. The core consists of an inner zone of graphite spheres, which serve as a neutron moderator and to provide heat capacity, and an outer annular zone that contains about 370,000 tennis ball size fuel pebbles. Helium flows downward around the fuel spheres. This reactor module would produce 110 MWe.

Each 60 mm fuel sphere, or pebble, is coated with a 5 mm thick graphite layer and contains about 15,000 coated TRISO fuel particles each 0.92 mm in diameter, as illustrated in Fig. 7.17. The central kernel of the TRISO particle contains UO_2 enriched up to 8%, surrounded by a porous carbon buffer layer to contain the gaseous fission products and then by two pyrolytic graphite layers and a SiC structural layer which prevent release of fission products.

Criticality is maintained during PBMR operation by removing depleted fuel pebbles from the bottom of the reactor and replenishing with fresh or slightly burned fuel pebbles at the top of the core. The burnup of fuel spheres leaving the core would be measured, those exceeding the reference burnup would be removed to storage and the others would be recycled (about 10 times).

The Gas Turbine-Modular Helium Reactor (GT-MHR) is based on an extension of the HTGR prismatic designs discussed in Section 7.5 to use coated particle TRISO fuel, which is projected to lead to fuel burnup > 100,000 MWd/t, coupled to a high efficiency direct Brayton cycle gas turbine. Each modular reactor unit would produce 600 MWth which would be converted at 48% efficiency to 286 MWe.

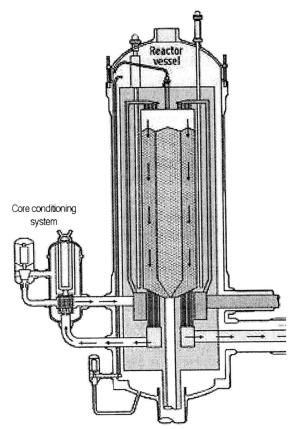

Fig. 7.16 Pebble Bed Modular Reactor.

In the GT-MHR, somewhat smaller TRISO fuel particles are mixed with a carbonaceous matrix and formed into cylindrical fuel compacts 13 mm in diameter and 51 mm long which are loaded into fuel channels in hexagonal graphite fuel elements 793 mm long by 360 mm across flats. The hexagonal fuel elements are stacked 10 elements high into 102 columns to form an annular core. Graphite reflector blocks are place inside and outside of the annular core to reflect and moderate neutrons and to provide adequate heat capacity to maintain fuel temperatures below damage limits during a loss-of-coolant-accident without the need for corrective operator actions.

7.10
Advanced Generation-IV Reactors

The ongoing (2006) international Generation-IV studies are intended to identify reactors and associated separations technologies that will make a major step to-

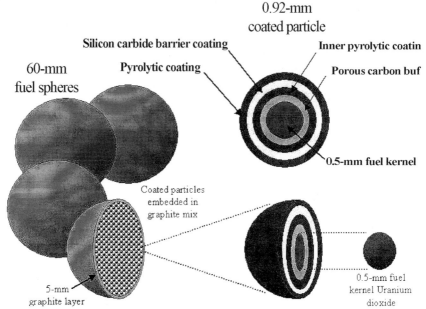

Fig. 7.17 Pebble Bed TRISO Fuel Sphere Cross Section.

wards closing the nuclear fuel cycle. The specific goals are: (1) long term availability of systems that realize effective uranium utilization; (2) minimization of the discharged nuclear waste; (3) safety, reliability, a very low likelihood and degree of core damage, and no need for off-site emergency response; (4) unattractive targets for the diversion of weapons-usable materials and acts of terrorism; and (5) a life-cycle cost advantage over and a level of financial risk comparable to other energy systems.

Six reactor systems were identified for the studies in order to provide some redundancy and complementarity in the capability to satisfy the three missions (electricity production, hydrogen and process heat production, and actinide management), to attain the performance objectives and to attract commercial deployment consistent with the national priorities of the countries involved. The six reactor systems chosen were (1) the Gas-cooled Fast Reactor (GFR), (2) the Lead-cooled Fast Reactor (LFR), (3) the Sodium-cooled Fast Reactor (SFR), (4) the Very High Temperature gas-cooled Reactor (VHTR), (5) the Super-Critical Water Reactor (SCWR), and (6) the Molten Salt Reactor (MSR).

The salient features and objectives of these reactors are given in Table 7.2.

Gas-Cooled Fast Reactors (GFR)

The GFR system features a fast neutron spectrum, helium coolant and a closed fuel cycle. The potential for high outlet helium temperature would enable the GFR to meet the hydrogen and electricity production missions, and the fast neutron spectrum would facilitate achieving the actinide fissioning mission. Core config-

Table 7.2 Features and Objectives of GEN-IV Reactor Designs

Reactor	Coolant	Neutron Spectrum	Electric Mission	H-Prod Mission	Actinide Mission	Earliest Deploy
GFR	Gas	Fast	X	X	X	2025
LFR	Lead	Fast	X	X	X	2025
MSR	Molten salt	Epi-thermal	X	X	X	2025
SCWR	Water	Thermal	X		(fast option)	2025
SFR	Sodium	Fast	X		X	2020
VHTR	Gas	Thermal		X		2020

urations may be based on assemblies of prismatic blocks (as for the HTGR and GT-MHR) or pin or plate fuel elements. Composite ceramic fuel, coated TRISO fuel particles and ceramic-clad dispersion fuel are being considered. The GFR can use a direct Brayton cycle helium turbine for electricity production or can use its process heat for the thermochemical production of hydrogen. By full recycling of actinides in a fast spectrum, the GFR would minimize the discharge of long-lived waste to HLW repositories and could maximize the utilization of the energy content of uranium (including depleted U) by transmutation of ^{238}U into fissionable transuranics. The current (2006) reference design produces 288 MWe.

Lead-Cooled Fast Reactors (LFR)

The LFR system features a fast neutron spectrum, liquid lead or lead-bismuth eutectic coolant and a closed fuel cycle. The potential for high outlet liquid metal temperature (up to 850°C) would enable the LFR to meet the hydrogen and electricity production missions, and the fast neutron spectrum would facilitate achieving the actinide fissioning mission. Metal and nitride-based dispersion fuels, containing uranium and transuranics, would be cooled by natural convection of the liquid metal coolant. By full recycling of actinides in a fast spectrum, the LFR would minimize the discharge of long-lived waste to HLW repositories and could maximize the utilization of the energy content of uranium (including depleted U) by transmutation of ^{238}U into fissionable transuranics. At present (2006), a range of plant sizes are being considered, including a 1200 MWe plant, a modular system rated at 300–400 MWe, and a "battery" rated at 50–150 MWe that features a 15–20 year refueling interval and a replaceable core cassette. By full recycling of actinides in a fast spectrum, the LFR would minimize the discharge of long-lived waste to HLW repositories and could maximize the utilization of the energy content of uranium (including depleted U) by transmutation of ^{238}U into fissionable transuranics.

Molten Salt Reactors (MSR)

The MSR system features an epi-thermal neutron spectrum, a circulating molten salt fuel-coolant mixture and a full actinide recycle fuel cycle. The potential for coolant outlet temperatures up to 800°C would enable the MSR to meet the electricity and hydrogen production missions, but the epi-thermal neutron spectrum would require multiple transmutation of some actinides to arrive at species with a significant fission cross-section in meeting the actinide fissioning mission. The fuel would be a circulating liquid mixture of sodium, zirconium, uranium and actinide fluorides flowing through channels in a graphite core structure.

Super-Critical Water Reactors (SCWR)

The SCWR system features a high-temperature, high-pressure water-cooled reactor that operates above the critical point of water to obtain a higher thermal efficiency than current LWRs. The balance of plant is simplified by the use of a direct cycle energy conversion system. The present (2006) reference is a 1700 MWe plant operating at 25 MPa and outlet temperature of 510°C. A second option with a fast neutron spectrum would have a closed fuel cycle with full actinide recycle and an aqueous processing facility.

Sodium-Cooled Fast Reactors (SFR)

The SFR system features a fast neutron spectrum, sodium coolant and a closed fuel cycle. With an outlet coolant temperature of 550°C, the present (2006) SFR designs can meet the electricity mission but not the hydrogen production mission, and the fast neutron spectrum would facilitate achieving the actinide fissioning mission. The fuel cycle includes full actinide recycle, with two major options: (1) a 150–500 MWe reactor with a uranium–plutonium–actinide–zirconium metal alloy fuel and a pyrometallurgical processing facility, based on IFR technology; and (2) a 500–1500 MWe reactor with uranium–plutonium–oxide dispersion fuel and an advanced aqueous processing facility, based on LMFBR technology. By full recycling of actinides in a fast spectrum, the SFR would minimize the discharge of long-lived waste to HLW repositories and could maximize the utilization of the energy content of uranium (including depleted U) by transmutation of ^{238}U into fissionable transuranics.

Very High Temperature Reactors (VHTR)

The VHTR system features a thermal neutron spectrum, helium coolant and a once-through uranium fuel cycle. With an outlet helium temperature of 1000°C, the VHTR is intended to be a high-efficiency system that provides process heat for non-electrical applications such as thermochemical hydrogen production, but it could include cogeneration of electricity. The reactor core can be prismatic (like the HTGR and GT-MHR) or pebble-bed (like the PBMR). The present (2006) reference design is a 600 MWth core connected to an intermediate heat exchanger to deliver process heat.

7.11
Advanced Sub-critical Reactors

Sub-critical operation would allow a much wider range of fuel cycle options for any of the above reactors because remaining critical would not constrain the fuel burn-up. Also, because the delayed neutron fraction for the actinides (e.g., $\beta = 0.0020$ for ^{239}Pu in a fast spectrum) is considerably less than for ^{235}U ($\beta = 0.0064$), operating a reactor with a substantial transuranic loading would significantly reduce the reactivity margin to prompt critical. Furthermore, the requirements to fission a substantial fraction of the fissile isotopes in the fuel before removing it from the reactor or to achieve a significant ^{238}U transmutation rate translate into requirements to compensate a substantial negative reactivity decrement. Since sub-critical operation provides an additional reactivity margin $\approx 1 - k_{sub}$ and an external neutron source S can be increased to offset any decrease in reactivity and hence to maintain constant the number of neutrons in the reactor ($N = (S\ell/(1 - k_{sub}))$, where ℓ is the neutron lifetime), achievement of both the uranium utilization and the actinide fissioning missions would be facilitated by sub-critical operation. Based on the studies to date, it is the preponderance of informed opinion that sub-critical operation will be required in at least some of the reactors in the international "fleet" in order to fully achieve the closed fuel cycle.

Two types of neutron sources of the size required are being developed on a time scale that should make them available shortly after the first deployment of advanced GEN-IV reactors, (1) the proton accelerator-spallation target neutron source which is the basis for the Spallation Neutron Source which recently (2006) became operational at Oak Ridge National Laboratory, and (2) the tokamak D–T fusion neutron source which is the basis of the international ITER fusion project that is beginning the construction phase this year (2006) in France.

Conceptual designs for an Accelerator Transmutation of Waste (ATW) reactor are based on a proton Linac (about 0.9 kilometer in length) accelerating protons to about 1000 MeV, at which point they would be directed downward through a vertical column in the center of the reactor into a Pb–Li target to produce via the spallation reaction copious neutrons ranging in energy up to 20 MeV that are incident onto the surrounding sub-critical reactor, as illustrated in Fig. 7.18. Projected neutron source rates are 0.1–1×10^{19}/s, and the source distribution would be highly localized about the target. The technology needed for the ATW will be tested on the Spallation Neutron Source.

Conceptual designs for a Fusion Transmutation of Waste (FTW) reactor are based on a D–T plasma confined in a toroidal confinement volume of major radius $R = 3$–4 m producing 14 MeV neutrons. The sub-critical reactor forms an annular ring (inner radius 4–5 m, height ≈ 3 m, width ≈ 1 m) located on the outboard side of the plasma chamber, as shown in Fig. 7.19. Projected neutron source rates are 1–10×10^{19}/s, and the source would be broadly distributed over the surface of the toroidal plasma chamber. The physics and technology needed for a fusion neutron

GAS COOLED ADS DEMO
SPALLATION TARGET

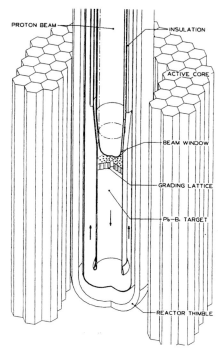

Fig. 7.18 Gas Cooled ADS Demo Spallation Target.

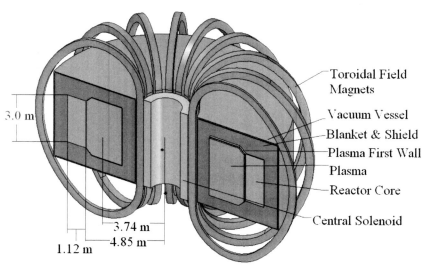

Fig. 7.19 Tokamak D–T Fusion Neutron Source-Driven Sub-Critical Reactor.

source will be tested in the ITER experiment which will begin operation in 2016–2019.

7.12
Nuclear Reactor Analysis

We now turn to a brief discussion of the application of the computational methods of reactor physics to analysis of the nuclear, or neutronics, performance of nuclear power reactors. More detailed discussions of reactor analysis procedures and a description of the various codes employed are given in Refs. 5 and 6. The advanced reactor physics calculational methods used in nuclear reactor analysis, in addition to those described in previous chapters, are described in Chapters 9 to 16.

Construction of Homogenized Multigroup Cross Sections

As the discussion of nuclear power reactors above illustrates, nuclear reactor cores are composed of tens of thousands of components of very different material properties, some of them highly absorbing fuel and control elements, with dimensions that are comparable to or smaller than the neutron diffusion length. Yet the major computational tool of nuclear reactor analysis is multigroup diffusion theory, which is rigorously valid only in weakly absorbing media at distances of a few diffusion lengths away from interfaces with strongly dissimilar media. Furthermore, many of the nuclear cross sections depend strongly on the details of the neutron energy distribution (e.g., resonances), which in turn are spatially dependent through the spatial distribution of materials. Thus the first major step of nuclear reactor analysis is to develop equivalent homogenized cross sections for the different fuel assemblies or fuel modules, which incorporate the effects of the detailed neutron distribution in space and energy, and to develop an equivalent representation of highly absorbing control elements. The word *equivalent* implies that these approximate representations would yield the same prediction of reaction rates as a detailed heterogeneous fine-energy calculation would, were it practical to perform the latter. Construction of such equivalent representations is a major and ongoing reactor physics activity.

The relative importance of the treatment of spatial and energy details differs among reactor types. For thermal reactors, in which most of the neutrons are absorbed in the thermal energy range where the neutron mean free path is small, treatment of the detailed spatial heterogeneity is paramount, and treatment of the details of the energy distribution is secondary but still important. On the other hand, for fast reactors, in which most of the neutron absorption takes place with fast neutrons with long mean free paths, treatment of the details of the energy distribution is paramount, and the spatial heterogeneity is secondary.

In a thermal reactor, the homogenization procedure starts at the pin-cell level of a fuel pin and the surrounding coolant, moderator, and structure. In a typical analysis, a volume-weighted homogenized fine group (30 to 60 fast groups, 15 to

172 thermal energy points or groups) pin-cell model is constructed, using integral transport theory to calculate heterogeneous resonance cross sections for the fuel nuclei. This model is used to calculate intermediate group cross sections to be used in a transport calculation of the heterogeneous pin cell. The spatially dependent intermediate group fluxes are used to construct volume-flux-weighted homogenized cross sections, usually with a smaller number of groups, for the pin cell. This is repeated for the various types of pin cells in a fuel assembly to obtain an intermediate group (5 to 15 groups) model for the fuel assembly which represents each fuel pin cell as an equivalent homogenized region. Then an intermediate group diffusion or transport calculation is performed for the fuel assembly or module, taking into account any water gaps, nearby structure or control elements, and so on. The intermediate group fluxes from the assembly transport calculation are then used to construct volume-flux-weighted few-group diffusion theory cross sections for the assembly. Usually, separate calculations are performed for the fast and thermal ($E < 1$ eV) energy regions. This process is repeated for the various fuel assemblies or modules that compose the core, resulting in equivalent few-group diffusion or transport theory cross sections which represent each homogenized fuel assembly or module. Supplemental transport calculations are used to construct effective few-group diffusion theory cross sections which represent the control elements in a diffusion theory model.

In a fast reactor the procedure is similar, but with more emphasis on treatment of the energy structure and of overlapping resonances and less on the treatment of the spatial structure (except as it affects the resonance treatment). In a typical analysis, an entire fuel assembly or group of similar fuel assemblies is homogenized on a volume-weighted basis to obtain an ultrafine-group ($\approx$2000) model, with integral transport calculations being used to construct heterogeneous resonance cross sections for the fuel nuclei. Ultrafine-group spectra are then calculated and used to construct fine-group cross sections for use in a multigroup (20 to 40 groups) diffusion or transport theory core calculation of the entire core.

Homogenized multigroup cross sections must be constructed for the variety of conditions encountered in subsequent applications because they depend on the details of the spatial and spectral flux distributions used in their construction. The presence or absence of a control element or a strong absorber such as xenon, the change in fuel composition with burnup, the buildup of plutonium and fission products, the different temperature and coolant densities encountered in a transient, and other factors, all affect the details of the spatial and spectral distributions and must be taken into account in the preparation of equivalent homogenized multi-group cross sections.

Criticality and Flux Distribution Calculations

The equivalent homogenized multigroup cross sections can be used to perform global diffusion or transport theory calculations of the reactor core, with the control rod positions adjusted to achieve criticality ($k = 1$) or with an eigenvalue k calculated. If three-dimensional finite-difference representations of the core are used for these calculations, the calculated fluxes are averaged global flux distributions. However, detailed pin-by-pin flux distributions are needed for the calculation of pin power limits and pin fuel depletion. The detailed local flux at the fuel pin-by-fuel pin level must be reconstructed by superimposing on this global average flux the detailed assembly and pin-cell transport flux distributions that were used in preparation of the homogenized multigroup cross sections, with the appropriate normalization.

Frequently, further approximations are made in the calculation of global flux distributions, in the interest of computational economy (e.g., nodal models that represent the global flux distribution within a fuel assembly or module with a few parameter polynomial). In such cases, the detailed local flux on a fuel pin-by-fuel pin level again must be reconstructed by superimposing on this representation of the global average flux the detailed assembly and pin-cell transport flux distributions that were used in the preparation of the homogenized multigroup cross sections. Care must be taken that the flux reconstruction procedure is consistent with the homogenization procedures and with the procedures used in the development of the approximate global flux calculation model.

Fuel Cycle Analyses

Calculation of the multigroup global flux distribution and critical control rod position and reconstruction of the flux distribution on a pin-by-pin basis is coupled with the calculation of fuel composition change and fission product buildup on a pin-by-pin basis in the multistep fuel cycle analysis calculation. The sequence of calculations is first to perform a number of flux calculations to establish the critical control rod position and corresponding flux distribution for the fresh fuel loading with and without equilibrium xenon and samarium, then solution of the fuel depletion and actinide/fission product buildup equations over a depletion time step using the initial equilibrium xenon and samarium flux distribution, then solution of neutron flux equations several times to establish the critical rod position and flux distribution corresponding to the new fuel composition and fission products, then the solution of the fuel depletion and actinide/fission product buildup equations over the next depletion time step using the newly calculated flux distribution with equilibrium xenon and samarium, and so on. Typical time steps might be 150, 350, 500, several 1000, and then 2000 MWd/T, the initial small time steps taken to build up equilibrium xenon and samarium and ^{239}Pu. Homogenized multigroup cross sections must be redetermined at each time step, either from a recomputation as described above or from interpolation in a table of fitted cross sections.

Efficient fuel management requires that the fuel cycle analysis described in the preceding paragraph be repeated many times. A series of such calculations will be made prior to fuel loading to determine the proper mixture and location of fresh and recycled fuel to achieve optimal fuel performance subject to a given set of assumptions about plant availability, power demand schedule, and refueling period. Then as the reactor operates and the assumptions are replaced by operating history, additional series of calculations are made to adjust the remaining operating plan and/or refueling date to achieve optimal fuel performance.

The large number of criticality and flux distribution calculations needed for fuel cycle analyses places a computational efficiency requirement on the neutron flux solution method. Approximate flux solution methods, such as the nodal model, are widely used. However, it should be noted that Monte Carlo codes capable of calculating fuel depletion on a point-by-point basis are available.

In fast breeder reactor calculations, there is a greater emphasis on determination of the initial plutonium concentration in the fuel and on the production and destruction of actinides with fuel depletion.

Transient Analyses

It is necessary to analyze a large number of planned operational transients (e.g., startup, power-level change) and potential transients that could result from off-normal or accident conditions (e.g., control rod ejection, loss of coolant flow), each subject to a variety of assumptions regarding the performance of control and reactor systems. Such calculations require solution of the time-dependent equations describing the neutron flux distribution and the reactor power level and distribution, heat conduction and the temperature distribution, the hydrodynamics and thermodynamics of the heat transport system, material expansion and movement, and in the case of extreme accident scenarios, the equations of state and the equations governing the hydrodynamics of melting and vaporizing fuel mixtures. Calculation of the neutron flux spatial distribution and level determines the reactivity, which is the driving function for any reactor transient, and the heating source level and distribution, which is the primary input to the other calculations.

In the simplest point kinetics model for neutron dynamics, the neutron flux distribution is assumed to be fixed and only the amplitude, or power-level, changes. The reactivity coefficients associated with fuel and moderator temperature and density changes, fuel and structure motion, and so on, are precomputed from a series of static neutron flux and criticality calculations (or from a few such calculations supplemented by perturbation theory estimates). Then, as changing temperatures and densities, fuel and structure motion, and so on, are calculated, the reactivity worth of these changes is incorporated into the power-level calculation using the reactivity coefficients.

Reference, or design basis, power distributions are often used in conjunction with point kinetics calculations to assess fuel integrity. Separate calculations are then performed to assure that the actual power distribution is not more limiting than the design basis power distribution. For certain accident simulations (e.g., those in which control elements are out of position), the design basis power distributions are inappropriate and new power distributions must be calculated based on the temperature, density, flow, and other information from the transient analysis calculation.

The reactivity feedback coefficients are determined for reference control rod positions and other core conditions. If the control rod positions or the core conditions are altered significantly, the reactivity coefficients, which depend on a flux-adjoint volume weighting of the perturbation, will be different because the flux and adjoint distributions will be different. The most important reactivity coefficients must be computed for conditions present during the most critical stages of the transient analysis.

The point-kinetics calculation cannot account for effects associated with changes in the spatial flux distribution, which may occur, for example, if there is a reduction of coolant flow only in one part of the reactor. Such changes in spatial flux distribution not only affect the local power distribution and heat source distribution but also produce changes in reactivity and in the reactivity coefficients. Thus there are situations in which calculation of the space- and time-dependent flux distribution is required. Such calculations require, in essence, a series of solutions for the spatial flux distribution, using at each step the most recent calculations of the temperature, density, and position of the materials in the reactor. Approximate flux solution methods, such as the nodal model, are normally used in such cases to make the computational requirements tractable.

Core Operating Data

Precalculated or on-line calculated values of various core physics parameters and responses must be available to the reactor operators to enable them to make core operational decisions, such as the control element insertion pattern, and to interpret instrument readings. Much of this information is developed in the course of fuel management and transient safety analyses, since the safety analysis considers a wide range of abnormal and normal conditions. Other information is provided by core operating data, although these are usually only for normal operating conditions. Additional power distribution and criticality calculations are necessary to fill in the database.

Criticality Safety Analysis

At various stages of the enrichment, fabrication, and transportation procedures prior to loading the fuel into the reactor, and at various stages of the temporary storage, processing, transportation, and permanent storage procedures for spent nuclear fuel, the nuclear fuel is distributed within a variety of configurations. Examples of such configurations are spent fuel assemblies stored in a swimming pool (to provide for decay heat removal) at the reactor site and barrels of processed fuel in liquid form arrayed on storage racks. Criticality safety requires a rigorous fuel management system to insure that the fuel inventories of each storage element is known and that the various configurations are well subcritical under all normal and conceivable off-normal conditions. Criticality calculations of the type discussed for the case when the fuel is loaded into the reactor must also be performed for these various ex-reactor configurations. While diffusion theory and the methodology discussed in previous chapters may suffice for certain of these configurations, the more rigorous transport methods of Chapter 9 are generally required for criticality safety analyses.

7.13
Interaction of Reactor Physics and Reactor Thermal Hydraulics

Power Distribution

More than 90% of the recoverable energy released in fission is in the form of kinetic energy of fission products and electrons, which is deposited in the fuel within millimeters of the site of the fission event, and somewhat less than 10% of the energy is in the form of energetic neutrons and gamma rays, which are deposited within about 10 cm around the fission site. Thus the heat deposition distribution is approximately the same as the fission rate distribution:

$$q'''(r) \approx \text{const.} \times \sum_f (r)\phi(r) \tag{7.1}$$

The requirement to remove this heat without violating constraints on maximum allowable values of materials temperature, heat flux from the fuel into the coolant, and so on, places limits on allowable neutron flux peaking factors, fuel element dimensions, coolant distribution, and so on. The neutron flux distribution affects the temperature in the fuel and coolant/moderator, the temperature of the fuel affects the fuel resonance cross section, and the temperature of the coolant/moderator affects the moderating power, both of which in turn affect the neutron flux distribution.

An increase in the local resonance absorption in the fuel when the local fuel temperature increases results because of the Doppler broadening of the resonances. This increase in local fuel absorption cross section will generally reduce the number of neutrons that reach thermal locally in LWRs, which will tend to reduce the local fission rate and compensate the original increase in fuel temperature. The increase in local fuel resonance absorption makes the fuel compete more effectively for local neutrons, which lends to make other nearby absorbers somewhat less effective (e.g., reduces the worth of nearby control rods).

The effect of coolant temperature on neutron moderation is also important. In most LWR cores, a local decrease in water density resulting from an increase in water temperature will cause a decrease in neutron moderation, which in turn causes a decrease in local power deposition. As the coolant passes up through the core, the cumulative heat input from the fuel elements causes the axial temperature distribution to increase with height: conversely, the axial density distribution decreases with height. This produces a power distribution peaked toward the bottom of the core, which is pronounced in BWRs, for which progressive coolant voiding occurs in the upper part of the core. Control rods are inserted from the bottom in BWRs to maximize rod worth and to avoid exacerbating this peaking in the axial neutron flux at the bottom of the core. The shift toward a harder spectrum associated with a local Na density decrease in a fast reactor results in an increase in local η, which increases the local heating. The coupling between reactor physics and thermal hydraulics is much weaker in gas-cooled reactors, in which the moderator is separate from the coolant.

Temperature Reactivity Effects

The general reactivity effects associated with changes in fuel, coolant/moderator, and structural temperatures and their effect on the reactor dynamics were discussed in Sections 5.7 to 5.12. The interaction of thermal-hydraulics and reactor physics phenomena to produce positive reactivity in the Three Mile Island and Chernobyl accidents is discussed in Section 8.4. The overall reactivity effect depends on the local changes in temperature and density in each zone of the reactor and the local neutron flux, weighted by the relative importance of these local reactivity contributions and summed over the reactor. The thermal-hydraulics characteristics of a reactor affect not only the local temperature and density changes in response to a change in the neutron flux distribution and magnitude, but also affect changes in the neutron flux distribution and magnitude in response to changes in local temperature and density.

Coupled Reactor Physics and Thermal-Hydraulics Calculations

It is clear from the discussion above that the power distribution and effective multiplication constant in a nuclear reactor depends not only on the distribution of material (fuel, coolant, structure, control) within a reactor core, but also on the temperature and density distribution within a reactor core. In the design process, it is necessary to determine a self-consistent material and temperature–density distribution that makes the reactor critical at operating conditions without violating thermal-hydraulics limits. The problem is further complicated by fuel depletion, which changes the materials in the fuel during the course of time; the distributions of materials and temperature–density must make the reactor critical over its entire lifetime without violating thermal-hydraulics limits. This is normally accomplished by trial and error, iterating between static neutron flux and thermal-hydraulics calculations until a self-consistent solution is found which can be made critical by adjusting control poison levels and which satisfies thermal-hydraulics and safety limits over the projected core lifetime.

Once the design is fixed, it is necessary to analyze a number of operational and off-normal transients to ensure that the reactor will operate without violation of thermal-hydraulics limits under normal conditions and that it will operate safely under off-normal conditions. The transient analyses codes usually solve for the neutron power amplitude and distribution and the corresponding temperature and density distributions, in some approximation.

References

1 "European Pressurized Water Reactor," *Nucl. Eng. Des. 187*, 1–142 **(1999)**.

2 D. C. WADE and R. N. HILL, "The Design Rationale of the IFR," *Prog. Nucl. Energy 31*, 13 **(1997)**; E. L. GLUEKLER, "U.S. Advanced Liquid Metal Reactor (ALMR)," *Prog. Nucl. Energy 31*, 43 **(1997)**.

3 R. A. KNIEF, *Nuclear Engineering*, Taylor & Francis, Washington, DC **(1992)**, Chaps. 8–12.

4 J. G. COLLIER and G. F. HEWITT, *Introduction to Nuclear Power*, Hemisphere Publishing, Washington, DC **(1987)**, Chaps. 2 and 3.

5 P. J. TURINSKY, "Thermal Reactor Calculations," in Y. Ronen, ed., *CRC Handbook of Nuclear Reactor Calculations*, CRC Press, Boca Raton, FL **(1986)**.

6 M. SALVATORES, "Fast Reactor Calculations," in Y. Ronen, ed., *CRC Handbook of Nuclear Reactor Calculations*, CRC Press, Boca Raton, FL **(1986)**.

7 R. H. SIMON and G. J. SCHLUETER, "From High-Temperature Gas-Cooled Reactors to Gas-Cooled Fast Breeder Reactors," *Nucl. Eng. Des. 4*, 1195 **(1974)**.

8 *Report on the Accident at the Chernobyl Nuclear Power Station*, NUREG-1250, U.S. Nuclear Regulatory Commission, Washington, DC **(1987)**.

Problems

7.1. Discuss the differences between 'thermal' and 'fast' reactors in terms of (a) neutron energy distribution, (b) capability to fission uranium and higher transuranic isotopes, (c) Doppler temperature coefficient of reactivity, (d) time constant for dynamic response to prompt supercritical reactivity insertions, (e) coolants that could be used, and (f) power density.

7.2. Discuss why fast reactors are necessary to close the nuclear fuel cycle.

8
Reactor Safety

A great deal of effort is devoted to ensuring that nuclear reactors operate safely. The fundamental objective of this effort is to ensure that radionuclides are not released to create a health hazard to the general public or operating personnel. Fundamental considerations of reactor safety, the methodology of safety analyses, reactor accidents, and the design approach to reactor safety are described in this chapter, with an emphasis on the role played by reactor physics.

8.1
Elements of Reactor Safety

Radionuclides of Greatest Concern

The radionuclides in a nuclear reactor that could most affect public health if released are the fission products and the actinides produced by neutron transmutation. For the most part, these radionuclides are harmful only if they are inhaled or ingested and concentrated chemically in a susceptible organ. As discussed in Chapter 6, the short-lived fission products constitute the major source of such radionuclides in an operating reactor. The most significant fission products and the organs they affect are identified in Table 8.1.

^{90}Sr and ^{137}Cs and the isotopes of iodine are of particular concern. Strontium has a high fission yield and behaves chemically like calcium and is deposited in bone tissue. Both ^{90}Sr and its daughter ^{90}Y produce a very high dose per unit activity, which is quite damaging to the blood cells produced in bone marrow. The iodine radioisotopes are concentrated in the thyroid gland, where they would produce tumors.

Multiple Barriers to Radionuclide Release

Multiple barriers against fission product (and actinide) release are a key safety feature of nuclear reactor design. The fission products in an operating reactor are contained within UO_2 pellets that are packed into clad fuel elements which are assembled within the reactor core. The reactor core is located within a pressure

Table 8.1 Significant Fission Products of Concern for Internal Doses in Reactor Accidents

Isotope	Radioactive Half-Life, $t_{1/2}$	Fission Yield (%)	Deposition Fraction*	Effective Half-Life	Internal Dose (mrem/μCi)	Reactor Inventory† (Ci/kWt) 400 Days	Equilibrium
Bone							
89Sr	50 d	4.8	0.28	50 d	413	43.4	43.6
90Sr–90Y	28 y	5.9	0.12	18 y	44,200	1.45	53.6
91Y	58 d	5.9	0.19	58 d	337	53.2	53.6
144Ce–144Pr	280 d	6.1	0.075	240 d	1,210	34.7	55.4
Thyroid							
131I	8.1 d	2.9	0.23	7.6 d	1,484	26.3	26.3
132I	2.4 h	4.4	0.23	2.4 h	54	40.0	40.0
133I	20 h	6.5	0.23	20 h	399	59.0	59.0
134I	52 m	7.6	0.23	52 m	25	69.0	69.0
135I	6.7 h	5.9	0.23	6.7 h	124	53.6	53.6
Kidney							
103Ru–103mRh	40 d	2.9	0.01	13 d	6.9	26.3	26.3
106Ru–106Rh	1.0 y	0.38	0.01	19 d	65	1.8	3.5
129mTe–129Te	34 d	1.0	0.02	10 d	46	9.1	9.1
Muscle							
137Cs–137mBa	33 y	5.9	0.36	17 d	8.6	1.2	53.6

Source: Data from Ref. 14.

* Fraction of inhaled material that deposits in the indicated tissue.

† A somewhat typical average residence time for fuel in an LWR is 400 full-power days; equilibrium inventories are achieved at times that are long compared to the radionuclide half-life.

vessel that in turn is located inside a containment building. Both the pressure vessel and the containment building are designed to withstand large overpressures. Thus the pellet, clad, pressure vessel, and containment building constitute four barriers against the release of fission products.

Defense in Depth

The first level of defense against fission product release is to design to *prevent* the occurrence of any event that could result in damage to the fuel or other reactor system. Negative reactivity coefficients that lead to inherently stable operating conditions, safety margins in design, reliable and known materials performance in structures and components, adequate instrumentation and control, and so on, are among the preventive measures employed in reactor design.

The second level of defense are *protective systems*, which are designed to halt or bring under control any transients resulting from operator error or component failure that may lead to fuel damage and fission product release within the pressure vessel. Reactor scram systems which inject control rods into the core for rapid shutdown upon being activated by any one of several signals being outside the tolerance range, pressure relief systems, and so on, constitute the reactor protective systems.

The third level of defense is provided by *mitigation systems*, which limit the consequences of accidents if they do occur. Emergency core cooling, emergency secondary coolant feedwater, emergency electrical power systems, systems for removing fission products that have been released into the reactor hall, and a reinforced containment building that can withstand high overpressure are elements of the mitigation system.

Energy Sources

The potential for the release of fission products is related directly to the amount of energy available. The primary energy source is the nuclear energy that is released in a positive reactivity insertion. However, there are other important energy sources that can play a role in an accident. The heat released in fission product decay is 7.5% of the operating power and constitutes a substantial heat source for some time after the reactor shuts down. There is thermal energy stored in the reactor materials which may become redistributed (e.g., the flashing of water to steam upon depressurization). There are several exoergic chemical reactions (Table 8.2) which may take place at elevated temperatures during the course of an accident, most of which produce hydrogen, which has an explosive potential.

8.2
Reactor Safety Analysis

All reasonably conceivable failures are postulated and analyzed to design reactor protective and mitigation systems, to prevent accidents, to prevent the release of

Table 8.2 Properties of Exoergic Reactions of Interest for Reactor Safety

Reactant (R)	Temperature (°C)	Oxide(s) Formed	Heat of Reaction* with: Oxygen (kcal/kg°R)	Water (kcal/kg°R)	Hydrogen Produced with Water (l/kg°R)
Zr (liquid)	1852[†]	ZrO_2	−2883	−1560	490
SS (liquid)	1370[†]	FeO, Cr_2O_3, NiO	−1330 to −1430	−144 to −253	440
Na (solid)	25	Na_2O	−2162	–	–
	25	NaOH	–	−1466	490
C (solid)	1000	CO	−2267	+2700	1870
	1000	CO_2	−7867	+2067	3740
H_2 (gas)	1000	H_2O	−29,560	–	–

Source: Data from Ref. 15; used with permission of MIT Press.

* Positive values indicate energy that must be added to initiate an endoergic reaction; negative values indicate energy released by exoergic reactions.

† Melting point.

fission products in the event of an accident, and to investigate the consequences of various accident scenarios for the release of radionuclides. The analyses are performed with sophisticated computer code systems that model the neutron dynamics and fission power production; the temperature, density, state, and location of materials within the reactor core and the reactivity worth of changes therein: the primary and secondary heat transport system[1]; the performance of the reactor safety protective and mitigation systems; the integrity of the fuel elements, pressure vessel and containment structure intended to prevent release of radionuclides; the dispersion of any released radionuclides; and a radiological assessment of resulting health effects in the population affected.

Accident scenarios are commonly classified by the initiating event, some of the major events being those discussed below.

Loss of Flow or Loss of Coolant

Loss-of-flow accidents (LOFAs) would be caused by failure of one or more pumps in the primary coolant system, which results in increased temperature and reduced density for the coolant. Loss-of-coolant accidents (LOCAs) can be caused by a rupture of the primary coolant line, failure of a primary coolant pump seal, inadvertent opening of a pressure relief or safety valve, and so on, and would result in increased temperature and decreased density of the coolant and possibly uncovering of the core. The negative coolant temperature reactivity coefficient of PWRs and BWRs, which would provide for an immediate power reduction, is an important feature in the early stages of such accidents.

Loss of Heat Sink

When steam flow in the secondary coolant system is decreased or lost due to a turbine trip (shutdown) or reduction or loss of feedwater in the secondary coolant system, an undercooling accident, or in an extreme case, a loss-of-heat sink accident (LOHA), would occur. Such an accident would result in the reduction or elimination of heat removal from the primary coolant system, causing the primary coolant temperature to increase and the density to decrease. Again, a negative coolant temperature coefficient of reactivity is an important feature in the early stages of such accidents.

Reactivity Insertion

Uncontrolled control rod withdrawal or ejection is the most common type of initiator for a reactivity insertion accident. However, there are other reactivity insertion

---

1) A PWR has a primary coolant system that removes heat from the reactor core and carries it to the steam generator, or heat exchanger, where the heat is transferred out of the primary coolant through tube walls to a cooler secondary coolant which is heated above the vaporization temperature to produce steam that is transported to turbines for the production of electricity.

mechanisms. The startup of an inactive primary coolant pump (or recirculation loop in a BWR), which injects cold water into the primary coolant system, would cause a positive reactivity insertion in reactors with a negative coolant reactivity coefficient. A steam line break in the secondary coolant system would result in increasing coolant flow in the secondary system, hence in increasing heat removal from the primary coolant, which would also result in a positive reactivity insertion in reactors with a negative coolant reactivity coefficient. The potential for such cold-water reactivity insertions places limits on the allowed magnitude of negative coolant reactivity coefficients.

Anticipated Transients without Scram

Anticipated transients without scram (ATWSs) are certain transient events which may occur once or twice in a reactor lifetime, on average, which can be handled by the protective system initiating a reactor scram. When the scram system is postulated to fail, such events may initiate an accident.

8.3
Quantitative Risk Assessment

Development and application of a methodology for quantification of the risk to public and worker safety associated with the occurrence of a reactor accident has provided a valuable basis for evaluating the relative safety of nuclear reactors. In broad terms, the (public safety) *risk* associated with a nuclear reactor may be characterized in terms of the various sequences of events, or *scenarios*, that could lead to the release of various quantities of radionuclides, the probabilities that each sequence of events could occur, and the public or worker health *consequences* of the release of various quantities of radionuclides.

Probabilistic Risk Assessment

Safety protective and mitigation systems are designed to minimize component damage and prevent radionuclide release for each of the potential accident-initiating events described in the preceding section (plus others), if the system works as designed. For a given initiating event (e.g., a loss-of-coolant accident), the success or failure of the hierarchy of relevant safety systems—electric power, emergency core cooling, fission product removal from the reactor hall, containment—are considered sequentially. The frequency of occurrence of the initiating event, λ, and the failure probabilities, P_i, for each safety system are first identified. Then an *event tree* is constructed, as shown in the upper portion of Fig. 8.1, tracing the various pathways that the accident could follow with respect to success or failure of the various safety systems. Since the conditional failure probabilities, P_i, are small, the overall probability of any given pathway is just the initiation frequency times the product of the P_i for the different failures in the pathway, if the failure probabilities

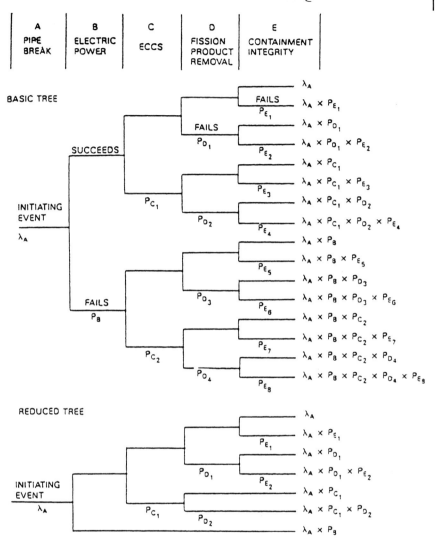

Fig. 8.1 Event tree logic diagram for a LOCA in an LWR. (From Ref. 13.)

are independent. However, the failure probabilities of the various safety systems are not independent (e.g., electrical power failure implies also failure of the emergency core cooling and fission product removal systems). Accounting for correlated failures reduces the event tree, as indicated in the lower portion of Fig. 8.1.

Quantification of the initiating event frequencies and of the safety system failure probabilities is, of course, the essential part of this methodology. A deductive technique known as *fault tree analysis* is employed for this purpose. A given safety system failure (e.g., loss of electrical power) requires failure of both the primary (off-site power supply) and the backup (on-site diesel generator) systems. Failure of the off-site power supplies requires failure of both the power sources on the lo-

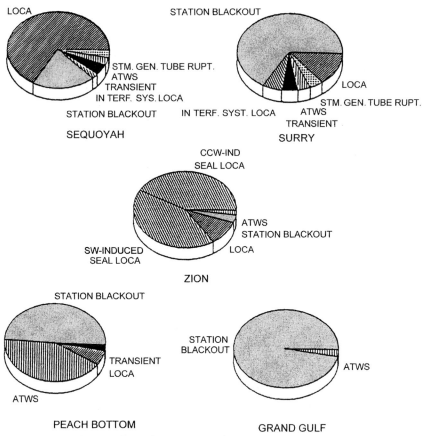

Fig. 8.2 Estimated principal contributions to core damage frequency for Sequoyah PWR, Surrey PWR, Zion PWR, Peach Bottom BWR, and Grand Gulf BWR. (From Ref. 5.)

cal grid and the tie-in with other power grids, or a failure of the local power grid. Each of these secondary failure possibilities can be related, in turn, to several possible tertiary causes, and so on. By tracing these possible failures back several levels, a fault tree can be constructed. By assigning failure probabilities at each level on the fault tree and combining these probabilities statistically, an overall failure probability for a given safety system can be constructed. Data used for fault tree analyses include component and system failure rate data, human error, maintenance and testing time when a system might not be available, and so on. Probabilistic risk assessment has proven to be a powerful tool for identifying the relative importance of various failure modes in a given plant. However, the results tend to be different for plants with different reactor types and containment systems, as shown in Fig. 8.2.

Radiological Assessment

The public health consequences of the release of a given inventory of radionuclides from a containment building depends on the dose received to various body organs by the affected population and the effect of that dose on those organs. The dispersion of radionuclide fallout from the release point depends on wind and weather conditions. The population that might be affected by this fallout depends on the population pattern of the fallout zone and any evacuation measures that would be taken. Calculation of radionuclide dispersion among the affected population is relatively straightforward. Most radionuclides must be inhaled or ingested to affect public health. Immediately following their release, breathing is the most likely pathway for radionuclides to enter the body. Over the longer term, there are many possible pathways, including breathing, drinking contaminated water, and eating contaminated food at any step in the food chain, that must be considered. Calculation of radionuclide uptake by the affected population is uncertain, and worst-case assumptions must be made when information is lacking.

Health effects of radiation exposure fall into three categories: early fatalities (acute), early illnesses, and latent effects. Early fatalities—defined as those that occur within a year of exposure—follow a linear dose–effect relationship varying from 0.01% fatality risk for 320 rad[2] to 99.99% fatality risk for 750 rad whole-body radiation exposure has been established from radiation effects data. Early illnesses are associated primarily with the respiratory tract and lung impairment in particular. A linear dose–effect relationship varying from 5% lung impairment for 3000 rad to 100% impairment for 6000 rad internal radiation exposure to the lung has been established from radiation effects data. Latent effects of radionuclide ingestion include cancer fatalities, thyroid nodules, and genetic damage, which generally occur 10 to 40 years after the accident. Linear dose–effect relationships can be established for significant levels of radiation exposure, but there are no radiation effects data at the low levels of exposure that would be encountered in trying to determine the latent effects of radionuclide ingestion following an accident. It is common practice to extrapolate the linear dose–effect relationship to zero dose in predicting latent health effects, but this practice is controversial because theoretical studies suggest that a threshold level of radiation energy deposition is required to cause cell damage. The predicted cancer fatality rate, using the linear extrapolation to zero dose, is about 100 per 10^6 person-rem exposure.

Reactor Risks

The estimated frequencies and public health and property damage consequences of possible PWR/BWR reactor accidents are given in Table 8.3. The most likely

2) Radiation doses are measured in a variety of units. The *rad* corresponds to the absorption of 100 ergs/g of material, and the *gray* (Gy) is equal to 100 rads. The *rem* is equal to the *rad* multiplied by a quality factor (1 for x-rays, gammas, and electrons; 10 for neutrons and protons; 20 for alpha particles), and the *sievert* (Sv) is 100 *rems*.

Table 8.3 Estimated Probabilities and Consequences of a Single Reactor Accident

Consequences	Chance per Reactor per Year					Normal incidence
	$1:2 \times 10^4$*	$1:10^6$	$1:10^7$	$1:10^8$	$1:10^9$	
Early fatalities	<1.0	<1.0	110	900	3,300	–
Early illness	<1.0	300	3,000	14,000	45,000	4.5×10^5
Latent cancer fatalities (per year)†	<1.0	170	460	860	1,500	17,000
Thyroid nodules (per year)†	<1.0	1,400	3,500	6,000	8,000	8,000
Genetic effects (per year)‡	<1.0	25	60	110	170	8,000
Total property damage ($10⁹)	<0.1	0.9	3	8	14	–
Decontamination area [km² (mi²)]	<0.3 (<0.1)	5,000 (2,000)	8,000 (3,200)	8,000 (3,200)	8,000 (3,200)	–
Relocation area [km² (mi²)]	<0.3 (<0.1)	340 (130)	650 (250)	750 (290)	750 (290)	–

Source: Data from Ref. 13.

* This is the predicted chance of core melt per reactor year.
† These rates would occur in approximately the 10 to 40-year period following a potential accident.
‡ This rate would apply to the first generation born after a potential accident. Subsequent generations would experience effects at a lower rate.

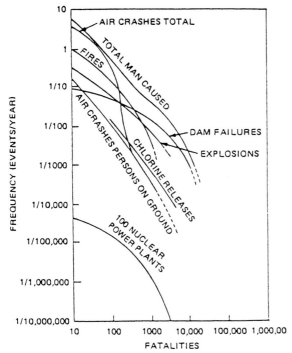

Fig. 8.3 Predicted frequency of fatality due to accidents from a number of technologies. (From Ref. 13.)

core meltdown accident, which has a probability of 5×10^{-5} per reactor-year of occurring, has rather modest consequences. The more serious accidents have lower probabilities of occurrence.

To put the risks of reactor accidents in perspective, the same methodology was applied to estimate the public health risks of other technological and natural phenomena to which the general public are exposed. As shown in Figs. 8.3 and 8.4, the risk to public health of the approximately 100 nuclear reactors operating in the United States is miniscule by comparison.

8.4
Reactor Accidents

There have been two major reactor accidents, at Three Mile Island and at Chernobyl. It is important to understand what went wrong. Examination of the causes provides a basis for the design of reactors with improved safety features and operating procedures for the future.

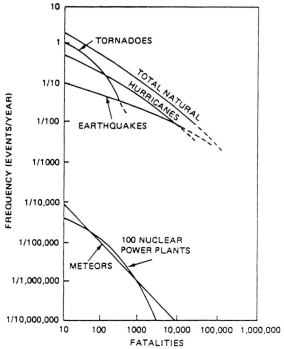

Fig. 8.4 Predicted frequency of fatalities due to nuclear reactor accidents and to a number of natural events. (From Ref. 13.)

Three Mile Island

On March 28, 1979, a series of events took place in unit 2 of the Three Mile Island plant near Harrisburg, Pennsylvania that resulted in the only major reactor accident in the history of commercial nuclear power in the United States. The TMI-2 unit was a standard PWR. Since this accident was associated primarily with the heat removal system, a simple diagram of a PWR heat removal system is shown in Fig. 8.5 to facilitate understanding of the sequence of events. The reactor was operating at about 97% of power, but with two valves on the emergency secondary coolant feedwater lines inadvertently closed, although the records available to the operators showed them to be open. The accident was apparently initiated by unsuccessful attempts to carry out a routine procedure of clearing a demineralizer line used to maintain secondary coolant purity, which apparently caused a condensate pump trip in the secondary cooling system. This led within a second to automatic trips in the main feedwater pumps for the secondary coolant system and the turbine. The loss of secondary coolant in the steam generators reduced the rate of heat removal from the primary coolant loop and the reactor core (a loss of heat sink accident).

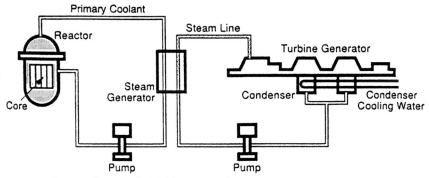

Fig. 8.5 Schematic diagram of a PWR heat removal system.

As the primary coolant became hotter and pressure increased, the overpressure relief valve in the pressurizer in the primary coolant system opened automatically when the (15.55 MPa) set point was exceeded, and 8 s into the accident the core protective system caused the control rods to be inserted in response to high coolant pressure signals. The primary system cooled following the control rod insertion, and the pressure dropped below the 15.21 MPa set point for closure of the overpressure relief valve at about 13 s into the accident, but the valve failed to close, although the solenoid deenergized, causing the primary coolant to be lost through the open valve into the drain tank at the bottom of the containment building, which reduced the pressure in the primary coolant system as well as the coolant level. At this point there was a loss of coolant accident, unbeknown to the operators. The control panel only indicated that the solenoid had deenergized, and primary coolant continued to be lost until the operators closed the blocked valve in the pressurizer drain line 142 min into the accident.

At 14 s into the accident, the emergency secondary coolant feedwater pumps reached full design pressure, but unbeknown to the operators, the two inadvertently closed valves in the emergency secondary system coolant lines prevented the emergency secondary coolant from reaching the steam generators. It was another 8 min before an operator noticed low pressure and water levels in the steam generators, discovered the closed valves, and opened them to restore secondary coolant to the steam generators.

At about 2 min into the accident, the primary system pressure dropped below the 11.31-MPa set point of the high-pressure injection system, which then started pumping borated water into the core. Because of the particular design, there was no direct relationship between the coolant levels in the reactor vessel and in the pressurizer. Even with continuing loss of primary coolant, the pressurizer signal indicated a filled system, which the operators had been trained to avoid because it prevented the pressurizer from fulfilling its function. Thus the operators turned off one of the pumps and throttled back the other pump in the high-pressure injection system, resulting in emergency coolant being added at a slower rate than primary coolant was being lost through the open pressurizer valve.

About 73 min into the accident, both primary coolant pumps in the loop to one of the two steam generators were shut down in response to indications of vibrations, low pressure, and low coolant flow. This was done to prevent destruction of seals, which the operators feared would have caused a loss of coolant accident, still being unaware that they already had one on their hands. At about 100 min into the accident, the primary coolant pumps in the other loop were shut down for similar reasons. The pump shutdown caused the steam and water in the primary coolant loop to separate and apparently prevented further coolant circulation through the steam generators. The remaining liquid did not cover the core, and decay heat caused continuing vaporization of the noncirculating coolant. At about 111 min into the accident, reactor outlet coolant temperatures rose rapidly to 325°C and remained there. As the core became uncovered, the clad temperatures became high enough that exoergic Zr–steam reactions occurred, adding energy to the system and producing hydrogen. The cladding, with a melting point of 2100 K, became molten and began to dissolve the UO_2 fuel.

The next 13 h was spent trying various means to reestablish core cooling, which was ultimately successful. The reactivation of the high-pressure injection at 200 min into the accident recovered the core and filled the reactor vessel. A major slumping of the molten core occurred at 224 min into the accident, resulting in molten debris being deposited onto the lower vessel head, where it was apparently quenched by the coolant. A sizable hydrogen bubble was created by the Zr–steam interactions involving about one-third of the zircaloy in the core, the concentration of which became large enough to support combustion, and hydrogen ignition occurred at about 9.5 h into the accident. However, the pressure was well within the design limits of the pressure vessel. The hydrogen was removed during the first week.

Reactor containment was successful in limiting radionuclide releases to less than 1% of total inventory, despite extensive core damage. Radiological assessments of the radionuclide release estimated average and maximum potential off-site doses of 0.015 and 0.83 mSv. As a point of reference, a dose of 1 mSv is estimated to result in a 1 in 50,000 chance of cancer, as contrasted with the 1 in 7 normal incidence of cancer in the population. The TMI-2 accident had no significant public health impact.

In hindsight, TMI-2 was a huge and costly but poorly instrumented safety experiment that provided a convincing demonstration of the safety of a properly engineered nuclear reactor. Two of the major credible accidents—loss of heat sink and loss of coolant—took place, while the operators, who were unaware of the state of the reactor, took about the worst possible actions for the actual situation in an attempt to deal with the situation they thought they had on their hands. Although the reactor was destroyed, no one got hurt. By the same token, TMI-2 exposed major deficiencies in reactor operating procedures, operator training, and exchange of safety-related operating information, which stimulated extensive subsequent improvements.

Chernobyl

In the early morning hours of April 26, 1986, a test was being performed on unit 4 of the Chernobyl nuclear power station about 130 km north of Kiev. The objective was to test the use of energy in the turbine during its post-trip coastdown as a source of emergency electrical power for cooling the reactor core following a scram, ironically to enhance the safety features of the reactor system. The test plan called for the power of the RBMK reactor to be reduced from the 3200-MWt full-power level to about 1000 to 700 MWt and for bypassing some safety systems that would have prevented the test conditions from being realized.

The test was initiated by inserting control rods to reduce power to about 1600 MWt, the emergency core cooling systems were shut off to prevent them from drawing power during the test, and the power reduction continued to the planned level. However, the operator failed to reprogram the computer to maintain power in the range 1000 to 700 MWt, and the power fell to 30 MWt. The majority of the control rods were withdrawn to compensate the buildup of xenon, causing the power to climb and stabilize briefly at about 200 MWt. At about 20 min into the test, all eight pumps were activated to ensure adequate post-test cooling. The normal scram trip on high flow level, which would have prevented this, was deactivated. The increase in coolant flow reduced coolant temperature and increased coolant density, which introduced negative reactivity due to increased neutron absorption in the coolant, requiring further control rod withdrawal. This increased coolant density also maximized the positive reactivity worth of coolant voiding. The combination of low power and high flow produced instability, which required numerous manual adjustments, causing the operators to deactivate other emergency shutdown signals.

At about 22 min into the test, the computer indicated excess reactivity. The operators blocked the last remaining trip signal just before it would have scrammed the reactor, in order to be able to complete the test. Power started to rise and coolant voiding in the pressure tubes occurred, leading to a positive reactivity input which enhanced the power rise. The operators began control rod insertion from the fully withdrawn position. However, the fully withdrawn control rods had graphite followers below the control poison (to enhance rod worth), and these entered the active core first, displacing neutron-absorbing water with graphite and thus adding further positive reactivity, which accelerated the power increase. The power surged to 100 times design full power in the next 4 s, then decreased momentarily. There then followed repeated power pulses, one of which may have reached 500 times design full power. The fuel disintegrated, breached the cladding, and entered the water coolant, causing a steam explosion that lifted the top shield of the reactor core, shearing all the coolant pipes and removing all the control rods. The explosion was well beyond the rather modest containment design basis and penetrated the concrete walls of the reactor building, dispersing burning fuel and graphite, and releasing a plume of radioactive gases and particles.

The accident resulted in 31 early fatalities. Over 1000 people received large doses of radiation. Many of the nearby population received doses greater that 0.25 Sv

(25 rem), with the most serious in the range 0.4 to 0.5 Sv (40 to 50 rem). As a reference, recommended annual dose limits by the International Council on Radiation Protection are 50 mSv (5 rem) whole-body radiation and 500 mSv (50 rem) for any body part other than the lens of the eye. The radioactivity released into the atmosphere fell out in measurable amounts over much of the world. Estimated individual whole-body doses immediately following the accident were on the order of 100 mGy (10 rad) in the immediate vicinity of the plant, 4 mGy (400 mrad) in Poland, 1 mGy (100 mrad) in the rest of Europe, and 0.01 mGy (1 mrad) in Japan and North America. The 24,000 evacuees who received an estimated average dose of 0.43 Sv were expected to incur an additional 26 fatal leukemias over the next decade, roughly doubling the natural incidence of leukemia fatalities in that population.

On a long-term basis, the predicted collective lifetime doses due to the fallout from the Chernobyl accident are 1.6×10^4 person-Gy for the evacuated population near the site, 4.7×10^5 person-Gy for the European part of the former USSR, 1.1×10^5 person-Gy for the Asian part of the former USSR, 5.8×10^5 person-Gy for Europe, 2.7×10^4 person-Gy for Asia, 1.1×10^3 person-Gy for the United States, and 1.2×10^6 person-Gy for the entire northern hemisphere. The increase in the estimated 50-year exposure doses in Europe, for example, varied from a fraction of the natural background to a few times the natural background. There is no scientific evidence on which to assess the effect, if any, of such small incremental doses. However, by extrapolating from higher dose levels, it is possible to estimate the long-term health effects of fallout from the Chernobyl accident. The estimated increase above natural incidence of fatal cancers in the respective populations due to the Chernobyl fallout is 2.4% for the evacuated population near the site, 0.12% for the European part of the former USSR, 0.01% for the Asian part of the former USSR, 0.02% for Europe, 0.00013% for Asia, and 0.00005% for the northern hemisphere.

Postaccident assessments identified design-related defects as (1) positive coolant void reactivity coefficient, (2) easy-to-block safety systems, (3) slow scram (15 to 20 s for full insertion, 5 s for effective negative reactivity), and (4) absence of containment and emergency fission product control systems. These design-related defects are uniquely applicable to the RBMK reactors, which are deployed only in the former Soviet Union. Technical fixes that have been implemented subsequently on other RBMK reactors include (1) maximum allowable control rod withdrawal limitations, (2) modifications to prevent operators from manually overriding safety systems, (3) reduction of the positive coolant void reactivity coefficient, and (4) development of an alternative shutdown capability.

Operator error and lax management were obviously at least partially responsible for the Chernobyl accident, and the government placed much of the blame there. Six members of plant management were subsequently tried and convicted for violation of safety rules, criminal negligence, and so on, and the station director, chief engineer, and deputy chief engineer were sentenced to 10 years in a labor camp. However, the positive coolant temperature coefficient and the absence of a containment building designed to withstand overpressure events were also major contributors to the accident.

8.5
Passive Safety

The experience of TMI-2 and Chernobyl has led to an emphasis on passive safety in the design of advanced reactors. Broadly speaking, the objectives of passive safety design are, to the extent possible, for the reactor to be able to maintain a balance between power production and heat removal, to shut itself down when an abnormal event occurs, and to remove decay heat, without requiring operator action or the functioning of engineered safety systems.

Pressurized Water Reactors

The AP-600 design features a passive emergency core cooling system consisting of water stored in large tanks above the core. During a loss-of-coolant accident, this water is injected into the core while the coolant system is still pressurized, and flows into the core under gravity when the system depressurizes, without requiring either pumps or electrical power. Decay heat, which is normally removed through the steam generators, would be removed by the natural circulation of water through the core into a large tank above the reactor vessel in the event that the steam generators were inoperable. The containment shell is cooled by gravity-driven water spray and the natural circulation of air. Because of reliance on passive safety, there are only half the number of large pumps as on a standard PWR.

The PIUS reactor vessel, pressurizer, and steam generators are all immersed in borated water. If a pump fails during normal operation, the hydrostatic pressure forces the borated water into the core, where it serves both as emergency coolant and a shutdown mechanism. The natural circulation between the core and the pool of borate water would remove decay heat.

Boiling Water Reactors

Main coolant flow for the boiling water reactor (SWBR) design is provided by natural circulation, eliminating the need for the recirculation pumps, valves, and associated controls of a standard boiling water reactor. In the event of a loss-of-coolant accident, steam is vented into a large suppression pool located above the core to depressurize the cooling system, which allows water from the pool to gravity flow down into the core to provide emergency core cooling. Decay heat can be removed to the suppression pool by natural circulation. The entire system is enclosed in a concrete containment structure that is cooled continuously by water flow downward from the suppression pool, the evaporation of which provides passive heat removal from the core to the atmosphere.

Integral Fast Reactors

The approach to safety embodied in the integral fast reactor (IFR) includes (1) large design margins between operating conditions and safety limits, (2) reliance on passive processes to hold power production in balance with heat removal, and (3) totally passive removal of decay heat. The IFR can be designed to achieve passive power regulation, even should equipment in the control and balance-of-plant systems fail, for anticipated transient without scram scenarios. The heat transport system that removes decay heat operates at ambient pressure, has large thermal inertia, is driven by natural convection, is contained along with the core in a double-wall top-entry coolant tank, is completely independent of the balance of plant equipment, and is always in operation.

The IFR system can be designed to have an inherent response that prevents release of radioactivity, even for accidents of extremely low probability far below the design basis level. Processes that are innate consequences of the materials and geometry cause dispersal of fuel early enough to avoid prompt criticality and the accompanying energy release and to ensure subcriticality and coolability inside an intact reactor vessel should significant fuel pin failures cause an accumulation of radioactive debris.

Passive Safety Demonstration

The passive safety features of the IFR have been demonstrated dramatically in a series of tests in the Experimental Breeder Reactor II (EBR-II), which has the same type of fuel and heat transport system as the IFR. It was demonstrated that the reactor operating at full power would be safely shut down by negative reactivity feedback, without benefit of the scram or any other safety system or of operator action, upon loss of forced coolant flow and upon loss of heat sink, two of the most demanding reactor accident scenarios. Transient temperatures during shutdown were measured to be below those of concern for fuel integrity and reactor safety.

In the first test, the coolant pumps were shut off while the reactor was operating at full power with the scram system deactivated (a separate emergency scram system was operable but not used). No operator action was taken. The response of EBR-II to the loss of coolant flow is shown in Fig. 8.6. The negative reactivity feedback associated with the increase in coolant temperature following the loss of coolant flow resulted in a rapid reduction in power, which reduced the coolant temperature. Because the metal fuel has a large heat conductivity and operates at a temperature only slightly greater than that of the coolant, there is a relatively small negative Doppler reactivity coefficient and consequently, relatively little positive reactivity addition when the coolant temperature decreases later in the transient.

In the second test, the ability of the system to reject heat from the primary coolant was eliminated while the reactor was at full power, with the scram system deactivated and no operator action taken. The response of EBR-II to the loss of heat sink is shown in Fig. 8.7. Again, negative reactivity feedback shut the reactor down without any danger to the plant.

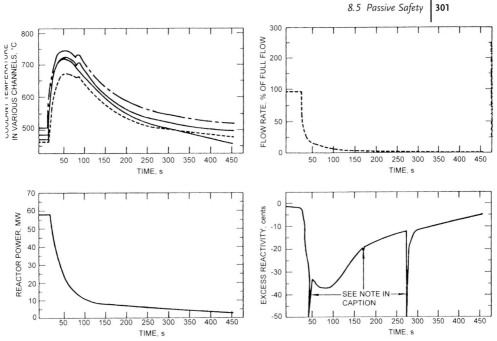

Fig. 8.6 Response of EBR-II to loss of flow (discontinuities in
reactivity are artificial). (From Ref. 6.)

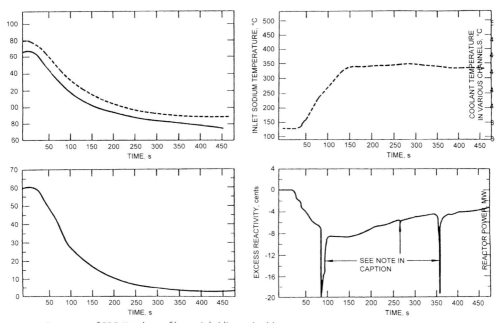

Fig. 8.7 Response of EBR-II to loss of heat sink (discontinuities
in reactivity are artificial). (From Ref. 6.)

References

1 D. C. WADE, R. A. WIGELAND, and D. J. HILL, "The Safety of the IFR," *Prog. Nucl. Energy. 31*, 63 (**1997**).

2 R. A. KNIEF, *Nuclear Engineering*, Taylor & Francis, Washington, DC (**1992**), Chaps. 13–16.

3 H. CEMBER, *Introduction to Health Physics*, 3rd ed., McGraw-Hill, New York (**1996**).

4 K. E. CARLSON et al., *RELAP5/MOD3 Code Manual*, Vols. I and II, EG&G Idaho report NUREG/CR-5535, U.S. Nuclear Regulatory Commission, Washington, DC (**1990**).

5 *Severe Accident Risks: An Assessment for Five U.S. Nuclear Plants*, NUREG-1150, U.S. Nuclear Regulatory Commission, Washington, DC (**1989**); *Nucl. Eng. Des. 135*, 1–135 (**1992**).

6 S. FISTEDIS, "The Experimental Breeder Reactor-II Inherent Safety Demonstration," *Nucl. Eng. Des. 101*, 1 (**1987**); J. I. SACKETT, "Operating and Test Experience with EBR-II, the IFR Prototype," *Prog. Nucl. Energy 31*, 111 (**1997**).

7 J. G. COLLIER and G. F. HEWITT, *Introduction to Nuclear Power*, Hemisphere Publishing, Washington, DC (**1987**), Chaps. 4–6.

8 "Special Issue: Chernobyl," *Nucl. Safety 28* (**1987**).

9 "Chernobyl: A Special Report," *Nucl. News, 29*, 87 (**1986**); "Chernobyl: The Soviet Report," *Nucl. News*, Special Report (**1986**).

10 *Report on the Accident at the Chernobyl Nuclear Power Station*, NUREG-1250, U.S. Nuclear Regulatory Commission, Washington, DC (**1987**).

11 "The Ordeal at Three Mile Island," *Nucl. News*, Special Report (**1979**).

12 *The TMI-2 Lessons Learned Task Force Final Report*, NUREG-0585, U.S. Nuclear Regulatory Commission, Washington, DC (**1979**).

13 WASH-1400, *Reactor Safety Study: An Assessment of Accident Risks in U.S. Commercial Nuclear Power Plants*, NUREG-74/014, U.S. Nuclear Regulatory Commission, Washington, DC (**1975**).

14 T. J. BURNETT, *Nucl. Sci. Eng. 2*, 382 (**1957**).

15 T. J. THOMPSON and J. G. BECKERLEY, eds., *The Technology of Nuclear Reactor Safety*, MIT Press, Cambridge, MA (**1964**).

Problems

8.1. Discuss the differences in reactivity feedback between the Three-Mile Island and Chernobyl accidents.

8.2. Discuss the important differences between a modern LWR and the Chernobyl RBMK reactors that make it extremely unlikely that a Chernobyl-type accident, with a large-scale release of radioactivity, could occur in a LWR.

8.3. What design changes would you make to the Three-Mile Island PWR reactor?

8.4. What design changes would you make to the Chernobyl RBMK reactor?

Part 2: Advanced Reactor Physics

9
Neutron Transport Theory

Calculation of the transport of neutrons and their interaction with matter are perhaps the fundamental topics of reactor physics. In this chapter, the major computational methods used for the transport of neutrons in nuclear reactors are described.

9.1
Neutron Transport Equation

The distribution of neutrons in space and angle is defined by the particle distribution function $N(\mathbf{r}, \mathbf{\Omega}, t)$, such that $N(\mathbf{r}, \mathbf{\Omega}, t) d\mathbf{r} d\mathbf{\Omega}$ is the number of neutrons in volume element $d\mathbf{r}$ at position r moving in the cone of directions $d\mathbf{\Omega}$ about direction $\mathbf{\Omega}$, as depicted in Fig. 9.1. An equation for $N(\mathbf{r}, \mathbf{\Omega}, t)$ can be derived by considering a balance on the differential cylindrical volume element of length $dl = \mathrm{v} dt$, where v is the neutron speed, and cross-section area dA surrounding the direction of neutron motion, as shown in Fig. 9.2. The rate of change of $N(\mathbf{r}, \mathbf{\Omega}, t)$ within this differential volume is equal to the rate at which neutrons with direction $\mathbf{\Omega}$ are flowing into the volume element (e.g., across the left face in Fig. 9.2) less the rate at which they are flowing out of the volume element (e.g., across the right face), plus the rate at which neutrons traveling in direction $\mathbf{\Omega}$ are being introduced into the volume element by scattering of neutrons within the volume element from different directions $\mathbf{\Omega}'$ and by fission, plus the rate at which neutrons are being introduced into the volume element by an external source S_{ex}, minus the rate at which neutrons within the volume element traveling in direction $\mathbf{\Omega}$ are being absorbed or being scattered into a different direction $\mathbf{\Omega}'$:

$$
\begin{aligned}
\frac{\partial N}{\partial t}&(\mathbf{r}, \mathbf{\Omega}, t) d\mathbf{r} d\mathbf{\Omega} \\
&= \mathrm{v}\big(N(\mathbf{r}, \mathbf{\Omega}, t) - N(\mathbf{r} + \mathbf{\Omega} dl, \mathbf{\Omega}, t)\big) dA d\mathbf{\Omega} \\
&\quad + \int_0^{4\pi} d\mathbf{\Omega}' \Sigma_s(\mathbf{r}, \mathbf{\Omega}' \rightarrow \mathbf{\Omega}) \mathrm{v} N(\mathbf{r}, \mathbf{\Omega}', t) d\mathbf{r} d\mathbf{\Omega} \\
&\quad + \frac{1}{4\pi} \int_0^{4\pi} d\mathbf{\Omega}' \nu \Sigma_f(\mathbf{r}) \mathrm{v} N(\mathbf{r}, \mathbf{\Omega}', t) d\mathbf{r} d\mathbf{\Omega} \\
&\quad + S_{\mathrm{ex}}(\mathbf{r}, \mathbf{\Omega}) d\mathbf{r} d\mathbf{\Omega} - \big(\Sigma_a(\mathbf{r}) + \Sigma_s(\mathbf{r})\big) \mathrm{v} N(\mathbf{r}, \mathbf{\Omega}, t) d\mathbf{r} d\mathbf{\Omega}
\end{aligned}
\tag{9.1}
$$

Nuclear Reactor Physics. Weston M. Stacey
Copyright © 2007 WILEY-VCH Verlag GmbH & Co. KGaA, Weinheim
ISBN: 978-3-527-40679-1

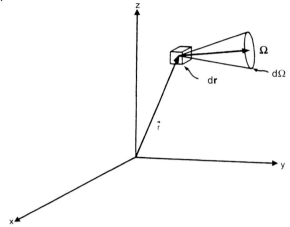

Fig. 9.1 Particles in $d\mathbf{r}$ at location $\mathbf{r}$ moving in the cone $d\mathbf{\Omega}$ about the direction $\mathbf{\Omega}$. (From Ref. 2; used with permission of Wiley.)

Making a Taylor's series expansion

$$N(\mathbf{r} + \mathbf{\Omega}\,dl, \mathbf{\Omega}, t) = N(\mathbf{r}, \mathbf{\Omega}, t) + \frac{\partial N(\mathbf{r}, \mathbf{\Omega}, t)}{\partial l} dl + \cdots$$

$$= N(\mathbf{r}, \mathbf{\Omega}, t) + \mathbf{\Omega} \cdot \nabla N(\mathbf{r}, \mathbf{\Omega}, t) + \cdots \tag{9.2}$$

to evaluate the streaming term, denning the directional flux distribution

$$\psi(\mathbf{r}, \mathbf{\Omega}, t) \equiv v N(\mathbf{r}, \mathbf{\Omega}, t) \tag{9.3}$$

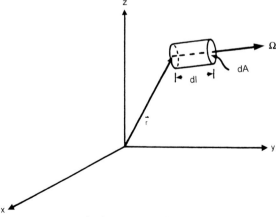

Fig. 9.2 Incremental volume element for particles at location $\mathbf{r}$ moving in the direction $\mathbf{\Omega}$ (From Ref. 2; used with permission of Wiley.)

and taking note of the fact that the scattering from $\mathbf{\Omega}'$ to $\mathbf{\Omega}$ depends only on $\mathbf{\Omega} \cdot \mathbf{\Omega} \equiv \mu_0$, so that

$$\Sigma_s(\mathbf{r}, \mathbf{\Omega}' \to \mathbf{\Omega}) = \frac{1}{2\pi} \Sigma_s(\mathbf{r}, \mathbf{\Omega} \cdot \mathbf{\Omega}') \equiv \frac{1}{2\pi} \Sigma_s(\mathbf{r}, \mu_0) \tag{9.4}$$

and writing $\Sigma_t = \Sigma_a + \Sigma_s$, leads to the neutron transport equation

$$\frac{1}{v} \frac{\partial \psi}{\partial t}(\mathbf{r}, \mathbf{\Omega}, t) + \mathbf{\Omega} \cdot \nabla \psi(\mathbf{r}, \mathbf{\Omega}, t) + \Sigma_t(\mathbf{r}) \psi(\mathbf{r}, \mathbf{\Omega}, t)$$

$$= \int_{-1}^{1} d\mu_0 \Sigma_s(\mathbf{r}, \mu_0) \psi(\mathbf{r}, \mathbf{\Omega}', t)$$

$$+ \frac{1}{4\pi} \int_0^{4\pi} d\mathbf{\Omega}' v \Sigma_f(\mathbf{r}) \psi(\mathbf{r}, \mathbf{\Omega}, t) + S_{ex}(\mathbf{r}, \mathbf{\Omega}) \equiv S(\mathbf{r}, \mathbf{\Omega}) \tag{9.5}$$

The representation of the neutron streaming operator, $\mathbf{\Omega} \cdot \nabla \psi$, in the common geometries is given in Table 9.1, and the respective coordinate systems are defined in Figs. 9.3 to 9.5.

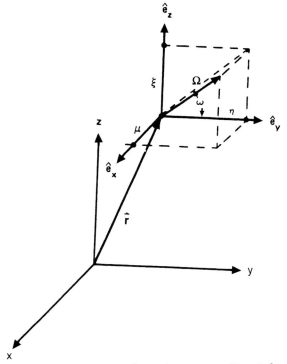

Fig. 9.3 Cartesian space–angle coordinate system. (From Ref. 2; used with permission of Wiley.)

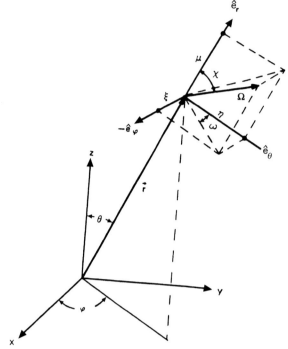

Fig. 9.4 Spherical space–angle coordinate system. (From Ref. 2; used with permission of Wiley.)

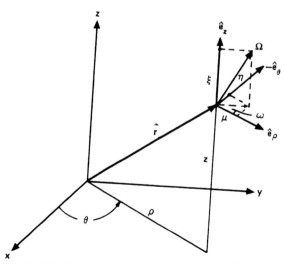

Fig. 9.5 Cylindrical space–angle coordinate system. (From Ref. 2; used with permission of Wiley.)

Table 9.1 Neutron Streaming Operator in Conservative Form

Spatial Variables	Angular Variables	$\Omega \cdot \nabla \psi$
Streaming Operator in Rectangular Coordinates		
x (one dimension)	μ	$\mu \dfrac{\partial \psi}{\partial x}$
x, y (two dimensions)	μ, η	$\mu \dfrac{\partial \psi}{\partial x} + \eta \dfrac{\partial \psi}{\partial y}$
x, y, z (three dimensions)	μ, η, ξ	$\mu \dfrac{\partial \psi}{\partial x} + \eta \dfrac{\partial \psi}{\partial y} + \xi \dfrac{\partial \psi}{\partial z}$
Streaming Operator in Cylindrical Coordinates in Conservation Form		
ρ (one dimension)	ω, ξ	$\dfrac{\mu}{\rho} \dfrac{\partial}{\partial \rho}(\rho \psi) - \dfrac{1}{\rho} \dfrac{\partial}{\partial \omega}(\eta \psi)$
ρ, θ (two dimensions)	ω, ξ	$\dfrac{\mu}{\rho} \dfrac{\partial}{\partial \rho}(\rho \psi) - \dfrac{\eta}{\rho} \dfrac{\partial \psi}{\partial \theta} - \dfrac{1}{\rho} \dfrac{\partial}{\partial \omega}(\eta \psi)$
ρ, z (three dimensions)	ω, ξ	$\dfrac{\mu}{\rho} \dfrac{\partial}{\partial \rho}(\rho \psi) + \xi \dfrac{\partial \psi}{\partial z} - \dfrac{1}{\rho} \dfrac{\partial}{\partial \omega}(\eta \psi)$
ρ, θ, z	ω, ξ	$\dfrac{\mu}{\rho} \dfrac{\partial}{\partial \rho}(\rho \psi) - \dfrac{\eta}{\rho} \dfrac{\partial \psi}{\partial \theta} + \xi \dfrac{\partial \psi}{\partial z} - \dfrac{1}{\rho} \dfrac{\partial}{\partial \omega}(\eta \psi), \quad \mu = (1 - \xi^2)^{1/2} \cos \omega; \ \eta = (1 - \xi^2)^{1/2} \sin \omega$
Streaming Operator in Spherical Coordinates in Conservation Form		
ρ	μ	$\dfrac{\mu}{\rho^2} \dfrac{\partial}{\partial \rho}(\rho^2 \psi) + \dfrac{1}{\rho} \dfrac{\partial}{\partial \mu}[(1 - \mu^2) \psi]$
ρ, θ	μ, ω	$\dfrac{\mu}{\rho^2} \dfrac{\partial}{\partial \rho}(\rho^2 \psi) + \dfrac{\eta}{\rho \sin \theta} \dfrac{\partial}{\partial \theta}(\sin \theta \psi) + \dfrac{1}{\rho} \dfrac{\partial}{\partial \mu}[(1 - \mu^2) \psi] - \dfrac{\cot \theta}{\rho} \dfrac{\partial}{\partial \omega}(\xi \psi)$
ρ, θ, φ	μ, ω	$\dfrac{\mu}{\rho^2} \dfrac{\partial}{\partial \rho}(\rho^2 \psi) + \dfrac{\eta}{\rho \sin \theta} \dfrac{\partial}{\partial \theta}(\sin \theta \psi) + \dfrac{\xi}{\rho \sin \theta} \dfrac{\partial \psi}{\partial \varphi} + \dfrac{1}{\rho} \dfrac{\partial}{\partial \mu}[(1 - \mu^2) \psi] - \dfrac{\cot \theta}{\rho} \dfrac{\partial}{\partial \omega}(\xi \psi)$
		$\eta = (1 - \mu^2)^{1/2} \cos \omega; \ \xi = (1 - \mu^2)^{1/2} \sin \omega$

Source: Data from Ref. 2; used with permission of Wiley.

Boundary Conditions

Boundary conditions for Eq. (9.5) are generally specified by the physical situation. For a left boundary at $\mathbf{r}_L$ with inward normal vector $\mathbf{n}$, such that $\mathbf{n} \cdot \boldsymbol{\Omega} > 0$ indicates inward, one of the following boundary conditions is usually appropriate:

$$
\begin{aligned}
&\text{Vacuum:} && \psi(\mathbf{r}_L, \boldsymbol{\Omega}) = 0, && \boldsymbol{\Omega} \cdot \mathbf{n} > 0 \\
&\text{Incident flux known:} && \psi(\mathbf{r}_L, \boldsymbol{\Omega}) = \psi_{\text{in}}(\mathbf{r}_L, \boldsymbol{\Omega}), && \boldsymbol{\Omega} \cdot \mathbf{n} > 0 \\
&\text{Reflection:} && \psi(\mathbf{r}_L, \boldsymbol{\Omega}) = \int_0^{4\pi} \alpha(\boldsymbol{\Omega}' \to \boldsymbol{\Omega}) \psi(\mathbf{r}_L, \boldsymbol{\Omega}') d\boldsymbol{\Omega}'
\end{aligned}
\tag{9.6}
$$

where α is a reflection or albedo function.

Scalar Flux and Current

The scalar flux is the product of the total number of neutrons in a differential volume, which is the integral over direction of the number of neutrons with direction within $d\boldsymbol{\Omega}$ about $\boldsymbol{\Omega}$, times the speed:

$$
\phi(\mathbf{r}) \equiv \int_0^{4\pi} d\boldsymbol{\Omega}\, \psi(\mathbf{r}, \boldsymbol{\Omega})
\tag{9.7}
$$

and the current with respect to the ξ-coordinate is the net flow of neutrons in the positive ξ-direction:

$$
\mathbf{J}_\xi(\mathbf{r}) \equiv \mathbf{n}_\xi \int_0^{4\pi} d\boldsymbol{\Omega}\, (\mathbf{n}_\xi \cdot \boldsymbol{\Omega}) \psi(\mathbf{r}, \boldsymbol{\Omega})
\tag{9.8}
$$

Partial Currents

The positive and negative partial currents, with respect to the ξ-direction, are the total neutron flows in the positive and negative ξ-directions, respectively:

$$
\begin{aligned}
\mathbf{J}_\xi^+(\mathbf{r}) &\equiv \mathbf{n}_\xi \int_0^{2\pi} d\phi \int_0^1 d\mu (\mathbf{n}_\xi \cdot \boldsymbol{\Omega}) \psi(\mathbf{r}, \boldsymbol{\Omega}) \\
\mathbf{J}_\xi^-(\mathbf{r}) &\equiv \mathbf{n}_\xi \int_0^{2\pi} d\phi \int_{-1}^0 d\mu (\mathbf{n}_\xi \cdot \boldsymbol{\Omega}) \psi(\mathbf{r}, \boldsymbol{\Omega})
\end{aligned}
\tag{9.9}
$$

9.2
Integral Transport Theory

The steady-state version of Eq. (9.5) may be written

$$
\frac{d}{dR} \psi(\mathbf{r}, \boldsymbol{\Omega}) dr\, d\boldsymbol{\Omega} + \Sigma_t(\mathbf{r}) \psi(\mathbf{r}, \boldsymbol{\Omega}) dr\, d\boldsymbol{\Omega} = S(\mathbf{r}, \boldsymbol{\Omega}) dr\, d\boldsymbol{\Omega}
\tag{9.10}
$$

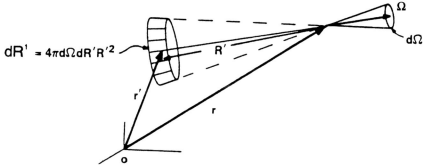

$$dR' = 4\pi d\Omega dR'R'^2$$

Fig. 9.6 Incremental volume subtended by cone $d\Omega$ at distance $R = |\mathbf{r} - \mathbf{r}'|$ from point $\mathbf{r}$. (From Ref. 2; used with permission of Wiley.)

where dR is the differential length along the direction $\mathbf{\Omega}$ (i.e., $\mathbf{\Omega} \cdot \nabla = d/dR$). This equation may be integrated along the direction $\mathbf{\Omega}$ from $\mathbf{r}_0$ to $\mathbf{r}$, to obtain

$$\psi(\mathbf{r}, \mathbf{\Omega})d\mathbf{r} = e^{-\alpha(\mathbf{r}_0, \mathbf{r})}\psi(\mathbf{r}_0, \mathbf{\Omega})d\mathbf{r}_0 + \int_{\mathbf{r}_0}^{\mathbf{r}} e^{-\alpha(\mathbf{r}', \mathbf{r})}S(\mathbf{r}', \mathbf{\Omega})d\mathbf{r}' \tag{9.11}$$

where $\alpha(\mathbf{r}', \mathbf{r})$ is the optical path length along the direction $\mathbf{\Omega}$ between $\mathbf{r}'$ and $\mathbf{r}$:

$$\alpha(\mathbf{r}', \mathbf{r}) \equiv \left| \int_{\mathbf{r}'}^{\mathbf{r}} \Sigma_t(R) \, dR \right| \tag{9.12}$$

Isotropic Point Source

For an isotropic point source of strength $S_0 (n/s)$ located at $\mathbf{r}_0$, the directional flux outward through the cone $d\Omega$ about direction $\mathbf{\Omega}$ is $S_0(d\Omega/4\pi)$. The volume element $d\mathbf{r}$ subtended by this cone at distance $R = |\mathbf{r} - \mathbf{r}'|$ away is $4\pi d\Omega R^2 dR$, as depicted in Fig. 9.6. From Eq. (9.11), the directional flux at $\mathbf{r}$ of uncollided neutrons from an isotropic point source at $\mathbf{r}'$ (such that the direction from $\mathbf{r}'$ to $\mathbf{r}$ is $\mathbf{\Omega}$) is given by

$$\psi_{\text{pt}}(R) = \psi\big(|(\mathbf{r} - \mathbf{r}')|, \mathbf{\Omega}\big) = \frac{S_0 e^{-\alpha(\mathbf{r}, \mathbf{r}')}}{4\pi |\mathbf{r} - \mathbf{r}'|^2} = \frac{S_0 e^{-\alpha(R,0)}}{4\pi R^2} \tag{9.13}$$

Isotropic Plane Source

The scalar flux of uncollided neutrons at a distance x from a uniform planar isotropic source can be constructed by treating each point in the plane as an isotropic point source and integrating over the plane, as indicated in Fig. 9.7, to obtain

$$\phi_{\text{pl}}(x, 0) = \int_0^{\infty} 2\pi \rho \psi_{\text{pt}}(R) \, d\rho = \int_x^{\infty} 2\pi R \psi_{\text{pt}}(R) \, dR$$

$$= \frac{S_0}{2} \int_x^{\infty} e^{-\alpha(R,0)} \frac{dR}{R} = \frac{1}{2} S_0 E_1\big(\alpha(x, 0)\big) \tag{9.14}$$

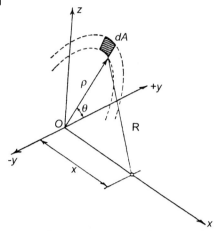

Fig. 9.7 Coordinate system for plane isotropic source calculation. (From Ref. 10; used with permission of McGraw-Hill.)

where the exponential integral function is defined as

$$E_n(y) \equiv \int_1^\infty du\, e^{-yu} u^{-n} = \int_0^1 d\mu\, e^{-y/\mu} \mu^{n-2} \tag{9.15}$$

The x-direction current of uncollided neutrons at a distance x from a uniform planar isotropic source can be constructed in a similar manner by noting that for a neutron originating on the plane with direction $\boldsymbol{\Omega}$, the quantity $\mu = \boldsymbol{\Omega} \cdot \mathbf{n}_x = x/R$:

$$
\begin{aligned}
J_{\mathrm{pl},x}(x,0) &= \int_0^\infty 2\pi\rho \frac{x}{R} \psi_{\mathrm{pt}}(R)\, d\rho = x \int_x^\infty 2\pi R \psi_{\mathrm{pt}}(R) \frac{dR}{R} \\
&= \frac{S_0}{2} \int_x^\infty e^{-\alpha(R,0)} \frac{dR}{R^2} = \frac{1}{2} S_0 E_2(\alpha(x,0))
\end{aligned}
\tag{9.16}
$$

A one-dimensional isotropic source distribution $S_0(x)$ in a slab of thickness a can be considered as a distribution of isotropic planar sources, and the uncollided scalar flux distribution can be constructed by integrating over the contributions from each planar source:

$$\phi(x) = \int_0^a S_0(x') \phi_{\mathrm{pl}}(x,x')\, dx' = \frac{1}{2} \int_0^a S_0(x') E_1(\alpha(x,x'))\, dx' \tag{9.17}$$

Anisotropic Plane Source

Using the relations $\mu = \cos\theta = x/R$ and $R^2 = x^2 + \rho^2$ and noting that all source neutrons in the annular region $2\pi\rho\, d\rho$ on the source plane will pass through a point at a distance x above the center of the annular region within $d\mu$ about the same value of μ, the directional flux of uncollided neutrons which results from an anisotropic planar source $S(\mu)$ can be constructed:

$$\psi(x,\mu)\,d\mu = \psi_{\text{pt}}\big(R(\rho)\big)2\pi\rho\,d\rho = \frac{S(\mu)e^{-\alpha(x,0)/\mu}}{\mu}d\mu \tag{9.18}$$

The scalar flux and current of uncollided neutrons at a distance x from an uniform anisotropic planar source $S(\mu)$ are

$$\phi(x) \equiv \int_{-1}^{1}\psi(x,\mu)\,d\mu = \int_{0}^{1}S(\mu)e^{-\alpha(x,0)/\mu}\frac{d\mu}{\mu} \tag{9.19}$$

$$J_x(x) \equiv \int_{-1}^{1}\mu\psi(x,\mu)\,d\mu = \int_{0}^{1}S(\mu)e^{-\alpha(x,0)/\mu}\,d\mu \tag{9.20}$$

It is convenient to expand the directional dependence of the source:

$$S(\mu) = \sum_{n=0}(2n+1)p_n^+(\mu)S_n \tag{9.21}$$

in half-range Legendre polynomials:

$$p_n^+(\mu) = P_n(2\mu-1)$$
$$p_0^+(\mu)=1,\quad p_1^+(\mu)=2\mu-1,\quad p_2^+(\mu)=6\mu^2-6\mu+1, \tag{9.22}$$
$$p_3^+(\mu)=20\mu^3-30\mu^2+12\mu-1,\quad \text{etc.}$$

which have the orthogonality properties

$$\int_{0}^{1}p_n^+(\mu)p_m^+(\mu)d\mu = \frac{\delta_{nm}}{2n+1} \tag{9.23}$$

With these orthogonality properties, it follows immediately that $S_n = \int_0^1 p_n^+(\mu) \times S(\mu)d\mu$.

Using this expansion in Eq. (9.19), the flux of uncollided neutrons at a distance x from an uniform anisotropic planar source is

$$\phi(x) = \sum_{n=0}(2n+1)S_n B_n^+(\alpha(x,0)) \tag{9.24}$$

where

$$B_n^+(\alpha(x,0)) \equiv \int_0^1 p_n^+(\mu)e^{-\alpha(x,0)/\mu}\frac{d\mu}{\mu}$$
$$B_0^+(\alpha(x,0)) = E_1(\alpha(x,0)) \tag{9.25}$$
$$B_1^+(\alpha(x,0)) = 2E_2(\alpha(x,0)) - E_1(\alpha(x,0)),\quad \text{etc.}$$

Similarly, the x-directed current of uncollided neutrons at a distance x from an uniform anisotropic planar source is

$$J_x(x) = \sum_{n=0}(2n+1)S_n L_n^+(\alpha(x,0)) \tag{9.26}$$

where

$$
L_n^+\big(\alpha(x,0)\big) = \int_0^1 p_n^+(\mu) e^{-\alpha(x,0)/\mu} d\mu
$$
$$
L_0^+\big(\alpha(x,0)\big) = E_2\big(\alpha(x,0)\big) \tag{9.27}
$$
$$
L_1^+\big(\alpha(x,0)\big) = 2E_3\big(\alpha(x,0)\big) - E_2\big(\alpha(x,0)\big), \quad \text{etc.}
$$

Transmission and Absorption Probabilities

As an example of an application of the formalism above, consider a purely absorbing slab of thickness a with an isotropic plane source of neutrons on one surface. The transmission probability for the slab is just the ratio of the exiting current on the opposite surface to the incident partial current on the other surface:

$$
T = \frac{J(a)}{J_{\text{in}}(0)} = \frac{S_0 L_0^+\big(\alpha(a,0)\big)}{S_0} = E_2\big(\alpha(a,0)\big) \tag{9.28}
$$

and the absorption probability is $A = 1 - T = 1 - E_2(\alpha(a,0))$.

Escape Probability

As another example, consider a uniform, purely absorbing slab of thickness a with an isotropic neutron source S_0 distributed uniformly throughout. Representing the source of neutrons at x within the slab as a plane isotropic source of strength $S_0/2$ to the right and $S_0/2$ to the left, the current of neutrons produced by the source at $x = x'$ which exit through the surface at $x = a$ is

$$
J_{\text{out}}(a:x') = \tfrac{1}{2} S_0 L_0^+\big(\alpha(a,x')\big) = \tfrac{1}{2} S_0 E_2\big(\alpha(a,x')\big) \tag{9.29}
$$

The total current of neutrons out through the surface at $x = a$ is found by integrating this expression over the slab:

$$
J_{\text{out}}(a) = \int_0^a dx' \, J_{\text{out}}(a:x') = \frac{1}{2} S_0 \int_0^a dx' E_2\big(\alpha(a,x')\big) \tag{9.30}
$$

Using the differentiation property of the exponential integral function

$$
\frac{dE_n}{dy} = -E_{n-1}(y), \quad n = 1, 2, 3, \ldots \tag{9.31}
$$

Eq. (9.30) may be evaluated:

$$
J_{\text{out}}(a) = -\frac{1}{2} S_0 \int_0^a \frac{dE_3(\alpha)}{d\alpha} dx' = -\frac{1}{2}\frac{S_0}{\Sigma_t} \int_0^{\alpha(a,0)} \frac{dE_3}{d\alpha} d\alpha
$$
$$
= \frac{1}{2}\frac{S_0}{\Sigma_t}\big[E_3(0) - E_3\big(\alpha(a,0)\big)\big] = \frac{1}{2}\frac{S_0}{\Sigma_t}\left[\frac{1}{2} - E_3\big(\alpha(a,0)\big)\right] \tag{9.32}
$$

By symmetry, the current out through the surface at $x = 0$ must be the same. The escape probability from the slab is the ratio of the total current out of the slab through both surfaces to the total neutron source rate $a S_0$ in the slab:

$$P_0 = \frac{J_{\text{out}}(a) + J_{\text{out}}(0)}{a S_0} = \frac{1}{a \Sigma_t} \left[\frac{1}{2} - E_3(a \Sigma_t) \right] \tag{9.33}$$

First-Collision Source for Diffusion Theory

As a further application, consider a medium with a surface source of neutrons, which is highly forward directed but almost isotropic within the forward-directional hemisphere, incident on one surface of a diffusing medium; that is, the forward-directed neutrons incident on the medium are nearly isotropic within the forward-directional hemisphere, but many more neutrons are moving forward into the medium than are moving backward out of it. Diffusion theory will not be accurate for treating these source neutrons, because diffusion theory is based on an implicit assumption that the neutron flux is nearly isotropic over the full angle (this is discussed in Section 9.6), even though diffusion theory may otherwise be sufficient for the analysis of neutrons once their direction is randomized by a scattering event within the medium. The first collision of the incident source neutrons can be calculated with integral transport theory and used as a distributed *first-collision source* for the diffusion theory calculation:

$$S_{\text{fc}}(x) = \Sigma_s(x)\phi(x) = S_0 \Sigma_s(x) E_1\big(\alpha(x, 0)\big) \tag{9.34}$$

If the distribution of incident source neutrons is more highly forward directed, so that it is anisotropic even over the forward-directional hemisphere, it may be represented by an anisotropic plane source and the first-collision source becomes

$$S_{\text{fc}}(x) = \Sigma_s(x) \sum_{n=0} (2n + 1) S_n B_n^+\big(\alpha(x, 0)\big) \tag{9.35}$$

Inclusion of Isotropic Scattering and Fission

Consider again the slab with a distributed isotropic source of neutrons, but now with isotropic elastic scattering and fission, as well as absorption represented explicitly. The flux of uncollided source neutrons is

$$\phi_0(x) = \frac{1}{2} \int_0^a S_0(x') E_1\big(\alpha(x, x')\big) dx' \tag{9.36}$$

If the first-collision rate at $x = x'$ is considered as a plane isotropic source of once-collided neutrons at x', the flux of once-collided neutrons due to the once-collided source at x' is

$$\phi_1(x : x') = \tfrac{1}{2}\big(\Sigma_s(x') + \nu \Sigma_f(x')\big)\phi_0(x') E_1\big(\alpha(x, x')\big) \tag{9.37}$$

and the total flux of once-collided neutrons at x is found by integrating over the distribution of first-collision sources:

$$\phi_1(x) = \int_0^a \phi_1(x : x')dx'$$

$$= \int_0^a \left[\frac{1}{2}\left(\Sigma_s(x') + \nu\Sigma_f(x') \right)\phi_0(x') \right] E_1\left(\alpha(x, x') \right)dx' \tag{9.38}$$

Continuing in this vein, the flux of n-collided neutrons is given by

$$\phi_n(x) = \int_0^a \left[\frac{1}{2}\left(\Sigma_s(x') + \nu\Sigma_f(x') \right)\phi_{n-1}(x') \right] \cdot E_1\left(\alpha(x, x') \right)dx',$$

$$n = 1, 2, 3, \ldots, \infty \tag{9.39}$$

The total neutron flux is the sum of the uncollided, once-collided, twice-collided, and so on, fluxes:

$$\phi(x) \equiv \phi_0(x) + \sum_{n=1}^{\infty} \phi_n(x)$$

$$= \int_0^a \frac{1}{2}\left[\Sigma_s(x') + \nu\Sigma_f(x') \right] \sum_{n=1}^{\infty} \phi_{n-1}(x') E_1\left(\alpha(x, x') \right)dx'$$

$$+ \frac{1}{2}\int_0^a \left(S_0(x') E_1\left(\alpha(x, x') \right) \right)dx'$$

$$= \int_0^a \frac{1}{2}\left(\Sigma_s(x') + \nu\Sigma_f(x') \right)\phi(x') E_1\left(\alpha(x, x') \right)dx'$$

$$+ \frac{1}{2}\int_0^a S_0(x') E_1\left(\alpha(x, x') \right)dx' \tag{9.40}$$

Thus we have found an integral equation for the neutron flux in a slab with isotropic scattering and fission, with a kernel $\frac{1}{2}[\Sigma_s(x') + \nu\Sigma_f(x')]E_1(\alpha(x, x'))$ and a first-collision source $\frac{1}{2}S_0 E_1(\alpha(x, 0))$.

Distributed Volumetric Sources in Arbitrary Geometry

The scalar flux of uncollided neutrons resulting from an arbitrary neutron source distribution can be constructed by treating each spatial location as a point source with strength given by the source distribution for that location. The uncollided directional flux at $\mathbf{r}$ arising from a point source at $\mathbf{r}'$ is given by Eq. (9.13). The total uncollided directional flux at $\mathbf{r}$ is obtained by integrating over all source points $\mathbf{r}'$, and the total uncollided scalar flux is then calculated by integrating over $\boldsymbol{\Omega}$:

$$\phi_{\text{un}}(\mathbf{r}) = \int d\mathbf{r}' \frac{S_0(\mathbf{r}')e^{-\alpha(\mathbf{r}, \mathbf{r}')}}{4\pi |\mathbf{r}' - \mathbf{r}|^2} \tag{9.41}$$

Following the same development as that leading to Eq. (9.40), an integral equation for the total neutron flux can be developed for the case of isotropic scattering:

$$\phi(r) = \int d\mathbf{r}' \frac{[\Sigma_s(\mathbf{r}') + \nu\Sigma_f(\mathbf{r}')]\phi(\mathbf{r}')e^{-\alpha(\mathbf{r},\mathbf{r}')}}{4\pi|\mathbf{r}' - \mathbf{r}|^2} + \phi_{un}(\mathbf{r}')$$

$$= \int d\mathbf{r}' \{[\Sigma_s(\mathbf{r}') + \nu\Sigma_f(\mathbf{r}')]\phi(\mathbf{r}') + S_0(\mathbf{r}')\} \frac{e^{-\alpha(\mathbf{r},\mathbf{r}')}}{4\pi|\mathbf{r}' - \mathbf{r}|^2} \qquad (9.42)$$

where $\exp[-\alpha(\mathbf{r}, \mathbf{r}')]/4\pi|\mathbf{r} - \mathbf{r}'|^2$ is the isotropic point source kernel and ϕ_{un} given by Eq. (9.41) is the uncollided flux contribution.

The derivations leading to Eqs. (9.40) and (9.42) did not explicitly take boundary conditions into account. Since scattering source rates integrated over the volume of the reactor were used to derive successive n-collided fluxes, the implicit assumption was that neutrons which escaped from the reactor did not return. Thus these equations are valid with vacuum boundary conditions, but not with reflective boundary conditions.

Flux from a Line Isotropic Source of Neutrons

Consider the situation illustrated in Fig. 9.8 of a line isotropic source of neutrons of strength $S_0(\text{n/cm} \cdot \text{s})$. The point source kernel can be used to construct the differential scalar flux at a point P located a distance t from the line source due to the differential element dz of the line source located at z:

$$d\phi(t) = \frac{S_0 \, dz \, e^{-\alpha(t,z)}}{4\pi R^2} \qquad (9.43)$$

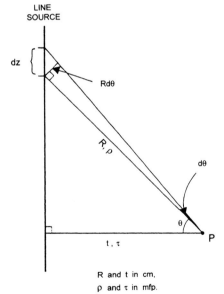

Fig. 9.8 Geometry for calculating flux at P from a line isotropic neutron source $[t = x, \tau = \alpha(x, 0)]$. (From Ref. 3; used with permission of Academic Press.)

where $\alpha(t, z)$ denotes the optical thickness along the path of length R from the source point at coordinate z to the point P a perpendicular distance t from the line source at $z = 0$. Noting that $R = t/\cos\theta$ and $dz = R\,d\theta/\cos\theta = t\,d\theta/\cos^2\theta$, the total flux at a point at a distance t can be found by integrating the differential flux contribution from all such differential elements dz:

$$\phi(t) = \int_{-\infty}^{\infty} \frac{S_0\,dz\,e^{-\alpha(t,z)}}{4\pi R^2} = S_0 \int_0^{\infty} \frac{dz\,e^{-\alpha(t,0)/\cos\theta}}{2\pi t^2/\cos^2\theta} \equiv \frac{Ki_1[\alpha(t,0)]}{2\pi t} S_0 \tag{9.44}$$

where $Ki_1(x)$ is the Bickley function of order one.

Bickley Functions

The general *Bickley function* is defined as

$$Ki_n(x) \equiv \int_0^{\pi/2} \cos^{n-1}\theta\,e^{-x/\cos\theta}\,d\theta = \int_0^{\infty} \frac{e^{-x\cosh(u)}}{\cosh^n(u)}\,du \tag{9.45}$$

These functions satisfy the following differential and integral laws:

$$\frac{dKi_n(x)}{dx} = -Ki_{n-1}(x) \tag{9.46}$$

$$Ki_n(x) = Ki_n(0) - \int_0^x Ki_{n-1}(x')\,dx' = \int_x^{\infty} Ki_{n-1}(x')\,dx' \tag{9.47}$$

and the recurrence relation

$$nKi_{n+1}(x) = (n-1)Ki_{n-1}(x) + x[Ki_{n-2}(x) - Ki_n(x)] \tag{9.48}$$

The Bickley functions must be evaluated numerically (e.g., Ref. 3).

Probability of Reaching a Distance t from a Line Isotropic Source without a Collision

With reference to Fig. 9.9, the probability P that a neutron emitted isotropically from point P on the line source is able to get a perpendicular distance t away from the line source without having a collision depends on the direction in which the neutron is traveling relative to the perpendicular to the line source. The uncollided differential neutron current arising from a point on the line source and passing through a differential surface area $dA = R\,d\theta t\,d\varphi = t^2\,d\theta\,d\varphi/\cos\theta$ normal to the R-direction at a perpendicular distance t from the line source is

$$dJ(t,\theta) = \frac{e^{-\alpha(t,z)}}{4\pi R^2}\,dA = \frac{e^{-\alpha(t,0)/\cos\theta}}{4\pi(t/\cos\theta)^2}\,\frac{t^2\,d\theta\,d\varphi}{\cos\theta} \tag{9.49}$$

where the optical thickness $\alpha(t, z)$ is taken along the path length R. Integrating over all possible values of the angles, the probability of a neutron emitted isotropically from a line source crossing the cylindrical surface at a distance t from the line source is

$$P(t) = \int_0^{2\pi} d\varphi \int_{-\pi/2}^{\pi/2} d\theta\,\frac{e^{-\alpha(t,0)/\cos\theta}(t^2/\cos\theta)}{4\pi(t/\cos\theta)^2} = Ki_2(\alpha(t,0)) \tag{9.50}$$

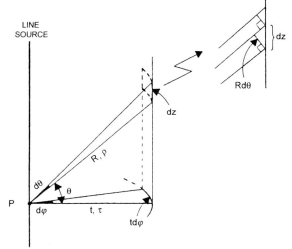

Fig. 9.9 Geometry for calculating probability that a neutron from an isotropic line source does not have a collision within perpendicular distance t from the line source [$t = x$, $\tau = \alpha(x, 0)$]. (From Ref. 3; used with permission of Academic Press.)

where now $\alpha(t, 0)$ is the optical path length perpendicular to the line source out to the cylindrical surface at distance t.

The Bickley and exponential integral functions arise because of the assumption of spatial symmetry. They take into account that the neutron flight path is always in three spatial dimensions, even though symmetry otherwise allows reduction in the dimensionality of the problem.

9.3
Collision Probability Methods

If the volume of the problem of interest is partitioned into discrete volumes, V_i, within each of which uniform average cross sections and a flat flux are assumed, Eq. (9.42) can be integrated over V_i, and the resulting equation can be divided by V_i to obtain

$$\phi_i = \sum_j T^{j \to i}[(\Sigma_{sj} + \nu \Sigma_{fj})\phi_j + S_{0_j}] \tag{9.51}$$

which relates the fluxes in the various volumes by the first-flight transmission probabilities $T^{j \to i}$:

$$T^{j \to i} \equiv \frac{1}{V_i} \int_{V_i} d\mathbf{r}_i \int_{V_j} d\mathbf{r}_j \frac{e^{-\alpha(\mathbf{r}_i, \mathbf{r}_j)}}{4\pi |\mathbf{r}_i - \mathbf{r}_j|^2} \tag{9.52}$$

Reciprocity Among Transmission and Collision Probabilities

Since $\alpha(\mathbf{r}_i, \mathbf{r}_j) = \alpha(\mathbf{r}_j, \mathbf{r}_i)$ (i.e., the optical path is the same no matter which way the neutron traverses the straight-line distance between $\mathbf{r}_i$ and $\mathbf{r}$) there is a reciprocity relation between the transmission probabilities:

$$V_i T^{j \to i} = V_j T^{i \to j} \tag{9.53}$$

Upon multiplication by $\Sigma_{ti} V_i$, Eq. (9.51) can be written

$$\Sigma_{ti} V_i \phi_i = \sum_j P^{ji} \frac{(\Sigma_{sj} + \nu \Sigma_{fj})\phi_j + S_{0j}}{\Sigma_{tj}} \tag{9.54}$$

where the collision rate in cell i is related to the neutrons introduced by scattering, fission, and an external source in all cells j by the collision probabilities

$$P^{ji} \equiv \Sigma_{tj} \Sigma_{ti} V_i T^{j \to i} = \Sigma_{tj} \Sigma_{ti} \int_{V_i} d\mathbf{r}_i \int_{V_j} d\mathbf{r}_j \frac{e^{-\alpha(\mathbf{r}_i, \mathbf{r}_j)}}{4\pi |\mathbf{r}_i - \mathbf{r}_j|^2} \tag{9.55}$$

Because $\alpha(\mathbf{r}_i, \mathbf{r}_j) = \alpha(\mathbf{r}_j, \mathbf{r}_i)$, there is reciprocity between the collision probabilities; that is,

$$P^{ij} = P^{ji} \tag{9.56}$$

Collision Probabilities for Slab Geometry

For a slab lattice the volumes, V_i, become the widths $\Delta_i \equiv x_{i+1/2} - x_{i-1/2}$ of the slab regions centered at x_i, and the slab kernel $E_1(\alpha(x', x))/2$ replaces the point source kernel in Eq. (9.55), which becomes

$$P^{ji} = \Sigma_{ti} \Sigma_{tj} \int_{\Delta_i} dx' \int_{\Delta_j} dx \frac{1}{2} E_1(\alpha(x', x)) \tag{9.57}$$

For $j \neq i$, the probability that a neutron introduced in cell j has its next collision in cell i is

$$P^{ji} = \tfrac{1}{2}\big[E_3(\alpha_{i+1/2, j+1/2}) - E_3(\alpha_{i-1/2, j+1/2}) - E_3(\alpha_{i+1/2, j-1/2}) + E_3(\alpha_{i-1/2, j-1/2})\big] \tag{9.58}$$

where $\alpha_{i,j} \equiv \alpha(x_i, x_j)$. For $j = i$, a similar development leads to an expression for the probability that a neutron introduced in cell j has its next collision in cell i is

$$P^{ii} = \Sigma_{ti} \Delta_i \left[1 - \frac{1}{2\Sigma_{ti} \Delta_i}(1 - 2E_3(\Sigma_{ti} \Delta_i))\right] \tag{9.59}$$

COLLISION PROBABILITY METHOD

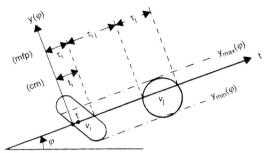

Fig. 9.10 Geometry for calculating collision probabilities in two-dimensional geometry (τ is the optical path length α over the indicated path). (From Ref. 3; used with permission of Academic Press.)

Collision Probabilities in Two-Dimensional Geometry

Consider the two-dimensional cross section shown in Fig. 9.10, in which the volumes V_i and V_j extend indefinitely in the direction perpendicular to the page. With respect to Fig. 9.9, a neutron emitted at point t defined by the angle φ and coordinate y in volume V_i in Fig. 9.10 and traveling in the direction defined by the angle φ which passes through volume V_j may be traveling at any angle $-\pi/2 \le \theta \le \pi/2$ with respect to the horizontal cross section shown in Fig. 9.10.

The probability that a neutron emitted at point t will reach some point on the line perpendicular to the page which passes through the page at point t' in volume V_j is, from Eq. (9.50), given by $Ki_2(\alpha(t', t))$, where $\alpha(t', t)$ is the optical path length in the horizontal plane of Fig. 9.10. With respect to Fig. 9.10, identify t_i and t_j as the points along the horizontal line between t and t' at which the line passes through the surfaces of volumes V_i and V_j, respectively. Thus $Ki_2(\Sigma_{ti}(t_i - t) + \alpha(t_j, t_i))$ is the probability that a neutron emitted from point t in volume V_i in direction φ reaches volume V_j, and $Ki_2(\Sigma_{ti}(t_i - t) + \alpha(t_j, t_i) + \alpha(t_j + \Delta t_j, t_j))$, with Δt_j being the distance in the horizontal plane across volume V_j, is the probability that the neutron not only reaches volume V_j but continues through volume V_j and emerges from the opposite side without a collision, both probabilities being averaged over an isotropic distribution of neutron directions with respect to the horizontal, as measured by the angle θ. The probability that neutrons emitted from point t in volume V_i with direction φ have their first collision in volume V_j is then $p_{ij}(t, \varphi, y) = -Ki_2(\Sigma_{ti}(t_i - t) + \alpha(t_j, t_i) + \alpha(t_j + \Delta t_j, t_j)) + Ki_2(\Sigma_{ti}(t_i - t) + \alpha(t_j, t_i))$. Averaging this probability over all source points along the line defined by angle φ within volume V_i and using the differential property of the Bickley functions given by Eq. (9.45) leads to

$$p_{ij}(\varphi, y) = \frac{1}{t_i} \int_0^{t_i} p_{ij}(t, \varphi, y)dt$$

$$= \frac{1}{\Sigma_{ti} t_i} \left[Ki_3\big(\alpha(t_j, t_i)\big) - Ki_3\big(\alpha(t_j, t_i) + \alpha(t_j + \Delta t_j, t_j)\big) \right.$$

$$- Ki_3\big(\alpha(t_j, t_i) + \alpha(t_i, 0)\big) + Ki_3\big(\alpha(t_j, t_i) + \alpha(t_j + \Delta t_j, t_i)\big)$$

$$\left. + \alpha(t_i, 0)\big) \right] \tag{9.60}$$

To obtain the average probability P^{ij} that a neutron introduced by an isotropic source uniformly distributed over volume V_i will have its first collision in volume V_j, this expression must be multiplied by the probability that an isotropically emitted neutron source will emit a neutron in the differential direction $d\varphi$ about φ, which is $d\varphi/2\pi$, and the probability that for a uniform source within V_i the neutron will be emitted from along the chord of length $t_i(y)$ at coordinate y, which is $t_i(y)dy/V_i$, and integrated over all relevant values of φ and y. Note that the volumes V_i and V_j are actually the respective areas within the planar cross section of Fig. 9.10. The result for the collision probability is

$$p^{ij} = \frac{1}{2\pi} \int_{\varphi_{min}}^{\varphi_{max}} d\varphi \int_{y_{min}(\varphi)}^{y_{max}(\varphi)} dy \big[Ki_3\big(\alpha(t_j, t_i)\big)$$

$$- Ki_3\big(\alpha(t_j, t_i) + \alpha(t_j + \Delta t_j, t_j)\big) - Ki_3\big(\alpha(t_j, t_i) + \alpha(t_i, 0)\big)$$

$$+ Ki_3\big(\alpha(t_j, t_i) + \alpha(t_j + \Delta t_j, t_j) + \alpha(t_i, 0)\big) \big] \tag{9.61}$$

A similar development leads to an expression for the probability that the next collision for a neutron introduced in volume V_i is within that same volume V_i:

$$P^{ii} = \Sigma_{ti} V_i - \frac{1}{2\pi} \int_{\varphi_{min}}^{\varphi_{max}} d\varphi \int_{y_{min}(\varphi)}^{y_{max}(\varphi)} dy \big[Ki_3(0) - Ki_3\big(\alpha(t_i, 0)\big) \big] \tag{9.62}$$

Collision Probabilities for Annular Geometry

The annular geometry of a fuel pin, its clad, and the surrounding moderator is of particular interest. For the annular geometry of Fig. 9.11, Eq. (9.61) specializes to

$$P^{ij} = \delta_{ij} \Sigma_{tj} V_j + 2(S_{i-1,j-1} - S_{i-1,j} - S_{i,j-1} + S_{i,j}) \tag{9.63}$$

where

$$S_{i,j} \equiv \int_0^{R_i} [Ki_3(\tau_{ij}^+) - Ki_3(\tau_{ij}^-)] dy \tag{9.64}$$

with the τ being optical path lengths α over the indicated chords in the horizontal plane in Fig. 9.11:

$$\tau_{ij}^{\pm} \equiv \left(\sqrt{R_j^2 - y^2} \mp \sqrt{R_i^2 - y^2} \right) \tag{9.65}$$

Methods for the numerical evaluation of these expressions are given in Ref. 3.

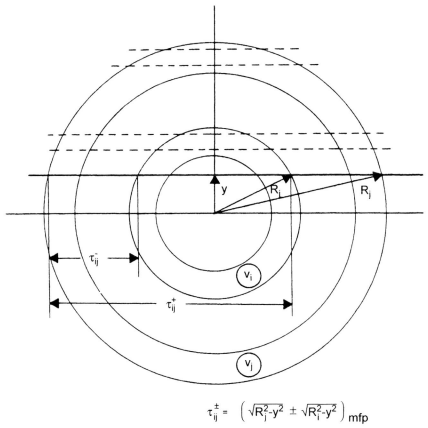

$$\tau_{ij}^{\pm} = \left(\sqrt{R_j^2 - y^2} \pm \sqrt{R_i^2 - y^2} \right) \text{ mfp}$$

Fig. 9.11 Annular geometry notation for calculation of collision probabilities (τ is the optical path length α over the indicated path). (From Ref. 3; used with permission of Academic Press.)

9.4
Interface Current Methods in Slab Geometry

Emergent Currents and Reaction Rates Due to Incident Currents

Consider the slab geometry configuration depicted in Fig. 9.12, in which a slab region i is bounded by surfaces i and $i+1$ with incident currents J_i^+ and J_{i+1}^- and emergent currents J_i^- and J_{i+1}^+. The angular flux of particles at x arising from an angular flux of neutrons at x' is

$$\psi(x, \mu) = e^{-\Sigma_{ti}(x-x')/\mu} \psi(x'\mu) \tag{9.66}$$

where it is assumed that the total cross section, Σ_t, is uniform over Δ_i, and μ is the cosine of the angle that the particle direction makes with the x-axis. Further assuming that the incident fluxes, ψ_i^+ and ψ_{i+1}^-, are isotropically distributed in angle

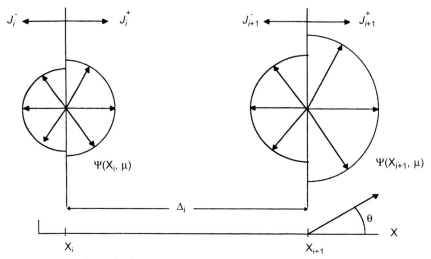

Fig. 9.12 Nomenclature for slab geometry interface current method.

over the incident hemisphere (i.e., a double P_0 approximation), the uncollided currents emergent from the opposite surface are given in terms of the incident partial currents ($J_i^+ = \frac{1}{2}\psi_i^+$, $J_{i+1}^- = \frac{1}{2}\psi_{i+1}^-$) by

$$\hat{J}_{un}^+(x_{i+1}) = 2J_i^+ \int_0^1 \mu\left(e^{-\Sigma_{ti}\Delta_i/\mu}\right)d\mu = 2E_3(\Delta_i\Sigma_{ti})J_i^+$$

$$\hat{J}_{un}^-(x_i) = 2J_{i+1}^- \int_{-1}^0 \mu\left(e^{+\Sigma_{ti}\Delta_i/\mu}\right)d\mu = 2E_3(\Delta_i\Sigma_{ti})J_{i+1}^-$$

(9.67)

where E_n is the exponential integral function given by

$$E_n(z) \equiv \int_0^1 \mu^{n-2}e^{-z/\mu}d\mu$$

(9.68)

The first collision rate for incident particles within Δ_i is given by

$$\hat{R}_{i1} = \Sigma_{ti}\left[2J_i^+ \int_0^1 d\mu \int_{x_i}^{x_{i+1}} dx\left(e^{-\Sigma_{ti}(x-x_i)/\mu}\right)\right.$$

$$\left. + 2J_{i+1}^- \int_{-1}^0 d\mu \int_{x_i}^{x_{i+1}} dx\left(e^{-\Sigma_{ti}(x-x_{i+1})/\mu}\right)\right]$$

$$= \left(J_i^+ + J_{i+1}^-\right)[1 - 2E_3(\Delta_i\Sigma_{ti})]$$

(9.69)

The fraction c_i of the collision rate that is due to scattering (i.e., to events that do not remove the particle) from the cohort under consideration (i.e., $c_i = (\Sigma_{si} + \nu\Sigma_{fi})/\Sigma_{ti}$) constitutes a source of once-collided particles, which we assume to be isotropic ($\frac{1}{2}$ emerge going to the right and $\frac{1}{2}$ to the left) and distributed uniformly over Δ_i. Treating these "scattered" neutrons as a distribution of plane isotropic sources, with the source at x' producing exiting uncollided fluxes

$$\left(\frac{1}{2}c_i\frac{\hat{R}_{i1}}{\Delta_i}\right)\frac{\exp(-\Sigma_{ti}(x_{i+1}-x')/\mu)}{\mu}$$

and

$$\left(\frac{1}{2}c_i\frac{\hat{R}_{i1}}{\Delta_i}\right)\frac{\exp(-\Sigma_{ti}(x_i-x')/\mu)}{\mu}$$

at x_{i+1} and x_i, respectively, the emergent currents of once-collided particles are

$$\hat{J}_1^+(x_{i+1}) = \int_{x_i}^{x_{i+1}} dx \int_0^1 \mu \left(\frac{1}{2}c_i\frac{\hat{R}_{i1}}{\Delta_i}\right)\frac{e^{-\Sigma_{ti}(x_{i+1}-x)/\mu}}{\mu}d\mu$$

$$= \frac{1}{2}\frac{c_i\hat{R}_{i1}}{\Delta_i\Sigma_{ti}}\left[\frac{1}{2}-E_3(\Delta_i\Sigma_{ti})\right] = \frac{1}{2}c_i P_{0i}\hat{R}_{i1} \qquad (9.70)$$

$$\hat{J}_1^-(x_i) = \tfrac{1}{2}c_i P_{0i}\hat{R}_{i1} = \tfrac{1}{2}c_i P_{0i}\left(J_i^+ + J_{i+1}^-\right)[1-2E_2(\Delta_i\Sigma_{ti})]$$

where the average first-flight escape probability for source particles distributed uniformly over Δ_i has been defined as

$$P_{0i} \equiv \frac{\frac{1}{2}}{\Delta_i}\int_{x_i}^{x_{i+1}} dx \int_0^1 d\mu\,\mu\frac{e^{-\Sigma_{ti}(x_{i+1}-x)/\mu}}{\mu}$$

$$+ \frac{\frac{1}{2}}{\Delta_i}\int_{x_i}^{x_{i+1}} dx \int_{-1}^0 d\mu\,\mu\frac{e^{-\Sigma_{ti}(x_i-x)/\mu}}{\mu}$$

$$= \frac{1}{\Delta_i\Sigma_{ti}}\left[\frac{1}{2}-E_3(\Delta_i\Sigma_{ti})\right] \qquad (9.71)$$

The collision rate for incident particles undergoing a second collision in Δ_i is

$$\hat{R}_{i2} = \Sigma_{ti}\frac{1}{2}c_i\frac{\hat{R}_{i1}}{\Delta_i}\left(\int_0^1 d\mu\int_{x_i}^{x_{i+1}} dx'\int_{x_i}^{x_{i+1}} dx\frac{e^{-\Sigma_{ti}(x-x')/\mu}}{\mu}\right.$$

$$\left.+ \int_{-1}^0 d\mu\int_{x_i}^{x_{i+1}} dx'\int_{x_i}^{x'} dx\frac{e^{-\Sigma_{ti}(x-x')/\mu}}{\mu}\right)$$

$$= c_i\hat{R}_{i1}(1-P_{0i}) = c_i(J_i^+ + J_{i+1}^-)[1-2E_3(\Delta_i\Sigma_{ti})](1-P_{0i}) \qquad (9.72)$$

As before, the fraction c_i of this collision rate constitutes a source of twice-collided particles which are assumed to be isotropic. The emergent currents of twice-collided particles are given by Eqs. (9.70) but with $\hat{R}_{i1}$ replaced by $\hat{R}_{i2}$:

$$\hat{J}_2^+(x_{i+1}) = \hat{J}_2^-(x_i) = \tfrac{1}{2}c_i\hat{R}_{i2}P_{0i}$$

$$= \tfrac{1}{2}c_i^2\left(J_i^+ + J_{i+1}^-\right)[1-2E_3(\Delta_i\Sigma_{ti})](1-P_{0i})P_{0i} \qquad (9.73)$$

Continuing this line of argument, we derive general expressions for the rate at which incident particles undergo their nth collision in Δ_i:

$$\hat{R}_{\text{in}} = c_i^{n-1}\left(J_i^+ + J_{i+1}^-\right)[1 - 2E_3(\Delta_i\Sigma_{ti})](1 - P_{0i})^{n-1} \tag{9.74}$$

and for the emergent currents of n-collided incident particles,

$$\hat{J}_n^+(x_{i+1}) = \hat{J}_n^-(x_i)$$
$$= \tfrac{1}{2}c_i^n\left(J_i^+ + J_{i+1}^-\right)[1 - 2E_3(\Delta_i\Sigma_{ti})](1 - P_{0i})^{n-1}P_{0i} \tag{9.75}$$

The total collision rate in Δ_i due to incident currents is obtained by summing Eqs. (9.74):

$$\hat{R}_i = \sum_{n=1}^{\infty} \hat{R}_{\text{in}} = \left(J_i^+ + J_{i+1}^-\right)[1 - 2E_3(\Delta_i\Sigma_{ti})]\sum_{n=0}^{\infty}[c_i(1 - P_{0i})]^n$$
$$= \frac{(J_i^+ + J_{i+1}^-)[1 - 2E_3(\Delta_i\Sigma_{ti})]}{1 - c_i(1 - P_{0i})} \tag{9.76}$$

and the total emergent currents due to incident currents are obtained by summing Eq. (9.75) and adding the uncollided contributions of Eqs. (9.67):

$$\hat{J}^+(x_{i+1}) = \left[\frac{\tfrac{1}{2}c_iP_{0i}[1 - 2E_3(\Delta_i\Sigma_{ti})]}{1 - c_i(1 - P_{0i})} + 2E_3(\Delta_i\Sigma_{ti})\right]J_i^+$$
$$+ \frac{\tfrac{1}{2}c_iP_{0i}[1 - 2E_3(\Delta_i\Sigma_{ti})]}{1 - c_i(1 - P_{0i})}J_{i+1}^-$$
$$\hat{J}^-(x_i) = \left[\frac{\tfrac{1}{2}c_iP_{0i}[1 - 2E_3(\Delta_i\Sigma_{ti})]}{1 - c_i(1 - P_{0i})} + 2E_3(\Delta_i\Sigma_{ti})\right]J_{i+1}^-$$
$$+ \frac{\tfrac{1}{2}c_iP_{0i}[1 - 2E_3(\Delta_i\Sigma_{ti})]}{1 - c_i(1 - P_{0i})}J_i^+ \tag{9.77}$$

Emergent Currents and Reaction Rates Due to Internal Sources

We consider a uniform distribution of particle sources within Δ_i of strength s_i/Δ_i per unit length. This source is allowed to be anisotropic, with a number s_i^+ emitted to the right and s_i^- emitted to the left. The emergent currents of uncollided source particles are

$$J_{\text{un},s}^+(x_{i+1}) = \frac{s_i^+}{\Delta_i}\int_{x_i}^{x_{i+1}}dx\int_0^1 d\mu\,\mu\,\frac{e^{-\Sigma_{ti}(x_{i+1}-x)/\mu}}{\mu} = s_i^+P_{0i}$$
$$J_{\text{un},s}^-(x_i) = \frac{s_i^-}{\Delta_i}\int_{x_i}^{x_{i+1}}dx\int_{-1}^0 d\mu\,\mu\,\frac{e^{-\Sigma_{ti}(x_i-x)/\mu}}{\mu} = s_i^-P_{0i} \tag{9.78}$$

The first collision rate of source particles within Δ_i is given by

$$\hat{R}_{i1,s} = \frac{s_i^+}{\Delta}\Sigma_{ti}\int_{x_i}^{x_{i+1}}dx'\int_{x'}^{x_{i+1}}dx\int_0^1 d\mu\,\frac{e^{-\Sigma_{ti}(x-x')/\mu}}{\mu}$$

$$+ \frac{s_i^-}{\Delta_i} \Sigma_{ti} \int_{x_i}^{x_{i+1}} dx' \int_{x_i}^{x'} dx \int_{-1}^{0} d\mu \frac{e^{-\Sigma_{ti}(x-x')/\mu}}{\mu}$$

$$= (s_i^+ + s_i^-) \left[1 - \frac{1}{\Delta_i \Sigma_{ti}} \left(\frac{1}{2} - E_3(\Sigma_{ti}\Delta_i) \right) \right] = s_i(1 - P_{0i}) \qquad (9.79)$$

As before, treating the fraction c_i of these particles that undergo scattering collisions as an isotropic plane source of once-collided particles, the emergent currents of once-collided source particles are given by

$$J_{1s}^+(x_{i+1}) = \int_0^1 d\mu\, \mu \int_{x_i}^{x_{i+1}} dx\, \frac{1}{2} c_i \frac{\hat{R}_{i1,s}}{\Delta_i} \frac{e^{-\Sigma_{ti}(x_{i+1}-x)/\mu}}{\mu}$$

$$= \frac{1}{2} c_i \hat{R}_{i1,s} P_{0i} = \frac{1}{2} c_i s_i (1 - P_{0i}) P_{0i}$$

$$(9.80)$$

$$J_{1s}^-(x_i) = \int_{-1}^0 d\mu\, \mu \int_{x_i}^{x_{i+1}} dx\, \frac{1}{2} c_i \frac{\hat{R}_{i1,s}}{\Delta_i} \frac{e^{-\Sigma_{ti}(x_i-x)/\mu}}{\mu}$$

$$= \frac{1}{2} c_i R_{i1,s} P_{0i} = \frac{1}{2} c_i s_i (1 - P_{0i}) P_{0i}$$

Continuing in this fashion, the general expression for the nth collision rate of source particles in Δ_i is

$$\hat{R}_{in,s} = c_i^{n-1} s_i (1 - P_{0i})^n \qquad (9.81)$$

and the general expressions for the emergent currents of n-collided source particles are

$$J_{ns}^+(x_{i+1}) = J_{ns}^-(x_i) = \tfrac{1}{2} s_i P_{0i} c_i^n (1 - P_{0i})^n \qquad (9.82)$$

The total collision rate of source particles within Δ_i is

$$\hat{R}_{i,s} = \sum_{n=1}^{\infty} \hat{R}_{in,s} = \frac{s_i(1 - P_{0i})}{1 - c_i(1 - P_{0i})} \qquad (9.83)$$

and the total emergent currents due to an anisotropic particle source within Δ_i are obtained by summing Eqs. (9.82) and adding Eqs. (9.78):

$$J_s^+(x_{i+1}) = \left(s_i^+ - \frac{1}{2} s_i \right) P_{0i} + \frac{\frac{1}{2} s_i P_{0i}}{1 - c_i(1 - P_{0i})}$$

$$(9.84)$$

$$J_s^-(x_i) = \left(s_i^- - \frac{1}{2} s_i \right) P_{0i} + \frac{\frac{1}{2} s_i P_{0i}}{1 - c_i(1 - P_{0i})}$$

Total Reaction Rates and Emergent Currents

The total reaction rate in Δ_i due to incident currents and to internal sources is obtained by adding Eqs. (9.76) and (9.83):

$$R_i = \frac{(J_i^+ + J_{i+1}^-)(1 - T_{0i}) + s_i(1 - P_{0i})}{1 - c_i(1 - P_{0i})} \qquad (9.85)$$

where the first-flight, or uncollided, transmission probability has been identified:

$$T_{0i} \equiv E_2(\Delta_i \Sigma_{ti}) \tag{9.86}$$

Further identifying the total escape probability,

$$P_i \equiv P_{0i} \sum_{n=0}^{\infty} [c_i(1 - P_{0i})]^n = \frac{P_{0i}}{1 - c_i(1 - P_{0i})} \tag{9.87}$$

the total reflection probability,

$$R_i \equiv \frac{\frac{1}{2} c_i P_{0i} [1 - 2E_3(\Delta_i \Sigma_{ti})]}{1 - c_i(1 - P_{0i})} = \frac{1}{2} c_i P_i(1 - T_{0i}) \tag{9.88}$$

and the total transmission probability,

$$T_i = T_{0i} + R_i = T_{0i} + \frac{1}{2} c_i P_i(1 - T_{0i}) \tag{9.89}$$

Eqs. (9.77) and (9.84) can be summed to obtain expressions for the total emergent currents due to incident currents and internal particle sources:

$$
\begin{aligned}
J_{i+1}^+ &= T_i J_i^+ + R_i J_{i+1}^- + \frac{1}{2} s_i P_i + \left(s_i^+ - \frac{1}{2} s_i\right) P_{0i} \\
J_i^- &= T_i J_{i+1}^- + R_i J_i^+ + \frac{1}{2} s_i P_i + \left(s_i^- - \frac{1}{2} s_i\right) P_{0i}
\end{aligned}
\tag{9.90}
$$

The inherent advantage of an interface current formulation of integral transport theory is evident from Eqs. (9.90). To solve for the currents across interface i, one needs only the currents at interface $i + 1$ and the source in the intervening region. This leads, in essence, to a "four-point" coupling of the unknowns, the partial currents at i and $i + 1$, and the evaluation of only one E_3 function for each region. By contrast, in the standard collision probabilities formulation of the preceding section, the fluxes in all other regions in the problem and the transition probabilities from all of these regions to the region in question are needed in order to solve for the flux in a given region, in essence coupling all regions in the problem. In both formulations, an iterative solution is needed.

As formulated above, the interface current method is based on the D-P_0 assumption of an isotropic angular flux distribution within the incident hemisphere at each interface, for the purpose of calculating the uncollided transmission across the region. This assumption is physically plausible for problems with scattering (and fission) rates comparable to or larger than absorption rates, because this tends to isotropize the flux exiting from a region. However, for problems with an incident neutron source on one boundary of an almost purely absorbing medium, the flux will become increasingly forward directed with distance into the region. In the limit of a purely absorbing region with an incident isotropic neutron source at $x = 0$, the current attenuation at a distance x from the source plane is exactly $E_2(\Sigma x)$. If this problem is modeled in the interface current formulation and the distance x is subdivided into N intervals Δ, the calculated current attenuation at x

is $\prod_{n=1}^{N} E_2(\Sigma \Delta)$, which differs from the exact answer $E_2(\Sigma n \Delta)$. Thus inaccuracies might be expected in highly absorbing multiregion problems.

It is informative to sum Eqs. (9.90) to obtain an intuitively obvious balance between incident and emergent currents and internal sources:

$$J_{i+1}^{+} + J_i^{-} = (T_i + R_i)(J_i^{+} + J_{i+1}^{-}) + s_i P_i$$

or

(9.91)

$$J_{\text{out}} = (T_{0i} + (1 - T_{0i})c_i P_i) J_{\text{in}} + s_i P_i$$

Solving the first of Eqs. (9.90) for J_i^{+} and using the result in the second equation leads to a matrix relation among the partial currents at adjacent surfaces:

$$\begin{bmatrix} J_i^{+} \\ J_i^{-} \end{bmatrix} = \begin{bmatrix} T_i^{-1} & -T_i^{-1} R_i \\ R_i T_i^{-1} & T_i - R_i T_i^{-1} R_i \end{bmatrix} \begin{bmatrix} J_{i+1}^{+} \\ J_{i+1}^{-} \end{bmatrix}$$

$$+ \frac{1}{2} s_i \left\{ P_i \begin{pmatrix} -T_i^{-1} \\ 1 - R_i T_i^{-1} \end{pmatrix} \right.$$

$$\left. + P_{0i} \begin{bmatrix} -T_i^{-1}(s_i^{+} - \frac{1}{2} s_i) \\ (s_i^{-} - \frac{1}{2} s_i) - R_i T_i^{-1}(s_i^{+} - \frac{1}{2} s_i) \end{bmatrix} \right\}$$

(9.92)

Equation (9.92) is well suited for numerical evaluation simply by marching from one boundary of the problem to the other.

Boundary Conditions

Boundary conditions take on a particularly simple form for an interface current formulation of integral transport. Let $x = 0$, $i = 0$ represent the leftmost surface of the transport medium. If a vacuum or nonscattering medium with no particle source exists for $x < 0$, then $J_0^{+} = 0$ is the appropriate boundary condition. If, on the other hand, a source-free scattering medium exists for $x < 0$, an albedo or reflection condition of the form $J_0^{+} = \beta J_0^{-}$, where β is the reflection coefficient or albedo, is appropriate. Finally, if a known current of particles J_{in} is incident upon the medium from the left at $x = 0$, the appropriate boundary condition is $J_0^{+} = J_{\text{in}}$.

Response Matrix

Suppressing internal sources, the matrix equation (9.92) may be written in more compact notation:

$$\boldsymbol{J}_i^{\pm} = \boldsymbol{R}_i \boldsymbol{J}_{i+1}^{\pm}$$

(9.93)

(where $\boldsymbol{J}$ indicates a column vector and $\boldsymbol{R}$ indicates a matrix) and applied successively to relate the incident and exiting currents on the left boundary, $\boldsymbol{J}_0^{\pm}$, to the incident and exiting currents on the right boundary, $\boldsymbol{J}_I^{\pm}$:

$$\boldsymbol{J}_0^{\pm} = [\boldsymbol{R}_0 \cdot \boldsymbol{R}_1 \cdot \boldsymbol{R}_2 \cdots \boldsymbol{R}_{I-2} \cdot \boldsymbol{R}_{I-1}] \boldsymbol{J}_I^{\pm} \equiv \boldsymbol{R} \boldsymbol{J}_I^{\pm}$$

(9.94)

where the matrix $\boldsymbol{R}$ is the matrix product of the matrices $\boldsymbol{R}_i$ for each slab Δ_i and has the form

$$\boldsymbol{R} = \begin{bmatrix} R^{11} & R^{12} \\ R^{21} & R^{22} \end{bmatrix} \tag{9.95}$$

in terms of which Eq. (9.94) may be written as the two equations

$$\begin{aligned} J_0^+ &= R^{11} J_I^+ + R^{12} J_I^- \\ J_0^- &= R^{21} J_I^+ + R^{22} J_I^- \end{aligned} \tag{9.96}$$

which may be solved to obtain the response matrix relationship between the incident currents, J_0^+ and J_I^-, and the exiting currents, J_0^- and J_I^+.

$$\begin{bmatrix} J_0^- \\ J_I^+ \end{bmatrix} = \begin{bmatrix} R^{22} - R^{21}(R^{11})^{-1}R^{12} & R^{21}(R^{11})^{-1} \\ -R^{21}(R^{11})^{-1}R^{12} & R^{21}(R^{11})^{-1} \end{bmatrix} \begin{bmatrix} J_I^- \\ J_0^+ \end{bmatrix} \tag{9.97}$$

or

$$\boldsymbol{J}_{\text{out}} = \boldsymbol{R}\boldsymbol{M}\boldsymbol{J}_{\text{in}}$$

Once the response matrix, $\boldsymbol{RM}$, is evaluated, the exiting currents can be computed rapidly for a given set of incident currents. This formalism can be extended in an obvious way to treat internal sources.

9.5
Multidimensional Interface Current Methods

Extension to Multidimension

The interface current formulation of integral transport theory can be extended to two and, in principle, three dimensions. First, for conceptual purposes, we rewrite Eqs. (9.90) by making the identification $J_i^+ = J_i^{\text{in}}$, $J_i^- = J_i^{\text{out}}$, $J_{i+1}^+ = J_{i+1}^{\text{out}}$, $J_{i+1}^- = J_{i+1}^{\text{in}}$, and

$$\begin{aligned} \Lambda_{i+1}^s s_i P_i &\equiv \tfrac{1}{2} s_i P_i + \left(s_i^+ - \tfrac{1}{2} s_i\right) P_{0i} \\ \Lambda_i^s s_i P_i &\equiv \tfrac{1}{2} s_i P_i + \left(s_i^- - \tfrac{1}{2} s_i\right) P_{0i} \end{aligned} \tag{9.98}$$

where Λ_i^s is the fraction of escaping source neutrons that escapes to the left across surface i and Λ_{i+1}^s is the fraction escaping to the right across surface $i + 1$. Then, using Eqs. (9.85) to (9.89), Eqs. (9.90) may be written

$$\begin{aligned} J_{i+1}^{\text{out}} &= T_{0i} J_i^{\text{in}} + (1 - T_{0i})\left(J_i^{\text{in}} + J_{i+1}^{\text{in}}\right) c_i P_i \Lambda_{i+1} + \Lambda_{i+1}^s s_i P_i \\ J_i^{\text{out}} &= T_{0i} J_{i+1}^{\text{in}} + (1 - T_{0i})\left(J_i^{\text{in}} + J_{i+1}^{\text{in}}\right) c_i P_i \Lambda_i + \Lambda_i^s s_i P_i \end{aligned} \tag{9.99}$$

where $\Lambda_i = \Lambda_{i+1} = \tfrac{1}{2}$ is the fraction of escaping scattered incident neutrons that escape across surfaces i and $i + 1$, respectively.

Uncollided flux

Collided flux

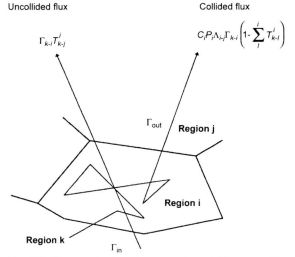

$$\Gamma_{k\text{-}i}T^i_{k\text{-}j}$$

$$C_iP_i\Lambda_{i\text{-}j}\Gamma_{k\text{-}i}\left(1\text{-}\sum_i T^i_{k\text{-}l}\right)$$

Γ_{out} **Region j**

Region i

Region k

Γ_{in}

Fig. 9.13 Planar projection of geometry for multidimensional interface current methods.

In this form, the terms in Eqs. (9.99) for the emergent currents have a direct physical interpretation which leads immediately to a generalization to multidimensions. The outward current across surface $i + 1$ consists of three terms: (1) the inward current across surface i times the probability T_{0i} that it is transmitted across region i without collision to surface $i + 1$; (2) the inward currents across all surfaces times the probability $(1 - T_{0i})$ that these currents are not transmitted across region i without collision, times the probability c_i that the first collision is a "scattering" event, times the probability P_i that the scattered neutrons subsequently escape from region i, times the probability Λ_{i+1} that escaping neutrons escape across surface $i + 1$; and (3) the total particle source s_i in region i times the probability P_i that these neutrons will escape from region i, times the probability Λ^s_{i+1} that escaping source neutrons escape across surface $i + 1$. Note that Λ_{i+1} and Λ^s_{i+1} can in principle differ because an anisotropic source is allowed [i.e., $\Lambda_{i+1} = \frac{1}{2}$ and Λ^s_{i+1} is given by Eq. (9.98) for slab geometry].

Generalization to multidimensions is straightforward, in principle. Consider the two-dimensional configuration in Fig. 9.13. The current from region k into region i is denoted J_{ki} ($\Gamma_{k \to i}$ in the figure), the probability that the current entering region i from region k is transmitted across region i without collision to contribute to the current from region i into region j is denoted T^{kj}_{0i}, and the probability that a collided or source neutron escaping from region i escapes into region j is denoted Λ_{ij}. The generalization of Eqs. (9.99) to two-dimensions is then

$$J_{ij} = \sum_k^i T^{kj}_{0i} J_{ki} + \sum_k^i \left(1 - \sum_l^i T^{kl}_{0i}\right) J_{ki} c_i P_i \Lambda_{ij} + \Lambda^s_{ij} s_i P_i \qquad (9.100)$$

where the summation $\sum_k^i$ is over all regions k that are contiguous to region i. The three terms in Eq. (9.100) correspond physically to (1) the sum of the currents incident into region i from all contiguous regions times the probability that each is transmitted across region i without collision to exit into region j (note that the possibility of concave surfaces is allowed by including uncollided transmission from region j across region i back into region j); (2) the sum of the currents incident into region i from all contiguous regions times the probability that each is not transmitted without collision across region i to any of the contiguous regions, times the probability that the first collision is a "scattering" event, times the probability that the scattered neutron eventually escapes from region i into region j; and (3) the source of neutrons in region i times the probability that a source neutron in region i eventually escapes into region j.

Evaluation of Transmission and Escape Probabilities

The general form for the evaluation of transmission and escape probabilities can be developed using the point kernel discussed previously. We treat the case of incident fluxes that are distributed isotropically in the incident hemisphere of directions and volumetric neutron sources (scattering, fission, external) which are uniformly distributed over volume and emitted isotropically in direction. These results can be extended to anisotropic incident fluxes and nonuniform and anisotropic volumetric source distributions by extending the procedures indicated below.

The probability that a neutron introduced isotropically at location $\mathbf{r}_i$ within volume V_i escapes without collision across the surface S_{ki} that defines the interface between volume V_i and contiguous volume V_k is the probability $d\mathbf{\Omega}/4\pi|\mathbf{r}_{S_{ki}} - \mathbf{r}_i|^2$ that the neutron is traveling within a cone of directions $d\mathbf{\Omega}$ which intersects that surface, times the probability $\exp[-\alpha(\mathbf{r}_{S_{ki}}, \mathbf{r}_i)]$ that the neutron reaches the surface at location $\mathbf{r}_{S_{ki}}$ along the direction $\mathbf{\Omega}$ from $\mathbf{r}_i$ without a collision, integrated over all $\mathbf{\Omega}$ that intersect the surface S_{ki} from point $\mathbf{r}_i$. This probability is then averaged over all points $\mathbf{r}_i$ within volume V_i to obtain

$$P_{0i}\Lambda_k = \frac{1}{4\pi V_i} \int_{V_i} d\mathbf{r}_i \int_{S_{ki}} dS \frac{e^{-\alpha(\mathbf{r}_{S_{ki}}, \mathbf{r}_i)}}{|\mathbf{r}_{S_{ki}} - \mathbf{r}_i|^2} \tag{9.101}$$

Extension of this expression to treat anisotropic neutron emission would be accomplished by including a function $f(\mathbf{r}_{S_{ki}}, \mathbf{r}_i)$ under the integral to represent any directional dependence of neutron emission. Extension to include a spatial distribution $g(\mathbf{r}_i)$ of neutron sources would be accomplished by including this function in the integrand.

The probability that an incident unit neutron flux which is isotropically distributed over the inward hemisphere of directions entering volume V_i from volume V_k across surface S_{ki} is transmitted without collision across volume V_i to the surface S_{ji} which forms the interface with contiguous volume V_j is the product of the probability $\mathbf{n}_{S_{ki}} \cdot d\mathbf{\Omega}/2\pi|\mathbf{r}_{S_{ki}} - \mathbf{r}_{S_{ji}}|^2 = (\mathbf{n}_{S_{ki}} \cdot \mathbf{\Omega})d\mathbf{\Omega}/2\pi|\mathbf{r}_{S_{ki}} - \mathbf{r}_{S_{ji}}|^2$ that a neutron incident across S_{ki} is traveling within a cone of directions $d\mathbf{\Omega}$ which intersects the surface S_{ji}, times the probability $\exp[-\alpha(\mathbf{r}_{S_{ji}}, \mathbf{r}_{S_{ki}})]$ that the neutron reaches the surface at location $\mathbf{r}_{S_{ji}}$ along the direction $\mathbf{\Omega}$ from $\mathbf{r}_{S_{ki}}$ without a collision, integrated over all $\mathbf{\Omega}$ that intersect the surface S_{ji} from point $\mathbf{r}_{S_{ki}}$. The quantity $\mathbf{n}_{S_{ki}}$

is the unit vector normal to the surface S_{ki} in the direction from volume V_k into volume V_i. This probability is then averaged over all points $\mathbf{r}_{S_{ki}}$ on S_{ki}, to obtain

$$T_{0i}^{kj} = \frac{\int_{S_{ki}} dS \int_{S_{ji}} dS [e^{-\alpha(\mathbf{r}_{S_{ji}}, \mathbf{r}_{S_{ki}})} (\mathbf{n}_{S_{ki}} \cdot \mathbf{\Omega})/|\mathbf{r}_{S_{ji}} - \mathbf{r}_{S_{ki}}|^2]}{\int_{\mathbf{n}_{S_{ki}} \cdot \mathbf{\Omega} > 0} d\mathbf{\Omega} \int_{S_{ki}} dS} \tag{9.102}$$

Extension of this expression to include an anisotropic incident neutron flux would be accomplished by including a function $f(\mathbf{r}_{S_{ki}}, \mathbf{r}_{S_{ji}})$ in the integrand.

Transmission Probabilities in Two-Dimensional Geometries

To develop computational algorithms, we consider geometries with symmetry in one direction, which are conventionally known as two-dimensional geometries. It is important to keep in mind, however, that neutron flight paths take place in three dimensions. Consider a volume V_i that is symmetric in the axial direction and bounded by flat vertical surfaces, so that a horizontal $(x-y)$ planar slice is as shown in Fig. 9.14, with the vertical dimension normal to the page. We want to calculate the transmission coefficient from volume 1 through the volume i into volume 3. A three-dimensional projection and a vertical projection are shown in Fig. 9.15. The points ξ_1 and ξ_3 in Fig. 9.14 are the projection onto the horizontal plane of the vertical axes shown in Fig. 9.15. The differential solid angle in this coordinate system is

$$d\mathbf{\Omega} = \frac{1}{4\pi} \sin\theta' d\theta' d\phi = -\frac{1}{4\pi} \cos\theta \, d\theta \, d\phi \tag{9.103}$$

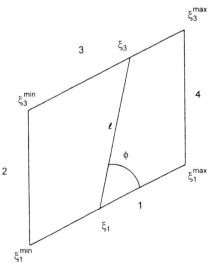

Fig. 9.14 Planar projection of geometry for transmission probability calculation in two-dimensions.

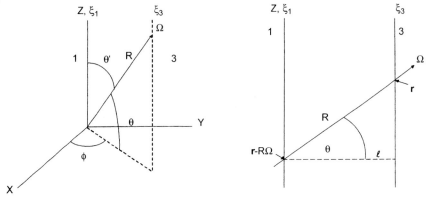

Fig. 9.15 Three-dimensional and axial projection of geometry for transmission probability calculation in two-dimensions.

The incident directional flux from volume 1 at point ξ_1, $\psi(\mathbf{r} - R\mathbf{\Omega}, \mathbf{\Omega})$ is attenuated when it traverses the distance R to reach the point ξ_3 and enter volume 3:

$$\psi(\mathbf{r}, \mathbf{\Omega}) = \psi(\mathbf{r} - R\mathbf{\Omega}, \mathbf{\Omega})e^{-\Sigma R} \qquad (9.104)$$

The incident partial current density ($\text{n/cm}^2 \cdot \text{s}$) from volume 1 at point ξ_1 is

$$
\begin{aligned}
j_{\text{in}}(\xi_1) &= \int_{\mathbf{n}_{\text{in}} \cdot \mathbf{\Omega} > 0} d\mathbf{\Omega}(\mathbf{n}_{\text{in}} \cdot \mathbf{\Omega})\psi(\mathbf{r} - R\mathbf{\Omega}, \mathbf{\Omega})d\mathbf{\Omega} \\
&= -\frac{1}{4\pi} \int_0^\pi d\phi \int_{-\pi/2}^{\pi/2} d\theta \cos^2\theta \sin\phi\,\psi(\mathbf{r} - R\mathbf{\Omega}, \mathbf{\Omega}) \qquad (9.105)
\end{aligned}
$$

where $\mathbf{n}_{\text{in}} \cdot \mathbf{\Omega} = \cos\theta\sin\phi$ has been used. When the incident flux is isotropic in the incident hemisphere (double-P_0 approximation), this becomes

$$j_{\text{in}}^{\text{iso}}(\xi_1) = \tfrac{1}{4}\psi(\mathbf{r} - R\mathbf{\Omega}) \qquad (9.106)$$

The incident partial current (n/s) is obtained by multiplying by the (arbitrary) axial dimension H and integrating over $\xi_1^{\text{min}} \le \xi_1 \le \xi_1^{\text{max}}$

$$J_{\text{in}} = H \int_{\xi_1^{\text{min}}}^{\xi_1^{\text{max}}} d\xi_1\, j_{\text{in}}(\xi_1) \qquad (9.107)$$

The incident neutrons from volume 1 which enter volume V_i at ξ_1 within the solid angle subtended by volume 3 and traverse volume i without collision to enter volume 3 constitute an uncollided neutron current out of volume V_i into volume 3, and hence a contribution to the incident current into volume 3 from volume i. For the moment we write this contribution to the current into volume 3 as

$$J_{\text{out}} = H \int_{\xi_1^{\text{min}}}^{\xi_1^{\text{max}}} d\xi_1 \int_{\substack{\mathbf{\Omega} \cdot \bar{n}_{\text{out}} > 0 \\ \phi(\xi_1) \ni 3}} d\mathbf{\Omega}(\mathbf{\Omega} \cdot \mathbf{n}_{\text{out}})\psi(\mathbf{r} - R\mathbf{\Omega}, \mathbf{\Omega})e^{-\Sigma R}$$

$$= H \int_{\xi_1^{\min}}^{\xi_1^{\max}} d\xi_1 \int_{\phi_{\min}(\xi_1)}^{\phi_{\max}(\xi_1)} d\phi \int_{-\pi/2}^{\pi/2} d\theta \cos^2\theta \sin\phi e^{(-\Sigma l(\phi(\xi_1))/\cos\theta)}$$

$$\times \psi(\mathbf{r} - R\mathbf{\Omega}, \mathbf{\Omega}) \tag{9.108}$$

where $\mathbf{n}_{\text{out}} \cdot \mathbf{\Omega} = \cos\theta \sin\phi_{\text{out}}$ may differ from $\mathbf{n}_{\text{in}} \cdot \mathbf{\Omega} = \cos\theta \sin\phi$ if the interfaces with volumes 1 and 3 are not parallel, and $\phi(\xi_1) \ni 3$ indicates angles ϕ from a point ξ_1 which intersect the interface with region 3. When the incident flux from volume 1 is isotropic in the incident directional hemisphere, this becomes

$$J_{\text{out}}^{\text{iso}} = \frac{H}{2\pi} \int_{\xi_1^{\min}}^{\xi_1^{\max}} d\xi_1 \int_{\phi_{\min}(\xi_1)}^{\phi_{\max}(\xi_1)} d\phi \sin\phi_{\text{out}} Ki_3\left[\Sigma l\left(\phi(\xi_1)\right)\right]\psi(\mathbf{r} - R\mathbf{\Omega}) \tag{9.109}$$

The transmission probability for an isotropic incident flux distribution from volume 1 that is uniform over $\xi_1^{\min} \leq \xi_1 \leq \xi_1^{\max}$ can be written in a form that couples the contribution to the incident current into volume 3 with the incident current into volume i:

$$T_{0i}^{13} \equiv \frac{J_{\text{out}}^{\text{iso}}}{J_{\text{in}}^{\text{iso}}} = \frac{2}{\pi} \frac{\int_{\xi_1^{\min}}^{\xi_1^{\max}} d\xi_1 \int_{\phi_{\min}(\xi_1)}^{\phi_{\max}(\xi_1)} d\phi \sin\phi_{\text{out}} Ki_3[\Sigma l(\phi(\xi_1))]}{\xi_1^{\max} - \xi_1^{\min}} \tag{9.110}$$

When the incident and exiting surfaces (the interfaces of volumes 1 and i-and of volumes 3 and i in this example) are not parallel, there is a subtlety about the direction to take for $\mathbf{n}_{\text{out}}$ in the equations above. The incident current into volume i from volume 1 was calculated on the basis of a DP-0 angular flux approximation with respect to the orientation of the incident surface. The transport of the uncollided incident DP-0 angular flux across region i is properly calculated, and by using $\mathbf{n}_{\text{out}} = \mathbf{n}_{\text{in}}$, the exiting uncollided partial current in the direction normal to the incident surface is properly calculated. So the neutron flow into volume 3 is properly calculated, although the direction of this current exiting volume i is not normal to the exit surface. In constructing the incident current for region 3 from region i, this uncollided contribution from region 1 is added to the uncollided contribution from regions 2 and 4 and to the collided contribution, and the combination is assumed to have a DP-0 incident angular distribution into volume 3 with respect to the orientation of this incident interface of volume 3 (the exiting interface of volume i). Thus, in the equations above, $\mathbf{n}_{\text{out}} = \mathbf{n}_{\text{in}}$ should be used.

Escape Probabilities in Two-Dimensional Geometries

The neutron flux per unit surface area, dA, normal to the direction of neutron flight at a distance R away from an isotropic point source is $\exp(-\Sigma R)/4\pi R^2$, and with reference to Fig. 9.15, the surface area normal to the direction $\mathbf{\Omega}$ of neutron travel is $dA = R\,d\theta l\,d\phi = l^2\,d\theta\,d\phi/\cos\theta$. Thus, with reference to Fig. 9.16, an isotropic neutron source of unit strength per axial length located at $\mathbf{r}_i$ within volume V_i produces an outward current of uncollided neutrons over the surface labeled ξ_3

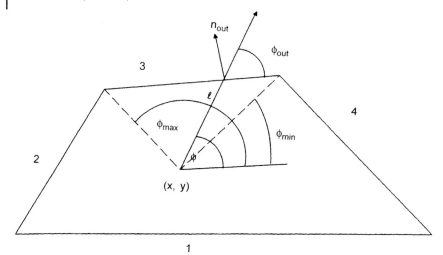

Fig. 9.16 Planar projection of geometry for escape probability calculation in two dimensions.

into volume 3 that is described by

$$
\begin{aligned}
J_{\text{out}}^3(\mathbf{r}_i) &= \int_{A \supset S_3} (\mathbf{n}_{\text{out}} \cdot \boldsymbol{\Omega}) \frac{e^{-\Sigma R} dA}{4\pi R^2} \\
&= \int_{\phi \supset S_3} d\phi \int_{-\pi/2}^{\pi/2} d\theta \, (\cos\theta \sin\phi_{\text{out}}) \frac{e^{-\Sigma l(\phi)/\cos\theta}(l^2/\cos\theta)}{4\pi (l/\cos\theta)^2} \\
&= \int_{\phi \supset S_3} d\phi \sin\phi_{\text{out}} \frac{1}{2\pi} \int_0^{\pi/2} d\theta \cos^2\theta \, e^{-\Sigma l(\phi)/\cos\theta} \\
&= \int_{\phi \supset S_3} d\phi \sin\phi_{\text{out}} \frac{Ki_3(\Sigma l(\phi))}{2\pi}
\end{aligned}
\tag{9.111}
$$

where $\phi \supset S_3$ indicates the range of $\phi_{\text{min}} < \phi < \phi_{\text{max}}$ subtended by side S_3 at location $\mathbf{r}_i$ within volume V_i.

The average value of $J_{\text{out}}^3(x, y)$ over the planar two-dimensional area A_i of volume V_i is just the probability that an isotropic, uniform neutron source s_i will produce an uncollided current $s_i \Lambda_{i3}^s P_{0i}$ from volume V_i into volume V_3:

$$
\begin{aligned}
\Lambda_{i3} P_{0i} &= \frac{1}{A_i} \int_{A_i} dx \, dy \, J_{\text{out}}^3(x, y) \\
&= \frac{1}{A_i} \int_{A_i} dx \, dy \int_{\phi \supset S_3} d\phi \sin\phi_{\text{out}} \frac{Ki_3(\Sigma l(\phi))}{2\pi}
\end{aligned}
\tag{9.112}
$$

The proper value of $\mathbf{n}_{\text{out}}$ is the outward normal to the surface in question, and ϕ_{out} is measured with respect to the orientation of that surface, whereas ϕ may be measured with respect to a fixed coordinate system, so that in general $\phi_{\text{out}} \neq \phi$, although it is convenient to orient the coordinate system so that $\phi_{\text{out}} = \phi$.

The total uncollided escape probability is obtained by summing Eq. (9.112) over all volumes V_k that are contiguous to volume V_i:

$$P_{0i} = \sum_k \Lambda_{ik} P_{0i} \tag{9.113}$$

and the directional escape fractions are calculated from

$$\Lambda_{ij} = \frac{\Lambda_{ij} P_{0i}}{P_{0i}} = \frac{\Lambda_{ij} P_{0i}}{\sum_k \Lambda_{ik} P_{0i}} \tag{9.114}$$

Using the same arguments as were made for the one-dimensional case in the preceding section, the total escape probability, including escape after zero, one, two, ... collisions can be calculated from

$$P_i = \frac{P_{0i}}{1 - c_i(1 - P_{0i})} \tag{9.115}$$

where $c_i = (\Sigma_{si} + \nu\Sigma_{fi})/\Sigma_{ti}$ is the number of secondary neutrons produced per collision.

Simple Approximations for the Escape Probability

Physical considerations lead to a simple approximation for the first-flight escape probability. In the limit that the average neutron path length $\langle l \rangle$ in a volume V is much less than the mean free path λ for a collision, the escape probability tends to unity. In the limit when $\langle l \rangle \gg \lambda$, a simple approximation for the first-flight escape probability is $1 - \exp(-\lambda/\langle l \rangle) \approx \lambda/\langle l \rangle$. If we associate the average neutron path length in the volume with the mean chord length $4V/S$, where S is the surface area of the volume V, a simple rational approximation for the escape probability, first proposed by Wigner and with which his name is associated, is

$$P_0 = \frac{1}{1 + \langle l \rangle/\lambda} = \frac{1}{1 + 4V/S\lambda} = \frac{1}{4V/S\lambda}\left[1 - \frac{1}{1 + (4V/S\lambda)}\right] \tag{9.116}$$

This Wigner rational approximation is known to underpredict the first-flight escape probability. However, extensive Monte Carlo calculations have confirmed that the first-flight escape probability depends only on the parameter $4V/S\lambda$, and improved rational approximations of the form

$$P_0 = \frac{1}{(4V/S\lambda)}\left\{1 - \frac{1}{[1 + (4V/S\lambda)/c]^c}\right\} \tag{9.117}$$

have been proposed. The Sauer approximation, developed from theoretical considerations for cylindrical geometry, corresponds to $c = 4.58$. The best fit to Monte Carlo calculations of first-flight escape probabilities for a uniform neutron source distribution in volumes with a wide range of geometries and values of the parameter $4V/S\lambda$ was found by using $c = 2.09$.

9.6
Spherical Harmonics (P_L) Methods in One-Dimensional Geometries

The spherical harmonics, or P_L, approximation is developed by expansion of the angular flux and the differential scattering cross section in Legendre polynomials.

Legendre Polynomials

The first few *Legendre polynomials* are

$$P_0(\mu) = 1, \quad P_2(\mu) = \tfrac{1}{2}(3\mu^2 - 1)$$
$$P_1(\mu) = \mu, \quad P_3(\mu) = \tfrac{1}{2}(5\mu^3 - 3\mu) \tag{9.118}$$

and higher-order polynomials can be generated from the recursion relation

$$(2n + 1)\mu P_n(\mu) = (n + 1)P_{n+1}(\mu) + n P_{n-1}(\mu) \tag{9.119}$$

The Legendre polynomials satisfy the orthogonality relation

$$\int_{-1}^{1} d\mu\, P_m(\mu) P_n(\mu) = \frac{2\delta_{mn}}{2n + 1} \tag{9.120}$$

With reference to Fig. 9.17, the Legendre polynomials of $\mu_0 = \cos\theta_0$, the cosine of the angle, between μ' and μ, can be expressed in terms of the Legendre polynomials of μ' and μ by the addition theorem

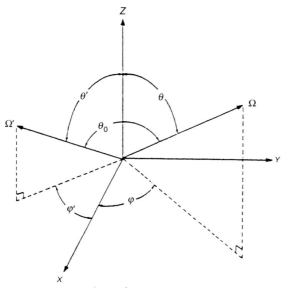

Fig. 9.17 Scattering from Ω' to Ω.

$$P_n(\mu_0) = P_n(\mu)P_n(\mu')$$

$$+ 2\sum_{m=1}^{n} \frac{(n-m)!}{(n+m)!} P_n^m(\mu')P_n^m(\mu)\cos m(\phi - \phi') \tag{9.121}$$

where the *associated Legendre functions* are defined by

$$P_n^m(\mu) \equiv \left(1 - \mu^2\right)^{m/2} \frac{d^m P_n(\mu)}{d\mu^m} \tag{9.122}$$

Neutron Transport Equation in Slab Geometry

Consider a situation in which there is symmetry in the y- and z-coordinate directions but a variation in properties in the x-coordinate direction. The steady-state neutron transport equation (9.5) in this case becomes

$$\mu \frac{\partial \psi(x,\mu)}{\partial x} + \Sigma_t(x,\mu)\psi(x,\mu)$$

$$= \int_{-1}^{1} \Sigma_s(x,\mu' \to \mu)\psi(x,\mu')d\mu' + S(x,\mu)$$

$$= \int_{-1}^{1} \Sigma_s(x,\mu_0)\psi(x,\mu')d\mu' + S(x,\mu) \tag{9.123}$$

where, with reference to Fig. 9.17, we take advantage of the fact that the scattering from a cone of directions $\mu' = \cos\theta'$ to a cone of directions $\mu = \cos\theta$ only depends on $\mu_0 = \cos\theta_0$, the cosine of the angle between μ' and μ, and not on the incident and exiting directions for the scattering event.

P_L **Equations**

The P_L equations are based on the approximation that the angular dependence of the neutron flux can be expanded in the first $L + 1$ Legendre polynomials:

$$\psi(x,\mu) = \sum_{l=0}^{L} \frac{2l+1}{2}\phi_l(x)P_l(\mu) \tag{9.124}$$

The angular dependence of the differential scattering cross section is also expanded in Legendre polynomials:

$$\Sigma_s(x,\mu_0) = \sum_{m=0}^{M} \frac{2m+1}{2}\Sigma_{sm}(x)P_m(\mu_0) \tag{9.125}$$

When these expansions are used in Eq. (9.123), the addition theorem of Eq. (9.121) is used to replace $P_m(\mu_0)$ with $P_m(\mu)$ and $P_m(\mu')$, the recursion relation of Eq. (9.119) is used to replace $\mu P_n(\mu)$ terms with $P_{n\pm1}(\mu)$ terms, the resulting equation is multiplied in turn by $P_k(\mu)$ ($k = 0$ to L) and integrated over

$-1 \leq \mu \leq 1$, and the orthogonality relation of Eq. (9.120) is used, the $L + 1$ P_L equations

$$\frac{d\phi_1(x)}{dx} + (\Sigma_t - \Sigma_{so})\phi_0(x) = S_0(x), \quad n = 0$$

$$\frac{n+1}{2n+1}\frac{d\phi_{n+1}(x)}{dx} + \frac{n}{2n+1}\frac{d\phi_{n-1}}{dx} + (\Sigma_t - \Sigma_{sn})\phi_n(x) = S_n(x), \quad (9.126)$$

$$n = 1, \ldots, L$$

are obtained. The n subscript indicates the nth Legendre moment of the angular dependent quantity:

$$\phi_n(x) \equiv \int_{-1}^{1} d\mu \, P_n(\mu)\psi(x, \mu)$$

$$S_n(x) \equiv \int_{-1}^{1} d\mu \, P_n(\mu)S(x, \mu) \qquad (9.127)$$

$$\Sigma_{sn}(x) \equiv \int_{-1}^{1} d\mu_0 \, P_n(\mu_0)\Sigma_s(x, \mu_0)$$

This set of $L + 1$ equations has a closure problem—they involve $L + 2$ unknowns. This problem is usually resolved by ignoring the term $d\phi_{L+1}/dx$ which appears in the $n = L$ equation.

Boundary and Interface Conditions

The true boundary condition at the left boundary x_L,

$$\psi(x_L, \mu) = \psi_{in}(x_L, \mu), \quad \mu > 0 \qquad (9.128)$$

where $\psi_{in}(x_L, \mu > 0)$ is a known incident flux [$\psi_{in}(x_L, \mu > 0) = 0$ is the vacuum boundary condition], cannot be satisfied exactly by the angular flux approximation of Eq. (9.124), for finite L. The most obvious way to develop approximate boundary conditions that are consistent with the flux approximation is to substitute Eq. (9.124) into the exact boundary condition of (9.128), multiply by $P_m(\mu)$, and integrate over $0 \leq \mu \leq 1$. Since it is the odd Legendre polynomials that represent directionality (i.e., are different for μ and $-\mu$), this procedure is repeated for all the odd Legendre polynomials $m = 1, 3, \ldots, L$ (or $L - 1$) as weighting functions to obtain, with the use of the orthogonality relation of Eq. (9.120), the *Marshak boundary conditions*

$$\int_{0}^{1} d\mu \, P_m(\mu) \sum_{n=0}^{N} \frac{2n+1}{2}\phi_n(x_L)P_n(\mu) \equiv \phi_m(x_L) = \int_{0}^{1} d\mu \, P_m(\mu)\psi_{in}(x_L, \mu),$$

$$m = 1, 3, \ldots, L \text{ (or } L - 1) \qquad (9.129)$$

Equations (9.129) constitute a set of $(L + 1)/2$ boundary conditions. An additional $(L + 1)/2$ boundary conditions are obtained similarly for the right boundary. The

Marshak boundary conditions ensure that the exact inward partial current at the boundary is incorporated into the solution; that is,

$$J^+(x_L) \equiv \int_0^1 d\mu\, P_1(\mu) \sum_{n=0}^N \frac{2n+1}{2} \phi_n(x_L) P_n(\mu)$$

$$\equiv \int_0^1 d\mu\, P_1(\mu) \psi_{\text{in}}(x_L, \mu) \equiv J_{\text{in}}^+(x_L) \tag{9.130}$$

A less intuitive set of *Mark boundary conditions* arises from requiring that the flux expansion of Eq. (9.124) satisfy the boundary condition

$$\sum_{n=0}^N \frac{2n+1}{2} \phi_n(x_L) P_n(\mu_i) = \psi_{\text{in}}(x_L, \mu_i), \quad \mu_i > 0 \tag{9.131}$$

for the $(L+1)/2$ discrete values of μ_i in the inward direction which are the positive roots of $P_{L+1}(\mu_i) = 0$. Another $(L+1)/2$ approximate boundary conditions are obtained at the other boundary by requiring that the flux expansion satisfy the true boundary condition for the $(L+1)/2$ discrete values of μ_i in the inward direction which are the negative roots of $P_{L+1}(\mu_i) = 0$. These Mark boundary conditions are justified by the fact that analytical solution of the P_L equations for a source-free, purely absorbing problem in a infinite half-space leads to these conditions. However, experience has shown that results obtained with the Mark boundary conditions are generally less accurate than results obtained with the Marshak boundary conditions.

A symmetry, or reflective, boundary condition $\psi(x_L, \mu) = \psi(x_L, -\mu)$ obviously requires that all odd moments of the flux vanish [i.e., $\phi_n(x_L) = 0$ for $n = 1, 3, \ldots$, odd].

The exact interface condition of continuity of angular flux

$$\psi(x_s - \varepsilon, \mu) = \psi(x_s + \varepsilon, \mu) \tag{9.132}$$

where ε is a vanishingly small distance, cannot, of course, be satisfied exactly by the flux approximation of Eq. (9.124), for finite L. Following the same procedure as for Marshak boundary conditions, we replace the exact flux with the expansion of Eq. (9.124) and require that the first $L+1$ Legendre moments of this relation be satisfied (i.e., multiply by P_m and integrate over $-1 \le \mu \le 1$, for $m = 0, \ldots, L$). Using the orthogonality relation of Eq. (9.120) then leads to the interface conditions of continuity of the moments:

$$\phi_n(x - \varepsilon) = \phi_n(x_s + \varepsilon), \quad n = 0, 1, 2, \ldots, L \tag{9.133}$$

There are some subtle reasons why this approximation is not appropriate for even-L approximations (see Ref. 6), but since odd-L approximations are almost always used, we will only raise a caution.

P_1 Equations and Diffusion Theory

Neglecting the $d\phi_2/dx$ term, the first two of Eqs. (9.126) constitute the P_1 equations

$$\frac{d\phi_1}{dx} + (\Sigma_t - \Sigma_{s0})\phi_0 = S_0$$

$$\frac{1}{3}\frac{d\phi_0}{dx} + (\Sigma_t - \Sigma_{s1})\phi_1 = S_1 \tag{9.134}$$

Noting that $\Sigma_{s0} = \Sigma_s$, the total scattering cross section, and that $\Sigma_{s1} = \bar{\mu}_0\Sigma_s$, where $\bar{\mu}_0$ is the average cosine of the scattering angle, and assuming that the source is isotropic (i.e., $S_1 = 0$), the second of the P_1 equations yields a Fick's law for neutron diffusion:

$$\phi_1(x) = \int_{-1}^{1} \mu\psi(x,\mu)\,d\mu \equiv J(x) = -\frac{1}{3(\Sigma_t - \bar{\mu}_0\Sigma_s)}\frac{d\phi_0}{dx} \tag{9.135}$$

which, when used in the first of the P_1 equations, yields the neutron diffusion equation

$$-\frac{d}{dx}\left[D_0(x)\frac{d\phi_0(x)}{dx}\right] + (\Sigma_t - \Sigma_s)\phi_0(x) = S_0(x) \tag{9.136}$$

where the diffusion coefficient and the transport cross section are defined by

$$D_0 \equiv \frac{1}{3(\Sigma_t - \bar{\mu}_0\Sigma_s)} \equiv \frac{1}{3\Sigma_{\text{tr}}} \tag{9.137}$$

The basic assumptions made in this derivation of diffusion theory are that the angular dependence of the neutron flux is linearly anisotropic:

$$\psi(x,\mu) \simeq \tfrac{1}{2}\phi_0(x) + \tfrac{3}{2}\mu\phi_1(x) \tag{9.138}$$

and that the neutron source is isotropic, or at least has no linearly anisotropic component ($S_1 = 0$). Diffusion theory should be a good approximation when these assumptions are valid (i.e., in media for which the distribution is almost isotropic because of the preponderance of randomizing scattering collisions, away from interfaces with dissimilar media, and in the absence of anisotropic sources).

The boundary conditions for diffusion theory follow directly from the Marshak condition (9.130):

$$J_{\text{in}}^+(x_L) = \int_0^1 d\mu\, P_1(\mu)\left[\frac{1}{2}\phi_0(x_L) + \frac{3}{2}\mu\phi_1(x_L)\right]$$

$$= \frac{1}{4}\phi_0(x_L) - \frac{1}{2}D\frac{d\phi_0(x_L)}{dx} \tag{9.139}$$

When the prescribed incident current, $J_{\text{in}}^+ = 0$, the vacuum boundary condition for diffusion theory can be constructed from a geometrical interpretation of the ratio of the flux gradient to the flux in this equation to obtain the condition that the extrapolated flux vanishes a distance λ_{ex} outside the boundary:

$$\phi(x_L - \lambda_{ex}) = 0, \qquad \lambda_{ex} = \frac{2}{3\Sigma_{tr}} \equiv \frac{2}{3}\lambda_{tr} \qquad (9.140)$$

The interface conditions of Eq. (9.133) become in the diffusion approximation

$$\phi_0(x_s + \varepsilon) = \phi_0(x_s - \varepsilon)$$

$$-D_0(x_s + \varepsilon)\frac{d\phi_0(x_s + \varepsilon)}{dx} = -D_0(x_s - \varepsilon)\frac{d\phi_0(x_s - \varepsilon)}{dx} \qquad (9.141)$$

Simplified P_L or Extended Diffusion Theory

The same procedure used to derive diffusion theory from the P_1 equations—solve the odd-order equation for the odd-order moment of the flux in terms of a gradient of the even-order flux moment and use the result to eliminate the odd-order flux—can be used to simplify odd-L P_L approximations of higher order. For example, in the P_3 approximation with an isotropic source and isotropic scattering, the following change of variables is made:

$$\begin{aligned} F_0 &= 2\phi_2 + \phi_0 \\ F_1 &= \phi_2 \end{aligned} \qquad (9.142)$$

to facilitate the reduction of the four coupled P_3 equations to the two coupled diffusion equations

$$-\frac{d}{dx}\left(D_0\frac{dF_0}{dx}\right) + (\Sigma_t - \Sigma_{s0})F_0 = S_0 + 2(\Sigma_t - \Sigma_{s0})F_1$$

$$-\frac{d}{dx}\left(D_1\frac{dF_1}{dx}\right) + \left[\frac{5}{3}(\Sigma_t - \Sigma_{s2}) + \frac{4}{3}(\Sigma_t - \Sigma_{s0})\right]F_1 \qquad (9.143)$$

$$= -\frac{2}{3}S_0 + \frac{2}{3}(\Sigma_t - \Sigma_{s0})F_0$$

where

$$D_1 \equiv \frac{3}{7(\Sigma_t - \Sigma_{s3})} \qquad (9.144)$$

The Marshak vacuum ($J_{in}^+ = 0$) boundary conditions of Eq. (9.129) become

$$\frac{1}{2}F_0(x_L) - \frac{3}{8}F_1(x_L) = D_0\frac{dF_0(x_L)}{dx}$$

$$-\frac{1}{8}F_0(x_L) + \frac{7}{8}F_1(x_L) = D_1\frac{dF_1(x_L)}{dx} \qquad (9.145)$$

This formulation of the P_L equations allows the powerful numerical solution techniques for diffusion theory to be used to solve a higher-order transport approximation.

P_L Equations in Spherical and Cylindrical Geometries

In the case of spherical symmetry, the neutron transport equation becomes

$$\mu \frac{\partial \psi(r, \mu)}{\partial r} + \frac{1 - \mu^2}{r} \frac{\partial \psi(r, \mu)}{\partial \mu} + \Sigma_t(r)\psi(r, \mu)$$

$$= \int_{-1}^{1} d\mu' \Sigma_s(r, \mu' \to \mu)\psi(r, \mu') + S(r, \mu) \tag{9.146}$$

where r is the magnitude of the radius vector $\mathbf{r}$ from the center of the spherical geometry and $\mu = \mathbf{\Omega} \cdot r$. Following the same procedure as above of expanding the angular dependence of the flux and differential scattering cross section according to Eqs. (9.124) and (9.125) and making use of the addition theorem, orthogonality relations, and the recursion relation

$$(1 - \mu^2) \frac{d P_m}{d \mu}(\mu) = (m + 1)[\mu P_m(\mu) - P_{m+1}(\mu)] \tag{9.147}$$

yields the P_L equations in spherical geometry:

$$\frac{d\phi_1}{dr} + \frac{2}{r}\phi_1 + (\Sigma_t - \Sigma_{s0})\phi_0 = S_0, \quad n = 0$$

$$\frac{n + 1}{2n + 1}\left(\frac{d\phi_{n+1}}{dr} + \frac{n + 2}{r}\phi_{n+1}\right) + \frac{n}{2n + 1}\left(\frac{d\phi_{n-1}}{dr} - \frac{n - 1}{r}\phi_{n-1}\right)$$

$$+ (\Sigma_t - \Sigma_{sn})\phi_n = S_n, \quad n = 1, \ldots, L \tag{9.148}$$

For cylindrical symmetry, the formalism becomes more complex because the angular flux depends on two components of the neutron direction vector $\mathbf{\Omega}$, instead of one as in the case of slab and spherical symmetry. With reference to Fig. 9.18, μ is defined with respect to the angle θ between $\mathbf{\Omega}$ and the cylindrical axis, which is taken in the z-direction, and φ is defined as the angle in the $x-y$ plane between the $x-y$ projection of $\mathbf{\Omega}$ and the radius vector $\mathbf{r}$, noting that $\mathbf{\Omega}_p / \sin\theta$ is a unit vector:

$$\mu = \cos\theta = \mathbf{\Omega} \cdot \mathbf{n}_z, \quad \varphi = \cos^{-1}\frac{\mathbf{r} \cdot \mathbf{\Omega}_p}{\sin\theta} \tag{9.149}$$

The neutron transport equation in systems with cylindrical symmetry becomes

$$\sin\theta\left[\cos\varphi \frac{\partial \psi(r, \mu, \varphi)}{\partial r} - \frac{\sin\theta}{4}\frac{\partial \psi(r, \mu, \varphi)}{\partial \varphi}\right] + \Sigma_t(r)\psi(r, \mu, \varphi)$$

$$= \int_0^{4\pi} d\mathbf{\Omega}\Sigma_s(r, \mathbf{\Omega} \cdot \mathbf{\Omega}')\psi(r, \mu', \varphi') + S(r, \mu, \varphi) \tag{9.150}$$

The expansion of the angular dependence of the differential scattering cross section is written in this coordinate system as

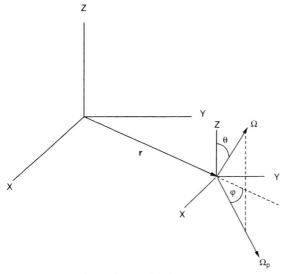

Fig. 9.18 Nomenclature for cylindrical geometry P_L equations.

$$\Sigma_s(r, \mathbf{\Omega} \cdot \mathbf{\Omega}')$$

$$= \frac{\Sigma_s(r, \mu_0)}{2\pi} = \frac{1}{2\pi} \sum_{l'=0}^{L} \frac{2l'+1}{2} \Sigma_{sl'}(r) P_{l'}(\mu_0)$$

$$= \frac{1}{2\pi} \sum_{l'=0}^{L} \frac{2l'+1}{2} \Sigma_{sl'}(r)$$

$$\times \left[P_{l'}(\mu) P_{l'}(\mu') + 2 \sum_{m=0}^{l'} \frac{(l'-m)!}{(l'+m)!} P_{l'}^m(\mu) P_{l'}^m(\mu') \cos m(\varphi - \varphi') \right]$$

$$(9.151)$$

where the addition theorem for Legendre polynomials has been used in the last step. Using this expansion in Eq. (9.149) and multiplying, in turn, by all functions $P_l^m(\mu) \cos(m\varphi)$ for which $l \leq L$, and making use of the recursion relations

$$(l + m - 1)(l + m) P_{l-1}^{m-1}(\mu) - (l - m + 1)(l - m + 2) P_{l+1}^{m-1}(\mu)$$

$$= (2l + 1)\sqrt{1 - \mu^2} P_l^m(\mu), \quad m \neq 0 \qquad (9.152)$$

$$P_{l+1}^{m+1}(\mu) - P_{l-1}^{m+1}(\mu) = (2l + 1)\sqrt{1 - \mu^2} P_l^m(\mu)$$

(where it is understood here and below that terms with negative super- or subscripts are to be omitted) and the orthogonality relations

$$\int_0^{2\pi} d\varphi \int_{-1}^1 d\mu \, P_l^m(\mu) P_{l'}^{m'}(\mu) \cos m\varphi \cos m'\varphi$$

$$= \begin{cases} \pi \dfrac{2(l+m)! \delta_{ll'} \delta_{mm'}}{(2l+1)(l-m)!}, & m \neq 0 \\[3mm] 2\pi \dfrac{2}{2l+1} \delta_{ll'} \delta_{mm'}, & m = 0 \end{cases} \tag{9.153}$$

leads to the P_L equations with an isotropic source in systems with cylindrical symmetry:

$$J_l^{m+1} + J_l^{m-1} + (\Sigma_t - \Sigma_{sl})\phi_l^m = S\delta_{l0}, \quad l = 0, \ldots, L-1; \; m = 1, \ldots, l$$

$$\frac{(1-\delta_{m0})(L+m-1)(L+m)}{2(2L+1)} \left[\frac{d}{dr}\phi_{L-1}^{m-1} - (m-1)\frac{\phi_{L-1}^{m-1}}{r} \right]$$

$$-\frac{1+\delta_{m0}}{2(2L+1)} \left[\frac{d}{dr}\phi_{L-1}^{m+1} + (m+1)\frac{\phi_{L-1}^{m+1}}{r} \right] + (\Sigma_t - \Sigma_{sL})\phi_L^m = 0, \tag{9.154}$$

$$l = L, \; m = 1, \ldots, L$$

where

$$J_l^{m+1} \equiv \frac{1}{2}(1+\delta_{m0}) \left\{ \left[\frac{d\phi_{l+1}^{m+1}}{dr} + (m+1)\frac{\phi_{l+1}^{m+1}}{r} \right] \right.$$

$$\left. - \left[\frac{d\phi_{l-1}^{m+1}}{dr} + (m+1)\frac{\phi_{l-1}^{m+1}}{r} \right] \right\} \Big/ (2l+1)$$

$$J_l^{m-1} \equiv \frac{1}{2}(1-\delta_{m0}) \left\{ (l+m-1)(l+m) \left[\frac{d\phi_{l-1}^{m-1}}{dr} - (m-1)\frac{\phi_{l-1}^{m-1}}{r} \right] \right. \tag{9.155}$$

$$- (l-m+1)(l-m+2)$$

$$\left. \times \left[\frac{d\phi_{l+1}^{m-1}}{dr} - (m-1)\frac{\phi_{l+1}^{m-1}}{r} \right] \right\} \Big/ (2l+1)$$

The P_L equations are equations for the $L+1$ flux moments

$$\phi_l^m(r) \equiv \int_{-1}^1 d\mu \, P_l^m(\mu) \int_0^{2\pi} d\varphi \cos m\varphi \, \psi(r, \mu, \varphi) \tag{9.156}$$

in terms of which the angular flux distribution is given by

$$\psi(r, \mu, \varphi) = \sum_{l=0}^L \frac{2l+1}{4\pi} \sum_{m=-l}^l \frac{(l-m)!}{(l+m)!} \phi_l^m(r) P_l^m(\mu) \cos m\varphi \tag{9.157}$$

Either the Mark or Marshak boundary conditions are applicable in spherical or cylindrical geometry, on the exterior boundary, but these provide only $(L+1)/2$ conditions. The other $(L+1)/2$ conditions are provided by the requirement for symmetry about the origin, which requires the odd flux moments to vanish at the origin. The Marshak form of the interface continuity conditions are also applicable in these geometries.

Diffusion Equations in One-Dimensional Geometry

The P_1 equations may be reduced to a diffusion theory form for spherical and cylindrical geometries. In general, letting r be the spatial coordinate upon which the flux distribution depends, the P_1 equations may be reduced to

$$-\frac{1}{r^n}\frac{d}{dr}\left[r^n D_0\frac{d\phi(r)}{dr}\right]+[\Sigma_t(r)-\Sigma_{s0}]\phi(r)=S_0(r) \tag{9.158}$$

where $n = 0$ for planar geometry, 1 for cylindrical geometry, and 2 for spherical geometry.

The reduction of the P_L equations to coupled diffusion equations that was discussed for slab geometry is not possible in spherical and cylindrical geometries because it is not possible to eliminate coupling terms containing spatial derivatives of the Legendre flux moments. Thus the efficient diffusion theory solution procedures cannot be employed with the spherical and cylindrical P_L equations, and other, generally less efficient iterative methods must be used (e.g., Ref. 6).

Half-Angle Legendre Polynomials

The efficacy of the P_L method depends on the validity of representing the angular dependence of the neutron flux as a low-order continuous polynomial expansion over $-1 \leq \mu \leq 1$. There are situations in which the flux may be highly directional and thus not well represented by a continuous polynomial expansion over both forward and backward directions, but in which the flux may be well represented by separate low-order polynomial expansions over the forward and backward directions. The half-angle Legendre polynomials have been developed for this purpose. The forward ($\mu > 0$) and backward ($\mu < 0$) half-angle Legendre polynomials are defined as

$$\begin{aligned}
p_l^+(\mu) &\equiv P_l(2\mu - 1), \quad \mu > 0 \\
p_l^-(\mu) &\equiv P_l(2\mu + 1), \quad \mu < 0
\end{aligned} \tag{9.159}$$

These polynomials clearly satisfy

$$\begin{aligned}
p_l^+(0) &\equiv p_l^-(-1) = P_l(-1) \\
p_l^+(1) &= p_l^-(0) = P_l(+1)
\end{aligned} \tag{9.160}$$

and may be shown from the orthogonality and recursion relations for the full-range Legendre polynomials to satisfy the orthogonality conditions

$$\int_0^1 p_l^+(\mu)p_m^+(\mu)d\mu = \int_{-1}^0 d\mu\, p_l^-(\mu)p_m^-(\mu) = \frac{\delta_{lm}}{2l+1} \tag{9.161}$$

and to have the recursion relations

$$\begin{aligned}
(l+1)p_{l+1}^+(\mu) + (2l+1)p_l^+(\mu) + lp_{l-1}^+(\mu) &= 2(2l+1)\mu p_l^+(\mu) \\
(l+1)p_{l+1}^-(\mu) - (2l+1)p_l^-(\mu) + lp_{l-1}^-(\mu) &= 2(2l+1)\mu p_l^-(\mu)
\end{aligned} \tag{9.162}$$

Double-P_L Theory

Expanding the flux separately in each half-space $0 \leq \mu \leq 1$ and $-1 \leq \mu \leq 0$,

$$\psi(x, \mu) \simeq \begin{cases} \sum_{l'=0}^{L}(2l'+1)\phi_{l'}^+(x)p_{l'}^+(\mu), & \mu > 0 \\ \sum_{l'=0}^{L}(2l'+1)\phi_{l'}^-(x)p_{l'}^-(\mu), & \mu < 0 \end{cases} \tag{9.163}$$

substituting into Eq. (9.123), weighting in turn by each p_l^+ ($l \leq L$) and integrating over $0 \leq \mu \leq 1$ and by each p_l^- ($l \leq L$) and integrating over $-1 \leq \mu \leq 0$, and making use of the orthogonality and recursion relations above yields a coupled set of $2(L+1)$ double-P_L, or D-P_L, equations:

$$\frac{l+1}{2(2l+1)}\frac{d\phi_{l+1}^+}{dx} + \frac{2l+1}{2(2l+1)}\frac{d\phi_l^+}{dx} + \frac{l}{2(2l+1)}\frac{d\phi_{l-1}^+}{dx} + \Sigma_t\phi_l^+$$

$$= \sum_{l'=l}^{2L+1}\frac{2l'+1}{2}C_{ll'}^+\Sigma_{sl'}\phi_{l'} + S_l^+$$

$$\frac{l+1}{2(2l+1)}\frac{d\phi_{l+1}^-}{dx} - \frac{2l+1}{2(2l+1)}\frac{d\phi_l^-}{dx} + \frac{l}{2(2l+1)}\frac{d\phi_{l-1}^-}{dx} + \Sigma_t\phi_l^- \tag{9.164}$$

$$= \sum_{l'=l}^{2L+1}\frac{2l'+1}{2}C_{ll'}^-\Sigma_{sl'}\phi_{l'} + S_l^-, \quad l = 0, 1, \ldots, L$$

where

$$C_{ll'}^+ \equiv \int_0^1 p_l^+(\mu)P_{l'}(\mu)\,d\mu, \qquad C_{ll'}^- \equiv \int_{-1}^0 d\mu\, p_l^-(\mu)P_{l'}(\mu)$$

$$S_l^+ \equiv \int_0^1 p_l^+(\mu)S(x, \mu)\,d\mu, \qquad S_l^- \equiv \int_{-1}^0 d\mu\, p_l^-(\mu)S(x, \mu) \tag{9.165}$$

The coupling between the forward (+) and backward (−) flux moment equations comes about because of the possibility of scattering from the interval $-1 \leq \mu \leq 0$ to the interval $0 \leq \mu \leq 1$, and vice versa, as indicated by the scattering sums on the right in Eqs. (9.164). The upper limits on these summations arise because the expansion of the differential scattering cross section was terminated at $2L+1$. These scattering terms contain full-range Legendre flux moments which must be represented in terms of the half-range moments by using the approximate representation of the full-range Legendre polynomials in terms of the half-range polynomials:

$$P_l(\mu) \simeq \begin{cases} \sum_{l'=0}^{L,l}(2l'+1)C_{l'l}^+p_{l'}^+(\mu), & \mu > 0 \\ \sum_{l'=0}^{L,l}(2l'+1)C_{l'l}^-p_{l'}^-(\mu), & \mu < 0 \end{cases} \tag{9.166}$$

where the summation extends to l or L, whichever is smaller. This representation leads to

$$\phi_l(x) \equiv \int_{-1}^1 d\mu\, P(\mu)\psi(x, \mu)dx$$

$$\simeq \sum_{l'=0}^{L,l} (2l'+1) \left[C_{l'l}^+ \phi_{l'}^+(x) + C_{l'l}^- \phi_{l'}^-(x) \right] \tag{9.167}$$

The final form of the D-P_L equations is

$$\frac{l+1}{2(2l+1)} \frac{d\phi_{l+1}^+}{dx} + \frac{2l+1}{2(2l+1)} \frac{d\phi_l^+}{dx} + \frac{l}{2(2l+1)} \frac{d\phi_{l-1}^+}{dx} + \Sigma_t \phi_l^+$$

$$= S_l^+ + \sum_{l'=l}^{2L+1} \frac{(2l'+1)}{2} C_{ll'}^+ \Sigma_{sl'} \sum_{l''=0}^{L,l'} \left[C_{l''l'}^+ \phi_{l''}^+ + C_{l''l'}^- \phi_{l''}^- \right]$$

$$\frac{l+1}{2(2l+1)} \frac{d\phi_{l+1}^-}{dx} - \frac{2l+1}{2(2l+1)} \frac{d\phi_l^-}{dx} + \frac{l}{2(2l+1)} \frac{d\phi_{l-1}^-}{dx} + \Sigma_t \phi_l^- \tag{9.168}$$

$$= S_l^- + \sum_{l'=l}^{2L+1} \frac{2l'+1}{2} C_{ll'}^- \Sigma_{sl'} \sum_{l''=0}^{L,l'} \left[C_{l''l'}^+ \phi_{l''}^+ + C_{l''l'}^- \phi_{l''}^- \right], \quad l=0,1,\ldots,L$$

Interface and boundary conditions for the D-P_L equations are straightforward extensions of the conditions derived for the P_L equations. All of the ϕ_l^+ and ϕ_l^- are continuous at interfaces. A vacuum boundary condition requires that the incoming flux moments be zero at that boundary [e.g., a vacuum condition on the left boundary requires that all $\phi_l^+(x_L) = 0$, and a vacuum condition on the right boundary requires that all $\phi_l^-(x_L) = 0$]. A symmetry, or reflective, boundary condition requires that $\phi_l^+(x_L) = \phi_l^-(x_L)$. Known incident flux conditions at the left boundary, $\psi_{in}(x_L, \mu > 0)$, or at the right boundary, $\psi_{in}(x_R, \mu < 0)$, lead to boundary conditions

$$\phi_l^+(x_L) = \int_0^1 d\mu \, p_l^+(\mu) \psi_{in}(x_L, \mu > 0)$$

$$\phi_l^-(x_R) = \int_{-1}^0 d\mu \, p_l^-(\mu) \psi_{in}(x_R, \mu < 0) \tag{9.169}$$

The D-P_L approximation results in $2(L+1)$ first-order ordinary differential equations to be solved for $2(L+1)$ unknowns, the flux moments ϕ_l^+ and ϕ_l^-. The same number of first-order ordinary differential equations and unknown flux moments ϕ_l are obtained in the P_{2L} approximation. In problems in which the difference in the number of neutrons moving in the forward and backward directional half-spaces is more important than the angular distribution per se, the D-P_L approximation is more accurate than the P_{2L} approximation with the same number of unknowns. Thus the D-P_L approximation is to be preferred for interface and boundary problems, whereas the P_{2L} approximation is to be preferred for deep penetration problems.

D-P_0 Equations

This simplest and most widely used of the D-P_L methods is obtained by setting $L=0$ in the equations above and noting that $C_{00}^\pm = 1$ and $C_{01}^\pm = \frac{1}{2}$:

$$\frac{1}{2}\frac{d\phi_0^+}{dx} + \left[\Sigma_t - \frac{1}{2}\Sigma_{s0}\left(1 + \frac{3}{4}\bar{\mu}_0\right)\right]\phi_0^+ = \frac{1}{2}\Sigma_{s0}\left(1 + \frac{3}{4}\bar{\mu}_0\right)\phi_0^- + S_0^+$$

$$-\frac{1}{2}\frac{d\phi_0^-}{dx} + \left[\Sigma_t - \frac{1}{2}\Sigma_{s0}\left(1 + \frac{3}{4}\bar{\mu}_0\right)\right]\phi_0^- = \frac{1}{2}\Sigma_{s0}\left(1 + \frac{3}{4}\bar{\mu}_0\right)\phi_0^+ + S_0^-$$

(9.170)

9.7
Multidimensional Spherical Harmonics (P_L) Transport Theory

Spherical Harmonics

The *spherical harmonics* are defined as (note that there are several normalizations in use)

$$Y_{lm}(\mu,\varphi) = \frac{\sqrt{(l-m)!}}{\sqrt{(l+m)!}} P_l^m(\mu)\exp(im\varphi)$$

(9.171)

in terms of the previously discussed associated Legendre functions. Denoting the complex conjugate by an asterisk, it follows that

$$Y_{l,-m}(\mu,\varphi) = (-1)^m Y_{lm}^*(\mu,\varphi)$$

(9.172)

The first few such functions are

$$Y_{00}(\mu,\varphi) = P_0^0(\mu) = P_0(\mu) = 1$$
$$Y_{10}(\mu,\varphi) = P_1^0(\mu) = P_1(\mu) = \mu$$
$$Y_{11}(\mu,\varphi) = -\frac{1}{\sqrt{2}}\sqrt{1-\mu^2}(\cos\varphi + i\sin\varphi)$$
$$Y_{1,-1}(\mu,\varphi) = \frac{1}{\sqrt{2}}\sqrt{1-\mu^2}(\cos\varphi - i\sin\varphi)$$

(9.173)

and the remaining spherical harmonics can be generated using the recursion relation for the associated Legendre functions defined by Eq. (9.122):

$$(2l+1)\mu P_l^m(\mu) = (l-m+1)P_{l+1}^m(\mu) + (l+m)P_{l-1}^m(\mu)$$

(9.174)

With respect to Fig. 9.19, the directional cosines along Cartesian coordinate axes are given in terms of the spherical harmonics by

$$\Omega_z \equiv \mathbf{\Omega}\cdot\mathbf{n}_z = \mu$$
$$\Omega_x \equiv \mathbf{\Omega}\cdot\mathbf{n}_x = \sqrt{1-\mu^2}\cos\varphi = -\frac{1}{\sqrt{2}}(Y_{11} - Y_{1,-1}) = \frac{1}{\sqrt{2}}(Y_{11}^* - Y_{1,-1}^*)$$
$$\Omega_y \equiv \mathbf{\Omega}\cdot\mathbf{n}_y = \sqrt{1-\mu^2}\sin\varphi = \frac{i}{\sqrt{2}}(Y_{11} + Y_{1,-1}) = \frac{-i}{\sqrt{2}}(Y_{11}^* + Y_{1,-1}^*)$$

(9.175)

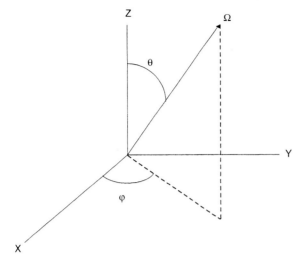

Fig. 9.19 Nomenclature for spherical harmonics.

The spherical harmonics satisfy the orthogonality relationship

$$\int_{-1}^{1} d\mu \int_{0}^{2\pi} d\varphi Y_{l'm'}^{*}(\mu, \varphi) Y_{lm}(\mu, \varphi) = \frac{4\pi}{2l+1} \delta_{ll'} \delta_{mm'} \qquad (9.176)$$

and in terms of the spherical harmonics the addition theorem for Legendre polynomials can be written

$$P_l(\mu_0) = \sum_{m=-l}^{l} Y_{lm}(\mu, \varphi) Y_{lm}^{*}(\mu', \varphi') \qquad (9.177)$$

Spherical Harmonics Transport Equations in Cartesian Coordinates

Expanding the angular dependence of the neutron flux

$$\psi(r, \boldsymbol{\Omega}) = \sum_{l=0}^{L} \frac{2l+1}{4\pi} \sum_{m=-l}^{l} \phi_{lm}(r) Y_{lm}(\mu, \varphi) \qquad (9.178)$$

and the differential scattering cross section

$$\Sigma_s(r, \mu_0) = \sum_{l'=0}^{L'} \frac{2l'+1}{4\pi} \Sigma_{sl'}(r) P_{l'}(\mu_0) \qquad (9.179)$$

in spherical harmonics, substituting the expansions into the neutron transport equation

$$\boldsymbol{\Omega} \cdot \nabla \psi(r, \boldsymbol{\Omega}) + \Sigma_t(r, \boldsymbol{\Omega}) \psi(r, \boldsymbol{\Omega})$$

$$= \frac{1}{4\pi} \int_{4\pi} d\boldsymbol{\Omega}' \nu \Sigma_f(r) \psi(r, \boldsymbol{\Omega}') + \int_{4\pi} d\boldsymbol{\Omega}' \Sigma_s(r)(r, \boldsymbol{\Omega} \cdot \boldsymbol{\Omega}') \psi(r, \boldsymbol{\Omega}') + S(r, \boldsymbol{\Omega})$$

$$(9.180)$$

multiplying by each Y_{lm}^* in turn, integrating over $d\Omega$, and making use of the orthogonality and recursion relations and the addition theorem yields the spherical harmonics equations for the flux moments ϕ_{lm}:

$$
\begin{aligned}
\frac{1}{2l+1}\bigg\{ &\frac{1}{2}\sqrt{(l+m+2)(l+m+1)}\left(-\frac{\partial}{\partial x}-i\frac{\partial}{\partial y}\right)\phi_{l+1,m+1} \\
&+\frac{1}{2}\sqrt{(l-m+2)(l-m+1)}\left(\frac{\partial}{\partial x}-i\frac{\partial}{\partial y}\right)\phi_{l+1,m-1} \\
&+\frac{1}{2}\sqrt{(l-m-1)(l-m)}\left(\frac{\partial}{\partial x}+i\frac{\partial}{\partial y}\right)\phi_{l-1,m+1} \\
&+\frac{1}{2}\sqrt{(l+m-1)(l+m)}\left(-\frac{\partial}{\partial x}+i\frac{\partial}{\partial y}\right)\phi_{l-1,m-1} \\
&+\sqrt{(l+m+1)(l+m-1)}\frac{\partial}{\partial z}\phi_{l+1,m} \\
&+\sqrt{(l+m)(l-m)}\frac{\partial}{\partial z}\phi_{l-1,m}\bigg\}+\Sigma_t\phi_{lm} \\
&=\Sigma_{sl}\phi_{lm}+\delta_{l0}\nu\Sigma_f\phi_{00}+Q_{lm}, \quad l=0,\dots,L,\ m=-l,\dots,l \quad (9.181)
\end{aligned}
$$

where Q_{lm} is the Y_{lm}^* moment of the external source and the other quantities have been discussed previously.

This formidable set of equations is rarely solved as is; however, it provides the basis for the development of a number of useful approximations. Note that the equation for each flux moment ϕ_{lm} contains scattering terms involving only that same flux moment, so that the coupling among equations for different flux moments is entirely through the streaming terms arising from the $\boldsymbol{\Omega}\cdot\nabla\psi$ term.

P_1 Equations in Cartesian Geometry

As was the case in one dimension, the spherical harmonics equations lack closure. When the spatial derivatives involving ϕ_{L+1} in the $l=L$ equation are set to zero, the three-dimensional P_L approximation is obtained. We consider the lowest-order P_1 approximation in more detail. Using Eqs. (9.173) and (9.175), it can be shown that the flux moments are related to the scalar flux and to the currents along the various coordinate axes:

$$
\begin{aligned}
\phi(r) &\equiv \int \psi(r,\boldsymbol{\Omega})d\Omega = \phi_{00} \\
J_x(r) &\equiv \int (\hat{\mathbf{n}}_x\cdot\boldsymbol{\Omega})\psi(r,\boldsymbol{\Omega})d\Omega = \frac{1}{\sqrt{2}}(\phi_{1,-1}-\phi_{11}) \\
J_y(r) &\equiv \int (\mathbf{n}_y\cdot\boldsymbol{\Omega})\psi(r,\boldsymbol{\Omega})d\Omega = \frac{-i}{\sqrt{2}}(\phi_{1,-1}+\phi_{11}) \\
J_z(r) &\equiv \int (\mathbf{n}_z\cdot\boldsymbol{\Omega})\psi(r,\boldsymbol{\Omega})d\Omega = \phi_{10}
\end{aligned}
\qquad (9.182)
$$

Using these relations to express the flux moments in terms of the scalar flux and the currents, Eq. (9.178) becomes (for $L = 1$)

$$\psi(r, \mathbf{\Omega}) = \frac{1}{4\pi}[\phi(r) + 3\mathbf{\Omega} \cdot \mathbf{J}(r)] \tag{9.183}$$

Using the flux moments calculated from Eqs. (9.182) in the ($l = 0$, $m = 0$) equation (9.181) yields the exact equation (i.e., it was not necessary to discard a derivative of a higher moment in this equation)

$$\nabla \cdot \mathbf{J}(r) + \Sigma_t(r)\phi(r) = \Sigma_{s0}(r)\phi(r) + \nu\Sigma_f(r)\phi(r) + Q_{00}(r) \tag{9.184}$$

Adding and subtracting the ($l = 1$, $m = 1$) and ($l = 1$, $m = -1$) equations (9.181) yields the approximate (i.e., it was necessary to discard a derivative of a higher moment in these equations) equations

$$\frac{1}{3}\frac{\partial\phi(r)}{\partial x} + \Sigma_t(r)J_x(r) - \Sigma_{s1}(r)J_x(r) = \int (\mathbf{\Omega} \cdot \mathbf{n}_x)Q\,d\mathbf{\Omega} \equiv Q_{1x}$$
$$\frac{1}{3}\frac{\partial\phi(r)}{\partial y} + \Sigma_t(r)J_y(r) - \Sigma_{s1}(r)J_y(r) = \int (\mathbf{\Omega} \cdot \mathbf{n}_y)Q\,d\mathbf{\Omega} \equiv Q_{1y} \tag{9.185}$$

and the ($l = 1$, $m = 0$) equation yields the approximate equation

$$\frac{1}{3}\frac{\partial\phi(r)}{\partial z} + \Sigma_t(r)J_z(r) - \Sigma_{s1}(r)J_z(r) = \int (\mathbf{\Omega} \cdot \mathbf{n}_z)Q\,d\mathbf{\Omega} \equiv Q_{1z} \tag{9.186}$$

Equations (9.184) to (9.186) are the three-dimensional P_1 equations in Cartesian geometry.

Diffusion Theory

The one-dimensional P_1 equations led to diffusion theory, and it is of some interest to see if the same is true in three dimensions. Equations (9.185) and (9.186) can be written as a Fick's law:

$$\mathbf{J}(r) = -\frac{1}{3[\Sigma_t(r) - \bar{\mu}_0\Sigma_s(r)]}\nabla\phi(r) \equiv -D\nabla\phi(r) \tag{9.187}$$

if the anisotropic source terms Q_1 vanish. Equation (9.187) can be used in Eq. (9.184) to obtain the three-dimensional diffusion equation in Cartesian coordinates

$$-\nabla \cdot D(r)\nabla\phi(r) + [\Sigma_t(r) - \Sigma_{s0}(r)]\phi(r) = \nu\Sigma_f(\mathbf{r})\phi(r) + Q_{00}(r) \tag{9.188}$$

Equation (9.187) and hence also the diffusion equation are thus based on two major assumptions: (1) spatial derivatives of higher flux moments ϕ_2 can be neglected; and (2) anisotropic neutron sources can be neglected. Had we carried out the development from the time-dependent transport equation, it would have also been necessary to assume that the time derivatives of the current could be neglected to obtain a Fick's law.

9.8
Discrete Ordinates Methods in One-Dimensional Slab Geometry

The discrete ordinate methods are based on a conceptually straightforward evaluation of the transport equation at a few discrete angular directions, or ordinates, and the use of quadrature relationships to replace scattering and fission neutron source integrals over angle with summations over ordinates. The essence of the methods are the choice of ordinates, quadrature weights, differencing schemes, and iterative solution procedures. In one dimension, the ordinates can be chosen such that the discrete ordinates methods are completely equivalent to the P_L and D-P_L methods discussed in Section 9.6, and in fact the use of discrete ordinates is probably the most effective way to solve the P_L and D-P_L equations in one dimension. This equivalence does not carry over into multidimensional geometries.

Making use of the spherical harmonics expansion of the differential scattering cross section of Eq. (9.125) and the addition theorem for Legendre polynomials of Eq. (9.121), the one-dimensional neutron transport equation (9.123) in slab geometry becomes

$$\mu \frac{d\psi}{dx}(x,\mu) + \Sigma_t(x)\psi(x,\mu)$$
$$= \sum_{l'=0} \frac{2l'+1}{2} \Sigma_{sl'}(x) P_{l'}(\mu) \int_{-1}^{1} d\mu' P_{l'}(\mu')\psi(x,\mu') + S(x,\mu) \tag{9.189}$$

where the source term includes an external source and, in the case of a multiplying medium such as a reactor core, a fission source. We first discuss the solution of the fixed external source problem (which implicitly assumes a subcritical reactor) and then return to the solution of the critical reactor problem, in which the solution of the fixed source problem constitutes part of the iteration strategy.

Defining N ordinate directions, μ_n, and corresponding quadrature weights, w_n, the integral over the angle in Eq. (9.189) can be replaced by

$$\phi_l(x) \equiv \int_{-1}^{1} d\mu\, P_l(\mu)\psi(x,\mu) \simeq \sum_n w_n P_l(\mu_n)\psi_n(x) \tag{9.190}$$

where $\psi_n \equiv \psi(\mu_n)$. The quadrature weights are normalized by

$$\sum_{n=1}^{N} w_n = 2 \tag{9.191}$$

It is convenient to choose ordinates and quadrature weights that are symmetric about $\mu = 0$, hence providing equal detail in the description of forward and backward neutron fluxes. This can be accomplished by choosing

$$\mu_{N+1-n} = -\mu_n, \qquad \mu_n > 0, \quad n = 1, 2, \ldots, N/2$$
$$w_{N+1-n} = w_n, \qquad w_n > 0, \quad n = 1, 2, \ldots, N/2 \tag{9.192}$$

With such even ordinates, reflective boundary conditions are simply prescribed:

$$\psi_n = \psi_{N+1-n}, \quad n = 1, 2, \ldots, N/2 \tag{9.193}$$

Known incident flux, $\psi_{in}(\mu)$, boundary conditions, including vacuum conditions when $\psi_{in}(\mu) = 0$, are

$$\psi_n = \psi_{in}(\mu_n), \quad n = 1, 2, \ldots, N/2 \tag{9.194}$$

Normally, an even number of ordinates is used ($N = $ even), because this results in the correct number of boundary conditions and avoids certain other problems encountered with $N = $ odd. Even with these restrictions, there remains considerable freedom in the choice of ordinates and weights.

P_L and D-P_L Ordinates

If the ordinates are chosen to be the N roots of the Legendre polynomial of order N,

$$P_N(\mu_i) = 0 \tag{9.195}$$

and the weights are chosen to integrate all Legendre polynomials correctly up to P_{N-1}

$$\int_{-1}^{1} P_l(\mu) d\mu = \sum_{n=1}^{N} w_n P_l(\mu_n) = 2\delta_{l0}, \quad l = 0, 1, \ldots, N - 1 \tag{9.196}$$

then the discrete ordinates equations with N ordinates are entirely equivalent to the P_{N-1} equations. To establish this, we multiply Eq. (9.189) by $w_n P_l(\mu_n)$ for $0 \le l \le N - 1$, in turn, and use the recursion relation of Eq. (9.119) to obtain

$$w_n \left[\frac{l+1}{2l+1} P_{l+1}(\mu_n) + \frac{l}{2l+1} P_{l-1}(\mu_n) \right] \frac{d\psi_n}{dx} + w_n \Sigma_t \psi_n$$

$$= \sum_{l'=0}^{N-1} \frac{2l'+1}{2} \Sigma_{sl'} w_n P_{l'}(\mu_n) P_l(\mu_n) \phi_{l'} + w_n P_l(\mu_n) S(\mu_n),$$

$$l = 0, \ldots, N - 1, \; n = 1, \ldots, N \tag{9.197}$$

Summing these equations over $1 \le n \le N$ yields

$$\frac{l+1}{2l+1} \frac{d\phi_{l+1}}{dx} + \frac{l}{2l+1} \frac{d\phi_{l-1}}{dx} + \Sigma_t \phi_l$$

$$= \sum_{l'=0}^{N-1} \frac{2l'+1}{2} \Sigma_{sl'} \phi_{l'} \left[\sum_{n=0}^{N} w_n P_{l'}(\mu_n) P_l(\mu_n) \right] + \sum_{n=1}^{N} w_n P_l(\mu_n) S(\mu_n),$$

$$l = 0, \ldots, N - 1 \tag{9.198}$$

Weights chosen to satisfy Eqs. (9.196) obviously correctly integrate all polynomials through order N (any polynomial of order n can be written as a sum of Legendre polynomials through order n), but fortuitously they also integrate correctly all polynomials through order less than $2N$. Thus the term in the scattering integral becomes

$$\sum_{n=1}^{N} w_n P_{l'}(\mu_n) P_l(\mu_n) = \int_{-1}^{1} P_{l'}(\mu) P_l(\mu) d\mu = \frac{2\delta_{ll'}}{2l+1} \tag{9.199}$$

and assuming that the angular dependence of the source term can be represented by a polynomial of order $< 2N$:

$$\sum_{n=1}^{N} w_n P_l(\mu_n) S(\mu_n) = \int_{-1}^{1} P_l(\mu) S(\mu) d\mu = \frac{2S_l}{2l+1} \tag{9.200}$$

where S_l is the Legendre moment of the source given by Eq. (9.127).

Using Eqs. (9.199) and (9.200), Eqs. (9.198) become

$$\frac{l+1}{2l+1} \frac{d\phi_{l+1}}{dx} + \frac{l}{2l+1} \frac{d\phi_{l-1}}{dx} + (\Sigma_t - \Sigma_{sl})\phi_l = S_l, \qquad l = 0, \dots, N-2$$
$$\frac{N-1}{2(N-1)+1} \frac{d\phi_{(N-1)-1}}{dx} + (\Sigma_t - \Sigma_{s,N-1})\phi_{N-1} = S_{N-1}, \quad l = N-1 \tag{9.201}$$

which, when ϕ_{-1} is set to zero, are identically the P_L equations (9.126) for $L = N-1$. These P_L ordinates and weights are given in Table 9.2.

The D-P_L ordinates are the roots of the half-angle Legendre polynomials for $L = N/2 - 1$:

$$P_{N/2}(2\mu_n + 1) = 0, \quad n = 1, 2, \dots, \frac{N}{2}$$
$$P_{N/2}(2\mu_n - 1) = 0, \quad n = \frac{N}{2} + 1, \dots, N \tag{9.202}$$

and the corresponding weights are determined from

$$\sum_{n=1}^{N/2} w_n P_l(2\mu_n + 1) = \delta_{l0}, \qquad l = 0, \dots, \frac{N-2}{2}$$
$$\sum_{n=(N+2)/2}^{N} w_n P_l(2\mu_n - 1) = \delta_{l0}, \quad l = 0, \dots, \frac{N-2}{2} \tag{9.203}$$

These ordinates and weights may be evaluated from the data in Table 9.2.

The P_L ordinates and weights are preferable to the D-P_L ordinates and weights for deep penetration problems in heterogeneous media and for problems in which anisotropic scattering is important, for both of which the correct calculation of a large number of Legendre moments of the flux are required. Conversely, for the

Table 9.2 P_{N-1} Ordinates and Weights

$\pm\mu_n$	w_n
$N = 2$	
0.57735	1.00000
$N = 4$	
0.33998	0.65215
0.86114	0.34785
$N = 6$	
0.23862	0.46791
0.66121	0.36076
0.93247	0.17132
$N = 8$	
0.18343	0.36268
0.52553	0.31371
0.79667	0.22238
0.96029	0.10123
$N = 10$	
0.14887	0.29552
0.43340	0.26927
0.67941	0.21909
0.86506	0.14945
0.97391	0.06667
$N = 12$	
0.12523	0.24915
0.36783	0.23349
0.58732	0.20317
0.76990	0.16008
0.90412	0.10694
0.98156	0.04718

Source: Data from Ref. 2; used with permission of Wiley.

calculation of highly anisotropic neutron fluxes near boundaries, the D-P_L ordinates and weights are preferable. With either set of ordinates and weights, the discrete ordinates method in one dimension is essentially a numerical method for solving the P_L or D-P_L equations. Other choices of weights and ordinates can be made to specialize the discrete ordinates method to the problem to be solved (e.g., bunching ordinates to emphasize an accurate calculation of the neutron flux in a certain direction). However, care must be exercised when choosing ordinates and weights that do not correctly integrate the low-order angular polynomials, because surprising results sometimes turn up.

Spatial Differencing and Iterative Solution

Defining cross sections to be constant over $x_{i-1/2} < x < x_{i+1/2}$, Eq. (9.189), for each ordinate, can be integrated over $x_{i-1/2} < x < x_{i+1/2}$ to obtain

$$\mu_n\left(\psi_n^{i+1/2} - \psi_n^{i-1/2}\right) + \Sigma_t^i \psi_n^i \Delta_i$$

$$= \Delta_i Q_n^i \equiv \Delta_i \left[\sum_{l'=0}^{L} \frac{2l'+1}{2} \Sigma_{sl'}^i P_{l'}(\mu_n)\phi_{l'}^i + S^i(\mu_n) \right] \tag{9.204}$$

where $\psi_n^i \equiv \psi(x_i, \mu_n)$, and so on, and $\Delta_i = x_{i+1/2} - x_{i-1/2}$. Using the *diamond difference relation*

$$\psi_n^i = \tfrac{1}{2}\left(\psi_n^{i+1/2} + \psi_n^{i-1/2}\right) \tag{9.205}$$

algorithms for sweeping to the right in the direction of neutrons traveling with $\mu_n > 0$,

$$\psi_n^i = \left(1 + \frac{\Sigma_t^i \Delta_i}{2|\mu_n|}\right)^{-1} \left(\psi_n^{i-1/2} + \frac{\Delta_i Q_n^i}{2|\mu_n|}\right)$$

$$\psi_n^{i+1/2} = 2\psi_n^i - \psi_n^{i-1/2} \tag{9.206}$$

and for sweeping to the left in the direction of neutrons traveling with $\mu_n < 0$,

$$\psi_n^i = \left(1 + \frac{\Sigma_t^i \Delta_i}{2|\mu_n|}\right)^{-1} \left(\psi_n^{i+1/2} + \frac{\Delta_i Q_n^i}{2|\mu_n|}\right)$$

$$\psi_n^{i-1/2} = 2\psi_n^i - \psi_n^{i+1/2} \tag{9.207}$$

are specified.

The boundary conditions at the left boundary (for incident flux or vacuum conditions) are specified for the positive-direction ordinates by Eqs. (9.194) (e.g., $\psi_n^{1/2} = 0$, $\mu_n > 0$ for a vacuum condition). Note that the physical boundaries are located at $x_{1/2}$ and $x_{I+1/2}$. Equations (9.206) are then used to sweep the solutions for ordinates $\mu_n > 0$ to the right boundary, where conditions similar to Eqs. (9.194) specify the boundary conditions (e.g., $\psi_n^{I+1/2} = 0$, $\mu_n < 0$ for a vacuum condition) for the ordinates with $\mu_n < 0$, and Eqs. (9.207) are used to sweep the solutions for $\mu_n < 0$ from the right to the left boundary. If there were no scattering or fission sources in Q_n^i, the solution would be complete. However, there are, and this iterate of the fluxes must be used to update the Q_n^i and the double-sweep repeated until convergence. If there is a reflective boundary, say on the right, the condition $\psi_{N+1-n}^{I+1/2} = \psi_n^{I+1/2}$ is used for the return sweep (the problem should be stated so that the reflective boundary is on the right). If there are reflective conditions on both boundaries, the boundary conditions on the left must be initially guessed, then updated following a double-sweep, and so on, which, of course, slows convergence.

Limitations on Spatial Mesh Size

Truncation error determines the allowable spatial mesh size. Consider Eq. (9.189), for a given ordinate, but without the source term:

$$\mu_n \frac{d\psi_n(x)}{dx} + \Sigma_t(x)\psi_n(x) = 0 \tag{9.208}$$

The exact solution for the flux at $x_{i+1/2}$ in terms of the flux at $x_{i-1/2}$ is

$$\psi_n^{i+1/2} = \exp\left(\frac{-\Sigma_t^i \Delta_i}{|\mu_n|}\right) \psi_n^{i-1/2} \tag{9.209}$$

The finite difference solution is found by using Eq. (9.205) to eliminate ψ_n^i in Eq. (9.204) with $Q_n^i = 0$:

$$\psi_n^{i+1/2} = \frac{1 - \Sigma_t^i \Delta_i/2|\mu_n|}{1 + \Sigma_t^i \Delta_i/2|\mu_n|} \psi_n^{i-1/2} \tag{9.210}$$

The error in the approximate solution is $O((\Sigma_t^i \Delta_i/2|\mu_n|)^2)$. The allowable mesh spacing is determined by the accuracy required and the smallest value of $|\mu_n|$.

Negative fluxes will occur if $\Delta_i > 2|\mu_n|/\Sigma_t^i$. Negative flux fix-up schemes have been developed, which amount to setting negative fluxes to zero when they occur in the iteration, but this introduces difficulties. This problem is sufficiently serious to have motivated the development of a number of alternative difference schemes, but variants of the diamond differencing scheme remain the most commonly used.

9.9
Discrete Ordinates Methods in One-Dimensional Spherical Geometry

The angles that specify the neutron direction in curvilinear geometry change as the neutron moves, as shown in Fig. 9.20. This leads to angular derivatives in the neutron streaming operator, making curvilinear geometries qualitatively different from slab geometry. The conservative form of the neutron transport equation in spherical geometry is

$$\frac{\mu}{\rho^2}\frac{\partial}{\partial\rho}\left[\rho^2\psi(\rho,\mu)\right] + \frac{1}{\rho}\frac{\partial}{\partial\mu}\left[(1-\mu^2)\psi(\rho,\mu)\right] + \Sigma_t(\rho)\psi(\rho,\mu) = Q(\rho,\mu) \tag{9.211}$$

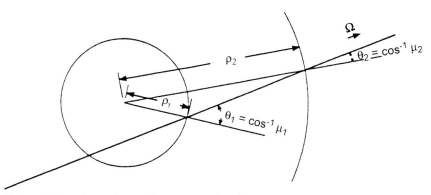

Fig. 9.20 Change in angular coordinate $\mu = \cos\theta$ as the neutron moves. (From Ref. 2; used with permission of Wiley.)

Representation of Angular Derivative

The difference scheme for the angular derivative is determined by the requirement that the sum of the angular and radial streaming terms (the first two terms in the equation above) satisfy the physical constraint of vanishing for an uniform, isotropic flux in an infinite medium. Approximating the angular derivative as

$$\frac{1}{\rho}\frac{\partial}{\partial\mu}\left[(1-\mu^2)\psi(\rho,\mu)\right] \simeq \frac{2}{\rho w_n}(\alpha_{n+1/2}\psi_{n+1/2} - \alpha_{n-1/2}\psi_{n-1/2}) \qquad (9.212)$$

and noting that for an uniform medium and an isotropic flux that $\psi_n = \psi_{n\pm1} = \phi_n/2$, the scalar flux, the requirement that the spatial plus angular derivative terms vanish is

$$\alpha_{n+1/2} = \alpha_{n-1/2} - \mu_n w_n \qquad (9.213)$$

which is an algorithm for determining the $\alpha_{n+1/2}$ once $\alpha_{1/2}$ is known. By choosing $\alpha_{1/2} = 0$ and N even, Eq. (9.213) yields $\alpha_{N+1/2} = 0$, which leads to closure in the angular differencing algorithm.

Using this form for the angular derivative and an angular diamond difference relation

$$\psi_n = \tfrac{1}{2}(\psi_{n+1/2} + \psi_{n-1/2}) \qquad (9.214)$$

in Eq. (9.211) yields

$$\frac{\mu_n}{\rho^2}\frac{\partial}{\partial\rho}\left(\rho^2\psi_n\right)$$

$$+ \frac{2}{\rho w_n}[2\alpha_{n+1/2}\psi_n - (\alpha_{n+1/2} + \alpha_{n-1/2})\psi_{n-1/2}] + \Sigma_t\psi_n = Q_n \qquad (9.215)$$

The spatial differencing proceeds as for the slab case, but taking into account the variation of differential area and volume with radius.

Iterative Solution Procedure

The equations are solved by sweeping in the direction of neutron travel. With reference to Fig. 9.21, for an S_4 ($N = 4$) calculation, the calculation is started on the outer surface of the sphere for the direction $n = \tfrac{1}{2}$.

A known incident flux (including vacuum) boundary condition

$$\psi_n^{I+1/2} = \psi_{in}(\mu_n), \quad n = 1, \ldots, N/2 \qquad (9.216)$$

provides a starting value for $\psi_{1/2}^{I+1/2}$. The calculation sweeps inward (decreasing i) for $n = 1/2$ ($\mu = -1$) using

$$\psi_n^{i-1/2} = 2\psi_n^i - \psi_n^{i+1/2} \qquad (9.217)$$

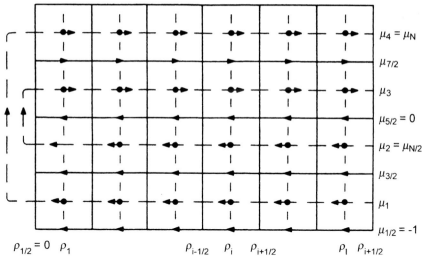

Fig. 9.21 Sweep of the space–angle mesh for one-dimensional spherical geometry. (From Ref. 2; used with permission of Wiley.)

$$\psi_{1/2}^{i} = \frac{2\psi_{1/2}^{i+1/2} + (\rho_{i+1/2} - \rho_{i-1/2})Q_{1/2}^{i}}{2 + \Sigma_{t}^{i}(\rho_{i+1/2} - \rho_{i-1/2})}, \quad \mu < 0 \tag{9.218}$$

Next, the ψ_1^i row is calculated using the starting value from Eq. (9.216) and using Eq. (9.217) and

$$\psi_n^i = \left[2|\mu_n|A_{i-1/2} + \frac{2}{w_n}(A_{i+1/2} - A_{i-1/2})\alpha_{n+1/2} + V_i \Sigma_t^i \right]^{-1}$$

$$\times \left[|\mu_n|(A_{i+1/2} + A_{i-1/2})\psi_n^{i+1/2} \right.$$

$$\left. + \frac{1}{w_n}(A_{i+1/2} - A_{i-1/2})(\alpha_{n+1/2} + \alpha_{n-1/2})\psi_{n-1/2}^i + V_i Q_n^i \right] \tag{9.219}$$

to sweep the solution inward. Then the $\psi_{3/2}^i$ are calculated from the angular diamond difference relation

$$\psi_{n+1/2}^i = 2\psi_n^i - \psi_{n-1/2}^i \tag{9.220}$$

These inward sweeps are continued, using, alternatively, Eqs. (9.217) and (9.219) for the ψ_n^i and Eqs. (9.220) for the $\psi_{n+1/2}^i$ until all the inward ($\mu_n < 0$, $n \leq N/2$) fluxes are calculated.

The starting fluxes at the center of the sphere ($i = 1/2$) for the outward ($\mu_n > 0$, $n > N/2$) calculation are determined from the symmetry condition at the center of the sphere:

$$\psi_{N+1-n}^{1/2} = \psi_n^{1/2}, \quad n = 1, 2, \ldots, N/2 \tag{9.221}$$

Then the calculation is swept outward (increasing i) using for ψ_n^i

$$\psi_n^{i+1/2} = 2\psi_n^i - \psi_n^{i-1/2} \tag{9.222}$$

$$\psi_n^i = \left[2|\mu_n|A_{i+1/2} + \frac{2}{w_n}(A_{i+1/2} - A_{i-1/2})\alpha_{n+1/2} + V_i\Sigma_t^i \right]^{-1}$$
$$\times \left[|\mu_n|(A_{i+1/2} + A_{i-1/2})\psi_n^{i-1/2} \right.$$
$$\left. + \frac{1}{w_n}(A_{i+1/2} - A_{i-1/2})(\alpha_{n+1/2} + \alpha_{n-1/2})\psi_{n-1/2}^i + V_i Q_n^i \right] \tag{9.223}$$

and the angular diamond difference relation for $\psi_{n+1/2}^i$:

$$\psi_{n+1/2}^i = 2\psi_n^i - \psi_{n-1/2}^i \tag{9.224}$$

The A's and V's in the equations above are the shell areas and differential volume elements at the radii indicated:

$$A_{i+1/2} = 4\pi\rho_{i+1/2}^2, \qquad V_i = \frac{4\pi}{3}\left(\rho_{i+1/2}^3 - \rho_{i-1/2}^3\right) \tag{9.225}$$

From these directional fluxes the scalar flux is calculated and the scattering and fission source terms in Q are updated for the next iteration.

Acceleration of Convergence

The numerical solution for the fluxes ψ_n^i on each double sweep is exact for the given scattering and fission source guess Q. The rate of convergence of the solution depends on the rate of convergence of these sources. Note from Eq. (9.204) that these sources depend only on the Legendre flux moments defined by Eq. (9.190) as a weighted sum over the ordinates of the ψ_n^i. This suggests that the iterative solution for the ψ_n^i can be accelerated by advancing the solution for the ϕ_l^i in a low order (e.g., diffusion theory) approximation at intermittent steps during the iteration, which is the basis of the *synthetic method*.

Another acceleration technique—*coarse mesh rebalance*—makes use of the fact that the converged solution for the ψ_n^i must satisfy neutron balance. Imposing this condition on the unconverged solution over coarse mesh regions that include a number of spatial mesh points at intermittent steps in the iteration provides a means for accelerating the solution. Both acceleration methods, which are discussed in detail in Ref. 2, may become unstable if the spatial mesh spacing is not sufficiently small. The synthetic method may even become unstable with small mesh spacing. Other acceleration methods, such as Chebychev acceleration, may also be applied to accelerate the discrete ordinates solution.

Calculation of Criticality

Up to this point, we have discussed solving the discrete ordinates equations for a fixed external source. We now consider the critical reactor problem, in which there is no external source. In this case the equations above would be modified by the inclusion of an effective multiplication constant, k^{-1}, as an eigenvalue in the fission term. A value k_0 and an initial flux guess $\psi^{(0)}$ would be used to evaluate the fission $S_f^{(0)}$ and scattering $S_s^{(0)}$ sources, and the solution above would be carried out to obtain a first iterate flux solution $\psi^{(1)}$. An improved fission source $S_f^{(1)}(\psi^{(1)}/k_1)$, an improved eigenvalue guess $k_1 = k_0 S_f^{(1)}/S_f^{(0)}$, and an improved scattering source $S_s^{(1)}(\psi^{(1)}/k_1)$ would be constructed, and the solution would be repeated to obtain $\psi^{(2)}$, and so on, until the eigenvalues obtained on successive iterates converged to within a specified tolerance. There are also techniques for accelerating this *power iteration* procedure.

9.10
Multidimensional Discrete Ordinates Methods

Ordinates and Quadrature Sets

Two angular coordinates are required to specify the direction of motion in multidimensional geometries. With reference to Fig. 9.22, denote the direction cosines of the neutron direction $\mathbf{\Omega}$ with respect to the x_1-, x_2-, and x_3-coordinate axes as μ, η, and ξ, respectively. Only two of these direction cosines are independent, and since $\mathbf{\Omega}$ is a unit vector, $\mu^2 + \eta^2 + \xi^2 = 1$.

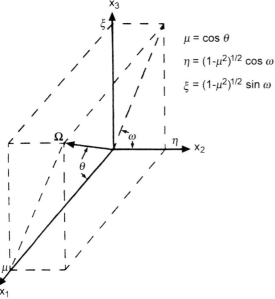

$$\mu = \cos\theta$$
$$\eta = (1-\mu^2)^{1/2}\cos\omega$$
$$\xi = (1-\mu^2)^{1/2}\sin\omega$$

Fig. 9.22 Coordinate system for multidimensional discrete ordinates. (From Ref. 2; used with permission of Wiley.)

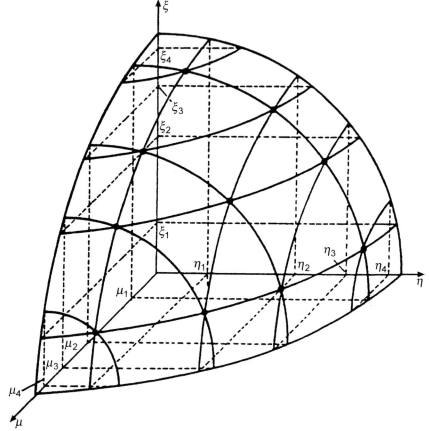

Fig. 9.23 Level symmetric S_8 discrete ordinates quadrature set.
(From Ref. 2; used with permission of Wiley.)

In three-dimensional problems, the flux must be determined in all eight octants of the unit sphere over which $\boldsymbol{\Omega}$ varies. In two-dimensional geometries, there is an assumption of symmetry in one of the coordinate directions, which reduces to four the number of octants over which the flux must be determined. (In one-dimensional geometries, there is an assumption of symmetry in two of the coordinate directions, and the flux must be determined only within two of the octants.) It is convenient to use a set of ordinates that are symmetric in the eight octants (i.e., can satisfy reflective conditions across surfaces in the x_1–x_2 plane, the x_2–x_3 plane, and the x_3–x_1 plane). Then, if the ordinates and weights are constructed for a set of direction cosines satisfying $\mu_n^2 + \eta_n^2 + \xi_n^2 = 1$ in one octant, the ordinates and weights for the other octants with direction cosine sets $(-\mu_n, \eta_n, \xi_n)$, $(\mu_n, -\eta_n, \xi_n)$, $(\mu_n, \eta_n, -\xi_n)$, $(-\mu_n, -\eta_n, \xi_n)$, $(-\mu_n, \eta_n, -\xi_n)$, $(\mu_n, -\eta_n, -\xi_n)$, and $(-\mu_n, -\eta_n, -\xi_n)$ are obtained simply by changing the signs of one or more direction cosines.

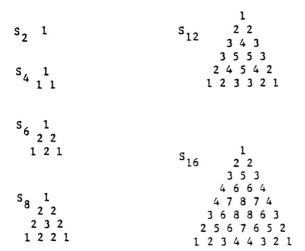

S_2 1

S_4 1
    1 1

S_6 1
   2 2
  1 2 1

S_8 1
   2 2
  2 3 2
 1 2 2 1

S_{12}
      1
     2 2
    3 4 3
   3 5 5 3
  2 4 5 4 2
 1 2 3 3 2 1

S_{16}
       1
      2 2
     3 5 3
    4 6 6 4
   4 7 8 7 4
  3 6 8 8 6 3
 2 5 6 7 6 5 2
1 2 3 4 4 3 2 1

Fig. 9.24 Equal-weighted ordinates for one octant in the S_N quadrature. (From Ref. 2; used with permission of Wiley.)

The *level symmetric quadratures* shown in Fig. 9.23 use the same set of $N/2$ positive values of the direction cosines with respect to each of the three axes (i.e., $\mu_n = \eta_n = \xi_n$, $n = 1, \ldots, N/2$). Use of such a quadrature set strictly defines the S_N method, although the term S_N is loosely used more widely as a synonym for discrete ordinates. The rotational symmetry of the level symmetric quadrature set and the requirement $\mu_n^2 + \eta_n^2 + \xi_n^2 = 1$ determines all the direction cosines except one. Once μ_1 is chosen, the other μ_n are calculated from

$$\mu_n^2 = \mu_1^2 + 2(n-1)\frac{1 - 3\mu_1^2}{N-2} \tag{9.226}$$

and the $\eta_n = \xi_n = \mu_n$. For the S_2 approximation, with only one direction cosine, satisfaction of $\mu_1^2 + \eta_1^2 + \xi_1^2 = 1$ uniquely specifies $\eta_1 = \xi_1 = \mu_1 = \sqrt{1/3}$, and there are no degrees of freedom in the choice of ordinates.

The weights in each octant are normalized by

$$\sum_{n=1}^{N(N+2)/8} w_n = 1 \tag{9.227}$$

where the index n runs over all the (μ_i, η_j, η_k), $i, j, k = 1, \ldots, N/2$ ordinate combinations in the octant. For the S_2 approximation, with only one ordinate per octant, $w_1 = 1$. For other S_N approximations the level symmetry condition $\mu_n = \eta_n = \xi_n$ requires that the weights be equal for ordinates obtained by permuting the direction cosines, as shown in Fig. 9.24, where the same value of w_n is assigned to all the ordinates indicated by the same number.

Note that unlike the situation in one dimension, this level symmetric quadrature set does not integrate Legendre polynomials to any given order accurately. However, even within the restrictions discussed above, there remain a few degrees of

Table 9.3 Level Symmetric S_N Quadrature Set

S_N	n	μ_n	w_n
S_4	1	0.35002	0.33333
	2	0.86889	–
S_6	1	0.26664	0.17613
	2	0.68150	0.15721
	3	0.92618	–
S_8	1	0.21822	0.12099
	2	0.57735	0.09074
	3	0.78680	0.09259
	4	0.95119	–
S_{12}	1	0.16721	0.07076
	2	0.45955	0.05588
	3	0.62802	0.03734
	4	0.76002	0.05028
	5	0.87227	0.02585
	6	0.97164	–

Source: Data from Ref. 2; used with permission of Wiley.

freedom, and these may be chosen for the purpose of correctly integrating the maximum number of Legendre polynomials in each of the angular variables consistent with the number of degrees of freedom. A quadrature set so constructed is given in Table 9.3.

S_N Method in Two-Dimensional $x-y$ Geometry

The discrete ordinates equations in two-dimensional $x-y$ geometry are

$$\mu_n \frac{\partial \psi(\mathbf{\Omega}_n)}{\partial x} + \eta_n \frac{\partial \psi(\mathbf{\Omega}_n)}{\partial y} + \Sigma_t \psi(\mathbf{\Omega}_n) = Q(\mathbf{\Omega}_n) \tag{9.228}$$

where the spatial dependence has been suppressed, $\mathbf{\Omega}_n = \mathbf{\Omega}(\mu_n, \eta_n)$, and the source Q includes a spherical harmonics representation of the scattering source plus a fission and external source S:

$$Q(\mathbf{\Omega}_n) = \sum_{l=0}^{L} \sum_{m=0}^{l} (2 - \delta_{m0}) Y_{lm}(\mathbf{\Omega}_n) \Sigma_{sl} \phi_l^m + S(\mathbf{\Omega}_n) \tag{9.229}$$

and the discrete ordinates approximation for the flux moments are

$$\phi_l^m = \frac{1}{4} \sum_{n=1}^{N(N+2)/2} w_n Y_{lm}(\mathbf{\Omega}_n) \psi(\mathbf{\Omega}_n) \tag{9.230}$$

Dividing the $x-y$ domain of the problem into mesh boxes $x_{i-1/2} < x < x_{i+1/2}$, $y_{j-1/2} < y < y_{j+1/2}$ centered at (x_i, y_j) with constant cross sections within each mesh box, integrating Eq. (9.228) over a mesh box, and defining volume-averaged quantities

$$\psi_n^{ij} \equiv \frac{1}{\Delta x_i \Delta y_j} \int_i dx \int_j dy\, \psi_n(x,y) \tag{9.231}$$

$$Q_n^{ij} \equiv \frac{1}{\Delta x_i \Delta y_j} \int_i dx \int_j dy\, Q_n(x,y) \tag{9.232}$$

and surface-averaged fluxes

$$\psi_n^{i+1/2,j} \equiv \frac{1}{\Delta y_j} \int_j dy\, \psi_n(x_{i+1/2},y) \tag{9.233}$$

$$\psi_n^{i,j+1/2} \equiv \frac{1}{\Delta x_i} \int_i dx\, \psi_n(x,y_{j+1/2}) \tag{9.234}$$

yields the neutron balance equation on a mesh box:

$$\frac{\mu_n}{\Delta x_i}\left(\psi_n^{i+1/2,j} - \psi_n^{i-1/2,j}\right) + \frac{\eta_n}{\Delta y_j}\left(\psi_n^{i,j+1/2} - \psi_n^{i,j-1/2}\right) + \Sigma_t^{ij}\psi_n^{ij} = Q_n^{ij} \tag{9.235}$$

It is necessary to relate the volume-averaged flux to the surface-averaged fluxes for each mesh box. There are several methods for doing this, the most common of which are the diamond difference method, which is used here, and the theta-weighted method. The volume- and surface-averaged fluxes are related in the diamond difference method by

$$\psi_n^{ij} = \tfrac{1}{2}\left(\psi_n^{i+1/2,j} + \psi_n^{i-1/2,j}\right) = \tfrac{1}{2}\left(\psi_n^{i,j+1/2} + \psi_n^{i,j-1/2}\right) \tag{9.236}$$

These equations are solved by sweeping the two-dimensional mesh grid in the direction of neutron travel. With respect to Fig. 9.25, each iteration (on the scattering source) consists of four sweeps through the grid corresponding to the four octants. For the octant with ($\mu_n > 0$, $\eta_n > 0$), the sweep is left to right, bottom to top; for the octant with ($\mu_n < 0$, $\eta_n > 0$), the sweep is right to left, bottom to top; for the octant with ($\mu_n > 0$, $\eta_n < 0$), the sweep is left to right, top to bottom; and for the octant with ($\mu_n < 0$, $\eta_n < 0$), the sweep is right to left, top to bottom.

For the octant with ($\mu_n > 0$, $\eta_n > 0$), Eqs. (9.236) can be used to write Eqs. (9.235) as

$$\psi_n^{ij} = \left(\Sigma_t^{ij} + \frac{2\mu_n}{\Delta x_i} + \frac{2\eta_n}{\Delta y_i}\right)^{-1}$$

$$\times \left(\frac{2\mu_n}{\Delta x_i}\psi_n^{i-1/2,j} + \frac{2\eta_n}{\Delta y_j}\psi_n^{i,j-1/2} + Q_n^{ij}\right) \tag{9.237}$$

Starting with known incident flux (including vacuum) conditions $\psi_n^{1/2,j} = \psi_{in}(x_L, \mu_n > 0)$, $j = 1, \dots, J$ and $\psi_n^{i,1/2} = \psi_{in}(y_B, \eta_n > 0)$, $i = 1, \dots, I$, where x_L refers

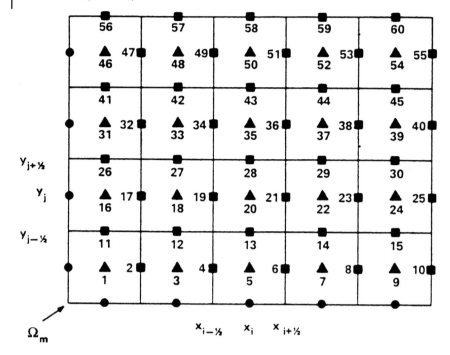

• **Known boundary flux**
■ **Calculated flux at cell interface**
▲ **Calculated flux at cell center**

Fig. 9.25 Order of sweeping the two-dimensional $(x-y)$ mesh grid for the octant with $(\mu_n > 0, \eta_n > 0)$. (From Ref. 2; used with permission of Wiley.)

to the left boundary and y_B refers to the bottom boundary, the flux ψ_n^{11} can be calculated with Eq. (9.237). The solution is then swept to the right using, alternatively, Eq. (9.235) and (9.236) to calculate $\psi_n^{3/2,1}, \psi_n^{2,1}, \ldots, \psi_n^{I+1/2,1}$. Then Eq. (9.236) is used to calculate $\psi_n^{1,3/2}, \psi_n^{2,3/2}, \ldots, \psi_n^{I,3/2}$. Using the boundary conditions $\psi_n^{1/2,2} = \psi_{\text{in}}(x_L, \mu_n > 0)$, Eqs. (9.235) and (9.236) can be used alternatively to sweep to the right across the $j = 2$ row, and then Eq. (9.236) can be used to sweep to the right across the $j = 2\frac{1}{2}$ row, and so on, until all the outgoing fluxes are calculated. Sweeps through the other three octants are carried out in a similar manner but in the order indicated above and with Eqs. (9.235) and (9.236) combined in such a way as to obtain an algorithm like Eq. (9.237) appropriate for that octant. The scalar flux

$$\phi^{ij} = \frac{1}{4} \sum_{n=1}^{N(N+2)/2} w_n \psi_n^{ij} \qquad (9.238)$$

and the Legendre moments

$$\phi_{lm}^{ij} = \frac{1}{4} \sum_{n=1}^{N(N+2)/2} w_n Y_{lm}(\mathbf{\Omega}_n) \psi_n^{ij} \tag{9.239}$$

are then constructed and used to evaluate the scattering and fission source terms. The process is repeated until source convergence on successive iterations is within a specified tolerance.

Further Discussion

The discrete ordinates method in multidimensional geometries is highly geometry dependent. Because of the coupling of spatial and angular mesh intervals, the methodology was initially limited to the regular geometries: parallelepipeds, cylinders, and spheres. However, the development of triangular spatial mesh techniques enables a variety of geometries to be approximated. A number of other ordinate and weight quadrature sets have been devised for special purposes (e.g., to emphasize a given direction in a deep penetration problem). The acceleration methods discussed for the one-dimensional discrete ordinates methods are also used for multidimensional discrete ordinates solutions, but the higher dimensionality introduces complications that diminish their efficacy. In problems with optically thick regions in which the scattering cross section (within-group scattering cross section in multigroup applications) is much larger than the absorption cross section, the source convergence can become intolerably slow. In problems with very little scattering and localized neutron sources, unphysical oscillations in the angular distribution, known as *ray effects*, arise because of discrete directions in which the solution is calculated. There are special remedies for these ray effects, such as a semianalytical calculation of a first collision source to be used in a subsequent discrete ordinates calculation. These difficulties notwithstanding, the discrete ordinates method provides a powerful means for calculating the neutron flux distribution in a nuclear reactor core and the surrounding shield and structure, and is widely used for problems in which diffusion theory is inadequate. Detailed discussions of discrete ordinates methods can be found in Refs. 2 and 5.

9.11
Even-Parity Transport Formulation

The one-group, or within-group, transport equation can be written in the case of isotropic sources and isotropic scattering:

$$\mathbf{\Omega} \cdot \nabla \psi(\mathbf{r}, \mathbf{\Omega}) + \Sigma_t(\mathbf{r}, \mathbf{\Omega}) \psi(\mathbf{r}, \mathbf{\Omega}) = \Sigma_s(\mathbf{r}) \phi(\mathbf{r}) + S(\mathbf{r}) \tag{9.240}$$

Defining the (+) even- and (−−) odd-parity components of the angular flux

$$\psi^{\pm}(\mathbf{r}, \mathbf{\Omega}) = \frac{1}{2}[\psi(\mathbf{r}, \mathbf{\Omega}) \pm \psi(\mathbf{r}, -\mathbf{\Omega})] \tag{9.241}$$

results in the following identities

$$\psi(\mathbf{r}, \mathbf{\Omega}) = [\psi^+(\mathbf{r}, \mathbf{\Omega}) + \psi^-(\mathbf{r}, \mathbf{\Omega})]$$
$$\psi^+(\mathbf{r}, \mathbf{\Omega}) = \psi^+(\mathbf{r}, -\mathbf{\Omega}) \tag{9.242}$$
$$\psi^-(\mathbf{r}, \mathbf{\Omega}) = -\psi^-(\mathbf{r}, -\mathbf{\Omega})$$

which can be used to demonstrate that the scalar flux and current can be written in terms of the even and odd, respectively, components

$$\psi(\mathbf{r}) \equiv \int d\mathbf{\Omega} \big(\psi^+(\mathbf{r}, \mathbf{\Omega}) + \psi^-(\mathbf{r}, \mathbf{\Omega})\big) = \int d\mathbf{\Omega}\, \psi^+(\mathbf{r}, \mathbf{\Omega})$$
$$\mathbf{J}(\mathbf{r}) \equiv \int d\mathbf{\Omega}\, \mathbf{\Omega} \big(\psi^+(\mathbf{r}, \mathbf{\Omega}) + \psi^-(\mathbf{r}, \mathbf{\Omega})\big) = \int d\mathbf{\Omega}\, \mathbf{\Omega} \psi^-(\mathbf{r}, \mathbf{\Omega}) \tag{9.243}$$

Adding Eq. (9.240) written for $-\mathbf{\Omega}$ to the same equation written for $\mathbf{\Omega}$ and using Eq. (9.241) yields

$$\mathbf{\Omega} \cdot \nabla \psi^-(\mathbf{r}, \mathbf{\Omega}) + \Sigma_t(\mathbf{r})\psi^+(\mathbf{r}, \mathbf{\Omega}) = \Sigma_s(\mathbf{r})\phi(\mathbf{r}) + S(\mathbf{r}) \tag{9.244}$$

and subtracting the same two equations yields

$$\mathbf{\Omega} \cdot \nabla \psi^+(\mathbf{r}, \mathbf{\Omega}) + \Sigma_t(\mathbf{r})\psi^-(\mathbf{r}, \mathbf{\Omega}) = 0 \tag{9.245}$$

The second of these equations may be used in the first to eliminate the odd-parity flux component, resulting in an equation for the even-parity flux:

$$-\mathbf{\Omega} \cdot \nabla \left[\frac{1}{\Sigma_t(\mathbf{r})} \mathbf{\Omega} \cdot \nabla \psi^+(\mathbf{r}, \mathbf{\Omega}) \right] + \Sigma(\mathbf{r})\psi^+(\mathbf{r}, \mathbf{\Omega}) = \Sigma_s(\mathbf{r})\phi(\mathbf{r}) + S(\mathbf{r}) \tag{9.246}$$

and Eq. (9.245) may be used to write the current in terms of the even-parity component:

$$\mathbf{J}(\mathbf{r}) = -\int d\mathbf{\Omega}\, \mathbf{\Omega} \frac{1}{\Sigma_t(\mathbf{r})} \mathbf{\Omega} \cdot \nabla \psi^+(\mathbf{r}, \mathbf{\Omega}) \tag{9.247}$$

The vacuum boundary condition becomes [from Eqs. (9.242) and (9.245)]

$$0 = \psi(\mathbf{r}_s, \mathbf{\Omega}) = \mathbf{\Omega} \cdot \nabla \psi^+(\mathbf{r}_s, \mathbf{\Omega}) \pm \Sigma_t(\mathbf{r}_s)\psi^+(\mathbf{r}_s, \mathbf{\Omega}), \quad \mathbf{\Omega} \cdot \mathbf{n}_s \gtrless 0 \tag{9.248}$$

and the reflection boundary condition is

$$\psi^+(\mathbf{r}_s, \mathbf{\Omega}) = \psi^+(\mathbf{r}_s, \mathbf{\Omega}') \tag{9.249}$$

where $\mathbf{\Omega}$ is the direction of spectral reflection relative to incident direction $\mathbf{\Omega}$.

9.12
Monte Carlo Methods

At a fundamental level, neutron transport through matter is formulated as an essentially stochastic process. The total cross section is a probability (per unit path length and unit atom density), but not a certainty, that a neutron will have a collision while traversing a certain spatial interval. If the neutron does have a collision, the cross sections for the various processes are probabilities, but not certainties, that the collision will be a scattering, radiative capture, fission, and so on, event. The neutron flux that we have discussed earlier in the chapter is actually the mean, or expectation, value of the neutron distribution function. The Monte Carlo method directly simulates neutron transport as a stochastic process.

Probability Distribution Functions

Let us postulate that variable x may take on various values over the interval $a \leq x \leq b$ and that there exists a probability distribution function (pdf), $f(x)$, such that $f(x)dx$ is the probability that a variable takes on a value within dx about x. The normalization is chosen such that

$$\int_a^b f(x)dx = 1 \tag{9.250}$$

In general, $f(x) \geq 0$ will not be a monotonically increasing function of x, which means that a given value for f does not correspond to a unique value of x.

A more useful quantity is the cumulative probability distribution function (cdf), $F(x)$, defined as the probability that the variable x takes on a value less than or equal to x:

$$F(x) = \int_a^x f(x')dx' \tag{9.251}$$

which is a monotonically increasing function of x. Thus the probability of a neutron having a value of x between x and $x + dx$ is $F(x + dx) - F(x) = f(x)dx$. If κ is a random number distributed between 0 and 1, the values of x determined from $F(x) = \kappa$ will be distributed as $f(x)$. In some cases, it is possible to solve directly for $x = F^{-1}(\kappa)$. In other cases, the cumulative distribution function may be known as a large table of $F(x_i)$ and the value of x determined by interpolation; for example, if $F(x_j) < \kappa < F(x_{j-1})$ linear interpolation yields

$$x = x_j - \frac{F(x_j) - \kappa}{F(x_j) - F(x_{j-1})}(x_j - x_{j-1}) \tag{9.252}$$

There are also methods of selection from the pdf, but it is generally preferable to select from the cdf.

Analog Simulation of Neutron Transport

By tracing the path of an individual neutron as it traverses matter and considering the various processes that may determine its history, we can understand how a Monte Carlo calculation simulates the stochastic nature of neutron transport through matter. We begin with the source of neutrons in a nuclear reactor, which is predominantly if not entirely the fission source. The fission source has a distribution in space (we discuss calculation of the fission source distribution in Monte Carlo later), a distribution in energy given by the fission spectrum, and a distribution in direction that is isotropic. Each of these distributions may be characterized by a pdf and a cdf. Generating a random number and selecting from the cdf for the spatial fission distribution defines a location in space for the source particle. Generating another random number and selecting from the cdf for the fission spectrum determines the energy of the source particle. Generating third and fourth random numbers and selecting from the cdf's for the two independent angular variables (say $\mu = \cos\theta$ and φ) defines the direction of the source neutron.

Once launched, the source neutron will travel in a straight line until it has a collision. The probability that a neutron has a collision at a distance s along the flight path is

$$T(s) = \Sigma_t(s)\exp\left[-\int_0^s \Sigma_t(s')ds'\right] \tag{9.253}$$

which is the pdf for the collision distance s. Generating a random number λ and selecting s from the cdf

$$-\ln\lambda = \int_0^s \Sigma_t(s')ds' \tag{9.254}$$

locates the position of the first collision, in principle. In fact, the process is considerably complicated by the nonuniform geometry. It is necessary to know the composition at the point of the first collision. We treat the medium as piecewise homogeneous and define the lengths of each uniform segment of the straightline flight path as s_j. If

$$\sum_{j=1}^{n-1}\Sigma_{tj}s_j \le -\ln\lambda < \sum_{j=1}^{n}\Sigma_{tj}s_j \tag{9.255}$$

the collision occurs in the nth region at a distance

$$s'_n = \frac{1}{\Sigma_{tn}}\left(-\ln\lambda - \sum_{j=1}^{n}\Sigma_{tj}s_j\right) \tag{9.256}$$

beyond the entrance of the flight path into region n. The actual procedures for treating flight paths in complex geometries are quite involved but highly developed. Modern Monte Carlo codes can essentially model any geometry exactly, which is a great strength of the method.

Having determined that a collision occurred at a distance s_n into region n on the original flight path, it is now necessary to determine what type of nuclide and what type of reaction are involved. The probability for a reaction of type x with a nuclide of species i is

$$p_{ix} = \frac{N_i \sigma_{ix}}{\sum_{i,x} N_i \sigma_{ix}} \tag{9.257}$$

where N_i is the number density of nuclide i in region n, σ_{ix} is the microscopic cross section for reaction x for nuclide i at the energy of the neutron. Constructing a pdf and a cdf, generating a random number η, and selecting the nuclide and reaction type by equating η and the cdf [probably involving table interpolation per Eq. (9.252)], the nuclide and reaction type can be determined.

If the reaction type is absorption, the neutron history is terminated, the energy and location of the absorbed neutron are recorded, and another history is started. If the reaction type is elastic scattering, another random number is generated and equated to the cdf for the cosine of the scattering angle in the center of mass (CM) to obtain μ_{cm} (it is convenient to work in the CM because the scattering is isotropic except for high-energy neutrons scattering from heavy mass nuclei, and the pdf and cdf are simple) and by transformation to obtain the scattering angle in the lab. For energies above thermal, the energy of the scattered neutron is uniquely correlated to μ_{cm} from the scattering kinematics:

$$E' = \frac{E(A^2 + 2A\mu_{cm} + 1)}{(A+1)^2} \tag{9.258}$$

Knowing E', the cosine of the scattering angle in the lab can be determined from

$$\mu = \cos\theta = \frac{1}{2}(A+1)\sqrt{\frac{E}{E'}} + \frac{1}{2}(A-1)\sqrt{\frac{E'}{E}} \tag{9.259}$$

When inelastic scattering or elastic scattering of thermal neutrons from bound lattice atoms is involved, the cdf's are more complicated. Generating another random number and equating it to the cdf for the azimuthal angle φ, the direction of the scattered neutron can be determined. The scattered neutron is treated as described above for a fission source neutron, and the calculation is repeated until the neutron either leaks from the system or is absorbed.

Statistical Estimation

The mean, or expectation, value of a function $h(x)$ of x is defined in terms of the pdf for x by

$$\langle h \rangle = \int_a^b dx\, h(x) f(x) \tag{9.260}$$

and the standard deviation, σ, and the variance, V, are defined:

$$\sigma(h) = \sqrt{V(h)} = \left\{ \int_a^b dx [h(x) - \langle h \rangle]^2 f(x) \right\}^{1/2}$$

$$= \left[\langle h^2 \rangle - \langle h \rangle^2 \right]^{1/2} \tag{9.261}$$

If N random values of the variable x are chosen from the cdf, as discussed above, a statistical estimate of the mean value $\langle h \rangle$ is

$$\bar{h} = \frac{1}{N} \sum_{n=1}^{N} h(x_n) \tag{9.262}$$

A bound for the error in an estimate of this type is given by the *central limit theorem*, which states that if many estimates $\bar{h}$ of $\langle h \rangle$ are obtained, each estimate involving N trials, the variable $\bar{h}$ is normally distributed about $\langle h \rangle$ to terms of accuracy $O(1/N^{1/2})$. In the limit $N \to$ infinity, this theorem takes the form

$$\text{Prob} \left\{ \langle h \rangle - \frac{M\sigma(h)}{\sqrt{N}} \leq \bar{h} \leq \langle h \rangle + \frac{M\sigma(h)}{\sqrt{N}} \right\} = \begin{cases} 0.6826, & M = 1 \\ 0.954, & M = 2 \\ 0.997, & M = 3 \end{cases} \tag{9.263}$$

[i.e., the probability that the statistical estimate of the mean value of Eq. (9.262) is within $\pm M\sigma/N^{1/2}$ of the exact value $\langle h \rangle$ is 68.3% for $M = 1$, 95.4% for $M = 2$, 99.7% for $M = 3$, etc.].

In general, the first and second moments of $h(x)$ are unknown. The statistical data can be used to construct approximations to these moments. The expectation value of $\bar{h}$ is

$$\langle \bar{h} \rangle = \frac{1}{N} \sum_{n=1}^{N} \langle h(x_n) \rangle = \frac{1}{N} \sum_{n=1}^{N} \int_a^b dx_n \, f(x_n) h(x_n)$$

$$= \frac{1}{N} \sum_{n=1}^{N} \int_a^b dx \, f(x) h(x) = \frac{1}{N} \sum_{n=1}^{N} \langle h \rangle = \langle h \rangle \tag{9.264}$$

(i.e., the statistical estimate $\bar{h}$ is an unbiased estimate of $\langle h \rangle$ since $\langle \hat{h} \rangle = \langle h \rangle$. The expected value of $\bar{h}^2$ is

$$\langle \bar{h}^2 \rangle = \frac{1}{N^2} \left\langle \sum_{n=1}^{N} h(x_n) \sum_{m=1}^{N} h(x_m) \right\rangle = \frac{1}{N^2} \left\langle \sum_{n=1}^{N} h^2(x_n) + \sum_{n=1}^{N} h(x_m) \sum_{m \neq n}^{N} h(x_m) \right\rangle$$

$$= \frac{1}{N^2} \left[N \langle h^2 \rangle + N(N-1) \langle h \rangle^2 \right] = \frac{\langle h^2 \rangle}{N} - \frac{N-1}{N} \langle h \rangle^2 \tag{9.265}$$

(i.e., the statistical estimate $\overline{h^2}$ is a biased estimate of $\langle h^2 \rangle$ since $\langle \bar{h}^2 \rangle \neq \langle h^2 \rangle$.

Since $\langle \bar{h}^2 \rangle = \langle h^2 \rangle$, the variance in the statistical estimate of $\bar{h}$ (h-bar) can be approximated:

$$V(\bar{h}) = \frac{1}{N} \left(\langle h^2 \rangle - \langle h \rangle^2 \right) = \frac{V(h)}{N}$$

$$\simeq \frac{1}{N-1} \left(\overline{h^2} - \bar{h}^2 \right) \tag{9.266}$$

and the mean squared fractional error associated with the statistical estimate of $\bar{h}$ is

$$\varepsilon^2 = \frac{1}{N}\left(\frac{\langle h^2 \rangle}{\langle h \rangle^2} - 1\right) \simeq \frac{1}{N-1}\left(\frac{\overline{h^2}}{\bar{h}^2} - 1\right) \tag{9.267}$$

Variance Reduction

It is clearly important to reduce the mean-squared error in order to increase confidence in the Monte Carlo calculation of the mean value of a quantity $h(x)$ based on a random sampling of the variable x. From Eq. (9.267), this can be accomplished by just running more histories, but that involves longer computational times. There are other methods of reducing the mean-squared error, or the related variance.

We now discuss a number of such variance reduction methods.

The basic idea of *importance sampling* is to select from a modified distribution function that yields the same mean value but a smaller variance. Suppose that instead of evaluating $\langle h \rangle$ and the statistical estimate from Eqs. (9.260) and (9.262), we evaluate them from

$$\langle h_2 \rangle = \int_a^b dx\, h(x) \frac{f(x)}{f^*(x)} f^*(x) \tag{9.268}$$

$$\bar{h}_2 = \frac{1}{N}\sum_{n=1}^{N} h(x_n)\frac{f(x_n)}{f^*(x_n)} \equiv \frac{1}{N}\sum_{n=1}^{N} h(x_n)w(x_n) \tag{9.269}$$

where the values of x_n are now selected from the distribution $f^*(x)$ according to the procedures described previously. The quantity $w(x_n) \equiv f(x_n)/f^*(x_n)$ is known as a *weight function*. Obviously, the mean value $\langle h_1 \rangle$ computed from Eq. (9.260) and the mean value $\langle h_2 \rangle$ computed from Eq. (9.268) are the same. The statistical estimate $\bar{h}_2$ of Eq. (9.269) and the statistical estimate $\bar{h}_1$ of Eq. (9.262) both have the expectation value $\langle h \rangle$. However, the variances are different, and this is the point. The variances computed by the two sampling procedures are

$$V_1(\bar{h}_1) = \frac{1}{N}\int_a^b dx\left(h(x) - \langle h \rangle\right)^2 f(x)$$

$$= \frac{1}{N}\left[\int_a^b dx\, h^2(x) f(x) - \langle h \rangle^2\right] \tag{9.270}$$

$$V_2(\bar{h}_2) = \frac{1}{N}\int_a^b dx\left(\frac{h(x) f(x)}{f^*(x)} - \langle h \rangle\right)^2 f^*(x)$$

$$= \frac{1}{N}\left[\int_a^b dx\, \frac{h^2(x) f^2(x)}{f^*(x)} - \langle h \rangle^2\right] \tag{9.271}$$

The objective is to choose $f^*(x)$ so that $V_2 < V_1$. If the distribution $h(x)$ and its expectation value were known, the optimum choice of $f^*(x)$ would be

$$f^*(x) = \frac{h(x) f(x)}{\langle x \rangle} \tag{9.272}$$

for which $V_2 = 0$. This suggests that a good estimate of $f^*(x)$ could reduce the variance significantly. The function $f^*(x)$ should be chosen to emphasize those neutrons which in some sense are the most important to the quantity that is being estimated, $\langle h \rangle$, which suggests that it is an importance or adjoint function (Chapter 13). However, the variance reduction techniques which are in common use are schemes for emphasizing neutrons which are most likely to contribute to the tally for the quantity of interest, $\langle h \rangle$, based on experience and intuition. Nonanalog variance reduction schemes are implemented by adjusting the neutron weight at each event in its history. An event may be a collision, crossing a boundary into a different region, and so on.

An *exponential transformation* is useful in penetration problems to increase the number of neutron histories which penetrate deeply to contribute to the event of interest (e.g., penetration of a shield, penetration into a control rod). If the event of interest depends primarily on neutrons moving in the positive x-direction, the cross section can be artificially reduced in the x-direction to enhance penetration:

$$\Sigma_t^{ex} = \Sigma_t(1 - p\mathbf{\Omega} \cdot \mathbf{n}_x) \tag{9.273}$$

where $0 \le p \le 1$. At a collision, the particle weight must be multiplied by a weight w_{ex} to preserve the expected weight of the collided neutron; that is,

$$\Sigma_t e^{-\Sigma_t s}ds = w_{ex}\Sigma_t^{ex}e^{-\Sigma_t^{ex}s}ds \tag{9.274}$$

must be satisfied, which defines the weight

$$w_{ex} = \frac{\exp[-p(\mathbf{\Omega} \cdot \mathbf{n}_x)s]}{1 - p(\mathbf{\Omega} \cdot \mathbf{n}_x)} \tag{9.275}$$

When a reaction rate is to be calculated over a small volume in which the collision probability is small, the artifice of *forced collisions* is useful. A neutron entering the volume with weight w which would have to travel a distance l to cross the volume is split into two neutrons, the first of which passes through the volume without collision and the second of which is forced to collide within the volume. Since the actual probability for the particle to cross the region without collision is $\exp(-\Sigma_t l)$, the collided and uncollided neutrons must be given weights $w_c = w[1 - \exp(-\Sigma_t l)]$ and $w_{un} = w\exp(-\Sigma_t l)$, respectively. The history of the uncollided particle with weight w_{un} is restarted on the exiting surface of the volume. A new history is started for the collided particle. The pdf for collision of this second particle within the volume is

$$f(s) = \frac{\Sigma_t e^{-\Sigma_t s}}{1 - e^{-\Sigma_t l}} \tag{9.276}$$

Generating a random number ξ ($0 \le \xi \le 1$), the distance into the volume at which the collision takes place is selected:

$$s = -\frac{1}{\Sigma_t} \ln\left[1 - \xi\left(1 - e^{-\Sigma_t l}\right)\right] \tag{9.277}$$

and the subsequent history of the collided particle with weight w_c is followed.

In some problems, the penetration of a neutron to a particular region may be of interest, and absorption in other regions may unduly reduce the number of neutrons that survive to do so. *Absorption weighting* can be used as an alternative to terminating a history by an absorption event. In a collision all outcomes are treated as scattering events, but the emerging neutron is given a weight

$$w_a = w\left(1 - \frac{\Sigma_a}{\Sigma_t}\right) \tag{9.278}$$

to preserve the survival probability.

Since continuing the computation of histories of neutrons with small weights is inefficient, *Russian roulette* can be used to either increase the neutron weight or terminate the history. A random number ξ ($0 \leq \xi \leq 1$) is generated and compared with an input number v typically between 2 and 10. If $\xi > 1/v$, the history is terminated; if $\xi < 1/v$, the history is continued with original neutron weight w increased to $w_{RR} = wv$.

Splitting can be used to increase the number of histories that penetrate in deep penetration problems. When a neutron with weight w crosses a fixed surface in the direction of penetration from a region with importance I_i into a region with importance I_{i+1}, the history is terminated and I_{i+1}/I_i new histories are started for neutrons with the same energy and direction and weights $w_s = wI_i/I_{i+1}$. Here *importance* refers to importance with respect to the quantity of interest $\langle h \rangle$. Russian roulette can be used in conjunction with splitting to terminate histories of particles with low weights moving across the surfaces away from the direction of penetration.

Tallying

The calculation of reaction rates in various regions, over various energies, and by various nuclides is accomplished straightforwardly by tallying each collision event. Neutron fluxes and currents can also be constructed by tallying events and surface crossings. By definition, the collision rate in a region is equal to the product of the cross section times the flux times the volume. Thus, by tallying the collision rate (CR), the flux can be calculated from

$$\phi = \frac{CR}{\Sigma_t V} \tag{9.279}$$

A shortcoming of this algorithm is that only particles which collide within the volume V will contribute to CR, hence to ϕ. Another definition of the scalar flux is the path length traversed by all particles passing through a volume per unit volume per unit time:

$$\bar{\phi} = \frac{\bar{l}}{V} = \frac{1}{V}\frac{1}{N}\sum_{n=1}^{N} l_n \tag{9.280}$$

where $\bar{l}$ is the track length per unit time in the volume in question of the nth history. Taking into account the weights of neutrons at various stages of their histories, this definition of flux becomes

$$\bar{\phi} = \frac{1}{V} \frac{1}{N} \sum_m w_n l_n \tag{9.281}$$

where w_n is the weight the neutron on the nth history had when it traversed the volume (note that a neutron history may traverse a given volume more than once, and it should be tallied each time). The variance in the flux estimate is given by

$$Var = \frac{N}{N-1} \left[\frac{1}{V^2 N} \sum_{n=1}^{n} (w_n l_n)^2 - \frac{1}{V^2 N^2} \left(\sum_{n=1}^{N} w_n l_n \right)^2 \right] \tag{9.282}$$

Currents across surfaces are also of interest. It is straightforward to tally the rate at which particles are passing through a given surface in the positive and negative directions, $p_n^{\pm}$ for history n. The total number of particles per unit time passing through the surface in the positive and negative directions can then be estimated from

$$p^{\pm} = \frac{1}{N} \sum_{n=1}^{N} w_n p_n^{\pm} \tag{9.283}$$

Here w_n is the weight that the neutron in the nth history had when it crossed through the surface to contribute to the tally (note that a neutron in a given history may cross through a surface more than once, and it should be tallied each time). The partial currents are obtained by dividing by the surface area A, and the net current is obtained by subtracting the partial currents:

$$J = J^{+} - J^{-} = \frac{p^{+} - p^{-}}{A} = \frac{1}{AN} \sum_{n=1}^{N} w_n \left(p_n^{+} - p_n^{-} \right) \tag{9.284}$$

If the Monte Carlo calculation is to be used to determine small differences, such as reactivity worths of perturbations, or reactivity coefficients, special methods must be used to avoid the small difference in two calculations being masked by statistical errors. The method of *correlated sampling* addresses this problem by using the same sequence of random numbers to generate the sequence of events that describes the histories in the two problems. If the system is unchanged, the two calculations must yield identical results. So any difference in results is due to the perturbation.

Criticality Problems

Monte Carlo can be used to calculate the multiplication constant and associated eigensolution for the flux distribution. The problem is started with an arbitrary spatial distribution of neutrons distributed in the fission energy spectrum and isotropically in direction. This initial spatial distribution can be uniform or a spatial distribution that is the result of a previous Monte Carlo calculation for a similar problem or of a deterministic transport (e.g., discrete ordinates) solution for the problem at hand. The history of a large number of neutrons in a given generation is followed in parallel to termination, thus obtaining a new fission distribution for the next generation of neutrons, and the process is repeated until the fission neutron spatial source distribution has settled down. The total number of neutrons in successive generations may be increased during the settling down period to obtain greater detail only after the solution has settled. Once the spatial fission neutron source distribution has settled down, the ratio of the total number of fission neutrons on successive generations is the statistical estimate of the multiplication constant. The computational effort in the period before the distribution settles down can be reduced by a number of techniques.

The fission source distribution is determined from generation to generation as follows. If w_n is the weight of the nth history neutron when it has an absorption event that terminates the history, then either I_n or $I_n + 1$ fission neutrons are produced at that location in the next generation. The selection is made by writing

$$w_n \frac{\nu \Sigma_f}{\Sigma_a} = I_n + R_n \tag{9.285}$$

where I_n is an integer and $0 < R_n < 1$. A random number ξ $(0 \leq \xi \leq 1)$ is generated. If $R_n > \xi$, then $I_n + 1$ neutrons are launched in the next generation; otherwise, I_n neutrons are launched. Track lengths can provide a second estimate of the total number of fission neutrons produced by history n:

$$\sum_i w_{ni} l_n \frac{\nu \Sigma_{fi}}{\Sigma_{ai}} = \text{total number of secondaries produced by history } n \tag{9.286}$$

where w_{ni} is the neutron weight as it crosses region i and l_{ni} is the total track length across region i.

One of the problems in criticality calculations is to prevent the total neutron population from increasing or decreasing too much, which it will do if the assembly is supercritical or subcritical, respectively. One technique is to change the neutron weight at each collision by multiplying the previous weight by the expected number of secondary neutrons. A second method is simply to start off each generation with the same number of neutrons by eliminating some of the next-generation neutrons if there are more neutrons than in the previous generation or using some of the neutrons twice if there are less neutrons than in the previous generation.

Source Problems

A number of reactor physics problems can be formulated as source problems. The most obvious is the shielding problem, where the reactor core can be considered as a fixed neutron source. The calculation of resonance absorption of neutrons from a slowing-down source in a heterogeneous lattice, the thermalization of neutrons from a slowing-down source into the thermal range, and the calculation of temperature coefficients of reactivity from a fixed fission source in a heterogeneous lattice are other problems which are treated as source problems.

The resonance cross sections in the resolved region can be represented by values at a very large number of energy points or calculated from the Doppler-broadened Breit–Wigner formula, and the resonance cross sections in the unresolved region can be selected from a pdf based on the statistics of the nuclear level spacing and width (Chapter 11). The neutron slowing down through the resonance region is then treated by sampling the uniform distribution of neutrons scattered at energy E over the interval E to αE, sampling the path length distribution to determine the point of collision, sampling the reaction-type distribution to determine whether the collision is absorption or scattering, and so on. Effective Doppler-broadened cross sections at different temperatures can be used in conjunction with correlated sampling to compute temperature coefficients of reactivity.

A source distribution in energy of neutrons slowing down into the thermal range in the moderator can be used to launch neutrons isotropically in the thermal energy region. The distribution of rotational–vibrational levels (Chapter 12) which affect inelastic scattering of neutrons from bound atoms and molecules can be used to construct pdf's for inelastic scattering. Then the histories of thermal neutrons can be traced until termination by absorption. Path length estimators at different energies can be used to estimate the thermal flux spectrum.

Random Numbers

Generation of random numbers is essential to a Monte Carlo calculation. There exist a number of random number generators—algorithms for generating random numbers—and there is a great deal of controversy about just how random they are. A discussion of random number generators and several FORTRAN routines for generating random numbers are given in Ref. 1.

References

1 W. H. Press et al., *Numerical Recipes*, Cambridge University Press, Cambridge (**1989**), Chap. 7.

2 E. E. Lewis and W. F. Miller, *Computational Methods of Neutron Transport*, Wiley-Interscience, New York (**1984**); reprinted by American Nuclear Society, La Grange Park, IL (**1993**).

3 R. J. J. Stamm'ler and M. J. Abbate, *Methods of Steady-State Reactor Physics in Nuclear Design*, Academic Press, London (**1983**), Chaps. IV and V.

4 S. O. Lindahl and Z. Weiss, "The Response Matrix Method," in J. Lewins and M. Becker, eds., *Adv. Nucl. Sci. Technol. 13* (**1981**).

5 B. G. Carlson and K. D. Lathrop, "Transport Theory: The Method of Discrete Ordinates," in H. Greenspan, C. N. Kelber, and D. Okrent, eds., *Computing Methods in Reactor Physics*, Gordon and Breach, New York (**1968**).

6 E. M. Gelbard, "Spherical Harmonics Methods: P_L and Double-P_L Approximations," in H. Greenspan, C. N. Kelber, and D. Okrent, eds., *Computing Methods in Reactor Physics*, Gordon and Breach, New York (**1968**).

7 M. H. Kalos, F. R. Nakache, and J. Celnik, "Monte Carlo Methods in Reactor Computations," in H. Greenspan, C. N. Kelber, and D. Okrent, eds., *Computing Methods in Reactor Physics*, Gordon and Breach, New York (**1968**).

8 J. Spanier and E. M. Gelbard, *Monte Carlo Principles and Neutron Transport Problems*, Addison-Wesley, Reading, MA (**1964**).

9 M. Clark and K. F. Hansen, *Numerical Methods of Reactor Analysis*, Academic Press, New York (**1964**).

10 R. V. Meghreblian and D. K. Holmes, *Reactor Analysis*, McGraw-Hill, New York, (**1960**), pp. 160–267 and 626–747.

11 B. Davison, *Neutron Transport Theory*, Oxford University Press, London (**1957**).

12 K. M. Case, F. de Hoffmann, and G. Placzek, *Introduction to the Theory of Neutron Diffusion*, Los Alamos National Laboratory, Los Alamos, NM (**1953**).

13 A. F. Henry, *Nuclear Reactor Analysis*, MIT Press, Cambridge, MA (**1975**), Chap. 6.

14 R. Sanchez, "Approximate Solution of the Two-Dimensional Integral Transport Equation by Collision Probabilities Methods," *Nucl. Sci. Engr.* 64, 384 (**1977**); "A Transport Multicell Method for Two-Dimensional Lattices of Hexagonal Cells," *Nucl. Sci. Engr.* 92, 247 (**1986**).

Problems

9.1. Rederive the transmission and absorption probabilities for a purely absorbing slab given by Eq. (9.28) for the situation in which the incident flux is linearly anisotropic (i.e., $\sim \mu$).

9.2. Use the orthogonality relation for Legendre polynomials to derive the orthogonality relation for half-angle polynomials given by Eq. (9.23).

9.3. Carry through the indicated steps to derive the integral equation (9.42).

9.4. Develop analytical expressions for the two-dimensional transmission and escape probabilities of Eqs. (9.110) and (9.112) for rectangular geometry with dimension a_x and a_y on a side. Evaluate these transmission and first-flight escape probabilities for $X = 4V/S\lambda$ varying over the range $0.1 < X < 10.0$.

9.5. Evaluate the first-flight escape probabilities given by Eq. (9.117) with $c = 2.09$ for $X = 4V/S\lambda$ varying over the range $0.1 < X < 10.0$ and compare with the results of Problem 9.4.

9.6. Carry through the indicated steps to derive the P_L equations (9.126).

9.7. Derive the simplified P_3 equations (9.143) from the P_3 equations and derive the boundary conditions of Eqs. (9.145).

9.8. Demonstrate that when the ordinates and weights given by Eqs. (9.202) and (9.203) are used, the discrete ordinates equations with N ordinates reduce identically to the D-P_{N-1} equations.

9.9. Write a code to solve the one-dimensional discrete ordinates equations in slab geometry. Solve for the flux in the S_2 approximation in a uniform slab 100 cm thick with vacuum boundary conditions, with $\Sigma_t = 0.25$ cm^{-1}, $\Sigma_{s0} = 0.15$ cm^{-1}, and an isotropic source $S_0 = 10^{14}$ n/cm·s distributed over $0 < x < 25$ cm. Repeat for the S_4, S_8, and S_{12} approximations.

9.10. Repeat Problem 9.9 including anisotropic scattering $\Sigma_{s1} = 0.01$ and $\Sigma_{s2} = 0.0025$.

9.11. Derive the spatial difference equations for the one-dimensional discrete ordinates equations in spherical geometry. Reconcile your results with the algorithms of Eqs. (9.219) and (9.223).

9.12. Write a code to solve the S_N equations in two-dimensional x–y geometry. Solve for the flux in the S_2 approximation in a uniform square 100 cm on a side with vacuum boundary conditions, with $\Sigma_t = 0.25$ cm^{-1}, $\Sigma_{s0} = 0.15$ cm^{-1}, $\Sigma_{s1} = 0.01$, and $\Sigma_{s2} = 0.0025$, and an isotropic source $S_0 = 10^{14}$ n/cm^2·s distributed over $0 < x < 25$ cm, $25 < y < 50$ cm. Repeat for the S_4, S_8, and S_{12} approximations.

9.13. The pdf for variable x is $f(x) = 4/\pi(1 + x^2)$ with $0 \le x \le 1$. Show that if a random number ξ $(0 \le \xi \le 1)$ is generated, the corresponding value of $x = \tan(\xi\pi/4)$.

9.14. Derive the simplified P_5 diffusion equations and associated Marshak boundary conditions from the P_5 equations. (*Hint*: Use $F_0 = 2\phi_2 + \phi_0$, $F_1 = \frac{4}{3}\phi_4 + \phi_2$, $F_2 = \phi_4$.)

9.15. Derive the diffusion theory equation (9.158) from the one-dimensional P_l equations in cylindrical and spherical geometries.

9.16. Derive the spherical harmonics approximation to the neutron transport equation in three-dimensional x–y–z geometry given by Eqs. (9.181).

9.17. Plot the cumulative distribution function corresponding to the fission spectrum given approximately by $\chi(E) = 0.453 \exp(-1.036E) \sinh \sqrt{2.29E}$ over the energy range 10^4 eV $\le E \le 10^7$ eV.

9.18. Calculate the maximum spatial mesh size that could be used in a one-dimensional S_2 calculation for a problem with $\Sigma_t = 0.3$ cm^{-1}. Repeat for the S_4 and S_8 approximations.

9.19. Plot the pdf and cdf for the cross-section distribution in a region with $\Sigma_a = 0.15$ cm^{-1}, and $\Sigma_s = 0.08$ cm^{-1}, and $\Sigma_f = 0.08$ cm^{-1}.

9.20. Write a Monte Carlo code to calculate the multiplication constant and flux distribution for one-speed neutrons in a slab reactor of thickness $a = 1.0$ m with isotropic scattering for which ($\Sigma_a = 0.12$ cm^{-1}, $\Sigma_s = 0.05$ cm^{-1}, $\nu\Sigma_f = 0.15$ cm^{-1}) over $0 < x < 50$ cm and ($\Sigma_a = 0.10$ cm^{-1}, $\Sigma_s = 0.05$ cm^{-1}, $\nu\Sigma_f = 0.12$) over $50 < x < 100$ cm.

9.21. A $S(\mu) \sim \mu^2$ neutron source is present on the left face of a slab of thickness a with absorption cross section Σ_a and isotropic scattering cross section Σ_s. Derive expressions for the uncollided and total neutron currents exiting from the right surface of the slab.

9.22. Derive the Boltzmann transport equation from particle balance considerations on a differential element of space–angle–energy phase space. Justify any assumptions.

10
Neutron Slowing Down

The methods used to calculate the slowing down of fast neutrons above the thermal energy range are treated in this chapter. We also introduce the lethargy as an alternative to the energy variable and develop the formalism in terms of lethargy.

10.1
Elastic Scattering Transfer Function

Lethargy

It is convenient in treating neutron slowing down to replace the energy variable with the neutron *lethargy*

$$u = \ln \frac{E_0}{E} \tag{10.1}$$

where E_0 is the maximum energy that a neutron might have in a nuclear reactor, say 10 MeV. The incremental lethargy interval, du, corresponding to the incremental energy interval, dE, is

$$du = \frac{du}{dE}dE = -\frac{dE}{E} \tag{10.2}$$

with the minus sign indicating that as the neutron energy decreases, its lethargy increases—hence the name.

The fact that the total neutron flux in an incremental lethargy interval physically is the same as the neutron flux in the corresponding incremental energy interval provides a correspondence between the flux per unit energy, $\phi(E)$, and the flux per unit lethargy, $\phi(u)$:

$$\phi(u)\,du = -\phi(E)\,dE \Rightarrow \phi(u) = E\phi(E) \tag{10.3}$$

Elastic Scattering Kinematics

The principal results obtained in Chapter 2 from the conservation of energy and momentum in an elastic scattering event were the correlation between the energy

Nuclear Reactor Physics. Weston M. Stacey
Copyright © 2007 WILEY-VCH Verlag GmbH & Co. KGaA, Weinheim
ISBN: 978-3-527-40679-1

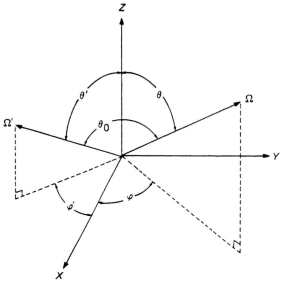

Fig. 10.1 Angles involved in a scattering event. (From Ref. 2; used with permission of MIT Press.)

change $E' \to E$ and the cosine of the scattering angle in the center-of-mass (CM) system $\mu_C = \cos\theta_C$:

$$\frac{E}{E'} = \frac{A^2 + 1 + 2A\mu_c}{(A+1)^2} = e^{u'-u} \equiv e^{-U} \tag{10.4}$$

and the relation between the cosine of the scattering angle in the lab system, $\mu_0 = \cos\theta_0$, and the cosine of the scattering angle in the CM system,

$$\mu_0 = \frac{1 + A\mu_c}{(1 + 2A\mu_c + A^2)^{1/2}} \tag{10.5}$$

which may be combined to express the correlation between the scattering angle in the lab system and the change in lethargy $U = u - u'$:

$$\mu_0(U) = \frac{1}{2}\left[(A+1)e^{-(1/2)U} - (A-1)e^{(1/2)U}\right] \tag{10.6}$$

Elastic Scattering Kernel

The general lethargy-angle scattering transfer function can be written

$$\sigma_s(\mu_0, u' \to u) = \sigma_s(u')p_0(u', \mu_0)g(\mu_0, u' \to u) \tag{10.7}$$

where $\mu_0 = \boldsymbol{\Omega}' \cdot \boldsymbol{\Omega}$ is the cosine of the angle in the lab system between the incident and exit directions of a neutron in a scattering collision, as shown in Fig. 10.1, $p_0(u', \mu_0)$ is the probability that a neutron of lethargy u' will scatter through an angle $\theta_0 = \cos^{-1} \mu_0$ and $g(\mu_0, u' \to u)$ is the probability that a neutron of lethargy u' which scatters through an angle $\theta_0 = \cos^{-1} \mu_0$ will have a final lethargy u. With the normalization

$$\int g(\mu_0, u' \to u)du = 1 \tag{10.8}$$

the angular transfer function for scattering through an angle $\theta_0 = \cos^{-1} \mu_0$ is

$$\sigma_s(\mu_0, u') \equiv \int \sigma_s(\mu_0, u' \to u)du = \sigma_s(u')p_0(u', \mu_0) \tag{10.9}$$

Writing the lethargy-angle transfer function as a function of $(u', U = u - u', \mu_0)$ and expanding in Legendre polynomials yields

$$\sigma_s(\mu_0, U, u') = \sum_{l'=0}^{\infty} \frac{1}{2}(2l' + 1)b_{l'}^0(u', U)P_{l'}(\mu_0) \tag{10.10}$$

where $P_l(\mu_0)$ is the lth Legendre polynomial of the argument of the cosine of the scattering angle in the lab system, and the orthogonality properties of the Legendre polynomials can be used to identify the Legendre coefficients of the scattering transfer function:

$$b_l^0(u', U) = \int_{-1}^{1} d\mu_0 P_l(\mu_0)\sigma_s(\mu_0, U, u') \tag{10.11}$$

For elastic scattering, there is a strict lethargy-angle correlation given by Eq. (10.6), which means that the probability for a scattering collision that produces a lethargy gain within dU about U is equal to the probability for scattering with a cosine of the scattering angle within $d\mu_0$ about μ_0 when U and μ_0 are related by Eq. (10.6) and is zero otherwise:

$$\sigma_s(\mu_0, U, u')dU = -\sigma_s(\mu_0, u')\delta(\mu_0 - \mu_0(U))d\mu_0(U) \tag{10.12}$$

where the minus sign reflects the fact that an increase in the cosine of the scattering angle corresponds to a decrease in the lethargy gain. Using Eq. (10.12) in Eq. (10.11) yields

$$b_l^0(u', U) = \sigma_s(\mu_0, u')P_l(\mu_0(U))\left[-\frac{d\mu_0(U)}{dU}\right] \tag{10.13}$$

Making use of the physical fact that the probabilities for scattering through a given scattering angle in the lab system to within $d\mu_0$ about μ_0 and for scattering through the corresponding [via Eq. (10.5)] scattering angle in the CM system to within $d\mu_C$ about μ_C must be equal:

$$\sigma_s(\mu_0, u')d\mu_0 = \sigma_s^c(\mu_c, u')d\mu_c \tag{10.14}$$

and making use of the observation that the experimental scattering data are well represented by a Legendre expansion in the cosine of the CM scattering angle $\theta_C = \cos^{-1} \mu_C$:

$$\sigma(\mu_0, u') = \sigma_s^c(\mu_c, u') \frac{d\mu_c}{d\mu_0} = \sum_{l'=0}^{\infty} \frac{1}{2}(2l'+1)b_{l'}^c(u')P_{l'}(\mu_c)\frac{d\mu_c}{d\mu_0} \qquad (10.15)$$

allows the Legendre moments of the lethargy gain, $b_l^0(u', U)$, to be related to the Legendre moments of the angular scattering distribution in the CM system, $b_{l'}^c(u')$, which are tabulated in the nuclear data files:

$$b_l^0(u', U) = \sum_{l'=0}^{\infty} \frac{1}{2}(2l'+1)b_{l'}^c(u')P_{l'}(\mu_c)\frac{d\mu_c}{d\mu_0}P_l(\mu_0(U))\left[-\frac{d\mu_0(U)}{dU}\right]$$

$$= \sum_{l'=0}^{\infty} \frac{1}{2}(2l'+1)b_{l'}^c(u')P_{l'}(\mu_c(U))P_l(\mu_0(U))\left[-\frac{d\mu_c(U)}{dU}\right]$$

$$\equiv \sum_{l'=0}^{\infty} T_{ll'}(U)b_{l'}^c(u') \qquad (10.16)$$

Using this result in Eq. (10.10) leads to

$$\sigma_s(\mu_0, U, u') = \sum_{l,l'} \frac{1}{2}(2l+1)T_{ll'}(U)P_l(\mu_0)b_{l'}^c(u') \qquad (10.17)$$

for the elastic scattering lethargy-angle transfer function. Integrating this result over angle yields the total probability for an elastic scattering event to cause a lethargy increase from u' to u:

$$\sigma_s(u' \to u) = \int_{-1}^{1} d\mu_0 \, \sigma_s(\mu_0, U, u') = \sum_{l'=0}^{\infty} T_{0l'}(U)b_{l'}^c(u') \qquad (10.18)$$

Isotropic Scattering in Center-of-Mass System

The angular distribution of elastic scattering in the CM system may be represented by an average value of the cosine of the CM scattering angle given by $\mu_c = 0.07A^{2/3}E$ (MeV), except near scattering resonances. Hence the elastic scattering distribution is essentially isotropic in the CM system, except for high-energy neutrons scattering from heavy mass nuclei. When the scattering is taken as spherically symmetric in the CM system, the Legendre moments of the angular scattering distribution in the CM system are

$$b_l^c(u') = \sigma_s(u')\delta_{l0} \qquad (10.19)$$

In this case, Eq. (10.18) becomes

$$\sigma_s^{iso}(u' \to u) = \sigma_s(u')T_{00}(U) = \frac{\sigma_s(u')e^{u'-u}}{1-\alpha} \qquad (10.20)$$

The average lethargy increase with isotropic scattering is

$$\xi^{iso} = \int_0^{\ln 1/\alpha} \frac{\sigma_s(u' \to u)U\,dU}{\sigma_s(u')} = \int_0^{\ln 1/\alpha} T_{00}(U)U\,du = 1 + \frac{\alpha \ln \alpha}{1-\alpha} \quad (10.21)$$

and the average cosine of the scattering angle in the lab system is

$$\langle \bar{\mu}_0^{iso} \rangle = \int_0^{\ln 1/\alpha} \frac{\sigma_s(u' \to u)\mu_0(U)\,dU}{\sigma_s(u')} = \int_0^{\ln 1/\alpha} T_{10}(U)\,dU = \frac{2}{3A} \quad (10.22)$$

where A is the atomic mass in amu of the scattering nuclei and $\alpha = [(A - 1)/(A + 1)]^2$. Both of these quantities are independent of lethargy for a given species of scattering nuclei. However, the composite values for a mixture, $\xi = \sum_j \sigma_{sj}(u)\xi_i / \sum_j \sigma_{sj}(u)$ and $\mu_0 = \sum_j \sigma_{sj}(u)\mu_{0j} / \sum_j \sigma_{sj}(u)$, may be lethargy dependent.

Linearly Anisotropic Scattering in Center-of-Mass System

When only the first two Legendre components of the scattering transfer function in the center of mass system are non-zero, Eq. (10.18) becomes

$$\sigma_s^{anis}(u' \to u) = T_{00}^{(U)} b_0^c(u') + T_{01}(U) b_1^c(u')$$

$$= \frac{\sigma_s(u')e^{u'-u}}{1-\alpha} \left\{ 1 + \bar{\mu}_c(u') \left[3 - \frac{6}{1-\alpha}(1 - e^{-u'-u}) \right] \right\} \quad (10.23)$$

In this case, the mean lethargy increase in an elastic scattering event,

$$\xi(u') = \frac{b_0^c(u')}{\sigma_s(u')} \int_0^{\ln 1/\alpha} dU\, U T_{00}(U) + \frac{b_1^c(u')}{\sigma_s(u')} \int_0^{\ln 1/\alpha} dU\, U T_{01}(U)$$

$$= \xi^{iso} - 3\frac{b_1^c(u')}{b_0^c(u')} \left[\frac{1}{4}\frac{A^2+1}{A} + \frac{1}{8}\frac{(A^2-1)^2}{A^2} \ln \frac{A-1}{A+1} \right]$$

$$\xrightarrow[\text{large } A]{} \xi^{iso} - \frac{2}{A}\frac{b_1^c(u')}{b_0^c(u')} = \xi^{iso}\left[1 - \bar{\mu}_c(u') \right] \quad (10.24)$$

is reduced by anisotropic scattering (i.e., the moderation in energy is reduced), and the average cosine of the scattering angle in the lab system,

$$\bar{\mu}_0(u') = \frac{b_0^c(u')}{\sigma_s(u')} \int_0^{\ln 1/\alpha} dU\, \mu_0(U)\left[T_{00}(U) + \frac{b_1^c(u')}{b_0^c(u')}T_{01}(U) \right]$$

$$= [\bar{\mu}_0(u')]^{iso} + \bar{\mu}_c(u')\left(1 - \frac{3}{5A^2} \right) \quad (10.25)$$

is increased by anisotropic scattering (i.e., the scattering is more forward directed). Both ξ and μ_0 become lethargy dependent with anisotropic scattering.

10.2
P_1 and B_1 Slowing-Down Equations

Derivation

The transport equation of Chapter 9 can immediately be generalized to include lethargy dependence by allowing for the scattering removal of neutrons from incremental interval du and for a scattering source of neutrons into du from other incremental intervals du' (in the slowing-down region above 1 eV, the in-scatter would only be from $u' \leq u$):

$$\mathbf{\Omega} \cdot \nabla \psi(\mathbf{r}, \mathbf{\Omega}, u) + \Sigma_t(\mathbf{r}, u)\psi(\mathbf{r}, \mathbf{\Omega}, u)$$

$$= \int_0^u du' \int_0^{4\pi} d\mathbf{\Omega}' \frac{\Sigma_s(\mathbf{r}, \mu_0, U, u')}{2\pi}\psi(\mathbf{r}, \mathbf{\Omega}', u')$$

$$+ \frac{1}{k}\chi(u) \int_0^{\infty} du' \int_0^{4\pi} d\mathbf{\Omega}' \frac{\nu\Sigma_f(\mathbf{r}, u')}{4\pi}\psi(\mathbf{r}, \mathbf{\Omega}', u')$$

$$\equiv \int_{u-\ln 1/\alpha}^u du' \int_0^{4\pi} d\mathbf{\Omega}' \frac{\Sigma_{sel}(\mathbf{r}, \mu_0, U, u')}{2\pi}\psi(\mathbf{r}, \mathbf{\Omega}', u') + S(\mathbf{r}, \mathbf{\Omega}, u) \quad (10.26)$$

where $\mu_0 = \mathbf{\Omega}' \cdot \mathbf{\Omega}$ is the cosine of the angle in the lab system between the incident and exit directions of a neutron in a scattering collision. In the last step, inelastically scattered and fission neutrons are grouped into a source term and the remaining scattering term includes only elastic scattering. The macroscopic elastic scattering transfer function is a sum over nuclear species of the density times the microscopic transfer function of Eq. (10.17):

$$\Sigma_{sel}(\mathbf{r}, \mu_0, U, u') = \sum_j N_j(r)\sigma_{sj}(\mu_0, U, u')$$

$$= \sum_j N_j(r) \sum_{l,l'} \frac{1}{2}(2l+1)T_{ll'}(U)P_l(\mu_0)b_{l'}^c(u') \quad (10.27)$$

and the lower limit of the in-scatter integral for each species is $1 - \ln(1/\alpha_j)$, but this is represented symbolically for notational convenience as a single $1 - \ln(1/\alpha)$.

The P_n equations were derived in Chapter 9 for one-dimensional geometry and one-speed neutrons by expanding the directional flux in a Legendre polynomial series, and this can immediately be generalized to the lethargy-dependent neutron flux

$$\psi(z, \mu, u) \simeq \frac{1}{2}\phi_0(z, u)P_0(\mu) + \frac{3}{2}\phi_1(z, \mu)P_1(\mu)$$

$$= \frac{1}{2}\phi_0(z, u) + \frac{3}{2}\mu J_z(z, u)$$

$$= \frac{1}{2}\phi_0(z, u) + \frac{3}{2}\Omega_z J_z(z, u) \quad (10.28)$$

where $\mu = \mathbf{\Omega} \cdot \mathbf{n}_z = \Omega_z = \cos\theta$ is the cosine of the angle made by the direction of neutron motion with the z-coordinate axis, as indicated in Fig. 10.2, and where the

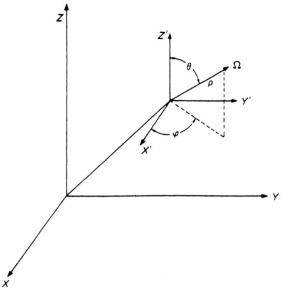

Fig. 10.2 Specification of the directional vector $\boldsymbol{\Omega}$ in a Cartesian coordinate system. (From Ref. 2; used with permission of MIT Press.)

current J_z has been associated with the $n = 1$ component of the flux expansion by using the orthogonality properties of the Legendre polynomials:

$$J_z(z, u) \equiv \phi_1(z, u) \equiv \int_{-1}^{1} P_1(\mu) \psi(z, \mu, u)\, d\mu \tag{10.29}$$

Further defining, $\boldsymbol{\Omega}_x = \boldsymbol{\Omega} \cdot \mathbf{n}_x = \sin\theta \cos\varphi$, $\boldsymbol{\Omega}_y = \boldsymbol{\Omega} \cdot \mathbf{n}_y = \sin\theta \sin\varphi$, and $d\boldsymbol{\Omega} = \sin\theta\, d\theta d\varphi / 4\pi$, the P_1 expansion of the directional neutron flux in three-dimensional geometry is

$$\psi(\mathbf{r}, \boldsymbol{\Omega}, u) \simeq \tfrac{1}{2}\phi_0(\mathbf{r}, u) + \tfrac{3}{2}\boldsymbol{\Omega} \cdot \mathbf{J}(\mathbf{r}, u) \tag{10.30}$$

In developing the P_1 equations in one dimension (Chapter 9), the expansion of Eq. (10.28) was substituted into the transport equation, and the resulting equation was weighted with $P_0 = 1$ and integrated over μ, and then weighted with $P_1(\mu = \boldsymbol{\Omega}_z) = \mu$ and integrated over μ, to obtain the two P_1 equations. We generalize this procedure to three dimensions by substituting Eq. (10.30) into Eq. (10.26) and weighting with 1, $\boldsymbol{\Omega}_x = \mu_x$, $\boldsymbol{\Omega}_y = \mu_y$, and $\boldsymbol{\Omega}_z = \mu_z$, that is, weight with 1 and $\boldsymbol{\Omega}$, and integrating over $\boldsymbol{\Omega}$ to obtain the P_1 equations in three-dimensional geometry:

$$\int_0^{4\pi} d\boldsymbol{\Omega} \left[\boldsymbol{\Omega} \cdot \nabla \left(\tfrac{1}{2}\phi_0 + \tfrac{3}{2}\boldsymbol{\Omega} \cdot \mathbf{J} \right) + \Sigma_t \left(\tfrac{1}{2}\phi_0 + \tfrac{3}{2}\boldsymbol{\Omega} \cdot \mathbf{J} \right) \right]$$

$$= \int_0^{4\pi} d\boldsymbol{\Omega} \int_0^{u} du' \int_0^{4\pi} d\boldsymbol{\Omega}' \frac{\Sigma_{\mathrm{sel}}(\mathbf{r}, \mu_0, U, u')}{2\pi} \left[\tfrac{1}{2}\phi_0(u') + \tfrac{3}{2}\boldsymbol{\Omega}' \cdot \mathbf{J}(u') \right]$$

$$+ S_0(\mathbf{r}, u) \tag{10.31}$$

$$\int_0^{4\pi} d\mathbf{\Omega}\,\mathbf{\Omega}\left[\mathbf{\Omega}\cdot\nabla\left(\frac{1}{2}\phi_0+\frac{3}{2}\mathbf{\Omega}\cdot\mathbf{J}\right)+\Sigma_t\left(\frac{1}{2}\phi_0+\frac{3}{2}\mathbf{\Omega}\cdot\mathbf{J}\right)\right]$$

$$=\int_0^{4\pi}d\mathbf{\Omega}\int_0^u du'\int_0^{4\pi}d\mathbf{\Omega}'\,\mathbf{\Omega}\frac{\Sigma_{\mathrm{sel}}(\mathbf{r},\mu_0,U,u')}{2\pi}\left[\frac{1}{2}\phi_0(u')+\frac{3}{2}\mathbf{\Omega}\cdot\mathbf{J}(u')\right]$$

$$+\mathbf{S}_1(\mathbf{r},u) \tag{10.32}$$

where

$$S_0(\mathbf{r},u)\equiv\int_0^{4\pi}d\mathbf{\Omega}\,S(\mathbf{r},\mathbf{\Omega},u)$$

$$\tag{10.33}$$

$$S_1(\mathbf{r},u)\equiv\int_0^{4\pi}d\mathbf{\Omega}\,\mathbf{\Omega}S(\mathbf{r},\mathbf{\Omega},u)$$

To simplify these P_1 equations, Eq. (10.27) is used for the scattering transfer function, and the addition theorem for Legendre polynomials,

$$P_n(\mu_0)=P_n(\mu')P_n(\mu)+2\sum_{m=1}^{n}\frac{(n-m)!}{(n+m)!}P_n^m(\mu')P_n^m(\mu)\cos m(\varphi'-\varphi) \tag{10.34}$$

is used to relate the cosine of the scattering angle $\mu_0=\cos\theta_0$ to the cosines of the angles that the incident and exiting neutron directions make with the z-axis, $\mu'=\cos\theta'$ and $\mu=\cos\theta$, respectively, as depicted in Fig. 10.1, where P_n^m is the associated Legendre function. Using the identities

$$\int d\mathbf{\Omega}=1,\qquad\int\mathbf{\Omega}_\xi\mathbf{\Omega}_\chi\,d\mathbf{\Omega}=\frac{1}{3}\delta_{\xi\chi}$$

$$\tag{10.35}$$

$$\int\mathbf{\Omega}_\xi\,d\mathbf{\Omega}=\int\mathbf{\Omega}_\xi^3\,d\mathbf{\Omega}=0,\quad\xi=x,y,z$$

Eqs. (10.32) and (10.33) then can be reduced to

$$\nabla\cdot\mathbf{J}(\mathbf{r},u)+\Sigma_t(\mathbf{r},u)\phi(\mathbf{r},u)=\int_{u-\ln 1/\alpha}^{u}du'\Sigma_{s0}(\mathbf{r},U,u')\phi(\mathbf{r},u')+S_0(\mathbf{r},u)$$

$$\tag{10.36}$$

$$\frac{1}{3}\nabla\phi(\mathbf{r},u)+\Sigma_t(\mathbf{r},u)\mathbf{J}(\mathbf{r},u)=\int_{u-\ln 1/\alpha}^{u}du'\,\Sigma_{s1}(\mathbf{r},U,u')\mathbf{J}(\mathbf{r},u')+\mathbf{S}_1(\mathbf{r},u)$$

$$\tag{10.37}$$

where the zero subscript on the flux has been dropped and the Legendre moments of the elastic scattering transfer functions are defined:

$$\Sigma_{sn}(\mathbf{r},U,u')\equiv\int_{-1}^{1}d\mu_0\,\Sigma_{\mathrm{sel}}(\mathbf{r},\mu_0,U,u')P_n(\mu_0)$$

$$=\sum_j N_j(r)\sum_{l'=0}T_{nl'}(U)b_{l'}^c(u')$$

$$\equiv\sum_j N_j(r)\sigma_{sn}^j(u'\to u) \tag{10.38}$$

In particular, the isotropic and linearly anisotropic lethargy change transfer functions are

$$
\sigma_{s0}^j(u' \to u) =
\begin{cases}
\sigma_s^j(u') \dfrac{e^{u'-u}}{1-\alpha_j}, & u - \ln \dfrac{1}{\alpha_j} < u' < u \\[2mm]
0, & \text{otherwise}
\end{cases}
\tag{10.39}
$$

$$
\sigma_{s1}^j(u' \to u)
$$

$$
=
\begin{cases}
\dfrac{\sigma_s^j(u')e^{u'-u}}{1-\alpha_j} \left[\dfrac{A+1}{2} e^{(1/2)(u'-u)} - \dfrac{A-1}{2} e^{-(1/2)(u'-u)} \right], & u - \ln \dfrac{1}{\alpha_j} < u' < u \\[2mm]
0, & \text{otherwise}
\end{cases}
\tag{10.40}
$$

The essential approximation that has been made in deriving Eqs. (10.36) and (10.37) is that the angular dependence of the neutron flux can be represented by only a linearly anisotropic dependence on the angular variable, as given by Eq. (10.30). This approximation should be good at more than a few mean free paths away from an interface between very dissimilar media (i.e., in the interior of large homogeneous regions) and more than a few mean free paths away from an anisotropic source.

Solution in Finite Uniform Medium

To solve Eqs. (10.36) and (10.37), it is assumed that the medium is uniform and that the spatial dependence of the flux and the current can both be represented by a simple buckling mode [i.e., $\phi(z, u) = \phi(u)\exp(iBz)$, $J(z, u) = J(u)\exp(iBz)$], so that these equations become

$$
iBJ(u) + \Sigma_t(u)\phi(u) = \int_{u-\ln 1/\alpha}^u du' \, \Sigma_{s0}(U, u')\phi(u') + S_0(u)
\tag{10.41}
$$

$$
\frac{1}{3}iB\phi(u) + \Sigma_t(u)J(u) = \int_{u-\ln 1/\alpha}^u du' \, \Sigma_{s1}(U, u')J(u') + S_1(u)
\tag{10.42}
$$

The parameter B may be considered to characterize the leakage from or into the medium. Note that this procedure is formally equivalent to Fourier transforming Eqs. (10.36) and (10.37).

These equations may be put in multigroup form by integrating over $\Delta u_g = u_g - u_{g-1}$ and defining

$$
\phi_g = \int_{u_{g-1}}^{u_g} du\, \phi(u), \qquad J_g \equiv \int_{u_{g-1}}^{u_g} du\, J(u)
$$

$$
\Sigma_t^g \equiv \frac{1}{\Delta u_g} \int_{u_{g-1}}^{u_g} du\, \Sigma_t, \qquad S_n^g \equiv \int_{u_{g-1}}^{u_g} du\, S_n(u)
\tag{10.43}
$$

$$
\Sigma_{sn}^{g' \to g} \equiv \frac{1}{\Delta u_{g'}} \int_{u_{g'-1}}^{u_{g'}} du' \int_{u_{g-1}}^{u_g} du\, \Sigma_{sn}(u' \to u), \quad n = 0, 1
$$

Here we have used the asymptotic flux solution $\phi(u) \sim 1$, corresponding to $\phi(E) \sim 1/E$, and assumed that $J(u) \sim 1$, also, in evaluating the total and scattering cross sections. The multigroup form of the P_1 equations is

$$iBJ_g + \Sigma_t^g \phi_g = \sum_{g' \leq g} \Sigma_{s0}^{g' \to g} \phi_{g'} + S_0^g, \qquad g = 1, \ldots, G \tag{10.44}$$

$$\frac{1}{3}iB\phi_g + \Sigma_t^g J_g = \sum_{g' \leq g} \Sigma_{s1}^{g' \to g} J_{g'} + S_1^g, \qquad g = 1, \ldots, G \tag{10.45}$$

B_1 Equations

The principal approximation involved in derivation of the P_1 equations is the assumption of linear anisotropy in the angular dependence of the neutron flux made in Eq. (10.28) or (10.30). There is an alternative formulation that avoids this approximation but instead makes the approximation that the angular dependence of the scattering can be represented by an isotropic plus a linearly anisotropic scattering transfer function. Returning to Eq. (10.26), but simplified to one-dimensional geometry,

$$\mu \frac{\partial \psi(z, \mu, u)}{\partial z} + \Sigma_t(z, u)\psi(z, \mu, u)$$

$$= \int_{-1}^{1} d\mu' \int_{u - \ln 1/\alpha}^{u} du' \Sigma_s(z, \mu_0, U, u')\psi(z, \mu', u') + S(z, \mu, u) \tag{10.46}$$

and making the same type of assumption about the spatial dependence [i.e., $\psi(z, \mu, u) = \psi(\mu, u)\exp(iBz)$] in a uniform medium leads to

$$[\Sigma_t(u) + iB\mu]\psi(\mu, u)$$

$$= \int_{-1}^{1} du' \int_{u - \ln 1/\alpha}^{u} du' \Sigma_s(\mu_0, U, u')\psi(\mu', u') + S(\mu, u) \tag{10.47}$$

Dividing by $(\Sigma_t + iB\mu)$ and assuming linearly anisotropic scattering yields

$$\psi(\mu, u) = (\Sigma_t(u) + iB\mu)^{-1} \left[\frac{1}{2} \int_{u - \ln 1/\alpha}^{u} du' \Sigma_{s0}(u' \to u)\phi(u') \right.$$

$$\left. + \frac{3}{2}\mu \int_{u - \ln 1/\alpha}^{u} du' \Sigma_{s1}(u' \to u)J(u') + S(\mu, u) \right] \tag{10.48}$$

The approximation of Eq. (10.28) has *not* been made in deriving this result; the quantities ϕ and J have been identified from the definitions

$$\phi(u) \equiv \int_{-1}^{1} d\mu\, \psi(\mu, u), \qquad J(u) \equiv \int_{-1}^{1} d\mu\, \mu\psi(\mu, u) \tag{10.49}$$

Now, Eq. (10.48) is multiplied by 1 and by μ and integrated over μ to obtain the two B_1 equations

$$i B J(u) + \Sigma_t(u)\phi(u) = \int_{u-\ln 1/\alpha}^{u} du' \Sigma_{s0}(u' \to u)\phi(u') + S_0(u)$$

$$\frac{1}{3} i B \phi(u) + \gamma(u)\Sigma_t(u) J(u) = \int_{u-\ln 1/\alpha}^{u} du' \Sigma_{s1}(u' \to u) J(u') + S_1(u) \tag{10.50}$$

where

$$\gamma(u) = \frac{[B/\Sigma_t(u)]^2 \tan^{-1}[B/\Sigma_t(u)]}{3\{B/\Sigma_t(u) - \tan^{-1}[B/\Sigma_t(u)]\}} \simeq 1 + \frac{4}{15}\frac{B}{\Sigma_t(u)} \tag{10.51}$$

The B_1 equations differ from the P_1 equations [Eqs. (10.44) and (10.45)] only by the factor γ. The essential B_1 approximation is a linearly anisotropic scattering transfer function; the essential P_1 approximation is linearly anisotropic neutron flux. The B_1 equations have been found to be somewhat more accurate for slab geometries, but clearly the two approximations will differ only when B is significant. The multigroup P_1 and B_1 equations are the basis of most multigroup fast spectrum codes (e.g., Refs. 4 and 10). Typical neutron energy distributions calculated for thermal (PWR) and fast (LMFBR) reactors are shown in Fig. 10.3.

Few-Group Constants

The usual procedure in reactor analysis is to solve the multigroup equations (with a large number of groups varying from 50 to 100 for thermal reactors to a few 1000 for fast reactors) for one or more large homogenized regions and then to develop few-group (2 to 4 for water-moderated thermal reactors, 5 to 10 for graphite-moderated thermal reactors, 20 to 30 for fast reactors) constants which can be used in a few-group diffusion theory calculation of the neutron diffusion during the slowing-down process. The few-group constants are constructed by using the fine-group fluxes to weight the fine-group constants over the fine groups contained within a few group. Denoting the fine groups with a g and the few groups with a k, the prescriptions for the few-group capture and fission cross sections

$$\sigma^k = \frac{\sum_{g \in k} \sigma^g \phi_g}{\sum_{g \in k} \phi_g} \tag{10.52}$$

and scattering transfer cross sections

$$\sigma_{sn}^{k' \to k} = \frac{\sum_{g' \in k'} \sum_{g \in k} \sigma_{sn}^{g' \to g} \phi_{g'}}{\sum_{g' \in k'} \phi_{g'}} \tag{10.53}$$

follow directly, where $g \in k$ indicates a sum over fine groups g within the lethargy interval of few group k.

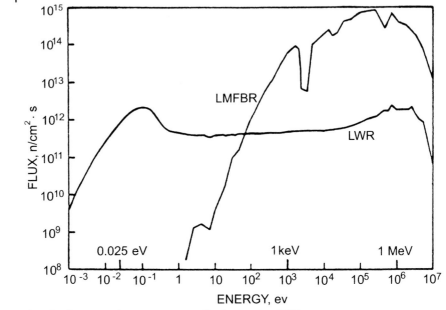

Fig. 10.3 Representative neutron energy distributions in a PWR and a LMFBR. (From Ref. 11; used with permission of Taylor & Francis.)

There is ambiguity about the definition of the few-group diffusion coefficient, as discussed in the following section. An appropriate definition is in terms of a few-group directional transport coefficient, defined as

$$\sigma_{tr,\xi}^k = \frac{\sum_{g\in k}\left(\sigma_t^g J_{g\xi} - \sum_{g'}\sigma_{s1}^{g'\to g}J_{g'\xi}\right)}{\sum_{g\in k}J_{g\xi}}, \quad \xi = x,y,z \tag{10.54}$$

where $J_{g\xi}$ is the fine-group current in the ξ-direction. The diffusion coefficient is related to the transport coefficient by $D^g = 1/3\Sigma_{tr}^g$. Many other prescriptions for the diffusion coefficient are found in practice.

10.3
Diffusion Theory

Lethargy-Dependent Diffusion Theory

It was shown in Chapter 9 that the one-speed P_1 equations led naturally to diffusion theory. Unfortunately, this is not the case for the lethargy-dependent P_1 equations of Eqs. (10.36) and (10.37). To derive from Eq. (10.37) a relationship of the form $\mathbf{J}(\mathbf{r}, E) = -D(\mathbf{r}, E)\nabla\phi(\mathbf{r}, E)$, it is necessary to require further (1) that $\Sigma_{s1}(u' \to u) \sim \Sigma_s(u')\delta(u' - u)$ or zero; and (2) that $\mathbf{J}(\mathbf{r}, E)$ and $\nabla\phi(\mathbf{r}, E)$ are parallel. Neither of these relations is satisfied in general, which gives rise to a number of ambiguities in defining the multigroup diffusion constants, in particular the diffusion coefficient. A common way to treat the anisotropic scattering difficulty is to make use of the one-speed results to approximate

$$\int_{u-\ln 1/\alpha}^{u} du' \Sigma_{s1}(\mathbf{r}, U, u') \mathbf{J}(\mathbf{r}, u') \simeq \Sigma_s(\mathbf{r}, u) \bar{\mu}_0 \mathbf{J}(\mathbf{r}, \mu) \tag{10.55}$$

which is equivalent to assuming no lethargy change in anisotropic elastic scattering. If, in addition, the anisotropic source that would arise from anisotropic inelastic scattering is assumed to vanish, then

$$\mathbf{J}(\mathbf{r}, u) = -\frac{1}{3(\sum_t(\mathbf{r}, u) - \sum_s(\mathbf{r}, u)\bar{\mu}_0)} \nabla \phi(\mathbf{r}, u)$$

$$= -\frac{1}{3\Sigma_{\mathrm{tr}}(\mathbf{r}, u)} \nabla \phi(\mathbf{r}, u) = -D(\mathbf{r}, u)\nabla \phi(\mathbf{r}, u) \tag{10.56}$$

is obtained from Eq. (10.37). This relation, a Fick's law, can be substituted into Eq. (10.36) to obtain lethargy-dependent diffusion theory:

$$-\nabla \cdot D(\mathbf{r}, u)\nabla \phi(\mathbf{r}, u) + \Sigma_t(\mathbf{r}, u)$$

$$= \int_{u-\ln 1/\alpha}^{u} du' \Sigma_{s0}(\mathbf{r}, U, u')\phi(\mathbf{r}, u') + S_0(r, u)$$

$$= \int_{u-\ln 1/\alpha}^{u} du' \Sigma_{s0}(\mathbf{r}, U, u')\phi(\mathbf{r}, u') + \int_{0}^{u} du' \Sigma_{\mathrm{in}}^0(\mathbf{r}, u' \to u)\phi(\mathbf{r}, u')$$

$$+ \frac{1}{k}\chi(u) \int_{0}^{\infty} du' \, v\Sigma_f(\mathbf{r}, u')\phi(\mathbf{r}, u') \tag{10.57}$$

where the inelastic and fission contributions to the isotropic source are shown explicitly in the last form.

Directional Diffusion Theory

In this derivation of lethargy-dependent diffusion theory from neutron transport theory, the lethargy change (energy change) in anisotropic scattering was neglected entirely. It is possible to formally include anisotropic lethargy change effects by defining

$$\Sigma_{\mathrm{tr}, \xi}(\mathbf{r}, u) \equiv \Sigma_t(\mathbf{r}, u) - \frac{\int_{u-\ln 1/\alpha}^{u} du' \Sigma_{s1}(\mathbf{r}, U, u') J_\xi(\mathbf{r}, u')}{J_\xi(\mathbf{r}, u)} \tag{10.58}$$

where J_ξ is the current in the ξ-direction. Since the lethargy dependence of the current could well be different for different ξ-directions, a different $\Sigma_{\mathrm{tr}, \xi}$ could be defined by Eq. (10.58) for each coordinate direction $\xi = x, y$, and z, giving rise to directional diffusion coefficients $D_\xi = 1/3\Sigma_{\mathrm{tr}, \xi}$ and to a directional diffusion equation

$$-\left[\frac{\partial}{\partial x}D_x(\mathbf{r},u)\frac{\partial}{\partial x}+\frac{\partial}{\partial y}D_y(\mathbf{r},u)\frac{\partial}{\partial y}+\frac{\partial}{\partial z}D_z(\mathbf{r},u)\frac{\partial}{\partial z}\right]$$

$$\times\phi(\mathbf{r},u)+\Sigma_t(\mathbf{r},u)\phi(\mathbf{r},u)$$

$$=\int_{u-\ln 1/\alpha}^{u}du'\Sigma_{s0}(\mathbf{r},U,u')\phi(\mathbf{r},u')+\int_{0}^{u}du'\Sigma_{\text{in}}^{0}(\mathbf{r},u'\to u)\phi(\mathbf{r},u')$$

$$+\frac{1}{k}\chi(u)\int_{0}^{\infty}du'\,\nu\Sigma_f(\mathbf{r},u')\phi(\mathbf{r},u') \tag{10.59}$$

Multigroup Diffusion Theory

Multigroup diffusion theory can be formally derived from the lethargy-dependent diffusion equations [Eqs. (10.57) or (10.59)] or directly from the lethargy-dependent P_1 equations [Eqs. (10.36) and (10.37)] by integrating over the lethargy interval of the group $\Delta u_g = u_g - u_{g-1}$. The definition of most of the group quantities is the same for all three procedures and is given by Eqs. (10.43), with fission and absorption cross sections evaluated similarly to the total cross section. However, the definition of the diffusion coefficient is different for the various derivations. In the derivation proceeding from Eq. (10.57), the multigroup diffusion term is formally defined by the replacement

$$-\nabla\cdot D^g(\mathbf{r})\nabla\phi_g(\mathbf{r})\equiv-\int_{u_{g-1}}^{u_g}du\,\nabla\cdot D(\mathbf{r},u)\nabla\phi(\mathbf{r},u) \tag{10.60}$$

but this leaves open how to define D_g. Since it is unlikely that lethargy-dependent flux gradients will be available, various heuristic definitions suggest themselves; for example,

$$D^g(\mathbf{r})=\frac{\int_{u_{g-1}}^{u_g}du\,D(\mathbf{r},u)}{\Delta u_g} \tag{10.61}$$

or

$$D^g(\mathbf{r})=\frac{1}{3\Sigma_{\text{tr}}^{g}}=\frac{1}{3(\Sigma_t^g-\bar{\mu}_0^g\Sigma_s^g)} \tag{10.62}$$

A similar ambiguity plagues the development of multigroup diffusion equations from Eq. (10.59).

The formal definition of ξ-direction diffusion coefficient that arises from the integration of Eq. (10.37) over Δu_g is

$$D_\xi^g(\mathbf{r})=\frac{J_{\xi,g}(\mathbf{r})}{3\int_{u_{g-1}}^{u_g}du\left[\Sigma_t(\mathbf{r},u)J_\xi(\mathbf{r},u)-\sum_{g'\leq g}\int_{u_{g'-1}}^{u_{g'}}du'\Sigma_{s1}(\mathbf{r},u'\to u)J_\xi(\mathbf{r},u')\right]} \tag{10.63}$$

The multigroup diffusion equations have the same form for all derivations:

$$-\frac{\partial}{\partial x}D_x^g(\mathbf{r})\frac{\partial \phi_g(\mathbf{r})}{\partial x} - \frac{\partial}{\partial y}D^g(\mathbf{r})\frac{\partial \phi_g(\mathbf{r})}{\partial y} - \frac{\partial}{\partial z}D_z^g(\mathbf{r})\frac{\partial \phi_g(\mathbf{r})}{\partial z} + \Sigma_t^g(\mathbf{r})\phi_g(\mathbf{r})$$

$$= \sum_{g' \leq g}\Sigma_s^{g' \rightarrow g}(\mathbf{r})\phi_{g'}(\mathbf{r}) + \frac{\chi^g}{k}\sum_{g'=1}^{G}\nu\Sigma_f^{g'}(\mathbf{r})\phi_{g'}(\mathbf{r}), \quad g = 1, \ldots, G \quad (10.64)$$

where the elastic and inelastic scattering terms have been combined into a single scattering term.

Boundary and Interface Conditions

The appropriate transport theory boundary conditions are zero return current at external boundaries (unless there is an external beam source):

$$\phi(\mathbf{r}_b, \mathbf{\Omega}, u) = 0, \qquad \hat{n}_b \cdot \mathbf{\Omega} < 0 \qquad (10.65)$$

where $\mathbf{n}_b$ is the outward unit vector to the external boundary at $\mathbf{r}_b$, and the appropriate interface condition is continuity of directional flux:

$$\psi(\mathbf{r}_i - \varepsilon, \mathbf{\Omega}, u) = \psi(\mathbf{r}_i + \varepsilon, \mathbf{\Omega}, u) \qquad (10.66)$$

where ε is a small quantity. These conditions obviously cannot be satisfied exactly by the diffusion theory approximation to the neutron flux.

The Marshak boundary conditions discussed in Chapter 9 generalize to

$$J_{\mathrm{in}} = -\mathbf{n}_b \cdot \mathbf{J}(\mathbf{r}_b, u) = -\mathbf{n}_b \cdot \int \mathbf{\Omega}\psi(\mathbf{r}_b, \mathbf{\Omega}, u)d\mathbf{\Omega} = 0 \qquad (10.67)$$

Making use of the partial currents and geometric interpretation discussed in Section 3.1, this condition can be interpreted as the vanishing of the flux at an extrapolated distance $0.71/\Sigma_{\mathrm{tr}}(u)$ outside the physical boundary. Given the ambiguity in defining $\Sigma_{\mathrm{tr}}(u)$, the computational difficulties that would ensue from an extrapolated boundary that varied with lethargy and the fact that the extrapolation distance is typically very small relative to the physical dimensions, the approximate boundary condition of vanishing flux on the physical boundary is appropriate as an approximation to Eq. (10.65):

$$\phi(\mathbf{r}_b, u) = 0, \qquad \phi_g(\mathbf{r}_b) = 0 \qquad (10.68)$$

As an approximation to the interface condition of Eq. (10.66), we require that the first two Legendre moments of this equation be satisfied:

$$\int \psi(\mathbf{r}_i - \varepsilon, \mathbf{\Omega}, u)d\mathbf{\Omega} = \int \psi(\mathbf{r}_i + \varepsilon, \mathbf{\Omega}, u)d\mathbf{\Omega}$$

$$\int \mathbf{n}_i \cdot \mathbf{\Omega}\psi(\mathbf{r}_i - \varepsilon, \mathbf{\Omega}, u)d\mathbf{\Omega} = \int \mathbf{n}_i \cdot \mathbf{\Omega}\psi(\mathbf{r}_i + \varepsilon, \mathbf{\Omega}, u)d\mathbf{\Omega}$$

$$(10.69)$$

Using the definitions of scalar flux and current as the first two Legendre moments of the angular flux, this may be written

$$\phi(\mathbf{r}_i - \varepsilon, u) = \phi(\mathbf{r}_i + \varepsilon, u)$$

$$\mathbf{n}_i \cdot \mathbf{J}(\mathbf{r}_i - \varepsilon, u) = \mathbf{n}_i \cdot \mathbf{J}(\mathbf{r}_i + \varepsilon, u) \tag{10.70}$$

and for multigroup diffusion theory

$$\phi_g(\mathbf{r}_i - \varepsilon) = \phi_g(\mathbf{r}_i + \varepsilon)$$

$$\mathbf{n}_i \cdot \mathbf{J}_g(\mathbf{r}_i - \varepsilon) = \mathbf{n}_i \cdot \mathbf{J}_g(\mathbf{r}_i + \varepsilon) \tag{10.71}$$

10.4
Continuous Slowing-Down Theory

Over much of the slowing-down range above (in lethargy) the fission spectrum and below the thermal range, neutron slowing down is due primarily to elastic scattering. Since there is no lethargy decrease in a scattering event below (in lethargy) the thermal range, the scatter-in integral is over lower lethargies only. It has been found convenient for computational purposes to replace the elastic scatter-in integral with a lethargy derivative of the associated elastic slowing-down density, which is computed in a coupled calculation, rather than evaluating the scatter-in integral directly. The various computational methods that have been developed for this purpose are known collectively as *continuous slowing-down theory*.

P_1 Equations in Slowing-Down Density Formulation

Generalizing the definition of slowing-down density introduced in Chapter 4 to include anisotropic scattering, the *isotropic slowing-down density* is defined as the number of neutrons slowing down past energy E, or lethargy u, by isotropic (in the lab system) elastic scattering:

$$q_0^i(x, u) \equiv \int_0^u du' \int_u^\infty du'' \Sigma_{s0}^i(x, u' \to u'') \phi(x, u') \tag{10.72}$$

and the linearly *anisotropic slowing-down density* is defined as the number of neutrons slowing down past lethargy u by linearly anisotropic scattering:

$$q_1^i(x, u) \equiv \int_0^u du' \int_u^\infty du'' \Sigma_{s1}^i(x, u' \to u'') J(x, u') \tag{10.73}$$

These two slowing-down densities are the zeroth and first Legendre components of the angular-dependent neutron slowing-down density.

Making use of Eq. (10.20), the first of these equations can be written explicitly as

$$q_0^i(x, u) \equiv \int_{u - \ln 1/\alpha_i}^u du' \int_u^{u + \ln 1/\alpha_i} \Sigma_s^i(u') \frac{e^{u' - u''}}{1 - \alpha_i} \phi(x, u')$$

$$= \int_{u - \ln 1/\alpha_i}^{u} du' \Sigma_s^i(u') \frac{e^{u'-u} - \alpha_i}{1 - \alpha_i} \phi(x, u') \tag{10.74}$$

and making use of Eq. (10.23), the second of these equations can be written explicitly as

$$q_1^i(x, u) = \int_{u - \ln 1/\alpha_i}^{u} du' \int_{u}^{u + \ln 1/\alpha_i} du'' \frac{\Sigma_s^i(u') e^{u'-u''}}{1 - \alpha_i}$$

$$\times \left[1 + 3\bar{\mu}_c(u') \left(1 - \frac{2(1 - e^{u'-u''})}{1 - \alpha_i} \right) \right] \tag{10.75}$$

where A_i is the atomic mass in amu of the scattering nuclei and $\alpha_i = [(A_i - 1)/(A_i + 1)]^2$.

These slowing-down densities can be related to the scatter-in integrals in the P_1 equations given by Eqs. (10.36) and (10.37):

$$\frac{\partial q_0^i}{\partial u} = \int_u^\infty du'' \Sigma_{s0}^i(x, u \to u'') \phi(x, u) - \int_0^u du' \Sigma_{s0}^i(x, u' \to u) \phi(x, u')$$

$$= \Sigma_s^i(x, u) \phi(x, u) - \int_0^u du' \Sigma_{s0}^i(x, u' \to u) \phi(x, u') \tag{10.76}$$

$$\frac{\partial q_1^i}{\partial u} = \int_u^\infty du'' \Sigma_{s1}^i(x, u \to u'') J(x, u) - \int_0^u du' \Sigma_{s1}^i(x, u' \to u) J(x, u')$$

$$= \bar{\mu}_0^i \Sigma_s^i(x, u) J(x, u) - \int_0^u du' \Sigma_{s1}^i(x, u' \to u) J(x, u') \tag{10.77}$$

Using Eqs. (10.76) and (10.77) to eliminate the scatter-in integrals in Eqs. (10.36) and (10.37) yields an equivalent form of the P_1 equations (written for one-dimensional slab geometry)

$$\frac{\partial J(x, u)}{\partial x} + \Sigma_{ne}(x, u) \phi(x, u) = - \sum_{i=1}^{I} \frac{\partial q_0^i}{\partial u}(x, u) + S_0(x, u) \tag{10.78}$$

$$\frac{1}{3} \frac{\partial \phi(x, u)}{\partial x} + \Sigma_{tr}(x, u) J(x, u) = - \sum_{i=1}^{I} \frac{\partial q_1^i(x, u)}{\partial u} + S_1(x, u) \tag{10.79}$$

where the nonelastic cross section

$$\Sigma_{ne}(x, u) \equiv \Sigma_t(x, u) - \sum_{i=1}^{I} \Sigma_s^i(x, u) \tag{10.80}$$

and the transport cross section

$$\Sigma_{tr}(x, u) \equiv \Sigma_t(x, u) - \sum_{i=1}^{I} \bar{\mu}_0^i \Sigma_s^i(x, u) \tag{10.81}$$

have been defined in a natural way.

Integrating these equations over $\Delta u_g = u_{g+1} - u_g$ leads to the multigroup P_1 equations in the slowing-down density formulation of elastic scattering:

$$\frac{\partial J_g(x)}{\partial x} + \Sigma_{\mathrm{ne}}^g(x)\phi_g(x) = -\sum_{i=1}^{I}\left[q_0^i(x, u_{g+1}) - q_0^i(x, u_g)\right] + S_0^g(x) \qquad (10.82)$$

$$\frac{1}{3}\frac{\partial \phi_g(x)}{\partial x} + \Sigma_{\mathrm{tr}}^g(x)J_g(x) = -\sum_{i=1}^{I}\left[q_1^i(x, u_{g+1}) - q_1^i(x, u_g)\right] + S_1^g(x)$$

$$g = 1, \ldots, G \qquad (10.83)$$

where the multigroup quantities are defined as

$$\phi_g(x) \equiv \int_{u_g}^{u_{g+1}} du\, \phi(x, u), \qquad J_g(x) \equiv \int_{u_g}^{u_{g+1}} du\, J(x, u)$$

$$S_n^g(x) \equiv \int_{u_g}^{u_{g+1}} du\, S_n(x, u)$$

$$\Sigma_{\mathrm{ne}}^g(x) \equiv \int_{u_g}^{u_{g+1}} du\, \Sigma_{\mathrm{ne}}(x, u)\frac{\phi(x, u)}{\phi_g(x)} \qquad (10.84)$$

$$\Sigma_{\mathrm{tr}}^g(x) \equiv \int_{u_g}^{u_{g+1}} du\, \Sigma_{\mathrm{tr}}(x, u)\frac{J(x, u)}{J_g(x)}$$

In this formulation, the natural definition of the group averaged transport equation is as a current averaged quantity.

The same type of difficulty encountered previously in reducing the energy-dependent P_1 equations to diffusion theory is present in Eq. (10.83); to obtain a Fick's law type of relationship $J = -D\, d\phi/dx$, it is necessary to require that the anisotropic source S_1^g vanish and that the anisotropic slowing-down density not change over the group, which would be the case if it was assumed to be identically zero. Making these assumptions, the multigroup diffusion equation in the slowing-down density formulation is

$$-\frac{\partial}{\partial x}\left[\frac{1}{3\Sigma_{\mathrm{tr}}^g(x)}\frac{\partial \phi_g(x)}{\partial x}\right] + \Sigma_{\mathrm{ne}}^g(x)\phi_g(x)$$

$$= \sum_{i=1}^{I}\left[q_0^i(x, u_g) - q_0^i(x, u_{g+1})\right] + S_0^g(x) \qquad (10.85)$$

with the diffusion coefficient being unambiguously defined in terms of a current spectrum-weighted group-averaged transport cross section, which contains some anisotropic effects—the average cosines of the scattering angle of the various nuclear species are embedded in the definition of the transport coefficient given by Eq. (10.81).

Making the approximation that the spatial dependence can be represented by a simple buckling [i.e., $\phi(x, u) = \phi(u) \exp(iBx)$, $J(x, u) = J(u) \exp(iBx)$] in Eqs. (10.82) and (10.83) reduces these equations to the forms that are found in various multigroup spectrum codes:

$$iBJ_g + \Sigma_{ne}^g \phi_g = \sum_{i=1}^{J} [q_0^i(u_g) - q_0^i(u_{g+1})] + S_0^g$$

$$\frac{1}{3}iB\phi_g + \Sigma_{tr}^g \phi_g = \sum_{i=1}^{J} [q_1^i(u_g) - q_1^i(u_{g+1})] + S_1^g, \quad g = 1, \dots, G \tag{10.86}$$

The asymptotic forms $\phi_{asy}(u) \sim 1$ and $J(u)_{asy} \sim 1$ are used in Eqs. (10.84) to define fine or ultrafine group constants. The few-group constants are then constructed from the solutions ϕ_g and J_g of the fine or ultrafine group calculation:

$$\Sigma_{ne}^k(x) = \frac{\sum_{g \in k} \Sigma_{ne}^g(x) \phi_g(x)}{\sum_{g \in k} \phi_g(x)}$$

$$\Sigma_{tr}^k(x) = \frac{\sum_{g \in k} \Sigma_{tr}^g(x) J_g(x)}{\sum_{g \in k} J_g(x)} \tag{10.87}$$

where $g \in k$ indicates that the sum is over fine groups g within few group k.

Slowing-Down Density in Hydrogen

The evaluation of the slowing-down densities in hydrogen is quite straightforward because a neutron can scatter from any lethargy to any greater lethargy in a single collision, which is implicit in the fact that $\alpha_H = 0$. This fact allows Eqs. (10.72) and (10.73) to be written

$$q_0^H(x, u) = \int_0^u du' \Sigma_s^H(u') e^{u'-u} \phi(x, u') \tag{10.88}$$

$$q_1(x, u) = \frac{2}{3} \int_0^u du' \Sigma_s^H(u') e^{3(u'-u)/2} J(x, u') \tag{10.89}$$

These equations may be differentiated to obtain

$$\frac{\partial q_0^H}{\partial u} + q_0^H(x, u) = \Sigma_s^H(u) \phi(x, u) \tag{10.90}$$

$$\frac{\partial q_1^H}{\partial n} + \frac{3}{2} q_1^H(x, u) = \frac{2}{3} \Sigma_s^H(u) J(x, u) \tag{10.91}$$

which may be put in multigroup form, to be used with Eqs. (10.86), by integration over Δu_g:

$$[q_0^H(u_{g+1}) - q_0^H(u_g)] + \frac{1}{2}[q_0^H(u_{g+1}) + q_0^H(u_g)]\Delta u_g = \Sigma_s^{H_g} \phi_g$$

$$[q_1^H(u_{g+1}) - q_1^H(u_g)] + \frac{3}{4}[q_1^H(u_{g+1}) + q_1^H(u_g)]\Delta u_g = \frac{2}{3}\Sigma_s^{H_g} J_g \tag{10.92}$$

Heavy Mass Scatterers

For moderators other than hydrogen, this procedure does not lead to such a simple differential equation for the slowing-down density, precisely because it is not possible for a neutron to lose all of its energy in a single collision, which means that the lower limits on the first integral in Eqs. (10.72) and (10.73) are $u - \ln(1/\alpha_i)$, not zero. At the other extreme from hydrogen are heavy mass nuclei for which the maximum lethargy gain in a scattering collision is quite small and it is reasonable to expand the integrands in Eqs. (10.72) and (10.73) in Taylor's series:

$$\Sigma_s^i(u')\phi(u') \simeq \Sigma_s^i(u)\phi(u) + (u' - u)\frac{\partial}{\partial u}\left[\Sigma_s^i(u)\phi(u)\right] + \cdots \tag{10.93}$$

$$\Sigma_s^i(u')J(u') \simeq \Sigma_s^i(u)J(u) + (u' - u)\frac{\partial}{\partial u}\left[\Sigma_s^i(u)J(u)\right] + \cdots \tag{10.94}$$

Various approximations result from keeping different terms in these expansions.

Age Approximation

The simplest such approximation, resulting from retaining only the first term in Eq. (10.93) and setting $q_1^i = 0$, is known as the *age approximation*:

$$q_0^i(x, u) \simeq \xi_i \Sigma_s^i(x, u)\phi(x, u) \tag{10.95}$$

where $\xi_i = \xi_i^{\text{iso}}$ given by Eq. (10.21). With these approximations for q_0^i and q_1^I, Eqs. (10.78) and (10.79) become the inconsistent (because of the neglect of q_1^i) P_1 equations:

$$\frac{\partial J}{\partial x}(x, u) + \Sigma_{\text{ne}}(x, u)\phi(x, u)$$
$$= -\sum_{i=1}^{I}\frac{\partial}{\partial u}\left[\xi_i \Sigma_s^i(x, u)\phi(x, u)\right] + S_0(x, u) \tag{10.96}$$

$$\frac{1}{3}\frac{\partial\phi(x, u)}{\partial x} + \Sigma_{\text{tr}}(x, u)J(x, u) = S_1(x, u)$$

which, with the additional assumption of zero anisotropic source ($S_1 = 0$), can be reduced to the age-diffusion equation

$$-\frac{\partial}{\partial x}\left[\frac{1}{3\Sigma_{\text{tr}}(x, u)}\frac{\partial\phi(x, u)}{\partial x}\right] + \Sigma_{\text{ne}}(x, u)\phi(x, u)$$
$$= -\sum_{i=1}^{I}\frac{\partial}{\partial u}\left[\xi_i \Sigma_s^i(x, u)\phi(x, u)\right] + S_0(x, u) \tag{10.97}$$

Selengut–Goertzel Approximation

The age approximation for the slowing-down density, and hence the inconsistent P_1 and age-diffusion equations, are restricted to heavy mass moderators for which the interval of the scatter-in integral, $\ln(1/\alpha_i)$, is quite small, and certainly would not be appropriate for hydrogen. For a mixture of hydrogen and heavy mass moderators, the hydrogen can be treated exactly and the age approximation can be used for the remaining nuclei, resulting in the *Selengut–Goertzel approximation*

$$\sum_{i=1}^{I} \frac{\partial q_0^i}{\partial u} = \frac{\partial q_0^H}{\partial u} + \sum_{i \neq H}^{I} \frac{\partial}{\partial u}(\xi_i \Sigma_s^i \phi) \tag{10.98}$$

Consistent P_1 Approximation

If, instead of setting $q_1^i = 0$, Eq. (10.73) is evaluated retaining the first term of the Taylor's series expansion of Eq. (10.94), to obtain

$$q_1^i(x,u) \simeq \xi_i^1 \Sigma_s^i(x,u) J(x,u) \tag{10.99}$$

where the first Legendre moment of the mean lethargy gain is defined as

$$\xi_i^1 = (A_i + 1)^2 \left\{ \frac{1 + 1/A_i}{9} \left[1 - \alpha_i^{3/2} \left(\frac{3}{2} \ln \frac{1}{\alpha_i} + 1 \right) \right] \right.$$
$$\left. - (1 - 1/A_i) \left[1 - \alpha_i^{1/2} \left(\frac{1}{2} \ln \frac{1}{\alpha_i} + 1 \right) \right] \right\} \tag{10.100}$$

the consistent P_1 equations (with the Selengut–Goertzel approximation) are obtained:

$$\frac{\partial J(x,u)}{\partial x} + \Sigma_{\text{ne}}(x,u)\phi(x,u)$$

$$= -\sum_{i \neq H}^{I} \frac{\partial}{\partial u} \left[\xi_i \Sigma_s^i(x,u)\phi(x,u) \right] - \frac{\partial q_0^H}{\partial u}(x,u) + S_0(x,u) \tag{10.101}$$

$$\frac{1}{3} \frac{\partial \phi(x,u)}{\partial x} + \left[\Sigma_{\text{tr}}(x,u) + \xi_i^1 \Sigma_s^i(x,u) \right] J(x,u) = S_1(x,u) \tag{10.102}$$

Extended Age Approximation

If the first two terms in the Taylor's series expansion of Eq. (10.93) are retained in evaluating Eq. (10.72), the result is

$$q_0^i(x,u) \simeq \xi_i \Sigma_s^i(x,u)\phi(x,u) + \frac{a_i}{\xi_i} \frac{\partial q_0(x,u)}{\partial u} \tag{10.103}$$

where

$$a_i = \int_{u - \ln 1/\alpha_i}^{u} du' \frac{e^{u'-u} - \alpha_i}{1 - \alpha_i}(u' - u) = \frac{\alpha_i [\ln(1/\alpha_i)]^2}{2(1 - \alpha_i)} - \xi_i \tag{10.104}$$

Using the balance equation for the elastic slowing-down density in a very large region (neglecting leakage)

$$\frac{\partial q_0(u)}{\partial u} = -\Sigma_{ne}(u)\phi(u) \tag{10.105}$$

allows Eq. (10.103) to be written

$$q_0(x, u) \simeq \sum_i \left[\xi_i \Sigma_s^i(x, u) - \frac{a_i}{\xi_i} \Sigma_{ne}(x, u) \right] \phi(x, u)$$

$$\simeq \xi \Sigma_t(x, u)\phi(x, u) \tag{10.106}$$

With this extended age approximation, the summation on the right in the first of Eqs. (10.101) is replaced by $-d(\xi \Sigma_t \phi)/du$ in the age-diffusion equations.

Grueling–Goertzel Approximation

The slowing-down density for hydrogen can be calculated exactly, and the slowing-down density for heavy mass nuclei can be well approximated by one of the variants of the age approximation given above. However, light nonhydrogen moderators are not well approximated by any of the age approximations above. Greater accuracy can obviously be obtained by retaining more terms in the Taylor's series expansions of Eqs. (10.93) and (10.94). In addition, it is possible to construct an approximate equation for the isotopic slowing-down densities which has the same form as the simple differential equation that describes the hydrogen slowing-down density and which reduces to the hydrogen equation when $A_i = 1$. Retaining three terms in the Taylor series expansion of Eq. (10.93) when used with Eq. (10.72) to evaluate $\lambda_0^i dq_0^i/du + q_0^i$ yields

$$\lambda_0^i \frac{\partial q_0^i}{\partial u} + q_0^i \simeq \lambda_0^i \left[\xi_i \frac{\partial}{\partial u} (\Sigma_s^i \phi) + a_i \frac{\partial^2 (\Sigma_s^i \phi)}{\partial u^2} \right] + \left[\xi_i \Sigma_s^i \phi + a_i \frac{\partial (\Sigma_s^i \phi)}{\partial u} \right] \tag{10.107}$$

The objective is to develop an equation for q_0^i which is like Eq. (10.90) for hydrogen. Neglecting $\partial^2 \phi / \partial u^2$ and choosing λ_0^i to make the $\partial \phi / \partial u$ term vanish leads to

$$\lambda_0^i \frac{\partial q_0^i}{\partial u}(x, u) + q_0^i(x, u) = \xi_i \Sigma_s^i(x, u)\phi(x, u) \tag{10.108}$$

which is of the same form as Eq. (10.90) for the hydrogen slowing-down density where

$$\lambda_0^i = \frac{1 - \alpha_i \{ 1 + \ln(1/\alpha_i) + \frac{1}{2}[\ln(1/\alpha_i)]^2 \}}{1 - \alpha_i [1 + \ln(1/\alpha_i)]} \tag{10.109}$$

Retaining the first three terms in the Taylor series expansion of Eq. (10.94) when used with Eq. (10.73) in a similar calculation leads to an equation similar to the hydrogen Eq. (10.91):

$$\lambda_1^i \frac{\partial q_1^i}{\partial u}(x, u) + q_1^i(x, u) = \xi_i^1 \Sigma_s^i(x, u)J(x, u) \tag{10.110}$$

where

$$\lambda_1^i = -\left[\frac{(1+1/A_i)^2}{4/A_i}\left(\frac{1+1/A_i}{3}\left\{\frac{8}{9} - \alpha_i^{3/2}\left[\left(\ln\frac{1}{2}\right)^2 - \frac{4}{3}\ln\frac{1}{\alpha_i} + \frac{8}{9}\right]\right\}\right)\right.$$

$$\left. - \left(1 - \frac{1}{A_i}\right)\left\{8 - \alpha_i^{1/2}\left[\left(\ln\frac{1}{\alpha_i}\right)^2 - 4\ln\frac{1}{\alpha_i} + 8\right]\right\}\right]/\xi_i^1 \quad (10.111)$$

again has been chosen to make $\partial\phi/\partial u$ terms vanish.

Summary of P_l Continuous Slowing-Down Theory

The P_l equations

$$\frac{\partial J(x,u)}{\partial x} + \Sigma_{ne}(x,u)\phi(x,u)$$

$$= -\frac{\partial q_0^H(x,u)}{\partial u} - \sum_{i\neq H}\frac{\partial q_0^i(x,u)}{\partial u} + S_0(x,u)$$

$$\frac{1}{3}\frac{\partial\phi(x,u)}{\partial x} + \Sigma_{tr}(x,u)J(x,u)$$

$$= -\frac{\partial q_1^H(x,u)}{\partial u} - \sum_{i\neq H}\frac{\partial q_1^i(x,u)}{\partial u} + S_1(x,u)$$

(10.112)

and the equations for the elastic slowing-down density, using the exact equations for hydrogen and the Grueling–Goertzel approximation for nonhydrogen nuclei,

$$\frac{\partial q_0^H}{\partial u}(x,u) + q_0^H(x,u) = \Sigma_s^H(x,u)\phi(x,u)$$

$$\frac{2}{3}\frac{\partial q_1^H}{\partial u}(x,u) + q_1^H(x,u) = \frac{4}{9}\Sigma_s^H(x,u)J(x,u)$$

$$\lambda_0^i\frac{\partial q_0^i}{\partial u}(x,u) + q_0^i(x,u) = \xi_i\Sigma_s^i(x,u)\phi(x,u)$$

$$\lambda_1^i\frac{\partial q_1^i}{\partial u}(x,u) + q_1^i(x,u) = \xi_i^1\Sigma_s^i(x,u)J(x,u)$$

(10.113)

represent the formulation usually referred to as P_l continuous slowing-down theory.

Inclusion of Anisotropic Scattering

In an ultrafine group calculation for which the group width is less than $\ln(1/\alpha_i)$ for some of the nuclei which contribute strongly to neutron moderation, it is necessary to retain a large number of Legendre moments to accurately represent the group transfer cross sections, which actually represent the probabilities for scattering to within relatively small angular intervals. (This situation is more likely to be found in a fast than in a thermal reactor.) The concept of slowing-down density can be extended to a higher order of anisotropy by defining the Legendre moments of the slowing-down density as the number of neutrons slowing down past lethargy u by the lth Legendre moment of the elastic scattering transfer function

$$q_l^i(u) \equiv \int_0^u du' \int_u^\infty du'' \Sigma_{sl}^i(u' \to u'') \phi_l(u') \tag{10.114}$$

where, recalling Eq. (10.38), we have

$$\Sigma_{sl}^i(u' \to u) = N_i \sigma_s^i(u') \sum_{l'=0} T_{ll'}^i(U) \frac{b_{l'i}^c(u')}{b_{0i}^c(u')} \equiv \sum_{l'=0} T_{ll'}^i(U) \Sigma_{sl'}^i(u') \tag{10.115}$$

Making a general Taylor series expansion about u in the integrand in Eq. (10.114),

$$\Sigma_{sl'}^i(u') \phi_{l'}(u') = \sum_{n=0}^\infty \frac{(u'-u)^n}{n!} \frac{d^n}{du^n} \left[\Sigma_{sl'}^i(u) \phi_{l'}(u) \right] \tag{10.116}$$

and using Eq. (10.115) yields

$$q_l^i(u) = -\sum_{n=0}^\infty \frac{d^n}{du^n} \left[G_{l,n+1}^i(u) \phi_l(u) \right] \tag{10.117}$$

where

$$G_{l,n}^i(u) = \sum_{l'=0} \Sigma_{sl'}^i(u) \frac{2l'+1}{2n!} \int_{u-\ln 1/\alpha_i}^u du' \int_u^{u+\ln 1/\alpha_i} du'' P_l[\mu_0(u'-u'')]$$

$$\times P_l[\mu_c(u'-u'')](u'-u)^n \frac{2e^{u'-u''}}{1-\alpha_i}$$

$$\equiv \sum_{l'=0}^\infty \Sigma_{sl'}^i(u) T_{ll',n}^i(u) \tag{10.118}$$

Extending the calculation that was described for the Grueling–Goertzel approximation yields an equation for each Legendre moment of the slowing-down density

$$\lambda_l(u) \frac{dq_l}{du}(u) + q_l(u)$$

$$= -\sum_i \left\{ G_{l,1}^i(u) \phi_l(u) \left[1 - \frac{d\lambda_l(u)}{du} \right] \right.$$

$$\left. + \sum_{n=2}^\infty \left[\frac{d^n[G_{l,n+1}^i(u)\phi_l(u)]}{du^n} + \lambda_l(u) \frac{d^n[G_{l,n}^i(u)\phi_l(u)]}{du^n} \right] \right\} \tag{10.119}$$

where

$$\lambda_l(u) \frac{\sum_i \sum_{l'=0}^{\infty} T_{ll',2}^i(u) \Sigma_{sl'}^i(u)}{\sum_i \sum_{l'=0}^{\infty} T_{ll',1}^i(u) \Sigma_{sl'}^i(u)} = \frac{\sum_i G_{l,2}^i(u)}{\sum_i G_{l,1}^i(u)} \qquad (10.120)$$

are chosen to eliminate first derivative terms involving ϕ_l.

The conventional Grueling–Goertzel theory is recovered by retaining only the $l = 0, 1$ slowing-down moments, neglecting terms $n \geq 2$ in Eq. (10.119) and identifying

$$\xi_i^l(u) = \frac{-G_{l,1}^i(u)}{\Sigma_s(u)}$$

$$a_l^i(u) = \frac{-G_{l,2}^i(u)}{\Sigma_s(u)} \qquad (10.121)$$

Inclusion of Scattering Resonances

An impractically large number of terms may have to be retained in the Taylor's series expansion to obtain an accurate approximation for the slowing-down density when resonance scattering nuclei are present in the mixture, because resonance scattering in nuclei j may cause ϕ, hence $\Sigma_s^i \phi$ for another nuclear species i, to be a rapidly varying quantity. In this case, a better approximation may be developed by expanding the total collision density in a Taylor's series:

$$\Sigma_t(u')\phi_l(u') = \sum_{n=0}^{\infty} \frac{(u'-u)^n}{n!} \frac{d^n[\Sigma_t(u)\phi_l(u)]}{du^n} \qquad (10.122)$$

Using this expansion to evaluate Eq. (10.114) yields

$$q_l(u) = -\sum_i \sum_{n=0}^{\infty} H_{l,n}^i(u) \frac{d^n[\Sigma_t(u)\phi_l(u)]}{du^n} \qquad (10.123)$$

where

$$H_{l,n}^i(u) \equiv \frac{1}{n!} \sum_{l'=0}^{\infty} \frac{1}{2}(2l'+1) \int_{u-\ln 1/\alpha_i}^{u} du' \int_{n}^{u+\ln 1/\alpha_i} du'' \, P_l\big[\mu_0(u'-u'')\big]$$

$$\times P_{l'}\big[\mu_c(u'-u'')\big] \frac{2e^{u'-u''}}{1-\alpha_i} \frac{\Sigma_{sl'}^i(u')}{\Sigma_t(u')}(u'-u)^n \qquad (10.124)$$

Differentiating Eq. (10.123),

$$\frac{dq_l(u)}{du} = -\sum_i \left[\sum_{n=0}^{\infty} H_{l,n}^i(u) \frac{d^{n+1}[\Sigma_t(u)\phi_l(u)]}{du^{n+1}} \right.$$

$$\left. + \sum_{n=0}^{\infty} \frac{dH_{l,n}^i(u)}{du} \cdot \frac{d^n[\Sigma_t(u)\phi_l(u)]}{du^n} \right] \qquad (10.125)$$

and carrying out a calculation similar to those described previously results in a hydrogen-like equation for the lth Legendre component of the slowing-down density:

$$
\hat{\lambda}_l(u)\frac{dq_l(u)}{du} + q_l(u) = -\sum_i \left\{ \left[H_{l,0}^i(u) + \hat{\lambda}_l(u)\frac{dH_{l,0}^i(u)}{du} \right] \Sigma_t(u)\phi_l(u) \right.
$$

$$
+ \sum_{n=2}^{\infty} \left[H_{l,n}^i(u)\frac{d^n[\Sigma_t(u)\phi_l(u)]}{du^n} \right.
$$

$$
+ \hat{\lambda}_l(u) H_{l,n-1}^i(u)\frac{d^n[\Sigma_t(u)\phi_l(u)]}{du^n}
$$

$$
\left. \left. + \hat{\lambda}_l(u)\frac{dH_{l,n}^i(u)}{du}\frac{d^n[\Sigma_t(u)\xi_l(u)]}{du^n} \right] \right\} \qquad (10.126)
$$

where

$$
\hat{\lambda}_l(u) = -\frac{\sum_i H_{l,1}^i(u)}{\sum_i[H_{l,0}^i(u) + dH_{l,1}^i(u)/du]} \qquad (10.127)
$$

has been chosen to eliminate first derivative terms involving ϕ_l.

P_l Continuous Slowing-Down Equations

The lethargy-dependent P_l equations are generalized from the one-speed P_l equations of Chapter 9 by including a scattering loss term and a scatter-in source of neutrons:

$$
\frac{l+1}{2l+1}\frac{\partial\phi_{l+1}(x,u)}{\partial x} + \frac{l}{2l+1}\frac{\partial\phi_{l-1}(x,u)}{\partial x} + \Sigma_t(x,u)\phi_l(x,u)
$$

$$
= \int_0^u du'\Sigma_{s,l}(x,u'\to u)\phi_l(x,u') + S_l(x,u), \quad l=1,\dots,L \qquad (10.128)
$$

The Legendre moments of the slowing-down density are related to the Legendre moments of the scatter-in integral. Differentiating Eq. (10.114) yields

$$
\frac{\partial q_l(u)}{\partial u} = \Sigma_{s,l}(u)\phi_l(u) - \int_0^u du'\Sigma_{s,l}(u'\to u)\phi_l(u') \qquad (10.129)
$$

Using this result to eliminate the elastic scatter-in integral in Eq. (10.128) leads to the P_l continuous slowing-down equations,

$$
\frac{l+1}{2l+1}\frac{\partial\phi_{l+1}(x,u)}{\partial x} + \frac{l}{2l+1}\frac{\partial\phi_{l-1}(x,u)}{\partial x} + \Sigma_{\text{ne}}^l(x,u)\phi_l(x,u)
$$

$$
= -\frac{\partial q_l(x,u)}{\partial u} + S_l(x,u), \quad l=1,\dots,L \qquad (10.130)
$$

where the nonelastic cross section is

$$\Sigma_{ne}^l(x, u) \equiv \Sigma_t(x, u) - \Sigma_{s,l}(x, u) \tag{10.131}$$

and the Legendre moments of the slowing-down density are calculated from Eqs. (10.126) for nonhydrogenic nuclei and Eqs. (10.90) and (10.91) and similarly derived higher Legendre moment equations for hydrogen.

10.5
Multigroup Discrete Ordinates Transport Theory

In situations in which a high degree of angular anisotropy in the neutron flux could be expected, the low-order P_1 and diffusion theory approximations might be inadequate to treat the combined slowing down and transport of neutrons. Such situations might arise in the treatment of slowing down in a highly heterogeneous lattice consisting of materials of very different properties or in the treatment of problems in which there is a highly directional flow of fast neutrons from one region to another. For such situations, the discrete ordinates methods of Chapter 9, extended to treat the neutron slowing down, are well suited. Generalizing the expansion of the differential (over scattering angle) elastic scattering cross section of Eq. (9.179) to an expansion of the double differential (over scattering angle and lethargy change) scattering cross section, and using the addition theorem for Legendre polynomials of Eq. (9.177) to relate the cosine of the scattering angle, μ_0, to the cosines of the incident, μ', and exiting, μ, directions for the scattering event, yields

$$\Sigma_s\left(r, \mathbf{\Omega}' \cdot \mathbf{\Omega}, u' \to u\right)$$

$$= \sum_{l'=0}^{L} \frac{2l'+1}{4\pi} \Sigma_{sl'}(\mathbf{r}, u' \to u) P_l(\mu_0)$$

$$= \sum_{l'=0}^{L} \frac{2l'+1}{4\pi} \Sigma_{sl'}(\mathbf{r}, u' \to u) \sum_{m=-l'}^{l'} Y_{l'm}(\mu, \varphi) Y_{l'm}^*(\mu', \varphi')$$

$$= \sum_{l'=0}^{L} \frac{2l'+1}{4\pi} \Sigma_{sl'}(\mathbf{r}, u' \to u)$$

$$\times \left[P_{l'}(\mu) P_{l'}(\mu') + 2 \sum_{m=1}^{l'} \frac{(l'-m)!}{(l'+m)!} P_{l'}^m(\mu) P_{l'}^m(\mu') \cos m(\varphi - \varphi') \right] \tag{10.132}$$

Using this representation of the double differential scattering cross section in the neutron transport equation (10.26) yields

$$\boldsymbol{\Omega} \cdot \nabla \psi(\mathbf{r}, \boldsymbol{\Omega}, u) + \Sigma_t(\mathbf{r}, u)\psi(\mathbf{r}, \boldsymbol{\Omega}, u)$$

$$= S_{ex}(\mathbf{r}, \boldsymbol{\Omega}, u) + \int_0^{\infty} du' \sum_{l'=0}^{L} Y_{l'm}(\boldsymbol{\Omega}) \Sigma_{sl'}(\mathbf{r}, u' \to u)\phi_{l'm}(\mathbf{r}, u')$$

$$+ \frac{\chi(u)}{4\pi} \int_0^{\infty} du' \nu\Sigma_f(\mathbf{r}, u')\phi(\mathbf{r}, u') \qquad (10.133)$$

where the Legendre moments of the angular flux, ϕ_{lm}, and the scalar flux, ϕ, are defined as

$$\phi_{lm}(\mathbf{r}, u') \equiv \int_{4\pi} d\boldsymbol{\Omega}' Y_{lm}^*(\boldsymbol{\Omega}')\psi(\mathbf{r}, \boldsymbol{\Omega}', u')$$

$$\phi(\mathbf{r}, u') = \phi_{00}(\mathbf{r}, u') \equiv \int_{4\pi} d\boldsymbol{\Omega}' \psi(\mathbf{r}, \boldsymbol{\Omega}', u') \qquad (10.134)$$

These equations may be reduced to a set of multigroup equations by integrating over the lethargy width $\Delta u_g = u_{g+1} - u_g$ of group g:

$$\boldsymbol{\Omega} \cdot \nabla \psi_g(\mathbf{r}, \boldsymbol{\Omega}) + \Sigma_t^g(\mathbf{r})\psi_g(\mathbf{r}, \boldsymbol{\Omega})$$

$$= S_{ex}^g(\mathbf{r}, \boldsymbol{\Omega}) + \sum_{g'=1}^{G} \sum_{l'=0}^{L} Y_{l'm}(\boldsymbol{\Omega}) \Sigma_{sl'}^{g' \to g}(\mathbf{r})\phi_{l'm}^{g'}(\mathbf{r})$$

$$+ \frac{\chi^g}{4\pi} \sum_{g'=1}^{G} \nu\Sigma_f^{g'}(\mathbf{r})\phi_{g'}(\mathbf{r}), \quad g = 1, \ldots, G \qquad (10.135)$$

where multigroup quantities have been defined

$$\psi_g(\mathbf{r}, \boldsymbol{\Omega}) \equiv \int_{\Delta u_g} du\, \psi(\mathbf{r}, \boldsymbol{\Omega}, u)$$

$$\phi_{lm}^g(\mathbf{r}) \equiv \int_{\Delta u_g} du\, \phi_{lm}(\mathbf{r}, u)$$

$$\phi_g(\mathbf{r}) \equiv \int_{\Delta u_g} du\, \phi(\mathbf{r}, u)$$

$$\chi^g \equiv \int_{\Delta u_g} du\, \chi(u), \qquad S_{ex}^g(\mathbf{r}, \boldsymbol{\Omega}) \equiv \int_{\Delta u_g} du\, S_{ex}^g(\mathbf{r}, \boldsymbol{\Omega}, u) \qquad (10.136)$$

$$\Sigma_t^g(\mathbf{r}) \equiv \frac{\int_0^{4\pi} d\boldsymbol{\Omega} \int_{\Delta u_g} du\, \Sigma_t(\mathbf{r}, u)\psi(\mathbf{r}, \boldsymbol{\Omega}, u)}{\int_0^{4\pi} d\boldsymbol{\Omega} \int_{\Delta u_g} du\, \psi(\mathbf{r}, \boldsymbol{\Omega}, u)} = \frac{\int_{\Delta u_g} du\, \Sigma_t(\mathbf{r}, u)\phi(\mathbf{r}, u)}{\int_{\Delta u_g} du\, \phi(\mathbf{r}, u)}$$

$$\nu\Sigma_f^g(\mathbf{r}) \equiv \frac{\int_{\Delta u_g} du\, \nu\Sigma_f(\mathbf{r}, u)\phi(\mathbf{r}, u)}{\int_{\Delta u_g} du\, \phi(\mathbf{r}, u)}$$

$$\Sigma_{sl}^{g' \to g}(\mathbf{r}) \equiv \frac{\int_{\Delta u_g} du \int_{\Delta u_{g'}} du'\, \Sigma_{sl}(\mathbf{r}, u' \to u)\phi_{lm}(\mathbf{r}, u')}{\int_{\Delta u_{g'}} du'\, \phi_{lm}(\mathbf{r}, u')}$$

Writing Eqs. (10.135) for each discrete ordinate, $\mathbf{\Omega}_n$, results in the set of multigroup discrete ordinates equations

$$\mathbf{\Omega}_n \cdot \nabla \psi_g(\mathbf{r}, \mathbf{\Omega}_n) + \Sigma_t^g \psi_g(\mathbf{r}, \mathbf{\Omega}_n) = Q^g(\mathbf{r}, \mathbf{\Omega}_n), \quad g = 1, \dots, G \qquad (10.137)$$

where the group scattering plus fission plus external source term is

$$Q^g(\mathbf{r}, \mathbf{\Omega}_n) \equiv \sum_{g'=1}^{G} \sum_{l'=0}^{L} Y_{l'm}(\mathbf{\Omega}_n) \Sigma_{sl'}^{g' \to g}(\mathbf{r}) \phi_{l'm}^{g'}(\mathbf{r})$$

$$+ \frac{\chi^g}{4\pi} \sum_{g'=1}^{G} \nu \Sigma_f^{g'}(\mathbf{r}) \phi_{g'}(\mathbf{r}) + S_{\text{ex}}^g(\mathbf{r}, \mathbf{\Omega}_n) \qquad (10.138)$$

Equation (10.137), for each group, is of the same form as the discrete ordinates equation discussed in Chapter 9. Thus the methods used to solve the discrete ordinates equations in Chapter 9 can be applied to solve the multigroup discrete ordinates equations, on a group-by-group basis. For a given fission and scattering source, the multigroup discrete ordinates equations are solved group by group using the methods of Chapter 9. Then on the source iteration, the new scattering and fission source for each group are constructed by summing over contributions from all groups, and the solution for the multigroup fluxes on a group-by-group basis is repeated until convergence. The power iteration procedure for criticality eigenvalue calculations is the same as discussed in Chapter 9, but now the fission source is summed over the contributions from all groups.

References

1 J. J. DUDERSTADT and L. J. HAMILTON, *Nuclear Reactor Analysis*, Wiley, New York (**1976**), pp. 347–369.

2 A. R. HENRY, *Nuclear Reactor Analysis*, MIT Press, Cambridge, MA (**1975**), pp. 359–367 and 386–423.

3 W. M. STACEY, "The Effect of Wide Scattering Resonances on Neutron Multigroup Cross Sections," *Nucl. Sci. Eng.* 47, 29 (**1972**); "The Effect of Anisotropic Scattering upon the Elastic Moderation of Fast Neutrons," *Nucl. Sci. Eng.* 44, 194 (**1971**); "Continuous Slowing Down Theory for Anisotropic Elastic Moderation in the P_n and B_n Representations," *Nucl. Sci. Eng.* 41, 457 (**1970**).

4 B. J. TOPPEL, A. L. RAGO, and D. M. O'SHEA, *MC²: A Code to Calculate Multigroup Cross Sections*, ANL-7318, Argonne National Laboratory, Argonne, IL (**1967**).

5 J. H. FERZIGER and P. F. ZWEIFEL, *The Theory of Neutron Slowing Down in Nuclear Reactors*, MIT Press, Cambridge, MA (**1966**).

6 M. M. R. WILLIAMS, *The Slowing Down and Thermalization of Neutrons*, North-Holland, Amsterdam (**1966**), pp. 317–516.

7 D. S. SELENGUT et al., "The Neutron Slowing Down Problem," in A. Radkowsky, ed., *Naval Reactors Physics Handbook*, U.S. Atomic Energy Commission, Washington, DC (**1964**).

8 G. GOERTZEL and E. GRUELING, "Approximate Method for Treating Neu-

tron Slowing Down," *Nucl. Sci. Eng. 7*, 69 (**1960**).

9 H. J. AMSTER, "Heavy Moderator Approximations in Neutron Transport Theory," *J. Appl. Phys. 29*, 623 (**1958**).

10 H. BOHL, JR., E. M. GELBARD, and G. H. RYAN, *MUFT-4: A Fast Neutron Spectrum Code*, WAPD-TM-22, Bettis Atomic Power Laboratory, West Miflin, PA (**1957**).

11 R. A. KNIEF, *Nuclear Engineering*, 2nd ed., Taylor & Francis, Washington, DC (**1992**).

Problems

10.1. Calculate the average cosine of the scattering angle in the CM system for neutrons at 1 MeV, 100 keV, and 1 keV colliding with uranium, iron, carbon, and hydrogen.

10.2. Calculate the values of the average lethargy increase, ξ, and the average cosine of the scattering angle in the lab system, μ_0, for the neutron energies and nuclei of Problem 10.1, for isotropic scattering and for linearly anisotropic scattering.

10.3. Carry through the steps in the derivation of the lethargy-dependent P_1 equations given by Eqs. (10.36) and (10.37).

10.4. Carry through the derivation of the isotropic and linearly anisotropic lethargy transfer functions of Eqs. (10.39) and (10.40).

10.5. Divide the energy interval 10 MeV $> E >$ 1 eV into 54 equal-lethargy intervals. Evaluate the multigroup scattering transfer functions $\Sigma_{s0}^{g' \rightarrow g}$ for carbon for $g' = 1, 10$, and 50.

10.6. Carry through the steps in the derivation of the lethargy-dependent B_1 equations given by Eqs. (10.50).

10.7. Solve for the lethargy-dependent neutron flux and current in an infinite medium, using the age approximation of Eqs. (10.96).

10.8. Derive a differential equation similar to Eqs. (10.90) and (10.91) for the higher Legendre moments of the slowing-down density in hydrogen.

10.9. Derive the multigroup approximation for the P_1 continuous slowing-down equations, Eqs. (10.112) and (10.113).

10.10. Write a computer code to solve the multigroup P_1 continuous slowing-down equations of Problem 10.9 for an assembly consisting of 3% enriched, zircalloy-clad UO_2 and water. The fuel pins are 1 cm in diameter with clad thickness of 0.05 cm in a square array with fuel pin center-to-center distance of 2.0 cm. Assume that spatial gradients can be neglected.

11
Resonance Absorption

11.1
Resonance Cross Sections

When the relative (center-of-mass) energy of an incident neutron and a nucleus plus the neutron binding energy match an energy level of the compound nucleus that would be formed upon neutron capture, the probability of neutron absorption is quite large. For the odd-mass fissionable fuel isotopes, resonances occur from a fraction of 1 eV up to a few thousand eV, and for the even-mass fuel isotopes, resonances occur from a few eV to about 10,000 eV, as shown in Figs. 11.1 to 11.4. At the lower energies the resonances are well separated, but at the higher energies the resonances overlap and become unresolvable experimentally. We first examine the widely spaced resonances at lower energy, where spatial self-shielding, as well as energy self-shielding, is important. At higher energies, spatial self-shielding becomes less important, but resonance overlap interference effects become important.

11.2
Widely Spaced Single-Level Resonances in a Heterogeneous Fuel–Moderator Lattice

Neutron Balance in Heterogeneous Fuel–Moderator Cell

At lower energies in the 10-eV range, the neutron mean free path becomes comparable to the fuel and moderator dimensions, and it is important to take into account the spatial heterogeneity of the fuel–moderator cell. The fuel assembly in a nuclear reactor generally consists of a repeating array of unit cells consisting of fuel, moderator/coolant, clad, and so on. For simplicity, we consider a two-region unit cell of fuel (F) and a separate moderator (M). We allow further for a moderator admixed with the fuel (e.g., the oxygen in UO_2 fuel). We return to the problem of calculating the absorption of neutrons in widely spaced resonances which was treated for a homogeneous mixture in Section 4.3, but now take into account the important spatial self-shielding effects that are present in a heterogeneous fuel–moderator

Nuclear Reactor Physics. Weston M. Stacey
Copyright © 2007 WILEY-VCH Verlag GmbH & Co. KGaA, Weinheim
ISBN: 978-3-527-40679-1

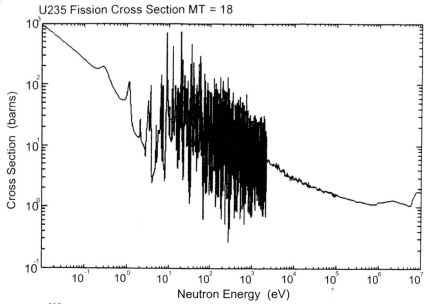

Fig. 11.1 ^{235}U fission cross section. (From *http://www.nndc.bnl.gov/.*)

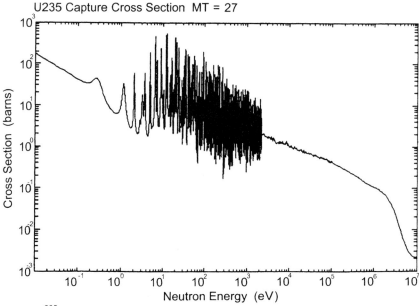

Fig. 11.2 ^{235}U capture cross section. (From *http://www.nndc.bnl.gov/.*)

U238 Capture Cross Section MT = 27

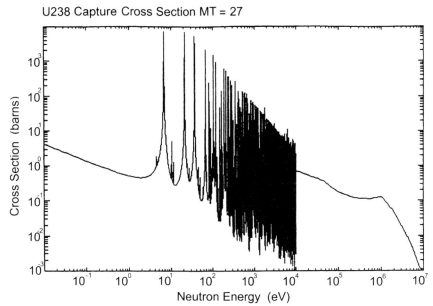

Fig. 11.3 ^{238}U capture cross section. (From *http://www.nndc.bnl.gov/.*)

U238 Elastic Scattering Cross Section MT = 2

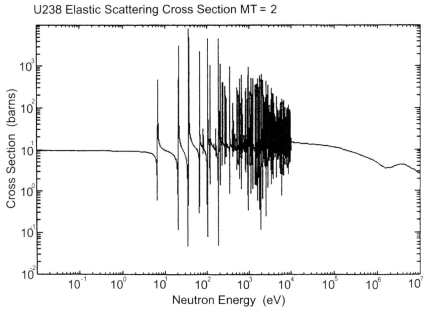

Fig. 11.4 ^{238}U elastic scattering cross section. (From *http://www.nndc.bnl.gov/.*)

lattice. Consider a repeating array of fuel–moderator cells with fuel volume V_F and moderator volume V_M. Define the first-flight escape probabilities

$P_{F0}(E) =$ probability that a neutron that slows down to energy E in the fuel will make its next collision in the moderator

$P_{M0}(E) =$ probability that a neutron that slows down to energy E in the moderator will make its next collision in the fuel

We assume that these probabilities are uniform over the fuel and moderator, respectively.

The neutron balance equation in the fuel can be written

$$
\left[\Sigma_m^F + \Sigma_t^F(E)\right]\phi_F(E)V_F
$$

$$
= V_F[1 - P_{F0}(E)]\left[\int_E^{E/\alpha_F} dE' \frac{\Sigma_S^F(E')\phi_F(E')}{(1-\alpha_F)E'} + \int_E^{E/\alpha_m} dE' \frac{\Sigma_m^F \phi_F(E')}{(1-\alpha_m)E'}\right]
$$

$$
+ V_M P_{M0}(E) \int_E^{E/\alpha_M} dE' \frac{\Sigma_S^M(E')\phi_M(E')}{(1-\alpha_M)E'} \tag{11.1}
$$

The left side of the equation is the total reaction rate of the fuel plus admixed moderator in the fuel volume. The first term on the right side is the source of neutrons scattering to energy E in the fuel (from scattering collisions with fuel and with admixed moderator nuclei) times the probability $(1 - P_{F0})$ that their next collision is in the fuel, and the second term is the source of neutrons scattering into energy E in the moderator times the probability P_{M0} that their next collision is in the fuel.

The practical width of an absorption resonance will generally be much smaller than the scattering-in interval of the moderator, $\Gamma_p \ll (1 - \alpha_M)E_0$, or of the admixed moderator, $\Gamma_p \ll (1 - \alpha_m)E_0$, which allows us to use the asymptotic form of the neutron flux above the resonance energy in the moderator and the fuel, $\phi_{\mathrm{asy}}(E) \sim 1/E$, to evaluate the moderator and admixed moderator scattering integrals in Eq. (11.1), leading to

$$
\left[\Sigma_m^F + \Sigma_t^F(E)\right]\phi_F(E)V_F
$$

$$
= V_F[1 - P_{F0}(E)]\left[\int_E^{E/\alpha_F} dE' \frac{\Sigma_S^F(E')\phi_F(E')}{(1-\alpha_F)E'} + \frac{\Sigma_m^F}{E}\right]
$$

$$
+ V_M P_{M0}(E)\frac{\Sigma_S^M(E)}{E} \tag{11.2}
$$

Reciprocity Relation

Define $G(r_F; r_M)$ as the probability that a neutron isotropically scattered to energy E at location r_F in the fuel travels without collision to location r_M in the moderator, and $G(r_M; r_F)$ as the probability that a neutron isotropically scattered to energy E at location r_M in the moderator travels without collision to location r_F in the fuel. For a uniformly distributed source of neutrons scattering to energy E in each region, the following identities must obtain:

$$V_F P_{F0}(E) \equiv \Sigma_t^M(E) \int_{V_F} dr_F \int_{V_M} G(r_F; r_M) \, dr_M$$

$$V_M P_{M0}(E) \equiv \left(\Sigma_m^F + \Sigma_t^F(E)\right) \int_{V_M} dr_M \int_{V_F} G(r_M; r_F) \, dr_F \tag{11.3}$$

Since $G(r_F; r_M)$ and $G(r_M; r_F)$ depend only on the collision probability along the path between r_F and r_M, and this probability is independent of the direction in which the neutron travels, $G(r_M; r_F) = G(r_F; r_M)$, Eqs. (11.3) may be combined to obtain the reciprocity relation between the two first-flight collision probabilities:

$$P_{F0}(E)\left[\Sigma_m^F + \Sigma_t^F(E)\right]V_F = P_{M0}(E)\Sigma_t^M(E)V_M \tag{11.4}$$

If we make the assumption that absorption is very small relative to scattering in the moderator, the reciprocity relation may be used to write Eq. (11.2) as

$$\left[\Sigma_m^F + \Sigma_t^F(E)\right]\phi_F(E)$$

$$= [1 - P_{F0}(E)]\left[\int_E^{E/\alpha_F} dE' \frac{\Sigma_S^F(E')\phi_F(E')}{(1-\alpha_F)E'} + \frac{\Sigma_m^F}{E}\right]$$

$$+ P_{F0}(E)\frac{\Sigma_m^F + \Sigma_t^F(E)}{E} \tag{11.5}$$

Narrow Resonance Approximation

If the practical width of the resonance is much smaller than the scattering-in interval of the resonance nucleus, $\Gamma_p \ll (1-\alpha_F)E_0$, the contribution of the resonance to the scattering-in integral in Eq. (11.5) can be neglected and the asymptotic flux in the fuel $\phi(E) \sim 1/E$ can be used to evaluate the integral to obtain

$$\phi_F^{\text{NR}}(E) = \frac{[1 - P_{F0}(E)]\Sigma_p^F + P_{F0}(E)\Sigma_t^F(E) + \Sigma_m^F}{[\Sigma_m^F + \Sigma_t^F(E)]E} \tag{11.6}$$

Using this form for the neutron flux to evaluate the capture resonance integral

$$I^\gamma \equiv \int dE\, \sigma_\gamma(E)\phi(E) \tag{11.7}$$

leads to the narrow resonance approximation for the heterogeneous resonance integral:

$$I_{\text{NR}}^\gamma = \int \frac{dE}{E} \sigma_\gamma(E) \frac{\sigma_p^F + \sigma_m^F + P_{F0}(E)[\sigma_t^F(E) - \sigma_p^F]}{\sigma_m^F + \sigma_t^F(E)}$$

$$= \int \frac{dE}{E} \sigma_\gamma(E) \frac{\sigma_p^F + \sigma_m^F + \sigma_e(E)}{\sigma_t^F(E) + \sigma_m^F + \sigma_e(E)} \tag{11.8}$$

where we have defined an escape cross section:

$$P_{F0}(E) \equiv \frac{\sigma_e(E)}{\sigma_e(E) + \sigma_t^F(E) + \sigma_m^F} \quad \Rightarrow \quad \sigma_e(E) = P_{F0}(E)\frac{\sigma_t^F(E) + \sigma_m^F}{1 - P_{F0}(E)}$$

(11.9)

and used the notation $\sigma_t^F(E)$ and σ_p^F for the total and potential scattering micro-scopic cross sections of the resonance absorber, and σ_m^F for the cross section of the admixed moderator per fuel nucleus.

Wide Resonance Approximation

If the practical width of the resonance is much larger than the scattering-in in-terval of the resonance nuclei, $\Gamma_p \gg (1 - \alpha_F)E_0$, the term $\Sigma_s^{\text{res}}(E')\phi(E')/E' \approx \Sigma_s^{\text{res}}(E)\phi(E)/E$ in the integrand of Eq. (11.5), leading to the wide resonance ap-proximation for the flux in the fuel region,

$$\phi_F^{\text{WR}}(E) = \frac{P_{F0}(E)\Sigma_t^F(E) + \Sigma_m^F}{[\Sigma_t^F(E) + \Sigma_m^F - (1 - P_{F0}(E))\Sigma_S^F(E)]E}$$

(11.10)

Using this result to evaluate the resonance integral of Eq. (11.7) leads to

$$I_{\text{WR}}^{\gamma} = \int \frac{dE}{E}\sigma_\gamma(E)\frac{P_{F0}(E)\sigma_t^F(E) + \sigma_m^F}{\sigma_t^F(E) + \sigma_m^F - [1 - P_{F0}(E)]\sigma_S^F(E)}$$

$$= \int \frac{dE}{E}\sigma_\gamma(E)\frac{\sigma_m^F + \sigma_e(E)}{\sigma_a^F(E) + \sigma_m^F + \sigma_e(E)}$$

(11.11)

where $\sigma_a^F = \sigma_\gamma^F + \sigma_f^F$ is the microscopic absorption cross section of the resonance absorber.

Evaluation of Resonance Integrals

Recalling from Section 4.3 the form of the single-level resonance cross section av-eraged over the thermal motion of the nuclei, the (n, γ) capture cross section or fission cross section averaged over the motion of the nucleus can be written

$$\sigma_q(E, T) = \sigma_0\frac{\Gamma_q}{\Gamma}\left(\frac{E_0}{E}\right)^{1/2}\psi(\xi, x), \quad q = \gamma, f$$

(11.12)

and the total scattering cross section, including resonance and potential scattering and interference between the two, can be written

$$\sigma_s(E, T) = \sigma_0\frac{\Gamma_n}{\Gamma}\psi(\xi, x) + \frac{\sigma_0 R}{\lambda_0}\chi(\xi, x) + 4\pi R^2$$

(11.13)

where R is the nuclear radius, λ_0 is the neutron De Broglie wavelength, and the functions

$$\psi(\xi, x) = \frac{\xi}{2\sqrt{\pi}}\int_{-\infty}^{\infty} e^{-(1/4)(x-y)^2\xi^2}\frac{dy}{1 + y^2}$$

(11.14)

$$\chi(\xi, x) = \frac{\xi}{\sqrt{\pi}} \int_{-\infty}^{\infty} e^{-(1/4)(x-y)^2 \xi^2} \frac{y \, dy}{1 + y^2} \tag{11.15}$$

are integrals over the relative motion of the neutron and nucleus, $x = 2(E_{cm} - E_0)/\Gamma$, assuming that the nuclear motion can be characterized by a Maxwellian distribution with temperature T, and E_{cm} is the energy of the neutron in the neutron–nucleus center-of-mass system. The parameters characterizing the resonance are σ_0, the peak value of the cross section; E_0, the neutron energy in the center-of-mass system at which it occurs; Γ, the resonance width; Γ_γ, the partial width for neutron capture, Γ_f, the partial width for fission; and Γ_n, die partial width for scattering. The resonance absorption cross section is symmetric about E_0, but the scattering cross section is asymmetric because the potential and resonance scattering interfere constructively for $E > E_0$ and destructively for $E < E_0$, as indicated in Fig. 11.4.

The temperature characterizing the nuclear motion is contained in the parameter

$$\xi = \frac{\Gamma}{(4E_0 k T / A)^{1/2}} \tag{11.16}$$

where A is the atomic mass (amu) and k is the Boltzmann constant.

Using these forms for the resonance cross sections in Eqs. (11.8) and (11.11), the resonance integrals become in the narrow resonance approximation (neglecting interference scattering)

$$
\begin{aligned}
I_{NR}^\gamma &= \frac{\Gamma_\gamma}{2E_0} \left(\sigma_p^F + \sigma_m^F + \sigma_e\right) \int_{-\infty}^{\infty} \frac{\psi(\xi, x) \, dx}{\psi(\xi, x) + \beta} \\
&\equiv \frac{\Gamma_\gamma}{E_0} \left(\sigma_p^F + \sigma_m^F + \sigma_e\right) J(\xi, \beta)
\end{aligned} \tag{11.17}
$$

and in the wide resonance approximation

$$I_{WR}^\gamma = \frac{\Gamma_\gamma}{2E_0} \left(\sigma_m^F + \sigma_e\right) \int_{-\infty}^{\infty} \frac{\psi(\xi, x) \, dx}{\psi(\xi, x) + \beta'} \equiv \frac{\Gamma_\gamma}{E_0} \left(\sigma_m^F + \sigma_e\right) J(\xi, \beta') \tag{11.18}$$

where

$$\beta \equiv \frac{\sigma_p^F + \sigma_m^F + \sigma_e}{\sigma_0}, \qquad \beta' \equiv \frac{\sigma_m^F + \sigma_e}{\sigma_0} \frac{\Gamma}{\Gamma_\gamma} \tag{11.19}$$

The $J(\xi, \beta)$ function is tabulated in Table 4.3. The properties of the moderator region do not appear explicitly in these expressions for the resonance integral because we have assumed that a neutron which escapes the fuel will have its next collision in the moderator and because we assumed that absorption in the moderator could be ignored in using the reciprocity relation.

Table 11.1 Infinite Dilution Total Resonance Integrals for Some Heavy Elements*

Isotope	RI(n, γ) (barns)	RI(n, f) (barns)
^{232}Th	84	–
^{233}U	138	774
^{233}Pa	864	–
^{234}U	631	7
^{235}U	133	278
^{236}U	346	8
^{237}Np	661	7
^{239}Np	445	–
^{239}Pu	181	302
^{241}Pu	180	573
^{241}Am	1305	14
^{238}U	278	2
^{240}Pu	8103	9
^{242}Pu	1130	6
^{242}Am	391	1258

* Calculated with ORIGEN (Ref. 14).

Infinite Dilution Resonance Integral

In the infinite dilution limit $\sigma_m^F + \sigma_e \gg \sigma_0$, all forms for the resonance integral approach the infinite dilution value:

$$I_\infty^\gamma = \frac{\pi}{2}\sigma_0\frac{\Gamma_\gamma}{E_0}, \qquad I_\infty^f = \frac{\pi}{2}\sigma_0\frac{\Gamma_f}{E_0} \qquad (11.20)$$

Infinite dilution resonance integrals for a number of fuel isotopes are given in Table 11.1. Actual resonance integrals will be smaller because of self-shielding effects.

Equivalence Relations

For a given resonance absorbing species, assemblies with the same values of $\sigma_m^F + \sigma_e$ have the same resonance integral. Furthermore, the heterogeneous assemblies with a given value of $\sigma_m^F + \sigma_e$ have the same resonance integrals as homogeneous assemblies which have moderator scattering cross section per resonance absorber nucleus $\sigma_s^M = \sigma_m^F + \sigma_e$. Equations (11.17) and (11.18) reduce to the homogeneous resonance integrals of Eqs. (4.68) and (4.71) when $\sigma_e \sim P_{F0} = 0$ (i.e., in the case of a resonance absorber with a homogeneously admixed moderator).

Heterogeneous Resonance Escape Probability

The resonance capture rate in a fuel–moderator cell with fuel volume V_F and moderator volume V_M is

$$R_\gamma = V_F N_F \int \sigma_\gamma^F (E) \phi_F(E) = V_F N_F I^\gamma \tag{11.21}$$

We have evaluated the resonance integral for an asymptotic flux above the resonance $\phi_{asy} = 1/E$, assumed uniform over the fuel and moderator. Using the asymptotic relationship between the slowing-down density, q, and the flux

$$q = \xi \Sigma_s E \phi_{asy} \tag{11.22}$$

where the average asymptotic moderating power of the cell is

$$\xi \Sigma_s = \frac{V_M \xi_M \Sigma_s^M + V_F (\xi_F \Sigma_p^F + \xi_m \Sigma_m^F)}{V_M + V_F} \tag{11.23}$$

indicates that an asymptotic flux of $1/E$ implies an asymptotic slowing-down density of $q = \xi \Sigma_s$. The resonance escape probability for the cell is unity minus the resonance absorption probability, and the latter is the resonance absorption rate divided by the total number of neutrons slowing down $q(V_F + V_M)$:

$$p = 1 - \frac{R_\gamma}{q(V_F + V_M)} = 1 - \frac{V_F N_F I^\gamma}{\xi \Sigma_s (V_F + V_M)} = 1 - \frac{I^\gamma}{\xi \sigma_s} \tag{11.24}$$

where $\xi \sigma_s$ is the cell moderating power per fuel nucleus. The total resonance escape probability over an energy group g containing several resonances is

$$P = \prod_{i \in g} p_i = \prod_{i \in g} \left(1 - \frac{I_i^\gamma}{\xi \sigma_s} \right) \simeq \exp \left(-\frac{1}{\xi \sigma_s} \sum_{i \in g} I_i^\gamma \right) \tag{11.25}$$

where $i \in g$ indicates all of the resonances within energy group g extending from E_g to E_{g-1}.

By lumping the fuel, the neutron flux in the fuel is reduced relative to the flux in the separate moderator—spatial self-shielding—and it is possible to decrease the resonance integral without decreasing the slowing-down power, thus increasing the resonance escape probability relative to the value for the same fuel and moderator distributed homogeneously. In fact, lumping the natural uranium fuel in the early graphite-moderated reactors was essential to achieving criticality—the resonance escape probability increased from about 0.7 in a homogeneous graphite–natural uranium assembly to about 0.88 in a heterogeneous assembly.

Homogenized Multigroup Resonance Cross Section

An effective multigroup cross section for the resonance absorber can be constructed by summing the resonance absorption rates over all of the resonances in the group, dividing by the fuel number density, N_F, dividing by the volume of the fuel–moderator cell, and dividing by the integral of the asymptotic flux over the energy interval of the group:

$$\sigma_g^{\text{res}} = \frac{\sum_{i \in g} R_\gamma^i / N_F}{(V_F + V_M) \int_{E_g}^{E_g - 1} \phi_{\text{asy}} \, dE} = \frac{V_F \sum_{i \in g} I_i^\gamma}{(V_F + V_M) \ln(E_{g-1}/E_g)} \tag{11.26}$$

Improved and Intermediate Resonance Approximations

The narrow $[\Gamma_p \ll (1 - \alpha_F) E_0]$ and wide $[\Gamma_p \gg (1 - \alpha_F) E_0]$ resonance approximations are limiting cases. For many resonances, the actual situation is intermediate to these extremes. It is possible to improve upon the narrow resonance and wide resonance approximations using the neutron balance equation to improve the flux solution iteratively:

$$\left[\sigma_m^F + \sigma_t^F(E) \right] \phi_F^{(n)}(E)$$

$$= [1 - P_{F0}(E)] \left[\int_E^{E/\alpha_F} dE' \frac{\sigma_s^F(E') \phi_F^{(n-1)}(E')}{(1 - \alpha_F) E'} + \frac{\sigma_m^F}{E} \right]$$

$$+ P_{F0}(E) \frac{\sigma_m^F + \sigma_t^F(E)}{E} \tag{11.27}$$

The initial flux guess can be the narrow or wide resonance approximation, or an intermediate resonance approximation which is suggested by comparison of the two:

$$\phi_{\text{NR}}^{(1)} = \frac{\sigma_m^F + \sigma_p^F + \sigma_e}{\sigma_m^F + \sigma_t^F + \sigma_e} \frac{1}{E}$$

$$\phi_{\text{WR}}^{(1)} = \frac{\sigma_m^F + \sigma_e}{\sigma_m^F + \sigma_a^F + \sigma_e} \frac{1}{E} \tag{11.28}$$

$$\phi_{\text{IR}}^{(1)} = \frac{\sigma_m^F + \lambda \sigma_p^F + \sigma_e}{\sigma_m^F + \lambda \sigma_s^F + \sigma_a^F + \sigma_e} \frac{1}{E}$$

where λ, which is in the range $0 < \lambda < 1$, is a parameter to be determined separately (Chapter 13). In practice, it is not practical to extend this procedure beyond a single iteration.

11.3
Calculation of First-Flight Escape Probabilities

To evaluate the resonance integrals of Section 11.2 it is first necessary to calculate the probability P_{F0} that a neutron reaching energy E in the fuel will have its next collision in the moderator. Although this can be done exactly with a Monte Carlo calculation, a large number of such calculations would be necessary, and a number of analytical approximations have been developed.

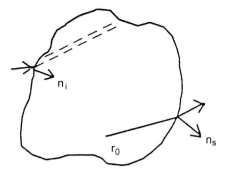

Fig. 11.5 Geometry notation for escape probability calculation.

Escape Probability for an Isolated Fuel Rod

For an isolated fuel rod surrounded by moderator, the probability P_{F0} that a neutron reaching energy E in the fuel will have its next collision in the moderator is just the probability that the neutron will escape from the fuel rod without a collision, P_0. For a uniform fuel rod, the probability that a neutron created isotropically at location r_0 within a fuel rod of arbitrary shape (Fig. 11.5) escapes from the fuel rod is

$$P_0'(r_0) = \int \frac{e^{-l/\lambda}}{4\pi l^2} (\mathbf{\Omega} \cdot \mathbf{n}_s)\, ds \tag{11.29}$$

where $\lambda = \Sigma_t^{-1}$ is the total mean free path, $l(r_0, \mathbf{\Omega})$ the distance from r_0 to the surface of the rod in the direction $\mathbf{\Omega}$, $\mathbf{n}_s$ the outward normal vector to the surface of the fuel rod, $(\mathbf{\Omega} \cdot \mathbf{n}_s)ds/4\pi l^2(r_0, \mathbf{\Omega})$ the solid angle that the surface element ds in the direction ω subtends, and $\exp(-l/\lambda)$ the probability that the neutron will reach the surface without collision.

If the neutrons are created isotropically, the average escape probability is

$$P_0 = \frac{1}{V} \int P_0'(r_0)\, dr_0 = \frac{1}{V} \int dV \int d\mathbf{\Omega} \frac{e^{-l/\lambda}}{4\pi} \tag{11.30}$$

If we represent the volume as tubular elements oriented in the $\mathbf{\Omega}$ direction with cross-sectional area $ds(\mathbf{n}_i \cdot \mathbf{\Omega})$ where $\mathbf{n}_i$ is the inward normal unit vector on the rod surface, the volume element is $dV = dl\, ds(\mathbf{n}_i \cdot \mathbf{\Omega})$, and Eq. (11.30) can be integrated over length l to obtain

$$P_0 = \frac{\lambda}{4\pi V} \int \int ds\, d\mathbf{\Omega}(\mathbf{\Omega} \cdot \mathbf{n}_i)\left(1 - e^{-l_s(\mathbf{\Omega})/\lambda}\right) \tag{11.31}$$

where $l_s(\mathbf{\Omega})$ is the chord length from surface to surface of the rod in direction $\mathbf{\Omega}$. For a long fuel plate of thickness a, this may be evaluated exactly:

$$P_0 = \frac{\lambda}{a}\left[\frac{1}{2} - E_3\left(\frac{a}{\lambda}\right)\right] \tag{11.32}$$

where E_3 is the exponential integral function. An approximate evaluation is possible for a sphere of radius a:

$$P_0 = \frac{\lambda}{R_0}\left[1 - \frac{8}{9(R_0/\lambda)^2} + \frac{4}{3(R_0/\lambda)}\left(1 + \frac{2}{3(R_0/\lambda)}\right)e^{-(3/2)(R_0/\lambda)}\right],$$

$$R_0 \equiv \frac{4a}{3} \tag{11.33}$$

and for a long cylindrical fuel rod of radius a,

$$P_0 = \frac{\lambda}{R_0}\left[1 - \frac{4}{\pi}\int_0^{\pi/2} d\beta \cos\beta\, Ki_3\left(\frac{R_0}{\lambda}\cos\beta\right)\right],$$

$$Ki_n(X) \equiv \int_0^\infty \frac{e^{-X\cosh u}\,du}{(\cosh u)^n}, \quad R_0 = 2a \tag{11.34}$$

A more general evaluation may be made by invoking the theory of chord distributions. The probability that the length of a chord lies between l_s and $l_s + dl_s$ is

$$\phi(l_s) = \frac{\int\left[\int_{l'_s = l_s}(\boldsymbol{\Omega}\cdot\mathbf{n}_i)\,d\boldsymbol{\Omega}\right]_{(\boldsymbol{\Omega}\cdot\mathbf{n}_i)>0}ds}{\iint(\boldsymbol{\Omega}\cdot\mathbf{n}_i)\,d\boldsymbol{\Omega}\,ds} \tag{11.35}$$

where the integral over $\boldsymbol{\Omega}$ in the numerator includes only those values of $\boldsymbol{\Omega}$ for which $l'_s = l_s$, (i.e., is a chord length of the fuel rod). The denominator is readily evaluated:

$$\iint(\boldsymbol{\Omega}\cdot\mathbf{n}_i)\,d\boldsymbol{\Omega}\,ds = 2\pi S\int_0^1 \mu\,d\mu = \pi S \tag{11.36}$$

where S is the surface area of the fuel rod.

In this representation the volume of the fuel rod is

$$V = \int l(\boldsymbol{\Omega}\cdot\mathbf{n}_i)\,ds, \quad (\boldsymbol{\Omega}\cdot\mathbf{n}_i) > 0 \tag{11.37}$$

and the mean chord length is

$$\bar{l}_s \equiv \int l_s\phi(l_s)\,dl_s = \frac{1}{\pi S}\int l_s\left[\iint_{l'_s=l_s}(\boldsymbol{\Omega}\cdot\mathbf{n}_i)\,d\boldsymbol{\Omega}\,ds\right]_{(\boldsymbol{\Omega}\cdot\mathbf{n}_i)>0}dl_s$$

$$= \frac{1}{\pi S}\iint l_s(\boldsymbol{\Omega}\cdot\mathbf{n}_i)\,d\boldsymbol{\Omega}\,ds = \frac{4V}{S}, \quad (\boldsymbol{\Omega}\cdot\mathbf{n}_i) > 0 \tag{11.38}$$

Hence

$$\frac{4\pi V}{\bar{l}_s}\phi(l_s)\,dl_s = \int\left[\int_{l'_s=l_s}(\boldsymbol{\Omega}\cdot\mathbf{n}_i)\,d\boldsymbol{\Omega}\right]_{(\boldsymbol{\Omega}\cdot\mathbf{n}_i)>0}ds \tag{11.39}$$

Using these results in Eq. (11.31) yields

$$P_0 = \frac{\lambda}{\bar{l}_s}\int\left(1 - e^{-l_s/\lambda}\right)\phi(l_s)\,dl_s = \frac{\lambda}{\bar{l}_s} - \int e^{-l_s/\lambda}\phi(l_s)\,dl_s \tag{11.40}$$

When the dimensions of the fuel rod are small compared to the mean free path, $l_s \ll \lambda$, this reduces to

$$P_0 \simeq 1 - \frac{1}{2} \frac{(\overline{l_s^2})}{\overline{l}_s \lambda} \simeq 1 \tag{11.41}$$

and when the dimensions are large compared to the mean free path, $l_s \gg \lambda$,

$$P_0 \simeq \frac{\lambda}{\overline{l}_s} \tag{11.42}$$

which suggests the rational approximation

$$P_0 = \frac{1}{1 + (\overline{l}_s/\lambda)} = \frac{1}{1 + (4V/\lambda S)} \tag{11.43}$$

The rational approximation is known to underestimate the escape probability.

An improved approximation for a long cylindrical rod has been obtained by integrating an empirical fit for the chord length distribution function:

$$P_0 = \frac{\lambda}{\overline{l}_s} \left[1 - \frac{1}{1 + (\overline{l}_s/4.58\lambda)^{4.58}} \right] \tag{11.44}$$

An improved rational approximation of the form

$$P_0 = \frac{1}{4V/\lambda s} \left[1 - \left(1 + \frac{4V/\lambda s}{c} \right)^{-c} \right] \tag{11.45}$$

with $c = 2.09$, has been determined empirically to be more accurate than the Wigner approximation and more accurate than the Sauer approximation for all geometries other than cylindrical. Note that the approximation of Eq. (11.45) reduces to the Wigner approximation for $c = 1$ and to the Sauer approximation for $c = 4.58$.

Closely Packed Lattices

In a lattice of closely packed fuel elements interspersed in moderating material a neutron escaping from a fuel element without collision may traverse a distance in moderator without collision and enter another fuel element, where it may have a collision with a fuel atom or may pass through uncollided to enter moderator again, and so on. In this situation, the probability of a neutron escaping from the fuel element in which it is scattered to energy $E(P_0)$ is not the same as the probability that the neutron will have its next collision in moderator (P_{F0}), but the two are related. Let $G_m^{(1)}$ be the probability that a neutron escaping from the original fuel element will traverse the line-of-sight distance of moderator separating the original fuel element from other fuel elements without a collision, $G_f^{(2)}$ be the probability that the neutron will traverse the second fuel element without collision to reenter moderator, and so on. Then we can write

$$P_{F0} = P_0\big[\big(1 - G_m^{(1)}\big) + G_m^{(1)}G_f^{(2)}\big(1 - G_m^{(2)}\big)$$
$$+ G_m^{(1)}G_f^{(2)}G_m^{(2)}G_f^{(3)}\big(1 - G_m^{(3)}\big) + \cdots\big] \tag{11.46}$$

The various $G_x^{(n)}$ depend on the lattice geometry and are not the same for successive flights in moderator or fuel (i.e., not equal for $n = 1$ and $n = 2$ or $n = 4$ and $n = 5$). However, if we make the approximation that the individual flight probabilities can be replaced by an average value, Eq. (11.46) can be summed:

$$P_{F0} = \frac{P_0(1 - G_f)(1 - G_m)}{1 - G_f G_m} \tag{11.47}$$

$G_{m,f}$ can be estimated by analogy with Eq. (11.43) or heuristically as

$$G_{m,f} = \frac{1}{1 + \bar{l}_{m,f}/\lambda_{m,f}} \quad \text{or} \quad G_{m,f} = e^{-\bar{l}_{m,f}/\lambda_{m,f}} \tag{11.48}$$

when $\bar{l}_m$ and λ_m are the mean chord length through the moderator between fuel elements and the mean free path of neutrons in the moderator, etc. Such corrections to P_0 are known as Dancoff corrections and allow P_{F0} to be written

$$P_{F0} = P_0(1 - \gamma) \tag{11.49}$$

where the factor γ accounts for the decrease in the probability that a neutron escaping a fuel element will first collide with a moderator nucleus because of the presence of other fuel elements. For fuel rods arranged in a square or hexagonal lattice structure,

$$\gamma = \frac{\exp(-\tau \bar{l}_s N_F \sigma_s^M)}{1 + (1 - \tau)\bar{l}_s N_F \sigma_s^M} \tag{11.50}$$

where

$$\tau = \begin{cases} \left[\frac{\sqrt{\pi}}{2}\big(1 + \frac{V_F}{V_M}\big)^{1/2} - 1\right]\frac{V_F}{V_M} - 0.08 & \text{square} \\ \left[\big(\frac{\pi}{2\sqrt{3}}\big)^{1/2}\big(1 + \frac{V_F}{V_M}\big)^{1/2} - 1\right]\frac{V_F}{V_M} - 0.12 & \text{hexagonal} \end{cases} \tag{11.51}$$

11.4
Unresolved Resonances

Unlike the case of the resonances in the resolved energy region (up to a few hundred eV or less) where parameters for each individual resonance can be evaluated explicitly from the high-resolution data, the evaluations of such parameters become increasingly more difficult as the Doppler and instrument resolution widths become much greater than the corresponding natural width in the relatively high energy range. Under such circumstances, it is not possible to deal with the physical quantities of interest as a function of energy in great detail. Instead, it is necessary to estimate the expectation values of these quantities on the basis of statistical theory. Two types of expectation values of particular interest in reactor applications are the reaction rate of a given process, denoted by $\langle \sigma_x \phi \rangle$, and the average flux, denoted by $\langle \phi \rangle$, in the energy interval where many resonances are present.

Since the NR-approximation described earlier is usually applicable in the unresolved energy range at relatively high energy, the extension of the J-integral approach is quite natural. The expectation values of interest can be expressed in terms of the population averages of an ensemble of resonance integrals with their resonance parameters determined by the known distribution functions from the statistical theory of spectra. In principle, these averages can be determined once the average resonance parameters are specified.

Two types of distributions are needed to characterize the statistical behavior of the resonance parameters. According to Porter–Thomas, the partial widths are theoretically expected to exhibit a chi-squared distribution with ν degrees of freedom about their mean value $\langle \Gamma_x \rangle$:

$$P_\nu(y)\,dy = \frac{\nu}{2\Gamma(\nu/2)} \left(\frac{\nu y}{2} \right)^{(\nu/2)-1} e^{-(\nu y/2)}\,dy \qquad (11.52)$$

where $y = \Gamma_x / \langle \Gamma_x \rangle$ and $F(\nu/2)$ is the gamma function of argument $\nu/2$. The degree of freedom ν is identifiable with the number of open channels for reaction process x.

The level spacing between two adjacent levels for a given spin state, $D = |E_k - E_{k+1}|$, is characterized by the Wigner distribution of the form

$$W(y)\,dy = \frac{\pi}{2} y \exp\left(-\frac{\pi}{4} y^2 \right) dy \qquad (11.53)$$

where $y = D/\langle D \rangle$. Physically, it signifies the tendency of repulsion between the adjacent levels of the same spin sequence. For the integral approach to be described, a level correlation function that specifies the probability of finding any level $E_{k'}$ at a distance $|E_{k'} - E_k|$ away from a given level E_k is also required. It is related to the Wigner distribution via the convolution integral equation, defined as follows:

$$\mathbf{\Omega}(y) = W(y) + \int_0^y W(t)\mathbf{\Omega}(y-t)\,dt \qquad (11.54)$$

For $E_{k'}$ belonging to a different spin sequence with respect to E_k, the levels are statistically independent, and consequently, the correlation function becomes unity.

With the specification of distributions, the averages can be determined once the average parameters are provided. For elastic scattering, in which the neutron width is explicitly energy dependent, one convenient average parameter usually used is the *strength function*. For neutrons with orbital angular momentum l, the strength function is

$$S_l = \frac{\sum_J g_J \langle \Gamma_{nlk}^{(0)} \rangle}{\langle D_k \rangle}, \quad k \in J, l \tag{11.55}$$

where $\langle \Gamma_{nlk}^{(0)} \rangle$ is the average reduced neutron width for given l and k, which is energy independent, and $\langle D_k \rangle$ is the average level spacing for the sequence in which level k occurs, with

$$g_J = \frac{2J + 1}{2(2I + 1)} \tag{11.56}$$

where J is the total spin of the neutron–nuclide system and I is the spin of the target nucleus. Statistical resonance parameters for some of the principal nuclides are given in Table 11.2.

The single-level Breit–Wigner formula, now generalized to include spin effects, is

$$\sigma_{xk} = \frac{\pi \lambda_0^2 g_J \Gamma_{xk} \Gamma_{nk}}{(E - E_k)^2 + (\Gamma_k/2)^2} \tag{11.57}$$

where F_{xk}, Γ_{nk}, and Γ_k are the capture ($x = \gamma$) or fission ($x = f$) width, the neutron width, and the total width, respectively, for a resonance at E_k, and λ_0 is the De Broglie neutron wavelength.

Multigroup Cross Sections for Isolated Resonances

Using a narrow resonance approximation for the neutron flux, $\phi \sim 1/\Sigma_t$, the effective multigroup cross section is

$$\sigma_{xk}^g = \frac{\int_{E_g}^{E_{g-1}} dE \, \sigma_{xk}(E)/\Sigma_t(E)}{\int_{E_g}^{E_{g-1}} dE/\Sigma_t(E)} = \frac{\Sigma_p \Gamma_{xk} J_k}{f \Delta E_g N_{\text{res}}} \tag{11.58}$$

where

$$f = \frac{\Sigma_p}{\Delta E_g} \int_{E_g}^{E_{g-1}} \frac{dE}{\Sigma_t(E)} \tag{11.59}$$

In the unresolved resonance region, statistical averages over the distribution functions of Eqs. (11.52) to (11.54) are used to construct an effective multigroup cross section for process x:

$$\langle \sigma_{xk}^g \rangle = \frac{\sigma_p}{f} \sum_{\substack{\text{spin} \\ \text{states}}} \frac{\langle \Gamma_{xk} J_k \rangle}{\langle D_k \rangle} \tag{11.60}$$

where $\sigma_p = \Sigma_p/N_{\text{res}}$ is the potential scattering cross section per resonance nucleus and $\langle \cdot \rangle$ indicates averages over statistical distributions of both widths and level spacings.

Table 11.2 Statistical Resonance Parameters for ^{235}U, ^{238}U, and ^{239}Pu

^{235}U	^{238}U	^{239}Pu
$S_0 = (0.915 \pm 0.5) \times 10^{-4}$	$S_0 = (0.90 \pm 0.10) \times 10^{-4}$	$S_0 = (1.07 \pm 0.1) \times 10^{-4}$
$(0 < E < 50 \text{ eV})$;		
$S_0^{J=3} \approx S_0^{J=4} \approx S_0$		
$S_1 = (2.0 \pm 0.3) \times 10^{-4}$	$S_1 = (2.5 \pm 0.5) \times 10^{-4}$	$S_1 = (2.5 \pm 0.5) \times 10^{-4}$
$\bar{D}_{\text{obs}}^{l=0} = 0.53 \pm 0.05$ eV	$\bar{D}_{\text{obs}}^{l=0} = D_{J=1/2}^{l=0,1} = 20.8 \pm 2.0$ eV	$\bar{D}_{\text{obs}}^{l=0} = 2.3 \pm 0.2$ eV
$\bar{D}_{J=2}^{l=1} = 1.23$ eV	$\bar{D}_{J=1/2}^{l=1} = 11.4 \pm 1.1$ eV	$\bar{D}_{J=0}^{l=0,1} = 8.78$ eV
$\bar{D}_{J=3}^{l=0,1} = D_{J=4}^{l=0,1} = 1.06$ eV	$\bar{\Gamma}_{\gamma}^{l=0} = 24.8 \pm 5.6$ meV	$\bar{D}_{J=1}^{l=0,1} = 3.12$ eV
$\bar{D}_{J=5}^{l=1} = 1.18$ eV	$\nu_\gamma = 39$	$\bar{D}_{J=2}^{l=1} = 2.12$ eV
$\bar{\Gamma}_{\gamma}^{l=0} = 47.9$ meV	$\sigma_{\text{pot}} = 10.6 \pm 0.2$ barns	$\bar{\Gamma}_{\gamma}^{l=0} = 38.7$ meV
$\Gamma^{l=0} = 65.1$ meV		
$(0 < E < 50 \text{ eV})$	$R = (9.18 \pm 0.13) \times 10^{-13}$ cm	$\bar{\Gamma}_{f\,J=0}^{l=0} = 2800$ meV
$\nu_f = 4$		$\bar{\Gamma}_{f\,J=1}^{l=0} = 57$ meV
		$(0 < E < 100 \text{ eV})$
$\nu_\gamma = 27$		$\nu_f = 2$ for $(1, J) = (0, 0)$
$\sigma_{\text{pot}} = 11.7 \pm 0.1$ barns		$\nu_f = 1$ for $(1, J) = (0, 1)$
$R = (9.65 \pm 0.05) \times 10^{-13}$ cm		$\nu_\gamma = 24$
		$\sigma_{\text{pot}} = 10.3 \pm 0.15$ barns
		$R = (9.05 \pm 0.11) \times 10^{-13}$ cm

Source: Data from Ref. 5; used with permission of American
Nuclear Society.

Self-Overlap Effects

The large Doppler width for high-energy neutrons and the small level spacing produce a high degree of self-overlap among the resonances for fissile isotopes, and significant but less self-overlap for the fertile isotopes. In fast reactor spectra, the self-overlap effect is not important for the fertile isotopes at operating temperatures, but does affect the temperature dependence of the Doppler effect above about 10 keV. The effect of the presence of other resonances on the effective cross section of resonance k arises from their effect on the flux, $\phi \sim 1/\Sigma_t$, and gives rise to a generalization of the J function:

$$J^*(\xi_k, \beta_k) = \frac{1}{2} \int_{-\infty}^{\infty} \frac{\psi_k\, dx_k}{\psi_k + \beta_k + \sum_{k' \neq k}(\sigma_{0k'}/\sigma_{0k})\psi_{k'}} = \frac{1}{2} \int_{-\infty}^{\infty} \frac{\psi_k\, dx_k}{\psi_k + \beta_k}$$

$$- \frac{1}{2} \int_{-\infty}^{\infty} \frac{\psi_k \sum_{k' \neq k}(\sigma_{0k'}/\sigma_{0k})\psi_{k'}\, dx_k}{(\beta_k + \psi_k)[\beta_k + \psi_k + \sum_{k' \neq k}(\sigma_{0k'}/\sigma_{0k})\psi_{k'}]}$$

$$\equiv J_k(\xi_k, \beta_k) + \sum_{k' \neq k} H_{kk'} \tag{11.61}$$

where ψ_k and $\psi_{k'}$ are evaluated for the respective resonance parameters of the resonances at E_k and $E_{k'}$. The evaluation of the second, overlap, term is quite complicated because of the statistical average over resonance parameters and level spacings, and useful approximations have been developed (Refs. 5 and 6).

The multigroup cross section then consists of a term like Eq. (11.58) plus a negative overlap correction term:

$$\langle \sigma_{xk}^g \rangle = \frac{\sigma_p}{f} \sum_{\substack{\text{spin} \\ \text{states}}} \frac{\langle \Gamma_{xk} J_k \rangle}{\langle D_k \rangle} + \frac{\sigma_p}{f} \sum_{k'} \sum_{\substack{\text{spin} \\ \text{states}}} \frac{\langle \Gamma_{xk} H_{kk'} \rangle}{\langle D_k \rangle} \tag{11.62}$$

Overlap Effects for Different Sequences

The spacings of resonances belonging to different J-spin states in the same isotope or to two different isotopes are not correlated. The most important case is the overlap of resonances in a fissile isotope by resonances in a fertile isotope. Neglecting self-overlap, for the moment, the generalized J function for a fissile isotope with a resonance sequence at energies E_k overlapped by a fertile isotope with a resonance sequence at energies E_i is

$$J_k^*(\xi_k, \beta_k) = \frac{1}{2} \int_{-\infty}^{\infty} \frac{\psi_k\, dx_k}{\beta_k + \psi_k + (\sigma_{0i}/\sigma_{0k})\psi_i} \tag{11.63}$$

Separating the generalized J-function into the normal J-function and an overlap term, as in Eq. (11.61), and making some further approximations, it is possible to write the effective multigroup resonance cross section as

$$\langle \sigma_{xk}^g \rangle = \frac{\sigma_p}{f} \sum_{\substack{\text{spin} \\ \text{states}}} \frac{\langle \Gamma_{xk} J_k \rangle}{\langle D_k \rangle} \left(1 - \frac{\langle \Gamma_i J_i \rangle}{\langle D_i \rangle} \right) \tag{11.64}$$

It can be shown that for a single spin state in the fissile isotope, the flux correction factor f of Eq. (11.59) can be written

$$f \simeq \left(1 - \frac{\langle \Gamma_k J_k \rangle}{\langle D_k \rangle} \right) \left(1 - \frac{\langle \Gamma_i J_i \rangle}{\langle D_i \rangle} \right) \tag{11.65}$$

so that the effective multigroup cross section for a fissile isotope overlapped by a fertile isotope can be written

$$\langle \sigma^g_{xk} \rangle = \sigma_p \frac{\langle \Gamma_{xk} J_k \rangle}{\langle D_k \rangle} \Big/ \left(1 - \frac{\langle \Gamma_k J_k \rangle}{\langle D_k \rangle} \right) \tag{11.66}$$

In this approximation, the effect of the resonance overlap is compensated by the corresponding change in flux that it produces, and the parameters of the overlapping fertile isotope sequence do not appear. With respect to Eq. (11.64), the effect of the overlapping sequence i enters via the $1/\Sigma_t$ in both the numerator and denominator and, to first order, these two effects cancel.

Combining the self-overlap and different sequence overlap results, the effective multigroup cross section for a fissile resonance sequence k with self-overlap and with overlap by a fertile isotope resonance sequence i is

$$\langle \sigma^g_{xk} \rangle = \frac{\sigma_p \sum_{\substack{\text{spin} \\ \text{states}}} [(\langle \Gamma_{xk} J_k \rangle / \langle D_k \rangle) + \sum_{k'} \langle \Gamma_{xk} H_{kk'} \rangle / \langle D_k \rangle]}{1 - \sum_{\substack{\text{spin} \\ \text{states}}} \langle \Gamma_k J_k \rangle / \langle D_k \rangle} \tag{11.67}$$

11.5
Multiband Treatment of Spatially Dependent Self-Shielding

Spatially Dependent Self-Shielding

Approximate methods for calculating effective multigroup cross sections for resonance absorbing isotopes have been discussed in Sections 4.3 and 11.2. It was found that the approximate flux to be used in evaluating the resonance integral was of the form $\phi(E) \sim f_{ss}(\Sigma_t(E)) \times M(E)$, where $M(E)$ is a spectral function with an energy dependence that would exist even in the absence of the resonance absorber and f_{ss} is a self-shielding factor that depends on the energy via the dependence of the total cross section on energy [e.g., $f_{ss} \sim 1/(\Sigma^{res}_t(E) + \Sigma^M_s)$ and $M(E) \sim 1/E$ in the narrow resonance approximation for a homogeneous mixture given by Eq. (4.65)]. This same general form persists in approximate treatments of heterogeneous resonance absorbers, as may be seen from Eqs. (11.6) and (11.10).

In the approximate treatment of heterogeneous resonance absorbers discussed in Section 11.2, the self-shielding factor, f_{ss}, and hence also the resulting multigroup cross section, was implicitly assumed to be spatially independent within the resonance absorber. However, simple physical considerations suggest that the self-shielding will be much more pronounced deep within a resonance absorber than on its surface, where the neutron spectrum is dominated by neutrons entering from the adjacent moderator and, furthermore, that the self-shielding near the surface will be different for neutrons entering from the moderator than for neutrons coming from deeper within the resonance absorber. Thus, even if accurate spatially constant multigroup cross sections that preserve volume-averaged reaction rates are obtained for a heterogeneous resonance absorber (e.g., a fuel pin), the spatial dependence of reaction rates within the resonance absorber will not be calculated properly, which will introduce an error into calculations of fuel depletion, fission heating distribution, and so on. Even if the spatial multigroup flux distributions within the resonance absorber are calculated with multigroup transport theory, there will remain an inaccuracy in calculating the spatial distribution of reaction rates because these spatially independent volume-averaged multigroup cross sections were used instead of spatially dependent cross sections which take into account the spatial dependence of the self-shielding.

The most straightforward way to solve this problem of spatially dependent within-group self-shielding might seem to be to further subdivide the normal multigroup structure (e.g., 20 to 50 groups) that would be used in a pin-cell transport calculation (see Section 14.4) into ultrafine groups in the resonance energy region. If the ultrafine groups could be sufficiently numerous that the within-group self-shielding term was almost unity (i.e., such that the variation of the cross section within any ultrafine group was small), the ultrafine-group cross sections could be accurately calculated as described previously, and the spatial self-shielding effect on the normal multigroup level would be calculated on the ultrafine-group level. However, this procedure is impractical except for special cases because each ultrafine-group width would have to be narrow compared with the width of the resonances at that energy, resulting in an enormous number of ultrafine groups to span the resonance energy region. This approach would not work at all for the unresolved resonances, of course.

Multiband Theory

In the multiband method, each normal group is further subdivided, not into finer energy intervals as in the ultrafine-group method discussed above, but into intervals of the total cross section magnitude which span the variation in the total cross section within the normal group. The multiband equations are derived by an extension to the derivation of the multigroup equations. Starting with the energy-dependent transport equation (with scattering and fission included in a general transfer function Σ_s),

$$\mathbf{\Omega} \cdot \nabla \Psi(\mathbf{r}, E, \mathbf{\Omega}) + \Sigma_t(\mathbf{r}, E)\Psi(\mathbf{r}, E, \mathbf{\Omega})$$
$$= \int_0^\infty dE' \int_0^{4\pi} d\mathbf{\Omega}' \Sigma_s\left(\mathbf{r}, E' \to E, \mathbf{\Omega}' \to \mathbf{\Omega}\right)\Psi\left(\mathbf{r}, E', \mathbf{\Omega}'\right) \quad (11.68)$$

the normal multigroup equations are formally derived by integrating over the energy interval $E_g \leq E \leq E_{g-1}$:

$$\mathbf{\Omega} \cdot \nabla \Psi_g(\mathbf{r}, \mathbf{\Omega}) + \Sigma_t^g(\mathbf{r}, \mathbf{\Omega})\Psi_g(\mathbf{r}, \mathbf{\Omega})$$
$$= \sum_{g'=1}^G \int_0^{4\pi} d\mathbf{\Omega}' \Sigma_s^{g' \to g}\left(\mathbf{r}, \mathbf{\Omega}' \to \mathbf{\Omega}\right)\Psi_{g'}(\mathbf{r}, \mathbf{\Omega}), \quad g = 1, \ldots, G \quad (11.69)$$

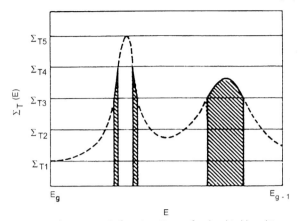

Fig. 11.6 The Heaviside function $H_{gb=3}$ for the third band in a four-band representation of the total cross section ($H_{sb=3} = 1.0$ in dark energy intervals and $= 0.0$ elsewhere). (From Ref. 11; used with permission of CRC Press.)

where

$$\Psi_g(\mathbf{r}, \mathbf{\Omega}) \equiv \int_{E_g}^{E_{g-1}} dE\, \Psi_g(\mathbf{r}, E, \mathbf{\Omega})$$

$$\Sigma_t^g(\mathbf{r}, \mathbf{\Omega}) \equiv \int_{E_g}^{E_{g-1}} dE\, \Sigma_t(\mathbf{r}, E)\Psi(\mathbf{r}, E, \mathbf{\Omega})/\Psi_g(\mathbf{r}, \mathbf{\Omega})$$

$$\Sigma_s^{g' \to g}\left(\mathbf{r}, \mathbf{\Omega}' \to \mathbf{\Omega}\right)$$
$$= \int_{E_g}^{E_{g-1}} dE \int_{E_{g'}}^{E_{g'-1}} dE'\, \Sigma_s\left(\mathbf{r}, E' \to E, \mathbf{\Omega}' \to \mathbf{\Omega}\right)\Psi\left(\mathbf{r}, E', \mathbf{\Omega}'\right)/\Psi_{g'}\left(\mathbf{r}, \mathbf{\Omega}'\right),$$
$$g, g' = 1, \ldots, G \tag{11.70}$$

The multiband equations are formally derived by a similar process, but now with each group (g) energy interval subdivided into B cross-section bands (g, b), which span the range of total cross section in group g, as depicted in Fig. 11.6. Defining a Heaviside function $H_{gb}(E)$ which is unity for those energy intervals for which the total cross section is within the band $\Sigma_{tb+1} \geq \Sigma_t(E) \geq \Sigma_{tb}$ and zero elsewhere, the multiband equations are derived by first multiplying Eq. (11.68) by $H_{gb}(E)$ and then integrating over both the energy interval $E_g \leq E \leq E_{g-1}$ of group g and over the total cross-section range $\Sigma_{tb+1} \geq \Sigma_t(E) \geq \Sigma_{tb}$ of band b:

$$\mathbf{\Omega} \cdot \nabla \Psi_{gb}(\mathbf{r}, \mathbf{\Omega}) + \Sigma_t^{gb}(\mathbf{r}, \mathbf{\Omega})\Psi_{gb}(\mathbf{r}, \mathbf{\Omega})$$
$$= \sum_{g'=1}^{G} \sum_{b'=1}^{B} \int_0^{4\pi} d\mathbf{\Omega}' \Sigma_s^{g'b' \to gb}\left(\mathbf{r}, \mathbf{\Omega}' \to \mathbf{\Omega}\right)\Psi_{g'b'}\left(\mathbf{r}, \mathbf{\Omega}'\right),$$
$$g, g' = 1, \ldots, G; \; b, b' = 1, \ldots, B \tag{11.71}$$

where the multiband parameters are given by

$$\Psi_{gb}(\mathbf{r}, \mathbf{\Omega}) \equiv \int_{E_g}^{E_{g-1}} dE \int_{\Sigma_{tgb}}^{\Sigma_{tgb+1}} d\Sigma_t^*(E) H_{gb}(E) \Psi(\mathbf{r}, E, \mathbf{\Omega})$$

$$\Sigma_t^{gb}(\mathbf{r}, \mathbf{\Omega}) \equiv \int_{E_g}^{E_{g-1}} dE \int_{\Sigma_{tgb}}^{\Sigma_{tgb+1}} d\Sigma_t^*(E) H_{gb}(E) \Sigma_t(\mathbf{r}, E) \Psi(\mathbf{r}, E, \mathbf{\Omega}) / \Psi_{gb}(\mathbf{r}, \mathbf{\Omega})$$

$$\Sigma_s^{g'b' \rightarrow gb}(\mathbf{r}, \mathbf{\Omega}' \rightarrow \mathbf{\Omega}) \equiv \int_{E_g}^{E_{g-1}} dE \int_{E_{g'}}^{E_{g'-1}} dE' \int_{\Sigma_{tb}}^{\Sigma_{tb+1}} d\Sigma_t^*(E) \qquad (11.72)$$

$$\times \int_{\Sigma_{tb'}}^{\Sigma_{tb'+1}} d\Sigma_t^*(E') H_{gb}(E) H_{g'b'}(E')$$

$$\times \Sigma_s(\mathbf{r}, E' \rightarrow E, \mathbf{\Omega}' \rightarrow \mathbf{\Omega}) \Psi(\mathbf{r}, E', \mathbf{\Omega}') / \Psi_{g'b'}(\mathbf{r}, \mathbf{\Omega}')$$

with the quantity Σ_t^* normalized such that

$$\sum_{b=1}^{B} \int_{E_{tb}}^{E_{tb+1}} d\Sigma_t^*(E) = 1 \qquad (11.73)$$

The normal multigroup quantities are related to the multiband quantities within the different groups as

$$\Psi_g(\mathbf{r}, \mathbf{\Omega}) = \sum_{b=1}^{B} \Psi_{gb}(\mathbf{r}, \mathbf{\Omega})$$

$$\Sigma_t^g(\mathbf{r}, \mathbf{\Omega}) = \sum_{b=1}^{B} \Sigma_t^{gb}(\mathbf{r}) \Psi_{gb}(\mathbf{r}, \mathbf{\Omega}) \Bigg/ \sum_{b=1}^{B} \Psi_{gb}(\mathbf{r}, \mathbf{\Omega}) \qquad (11.74)$$

Evaluation of Multiband Parameters

Direct evaluation of the multiband parameters from the relationships above (actually, from the relationships that result when some discrete representation of the angular dependence is invoked) is possible in principle, but these relationships may be recast into a form that can make use of existing self-shielded multigroup libraries. The definition of the normal multigroup cross section for process x as an integral over energy can be exactly transformed into an integral over total cross section:

$$\Sigma_x^g \equiv \frac{\int_{E_g}^{E_{g-1}} dE \, \Sigma_x(E) \Psi(E)}{\int_{E_g}^{E_{g-1}} dE \, \Psi(E)} = \frac{\int_{E_g}^{E_{g-1}} dE \, \Sigma_x(E) M(E) f_{ss}(E)}{\int_{E_g}^{E_{g-1}} dE \, M(E) f_{ss}(E)}$$

$$= \frac{\int_{E_g}^{E_{g-1}} dE \sum_{b=1}^{B} \int_{\Sigma_{tb}}^{\Sigma_{tb+1}} d\Sigma_t^*(E) H_{gb}(E) \Sigma_x(E) M(E) f_{ss}(E)}{\int_{E_g}^{E_{g-1}} dE \sum_{b=1}^{B} \int_{\Sigma_{tb}}^{\Sigma_{tb+1}} d\Sigma_t^*(E) H_{gb}(E) M(E) f_{ss}(E)} \qquad (11.75)$$

where use has been made of the approximate relationship $\phi(E) \sim M(E) f_{ss}(E)$ discussed previously. Performing the integration over energy first and defining

$$p\left(\Sigma_t^*\right) \equiv \frac{\int_{E_g}^{E_{g-1}} dE\, H_{gb}(E)M(E)}{\int_{E_g}^{E_{g-1}} dE\, M(E)} \tag{11.76}$$

$$\Sigma_x\left(\Sigma_t^*\right) \equiv \frac{\int_{E_g}^{E_{g-1}} dE\, H_{gb}(E)M(E)}{\int_{E_g}^{E_{g-1}} dE\, M(E)H_{gb}(E)} \tag{11.77}$$

leads to the equivalent definition of the normal multigroup cross section:

$$\Sigma_x^g = \frac{\sum_{b=1}^B \int_{\Sigma_{tb}}^{\Sigma_{tb+1}} d\Sigma_t^*\, \Sigma_x(\Sigma_t^*) f_{ss}(\Sigma_t^*) p(\Sigma_t^*)}{\sum_{b=1}^B \int_{\Sigma_{tb}}^{\Sigma_{tb+1}} d\Sigma_t^*\, f_{ss}(\Sigma_t^*) p(\Sigma_t^*)} \tag{11.78}$$

in terms of the total cross-section probability distribution function $p(\Sigma_t^*)$, defined such that $p(\Sigma_t^*)d\Sigma_t^*$ is just the normalized probability of the total cross section being within $d\Sigma_t^*$ of Σ_t^* within the energy interval $E_g \leq E \leq E_{g-1}$.

For the practical evaluation of Eq. (11.78), average values of the cross sections Σ_{xb} for process x in each band b are used to replace the integrals with quadratures:

$$\Sigma_x^g = \frac{\sum_{b=1}^B \Sigma_x^{gb} f_{ss}(\Sigma_{tb}^*) P_b}{\sum_{b=1}^B f_{ss}(\Sigma_{tb}^*) P_b} \tag{11.79}$$

where

$$P_b \equiv \int_{\Sigma_{tb}}^{\Sigma_{tb+1}} p\left(\Sigma_t^*\right) d\Sigma_t^* \tag{11.80}$$

is the band weight. The computational advantage of this approach relative to a direct quadrature approximation of the second form of Eq. (11.75) is that $\Sigma_x(\Sigma_t^*)$ is generally a much smoother function than is $\Sigma_x(E)$ over the resonance energy range, so it is much easier to define an appropriate quadrature. Once the total cross-section probability distribution is evaluated, integrals involving this distribution may be performed quite accurately and efficiently.

Calculation of Multiband Parameters

Although it would be most straightforward to choose the band structure (Σ_t) a priori and just evaluate the P_b and the various Σ_{tb}, it is more common to use a moments method to calculate these multiband parameters to reproduce the results obtained using certain limiting forms for the self-shielding. Using a generalized self-shielding factor of the form

$$f_{ss}(E) = \frac{1}{[\Sigma_t(E) + \Sigma_0]^n} \tag{11.81}$$

the band parameters can be calculated by requiring that the multiband expression agrees with the known results for various values of Σ_0 and n. As an example, for two bands, there are two weighting parameters (P_1, P_2) and for each reaction process x two group-band cross sections $(\Sigma_x^{g1}, \Sigma_x^{g2})$ in each group. A normal multigroup processing code will provide the unshielded $[f_{ss} = 1 = 1/(\Sigma_t + \Sigma_0)^0]$, the totally self-shielded flux-weighted $[f_{ss} = 1/(\Sigma_t + \Sigma_0)^1]$, and sometimes the totally self-shielded current-weighted $[f_{ss} = 1/(\Sigma_t + \Sigma_0)^2]$ values of the various cross sections in group g, $(\Sigma_x^g)_0$, $(\Sigma_x^g)_1$, and $(\Sigma_x^g)_2$, respectively. Requiring that Eq. (11.79) yield these three values of the total cross section and realizing that $P_1 + P_2 = 1$ yields four equations from which the band parameters can be calculated. It is necessary to introduce the ordering $\Sigma_{t1} < \Sigma_{t2}$ in order to obtain a unique solution, since the two bands are otherwise indistinguishable. The solutions are

$$P_{1/2} = \frac{1}{2} \pm \delta, \quad \Sigma_{t1/2} = \frac{1}{A \pm B} \tag{11.82}$$

where

$$\delta \equiv \frac{1 - A(\Sigma_t^g)_1}{2B(\Sigma_t^g)_1}$$

$$A \equiv \frac{1}{2(\Sigma_t^g)_2} \frac{(\Sigma_t^g)_0 - (\Sigma_t^g)_2}{(\Sigma_t^g)_0 - (\Sigma_t^g)_1} \tag{11.83}$$

$$B^2 \equiv \frac{1}{(\Sigma_t^g)_0 (\Sigma_t^g)_1} \left[1 - 2A(\Sigma_t^g)_1 + (\Sigma_t^g)_0 (\Sigma_t^g)_1 A^2 \right]$$

Having thus determined $(P_1, P_2, \Sigma_t^{g1}, \text{ and } \Sigma_t^{g2})$, the group-band cross sections for the individual processes x, $(\Sigma_x^{g1}, \Sigma_x^{g2})$ can then be determined by requiring that Eq. (11.79) yield the unshielded $[f_{ss} = 1 = 1/(\Sigma_t + \Sigma_0)^0]$ and the totally shielded flux-weighted $[f_{ss} = 1/(\Sigma_t + \Sigma_0)^1]$ cross sections $(\Sigma_x)_0$ and $(\Sigma_x^g)_1$, respectively, which yields

$$\Sigma_x^{g2/1} = (\Sigma_x^g)_0 \pm \frac{C}{P_1}, \quad C = \left[(\Sigma_x^g)_0 - (\Sigma_x^g)_1 \right] \frac{P_1/\Sigma_t^{g1} + P_2/\Sigma_t^{g2}}{1/\Sigma_t^{g1} - 1/\Sigma_t^{g2}} \tag{11.84}$$

This general procedure may be extended to more bands. In practice, it has been found that two to four bands are sufficient.

The scattering transfer rate from group g' to group g in the normal multigroup theory depends on the scattering cross section in group g', $\Sigma_s^{g'}$, and on the transfer probability, $T^{g' \to g}$, that a neutron scattered in group g' will have final energy in group g (i.e., $\Sigma_s^{g' \to g} = \Sigma_s^{g'} T^{g' \to g}$). The transfer probability does not depend on the scattering cross section in either group g' or group g. The usual procedure for constructing the group band $g'b'$ to group band gb scattering transfer rate is to replace $\Sigma_s^{g'}$ with $\Sigma_s^{g'b'}$ and to assume that the transfer probability from group band $g'b'$ to group band gb is just the group g' to group g transfer probability times the weight P_b^g of band b in group g (i.e., $\Sigma_s^{g'b' \to gb} = \Sigma_s^{g'b'} T^{g' \to g} P_b^g$). Various extensions of this definition of the group-band scattering transfer probability have been suggested, but its calculation remains heuristic.

Interface Conditions

The interface–boundary conditions for multiband transport theory remain somewhat heuristic as well. Clearly, the continuity of directional group-flux condition requires that

$$\sum_{b=1}^{B} \Psi_{gb}(-\mathbf{s}, \boldsymbol{\Omega}) \equiv \Psi_g(-\mathbf{s}, \boldsymbol{\Omega}) = \Psi_g(\mathbf{s}, \boldsymbol{\Omega}) \equiv \sum_{b=1}^{B} \Psi_{gb}(\mathbf{s}, \boldsymbol{\Omega}) \qquad (11.85)$$

where $\pm \mathbf{s}$ refers to the $+$ and $-$ sides of the interface at $\mathbf{r} = \mathbf{s}$. The argument that the cross sections are not correlated across an interface between dissimilar media is used to justify the distribution of directional group fluxes crossing an interface from $-$ to $+$ side according to the weights on the $+$ side:

$$\Psi_{gb}(+\mathbf{s}, \boldsymbol{\Omega}) = P_b^{g+} \Psi_g(-\mathbf{s}, \boldsymbol{\Omega}) = P_b^{g+} \sum_{b'=1}^{B} \Psi_{gb'}(-\mathbf{s}, \boldsymbol{\Omega}) \qquad (11.86)$$

11.6
Resonance Cross-Section Representations*

R-Matrix Representation

The quantum mechanical representation of reaction cross sections is given most generally by $\boldsymbol{R}$-matrix theory, in which the reaction cross section for any incident channel c and exit channel c' is generally expressed in terms of the collision matrix $U_{cc'}$:

$$\sigma_{cc'} = \pi \lambda^2 g_c |\delta_{cc'} - U_{cc'}|^2 \qquad (11.87)$$

where g_c and $\delta_{cc'}$ are the statistical factor and the Kronecker delta, respectively. The unitary property of $U_{cc'}$ leads to the expression of the total cross section as a linear function of $U_{cc'}$:

$$\sigma_t = \sum_{c'} \sigma_{cc'} = 2\pi \lambda^2 g_c (l - \mathrm{Re}\{U_{cc}\}) \qquad (11.88)$$

The collision matrix, in turn, can be expressed in terms of the resonance parameter matrix $\boldsymbol{R}$ according to Wigner and Eisenbud:

$$U_{cc'} = \exp[-i(\phi_c + \phi_{c'})]\{\delta_{cc'} + i\boldsymbol{P}_c^{1/2}[(\boldsymbol{I} - \boldsymbol{R}\boldsymbol{L}^0)^{-1}\boldsymbol{R}]_{cc'}\boldsymbol{P}_{c'}^{1/2}\} \qquad (11.89)$$

---

* This section was prepared with the extensive collaboration of R. N. Hwang.

Table 11.3 Momentum-Dependent Factors for Various *l*-States Defined at Channel Radius r_c ($\rho = kr_c$)

Factors	$l = 0$	$l = 1$	$l = 2$	$l = 3$
P_l	ρ	$\dfrac{\rho^3}{1+\rho^2}$	$\dfrac{\rho^5}{9+3\rho^2+\rho^4}$	$\dfrac{\rho^7}{225+45\rho^2+6\rho^4+\rho^6}$
S_l	0	$\dfrac{-1}{1+\rho^2}$	$\dfrac{-(18+3\rho^2)}{9+3\rho^2+\rho^4}$	$\dfrac{-(675+90\rho^2+6\rho^4)}{225+45\rho^2+6\rho^4+\rho^6}$
ϕ_l	ρ	$\rho - \tan^{-1}\rho$	$\rho - \tan^{-1}\dfrac{3\rho}{3-\rho^2}$	$\rho - \tan^{-1}\dfrac{\rho(15-\rho^2)}{15-6\rho^2}$

Source: Data from Ref. 12; used with permission of Institute for Nuclear Research and Nuclear Energy, Sofia.

where

$$R_{cc'} = \sum_{\lambda} \frac{\gamma_{\lambda c}\gamma_{\lambda c'}}{E_\lambda - E} \tag{11.90}$$

is a real symmetric matrix and

$$L^0_{cc'} = (S_c - B_c + i P_c)\delta_{cc'} \tag{11.91}$$

The energy-independent parameters E_λ, $\gamma_{\lambda c}$, and B_c denote the **R**-matrix state, reduced width amplitude and arbitrary boundary parameters, respectively. Of all parameters given above, ϕ_c, S_c, and P_c are momentum dependent for the elastic scattering channels only. ϕ_c, the hard-sphere phase shift factor, is related directly to the argument of the outgoing wavefunction at the channel radius, whereas S_c, the shift factor, and P_c, the penetration factor, reflect the real and imaginary parts of its logarithmic derivative, respectively, as defined in Table 11.3. These quantities, along with the matrix **R**, specify the explicit energy dependence of the cross section. It should be noted that the matrix **R** is primarily responsible for the sharp rise in cross section at the energy near the resonance energy, E_λ, while the other energy-dependent quantities are relatively smooth by comparison. All energy-independent quantities given above are, in principle, determined by the fitting of the experimental data.

An alternative expression for the collision matrix is the equivalent level matrix representation derived by Wigner, which can provide more analytical insight for the discussions to follow. It is given as

$$U_{cc'} = e^{-i(\phi_c + \phi_{c'})}\left(\delta_{cc'} + i\sum_{\lambda,\mu}\Gamma_{\lambda c}^{1/2}A_{\lambda\mu}\Gamma_{\mu c'}^{1/2}\right) \tag{11.92}$$

where the level matrix A is defined as

$$\left(A^{-1}\right)_{\lambda c} = (E_{\lambda} - E)\delta_{\lambda\mu} - \sum_{c} \gamma_{\lambda c} L_c^0 \gamma_{\mu c}, \qquad \Gamma_{\lambda c}^{1/2} = \sqrt{2P_c}\gamma_{\lambda c} \qquad (11.93)$$

This expression provides a clearer picture of the explicit energy dependence of the collision matrix.

Practical Formulations

Although the formal R-matrix representation is rigorous on theoretical grounds, it is quite obvious that simplifications are required before its deployment as the basis for nuclear data evaluations and subsequent use in reactor applications. In the current ENDF/B format, four major formalisms pertinent to the treatment of the resonance absorption are allowed: the single-level Breit–Wigner (SLBW), multilevel Breit–Wigner (MLBW), Adler–Adler (AA), and Reich–Moore (RM) formalisms. These formalisms are based on approximations of the formal R-matrix theory to various degrees of sophistication.

With exception of the Reich–Moore formalism, all these formalisms exhibit a similar form as a function of energy and can be considered as the consequences of various approximations of the Wigner level matrix. It is convenient to cast them into the pole expansion form in either the energy or momentum domain (k-plane). A simple generic form that is widely used in reactor physics is

$$\sigma_x = \frac{1}{E} \sum_{l,J} \sum_{\lambda} \mathrm{Re}\left\{\rho_{l,J,\lambda}^{(x)} \frac{-i\sqrt{E}}{d_{\lambda} - E}\right\} \quad \text{(energy domain)}$$

$$= \frac{1}{E} \sum_{l,J} \sum_{\lambda} \mathrm{Re}\left\{\frac{\rho_{l,J,\lambda}^{(x)}}{2} \left(\frac{-i}{\sqrt{d_{\lambda}} - \sqrt{E}} - \frac{-i}{\sqrt{d_{\lambda}} + \sqrt{E}}\right)\right\}$$

(momentum domain) $\qquad\qquad\qquad\qquad (11.94)$

where the superscript x denotes the type of reaction under consideration. The superscripts f, γ, and R will be used to denote fission, capture, and compound nucleus (or total resonance) cross sections, respectively. Physically, each term retains the general features of a Breit–Wigner resonance upon which the traditional resonance integral concept was based. The relationships between these pole and residue parameters and the traditional resonance parameters for three of the major formalism are tabulated in Table 11.4. The use of complex arithmetic here makes possible a direct comparison of these traditional formalisms to the rigorous pole representation to be discussed later. $\rho_{l,J,\lambda}^{(x)}$, however, are different and depend on the approximations assumed.

Single-Level Breit–Wigner Approximation (SLBW). The SLBW approximation represents the limiting case when the resonances are well separated from each other. Thus the level matrix A at a given E can be viewed as a matrix with only one element. In much of the previous discussion, the resonance integrals were also

Table 11.4 Poles and Residues for Traditional Formalism

Formalism	Poles, d_λ	Residues, $\rho^{(x)}_{l,j,\lambda}$
SLBW	$E_{0\lambda} - \dfrac{i\Gamma_{l\lambda}}{2}$	$C_{gJ}\dfrac{\Gamma_{x\lambda}\Gamma_{n\lambda}}{\Gamma_{l\lambda}/2}\dfrac{1}{\sqrt{E}}; x \in \{f,\gamma\}$
		$C_{gJ}2\Gamma_{n\lambda}\dfrac{1}{\sqrt{E}}\exp(-i2\phi_l); x \in R$
MLBW	Same as above	Same as above if $x \in \{f,\gamma\}$
		$C_{gJ}2\Gamma_{n\lambda}\dfrac{1}{\sqrt{E}}[\exp(-i2\phi_l) + W_{\lambda'}]; x \in R$
		where
		$W_{\lambda'} = \displaystyle\sum_{\lambda' \neq \lambda}\dfrac{i\Gamma_{n\lambda'}}{(E_\lambda - E_{\lambda'}) + i[(\Gamma_{t\lambda} + \Gamma_{t\lambda'})/2]}$
Adler–Adler	$\mu_\lambda - iv_\lambda$	$C_{gJ}[G^{(x)}_\lambda + iH^{(x)}_\lambda]; x \in \{f,\gamma\}$
		$C_{gJ}\exp(-i2\phi_l)[G^{(x)}_\lambda + iH^{(x)}]; x \in R$

Source: Data from Ref. 12; used with permission of Institute for Nuclear Research and Nuclear Energy, Sofia.

treated in this approximation. In reality, the resonance cross sections clearly cannot be taken as a disjoint set of isolated resonances in a rigorous treatment. Ambiguity can arise as to what constitutes the macroscopic cross sections in a detailed treatment of the neutron slowing-down problem over an energy span consisting of many resonances of more than one nuclide. For this reason, the single-level description used in practical applications such as that specified in the ENDF/B manual is often given in the context of Eq. (11.92) as a linear combination of Breit–Wigner terms supplemented by the tabulated pointwise smooth data so that the continuous nature of the cross sections, and thus the flux, can be preserved.

Multilevel Breit–Wigner Approximation (MLBW). The MLBW approximation corresponds to the situation in which the inverse of the level matrix is taken to be diagonal. One constraint for SLBW and MLBW approximations of practical interest is that all parameters must be positive. It is worth noting that poles and residues are energy dependent, although in many applications they are taken to be energy independent. Otherwise, additional terms in the $\sqrt{E}$-domain would result in all $l > 0$ sequences using SLBW and MLBW formalism.

It should be noted that, strictly speaking, the amplitude $\rho^{(x)}_{l,J,\lambda}$ and the pole d_λ are also energy dependent for the single-level Breit–Wigner and multilevel Breit–Wigner approximations if the explicit energy dependence of the penetration factor and the level-shift factor are considered for all $l > 1$ states. In reference to the

ENDF/B manual, the amplitude of an individual resonance is proportional to the penetration factor, while the real part of d_λ is identifiable as

$$d_\lambda = E'_\lambda = E_\lambda + \frac{S_l(|E_\lambda|) - S_l(E)}{2 P_l(|E_\lambda|)} \Gamma_{n\lambda}(|E_\lambda|)$$

The latter is equivalent to assuming that the boundary parameter is set to be $B_l = S_l(|E_\lambda|)$. Thus the rational function nature of $P_l(\rho)$ and $S_l(\rho)$ defined in Table 11.3 can lead to $2(l + 1)$ pole terms for each resonance in the momentum domain, instead of two terms. The absolute values of the additional $2l$ poles are generally large compared to those two, resulting from the line-shape function directly, so as to reflect the relatively smooth nature of the energy dependence of the penetration factor and the level shift factor. The inclusion of these secondary energy effects can readily be added within the context of the generalized pole representation to be described (Ref. 13).

Adler–Adler Approximation (A–A). The diagonalization of the inverse level matrix $\mathbf{A}^{-1}$ leads directly to the pole expansion defined by Eq. (11.92). The Adler–Adler approximation is equivalent to the Kapur–Peierls representation, in which the poles and residues are assumed to be energy independent. In the context of the forgoing discussion, it is equivalent to assuming the energy independence of L^0 of Eq. (11.91) when the inverse of $\mathbf{A}^{-1}$ is considered. The approximation is usually restricted to the s-wave sequences of the fissionable isotopes in the low-energy region, where the assumption is valid.

Reich–Moore Formalism. For practical applications, the formal $\mathbf{R}$-matrix representation is obviously difficult to use when many levels and channels are present. The problem has been simplified significantly by the method proposed by Reich and Moore. The only significant assumption required, in principle, is

$$\sum_{c \in \gamma} \gamma_{\lambda c} L^0_{cc'} \gamma_{\mu c'} = \delta_{\lambda\mu} \sum_{c \in \gamma} \gamma^2_{\lambda c} L^0_c \qquad (11.95)$$

which utilizes the presence of the large number of capture channels and the random sign of $\gamma_{\lambda c}$. It is consistent with the observed fact that the total capture width distribution is generally very narrow. If one partitions the collision matrix into a 2×2 block matrix arranged such that the upper and lower diagonal blocks consist of only noncapture and capture channels, respectively, and utilizes Eq. (11.95) as well as Wigner's identity between the channel matrix and the level matrix, the collision matrix can be reduced to the order of $m \times m$ where m is the total number of noncapture channels. The reduced collision matrix remains in the same form except that the real matrix $\mathbf{R}$ is replaced by a complex matrix $\mathbf{R}'$ with elements

$$R'_{cc'} = \sum_\lambda \frac{\gamma_{\lambda c} \gamma_{\lambda c'}}{E_\lambda - E - i\Gamma_{\lambda\gamma}/2} \qquad (11.96)$$

The substitution of the reduced $\boldsymbol{R}$-matrix into the original equation leads to the following general form of collision matrix expression for the Reich–Moore approximation in terms of the familiar notations commonly used in applications:

$$\boldsymbol{U}(E) = e^{-i\varphi}(\boldsymbol{I} + 2\boldsymbol{Y})e^{-i\varphi} \tag{11.97}$$

where

$$\boldsymbol{Y} = i\boldsymbol{P}^{1/2}(\boldsymbol{I} - \boldsymbol{R}\hat{\boldsymbol{L}})^{-1}\boldsymbol{R}\boldsymbol{P}^{1/2} = \boldsymbol{F}^{-1}\big[(\boldsymbol{I} - i\boldsymbol{F}\boldsymbol{K})^{-1} - \boldsymbol{I}\big] \tag{11.98}$$

$$\boldsymbol{K} = (\boldsymbol{L}^0)^{1/2}\boldsymbol{R}(\boldsymbol{L}^0)^{1/2}, \qquad \boldsymbol{F} = \boldsymbol{I} - i\hat{\boldsymbol{S}}\boldsymbol{P}^{-1} \tag{11.99}$$

and $\hat{\boldsymbol{S}} = \boldsymbol{S}(E) - \boldsymbol{B}$.

It should be noted that the traditional Reich–Moore representation currently specified by the ENDF/B manual was originally developed for applications in the relatively low energy regions. It is different from the general form given above because two additional assumptions were introduced. First, $\hat{\boldsymbol{S}}$ is taken to be zero. The rationale is based on the fact that $\lim S_l(E) = -l$, implied by the rational functions listed in Table 11.3. Thus, by taking $\lim_{E \to 0} B_c = -l$, the quantity $\hat{S}_l = 0$ and the level shift factor will not play a role in the low-energy region. Second, one elastic scattering channel is allowed in the channel matrix $\boldsymbol{K}$. Although the assumption simplifies the computation required, the issue may still arise for nuclides with odd atomic weight, for which the multiple elastic channels still may play a role. All evaluated resonance data given in the ENDF/B-VI to date are based on these assumptions.

One consequence of the Reich–Moore approximation is that the reduced collision matrix is no longer unitary because $R^l_{cc'}$ is complex. For practical applications, this presents no problem since the total cross section can be preserved if the capture cross section is defined as

$$\sigma_\gamma = \sigma_t - \sum_{c' \notin \gamma} \sigma_{cc'} \tag{11.100}$$

All parameters retain the physical as well as the statistical properties specified by the formal $\boldsymbol{R}$-matrix theory. The order of the channel matrix is usually no greater than 3×3. Hence the method is attractive for data evaluations, and in fact, Reich–Moore parameters have become widely available in the new ENDF/B-VI data.

However, unlike the other three formalisms, resonances defined by the Reich–Moore formalism can no longer be perceived in the context upon which the traditional resonance theory in reactor physics was based. The direct application of this formalism to reactor calculations not only requires the entry of excessive files of precomputed, numerically Doppler-broadened pointwise cross sections at various temperatures, but also renders useless many well-established methods based on the resonance integral concept. Hence there is strong motivation to seek remedies so that the newly released Reich–Moore parameters can be fully utilized within the framework of the existing methodologies.

Generalization of the Pole Representation

Although any given set of R-matrix parameters, including those in the Reich–Moore form, can be numerically converted into parameters of the Kapur–Peierls type, the parameters so obtained are implicitly energy dependent. With the exception of low-lying resonances of a few fissionable isotopes, such dependence is generally not negligible. Thus, from the practical point of view, the traditional pole expansion is not useful for most nuclides of interest. However, a desirable representation directly compatible with the traditional forms given by Eq. (11.94) can be derived if the pole expansion is cast into a somewhat different form.

Rigorous Pole Representation. One attractive means to preserve the rigor of the R-matrix description of cross sections is to perform the pole expansion in the k-plane (or momentum domain). Such a representation is natural for the SLBW, MLBW, and Adler–Adler approximations. The theoretical justification of such a representation is based on the rationale that the collision matrix must be single valued and meromorphic in the momentum domain. Any function that exhibits such properties must be a rational function according to a well-known theorem in complex analysis. The rational function characteristics are quite apparent if one examines the explicit $\sqrt{E}$-dependence of the collision matrix $U_{cc'}$ defined by Eq. (11.89), if the level matrix is expressed as the ratio of the cofactor and the determinant of its inverse. By substituting S_l and P_l into Eq. (11.89) or (11.92), the quantity $U_{cc'}$ is expressible in terms of a rational function of order $2(N + l)$, where N is the total number of resonances. This reflects the polynomial nature of the cofactor and the determinant of the inverse level matrix. Thus one obtains via partial fractions the similar pole representation for other approximations. A general expression that can be used with all cross-section representations is

$$\sigma_t = \sigma_p + \frac{1}{E} \sum_{l,J} \sum_{\lambda=1}^{M} \sum_{j=1}^{jj} \mathrm{Re}\left\{ \left[R^{(t)}_{l,J,j,\lambda} e^{-i2\phi_l} \right] \cdot \frac{-i}{p_\lambda^{(j)*} - \sqrt{E}} \right\} \tag{11.101}$$

and similarly,

$$\sigma_x = \frac{1}{E} \sum_{l,J} \sum_{\lambda=1}^{M} \sum_{j=1}^{jj} \mathrm{Re}\left\{ R^{(x)}_{l,J,j,\lambda} \frac{-i}{p_\lambda^{(j)*} - \sqrt{E}} \right\} \tag{11.102}$$

for the reaction cross section of process x, where $R^{(x)}_{l,J,j,\lambda}$ and $p_\lambda^{(j)}$ are pole and residue, respectively. Note that the complex conjugate $p_\lambda^{(j)*}$ is used here in order to cast the expressions into the form defined by Eq. (11.92). These equations can be viewed as the generalized pole expansion in which all parameters are truly energy independent and the energy dependence of the cross sections is specified explicitly by the rational terms alone.

The indices M and jj depend on the type of resonance parameters and assumptions used to generate these pole parameters:

- *Adler–Adler*: $M = N$ (total number of resonance); $jj = 2$. All pole parameters can be deduced directly via partial fractions.
- *SLBW and MLBW*: $M = N$; $jj = 2$ if penetration factor and level shift factor are taken to be energy independent, an assumption used in the traditional resonance integral approach. Otherwise, $M = N$; $jj = 2(l + 1)$ if all energy-dependent features are included.
- *Reich–Moore*: $M = N + l$; $jj = 2$ for both scenarios with $\hat{S}_l(E) = 0$ and $\hat{S}_l(E) \neq 0$ if Eq. (11.98) is used. Another possible scenario is to keep the traditional expression specified by the ENDF/B manual intact; that is, let $\boldsymbol{F} = \boldsymbol{I}$ in Eq. (11.98), but to introduce the level shift factor via replacing the resonance energy E_λ with

$$E'_\lambda = E_\lambda + \frac{S_l(|E_\lambda|) - S_l(E)}{2 P_l(|E_\lambda|)} \Gamma_{n\lambda}(|E_\lambda|)$$

the same as for the SLBW and MLBW approximations. The resulting number of poles becomes $M = N$; $jj = 2(l + 1)$.

By comparing Eqs. (11.94) and (11.101), one is led to the following observations: (1) For the s-wave, both the rigorous pole representation and the traditional formalism consist of an identical number of terms with the same functional form in the momentum domain. In particular, the Adler–Adler formalism for the s-wave can be considered as the special case of the former when $p_\lambda^{(1)} = -p_\lambda^{(2)}$ and $R_{l,J,1,\lambda}^{(x)} = R_{l,J,2,\lambda}^{(x)}$. (2) For higher angular momentum states, Eq. (11.101) consists of either $2l$ or $2lN$ more terms than those defined by Eq. (11.94). The difference, however, is only superficial. The same number of terms would have resulted if the detailed energy dependence of the penetration factor and the shift factor had been included in Eq. (11.94).

Equations (11.101) and (11.102) provide the basis whereby any given set of $\boldsymbol{R}$-matrix parameters, in principle, can be converted into pole parameters, although it may not be an easy task in practice. The recent availability of $\boldsymbol{R}$-matrix parameters in the Reich–Moore form greatly alleviates the numerical difficulties for such a conversion process. One obvious disadvantage of this method is that two to as many as $2(l + 1)$ terms must be considered for each resonance if the cross section is to be evaluated in the momentum domain. This is obviously undesirable for computing efficiency, storage requirement, and its amenability to the existing codes for reactor calculations.

Simplified Pole Representation. The $M \times jj$ poles for a given l and J defined in Eqs. (11.101) and (11.102) can be divided into two distinct classes. There are $2N$ s-wavelike poles with sharp peaks and distinct spacings, while the remaining $2l$ or $2lN$ poles are closely spaced and are characterized by their extremely large imaginary components (or widths). In fact, the contributions of the latter to the sums are practically without any resonance-like fluctuations, as if they were a smooth constituent. On the other hand, the s-wavelike poles always appear in pairs with opposite signs but not necessarily with the same magnitude. These characteristics provide a valuable basis for simplification.

Let $q_l^{(x)}(\sqrt{E})$ denote the contributions from those additional $2l$ or $2lN$ terms involving poles with giant width. Equation (11.102) can be cast into the same form as that of Humblet–Rosenfeld:

$$
\sigma_x = \frac{1}{E} \sum_l \mathrm{Re} \left\{ \sum_J \sum_{\lambda=1}^{N} \left[R_{l,J,1,\lambda}^{(x)} - \frac{2(-i)\sqrt{E}}{(p_\lambda^{(1)*})^2 - E} \right] + s_l^{(x)}(\sqrt{E}) \right.
$$

$$
\left. + q_l^{(x)}(\sqrt{E}) \cdot \delta_l \right\}, \quad \sqrt{E} > 0 \tag{11.103}
$$

where

$$
s_l^{(x)}(\sqrt{E}) = \sum_J \sum_{\lambda=1}^{N} \left[\frac{R_{l,J,2,\lambda}^{(x)}(-i)}{p_\lambda^{(2)*} - \sqrt{E}} - \frac{R_{l,J,1,\lambda}^{(x)}(-i)}{-p_\lambda^{(1)*} - \sqrt{E}} \right] \tag{11.104}
$$

and $\delta_0 = 0$ and $\delta_l = 1$ for $l > 0$. The quantity $s_l^{(x)}(\sqrt{E})$, physically signifying a measure of deviation from the Adler–Adler limit of $p_\lambda^{(2)*} = -p_\lambda^{(1)*}$, is usually not only small in magnitude but also smooth as a function of energy in the region where the calculations take place. Thus $s_l^{(x)}(\sqrt{E})$ and $q_l^{(x)}(\sqrt{E})$ can be construed as the energy-dependent smooth term in the Humblet–Rosenfeld representation with its energy dependence explicitly specified.

Hence, for a given range of practical interest, the rigorous pole representation can be viewed as a combination of fluctuating terms, consisting of N poles with $\mathrm{Re}\{p_\lambda^{(1)}\} > 0$ expressed in the energy domain consistent with the traditional formalism and two nonfluctuating (or background) terms attributed to the tails of outlying poles (in reference to the domain $\sqrt{E} > 0$, where calculations are to take place) with negative real component and the poles with extremely large width (or $|\mathrm{Im}\{p_\lambda^{(j)}\}|$) for $l > 0$ states, respectively. The striking behavior of the fluctuating and nonfluctuating components have been confirmed in recent calculations for all major nuclei specified by the Reich–Moore parameters in the ENDF/B VI files.

The smooth behavior of these terms clearly suggests that their energy dependence can be reproduced by other, simpler functions within the finite interval of practical interest. It is well known in numerical analysis that the rational functions are best suited to approximate a well-behaved function within a finite range. Hence the obvious choice is to set the approximate functions $\hat{s}_l^{(x)}(\sqrt{E})$ and $\hat{q}_l^{(x)}(\sqrt{E})$ to be rational functions of arbitrary order. Mathematically, they can be viewed as the analytic continuations of the original functions $\hat{s}_l^{(x)}(\sqrt{E})$ and $\hat{q}_l^{(x)}(\sqrt{E})$ within domain $\sqrt{E} > 0$. One attractive feature of the method proposed is that the rational functions so obtained can be again expressed in the form of a pole expansion via partial fraction, that is,

$$\hat{s}_l^{(x)}\left(\sqrt{E}\right) = \frac{P_{MM}\left(\sqrt{E}\right)}{Q_{NN}\left(\sqrt{E}\right)} = \sum_{\lambda=1}^{NN} \frac{r_\lambda^{(x)}(-i)}{\alpha_\lambda^* - \sqrt{E}} \tag{11.105}$$

$$\hat{q}_l^{(x)}\left(\sqrt{E}\right) = \sum_{\lambda=1}^{NN} \frac{b_\lambda^{(x)}(-i)}{\xi_\lambda^* - \sqrt{E}} \tag{11.106}$$

if $NN > MM$. α_λ^* and ξ_λ^* are the poles of the fitted rational functions (i.e., the ratio of the two low-order polynomials) for $\hat{s}_l^{(x)}(\sqrt{E})$ and $\hat{q}_l^{(x)}(\sqrt{E})$, respectively, while $r_\lambda^{(x)}$ and $\xi_\lambda^{(x)}$ are their corresponding residues.

Doppler Broadening of the Generalized Pole Representation

Either one of two approaches are usually taken, depending on the accuracy required. The rigorous broadening must be carried out in the momentum domain, whereas the simplified broadening is based on the approximate kernel in the energy domain. In the following discussions, the Doppler-broadened cross section based on the traditional formalism and the generalized pole representation are compared.

Exact Doppler Broadening. The Maxwell–Boltzmann kernel can be expressed rigorously as

$$S\left(\sqrt{E}, \sqrt{E'}\right) = \frac{\sqrt{E'}}{(\pi E)^{1/2}\Delta_m}\left\{ \exp\left[-\frac{(\sqrt{E} - \sqrt{E'})^2}{\Delta_m^2}\right]\right.$$
$$\left. - \exp\left[-\frac{(\sqrt{E} + \sqrt{E'})^2}{\Delta_m^2}\right]\right\} \tag{11.107}$$

where

$$\Delta_m = \sqrt{\frac{kT}{A}} = \text{Doppler width in momentum space} \tag{11.108}$$

The Doppler broadening of $\sqrt{E'}\sigma_x(\sqrt{E'})$ defined by Eq. (11.94) in momentum space and that defined by Eq. (11.102) lead immediately to:
- *Traditional representation*:

$$\sigma_x\left(\sqrt{E}, T\right) = \frac{1}{E}\sum_{l,J}\sum_{\lambda=1}^{N}\text{Re}\left\{ \frac{\rho_{l,J,\lambda}^{(x)}}{2} \frac{\sqrt{\pi}}{\Delta_m}\left[W\left(\frac{\sqrt{E} - \zeta_\lambda}{\Delta_m^2}\right)\right.\right.$$
$$\left.\left. - W^*\left(\frac{\sqrt{E} + \zeta_\lambda^*}{\Delta_m}\right)\right]\right\} \tag{11.109}$$

- *Generalized pole representation:*

$$\sigma_x\left(\sqrt{E},T\right) = \frac{1}{E}\sum_l \text{Re}\left\{ \sum_J \sum_{\lambda=1}^{N}\sum_{j=1}^{2} R_{l,J,\lambda,j}^{(x)}\left[\frac{\sqrt{\pi}}{\Delta_m}W\left(\frac{\sqrt{E}-p_\lambda^{(j)*}}{\Delta_m}\right)\right] \right.$$
$$\left. + \hat{s}_l^{(x)}\left(\sqrt{E},T\right) + \hat{q}_l^{(x)}\left(\sqrt{E}\right)\delta_l \right\} \qquad (11.110)$$

where $\hat{q}_l^{(x)}(\sqrt{E})$ is insensitive to Doppler broadening and

$$\hat{s}_l^{(x)}\left(\sqrt{E},T\right) = \sum_{k=1}^{NN} r_k^{(x)}\left[\frac{\sqrt{\pi}}{\Delta_m}W\left(\frac{\sqrt{E}-\alpha_k^*}{\Delta_m}\right)\right] \qquad (11.111)$$

and $W(z)$ is the complex probability integral and is directly related to the usual Doppler-broadened line shape functions via the relation

$$W(z) = \frac{i}{\pi}\int_{-\infty}^{\infty}\frac{e^{-t^2}}{z-t}dt \qquad (11.112)$$

$$\psi(x,y) + i\chi(x,y) = \sqrt{\pi}\,y\,W(z) \qquad (11.113)$$

and $z = x + iy$.

In the single-level limit, Eq. (11.109) is equivalent to the generalized form of the exact Doppler broadening defined by Ishiguro. Thus, except for the superficial difference leading to the smooth term $\hat{q}_l^{(x)}(\sqrt{E})\delta_l$, Eqs. (11.109) and (11.110) have the same functional form but are characterized by different parameters. From a practical point of view, the computational requirements for these equations are expected to be comparable if the smooth term is replaced by the approximation defined by Eqs. (11.100) and (11.101).

Approximate Doppler Broadening. For most of the existing codes based on the traditional formalism, the Doppler broadening is generally based on the approximate Gauss kernel defined in the energy domain

$$M(E_\lambda - E) = \frac{1}{\sqrt{\pi}\Delta_E}\exp\left[\frac{(E_\lambda - E)^2}{\Delta_E}\right] \qquad (11.114)$$

where $\Delta_E = \sqrt{4kTE/A}$ is the Doppler width in the energy domain.

The validity of such an approximation requires the criterion $E \gg \Delta_m$. It has been well established that the use of the Gauss kernel in the energy domain is generally satisfactory for $E > 1$ eV. The Doppler-broadened cross sections become:

- *Traditional formalism:*

$$\sigma_x(E,T) = \frac{1}{E}\sum_{l,J}\sum_{\lambda=1}^{N}\text{Re}\left\{ \sqrt{E}\rho_{l,J,\lambda}^{(x)}\frac{\sqrt{\pi}}{\Delta_E}W\left(\frac{E-d_\lambda}{\Delta_E}\right) \right\} \qquad (11.115)$$

- *Generalized pole representation after simplification:*

$$\sigma_x(E, T) = \frac{1}{E} \sum_l \text{Re} \left\{ \sum_J \sum_{\lambda=1}^{N} 2\sqrt{E} R_{l,J,1,\lambda}^{(x)} \frac{\sqrt{\pi}}{\Delta_E} W\left(\frac{E - \varepsilon_\lambda}{\Delta_E}\right) \right.$$

$$\left. + \hat{s}_l^{(x)}\left(\sqrt{E}, T\right) + \hat{q}_l^{(x)}(E)\delta_l \right\} \tag{11.116}$$

where

$$\varepsilon_\lambda = \left[p_\lambda^{(1)^*} \right]^2 \tag{11.117}$$

References

1 W. ROTHENSTEIN and M. SEGEV, "Unit Cell Calculations," in Y. Ronen, ed., *CRC Handbook of Nuclear Reactor Calculations I*, CRC Press, Boca Raton, FL (**1986**).

2 J. J. DUDERSTADT and L. J. HAMILTON, *Nuclear Reactor Analysis*, Wiley, New York (**1976**), Chaps. 2 and 8.

3 A. F. Henry, *Nuclear-Reactor Analysis*, MIT Press, Cambridge, MA (**1975**), Chap. 5.

4 G. I. BELL, *Nuclear Reactor Theory*, Van Nostrand Reinhold, New York (**1970**), Chap. 8.

5 H. H. HUMMEL and D. OKRENT, *Reactivity Coefficients in Large Fast Power Reactors*, American Nuclear Society, La Grange Park, IL (**1970**).

6 R. B. NICHOLSON and E. A. FISCHER, "The Doppler Effect in Fast Reactors," in *Advances in Nuclear Science and Technology*, Academic Press, New York (**1968**).

7 R. N. HWANG, "Doppler Effect Calculations with Interference Corrections," *Nucl. Sci. Eng. 21*, 523 (**1965**).

8 L. W. NORDHEIM, "The Doppler Coefficient," in T. J. Thompson and J. G. Beckerley, eds., *The Technology of Nuclear Reactor Safety*, MIT Press, Cambridge, MA (**1964**).

9 A. SAUER, "Approximate Escape Probabilities," *Nucl. Sci. Eng. 16*, 329 (**1963**).

10 L. DRESNER, *Resonance Absorption in Nuclear Reactors*, Pergamon Press, Elmsford, NY (**1960**).

11 D. E. CULLEN, "Nuclear Cross Section Preparation," in Y. Ronen, ed., *CRC Handbook of Nuclear Reactor Calculations I*, CRC Press, Boca Raton, FL (**1986**).

12 R. N. HWANG, "An Overview of the Current Status of Resonance Theory in Reactor Physics Applications," in W. Audrejtscheff and D. Elenkov, eds., *Proc. 11th Int. School Nuclear Physics, Neutron Physics, and Nuclear Energy*, Institute for Nuclear Research and Nuclear Energy, Sofia, Bulgaria (**1993**).

13 C. JAMMES and R. N. HWANG, "Conversion of Single- and Multi-Level Breit–Wigner Resonance Parameters to Pole Representation Parameters," *Nucl. Sci. Eng. 134*, 37 (**2000**).

14 A. G. CROFF, *ORIGEN2: A Revised and Updated Version of the Oak Ridge Isotope Generation and Depletion Code*, ORNL-5621, Oak Ridge National Laboratory, Oak Ridge, TN (**1980**).

Problems

11.1. Carry through the steps indicated to derive the narrow resonance approximation flux of Eq. (11.6) and the wide resonance approximation flux of Eq. (11.10).

11.2. A fuel assembly in a reactor consists of a uniform array of fuel pins 1 cm in diameter set in parallel rows such that the center-to-center separation between adjacent rows is 3 cm in both ways. The fuel is 2.8% enriched UO_2 operating at $800°C$. The moderator is H_2O at 0.85 g/cm^3. Calculate the heterogeneous resonance integral in the narrow resonance and the wide resonance approximations for the ^{238}U resonance at 36.8 eV, in the isolated fuel rod approximation.

11.3. Repeat Problem 11.2 taking into account the Dancoff correction for a closely packed square lattice.

11.4. Assume that the fuel and moderator in Problem 11.2 are mixed homogeneously together. Calculate the homogeneous resonance integral for the ^{238}U resonance at 36.8 eV at $800°C$.

11.5. Calculate the contribution of the ^{238}U resonance at 36.8 eV to the multigroup cross section of a group with $E_g = 10$ eV and $E_{g-1} = 100$ eV, for Problems 11.2 to 11.4.

11.6. Repeat Problem 11.2 for D_2O moderator. Calculate the contribution to the multigroup cross section over $E_g = 100$ eV to $E_{g-1} = 100$ eV.

11.7. Compare the calculated escape probability for a fuel plate immersed in water for values of λ/l_s in the range $0.1 \leq \lambda/l_s \leq 10.0$, using the exact expression of Eq. (11.32) and the rational approximation of Eq. (11.43).

11.8. Derive the first-flight escape probability for neutrons created uniformly over a slab of thickness a given by Eq. (11.32).

11.9. Write a code to calculate the unresolved ^{238}U multigroup capture cross section of Eq. (11.62) for a group extending from $E_g = 1$ keV to $E_{g-1} = 10$ keV.

11.10. Evaluate the two-band group absorption cross section for a group extending from $E_g = 10$ eV to $E_{g-1} = 100$ eV for a nuclide for which the absorption and total cross sections are $\Sigma_{a1} = 0.4$ cm^{-1} and $\Sigma_{t1} = 0.5$ cm^{-1} from $10 \leq E \leq 50$ eV and are $\Sigma_{a2} = 0.6$ cm^{-1} and $\Sigma_{t2} = 0.8$ cm^{-1} from 50 eV $\leq E \leq 100$ eV.

12
Neutron Thermalization

The thermalization of neutrons is complicated, relative to neutron slowing down, by the fact that the thermal energies of the target nuclei are comparable to the neutron energies, so that a neutron may gain or lose energy in a scattering collision, and by the fact that the nuclei are generally bound in a lattice or molecular structure, which considerably complicates both the calculation of a scattering cross section and the scattering kinematics. The objectives of neutron thermalization theory are first to calculate cross sections that characterize the thermal neutron scattering and energy transfer and then to use these cross sections in calculation of the thermal neutron spectrum. In this chapter we consider some approximate models of neutron thermalization that provide useful physical insights, discuss the construction of thermal neutron scattering kernels, and then discuss the analytical and numerical calculation of the neutron thermal energy spectrum in homogeneous media and heterogeneous reactor lattices.

12.1
Double Differential Scattering Cross Section for Thermal Neutrons

A quantum mechanical analysis of the scattering event in which an incident neutron interacts with an assembly of target nuclei of atomic mass A leads to an expression of the differential scattering cross section for scattering from energy E' to energy E and from direction $\mathbf{\Omega}'$ to direction $\mathbf{\Omega}$:

$$\Sigma_s(E' \to E, \mu_0) = \frac{\Sigma_b}{4\pi kT} \left(\frac{E}{E'} \right)^{1/2} \exp\left(-\frac{\beta}{2} \right) S(\alpha, \beta) \tag{12.1}$$

where $\mu_0 = \mathbf{\Omega}' \cdot \mathbf{\Omega}$, σ_b is the scattering cross section for a neutron incident on a bound nucleus,

$$\alpha \equiv \frac{E' + E - 2\mu_0\sqrt{E'E}}{AkT}, \qquad \beta \equiv \frac{E - E'}{kT} \tag{12.2}$$

Nuclear Reactor Physics. Weston M. Stacey
Copyright © 2007 WILEY-VCH Verlag GmbH & Co. KGaA, Weinheim
ISBN: 978-3-527-40679-1

and $S(\alpha, \beta)$ is a *scattering function* that depends in a complicated way on the detailed dynamics and structure of the scattering material. Hence

$$\Sigma_b = N\sigma_b = N\left(\frac{A+1}{A}\right)^2\sigma_f = \left(\frac{A+1}{A}\right)^2\Sigma_f$$

is the bound atom cross section, given in terms of the free atom cross section, Σ_f, and the ratio of scattering to neutron masses, A.

12.2
Neutron Scattering from a Monatomic Maxwellian Gas

Differential Scattering Cross Section

The simplest, but by no means simple, model of neutron thermalization is for neutrons scattering from a monatomic gas of unbound nuclei distributed in energy according to a Maxwellian distribution, for which the scattering function is

$$S(\alpha, \beta) = \frac{1}{2(\pi\alpha)^{1/2}}\exp\left(-\frac{\alpha^2+\beta^2}{4\alpha}\right) \tag{12.3}$$

which yields for the differential scattering cross section

$$\Sigma_s(E' \to E, \mu_0) = \left(1+\frac{1}{A}\right)^2\frac{\Sigma_f}{4\pi}\left(\frac{E}{E'}\right)^{1/2}\left(\frac{A}{2\pi kT\hbar^2\kappa^2}\right)^{1/2}$$

$$\times \exp\left[-\frac{A}{2kT\hbar^2\kappa^2}\left(\varepsilon - \frac{\hbar^2\kappa^2}{2A}\right)^2\right] \tag{12.4}$$

where A is the atomic mass (amu) of the target nuclei, σ_f is the total scattering cross section for a neutron incident on a free nucleus, and

$$\varepsilon \equiv E' - E, \qquad \hbar^2\kappa^2 = 2m\left(E' + E - 2\mu_0\sqrt{E'E}\right) \tag{12.5}$$

Integrating Eq. (12.4) over $-1 \le \mu_0 \le 1$ and using the relationship between μ_0 and (E, E') for elastic scattering,

$$\mu_0 = \frac{1}{2}\left[(A+1)\sqrt{\frac{E}{E'}} - (A-1)\sqrt{\frac{E'}{E}}\right] \tag{12.6}$$

yields for the zeroth Legendre moment of the scattering transfer function,

$$\Sigma_{s0}(E' \to E)$$

$$= \frac{\Sigma_f\theta^2}{2E'}\left\{\exp\left(\frac{E'}{kT} - \frac{E}{kT}\right)\right.$$

$$\times \left[\mathrm{erf}\left(\theta \sqrt{\frac{E'}{kT}} - \eta \sqrt{\frac{E}{kT}} \right) \pm \mathrm{erf}\left(\theta \sqrt{\frac{E'}{kT}} + \eta \sqrt{\frac{E}{kT}} \right) \right]$$

$$+ \mathrm{erf}\left(\theta \sqrt{\frac{E}{kT}} - \eta \sqrt{\frac{E'}{kT}} \right) \mp \mathrm{erf}\left(\theta \sqrt{\frac{E}{kT}} + \eta \sqrt{\frac{E'}{kT}} \right) \Bigg\} \qquad (12.7)$$

where $\mathrm{erf}(x)$ is the error function and

$$\theta \equiv \frac{A+1}{2\sqrt{A}}, \qquad \eta \equiv \frac{A-1}{2\sqrt{A}} \qquad (12.8)$$

The upper signs are used when $E' < E$, and the lower signs are used when $E' > E$.

Cold Target Limit

In the limit $T \to 0$, Eq. (12.7) reduces to the scattering transfer function for elastic scattering from a stationary target:

$$\Sigma_{s0}(E' \to E) = \begin{cases} \dfrac{\Sigma_f(E')}{E'} \dfrac{(A+1)^2}{4A}, & E < E' < \dfrac{E}{\alpha} \\ 0, & \text{otherwise} \end{cases} \qquad (12.9)$$

which was used in Chapter 10 for the treatment of neutron slowing down in the energy range well above thermal where the nuclear motion is negligible compared to the neutron motion.

Free-Hydrogen (Proton) Gas Model

Hydrogen, in the form of water molecules, is a dominant nuclear species for neutron thermalization in water-cooled nuclear reactors. The free-hydrogen gas model neglects the fact that hydrogen is present in molecular form and treats the thermalization of neutrons by a gas of free protons (hydrogen nuclei). For scattering from hydrogen nuclei ($A = 1$), the zeroth Legendre moment of the scattering energy transfer function of Eq. (12.7) simplifies to

$$\Sigma_{s0}(E' \to E) = \begin{cases} \dfrac{\Sigma_f^H(E')}{E'} \left[\exp\left(\dfrac{E'-E}{kT} \right) \mathrm{erf}\left(\sqrt{\dfrac{E'}{kT}} \right) \right], & E' < E \\ \dfrac{\Sigma_f^H(E')}{E'} \mathrm{erf}\left(\sqrt{\dfrac{E}{kT}} \right), & E' > E \end{cases} \qquad (12.10)$$

Radkowsky Model for Scattering from H_2O

The Radkowsky model uses the hydrogen gas model of Eq. (12.10) to describe the zeroth Legendre moment of the scattering transfer function, and uses

$$\Sigma_{s1}(E' \to E) = \Sigma_s(E) \bar{\mu}_0(E) \delta(E - E') \qquad (12.11)$$

to describe the first Legendre moment. The bound-state cross section, Σ_b, is related to the free-state cross section, Σ_f, by

$$\Sigma_s = \Sigma_b = \Sigma_f \frac{[(A+1)/A]^2}{(1+1/A_{mol})^2} = 4\Sigma_f \left(\frac{A_{mol}}{A_{mol}+1}\right)^2 \qquad (12.12)$$

where A is the mass of an atom bound in a molecule of mass A_{mol} and has been set to unity in the last step to represent the hydrogen bound in water molecules.

Application of this model is implemented by adjustment of A_{mol} until Σ_b agrees with an experimentally measured scattering cross section, as a function of energy. Then $\mu_0 = 2/(3A_{mol})$ is used to calculate $\Sigma_{tr}(E) = \Sigma_b(E)[1 - \mu_0(E)]$.

Heavy Gas Model

In the limit of large A, the scattering transfer function of Eq. (12.7) can be expanded in powers of A^{-1}. When only the leading term is retained, the result is

$$\Sigma_{s0}(E' \to E) = \Sigma_f \left\{ \delta(E - E') + \left(\frac{E}{E'}\right)^{1/2} \frac{E + E'}{A} \right.$$

$$\left. \times \left[-\delta'(E - E') + kT\delta''(E - E')\right]\right\} \qquad (12.13)$$

where δ' and δ'' are the first and second derivatives of the delta function with respect to x. Integrating this expression over E defines the total scattering cross section in this model:

$$\Sigma_{s0}(E) = \Sigma_f \left(1 + \frac{kT}{2AE}\right) \qquad (12.14)$$

Using Eq. (12.13) to evaluate the scatter-in integral yields

$$\int_0^\infty dE' \Sigma_{s0}(E' \to E)\phi(E')$$

$$= \frac{2\Sigma_f}{A}\left[kTE\frac{d^2\phi(E)}{dE^2} + E\frac{d\phi(E)}{dE} + \phi(E)\right] + \Sigma_{s0}(E)\phi(E) \qquad (12.15)$$

when the property of the derivatives of the delta functions,

$$\int dx\, f(x)\frac{d^n}{dx^n}\delta(x - a) = (-1)^n \frac{d^n f(x=a)}{dx^n} \qquad (12.16)$$

is taken into account. Substituting this expression for the scatter-in integral into the neutron balance equation

$$\Sigma_t(E)\phi(E) = \int_0^\infty \Sigma_{s0}(E' \to E)\phi(E')dE' \qquad (12.17)$$

yields

$$\Sigma_a(E)\phi(E) = 2\frac{\Sigma_f}{A}\left[EkT\frac{d^2\phi(E)}{dE^2} + E\frac{d\phi(E)}{dE} + \phi(E)\right] \tag{12.18}$$

which is the heavy gas model for the thermal neutron spectrum, $\phi(E)$.

12.3
Thermal Neutron Scattering from Bound Nuclei

The quantum mechanical theory for neutron scattering from a system of bound nuclei leads to an expression for the double differential scattering function for scattering from energy E' to energy E and from direction Ω' to direction Ω:

$$\Sigma_s\left(E' \to E, \Omega' \to \Omega\right) = \frac{\Sigma_{\text{coh}}}{4\pi\hbar}\sqrt{\frac{E}{E'}}\frac{1}{2\pi}\int_{-\infty}^{\infty}\int e^{i(\kappa\cdot\mathbf{r}-\varepsilon t/\hbar)}G(\mathbf{r},t)\,d\mathbf{r}\,dt$$

$$+ \frac{\Sigma_{\text{inc}}}{4\pi\hbar}\sqrt{\frac{E}{E'}}\frac{1}{2\pi}\int_{-\infty}^{\infty}\int e^{i(\kappa\cdot\mathbf{r}-\varepsilon t/\hbar)}G_s(\mathbf{r},t)\,d\mathbf{r}\,dt$$

$$\tag{12.19}$$

where $\hbar$ is the reduced Planck's constant, $\hbar\kappa = m(\mathbf{v}' - \mathbf{v})$ is the neutron momentum exchange vector, $\varepsilon = E' - E$ is the neutron energy change, and Σ_{coh} and Σ_{inc} are the bound coherent and incoherent macroscopic cross sections. The coherent scattering takes into account the interference of neutrons scattering from different nuclei, which is important when the neutron wavelength λ (cm) $= 2.86 \times 10^{-9}/[E(\text{eV})]^{1/2}$ is comparable with the spacing between atoms in a crystal or molecule, and the incoherent scattering takes into account the scattering of neutrons from isolated nuclei.

Pair Distribution Functions and Scattering Functions

The functions $G(\mathbf{r},t)$ and $G_s(\mathbf{r},t)$ are *pair distribution functions*. If a scattering target atom is at the origin $\mathbf{r} = 0$ at time $t = 0$, then $G(\mathbf{r},t)$ is the probability that an atom will be present within unit volume $d\mathbf{r}$ about $\mathbf{r}$ at time t. $G(\mathbf{r},t) = G_s(\mathbf{r},t) + G_d(\mathbf{r},t)$, where $G_s(\mathbf{r},t)$ is the probability that the atom present in $d\mathbf{r}$ about $\mathbf{r}$ at time t is the same atom that was present at $\mathbf{r} = 0$ at time $t = 0$, and $G_d(\mathbf{r},t)$ is the probability that a different atom is present in $d\mathbf{r}$ about $\mathbf{r}$ at time t. The integrals involving the pair distribution functions in Eq. (12.19) are defined as the scattering functions

$$S(\kappa, G) = \frac{1}{2\pi}\int_{-\infty}^{\infty}\int e^{i(\kappa\cdot\mathbf{r}-\varepsilon t/\hbar)}G(\mathbf{r},t)\,d\mathbf{r}\,dt \tag{12.20}$$

with a similar definition for S_s in terms of G_s.

The principle of detailed balance requires that

$$M(E,T)\sqrt{\frac{E'}{E}}\left[\Sigma_{\text{coh}}S(-\kappa,-\varepsilon)+\Sigma_{\text{inc}}S_s(-\kappa,-\varepsilon)\right]$$

$$= M(E',T)\sqrt{\frac{E}{E'}}\left[\Sigma_{\text{coh}}S(\kappa,\varepsilon)+\Sigma_{\text{inc}}S_s(\kappa,\varepsilon)\right] \tag{12.21}$$

be satisfied separately for the incoherent and coherent contributions. Recalling that

$$M(E,T)=\frac{2\pi\sqrt{E}}{(\pi kT)^{3/2}}\exp\left(-\frac{E}{kT}\right) \tag{12.22}$$

this detailed balance requirement may be written

$$e^{-\varepsilon/2kT}S(\kappa,\varepsilon)=e^{\varepsilon/2kT}S(-\kappa,-\varepsilon) \tag{12.23}$$

with a similar requirement for S_s, which requires that both $S(\kappa,\varepsilon)$ and $S_s(\kappa,\varepsilon)$ be even functions of ε.

In many scattering models, $S(\kappa,\varepsilon)$ is a function of κ^2, and an equivalent scattering function can be defined:

$$S(\alpha,\beta)=kTe^{\beta/2}S(\kappa,\varepsilon)=\frac{1}{2(\pi\alpha)^{1/2}}\exp\left(-\frac{\alpha^2+\beta^2}{4\alpha}\right) \tag{12.24}$$

where α and β are defined by Eq. (12.2). Using this scattering function, the double differential scattering transfer function can be represented as

$$\Sigma_s(E'\rightarrow E,\boldsymbol{\Omega}'\rightarrow\boldsymbol{\Omega})=\frac{1}{4\pi\hbar kT}\sqrt{\frac{E}{E'}}e^{-\beta/2}\left[\Sigma_{\text{coh}}S(\alpha,\beta)+\Sigma_{\text{inc}}S_s(\alpha,\beta)\right] \tag{12.25}$$

Intermediate Scattering Functions

An equivalent representation of the double differential scattering transfer function is

$$\Sigma_s(E'\rightarrow E,\boldsymbol{\Omega}'\rightarrow\boldsymbol{\Omega})$$

$$=\frac{1}{4\pi\hbar}\sqrt{\frac{E}{E'}}\frac{1}{2\pi}\int_{-\infty}^{\infty}e^{-i\varepsilon t/\hbar}\left[\chi_{\text{coh}}(\kappa,t)+\chi_{\text{inc}}(\kappa,t)\right]dt \tag{12.26}$$

where the *intermediate scattering functions* are defined:

$$\chi_{\text{coh}}(\kappa,t)\equiv\int e^{i\kappa\cdot\mathbf{r}}G(\mathbf{r},t)d\mathbf{r}$$

$$\tag{12.27}$$

$$\chi_{\text{inc}}(\kappa,t)\equiv\int e^{i\kappa\cdot\mathbf{r}}G_s(\mathbf{r},t)d\mathbf{r}$$

Incoherent Approximation

The interference effects, which are contained entirely in the pair distribution function G_d, are important in elastic scattering, but are less important in inelastic scattering, particularly in liquids and polycrystalline solids. This observation leads to the incoherent approximation, obtained by setting $G_d = 0$ in Eq. (12.19):

$$\Sigma_s (E' \rightarrow E, \mathbf{\Omega}' \rightarrow \mathbf{\Omega}) = \frac{\Sigma_{\text{coh}} + \Sigma_{\text{inc}}}{4\pi\hbar} \sqrt{\frac{E}{E'}} \frac{1}{2\pi} \int_{-\infty}^{\infty} e^{i(\boldsymbol{\kappa}\cdot\mathbf{r} - \varepsilon t/\hbar)} G_s (\mathbf{r}, t) \, d\mathbf{r} \, dt$$

(12.28)

Note that this approximation retains the coherent scattering cross section, Σ_{coh}. With the incoherent approximation, $S(\alpha, \beta) = S_s(\alpha, \beta)$ in Eq. (12.25) and $\chi_{\text{coh}} = \chi_{\text{inc}}$ in Eq. (12.26).

Gaussian Representation of Scattering

In the incoherent approximation, the intermediate scattering function has a Gaussian form in many important cases:

$$\chi_s (\boldsymbol{\kappa}, t) = \exp\left[-\frac{1}{2}\kappa^2 \Lambda(t) \right]$$

(12.29)

where

$$\Lambda\left(t + \frac{i}{2T}\right) = \int_0^{\infty} g(\omega) \left[\coth\frac{\omega}{2kT} - \frac{\cos\omega t}{\sinh(\omega/2kT)} \right] \frac{d\omega}{\omega}$$

(12.30)

The properties of a particular moderator are represented in the frequency distribution function, $g(\omega)$. For crystals, $g(\omega)$ is a true phonon frequency spectrum. For liquids and molecules, $g(\omega)$ contains the diffusive and vibrational characteristics and may be temperature dependent. Representative frequency distribution functions are:

Perfect gas: $\qquad g(\omega) = \frac{1}{A}\delta(\omega)$

Debye crystal: $\qquad g(\omega) = \frac{3\omega^2}{A\theta^3}$

Einstein crystal: $\quad g(\omega) = \frac{1}{A}\delta(\omega - \theta)$

Molecular liquid: $\quad g(\omega) = \frac{1}{A}f_d(\omega) + \sum_i \gamma_i \delta(\omega - \omega_i)$

(12.31)

where $\theta = h\nu_m/2\pi k \equiv \hbar\nu_m/k$, ν_m is the maximum allowed frequency, $f_d(\omega)$ describes the frequency distribution associated with diffusive motion of the molecule, and the $\gamma_i \delta(\omega - \omega_i)$ describe the internal vibrations with frequency ω_i of the individual atoms of which the molecular fluid is composed.

The corresponding scattering functions in the Gaussian representation are

$$S_s(\alpha, \beta) = \frac{1}{2\pi} \int_{-\infty}^{\infty} dt \, e^{i\beta t} \exp\left[-\alpha W^2(t)\right] \tag{12.32}$$

where

$$W^2(t) = \int_{-\infty}^{\infty} \frac{g(\beta)[\cosh(\beta/2) - \cos(kT\beta t/\hbar)]d\beta}{\beta \sinh(\beta/2)} \tag{12.33}$$

Measurement of the Scattering Function

The scattering transfer function can be determined from Eq. (12.25), in the incoherent approximation, by measuring $\Sigma_s(E' \to E, \mathbf{\Omega}' \to \mathbf{\Omega})$. For small values of neutron momentum transfer, κ^2, and energy transfer, the exponential in Eq. (12.32) can be expanded to obtain a relation between the frequency distribution function and the measured $S_s(\alpha, \beta)$:

$$f(\beta) = 2\beta \sinh \frac{\beta}{2} \lim_{\alpha \to 0} \frac{S_s(\alpha, \beta)}{\alpha} \tag{12.34}$$

and noting that $\hbar\omega/2\pi = E' - E = \beta kT$. Thus, by measuring the scattering double differential cross section for small momentum and energy transfer events, the scattering function S_s can be inferred and related to the frequency distribution. This enables the experimental determination of $g(\omega)$, which can then be extrapolated and used to calculate scattering transfer functions for larger energy and momentum transfers.

Applications to Neutron Moderating Media

The double-differential scattering transfer function for water has been calculated with a molecular liquid model in which the frequency distribution function is given by

$$g(\omega) = \frac{1}{18} + \sum_{i=2}^{4} \frac{1}{A_i} \delta(\omega - \omega_i) \tag{12.35}$$

where the first term represents the translational (diffusive) motion of free gas molecules, the second term represents hindered rotation ($A_2 = 2.32$, $\hbar\omega/2\pi = 0.06$ eV), and the third and fourth terms represent vibrational modes with ($A_3 = 5.84$, $\hbar\omega/2\pi = 0.205$ eV) and ($A_4 = 2.92$, $\hbar\omega/2\pi = 0.481$ eV). This Nelkin distribution function was used to evaluate the scattering function of Eq. (12.32), which was then used to evaluate the double differential scattering transfer function of Eq. (12.1). The results are compared with experimental measurements of the double differential scattering transfer function, for different incident neutron energies, in Fig. 12.1. Also shown are results calculated with the free-hydrogen gas model of Eq. (12.4).

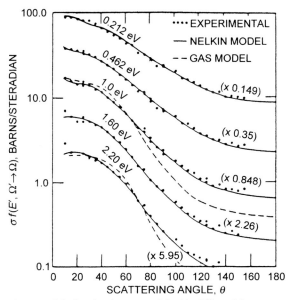

Fig. 12.1 Calculated and measured double differential scattering transfer functions in liquid water at various incident neutron energies. (From Ref. 3; used with permission of Wiley.)

The phonon frequency spectrum for graphite, based on two slightly different models, is shown in Fig. 12.2. Specializing the incoherent approximation for the double differential scattering transfer function of Eq. (12.28) to a crystal lattice with cubic symmetry and harmonic interatomic forces yields

$$\Sigma_s\left(E' \rightarrow E, \boldsymbol{\Omega}' \rightarrow \boldsymbol{\Omega}\right)$$

$$= \frac{\Sigma_b}{4\pi\hbar} \sqrt{\frac{E}{E'}} \frac{1}{2} \int_{-\infty}^{\infty} e^{-i\varepsilon t/\hbar}$$

$$\times \exp\left[\frac{\hbar\kappa^2}{2Am} \int_{-\infty}^{\infty} \frac{f(\omega)e^{-\hbar\omega/2kT}}{2\omega\sinh(\hbar\omega/2kT)} \left(e^{-i\omega t} - 1\right)d\omega\right]dt \qquad (12.36)$$

Using the Young–Koppel frequency distribution shown in Fig. 12.2 to evaluate Eq. (12.36) yields the inelastic cross section shown in Fig. 12.3. Adding to this the absorption cross section of graphite and an elastic scattering cross section calculated without making the incoherent approximation, the total calculated cross section for graphite is compared to measured values in Fig. 12.3. Here m is the neutron mass.

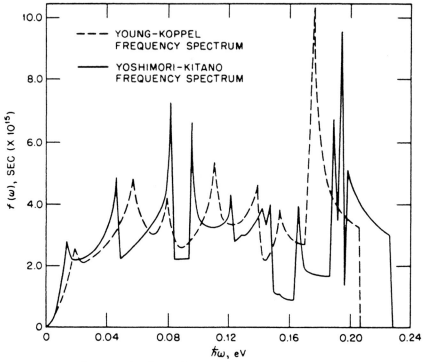

Fig. 12.2 Phonon frequency distribution functions for graphite derived from two different models. (From Ref. 3; used with permission of Wiley.)

12.4
Calculation of the Thermal Neutron Spectra in Homogeneous Media

Turning now to the calculation of the neutron energy spectrum, the neutron balance equation for thermal neutrons, neglecting leakage, is

$$[\Sigma_a(E) + \Sigma_s(E)]v(E)n(E) = \int_0^\infty dE' \Sigma_{s0}(E' \to E)v(E')n(E') \quad (12.37)$$

The principle of detailed balance for a neutron distribution in equilibrium,

$$M(E', T)\Sigma_{s0}(E' \to E) = M(E, T)\Sigma_{s0}(E \to E') \quad (12.38)$$

where $M(E, T)$ is the Maxwellian neutron particle distribution at temperature T,

$$M(E, T) = \frac{2\pi}{(\pi kT)^{3/2}} \sqrt{E} \exp\left(-\frac{E}{kT}\right) \quad (12.39)$$

is quite important in developing solutions for the thermal neutron distribution.

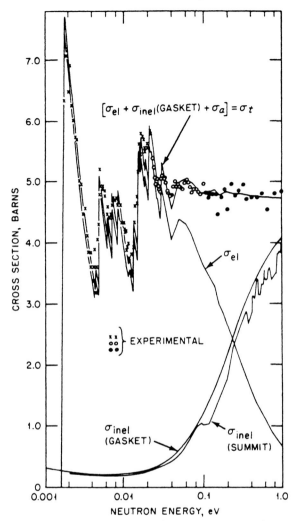

Fig. 12.3 Calculated and measured cross sections in graphite (GASKET and SUMMIT refer to codes). (From Ref. 3; used with permission of Wiley.)

Wigner–Wilkins Proton Gas Model

The zeroth Legendre moment of the scattering energy transfer function for neutron scattering from a free gas of hydrogen nuclei with a Maxwellian distribution is given by Eq. (12.10). It is convenient to define the dimensionless variable $x = (E/kT)^{1/2}$ and to symmetrize the scattering transfer function:

$$S(x' \rightarrow x) \equiv \frac{1}{\Sigma_f^H} \sqrt{\frac{M(x')}{M(x)}} x' \Sigma_{s0}(x' \rightarrow x) \tag{12.40}$$

In terms of the reduced density,

$$\chi(x) \equiv \frac{n(x)}{\sqrt{M(x)}} \tag{12.41}$$

Eq. (12.37) can be written

$$\left[\frac{x\Sigma_s(x)}{\Sigma_f} + \frac{\Sigma_{a0}}{\Sigma_f} \right] \chi(x) = \int_0^\infty S(x' \to x)\chi(x')dx' \tag{12.42}$$

or more explicitly,

$$\left[\left(x + \frac{1}{2x} \right) \mathrm{erf}(x) + \frac{\exp(-x^2)}{\sqrt{\pi}} + \frac{\Sigma_{a0}}{\Sigma_f} \right] \chi(x)$$

$$= 2\exp\left(-\frac{1}{2}x^2 \right) \int_0^x \chi(x') \exp\left[-\frac{1}{2}(x')^2 \right] \mathrm{erf}(x')dx'$$

$$+ 2\exp\left(\frac{1}{2}x^2 \right) \mathrm{erf}(x) \int_x^\infty \exp\left[-\frac{1}{2}(x')^2 \right] \chi(x')dx' \tag{12.43}$$

where $\mathrm{erf}(x)$ is the error function of argument x, $1/v$ absorption has been assumed, and $\Sigma_{a0} = \Sigma_a(v_0 = 2200 \text{ m/s})$ is the 2200-m/s macroscopic absorption cross section.

Equation (12.42) can be transformed into a second-order differential equation by defining a second-order differential operator which when applied to either $\mathrm{erf}(x)\exp(x^2/2)$ or $\exp(-x^2/2)$ yields zero. Such an operator is

$$L = \frac{d^2}{dx^2} + a(x)\frac{d}{dx} + b(x) \tag{12.44}$$

with

$$a(x) = \frac{-\sqrt{\pi}\,\mathrm{erf}(x)}{\exp(-x^2) + x\sqrt{\pi}\,\mathrm{erf}(x)}, \qquad b(x) = \frac{\exp(-x^2)}{\exp(-x^2) + \sqrt{\pi}x\,\mathrm{erf}(x)} - x^2 \tag{12.45}$$

When this operator is divided by

$$P(x) = \exp(-x^2) + \sqrt{\pi}x\,\mathrm{erf}(x) \tag{12.46}$$

and then applied to Eq. (12.42), the Wigner–Wilkins equation results:

$$-\frac{d}{dx}\left\{ \frac{1}{P(x)}\frac{d}{dx}\left[V(x) + \frac{\Sigma_{a0}}{\Sigma_f} \right]\chi(x) \right\}$$

$$+ \left\{ W(x)\left[V(x) + \frac{\Sigma_{a0}}{\Sigma_f} \right] - \frac{4}{\sqrt{\pi}} \right\}\chi(x) = 0 \tag{12.47}$$

where

$$W(x) = \frac{x^2}{P(x)} - \frac{\exp(-x^2)}{P^2(x)}, \qquad V(x) = x\Sigma_s(x) \tag{12.48}$$

Appropriate low-energy boundary conditions can be deduced from setting $x = 0$ in Eq. (12.43), which leads to the low-energy boundary condition $\chi(0) = 0$. The two solutions of Eq. (12.43) near $x = 0$ are a constant and a solution that varies like x; and only the latter can satisfy the boundary condition $\chi(0) = 0$. The other boundary condition follows from the requirement that the flux take on the asymptotic $1/E \sim 1/x^2$ form from the slowing-down region at high energies in the thermal range.

Defining

$$\mu(x) \equiv \left[V(x) + \frac{\Sigma_{a0}}{\Sigma_f} \right] \chi(x), \qquad y(x) \equiv \frac{1}{\mu(x)P(x)} \frac{d\mu(x)}{dx} \qquad (12.49)$$

Eq. (12.47) can be reduced to a Ricatti equation:

$$\frac{dy(x)}{dx} = W(x) - \frac{4}{\sqrt{\pi}} \left[V(x) + \frac{\Sigma_{a0}}{\Sigma_f} \right]^{-1} - P(x)y^2(x) \qquad (12.50)$$

At low energies (small x), Eq. (12.50) has a power series solution

$$y(x) \simeq \frac{a_1}{x} + a_2 x + a_3 x^3 + a_4 x^4 + \cdots \qquad (12.51)$$

Defining

$$\delta \equiv \frac{\sqrt{\pi}}{4} \frac{\Sigma_a(x)}{\Sigma_f} \simeq \frac{\sqrt{\pi}}{4} \frac{\Sigma_{a0}}{\Sigma_f} \qquad (12.52)$$

the coefficients are

$$a_1 = 1, \qquad a_2 = -\frac{4}{3} \left(\frac{1+\delta}{1+2\delta} \right), \qquad a_3 = \frac{103 + 380\delta + 364\delta^2}{90(1+2\delta)^2} \qquad (12.53)$$

The solution can be extended numerically to higher energies (larger x) by fitting a polynomial to values for which the power series is valid, say up to x_n, to define the polynomials

$$W(x) = \prod_{k=1}^{K} (x - x_k), \qquad q(x) = \sum_{k=1}^{K} \frac{W(x)}{x - x_k} \frac{dV(x_k)/dx}{dW(x_k)/dx} \qquad (12.54)$$

which can be used to extrapolate the solution to higher $x > x_n$:

$$y(x_{n+1}) = y(y_n) + \int_{x_n}^{x_{n+1}} q(x)dx \qquad (12.55)$$

These algorithms can be used as a predictor coupled with Eq. (12.50) in a predictor–corrector type of solution.

The boundary condition $\mu(0) = 0$, together with $\mu(x) \neq 0$, implies that

$$\lim_{x \to 0} \left\{ y(x) - \frac{1}{x} \right\} = 0 \qquad (12.56)$$

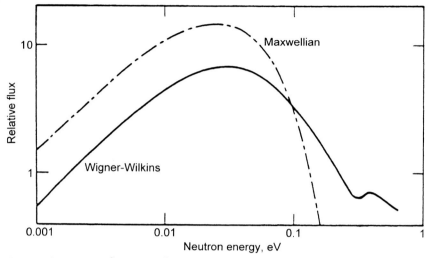

Fig. 12.4 Comparison of Wigner–Wilkins and Maxwellian thermal neutron spectra for a typical PWR composition. (From Ref. 2; used with permission of Wiley.)

which in turn implies that

$$\mu(x) = \frac{x \, d\mu(0)}{dx} \exp\left\{ \int_0^x \left[P(x')y(x') - \frac{1}{x'} \right] dx' \right\} \tag{12.57}$$

Numerical integration of the exponent then allows the density to be constructed from

$$n(x) = \frac{4}{\pi^{3/4}} \frac{x \sqrt{M(x)}}{V(x) + \Sigma_{a0}/\Sigma_f} \exp\left\{ \int_0^x \left[y(x')P(x') - \frac{1}{x'} \right] dx' \right\} \tag{12.58}$$

The development can be extended to include non-1/v absorbers and leakage by the replacement

$$\Sigma_{a0} \rightarrow \Sigma_a(E) + \frac{B^2}{3[\Sigma_a(E) + \Sigma_s(E)(1 - \bar{\mu}_0)]} \tag{12.59}$$

where B characterizes a simple buckling mode.

A thermal spectrum calculated for a 1/v absorber and with a thermal resonance, and matched to a $1/E$ slowing-down source upper boundary condition, is compared with a Maxwellian in Fig. 12.4. The spectrum hardening effects of the 1/v absorber in preferentially absorbing the lower-energy neutrons and of the $1/E$ slowing-down source in increasing the higher-energy neutron population are apparent, as is the flux depression in the vicinity of the resonance.

Heavy Gas Model

The heavy gas model given by Eq. (12.18) is a second-order differential equation for the thermal neutron flux. It is instructive to rederive that result before looking for a solution. We take advantage of the fact that the thermal neutron spectrum is expected to be similar to a Maxwellian for small absorption to look for a solution of Eq. (12.37) of the form

$$\phi(E) = M(E)\psi(E) \tag{12.60}$$

and then make use of the detailed balanced condition of Eq. (12.38) to rewrite Eq. (12.37):

$$\Sigma_t(E)M(E)\psi(E) = M(E)\int_0^\infty \Sigma_{s0}(E \to E')\psi(E')dE' \tag{12.61}$$

Assuming that ψ is a slowly varying function of E, we make a Taylor's series expansion

$$\psi(E') = \psi(E) + (E' - E)\frac{d\psi(E)}{dE} + \frac{1}{2!}(E' - E)^2\frac{d^2\psi(E)}{dE^2} + \cdots \tag{12.62}$$

of $\psi(E')$ in the scatter-in integral, to obtain

$$\Sigma_a(E)M(E)\psi(E) = M(E)\sum_{n=1}^\infty \frac{1}{n!}A_n(E)\frac{d^n\psi(E)}{dE_n} \tag{12.63}$$

where the energy moments of the scattering energy transfer function are

$$A_n(E) = \int_0^\infty (E' - E)^n \Sigma_{s0}(E \to E')dE' \tag{12.64}$$

and where the first term in the expansion has canceled with the scattering contribution to the total cross section on the left side of the equation. This expansion is valid for any scattering transfer function, but its utility depends on rapid convergence of the Taylor series, which requires that $\Sigma_{s0}(E \to E')$ is strongly peaked about $E' = E$ (i.e., for heavy mass moderators which cannot produce a large energy change). Making a $1/A$ expansion of the gas scattering transfer function of Eq. (12.7) and using the result to evaluate Eq. (12.64) yields

$$A_1(E) = \left(\frac{A}{A+1}\right)^2 \frac{2\Sigma_f}{A}(2kT - E) + O\left(\frac{1}{A^2}\right)$$

$$A_2(E) = \left(\frac{A}{A+1}\right)^2 \frac{4\Sigma_f}{A}EkT + O\left(\frac{1}{A^2}\right) \tag{12.65}$$

$$A_n(E) = O\left(\frac{1}{A^2}\right), \quad n \geq 3$$

If only terms through $n = 2$ are retained in Eq. (12.63), the resulting equation is identical to Eq. (12.18) to within a factor $[A/(1 + A)]^2$, which approaches unity for large A.

It is convenient to rewrite Eq. (12.63) in terms of the variable $x = (E/kT)^{1/2}$:

$$x \frac{d^2 n(x)}{dx^2} + (2x^2 - 1)\frac{dn(x)}{dx} + (4x - \Delta)n(x) = 0 \tag{12.66}$$

where $1/v$ absorption has been assumed and the absorption parameter is

$$\Delta \equiv \frac{2A\Sigma_{a0}}{\Sigma_f} \tag{12.67}$$

This equation can be solved exactly in the case of zero absorption ($\Delta = 0$):

$$n(x) = a_1 x^2 e^{-x^2} + a_2 \left[x^2 e^{-x^2} E_1(x^2) - 1 \right] \tag{12.68}$$

where E_1 is the exponential integral function. Since the second term is negative at $x = 0$ and positive for large x, a_2 must be zero.

When absorption is present, Eq. (12.66) can be integrated once to obtain

$$x \frac{dn(x)}{dx} + 2(x^2 - 1)n(x) = \Delta \int_0^x n(x')dx' = \frac{2A}{\Sigma_f} q(x) \tag{12.69}$$

where we have used the fact that all of the neutrons slowing down below x—the slowing down density $q(x)$—must be absorbed in the interval $x < x' < 0$. Integrating a second time yields an integral equation:

$$n(x) = x^2 e^{-x^2} \left[\frac{4}{\sqrt{\pi}} + \Delta \int_0^x \frac{e^{u^2}}{u^3} \int_0^u n(u')du' du \right] \tag{12.70}$$

that is well suited to solution by iteration. The asymptotic form for the neutron flux $\phi = nv$ at large values of x is

$$\phi(E) = \frac{A}{2\Sigma_f E} \left[1 - \frac{1}{2}\Delta\left(\frac{kT}{E}\right)^{1/2} + \frac{1}{8}(\Delta^2 + 16)\frac{kT}{E} + \cdots \right] \tag{12.71}$$

Equation (12.70) can be solved numerically to obtain the thermal neutron spectrum, $\phi(E)$. The solution is shown in Fig. 12.5 for different values of the parameter $\Gamma = \Sigma_{a0}/\Sigma_f$.

Numerical Solution

Neutron scattering kernels are frequently so complicated that analytical or even semianalytical solutions are impractical, in which case direct numerical solution of the governing equation is the method of choice. A general numerical solution method, applicable to any scattering kernel, is illustrated for the proton gas model, for which Eq. (12.37) may be rewritten

$$[V(x) + \Gamma]N(x) = \int_0^{x_c} dx' G(x' \to x)N(x') + \int_{x_c}^{\infty} dx' G(x' \to x)N_{\text{asym}}(x') \tag{12.72}$$

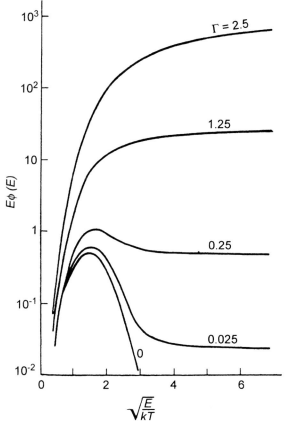

Fig. 12.5 Neutron spectrum predicted by the heavy gas model for a 1/v absorber and different values of $\Gamma = \Sigma_{a0}/\Sigma_f$. (From Ref. 2; used with permission of Wiley.)

where

$$V(x) = \frac{x\,\Sigma_s(x)}{\Sigma_f}, \qquad \Gamma = \frac{\Sigma_{a0}}{\Sigma_f}, \qquad G(x' \to x) = x'\Sigma(x' \to x) \qquad (12.73)$$

and x_c has been chosen so that the asymptotic form N_{asym} from the slowing-down range may be used for $x' > x_c$. In this case, the last term in Eq. (12.72) may be written $cx\,\mathrm{erf}(x)/(x_c + \Gamma)^2$, where $\mathrm{erf}(x)$ is the error function.

Dividing the thermal energy range ($0 \leq x \leq x_c$) into I intervals and using the trapezoidal rule, the right side of Eq. (12.72) may be approximated:

$$\sum_{j=1}^{i-1} G(x_j \to x_i)N(x_j)\Delta_j + G(x_i \to x_i)N(x_i)\Delta_i + \frac{2cx_i\,\mathrm{erf}(x_i)}{(x_c + \Gamma)^2} \qquad (12.74)$$

Equations (12.72) now may be solved directly by matrix inversion or by iteration. For the iterative solution, the equations are rearranged to obtain

$$N(x_i) = \frac{1}{(V(x_i) + \Gamma)} \left[\sum_{j=1}^{I} G(x_j \to x_i) N(x_j) \Delta_j + \frac{2cx_i \, \mathrm{erf}(x_i)}{(x_c + \Gamma)^2} \right]$$

$$i = 1, \dots, I \tag{12.75}$$

The iterative solution of Eqs. (12.75) proceeds by guessing $N^{(0)}(x_j)$, evaluating the right-hand side, calculating $N^{(1)}(x_i)$, and so on. A convenient starting guess is $N^{(0)}(x_i) = N_{\mathrm{asym}}(x_i)$. It is important to enforce neutron conservation during the iteration, which is done by adjusting c.

Moments Expansion Solution

The continuous slowing-down, or moments expansion, methodology that was applied in Chapter 11 to the neutron slowing-down problem is also applicable to the neutron thermalization problem. For heavy elements, the development is similar to that of age theory. Defining

$$u \equiv \ln \frac{T}{E} \tag{12.76}$$

and changing to the lethargy variable, Eq. (12.37) may be written

$$(\Sigma_a + \Sigma_s)\phi(u) \equiv \int_{-\infty}^{\infty} \Sigma(u' \to u)\phi(u') \, du' \tag{12.77}$$

where

$$\phi(u) \equiv En(E)v(E), \qquad \Sigma(u' \to u) \equiv E\Sigma(E' \to E)$$

Since $\phi(u)$ is approximately constant in the slowing-down range above thermal, $\phi(u')$ is expanded in a Taylor's series about u to obtain

$$\left[\Sigma_a(u) + \Sigma_s(u) - \langle \xi^0 \Sigma \rangle \right] \phi(u)$$

$$= -\langle \xi \Sigma \rangle \frac{d\phi(u)}{du} + \frac{1}{2!} \langle \xi^2 \Sigma \rangle \frac{d^2\phi(u)}{du^2} + \cdots + \frac{(-1)^n}{n!} \langle \xi^n \Sigma \rangle \frac{d^n\phi(u)}{du^n} + \cdots \tag{12.78}$$

where

$$\langle \xi^n \Sigma \rangle \equiv (-1)^n \int_{-\infty}^{\infty} (u' - u)^n \Sigma(u' \to u) \, du' \tag{12.79}$$

Noting that for energies above thermal (no upscattering) the nth term in Eq. (12.78) is of order $(\xi_0)^{n-1}$ relative to $\langle \xi^n \Sigma \rangle d\phi/du$, where ξ_0 is the average logarithmic energy loss for scattering by free atoms at rest [$\xi_0 \equiv \xi^{\mathrm{iso}} = 1 +$

$\alpha \ln \alpha / (1 - \alpha)$]. Hence, for scattering from atoms other than hydrogen and deuterium, Eq. (12.38) can be truncated after a few terms with little loss in accuracy.

Differentiating Eq. (12.38), truncating terms higher than $d^2\phi/du^2$, solving for $\langle \xi \Sigma \rangle d^2\phi/du^2$, using this result in Eq. (12.38), and neglecting terms of order (ξ_0^3) and higher yields

$$\left[\Sigma_a(u) + \Sigma_s(u) - \langle \xi^0 \Sigma \rangle \right] \phi(u)$$

$$= -\left\{ \Sigma_s(u)\xi(u) + \gamma(u)\left[\Sigma_a(u) + \Sigma_s(u) - \langle \xi^0 \Sigma_s \rangle \right] + \gamma(u)\frac{d}{du}\langle \xi \Sigma \rangle \right\} \frac{d\phi(u)}{du}$$

$$(12.80)$$

which can be integrated to obtain

$$\phi(u) = \frac{K(u)\exp\{-\int_{-\infty}^{u}[\hat{\Sigma}_a(u')/(\xi(u')\Sigma_s(u') + \gamma(u')\hat{\Sigma}_a(u'))]du'\}}{\xi(u)\Sigma_s(u) + \gamma(u)\hat{\Sigma}_a(u)} \quad (12.81)$$

where

$$K(u) \equiv \exp\left\{ \int_{-\infty}^{u} \left[\frac{\hat{\Sigma}_a(u')(d\gamma(u')/du') + (d[\xi(u')\Sigma_s(u')]/du')}{\xi(u')\Sigma_s(u') + \gamma(u')\hat{\Sigma}_a(u')} \right. \right.$$

$$\left. \left. - \frac{g(u')[\Sigma_a(u') + \gamma(u')(d\hat{\Sigma}_a(u')/du')]}{\xi(u')\Sigma_s(u') + \gamma(u')\hat{\Sigma}_a(u')} \right] du' \right\}$$

$$g(u) \equiv \frac{\xi(u)\Sigma_s(u) + \gamma(u)\hat{\Sigma}_a(u)}{\Sigma_s(u)\xi(u) + \gamma(u)\hat{\Sigma}_a(u) + \gamma(u)(d\langle \xi \Sigma \rangle/du)} - 1 \qquad (12.82)$$

$$\hat{\Sigma}_a(u) \equiv \Sigma_a(u) + \Sigma_s(u) - \langle \xi^0 \Sigma \rangle$$

$$\xi(u) \equiv \frac{\langle \xi \Sigma(u) \rangle}{\Sigma_s(u)}$$

$$\gamma(u) \equiv \frac{1}{2}\frac{\langle \xi^2 \Sigma(u) \rangle}{\langle \xi \Sigma(u) \rangle}$$

The moments of the scattering kernel are given by

$$\langle \xi^n \Sigma \rangle = (-1)^{n+1}\Sigma_f \left\{ \frac{\alpha}{1-\alpha}\left[(\ln \alpha)^n - n(\ln \alpha)^{n-1} + n(n-1)(\ln \alpha)^{n-2} \right. \right.$$

$$\left. + \cdots + (-1)^n n! \right] + \frac{(-1)^{n+1}n!}{1-\alpha}$$

$$- \frac{(1+\mu)^2}{4\mu}\left[\left(\frac{2\mu}{1+\mu} - \frac{1}{2}\mu(1-\alpha) - \frac{1-\mu}{1+\mu} \right)\alpha(\ln \alpha)^n \right.$$

$$\left. \left. - (1-\alpha)\alpha(\ln 2)^{n-1}(2\ln \alpha + n) \right] \frac{\bar{T}}{E} \right\} + O\left(\frac{\mu}{E^2} \right)$$

$$(12.83)$$

Table 12.1 Thermalization Parameters for Carbon

T (K)	Graphite			Free Gas		
	$\bar{T}/T$	$(K^2)_{av}/T^2$	B_{av}/T^2	$\bar{T}/T$	$(K^2)_{av}/T^2$	B_{av}/T^2
300	2.363	21.63	25	1	15/4	0
600	1.432	7.794	25/4	1	15/4	0

where $\mu \equiv m/M$, the ratio of the masses of the neutron and the scattering atom. For scattering by unbound atoms at rest, $K(u) \to 1$ and Eq. (12.81) is identical to the Grueling–Goertzel approximation of Chapter 10. For $\gamma = 0$, Eq. (12.81) reduces to Fermi age theory. The presence of $\gamma \neq 0$ accounts for upscattering in the thermal range of energies. It is the decrease in $\hat{\Sigma}_a = \Sigma_a + \Sigma_s - \langle \xi^0 \Sigma \rangle$, rather than the decrease in $\xi \Sigma_s$, that is the dominant effect of the chemical binding.

For neutron thermalization by graphite, an explicit expression for the thermal spectrum is given by

$$\xi \Sigma_f E \phi(E) = 1 - 1.1138 \left(\frac{1}{2} \Delta \right) z + \left[0.6526 \left(\frac{1}{2} \Delta \right) + 1.913 \frac{\bar{T}}{T} \right] z^2$$

$$- \left[0.2673 \left(\frac{1}{2} \Delta \right)^3 + 3.313 \frac{\bar{T}}{T} \right] z^3$$

$$+ \left[0.08596 \left(\frac{1}{2} \Delta \right)^4 + 2.752 \left(\frac{1}{2} \Delta \right)^2 \frac{\bar{T}}{T} + 4.935 \left(\frac{\bar{T}}{T} \right)^2 \right.$$

$$\left. + 0.201 \frac{(K^2)_{av}}{T^2} - 0.6204 \frac{B_{av}}{T^2} \right] z^4 + O(z^5) \tag{12.84}$$

where $z \equiv (T/E)^{1/2}$, $\Delta \equiv 2\Sigma_a(T)/\mu \Sigma_f$, and the other parameters are defined in terms of the crystal vibration spectra for perpendicular, $\rho_1(\omega)$, and parallel, $\rho_2(\omega)$, vibrations:

$$\bar{T} \equiv \tfrac{1}{3} T_1 + \tfrac{2}{3} T_2$$

$$T_i \equiv \frac{1}{2} \int_0^{\theta_i} \omega \rho_i(\omega) \coth \frac{\omega}{2T} \, d\omega$$

$$(K^2)_{av} \equiv \tfrac{3}{4} T_1^2 + T_1 T_2 + 2 T_2^2 \tag{12.85}$$

$$B_{av} \equiv \frac{1}{3} \int_0^{\theta_1} \omega^2 \rho_1(\omega) \, d\omega + \frac{2}{3} \int_0^{\theta_2} \omega^2 \rho_2(\omega) \, d\omega$$

where the θ_i are the cutoff frequencies for the respective crystal vibration modes. These parameters are given for graphite and a free carbon gas in a Maxwellian distribution in Table 12.1.

For hydrogenous atoms, it is not possible to truncate Eq. (12.78) as described above for heavy mass scattering atoms. However, noting that

$$\langle \xi^n \Sigma \rangle = (-1)^n n! \Sigma_f + O\left(\frac{1}{E^2}\right) \tag{12.86}$$

for hydrogen, it is possible to obtain a solution $\phi(u)$ accurate to $O(1/E^2)$ by neglecting terms of order $1/E^2$ and higher in Eq. (12.78), which enables this equation to be written

$$\frac{\hat{\Sigma}_a(u)}{\Sigma_f}\phi(u) = -\frac{d\phi(u)}{du} + \frac{d^2\phi(u)}{du^2} - \cdots + (-1)^n \frac{d\phi^n(u)}{du^n} + \cdots \tag{12.87}$$

Operating on Eq. (12.87) with $1 + d/du$ and integrating then yields

$$\phi(u) = \frac{1}{\hat{\Sigma}_a(u) + \Sigma_f} \exp\left[-\int_E^\infty \frac{\hat{\Sigma}_a(E')}{\hat{\Sigma}_a(E') + \Sigma_f}\frac{dE'}{E'}\right] \tag{12.88}$$

Expanding Eq. (12.88) in inverse powers of $(E/T)^{1/2}$ yields

$$
\begin{aligned}
E\phi E = 1 &- 3\left(\frac{1}{2}\Delta\right)z + \left[6\left(\frac{1}{2}\Delta\right)^2 + \frac{\bar{T}}{T}\right]z^2 \\
&- \left[10\left(\frac{1}{2}\Delta\right)^3 + \frac{25}{6}\left(\frac{1}{2}\Delta\right)\frac{\bar{T}}{T}\right]z^3 \\
&+ \left[15\left(\frac{1}{2}\Delta\right)^4 + \frac{43}{4}\left(\frac{1}{2}\Delta\right)^2\left(\frac{\bar{T}}{T}\right) + \frac{3}{4}\left(\frac{\bar{T}}{T}\right)^2 + \frac{4}{5}\frac{(K^2)_{av}}{T^2}\right. \\
&\left. - \frac{1}{2}\frac{B_{av}}{T^2}\right]z^4 + O\left(z^5\right)
\end{aligned}
\tag{12.89}
$$

For hydrogen bound in water molecules at 293 K, the thermalization parameters are $\bar{T}/T = 4.345$, $B_{av}/T^2 = 126.90$, and $(K^2)_{av}/T^2 = 53.63$.

Multigroup Calculation

The thermal neutron scattering transfer function discussed in the preceding sections can be used in a multigroup calculation of the thermal neutron energy spectrum. The group-to-group scattering transfer term is defined as

$$\Sigma^{g' \to g} = \frac{\int_{E_g}^{E_{g-1}} dE \int_{E_{g'}}^{E_{g'-1}} dE' \Sigma_{s0}(E' \to E)\phi(E')}{\int_{E_{g'}}^{E_{g'-1}} dE' \phi(E')} \tag{12.90}$$

Evaluation of Eq. (12.90) requires an approximation for the energy dependence of the thermal neutron flux over the energy interval $E_g < E < E_{g-1}$. One of the approximate thermal neutron spectra above can be used for this purpose, or if the interval is sufficiently small, $\phi = $ constant can be used.

The multigroup thermal neutron flux balance equation, neglecting leakage, is

$$\Sigma_a^g \phi_g = \sum_{\substack{g'=g}}^{G} \Sigma_s^{g' \to g} \phi_{g'} + S_g, \quad g = 1, \ldots, G \tag{12.91}$$

where S_g is the slowing-down source to the upper groups in the thermal energy range.

Applications to Moderators

The thermal neutron flux distribution has been calculated numerically for water with various amounts of admixed cadmium absorber, using both the free gas and Nelkin models to calculate the scattering transfer cross section. Results of the calculations are compared with experiment in Fig. 12.6.

The thermal neutron flux distribution has also been calculated numerically for a large graphite block poisoned with boron, using both the crystal model of Eq. (12.36) and the heavy gas model of Eq. (12.13) to evaluate the scattering transfer cross section. The results are compared with experiment in Fig. 12.7.

12.5
Calculation of Thermal Neutron Energy Spectra in Heterogeneous Lattices

The transport equation for neutrons in the thermal energy region $E < E_{th} \sim 1$ eV is

$$\mathbf{\Omega} \cdot \nabla \psi(\mathbf{r}, E, \mathbf{\Omega}) = \int_0^{E_{th}} dE' \int_0^{4\pi} d\mathbf{\Omega}' \Sigma_s(\mathbf{r}, \mathbf{\Omega}' \to \mathbf{\Omega}, E' \to E) \psi(\mathbf{r}, E', \mathbf{\Omega}')$$

$$+ \frac{S(\mathbf{r}, E)}{4\pi}, \quad 0 < E < E_{th} \tag{12.92}$$

where

$$S(\mathbf{r}, E) = \int_{E_{th}}^{\infty} dE' \Sigma_s(\mathbf{r}, E' \to E) \phi(\mathbf{r}, E') \tag{12.93}$$

is the source of neutron scattering into the thermal region from the slowing-down region. With reference to Section 9.2, this equation can be converted into an integral equation for the scalar neutron flux, which for the case of isotropic scattering may be written

$$\phi(\mathbf{r}, E) = \int d\mathbf{r}' \frac{e^{-\alpha(\mathbf{r}, \mathbf{r}')}}{4\pi |\mathbf{r} - \mathbf{r}'|^2} \left[\int_0^{E_{th}} dE' \Sigma_{s0}(\mathbf{r}, E' \to E) \phi(\mathbf{r}, E') + S(\mathbf{r}, E') \right] \tag{12.94}$$

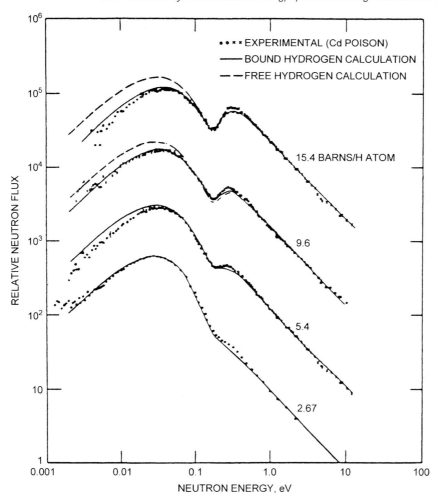

Fig. 12.6 Experimental and calculated neutron energy spectrum in water with cadmium poisons. (From Ref. 3; used with permission of Wiley.)

Dividing the problem of interest (e.g., a fuel assembly) up into I spatial regions, integrating Eq. (12.94) over the volume V_i of region i, and defining [by analogy with Eq. (9.52)]

$$T^{j \to i}(E) \equiv \frac{1}{V_i} \int_{V_i} d\mathbf{r}_i \int_{V_j} d\mathbf{r}_j \frac{e^{-\alpha(\mathbf{r}_i, \mathbf{r}_j)}}{4\pi |\mathbf{r}_i - \mathbf{r}_j|^2} \qquad (12.95)$$

leads to a coupled set of equations for the group fluxes ϕ_i in each region:

$$\phi_i(E) = \sum_{j=1}^{I} T^{j \to i}(E) \left[\int_0^{E_{\text{th}}} dE' \Sigma_{sj}(\mathbf{r}', E' \to E)\phi_j(E') + S_j(E') \right] \qquad (12.96)$$

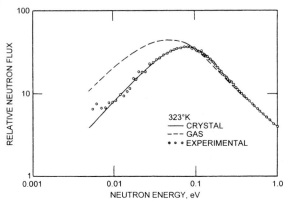

Fig. 12.7 Experimental and calculated neutron energy spectrum in graphite at 323 K. (From Ref. 3; used with permission of Wiley.)

Dividing the thermal energy range into G groups and using an appropriate differential scattering cross section and weighting spectrum to calculate

$$\Sigma_{sj}^{g' \to g} \equiv \frac{\int_{E_g}^{E_{g-1}} \int_{E_{g'}}^{E_{g'-1}} \Sigma_{sj}(E' \to E) w(E') dE' dE}{\int_{E_g}^{E_{g-1}} w(E') dE'} \tag{12.97}$$

Eqs. (12.96) can be integrated over $E_g < E < E_{g-1}$ to obtain the set of multigroup equations

$$\phi_i^g = \sum_{j=1}^{N} T_g^{j \to i} \left(\sum_{g'=1}^{G} \Sigma_{sj}^{g' \to g} \phi_j^{g'} + S_j^g \right), \quad g = 1, \dots, G \tag{12.98}$$

Following Section 9.3, define the *collision probability*

$$P_g^{ji} \equiv V_i \Sigma_{ti}^g \Sigma_{tj}^g T_g^{j \to i} \tag{12.99}$$

in terms of which Eqs. (12.98) can be written

$$\Sigma_{ti}^g V_i \phi_i^g = \sum_{j=1}^{N} P_g^{ji} \frac{\sum_{g'=1}^{G} \Sigma_{sj}^{g' \to g} \phi_i^{g'} + S_j^g}{\Sigma_{tj}^g}, \quad g = 1, \dots, G; \ i = 1, \dots, I \tag{12.100}$$

The collision probabilities can be calculated by the methods of Section 9.3. The multigroup scattering transfer cross sections can be calculated using one of the differential scattering cross sections and a plausible weighting function, as discussed in this chapter. Then the set of $I \times G$ Eqs. (12.100) can be solved for the group fluxes in each region. Such methods are widely employed for practical calculations of the thermal spectra in heterogeneous reactor fuel assemblies.

12.6
Pulsed Neutron Thermalization

Spatial Eigenfunction Expansion

The time-dependent diffusion equation that describes the neutron flux distribution following the introduction of a pulse Q of neutrons with energy E_0 at time $t = 0$ into a uniform but finite nonmultiplying medium is

$$\frac{1}{v}\frac{\partial \phi}{\partial t}(r, E, t) - D(E)\nabla^2 \phi(r, E, t) + \Sigma_a(E)\phi(r, E, t)$$

$$= \int_0^\infty \Sigma(E' \to E)\phi(r, E', t)dE' - \Sigma_s(E)\phi(r, E, t)$$

$$+ \delta(t)Q(r)\delta(E - E_0) \tag{12.101}$$

Assuming that the spatial eigenfunctions satisfying

$$\nabla^2 G_n(r) + B_n^2 G_n(r) = 0 \tag{12.102}$$

and the physical boundary conditions form a complete set, the solution of Eq. (12.101) can be expanded:

$$\phi(r, E, t) = \sum_n \phi_n(E, t)G_n(r) \tag{12.103}$$

and the general orthogonality property

$$\int G_n(r)G_m(r)dr = \delta_{nm} \tag{12.104}$$

can be used to reduce Eq. (12.101) to a coupled set of equations for the $\phi_n(E, t)$:

$$\frac{1}{v}\frac{\partial \phi_n}{\partial t}(E, t) + D(E)B_n^2 \phi_n(E, t) + \Sigma_a(E)\phi_n(E, t)$$

$$= \int_0^\infty \Sigma(E' \to E)\phi_n(E', t)dE' + \delta(E - E_0)\delta(t)Q_n \tag{12.105}$$

where $Q_n \equiv \int dr\, G_n(r)Q(r)$.

Energy Eigenfunctions of the Scattering Operator

The scattering operator S_0 defined by

$$S_0\phi(E) \equiv \int_0^\infty \Sigma(E' \to E)\phi(E')dE' - \Sigma_s(E)\phi(E) \tag{12.106}$$

possesses an eigenvalue spectrum and a set of eigenfunctions in terms of which the energy dependence of the neutron spectrum may be expanded. The general eigenvalue problem is

$$\kappa \chi(E) = \Sigma_s(E)\chi(E) - \int_0^\infty \Sigma(E' \to E)\chi(E')dE' \tag{12.107a}$$

or

$$\kappa \chi(E) = -S_0 \chi(E) \tag{12.107b}$$

The adjoint operator S_0^+ defined (see Chapter 13) by

$$\int \chi^+(E) S_0 \chi(E) dE \equiv \int \chi(E) S_0^+ \chi^+(E) dE \tag{12.108}$$

is

$$S_0^+ \chi^+(E) = \int_0^\infty dE' \Sigma(E \to E') \chi^+(E') - \Sigma_s(E) \chi^+(E) \tag{12.109}$$

The principle of detailed balance,

$$\Sigma(E' \to E) M(E') = \Sigma(E \to E') M(E) \tag{12.110}$$

requires that

$$\chi(E) = M(E) \chi^+(E) \tag{12.111}$$

where $M(E)$ is the Maxwellian distribution. Thus the principle of detailed balance ensures that the lowest eigenvalue $\kappa_0 = 0$ and eigenfunction $\chi_0(E) = M(E)$, independent of scattering model.

As an example, consider the heavy gas model of Section 12.2. From Eq. (12.15),

$$S_0 \chi(E) = \frac{2\Sigma_f}{A} \left[kTE \frac{d^2 \chi(E)}{dE^2} + E \frac{d\chi(E)}{dE} + \chi(E) \right] \tag{12.112}$$

and from Eq. (12.108),

$$S_0^+ \chi^+(E) = \frac{2\Sigma_f}{A} \left[kTE \frac{d^2 \chi^+(E)}{dE^2} + (2kT - E) \frac{d\chi^+(E)}{dE} \right] \tag{12.113}$$

The eigenvalues of the direct and adjoint eigenvalue problems of Eqs. (12.107) and (12.109) are identical (Chapter 13). Substitution of

$$\chi^+(E) = \sum_{n=0}^\infty a_n E^n \tag{12.114}$$

into the adjoint eigenvalue problem

$$S_0^+ \chi^+(E) + \kappa \chi^+(E) = 0 \tag{12.115}$$

and working use of Eq. (12.113) reveals that the eigenvalue spectrum is discrete:

$$\kappa_n = \frac{2\Sigma_f}{A} n, \quad n = 0, 1, 2, \ldots \tag{12.116}$$

The associated eigenfunctions are the Laguerre polynomials of order unity

$$\chi_n^+(E) = \frac{L_n^{(1)}(E)}{\sqrt{n+1}} \qquad (12.117)$$

where

$$L_0^{(1)}(E) = 1, \quad L_1^{(1)}(E) = 2 - E, \quad L_2^{(1)}(E) = 3 - 3E + \tfrac{1}{2}E^2, \quad \dots (12.118)$$

These polynomials constitute a complete set, so any arbitrary function can be expanded in them.

Expansion in Energy Eigenfunctions of the Scattering Operator

Assuming that the function $\phi_n(E, t)$ can be represented as

$$\phi_n(E, t) \sim \phi_n(E)e^{-\lambda_n t} \qquad (12.119)$$

the homogeneous part of Eq. (12.105) reduces to the eigenvalue problem

$$\left[-\frac{\lambda_n}{v} + D(E)B_n^2 \right]\phi_n(E) = S_0\phi_n(E) \qquad (12.120)$$

Expanding each $\phi_n(E)$ in the eigenfunctions of the scattering operator $\chi_m(E)$,

$$\phi_n(E) = \sum_m C_{mn}\chi_m(E) \qquad (12.121)$$

substituting into Eq. (12.120), multiplying by $\chi_p^+(E)$, and integrating over energy yields

$$\sum_m \left(-\lambda_n V_{mp} + D_{mp}B_n^2 + \kappa_m\delta_{mp} \right)C_{mn} = 0 \qquad (12.122)$$

where

$$V_{mp} = \int_0^\infty \frac{1}{v}\chi_m^{(E)}\chi_p^+(E)dE$$

$$\qquad (12.123)$$

$$D_{mp} = \int_0^\infty D(E)\chi_m(E)\chi_p^+(E)dE$$

The set of Eqs. (12.122) formed by multiplying by each $\chi_p^+(E)$ must simultaneously vanish, which by Cramer's rule requires that

$$\det\left(-\lambda_n V_{mp} + D_{mp}B_n^2 + \kappa_m\delta_{mp} \right) = 0 \qquad (12.124)$$

This is the eigenvalue condition from which the κ_m are determined.

The spatial harmonics $n > 0$ will decay more rapidly than the $n = 0$ modes, because $\lambda_{n>0} > \lambda_0$, due to the larger B_n^2. When all the higher spatial harmonics have become negligible, the neutron pulse will decay as a series of energy harmonics of the fundamental spatial mode:

$$\phi(r, E, t) = G_0(r) \sum_p A_p \phi_{p,0}(E) e^{-\lambda_{p,0} t} \tag{12.125}$$

At long times,

$$\phi(r, E, t) \rightarrow G_0(r) \phi_{0,0}(E) e^{-\lambda_{0,0} t} \tag{12.126}$$

since $\lambda_{p,0} < \lambda_{p,n}$ for $n > 0$. If Eq. (12.121) is truncated at one term (i.e., only the fundamental energy eigenfunction is retained), then Eq. (12.124) yields

$$\lambda_{0,0} = \frac{D_{00}}{V_{00}} B_0^2 \equiv \bar{D}_0 B_0^2 \tag{12.127}$$

If the first two terms are retained in Eq. (12.121), then

$$\lambda_{0,0} = \bar{D}_0 B_0^2 - \frac{1}{V_{00} \kappa_1} \left(D_{01} - \bar{D}_0 V_{01} \right)^2 B_0^4 \tag{12.128}$$

Thus measurement of the time decay of the neutron pulse yields information about the Maxwellian average diffusion coefficient, D_{00}. The second term in Eq. (12.128), which is known as the *diffusion cooling term*, depends explicitly on the thermalizing properties of the medium.

References

1 W. Rothenstein and M. Segev, "Unit Cell Calculations," in Y. Ronen, ed., *CRC Handbook of Nuclear Reactor Calculations I*, CRC Press, Boca Raton, FL (**1986**).

2 J. J. Duderstadt and L. J. Hamilton, *Nuclear Reactor Analysis*, Wiley, New York (**1976**), Chap. 9.

3 G. I. Bell and S. Glasstone, *Nuclear Reactor Theory*, Wiley (Van Nostrand Reinhold), New York (**1970**), Chap. 7.

4 D. E. Parks, M. S. Nelkin, N. F. Wikner, and J. R. Beyster, *Slow Neutron Scattering and Thermalization with Reactor Applications*, W. A. Benjamin, New York (**1970**).

5 *Neutron Thermalization and Reactor Spectra*, STI/PUB/160, International Atomic Energy Agency, Vienna (**1968**).

6 I. I. Gurevich and L. V. Tarasov, *Low Energy Neutron Physics*, Wiley, New York (**1968**).

7 M. M. R. Williams, *The Slowing Down and Thermalization of Neutrons*, Wiley-Interscience, New York (**1966**).

8 R. J. Breen et al., "The Neutron Thermalization Problem," in A. Radkowsky, ed., *Naval Reactors Physics Handbook*, U.S. Atomic Energy Commission, Washington, DC (**1964**).

9 K. H. Beckhurts and K. Wirtz, *Neutron Physics*, Springer-Verlag, Berlin (**1964**).

10 T.-Y. Wu and T. Ohmura, *Quantum Theory of Scattering*, Prentice Hall, Englewood Cliffs, NJ (**1962**).

11 H. C. Honeck, "Thermos, A Thermalization Transport Theory Code for Reactor Lattice Calculations," USAEC report BNL-5826, Brookhaven National Laboratory, Upton, NY (**1961**).

12 J. R. Beyster, N. Corngold, H. C. Honeck, G. D. Joanou, and D. E.

PARKS, in *Third U.N. Conference on Peaceful Uses of Atomic Energy* (**1964**), p. 258.

13 E. P. WIGNER and J. E. WILKINS, "Effect of the Temperature of the Mod-

erator on the Velocity Distribution of Neutrons with Numerical Calculations for Hydrogen as Moderator," USAEC report AECD-2275 (**1944**).

Problems

12.1. Use the proton gas model of Eqs. (12.51) to (12.53) and (12.58) to calculate the low-energy neutron flux distribution in water at 300 K. Use $\delta_s^H = 38$ barns, $\sigma_a^{H_2O} = 0.66$ barn, and $\sigma_s^O = 4.2$ barns.

12.2. Use the effective neutron temperature model of Eqs. (4.30) and (4.31) to calculate the thermal neutron spectrum in water at 300 K and compare with the results of Problem 12.1.

12.3. Repeat Problems 12.1 and 12.2 including a $1/v$ absorber with $\sigma_{a0} = 25$ barns and $N_a/N_{H_2O} = 0.1$.

12.4. An H_2O-moderated reactor has a thermal flux of 2.5×10^{14} n/cm$^2 \cdot$ s. Compute the absorption rate density in water at density 0.75 g/cm^3.

12.5. Evaluate the heavy gas model expression for the neutron flux in the limit $E \gg kT$ [Eq. (12.71)] for neutron moderation in graphite at 500 K. Use $\sigma_s^c = 4.8$ barns and $\sigma_a^c = 0.004$ barn.

12.6. Repeat the calculation of Problem 12.5 for an admixture of $1/v$ absorber with $\sigma_{a0} = 0.5$ barn per carbon atom.

12.7. Write a computer code to integrate the nonlinear differential equation (12.50) describing neutron thermalization in a free proton gas. Use an energy mesh of $\Delta E = 0.01$ eV. Calculate the neutron spectrum in water at 300 K. Use $\sigma_s^H = 38$ barns, $\sigma_a^{H_2O} = 0.66$ barn, and $\sigma_a^O = 4.2$ barns.

12.8. Solve the problem of neutron thermalization in a free proton gas model of water at 300 K by direct numerical solution. Compare your results with the results of Problem 12.7.

12.9. Calculate and plot the thermal energy spectrum of neutrons thermalizing in graphite and in a Maxwellian gas of carbon atoms of the same density at 300 K. Use Eq. (12.84).

12.10. Calculate and plot the thermal energy spectrum of neutrons thermalizing in water at 293 K from Eq. (12.89).

13
Perturbation and Variational Methods

In many situations it is necessary to estimate the effect of numerous individual perturbations in the materials properties of the reactor on the multiplication constant or on a reaction rate in a reactor. Perturbation theory provides a means for obtaining an estimate of the change in multiplication constant or reaction rate, neglecting the effect of any change in the neutron flux distribution caused by the perturbation. Generalized perturbation estimates and variational estimates provide a means for taking into account the change in the neutron flux distribution caused by the perturbation, without actually having to calculate it, thus providing a powerful methodology for calculating reactivity coefficients and for performing sensitivity studies. Variational methods also have a much wider application in reactor physics in the development of approximations, and several of these are described.

13.1
Perturbation Theory Reactivity Estimate

Multigroup Diffusion Perturbation Theory

Let us return to the question of estimating the *reactivity worth* of a small change made in a critical reactor described by multigroup diffusion theory:

$$-\nabla \cdot D_0^g(r)\nabla \phi_0^g(r) + \Sigma_{t0}^g(r)\phi_0^g(r)$$

$$= \sum_{g'=1}^{G} \Sigma_0^{g' \to g}(r)\phi_0^{g'}(r) + \frac{1}{k_0}\chi^g \sum_{g'=1}^{G} \nu\Sigma_f^{g'}(r)\phi_0^{g'}(r), \quad g = 1,\ldots,G \quad (13.1)$$

Assume a change in microscopic cross section, density, or geometry such that $D_0 \to D_0 + \Delta D$, $\Sigma_0 \to \Sigma_0 + \Delta\Sigma$. This change will produce a change in the flux $\phi_0 \to \phi_0 + \delta\phi$ and a change in effective multiplication constant $k_0 \to k_0 + \Delta k$, such that the perturbed system is described by

$$-\nabla \cdot \left(D_0^g + \Delta D^g\right)\nabla\left(\phi_0^g + \delta\phi^g\right) + \left(\Sigma_{t0}^g + \Delta\Sigma_t^g\right)\left(\phi_0^g + \delta\phi^g\right)$$

Nuclear Reactor Physics. Weston M. Stacey
Copyright © 2007 WILEY-VCH Verlag GmbH & Co. KGaA, Weinheim
ISBN: 978-3-527-40679-1

$$= \sum_{g'=1}^{G} \left(\Sigma_0^{g' \to g} + \Delta \Sigma^{g' \to g} \right) \left(\phi_0^{g'} + \delta\phi^{g'} \right) + \frac{\chi^g}{k_0 + \Delta k} \sum_{g'=1}^{G} \left(\nu\Sigma_{f0}^{g'} + \Delta \left(\nu\Sigma_f^{g'} \right) \right)$$

$$\times \left(\phi_0^{g'} + \delta\phi^{g'} \right), \quad g = 1, \ldots, G \tag{13.2}$$

Equation (13.2) can, in principle, be solved to determine Δk using the methods described previously. However, in some applications (e.g., the calculation of reactivity coefficients associated with many different possible changes or the evaluation of the sensitivity of the multiplication constant to cross-section uncertainties) this would be impractical because of the large number of such calculations that would be involved. The objective of perturbation theory is to provide an estimate of Δk without requiring a calculation of the perturbed configuration (i.e., without calculating $\delta\phi$).

Using Eqs. (13.1) to eliminate certain terms in Eqs. (13.2), multiplying the resulting equation in each group by an arbitrary (at this point) spatially dependent function ϕ_g^+, integrating over the reactor and summing over groups, we obtain an exact expression for Δk:

$$\int dr \sum_{g=1}^{G} \phi_g^+ \left\{ \left[-\nabla \cdot D_0^g \nabla(\delta\phi^g) + \Sigma_{t0}^g \delta\phi^g - \sum_{g'=1}^{G} \Sigma_0^{g' \to g} \delta\phi^{g'} \right.\right.$$

$$\left. - \frac{\chi^g}{k_0 + \Delta k} \sum_{g'=1}^{G} \nu\Sigma_{f0}^{g'} \delta\phi^{g'} \right]$$

$$+ \left[-\nabla \cdot \Delta D^g \nabla \phi_0^g + \Delta\Sigma_t^g \phi_0^g - \sum_{g'=1}^{G} \Delta\Sigma^{g' \to g} \phi_0^{g'} \right.$$

$$\left. - \frac{\chi^g}{k_0 + \Delta k} \sum_{g'=1}^{G} \Delta \left(\nu\Sigma_f^{g'} \right) \phi_0^{g'} \right]$$

$$+ \left[-\nabla \cdot \Delta D^g \nabla(\delta\phi^g) + \Delta\Sigma_t^g \delta\phi^g - \sum_{g'=1}^{G} \Delta\Sigma^{g' \to g} \delta\phi^{g'} \right.$$

$$\left.\left. - \frac{\chi^g}{k_0 + \Delta k} \sum_{g'=1}^{G} \Delta \left(\nu\Sigma_f^{g'} \delta\phi^{g'} \right) \right] \right\}$$

$$= \left(\frac{1}{k_0 + \Delta k} - \frac{1}{k_0} \right) \int dr \sum_{g=1}^{G} \phi_g^+ \chi^g \sum_{g'=1}^{G} \nu\Sigma_{f0}^{g'} \phi_0^{g'}$$

$$= -\frac{\Delta k}{k_0(\Delta k + k_0)} \int dr \sum_{g=1}^{G} \phi_g^+ \chi^g \sum_{g'=1}^{G} \nu\Sigma_{f0}^{g'} \phi_0^{g'} \tag{13.3}$$

if we were willing to calculate $\delta\phi^g$ in order to evaluate it. However, we wish to neglect $\delta\phi$, which appears in two of the [·] terms on the left side. We might argue (for the moment) that since the third [·] term is the product of $\delta\phi$ and $\Delta\Sigma$ or ΔD, we can neglect it as being of second order in small quantities. However, we cannot make this argument for the first [·] term, which is of first order in small quantities, as is the second [·] term which we wish to evaluate to obtain the perturbation estimate of Δk. Thus are we motivated to choose the ϕ_g^+ to cause the first [·] term in Eq. (13.3) to vanish for arbitrary $\delta\phi^g$. To determine the equation that must be satisfied by ϕ_g^+, it is necessary to twice integrate by parts the gradient part of the first [·] term and use the divergence theorem:

$$
-\int dr\, \phi_g^+ \nabla \cdot D_0^g \nabla(\delta\phi^g) = -\int dr\, \nabla \cdot \left(\phi_g^+ D_0^g \nabla(\delta\phi^g)\right)
$$

$$
+ \int dr\, D_0^g \nabla\phi_g^+ \cdot \nabla(\delta\phi^g) = -\int_s ds\, \phi_g^+ D_0^g \nabla(\delta\phi^g) \cdot \mathbf{n}_s
$$

$$
+ \int dr\, \nabla \cdot \left(\delta\phi^g D_0^g \nabla\phi_g^+\right) - \int dr\, \delta\phi^g \nabla \cdot \left(D_0^g \nabla\phi_g^+\right)
$$

$$
= -\int_s ds\, \phi_g^+ D_0^g \nabla(\delta\phi^g) \cdot \mathbf{n}_s + \int_s ds\, \delta\phi^g D_0^g \nabla\phi_g^+ \cdot \mathbf{n}_s
$$

$$
- \int dr\, \delta\phi^g \nabla \cdot \left(D_0^g \nabla\phi_g^+\right) \tag{13.4}
$$

where $\mathbf{n}_s$ is the outward normal unit vector to the surface of the reactor and the integrals over s are surface integrals. $\delta\phi^g$, which must satisfy the same boundary conditions as ϕ^g, vanishes on the surface of the reactor, which causes the second term on the right in the final form of Eq. (13.4) to vanish. If we choose a boundary condition $\phi_g^+(\mathbf{r}_s) = 0$ (i.e., ϕ_g^+ vanishes on the surface of the reactor), the first term on the right in Eq. (13.4) also vanishes. Using this result in Eq. (13.3) and interchanging the dummy g and g' indices, the vanishing of the first [·] term requires that

$$
\int dr \sum_{g=1}^{G} \delta\phi^g \left[-\nabla \cdot D_0^g \nabla\phi_g^+ + \Sigma_{t0}^g \phi_g^+ \right.
$$

$$
\left. - \sum_{g'=1}^{G} \Sigma_0^{g \to g'} \phi_{g'}^+ - \frac{\nu\Sigma_{f0}^g}{k_0 + \Delta k} \sum_{g'=1}^{G} \chi^{g'} \phi_{g'}^+ \right] = 0 \tag{13.5}
$$

which is satisfied for arbitrary $\delta\phi^g$ if ϕ_g^+ satisfies

$$
-\nabla \cdot D_0^g \nabla\phi_g^+ + \Sigma_{t0}^g \phi_g^+ = \sum_{g'=1}^{G} \Sigma_0^{g \to g'} \phi_{g'}^+ + \frac{\nu\Sigma_{f0}^g}{k_0 + \Delta k} \sum_{g'=1}^{G} \chi^{g'} \phi_{g'}^+,
$$

$$
g = 1, \ldots, G \tag{13.6}
$$

and vanishes on the surface of the reactor:

$$
\phi_{g'}^+(\mathbf{r}_s) = 0, \quad g = 1, \ldots, G \tag{13.7}
$$

With the function ϕ_g^+ which satisfies Eqs. (13.6) and (13.7), with neglect of the third [·] term on the left, and with the approximation $k_0(k_0 + \Delta k) \to k_0$, Eq. (13.3) reduces to the perturbation theory expression for the reactivity worth:

$$\rho_{\text{pert}} \equiv \frac{\Delta k}{k_0} = \int dr \sum_{g=1}^{G} \phi_g^+ \left[\nabla \cdot \Delta D^g \nabla \phi_0^g - \Delta \Sigma_t^g \phi_0^g + \sum_{g'=1}^{G} \Delta \Sigma^{g' \to g} \phi_0^{g'} \right.$$

$$\left. + \frac{\chi^g}{k_0} \sum_{g'=1}^{G} \Delta \left(\nu \Sigma_{f0}^{g'} \right) \phi_0^{g'} \right]$$

$$\div \int dr \sum_{g=1}^{G} \phi_g^+ \chi^g \sum_{g'=1}^{G} \nu \Sigma_{f0}^{g'} \phi_0^{g'} + O(\delta\phi) \tag{13.8}$$

where we indicate by $O(\delta\phi)$ that the neglected third [·] term in Eq. (13.3) introduces an error of order $\delta\phi$ (into a term which itself is of order $\Delta\Sigma$).

13.2
Adjoint Operators and Importance Function

Adjoint Operators

Equation (13.6) is mathematically adjoint to Eq. (13.1), when $k_0 + \Delta k \to k_0$, and the function ϕ_g^+ is called the *adjoint function*. Comparing Eqs. (13.1) and (13.6) term by term identifies the direct and adjoint operators of multigroup diffusion theory, which are denoted symbolically as

Direct	*Adjoint*	
$[D(\phi)]_g \equiv -\nabla \cdot D^g \nabla \phi^g$	$\left[D^+(\phi^+) \right]_g \equiv -\nabla \cdot D^g \nabla \phi_g^+$	
$[\Sigma(\phi)]_g \equiv \Sigma_t^g \phi^g$	$\left[\Sigma^+(\phi^+) \right]_g \equiv \Sigma_t^g \phi_g^+$	
$[S(\phi)]_g \equiv \sum_{g'=1}^{G} \Sigma^{g' \to g} \phi^{g'}$	$\left[S^+(\phi^+) \right]_g \equiv \sum_{g'=1}^{G} \Sigma^{g \to g'} \phi_{g'}^+$	(13.9)
$[F(\phi)]_g \equiv \chi^g \sum_{g'=1}^{G} \nu \Sigma_f^{g'} \phi^{g'}$	$\left[F^+(\phi^+) \right]_g \equiv \nu \Sigma_f^g \sum_{g'=1}^{G} \chi^{g'} \phi_{g'}^+$	

The direct and adjoint operators for group diffusion and group absorption are identical; these operators are said to be *self-adjoint*. On the other hand, the adjoint group scattering and fission operators differ from the direct operators. Note that there is an adjoint boundary condition [Eq. (13.7)] associated with the definition of the adjoint group diffusion operator. In terms of these operators, Eq. (13.8) for the perturbation theory estimate of the reactivity worth of a change in reactor properties becomes

$$
\rho_{\text{pert}} = \frac{\int dr \sum_{g=1}^{G} \phi_g^+ \left\{ -[\Delta D(\phi_0)]_g - [\Delta \Sigma(\phi_0)]_g + [\Delta S(\phi_0)]_g + (1/k_0)[\Delta F(\phi_0)]_g \right\}}{\int dr \sum_{g=1}^{G} \phi_g^+ [F_0(\phi_0)]_g}
$$

$$
\equiv \frac{\int dr \sum_{g=1}^{G} \phi_g^+ \left\{ -[\Delta A(\phi_0)]_g + (1/k_0)[\Delta F \phi_0]_g \right\}}{\int dr \sum_{g=1}^{G} \phi_g^+ [F_0(\phi_0)]_g} + O(\delta\phi) \tag{13.10}
$$

It is clear from the derivation above that the adjoint operators were defined by the requirement

$$
\int dr \sum_{g=1}^{G} \phi_g^+ [B(\phi)]_g \equiv \int dr \sum_{g=1}^{G} \phi^g \left[B^+(\phi^+) \right]_g \tag{13.11}
$$

where $[B(\phi)]_g$ represents any one of the operators in Eq. (13.9). This definition of adjoint group operator is quite general and provides for the immediate generalization of perturbation theory to multigroup transport theory by replacement of $[D(\phi)]_g$ with the appropriate transport group operator $[T(\phi)]_g$.

This formalism may be generalized immediately from multigroup to energy-dependent diffusion or transport theory by replacing the sum over groups by an integral over energy. At this point, we introduce the notation

$$
\langle B\phi \rangle \equiv \int dr \sum_{g=1}^{G} [B(\phi)]_g \quad \text{or} \quad \int dr \int_0^\infty dE \, B(\phi) \tag{13.12}
$$

which allows the compact expression of the perturbation theory estimate of reactivity worth:

$$
\rho_{\text{pert}} = \frac{\langle \phi^+, (\Delta F - \Delta A)\phi_0 \rangle}{\langle \phi^+, F_0\phi_0 \rangle} + O(\delta\phi) \tag{13.13}
$$

In this notation, the definition of the adjoint operator becomes

$$
\langle \phi^+, B\phi \rangle \equiv \langle \phi, B^+\phi^+ \rangle \tag{13.14}
$$

Importance Interpretation of the Adjoint Function

The mathematical definition of the adjoint function (in the diffusion theory approximation) is simply the function that satisfies Eq. (13.6). The physical interpretation is somewhat subtle and involves the concept of neutron importance, more specifically the importance of a neutron introduced into a reactor to a specific observable physical quantity. For example, an importance may be defined as the expected number of counts that will be produced at all subsequent times by a neutron introduced into a reactor with a given position, energy and direction or by the secondary, tertiary, etc. neutrons produced at other energies and directions as a result of fission and scattering events due to the original neutron. For a critical reactor, it is conventional to define the neutron 'importance,' $\psi^+(\mathbf{r}, \mathbf{\Omega}, E)$, as the asymptotic increase in neutron population in the reactor due to a single neutron introduced into the reactor with a given energy E, with a given direction $\mathbf{\Omega}$ and at a given location $\mathbf{r}$. (Actually, we need to speak of neutrons introduced within dE about E, $d\mathbf{r}$ about $\mathbf{r}$, and $d\mathbf{\Omega}$ about $\mathbf{\Omega}$, but we will leave this cumbersome terminology to be understood.) Neutrons introduced with a given energy and direction at a given location can (1) move to another location $\mathbf{r} + d\mathbf{r}$ where the importance is different; (2) be captured, which causes the importance to become zero; (3) be scattered into a different energy E' and direction $\mathbf{\Omega}'$ where the importance is different; or (4) produce fission, which causes the importance of the original neutron to become zero, but which produces ν new neutrons distributed in energy E' and distributed isotropically in direction $\mathbf{\Omega}'$ with different importances. In a critical reactor, the importance must be conserved as the N neutrons move about and undergo these various reactions, which can be expressed as

$$
N\Big[\big[\psi^+(\mathbf{r}+d\mathbf{r}, \mathbf{\Omega}, E) - \psi^+(\mathbf{r}, \mathbf{\Omega}, E)\big] - \big[\Sigma_a(\mathbf{r}, E) + \Sigma_s(\mathbf{r}, E)\big]\psi^+(\mathbf{r}, \mathbf{\Omega}, E)
$$

$$
+ \int_0^\infty dE' \int_0^{4\pi} d\mathbf{\Omega}' \Sigma_s(\mathbf{r}, E \to E', \mathbf{\Omega} \to \mathbf{\Omega}')\psi^+(\mathbf{r}, \mathbf{\Omega}', E')
$$

$$
+ \frac{\nu\Sigma_f(\mathbf{r}, E)}{k} \int_0^\infty dE' \int_0^{4\pi} d\mathbf{\Omega}' \chi(E')\psi^+(\mathbf{r}, \mathbf{\Omega}, E')\Big] = 0 \tag{13.15}
$$

Making a Taylor's series expansion

$$
\psi^+(\mathbf{r}+d\mathbf{r}, \mathbf{\Omega}, E) \simeq \psi^+(\mathbf{r}, \mathbf{\Omega}, E) + \mathbf{\Omega}\cdot\nabla\psi^+(\mathbf{r}, \mathbf{\Omega}, E) \tag{13.16}
$$

in Eq. (13.15) leads to the transport equation satisfied by the neutron importance:

$$
\mathbf{\Omega}\cdot\nabla\psi^+(\mathbf{r}, \mathbf{\Omega}, E) - \Sigma_t(\mathbf{r}, E)\psi^+(\mathbf{r}, \mathbf{\Omega}, E)
$$

$$
+ \int_0^\infty dE' \int_0^{4\pi} d\mathbf{\Omega}' \Sigma_s\big(\mathbf{r}, E \to E', \mathbf{\Omega} \to \mathbf{\Omega}'\big)\psi^+\big(\mathbf{r}, \mathbf{\Omega}', E'\big)
$$

$$
+ \frac{\nu\Sigma_f(\mathbf{r}, E)}{k} \int_0^\infty dE' \int_0^{4\pi} d\mathbf{\Omega}' \chi(E')\psi^+\big(\mathbf{r}, \mathbf{\Omega}', E'\big) = 0 \tag{13.17}
$$

The importance of neutrons leaving the reactor is zero, which provides a boundary condition for the neutron importance,

$$
\psi^+(\mathbf{r}_s, \mathbf{\Omega}, E) = 0, \qquad \mathbf{n}_s\cdot\mathbf{\Omega} > 0 \tag{13.18}
$$

where $\mathbf{n}_s$ is the outward normal unit vector to the surface of the reactor.

Compare these equations with the neutron transport equation and surface boundary condition derived in Chapter 9:

$$-\mathbf{\Omega} \cdot \nabla \psi (\mathbf{r}, \mathbf{\Omega}, E) - \Sigma_t (\mathbf{r}, E) \psi (\mathbf{r}, \mathbf{\Omega}, E)$$

$$+ \int_0^\infty dE' \int_0^{4\pi} d\mathbf{\Omega}' \Sigma_s \left(\mathbf{r}, E' \to E, \mathbf{\Omega}' \to \mathbf{\Omega} \right) \psi \left(\mathbf{r}, \mathbf{\Omega}', E' \right)$$

$$+ \frac{\chi (E)}{k} \int_0^\infty dE' \int_0^{4\pi} d\mathbf{\Omega}' \nu \Sigma_f \left(\mathbf{r}, E' \right) \psi \left(\mathbf{r}, \mathbf{\Omega}, E' \right) = 0 \tag{13.19}$$

and

$$\psi (\mathbf{r}_s, \mathbf{\Omega}, E) = 0, \qquad \mathbf{n}_s \cdot \mathbf{\Omega} < 0 \tag{13.20}$$

The neutron transport equation is based on a backward balance of neutrons among those neutrons that scattered or were produced in fission or moved from a nearby location in the immediate past (i.e., in the interval $t - \Delta t$ to t) and those neutrons that are undergoing absorption and scattering now (i.e., at time t). The importance equation is based on a forward balance of the importance among those neutrons that are being absorbed or scattered now (i.e., at time t) and the importance of those neutrons that will move to a nearby location or be scattered into a different energy and direction or produce fission neutrons with different energy and direction in the immediate future (i.e., in the interval t to $t + \Delta t$).

Eigenvalues of the Adjoint Equation

In the foregoing development of Eq. (13.17) from physical arguments, the same effective multiplication constant was used to achieve a steady-state importance balance equation as was used to achieve a steady-state neutron balance equation. We now establish formally that the eigenvalues of the neutron balance equation

$$(A - \lambda F)\phi = 0 \tag{13.21}$$

and of the adjoint equation

$$\left(A^+ - \lambda^+ F^+ \right) \phi^+ = 0 \tag{13.22}$$

are identical when the adjoint operators are related to the direct operators by Eq. (13.14). Multiplying Eq. (13.21) by ϕ^+ and integrating over space, direction, and energy, multiplying Eq. (13.22) by ϕ and integrating, and making use of Eq. (13.14) yields

$$\lambda = \frac{\langle \phi^+, A\phi \rangle}{\langle \phi^+, F\phi \rangle} = \frac{\langle \phi, A^+\phi^+ \rangle}{\langle \phi, F^+\phi^+ \rangle} = \lambda^+ \tag{13.23}$$

13.3
Variational/Generalized Perturbation Reactivity Estimate

In many practical applications, a perturbation to the properties of the reactor will cause a change in the neutron flux distribution which has a significant effect on the reactivity worth of the perturbation (i.e., the neglected third [·] term in Eq. (13.3) is important). The perturbation theory of Section 13.1 can be extended to take into account the change in the flux distribution without actually requiring its calculation. Such extensions can be developed within the context of variational theory or simply as a heuristic extension of perturbation theory; the results are the same except for minor differences. This extended perturbation theory is widely used in reactor physics in the calculation of reactivity worths (and reaction rate ratios—next section) and for the performance of sensitivity studies. Since the variational theory is more systematic and has broader applications in reactor physics, we follow the variational development of an extended perturbation theory for estimating reactivity worths.

One-Speed Diffusion Theory

Consider a critical reactor described by the one-speed diffusion equation

$$-\nabla \cdot D_0 \nabla \phi_0 + \Sigma_{a0}\phi_0 - \lambda_0 \nu \Sigma_{f0}\phi_0 = 0 \tag{13.24}$$

where, for convenience of notation, we set $\lambda \equiv k^{-1}$. Making use of the definition of adjoint operator given by Eq. (13.11) with $G = 1$, the one-speed diffusion theory adjoint equation satisfies

$$-\nabla \cdot D_0 \nabla \phi_0^+ + \Sigma_{a0}\phi_0^+ - \lambda_0 \nu \Sigma_{f0}\phi_0^+ = 0 \tag{13.25}$$

Thus the one-speed diffusion equation is self-adjoint and $\phi^+ = \phi$.

Now consider perturbations $D_0 \rightarrow D = D_0 + \Delta D$ and $\Sigma_0 \rightarrow \Sigma = \Sigma_0 + \Delta\Sigma$, which cause $\phi_0 \rightarrow \phi_{\text{ex}} = \phi_0 + \delta\phi$ and $\lambda_0 \rightarrow \lambda = \lambda_0 + \Delta\lambda$. The perturbed system satisfies

$$-\nabla \cdot D \nabla \phi_{\text{ex}} + \Sigma_a \phi_{\text{ex}} - \lambda \nu \Sigma_f \phi_{\text{ex}} = 0 \tag{13.26}$$

Multiplying Eq. (13.26) by ϕ_0^+, multiplying Eq. (13.25) by ϕ_{ex}, integrating over volume, subtracting, and rearranging yields an exact expression for the reactivity worth of the perturbation:

$$\rho_{\text{ex}}\{\phi_0^+, \phi_{\text{ex}}\} = -\Delta\lambda$$

$$= \frac{\langle \phi_0^+, (\lambda_0 \Delta(\nu\Sigma_f)\phi_{\text{ex}} + \nabla \cdot \Delta D \nabla \phi_{\text{ex}} - \Delta\Sigma_a \phi_{\text{ex}})\rangle}{\langle \phi_0^+, \nu\Sigma_f \phi_{\text{ex}}\rangle} \tag{13.27}$$

If we used the approximation $\phi_{\text{ex}} \approx \phi_0$ to evaluate Eq. (13.27), we would obtain the perturbation theory estimate of the reactivity worth of the change, in one-speed diffusion theory:

$$\rho_{\text{pert}}\{\phi_0^+, \phi_0\} = \frac{\langle \phi_0^+, (\lambda_0 \Delta(\nu\Sigma_f)\phi_0 + \nabla \cdot \Delta D \nabla \phi_0 - \Delta\Sigma_a \phi_0)\rangle}{\langle \phi_0^+, \nu\Sigma_f \phi_0\rangle}$$

$$+ O(\delta\phi) \tag{13.28}$$

which is accurate to first order in $\delta\phi$.

Variational or generalized perturbation theory allows us to obtain an estimate that is accurate to second order in $\delta\phi$. Note that Eq. (13.27) defines a number that is evaluated by performing integrals over space (more generally over space and energy) involving the functions ϕ_0^+ and ϕ_{ex}. Such a function of functions is known as a *functional*. The idea behind variational theory is to construct an equivalent variational functional $\rho_{var}\{\phi^+, \phi, \Gamma^+\}$ which has the properties: (1) $\rho_{var}\{\phi_0^+, \phi_{ex}, \Gamma_{ex}^+\}$ has the same value as the functional $\rho_{ex}\{\phi_0^+, \phi_{ex}\}$ if ϕ_0^+ and ϕ_{ex} are used to evaluate ρ_{var}, and (2) $\rho_{var}\{\phi_0^+, \phi, \Gamma^+\}$ evaluated with functions ϕ_0^+ and $\phi = \phi_{ex} + \delta\phi$ yields a value that differs from $\rho_{ex}\{\phi_0^+, \phi_{ex}\}$ by $O(\delta\phi^2, \delta\phi\delta\Gamma^+)$. In particular, $\rho_{var}\{\phi_0^+, \phi_0, \Gamma^+\} = \rho_{ex}\{\phi_0^+, \phi_{ex}\} + O(\delta\phi^2)$ when $\phi_{ex} = \phi_0 + \delta\phi$.

We construct

$$\rho_{var}\{\phi_0^+, \phi, \Gamma^+\} = \frac{\langle \phi_0^+, (\lambda_0 \Delta(\nu\Sigma_f)\phi + \nabla \cdot \Delta D\nabla\phi - \Delta\Sigma_a\phi)\rangle}{\langle \phi_0^+, \nu\Sigma_F\phi\rangle}$$
$$\times \left[1 - \langle \Gamma^+, (-\nabla \cdot D\nabla\phi + \Sigma_a\phi - \lambda\nu\Sigma_f\phi)\rangle \right] \quad (13.29)$$

by taking the exact functional of Eq. (13.27) and multiplying it by 1 minus a correction functional constructed by premultiplying the exact Eq. (13.26) by Γ^+ and integrating over space (space and energy in general). This functional obviously satisfies the first of the properties of the variational functional above, because when $\phi = \phi_{ex}$, the correction functional vanishes and the first term reduces identically to $\rho_{ex}\{\phi_0^+, \phi_{ex}\}$. Subtracting yields

$$\rho_{var}\{\phi_0^+, \phi_{ex}, \Gamma^+\} - \rho_{var}\{\phi_0^+, \phi_{ex} - \delta\phi, \Gamma^+\}$$
$$= -\frac{\langle \phi_0^+, (\lambda_0\Delta(\nu\Sigma_f)\phi_{ex} + \nabla \cdot \Delta D\nabla\phi_{ex} - \Delta\Sigma_a\phi_{ex})\rangle}{\langle \phi_0^+, \nu\Sigma_f\phi_{ex}\rangle}$$
$$\times \left[\langle \Gamma^+, (-\nabla \cdot D\nabla\delta\phi + \Sigma_a\delta\phi - \lambda\nu\Sigma_f\delta\phi)\rangle + \frac{\langle \phi_0^+, \nu\Sigma_f\delta\phi\rangle}{\langle \phi_0^+, \nu\Sigma_f\phi_{ex}\rangle} \right]$$
$$+ \frac{\langle \phi_0^+, (\lambda_0\Delta(\nu\Sigma_f)\delta\phi + \nabla \cdot \Delta D\nabla\delta\phi - \Delta\Sigma_a\delta\phi)\rangle}{\langle \phi_0^+, \nu\Sigma_f\phi_{ex}\rangle} + O(\delta\phi^2) \quad (13.30)$$

The explicit terms on the right in this expression will vanish for arbitrary $\delta\phi$ if Γ^+ is chosen to satisfy

$$-\nabla \cdot D\nabla\Gamma_{ex}^+ + \Sigma_a\Gamma_{ex}^+ - \lambda\nu\Sigma_f\Gamma_{ex}^+$$
$$= \frac{-\nabla \cdot \Delta D\nabla\phi_0^+ + \Delta\Sigma_a\phi_0^+ - \lambda_0\Delta(\nu\Sigma_f)\phi_0^+}{\langle \phi_0^+, (-\nabla \cdot \Delta D\nabla\phi_{ex} + \Delta\Sigma_a\phi_{ex} - \lambda_0\Delta(\nu\Sigma_f)\phi_{ex})\rangle} - \frac{\nu\Sigma_f\phi_0^+}{\langle \phi_0^+, \nu\Sigma_f\phi_{ex}\rangle}$$
$$(13.31)$$

so that $\rho_{var}\{\phi_0^+, \phi, \Gamma_{ex}^+\} = \rho_{ex}\{\phi_0^+, \phi_{ex}\} + O(\delta\phi^2)$. Thus evaluation of the variational functional of Eq. (13.29) using the functions ϕ_0^+ given by Eq. (13.25), Γ_{ex}^+ given by Eq. (13.31), and any function $\phi = \phi_{ex} + \delta\phi$ yields an estimate of the reactivity worth of the change which is accurate to $O(\delta\phi^2)$.

Unfortunately, solving Eq. (13.31) requires a knowledge of ϕ_{ex}, avoidance of the calculation of which is the purpose of this development. If, instead of Eq. (13.31), we use the equation obtained by changing $\phi_{ex} \rightarrow \phi_0$

$$-\nabla \cdot D_0 \nabla \Gamma_0^+ + \Sigma_{a0} \Gamma_0^+ - \lambda_0 \nu \Sigma_{f0} \Gamma_0^+$$

$$= \frac{-\nabla \cdot \Delta D \nabla \phi_0^+ - \lambda_0 \Delta (\nu \Sigma_f) \phi_0^+ + \Delta \Sigma_a \phi_0^+}{\langle \phi_0^+, (-\nabla \cdot \Delta D \nabla \phi_0 + \Delta \Sigma_a \phi_0 - \lambda_0 \Delta (\nu \Sigma_f) \phi_0)) \rangle} - \frac{\nu \Sigma_f \phi_0^+}{\langle \phi_0^+, \nu \Sigma_f \phi_0 \rangle}$$

(13.32)

it can be shown that $\rho_{var}\{\phi_0^+, \phi, \Gamma_0^+\} = \rho_{ex}\{\phi_0^+, \phi_{ex}\} + O(\delta\phi^2, \delta\phi\delta\Gamma^+)$, where $\Gamma_{ex}^+ = \Gamma_0^+ + \delta\Gamma^+$. Thus the variational estimate $\rho_{var}\{\phi_0^+, \phi, \Gamma_0^+\}$ is accurate to second order in the (presumably) small quantities $\delta\phi$ and $\delta\Gamma$.

The function Γ^+ is related to the flux change, $\delta\phi$, caused by the perturbation. The equation satisfied by $\delta\phi$ is obtained by using $D = D_0 + \Delta D$, $\Sigma = \Sigma_0 + \Delta\Sigma$, and $\phi_{ex} = \phi_0 + \delta\phi$ in Eq. (13.26) and making use of Eq. (13.24):

$$-\nabla \cdot D \nabla (\delta\phi) + \Sigma_a (\delta\phi) - \lambda \nu \Sigma_f (\delta\phi)$$

$$= -[-\nabla \cdot \Delta D \nabla \phi_0 + \Delta \Sigma_a \phi_0 - \Delta(\lambda \nu \Sigma_f) \phi_0]$$

(13.33)

Comparing this equation with Eq. (13.31), it is apparent that $\Gamma^+ \sim -\delta\phi$, since $\phi_0 = \phi_0^+$ for one group. A similar relationship may be established for multigroup theory.

Defining the variational flux correction factor

$$f_{var}\{\phi_0, \Gamma_0^+\} = \langle \Gamma_0^+, (-\nabla \cdot \Delta D \nabla \phi_0 + \Delta \Sigma_a \phi_0 - \lambda_0 \Delta (\nu \Sigma_f) \phi_0$$

$$- \Delta\lambda \Delta(\nu \Sigma_f) \phi_0) \rangle$$

(13.34)

the variational estimate for the reactivity worth of a change in reactor properties may be written

$$\rho_{var}\{\phi_0^+, \phi_0, \Gamma_0^+\} = \rho_{pert}\{\phi_0^+, \phi_0\}[1 - f_{var}\{\phi_0, \Gamma_0^+\}]$$

(13.35)

as the perturbation theory estimate times a flux correction factor.

The calculations required for the variational estimate include the solution for the three spatial functions ϕ_0^+, ϕ_0, and Γ_0^+ for the parameters of the critical reactor and the evaluation of the indicated spatial integrals in Eq. (13.29). The left side of Eq. (13.32) is identical with the homogeneous Eq. (13.24). However, the useful biorthogonality property $\langle \Gamma_0^+, F_0 \phi_0 \rangle = 0$ can be demonstrated (Ref. 13), which assures the existence of a solution. Note that the source term on the right of Eq. (13.32) will in general be the same for all perturbations taking place within a given spatial domain, since the magnitude of the perturbations appear in the numerator and denominator, implying that the calculation of one such Γ_0^+ for each distinct spatial domain of interest will allow the evaluation of the reactivity worths of a large number of perturbations of different types and magnitudes within that spatial domain.

The reactivity estimate of Eq. (13.35) has been found to be quite accurate when the change in properties is such as to produce a positive reactivity ($f_{var} < 0$) or a small negative reactivity for which $0 < f_{var} \ll 1$. However, for large negative reactivities such that $f_{var} \sim 1$, the $(1 - f_{var})$ term becomes inaccurate. In such cases it is more accurate to use $\rho_{var} = \rho_{pert}[1 - f_{var}/(1 + f_{var})]$, which form can be derived from consideration (Ref. 1) of the exact functional of Eq. (13.27). Thus a better variational estimate for the reactivity worth is

$$
\rho_{var}\{\phi_0^+, \phi_0, \Gamma_0^+\} = \rho_{pert}\{\phi_0^+, \phi_0\} \begin{cases} 1 - f_{var}\{\phi_0, \Gamma_0^+\}, & \rho_0 > 0 \\[2ex] 1 - \dfrac{f_{var}\{\phi_0, \Gamma_0^+\}}{1 + f_{var}\{\phi_0, \Gamma_0^+\}}, & \rho_0 < 0 \end{cases} \tag{13.36}
$$

Other Transport Models

This formalism can be generalized immediately to other representations of neutron transport (e.g., multigroup diffusion or transport theory). Let the operator A represent the transport, absorption and scattering and the operator F represent the fission. Then Eqs. (13.24) and (13.25) for the flux and adjoint in the critical reactor generalize to

$$
(A_0 - \lambda_0 F_0)\phi_0 = 0 \tag{13.37}
$$

$$
(A_0^+ - \lambda_0 F_0^+)\phi_0^+ = 0 \tag{13.38}
$$

and Eq. (13.26) for the flux in the perturbed reactor generalizes to

$$
(A - \lambda F)\phi_{ex} = 0 \tag{13.39}
$$

The exact value and perturbation theory estimate of the reactivity worth of the perturbation of Eqs. (13.27) and (13.28) become

$$
\rho_{ex}\{\phi_0^+, \phi_{ex}\} = \frac{\langle \phi_0^+, (\lambda_0 \Delta F - \Delta A)\phi_{ex}\rangle}{\langle \phi_0^+, F\phi_{ex}\rangle} \tag{13.40}
$$

$$
\rho_{pert}\{\phi_0^+, \phi_0\} = \frac{\langle \phi_0^+, (\lambda_0 \Delta F - \Delta A)\phi_0\rangle}{\langle \phi_0^+, F\phi_0\rangle} \tag{13.41}
$$

Equation (13.32) for the generalized adjoint function Γ_0^+ becomes

$$
(A_0 - \lambda_0 F_0)\Gamma_0^+ = \frac{(\Delta A^+ - \lambda_0 \Delta F^+)\phi_0^+}{\langle \phi_0^+, (\Delta A - \lambda_0 \Delta F)\phi_0\rangle} - \frac{F_0^+ \phi_0^+}{\langle \phi_0^+, F_0\phi_0\rangle} \tag{13.42}
$$

The variational estimate for the reactivity worth of the perturbation is still given by Eq. (13.36), where now ρ_{pert} is given by Eq. (13.41) and the flux correction factor is given by

$$
f_{var}\{\phi_0^+, \phi_0, \Gamma_0^+\} = \langle \Gamma_0^+, (\Delta A - \lambda_0 \Delta F - \Delta\lambda\Delta F)\phi_0\rangle \tag{13.43}
$$

Reactivity Worth of Localized Perturbations in a Large PWR Core Model

Exact, perturbation theory, and variational calculations were made of the reactivity worth of a change in the thermal group absorption cross section in
a two-group model of a large (about 40 migration lengths) slab model of a
PWR core. The perturbations were made in the left quarter of the core model.
Small cross-section changes produced small reactivity changes that were well estimated by both perturbation and variational methods because the associated flux
change was small. Larger cross-section changes, which produced larger reactivity worths and significant flux changes were poorly predicted by perturbation
theory, but the variational flux correction resulted in quite accurate predictions
even for flux tilts on the order of 100%. The reactivity predictions are shown in
Fig. 13.1, and associated flux shapes for the unperturbed core and for the core
with two of the perturbations are shown in Fig. 13.2. The unit of reactivity is
$\mathrm{pcm} = 10^{-5}$.

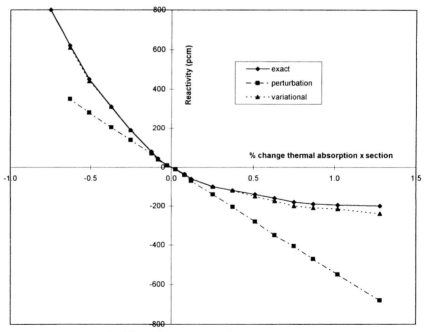

Fig. 13.1 Reactivity worth of thermal cross-section changes
over the left quarter of a slab PWR model in two-group
diffusion theory: comparison of exact, perturbation theory, and
variational calculations. (From Ref. 1; used with permission of
American Nuclear Society.)

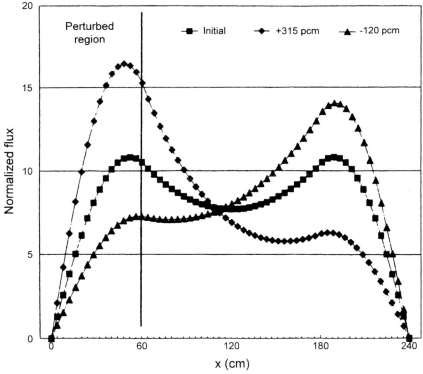

Fig. 13.2 Thermal flux distributions for the unperturbed and two perturbed conditions in a two-group slab PWR model. (From Ref. 1; used with permission of American Nuclear Society.)

Higher-Order Variational Estimates

A variational formalism (Refs. 17 and 20) has been developed for making reactivity estimates that are accurate to higher order in $\delta\phi = \phi_{ex} - \phi_0$. However, the complexity of such estimates has limited their practical application.

13.4
Variational/Generalized Perturbation Theory Estimates of Reaction Rate Ratios in Critical Reactors

Consider the problem of estimating the reaction rate ratio

$$\mathrm{RR}\{\phi_{ex}\} = \frac{\langle \Sigma_i \phi_{ex} \rangle}{\langle \Sigma_j \phi_{ex} \rangle} \tag{13.44}$$

in a perturbed system in which the exact flux is given by Eq. (13.39), without solving Eq. (13.39) for ϕ_{ex}. The unperturbed critical reactor is described by Eq. (13.37).

As before, the operators A and F represent transport-absorption-scattering and fission, respectively, in whatever theory is chosen to describe the neutron distribution in the reactor (e.g., multigroup diffusion, multigroup S_n). Clearly, a perturbation theory estimate

$$\text{RR}_{\text{pert}}\{\phi_0\} = \frac{\langle \Sigma_i \phi_0 \rangle}{\langle \Sigma_j \phi_0 \rangle} \qquad (13.45)$$

has errors of $O(\delta\phi = \phi_{\text{ex}} - \phi_0)$.

Defining the generalized adjoint function, Γ_R^+, by

$$\left(A^+ - \lambda F^+\right)\Gamma_R^+ = \frac{\Sigma_i}{\langle \Sigma_i \phi_{\text{ex}} \rangle} - \frac{\Sigma_j}{\langle \Sigma_j \phi_{\text{ex}} \rangle} \qquad (13.46)$$

the variational estimate

$$\text{RR}_{\text{var}}\left\{\phi_0, \Gamma_{R0}^+\right\} = \frac{\langle \Sigma_i \phi_0 \rangle}{\langle \Sigma_j \phi_0 \rangle}\left[1 - \left\langle \Gamma_{R0}^+, (\Delta A - \Delta(\lambda F))\phi_0 \right\rangle\right] \qquad (13.47)$$

where Γ_{R0}^+ is calculated from Eq. (13.46) but with $A \to A_0$, $F \to F_0$ and $\phi_{\text{ex}} \to \phi_0$, will have a second-order error $O(\delta\Gamma^+\delta\phi)$, as may be demonstrated by evaluating

$$\text{RR}_{\text{ex}}\{\phi_{\text{ex}}\} - \text{RR}_{\text{var}}\left\{\phi_0, \Gamma_{R0}^+\right\} = O\left(\delta\Gamma_R^+\delta\phi\right) \qquad (13.48)$$

Several reaction rate ratios calculated for a multigroup diffusion theory model of the spherical ZEBRA fast reactor critical assembly are given in Table 13.1. The breeding ratio is the ratio of the ^{238}U capture rate integrated over the region to the ^{239}Pu fission rate integrated over the region. The reference assembly composition is given in Table 13.2. It is clear that the flux correction provided by the variational (generalized perturbation theory) calculation is important in achieving an accurate estimate.

Table 13.1 Table Perturbed Reaction Rate Ratios

Ratio	Reference Value	Perturbation	RR$_{\text{exact}}$	RR$_{\text{pert}}$	RR$_{\text{var}}$
Central $\sigma_c^{28}/\sigma_f^{49}$	0.09866	Add 0.01 at/cm^3 Na 0 $\to$ 9.45 cm	0.10241	0.09866	0.10225
	0.09866	Increase σ_f^{49} 10%	0.08964	0.08969	0.08964
	0.09866	Add 0.0015 at/cm^3 Pu 9.45 $\to$ 22.95 cm	0.09887	0.09866	0.09884
Core breeding ratio	0.80040	Add 0.01 at/cm^3 Na 0 $\to$ 9.45 cm	0.80554	0.80040	0.80549
Assembly breeding ratio	2.1844	Increase σ_f^{49} 10%	1.9939	2.0038	1.9937
	2.1844	Add 0.0015 at/cm^3 Pu 9.45 $\to$ 22.95 cm	1.6034	1.6446	1.6049

Source: Data from Ref. 13; used with permission of Academic Press.

Table 13.2 Composition of Spherical Computational Model of ZEBRA Critical Assembly

Isotope	Core (0 → 22.95 cm)	Blanket (22.95 → 49.95 cm)
^{239}Pu	0.00371	0.0003
^{238}U	0.03174	0.04099
^{56}Fe	0.005698	0.00477

Source: Data from Ref. 13; used with permission of Academic Press.

13.5
Variational/Generalized Perturbation Theory Estimates of Reaction Rates

Many problems in reactor physics can be formulated as fixed source problems described by

$$A_0\phi_0 = S \tag{13.49}$$

where the operator A represents transport, absorption, scattering, and if present in the particular problem, fission. Let us imagine that Eq. (13.49) has been solved for ϕ_0 and then the reactor is perturbed, so that the flux now satisfies

$$A\phi_{\text{ex}} = S \tag{13.50}$$

and we wish to evaluate the reaction rate

$$R\{\phi_{\text{ex}}\} = \langle \Sigma \phi_{\text{ex}} \rangle \tag{13.51}$$

without calculating ϕ_{ex}. The perturbation theory estimate $R_{\text{pert}}\{\phi_0\} = \langle \Sigma \phi_0 \rangle + O(\delta\phi)$ obviously is only accurate to zero order in the flux perturbation that is caused by the perturbation in the reactor properties.

Defining an adjoint function, ϕ^+_{R0}, by

$$A^+\phi^+_{R0} = \Sigma \tag{13.52}$$

it is easy to show that the variational estimate

$$R_{\text{var}}\{\phi_0, \phi^+_{R0}\} = \langle \Sigma \phi_0 \rangle - \langle \phi^+_{R0}, (S - A_0\phi_0) \rangle \tag{13.53}$$

differs from the exact calculation of the reaction rate in the perturbed reactor by a second-order term,

$$R_{\text{ex}}\{\phi_{\text{ex}}\} - R_{\text{var}}\{\phi_0, \phi^+_{R0}\} = O(\delta\phi\delta\phi^+_R) \tag{13.54}$$

where ϕ^+_{R0} is calculated from Eq. (13.52) with $A \to A_0$.

By making use of the definition of the adjoint operator, it follows that

$$\langle \Sigma \phi \rangle = \langle A^+ \phi_R^+, \phi \rangle = \langle \phi_R^+, A\phi \rangle = \langle \phi_R^+ S \rangle \tag{13.55}$$

implying that the reaction rate can also be calculated by integrating the product of the source distribution S and the generalized adjoint function ϕ_R^+ over the volume of the reactor. This result suggests the interpretation of ϕ_R^+ as an importance function for a source neutron to produce the reaction represented by Σ.

13.6
Variational Theory

Stationarity

We have constructed variational extensions of perturbation theory by establishing functionals which when evaluated with the exact solutions of the governing equations yielded the exact value of a quantity of interest (e.g., reactivity worth, reaction rate) and which when evaluated with approximate solutions of the governing equations (or exact solutions of equations that approximated the governing equations) differed from the exact result by terms of second order in the difference between the approximate solutions and the exact solutions. In other words, terms involving first-order variations between the exact and approximate solutions vanished when the approximate solutions were used in the variational functionals. This property is described by stating that the variational functionals are *stationary* about the exact solutions of the governing equations (i.e., the first variations vanish), and the functions that make the variational functional stationary (by satisfying the governing equations) are known as the *stationary functions*. This means that the same value of the variational functional will be obtained when evaluated with two different functions that differ infinitesimally, if one of these functions exactly satisfies the governing equations (i.e., is the stationary function of the variational functional).

Minimum principles of various sorts are usually represented by variational functionals, and the minimum property of the variational functional is a form of stationarity condition. However, with a minimum principle or minimum variational functional, the value of the variational functional will increase when evaluated with any function which differs sufficiently from the stationary function that $\delta \phi^2$ is significant, whereas the value of a stationary variational functional may be greater or less than the stationary value when evaluated with a function that differs sufficiently from the stationary function.

Roussopolos Variational Functional

Consider again the variational functional of Eq. (13.53), which we now write in the more general form known as the *Roussopolos functional*:

$$R_{\text{var}}\{\phi, \phi_R^+\} = \langle \Sigma \phi \rangle - \langle \phi_R^+, A\phi - S \rangle \tag{13.56}$$

The stationarity condition is

$$\delta R_{\text{var}} \equiv R_{\text{var}}\{\phi + \delta\phi, \phi_R^+ + \delta\phi_R^+\} - R_{\text{var}}\{\phi, \phi_R^+\}$$
$$= \langle \Sigma \delta\phi \rangle - \langle \delta\phi_R^+ (A\phi - S) \rangle + \langle A^+\phi_R^+, \delta\phi \rangle = 0 \tag{13.57}$$

For arbitrary and independent variations $\delta\phi$ and $\delta\phi_R^+$, this requires that

$$\delta\phi: \ \Sigma - A^+\phi_{Rs}^+ = 0, \qquad \delta\phi^+: \ S - A\phi_s = 0 \tag{13.58}$$

where the subscript s indicates the stationary solution. When the stationary solutions are used to evaluate the functional of Eq. (13.56), the exact value $\langle \Sigma\phi_s \rangle$ is obtained. When approximate functions—trial functions—$\phi = \phi_s + \delta\phi$ and $\phi_R^+ = \phi_{Rs}^+ + \delta\phi_R^+$ are used to evaluate the functional of Eq. (13.56), the value obtained differs from the exact value by a term of order $(\delta\phi\delta\phi_R^+)$.

Schwinger Variational Functional

The estimate of the reaction rate provided by Eq. (13.56) or (13.53) is obviously sensitive to the normalization of the trial functions. The stationarity of the variational functional can be used to choose the best normalization. Write $\chi^+ = c^+\phi_R^+$ and $\chi = c\phi$. Substitute these trial functions into the variational functional of Eq. (13.56) and require stationarity with respect to arbitrary and independent variations δc^+ and δc:

$$\delta R_{\text{var}} = \langle \Sigma\phi \rangle \delta c + \langle \phi_R^+, (S - Ac\phi) \rangle \delta c^+ - \langle A^+ c^+ \phi_R^+, \phi \rangle \delta c = 0 \tag{13.59}$$

which is satisfied for arbitrary δc and δc^+ only if

$$c^+ = \frac{\langle \Sigma\phi \rangle}{\langle \phi_R^+, A\phi \rangle}, \qquad c = \frac{\langle \phi_R^+ S \rangle}{\langle \phi_R^+, A\phi \rangle} \tag{13.60}$$

Using these normalizations in Eq. (13.56) yields the equivalent Schwinger variational principle:

$$J\{\phi, \phi_R^+\} = \frac{\langle \Sigma\phi \rangle \langle \phi_R^+ S \rangle}{\langle \phi_R^+, A\phi \rangle} \tag{13.61}$$

the value of which is independent of the normalization of the trial functions.

Rayleigh Quotient

Consider the critical reactor eigenvalue problem described by the transport and adjoint equations

$$(A - \lambda F)\phi = 0, \qquad (A^+ - \lambda F^+)\phi^+ = 0 \tag{13.62}$$

The Rayleigh quotient

$$\lambda\{\phi^+, \phi\} = \frac{\langle \phi^+, A\phi \rangle}{\langle \phi^+, F\phi \rangle} \tag{13.63}$$

is a variational functional for the eigenvalue. The value of Eq. (13.63) when the exact solution of the first of Eqs. (13.62) is used in its evaluation is clearly the exact eigenvalue. The requirement that the first variation of the Rayleigh quotient vanish,

$$\delta\lambda = \frac{\langle \delta\phi^+, A\phi \rangle + \langle \phi^+, A\delta\phi \rangle}{\langle \phi^+, F\phi \rangle} - \frac{\langle \phi^+, A\phi \rangle}{\langle \phi^+, F\phi \rangle} \frac{\langle \delta\phi^+, F\phi \rangle + \langle \phi^+, F\delta\phi \rangle}{\langle \phi^+, F\phi \rangle} = 0 \tag{13.64}$$

for arbitrary and independent variations $\delta\phi^+$ and $\delta\phi$ requires that the stationary functions ϕ_s and ϕ_s^+ satisfy Eqs. (13.62).

Construction of Variational Functionals

Although the construction of variational functionals is usually done by trial and error, there is a systematic procedure that can guide the process. The basic idea is to add the inner product of some function ϕ^+ or Γ^+ with the governing equation for ϕ to the quantity of interest and then use the stationarity requirement to determine the equation satisfied by ϕ^+ or Γ^+. For example, if we want to estimate a reaction rate $\langle \Sigma\phi \rangle$ and ϕ is determined by $A\phi = S$, we construct the Roussopolos functional $R_{var}\{\phi^+, \phi\} = \langle \Sigma\phi \rangle - \langle \phi^+, (S - A\phi) \rangle$ of Eq. (13.56), and find from the stationarity requirement that ϕ^+ must satisfy $A^+\phi^+ = \Sigma$. As another example, if we want to estimate the reactivity worth $\langle \phi_0^+, (\lambda_0 \Delta F - \Delta A)\phi \rangle / \langle \phi_0^+, F\phi \rangle$ of changes ΔF and ΔA leading from $(A_0 - \lambda_0 F_0)\phi_0 = 0$ to $(A - \lambda F)\phi = 0$, we construct $\rho_{var}\{\phi_0^+, \phi, \Gamma^+\} = \langle \phi_0^+, (\lambda_0 \Delta F - \Delta A)\phi \rangle / \langle \phi_0^+, F\phi \rangle [1 - \langle \Gamma^+, (A - \lambda F)\phi \rangle]$ of Eq. (13.29).

13.7
Variational Estimate of Intermediate Resonance Integral

Consider, as an application, the elastic slowing down of neutrons in the presence of a resonance absorber and a moderator (m), which is described by

$$[\sigma_m + \sigma(u)]\phi(u) = \int_{u-\Delta_m}^{u} du' \frac{e^{(u'-u)}}{1-\alpha_m} \sigma_m \phi(u')$$

$$+ \int_{u-\Delta}^{u} du' \frac{e^{(u'-u)}}{1-\alpha} \sigma_s(u')\phi(u')$$

$$\simeq \sigma_m + \int_{u-\Delta}^{u} du' \frac{e^{(u'-u)}}{1-\alpha} \sigma_s(u')\phi(u') \tag{13.65}$$

where σ_m, σ, and σ_s are moderator scattering cross section per atom of resonance absorber and the total and scattering microscopic cross sections of the resonance absorber, respectively. It has been assumed that the moderator in-scatter integral can be evaluated using the asymptotic flux, which is constant in lethargy, and the constant has been chosen as unity, in writing the second form of the equation. This equation corresponds to the second of Eqs. (13.58).

The quantity of physical interest is the resonance integral

$$I = \int \sigma_a(u)\phi(u)\,du = \langle \sigma_a \phi \rangle \tag{13.66}$$

Using the definition of adjoint operator given by Eq. (13.14), where $\langle \cdot \rangle$ now indicates an integral over lethargy, the first of Eqs. (13.58)—the adjoint equation—for this problem is

$$[\sigma_m + \sigma(u)]\phi_R^+(u) - \int_u^{u+\Delta} du' \frac{e^{(u-u')}}{1-\alpha}\sigma_s(u)\phi_R^+(u') = \sigma_a(u) \tag{13.67}$$

and the Schwinger variational functional of Eq. (13.61) becomes

$$J\{\phi, \phi_R^+\} = \frac{\langle \Sigma \phi \rangle \langle \phi_R^+ S \rangle}{\langle \phi_R^+, A\phi \rangle}$$

$$= \frac{[\int_0^\infty du\, \sigma_a(u)\phi(u)][\int_0^\infty du\, \phi_R^+(u)\sigma_m]}{\int_0^\infty du\, \phi_R^+(u)\{[\sigma_m + \sigma(u)]\phi(u) - \int_{u-\Delta}^u du'\, [e^{(u'-u)}/(1-\alpha)]\sigma_s(u')\phi(u')\}} \tag{13.68}$$

In choosing trial functions, we recall the narrow resonance and wide resonance approximations of Chapter 4:

$$\phi_{NR}(u) = \frac{\sigma_m + \sigma_p}{\sigma_m + \sigma(u)}, \qquad \phi_{WR}(u) = \frac{\sigma_m}{\sigma_m + \sigma_a(u)} \tag{13.69}$$

where σ_p is the background scattering cross section of the resonance absorber. Making similar approximations in Eq. (13.67) as were made in deriving Eqs. (13.69), we can derive approximate adjoint functions. For wide resonances, $\sigma_s\phi_R^+$ is approximately constant over the scattering interval and can be removed from the integral in Eq. (13.67), yielding

$$\phi_{WR}^+(u) = \frac{\sigma_a(u)}{\sigma_m + \sigma_a(u)} \tag{13.70}$$

In the limit of very narrow resonances the off-resonance form for $\sigma_s\phi_R^+$ can be used to evaluate the scattering integral to obtain

$$\phi_{NR}^+(u) = \frac{\sigma_a(u)}{\sigma_m} \frac{\sigma_m + \sigma_p}{\sigma_m + \sigma(u)} \tag{13.71}$$

These results suggest the trial functions

$$\phi_\lambda(u) = \frac{\sigma_m + \lambda\sigma_p}{\sigma_m + \sigma_a(u) + \lambda\sigma_s(u)}$$

$$\phi_\kappa^+(u) = \frac{\sigma_m + \kappa\sigma_p}{\sigma_m} \frac{\sigma_a(u)}{\sigma_m + \sigma_a(u) + \kappa\sigma_s(u)} \tag{13.72}$$

which contain arbitrary constants λ and κ that are determined by using Eqs. (13.72) in the variational functional of Eq. (13.63) and requiring stationarity with respect to arbitrary and independent variations $\delta\lambda$ and $\delta\kappa$, which leads to the transcendental equations

$$\lambda = \frac{\chi_{\kappa\lambda}^2}{1 + \chi_{\kappa\lambda}^2}, \qquad \beta_\kappa = \beta_\lambda \frac{1 + 2\sigma_p(1 - Y_{\kappa\lambda})}{\sigma_m} \tag{13.73}$$

which must be solved for $\chi_{\kappa\lambda}$ and $Y_{\kappa\lambda}$, where

$$\beta_i^2 = 1 + \frac{\sigma_0}{\sigma_m + i\sigma_p} \frac{\Gamma_\gamma + i\Gamma_n}{\Gamma}, \qquad i = \lambda, \kappa, 0, 1$$

$$\chi_{\kappa\lambda} = \frac{2E_0(1 - \alpha)}{\Gamma(\beta_\kappa + \beta_\lambda)} \tag{13.74}$$

$$Y_{\kappa\lambda} = \arctan \chi_{\kappa\lambda} / \chi_{\kappa\lambda}$$

where the Γ's are the resonance widths, and σ_0 and E_0 are the peak resonance cross section and the energy at which it occurs.

The variational estimate of the resonance integral is

$$J\{\phi_\lambda, \phi_\kappa^+\}$$

$$= \frac{\pi\sigma_0\Gamma_\gamma/2E_0}{\beta_\lambda + ((\sigma_m + \sigma_p)(\beta_1^2 - \beta_0^2)/(\sigma_m + \lambda\sigma_p)(\beta_\kappa + \beta_\lambda))(1 - \lambda - Y_{\kappa\lambda})} \tag{13.75}$$

which has been shown to provide a more accurate estimate than either the narrow-resonance or wide-resonance approximations to the resonance integral for resonances of 'intermediate' width.

13.8
Heterogeneity Reactivity Effects

As an application of the Raleigh quotient, consider a heterogeneous lattice described by collision probability integral transport theory. Equations (12.100) become

$$\mu_t^{ng}\phi_n^g = \sum_{g'n'} P_{n'n}^g \left(\mu_s^{n'g'\to g} + \lambda\nu\mu_f^{n'g'}\chi^g\right)\phi_n^{g'},$$

$$n, n' = 1, \ldots, N; \ g, g' = 1, \ldots, G$$

and the corresponding adjoint equations are

$$\mu_t^{ng}\phi_n^{+g} = \sum_{g'n'} P_{nn'}^{g'} \left(\mu_s^{ng\to g'} + \lambda\nu\mu_f^{ng}\chi^{g'}\right)\phi_n^{+g'},$$

$$n, n' = 1, \ldots, N; \ g, g' = 1, \ldots, G \tag{13.76}$$

where n and g refer to spatial region and group, $P^g_{n'n}$ is the probability that a neutron in group g and region n' has its next collision in region n, and μ^{ng} is the cross section in group g and region n times the volume of region n divided by the total volume of all regions.

The Rayleigh quotient of Eq. (13.63) becomes

$$\lambda\{\phi^+, \phi\} = \frac{\sum_{gn} \phi_n^{+g}\left(\mu_t\phi_n^g - \sum_{g'n'} P^g_{n'n}\mu_s^{n'g'\to g}\phi_{n'}^{g'}\right)}{\sum_{gn} \phi_n^{+g}\chi^g \sum_{g'n'} P^g_{n'n}\nu\mu_f^{n'g'}\phi_{n'}^{g'}} \tag{13.77}$$

This expression can be used in a number of ways. For example, approximate flux and adjoint distributions (even one based on a homogenized model) can be used as trial functions in Eq. (13.77) to obtain a more accurate estimate of the infinite multiplication factor in a heterogeneous lattice.

13.9
Variational Derivation of Approximate Equations

The requirement that a variational functional be stationary about the function ϕ_s which causes the first variation of the functional to vanish is entirely equivalent to requiring that the function ϕ_s satisfy the governing equation for ϕ if the variational functional is constructed so that satisfaction of this governing equation is the stationarity condition. Thus the equations of reactor physics can be stated equivalently as stationary variational functionals, just as the equations of particle dynamics can be equivalently stated in terms of a Hamiltonian. For example, the statement that ϕ_s^+ and ϕ_s make the Raleigh quotient of Eq. (13.63) stationary is entirely equivalent to the statement that ϕ_s and ϕ_s^+ satisfy Eqs. (13.62) and the associated boundary conditions. This equivalence provides a basis for the variational derivation of approximate equations.

As an example, consider a reactor described by one-speed diffusion theory in two dimensions:

$$-\frac{\partial}{\partial x}\left[D(x, y)\frac{\partial\phi(x, y)}{\partial x}\right] - \frac{\partial}{\partial y}\left[D(x, y)\frac{\partial\phi(x, y)}{\partial y}\right]$$
$$+ \left[\Sigma_a(x, y) - \frac{\nu\Sigma_f(x, y)}{k}\right]\phi(x, y) = 0 \tag{13.78}$$

An equivalent variational description is the stationarity requirement for the variational functional:

$$F\{\phi^+, \phi\} = \int dx\, dy\, \phi^+(x, y)$$
$$\times\left[-\frac{\partial}{\partial x}D\frac{\partial\phi}{\partial x} - \frac{\partial}{\partial y}D\frac{\partial\phi}{\partial y} + \left(\Sigma_a - \frac{\nu\Sigma_f}{k}\right)\phi\right] \tag{13.79}$$

Recalling that the one-speed diffusion equation is self-adjoint, we look for a separable solution:

$$\phi^+(x, y) = \phi(x, y) = \phi_x(x)\phi_y(y) \tag{13.80}$$

consisting of a known function $\phi_y(y)$, perhaps obtained from a one-dimensional calculation, and an unknown function $\phi_x(x)$. Substituting Eq. (13.80) into Eq. (13.79) and requiring stationarity with respect to arbitrary variations $\delta\phi_x$ (ϕ_y is specified and hence does not allow arbitrary variations) leads to a one-dimensional equation for the unknown $\phi_x(x)$:

$$-\frac{d}{dx}\left[D_x(x)\frac{d\phi_x(x)}{dx}\right] + \left[\Sigma_{Rx}(x) + D_x(x)B_g^2(x) - \frac{\nu\Sigma_{fx}}{k}\right]\phi_x(x) = 0 \tag{13.81}$$

where the effective y-independent constants are defined as weighted integrals over y:

$$D_x(x) \equiv \int dy\, \phi_y^2(y)D(x, y)$$

$$D_x(x)B_g^2(x) \equiv -\int dy\, \phi_y(y)\frac{\partial}{\partial y}\left[D(x, y)\frac{\partial\phi_y(y)}{\partial y}\right] \tag{13.82}$$

$$\Sigma_x(x) \equiv \int dy\, \phi_y^2(y)\Sigma(x, y)$$

This procedure is referred to as *variational synthesis* and is described more fully in Chapter 15.

Inclusion of Interface and Boundary Terms
In deriving Eqs. (13.81) and (13.82) it was implicitly assumed that the known function $\phi_y(y)$ is continuous over all y, which limits the approximation to trial functions ϕ_y which are continuous in y. This limitation can be removed if the variational functional is modified so that stationarity requires not only satisfaction of Eq. (13.78) but also continuity of flux and current across an interface at $y = y_i$. Stationarity of the modified functional

$$F_{\mathrm{dis}}\{\phi^+, \phi\} = \int dx\, dy\, \phi^+(x, y)$$

$$\times\left[-\frac{\partial}{\partial x}\left(D\frac{\partial\phi}{\partial x}\right) - \frac{\partial}{\partial y}\left(D\frac{\partial\phi}{\partial y}\right) + \left(\Sigma_a - \frac{\nu\Sigma_f}{k}\right)\phi\right]$$

$$+ \int dx\, \phi_i^+(x, y_i)[\phi(x, y_i + \varepsilon) - \phi(x, y_i - \varepsilon)]$$

$$+ \int dx\, J_i^+(x, y_i)\left[-D(x, y_i + \varepsilon)\frac{\partial\phi(x, y_i + \varepsilon)}{\partial y}\right.$$

$$\left.+ D(x, y_i - \varepsilon)\frac{\partial\phi(x, y_i - \varepsilon)}{\partial y}\right] \tag{13.83}$$

with respect to arbitrary and independent variations $\delta\phi_x$ over the volume and variations $\delta\phi_i^+$ and δJ_i^+ on the interface at $y = y_i$ requires both that Eq. (13.78) be satisfied everywhere in the reactor except on the interface and that continuity of flux and current be satisfied at the interface:

$$\phi(x, y_i + \varepsilon) = \phi(x, y_i - \varepsilon)$$

$$-D(x, y_i + \varepsilon)\frac{\partial\phi(x, y_i + \varepsilon)}{\partial y} = -D(x, y_i - \varepsilon)\frac{\partial\phi(x, y_i - \varepsilon)}{\partial y} \qquad (13.84)$$

Boundary terms can be included in a similar fashion, leading to variational functionals which admit trial functions that do not satisfy the boundary conditions.

Inclusion of interface and boundary terms is important for the development of synthesis and nodal approximations and is discussed in greater detail in Chapter 15 as well as in Section 13.11.

13.10
Variational Even-Parity Transport Approximations

Variational Principle for the Even-Parity Transport Equation

The even-parity form of the transport equation introduced in Section 9.11 is convenient for the development of approximate transport equations when the scattering and source are isotropic. A variational functional for the even-parity component of the angular flux, which is self-adjoint, may be written

$$J\{\psi^+, \phi\} = \int_V d\mathbf{r}\left\{\int d\mathbf{\Omega}\left[\frac{1}{\Sigma}(\mathbf{\Omega}\cdot\nabla\psi^+)^2 + \Sigma_t(\psi^+)^2\right] - \Sigma_s\phi^2 - 2\phi S\right\}$$
$$+ \int_S ds \int d\mathbf{\Omega}|\mathbf{n}\cdot\mathbf{\Omega}|[\psi^+(\mathbf{r}_s)]^2 \qquad (13.85)$$

where the dependence on $(\mathbf{r}, \mathbf{\Omega})$ has been suppressed, and the two integrals are over the volume V and the bounding surface S, with $\mathbf{n}$ being the outward normal to the surface. Note that here ψ^+ refers to the even component of the angular flux, not an adjoint function. Taking the variation of the functional J with respect to arbitrary but dependent (since ϕ depends on ψ^+) variations $\delta\psi^+$ and $\delta\phi$ about some reference functions ψ_0^+ and ϕ_0 yields

$$\delta J \equiv J\{\psi_0^+ + \delta\psi^+, \phi + \delta\phi\} - J\{\psi_0^+, \phi\}$$
$$= 2\int_V d\mathbf{r}\int d\mathbf{\Omega}\left[\frac{1}{\Sigma(r)}(\mathbf{\Omega}\cdot\nabla\delta\psi^+)(\mathbf{\Omega}\cdot\nabla\psi_0^+)\right.$$
$$\left. + \Sigma_t\psi_0^+\delta\psi^+ - \delta\phi(\Sigma_s\phi_0 + S)\right]$$
$$+ 2\int_S ds \int d\mathbf{\Omega}|\mathbf{n}_s\cdot\mathbf{\Omega}|\psi_0^+\delta\psi^+ + O\big((\delta\psi^+)^2, (\delta\phi)^2\big)$$

$$= 2 \int_V d\mathbf{r} \int d\mathbf{\Omega} \delta \psi^+ \left[-\mathbf{\Omega} \cdot \nabla \left(\frac{1}{\Sigma_t} \mathbf{\Omega} \cdot \nabla \psi_0^+ \right) + \Sigma_t \psi_0^+ - \Sigma_s \phi_0 - S \right]$$

$$+ 2 \int_S ds \int d\mathbf{\Omega} \delta \psi^+ \left[|\mathbf{n} \cdot \mathbf{\Omega}| \psi_0^+ + \mathbf{n} \cdot \mathbf{\Omega} \left(\frac{1}{\Sigma_t} \mathbf{\Omega} \cdot \nabla \psi_0^+ \right) \right] + O \left((\delta \psi^+)^2 \right)$$

$$\tag{13.86}$$

where integration by parts and the divergence theorem have been used to obtain the final form. The requirements that the volume and surface integrals vanish for arbitrary and independent variations $\delta \psi^+$ in the volume and on the surface are just the transport equation for the one-speed (or within-group) even-parity transport equation:

$$-\mathbf{\Omega} \cdot \nabla \left[\frac{1}{\Sigma_t(\mathbf{r})} \mathbf{\Omega} \cdot \nabla \psi_0^+(\mathbf{r}, \mathbf{\Omega}) \right] + \Sigma_t(\mathbf{r}) \psi_0^+(\mathbf{r}, \mathbf{\Omega}) - \Sigma_s(\mathbf{r}) \phi_0(\mathbf{r}) - S(\mathbf{r}) = 0$$

$$\tag{13.87}$$

and the vacuum boundary condition satisfied by the even-parity flux component:

$$\mathbf{\Omega} \cdot \nabla \psi^+(\mathbf{r}_s, \mathbf{\Omega}) \pm \Sigma_t(\mathbf{r}_s) \psi^+(\mathbf{r}_s, \mathbf{\Omega}) = 0, \qquad \mathbf{\Omega} \cdot \mathbf{n}_s \gtrless 0 \tag{13.88}$$

Ritz Procedure

This is a procedure for constructing an improved approximate solution by combining several plausible approximate solutions, each of which perhaps represents some feature expected in the exact solution, that is, by approximating the even-parity flux by an expansion in known functions $\chi_i(\mathbf{r}, \mathbf{\Omega})$:

$$\psi^+(\mathbf{r}, \mathbf{\Omega}) \simeq \sum_i a_i \chi_i(\mathbf{r}, \mathbf{\Omega}) \tag{13.89}$$

The general Ritz method proceeds by substituting this expansion into the variational functional describing the system of interest, Eq. (13.85) in our case, and requiring stationarity (vanishing of first variations) for arbitrary and independent variations of the combining coefficients, a_i:

$$\delta J\{a\} = 0 = 2\delta a^{\mathrm{T}} [Aa - S] \tag{13.90}$$

where a is a column vector of the a_i, a^{T} is the transposed row vector, A is a matrix with elements

$$A_{ij} = \int_V d\mathbf{r} \left\{ \int d\mathbf{\Omega} \left[\frac{1}{\Sigma_t} (\mathbf{\Omega} \cdot \nabla \chi_i)(\mathbf{\Omega} \cdot \nabla \chi_j) - \Sigma_t \chi_i \chi_j \right] \right.$$

$$\left. - \Sigma_s \int d\mathbf{\Omega}' \chi_i \int d\mathbf{\Omega} \chi_j \right\} + \int_S ds \int d\mathbf{\Omega} |\mathbf{n}_s \cdot \mathbf{\Omega}| \chi_i \chi_j \tag{13.91}$$

and

$$S_i = \int_V d\mathbf{r} \int d\mathbf{\Omega} \chi_i S(\mathbf{r}) \tag{13.92}$$

Thus the requirement for stationarity of the variational principle defines the a_i as the solution of

$$Aa = S \tag{13.93}$$

Diffusion Approximation

The diffusion approximation was shown in Chapter 9 to follow from a representation of the angular flux of the form

$$\psi(\mathbf{r}, \boldsymbol{\Omega}) \simeq \phi(\mathbf{r}) + 3\boldsymbol{\Omega} \cdot \mathbf{J}(\mathbf{r}) \tag{13.94}$$

With this representation, the even-parity component of the angular flux is just the scalar flux,

$$\psi^+(\mathbf{r}, \boldsymbol{\Omega}) \equiv \frac{1}{2}[\psi(\mathbf{r}, \boldsymbol{\Omega}) + \psi(\mathbf{r}, -\boldsymbol{\Omega})] = \phi(\mathbf{r}) \tag{13.95}$$

Using this representation for the even-parity flux in the variational principle of Eq. (13.85) leads to

$$
\begin{aligned}
J\{\phi\} &= \int_V d\mathbf{r} \int d\boldsymbol{\Omega} \left[\frac{1}{\Sigma_t} (\boldsymbol{\Omega} \cdot \nabla\phi)^2 + \Sigma_t \phi^2 - \Sigma_s \phi^2 - 2\phi S \right] \\
&\quad + \int_S ds \, \phi^2 \int d\boldsymbol{\Omega} |\boldsymbol{\Omega} \cdot \mathbf{n}_s| \\
&= \int_V d\mathbf{r} \left[\frac{1}{3\Sigma_t} (\nabla\phi)^2 + (\Sigma_t - \Sigma_s)\phi^2 - 2\phi S \right] + \frac{1}{2} \int_S ds \, \phi^2
\end{aligned} \tag{13.96}
$$

Requiring stationarity with respect to arbitrary and independent variations $\delta\phi$ in the volume and on the surface leads to the equation

$$-\nabla \cdot \left[\frac{1}{3\Sigma_t(\mathbf{r})} \nabla\phi_0(\mathbf{r}) \right] + [\Sigma_t(\mathbf{r}) - \Sigma_s(\mathbf{r})]\phi_0(\mathbf{r}) = S(\mathbf{r}) \tag{13.97}$$

and the boundary condition

$$-\frac{2}{3\Sigma_t(\mathbf{r}_s)} \mathbf{n}_s \cdot \nabla\phi_0(\mathbf{r}_s) + \phi_0(\mathbf{r}_s) = 0 \tag{13.98}$$

Equation (13.97) differs from the previous diffusion equation only by the Σ_t^{-1} rather than $(\Sigma_t - \mu_0\Sigma_s)^{-1}$ in the first term, and had we neglected anisotropic scattering ($\to \mu_0 = 0$) in Chapter 9 as we have here, the two would be identical. Equation (13.98) specifies that the flux extrapolate to zero a distance $2/3\Sigma_t$ outside the boundary, which is the same result (for isotropic scattering) that was obtained from P_1 theory in Chapter 9.

One-Dimensional Slab Transport Equation

In a slab varying from $x = 0$ to $x = a$, the variational principle of Eq. (13.85) becomes

$$J\{\psi^+\} = \int_0^a dx \left\{ \int_{-1}^1 \frac{d\mu}{2} \left[\frac{\mu^2}{\Sigma_t} \left(\frac{\partial \psi^+}{\partial x} \right)^2 + \Sigma_t(\psi^+)^2 \right] - \Sigma_s \phi^2 - 2\phi S \right\}$$

$$+ \int_{-1}^1 \frac{d\mu}{2} |\mu| (\psi^+)^2 \big|_{x=0} + \int_{-1}^1 \frac{d\mu}{2} |\mu| (\psi^+)^2 \big|_{x=a} \qquad (13.99)$$

Requiring stationarity with respect to arbitrary and independent variations $\delta \psi^+$ within $0 < x < a$ and at $x = 0$ and $x = a$ yields a one-dimensional transport equation for the even-parity flux component:

$$-\mu^2 \frac{\partial}{\partial x} \left[\frac{1}{\Sigma_t(x)} \frac{\partial}{\partial x} \psi^+(x, \mu) \right] + \Sigma_t(x) \psi^+(x, \mu)$$

$$= \Sigma_s(x) \phi(x) + S(x) \qquad (13.100)$$

and a pair of extrapolated vacuum boundary conditions

$$\psi^+(a, \mu) + \frac{1}{\Sigma_t(a)} \frac{\partial}{\partial x} \psi^+(a, \mu) = 0$$

$$\psi^+(0, \mu) - \frac{1}{\Sigma_t(a)} \frac{\partial}{\partial x} \psi^+(0, \mu) = 0 \qquad (13.101)$$

13.11
Boundary Perturbation Theory

Consider a reactor described by the multigroup diffusion equations, which are written in operator notation as

$$A_0(\mathbf{r}) \phi_0(\mathbf{r}) = \lambda_0 F_0(\mathbf{r}) \phi_0(\mathbf{r}) \qquad (13.102)$$

with general boundary conditions given by

$$a_0 \mathbf{n} \cdot \nabla \phi_0(\mathbf{r}_s) + b_0 \phi_0(\mathbf{r}_s) = 0 \qquad (13.103)$$

where a_0 and b_0 are group-dependent operators which may vary with position on the surface $\mathbf{r}_s$.

The adjoint equation is

$$A_0^+(\mathbf{r}) \phi_0^+(\mathbf{r}) = \lambda_0 F_0^+(\mathbf{r}) \phi_0^+(\mathbf{r}) \qquad (13.104)$$

where the definition (13.11) or (13.14) of adjoint operator has been used. The double integration by parts of the spatial derivative term in the diffusion operator yields

$$-\int_V d\mathbf{r}\,\phi_0^2(\nabla \cdot D_0\nabla\phi_0)$$

$$= -\int_V d\mathbf{r}\,\phi_0(\nabla \cdot D_0\nabla\phi_0^+) - \int_S ds\,\mathbf{n} \cdot (\phi_0^+ D_0\nabla\phi_0 - \phi_0 D_0\nabla\phi_0^+) \quad (13.105)$$

Using the boundary condition of Eq. (13.103) to evaluate the $\mathbf{n} \cdot \nabla\phi_0$ term in the surface integral reveals that the natural adjoint boundary condition (the condition that leads to vanishing of the surface integral) is

$$a_0\mathbf{n} \cdot \nabla\phi_0^+(\mathbf{r}_s) + b_0\phi_0^+(\mathbf{r}_s) = 0 \qquad (13.106)$$

Now let the boundary condition be changed by perturbing b_0 to $b_0 + b_1$:

$$a_0\mathbf{n} \cdot \nabla\phi^+(\mathbf{r}_s) + (b_0 + b_1)\phi^+(\mathbf{r}_s) = 0 \quad \text{and}$$
$$a_0\mathbf{n} \cdot \nabla\phi(\mathbf{r}_s) + (b_0 + b_1)\phi(\mathbf{r}_s) = 0 \qquad (13.107)$$

where $|b_1/b_0| \equiv \varepsilon \ll 1$. The perturbed flux, which must satisfy a different boundary condition and is associated with a different eigenvalue as a consequence, now satisfies

$$A_0(\mathbf{r})\phi(r) = \lambda F_0(\mathbf{r})\phi(\mathbf{r}) \qquad (13.108)$$

Expanding the perturbed flux and eigenvalue

$$\phi = \phi_0 + \phi_1 + \phi_2 + \cdots \qquad (13.109)$$
$$\lambda = \lambda_0 + \lambda_1 + \lambda_2 + \cdots \qquad (13.110)$$

where the subscript indicates the order of the term with respect to the small parameter $|b_1/b_0| \equiv \varepsilon \ll 1$, and substituting into Eqs. (13.107) and (13.108) results in the following hierarchy of perturbation equations and boundary conditions:

- *Order ε^0:*

$$A_0(\mathbf{r})\phi_0(\mathbf{r}) = \lambda_0 F_0(\mathbf{r})\phi_0(\mathbf{r}) \qquad (13.111)$$
$$a_0\mathbf{n} \cdot \nabla\phi_0(\mathbf{r}_s) + b_0\phi_0(\mathbf{r}_s) = 0 \qquad (13.112)$$

- *Order ε^1:*

$$[A_0(\mathbf{r}) - \lambda_0 F_0(\mathbf{r})]\phi_1(\mathbf{r}) = \lambda_1 F_0(\mathbf{r})\phi_0(\mathbf{r}) \qquad (13.113)$$
$$a_0\mathbf{n} \cdot \nabla\phi_1(\mathbf{r}_s) + b_0\phi_1(\mathbf{r}_s) + b_1\phi_0(\mathbf{r}_s) = 0 \qquad (13.114)$$

- *Order ε^2:*

$$\big(A_0(\mathbf{r}) - \lambda_0 F_0(\mathbf{r})\big)\phi_2(\mathbf{r}) = \lambda_1 F_0(\mathbf{r})\phi_1(\mathbf{r}) + \lambda_2 F_0(\mathbf{r})\phi_0(\mathbf{r}) \qquad (13.115)$$
$$a_0\mathbf{n} \cdot \nabla\phi_2(\mathbf{r}_s) + b_0\phi_2(\mathbf{r}_s) + b_1\phi_1(\mathbf{r}_s) = 0 \qquad (13.116)$$

The leading-order estimate of the eigenvalue is obtained by multiplying Eq. (13.111) by ϕ_0^+ and integrating over space and summing over groups (indicated by $\langle \cdot \rangle$):

$$\lambda_0 = \frac{\langle \phi_0^+, A_0 \phi_0 \rangle}{\langle \phi_0^+, F_0 \phi_0 \rangle} \tag{13.117}$$

The first-order correction to the eigenvalue is obtained by multiplying Eq. (13.113) by ϕ_0^+ and integrating over space and summing over groups, integrating the derivative term by parts twice, and using the boundary conditions of Eqs. (13.106) and (13.114):

$$\lambda_1 = \frac{\langle \phi_0^+, D_0 a_0^{-1} b_1 \phi_0 \rangle_S}{\langle \phi_0^+, F_0 \phi_0 \rangle} \tag{13.118}$$

where $\langle \cdot \rangle_S$ indicates an integral over the surface and a sum over groups. The second-order correction to the eigenvalue is obtained by multiplying Eq. (13.115) by ϕ_0^+ and integrating over space and summing over groups, integrating the derivative term by parts twice and using the boundary conditions of Eqs. (13.106) and (13.114):

$$\lambda_2 = -\lambda_1 \frac{\langle \phi_0^+, F_0 \phi_1 \rangle}{\langle \phi_0^+, F_0 \phi_0 \rangle} + \frac{\langle \phi_0^+, D_0 a_0^{-1} b_1 \phi_1 \rangle_S}{\langle \phi_0^+, F_0 \phi_0 \rangle} \tag{13.119}$$

The perturbation theory estimate, through second order, is

$$\lambda \simeq \lambda_0 + \lambda_1 + \lambda_2$$
$$= \frac{\langle \phi_0^+, A_0 \phi_0 \rangle + (1 - \langle \phi_0^+, F_0 \phi_1 \rangle / \langle \phi_0^+, F_0 \phi_0 \rangle) \langle \phi_0^+, D_0 a_0^{-1} b_1 \phi_0 \rangle_S + \langle \phi_0^+, D_0 a_0^{-1} b_1 \phi_1 \rangle_S}{\langle \phi_0^+, F_0 \phi_0 \rangle}$$
$$\tag{13.120}$$

To evaluate this second-order estimate, it is necessary to solve Eqs. (13.104), (13.111), and (13.113), with the associated boundary conditions. The first two equations, for ϕ_0^+ and ϕ_0, and their boundary conditions are independent of the boundary perturbation b_1. Upon using Eq. (13.118), Eq. (13.113) for ϕ_1 can be written

$$[A_0(\mathbf{r}) - \lambda_0 F_0(\mathbf{r})]\phi_1(\mathbf{r}) = \frac{\langle \phi_0^+, D_0 a_0^{-1} b_1 \phi_0 \rangle_S \, F_0(\mathbf{r})\phi_0(\mathbf{r})}{\langle \phi_0^+, F_0 \phi_0 \rangle} \tag{13.121}$$

showing that the amplitude of ϕ_1 depends on the magnitude of the perturbation in boundary condition b_1. The first-order perturbation theory estimate is $\lambda = \lambda_0 + \lambda_1$ and corresponds to omitting the ϕ_1 terms in Eq. (13.120), which obviates the necessity of calculating ϕ_1.

References

1 W. M. STACEY and J. A. FAVORITE, "Variational Reactivity Estimates," *Joint Int. Conf. Mathematical Methods and Supercomputing for Nuclear Applications I*, American Nuclear Society, La Grange Park, IL (**1997**), pp. 900–909.

2 K. F. LAURIN-KOVITZ and E. E. LEWIS, "Solution of the Mathematical Adjoint Equations for an Interface Current Nodal Formulation," *Nucl. Sci. Eng. 123*, 369 (**1996**).

3 Y. RONEN, ed., *Uncertainty Analysis*, CRC Press, Boca Raton, FL (**1988**).

4 A. GANDINI, "Generalized Perturbation Theory (GPT) Methods: A Heuristic Approach," in J. Lewins and M. Becker, eds., *Advances in Nuclear Science and Technology*, Vol. 19, Plenum Press, New York (**1987**).

5 T. A. TAIWO and A. F. HENRY, "Perturbation Theory Based on a Nodal Model," *Nucl. Sci. Eng. 92*, 34 (**1986**).

6 M. L. WILLIAMS, "Perturbation Theory for Nuclear Reactor Analysis," in Y. Ronen, ed., *CRC Handbook of Nuclear Reactor Calculations I*, CRC Press, Boca Raton, FL (**1986**).

7 F. RAHNEMA and G. C. POMRANING, "Boundary Perturbation Theory for Inhomogeneous Transport Equations," *Nucl. Sci. Eng. 84*, 313 (**1983**).

8 E. W. LARSEN and G. C. POMRANING, "Boundary Perturbation Theory," *Nucl. Sci. Eng. 77*, 415 (**1981**).

9 D. G. CACUCI, "Sensitivity Theory for Nonlinear Systems," *J. Math. Phys. 22*, 2794 and 2803 (**1981**).

10 D. G. CACUCI, C. F. WEBER, E. M. OBLOW, and J. H. MARABLE, "Sensitivity Theory for General Systems of Nonlinear Equations," *Nucl. Sci. Eng. 75*, 88 (**1980**).

11 M. KOMATA, "Generalized Perturbation Theory Applicable to Reactor Boundary Changes," *Nucl. Sci. Eng. 64*, 811 (**1977**).

12 E. GREENSPAN, "Developments in Perturbation Theory," in J. Lewins and M. Becker, eds., *Advances in Nuclear Science and Technology*, Vol. 9, Plenum Press, New York (**1976**).

13 W. M. STACEY, *Variational Methods in Nuclear Reactor Physics*, Academic Press, New York (**1974**).

14 W. M. STACEY, "Variational Estimates of Reactivity Worths and Reaction Rate Ratios in Critical Systems," *Nucl. Sci. Eng. 48*, 444 (**1972**).

15 W. M. STACEY, "Variational Estimates and Generalized Perturbation Theory for Ratios of Linear and Bilinear Functionals," *J. Math. Phys. 13*, 1119 (**1972**).

16 G. I. BELL and S. GLASSTONE, *Nuclear Reactor Theory*, Van Nostrand Reinhold, New York (**1970**), Chap. 6.

17 J. DEVOOGHT, "Higher-Order Variational Principles and Iterative Processes," *Nucl. Sci. Eng. 41*, 399 (**1970**).

18 A. GANDINI, "A Generalized Perturbation Method for Bilinear Functionals of the Real and Adjoint Neutron Fluxes," *J. Nucl. Energy Part A/B 21*, 755 (**1967**).

19 G. C. POMRANING, "The Calculation of Ratios in Critical Systems," *J. Nucl. Energy Part A/B 21*, 285 (**1967**).

20 G. C. POMRANING, "Generalized Variational Principles for Reactor Analysis," *Proc. Int. Conf. Utilization of Research Reactors and Mathematics and Computation*, Mexico, D.F. (**1966**), p. 250.

21 G. C. POMRANING, "Variational Principles for Eigenvalue Equations," *J. Math. Phys. 8*, 149 (**1967**).

22 G. C. POMRANING, "A Derivation of Variational Principles for Inhomogeneous Equations," *Nucl. Sci. Eng. 29*, 220 (**1967**).

23 J. LEWINS, "A Variational Principle for Ratios in Critical Systems," *J. Nucl. Energy Part A/B 20*, 141 (**1966**).

24 G. C. POMRANING, "A Variational Description of Dissipative Processes," *J. Nucl. Energy Part A/B 20*, 617 (**1966**).

25 J. Lewins, *Importance: The Adjoint Function*, Pergamon Press, Oxford (1965).

26 L. N. Usachev, "Perturbation Theory for the Breeding Ratio and Other Number Ratios Pertaining to Various Reactor Processes," *J. Nucl. Energy Part A/B 18*, 571 (1964).

27 D. S. Selengut, *Variational Analysis of Multidimensional Systems*, Hanford Engineering Laboratory report HW-59126 (1959).

28 N. C. Francis, J. C. Stewart, L. S. Bohl, and T. J. Krieger, "Variational Solutions of the Transport Equation," *Prog. Nucl. Energy Ser. I 3*, 360 (1958).

Problems

13.1. Use one-speed diffusion theory and perturbation theory to estimate the reactivity worth of a 0.25% increase in the fission cross section over the left half of a critical slab reactor of 1-m thickness.

13.2. Use two-group diffusion theory perturbation theory to estimate the reactivity worth of a 0.5% change in the thermal absorption cross section of a very large core described by: group 1—$D = 1.2$ cm, $\Sigma_a = 0.012$ cm^{-1}, $\Sigma^{1\rightarrow2} = 0.018$ cm^{-1}, $\nu\Sigma_f = 0.006$ cm^{-1}; group 2— $D = 0.40$ cm, $\Sigma_a = 0.120$ cm^{-1}, $\nu\Sigma_f = 0.150$ cm^{-1}.

13.3. Prove that each term in the importance equation [Eq. (13.17)] is mathematically adjoint to the corresponding term in the neutron transport equation [Eq. (13.19)].

13.4. Derive the multigroup discrete ordinates adjoint equation for a critical reactor (a) directly from the discrete ordinates equations, and (b) by making the discrete ordinates approximation of the adjoint transport equation.

13.5. Derive an explicit expression for the perturbation theory reactivity estimate in the multigroup discrete ordinates representation of neutron transport.

13.6. Solve for the infinite medium neutron flux and adjoint energy distributions in a three-group representation: group 1—$\Sigma_a = 0.030$ cm^{-1}, $\Sigma^{1\rightarrow2} = 0.060$ cm^{-1}, $\nu\Sigma_f = 0.004$ cm^{-1}; group 2—$\Sigma_a = 0.031$ cm^{-1}, $\Sigma^{2\rightarrow3} = 0.088$ cm^{-1}, $\nu\Sigma_f = 0.018$ cm^{-1}; group 3—$\Sigma_a = 0.120$ cm^{-1}, $\nu\Sigma_f = 0.180$ cm^{-1}.

13.7. Carry through the derivation to show that $\rho_{var}\{\phi_0^+, \phi, \Gamma_0^+\} = \rho_{ex}\{\phi_0^+, \phi_{ex}\} + O(\delta\phi^2, \delta\phi\delta\Gamma)$, where $\Gamma_{ex}^+ = \Gamma_0^+ + \delta\Gamma^+$ and Γ_0^+ is obtained by solving Eq. (13.32).

13.8. Consider a critical slab reactor with one-speed diffusion theory constants $D = 1.0$ cm, $\Sigma_a = 0.15$ cm^{-1}, and $\nu\Sigma_f = 0.16$ cm^{-1}. Calculate the flux correction function, Γ_0^+, from Eq. (13.32) for a 1% change in the absorption cross section in the left one-fourth of the critical slab. (*Hint:* Note that Γ_0^+ is orthogonal to $\phi_0^+ = \phi_0$ and expand Γ_0^+ in the higher harmonics of the critical reactor eigenfunctions.)

13.9. Evaluate the variational/generalized perturbation reactivity estimate of Eq. (13.36) for Problem 13.8.

13.10. Carry out the missing steps of the derivation in Section 13.4 to show that $\mathrm{RR}_{\mathrm{var}}\{\phi_0, \Gamma_{R0}^+\} - \mathrm{RR}_{\mathrm{ex}}\{\phi_{\mathrm{ex}}\} = O(\delta\Gamma_R^+\delta\phi)$.

13.11. Consider a critical bare slab reactor described by one-speed diffusion theory with $D = 1.0$ cm, $\Sigma_a = 0.15$ cm^{-1}, and $\nu\Sigma_f = 0.16$ cm^{-1}. Use Eq. (13.47) to evaluate the variational estimate for a 1% increase in absorption cross section on the absorption-to-fission rate ratio in the right one-tenth of the slab core.

13.12. Use the Rayleigh quotient to estimate the effective multiplication constant for a bare cylindrical core with $H/D = 1$, $H = 2$ m and one-speed diffusion theory parameters $D = 1.0$ cm, $\Sigma_a = 0.15$ cm^{-1}, and $\nu\Sigma_f = 0.16$ cm^{-1}.

13.13. In the window-shade model, a control rod bank can be represented by a 10% increase in Σ_a for Problem 13.11. Use the Rayleigh quotient to estimate the effective multiplication constant when the control rod bank is inserted halfway, using the flux and adjoint distributions calculated in Problem 13.11. Recalculate the effective multiplication constant directly (i.e., solve the two-region diffusion theory problem) for the control rod bank inserted halfway and compare with the variational estimate.

13.14. Consider a uniform slab nonfissioning assembly of width 50 cm in which there is a uniform source S_f of fast neutrons in the left half. Calculate the thermal absorption rate in the right half (**a**) directly and (**b**) using the Schwinger variational estimate evaluated trial functions, obtain from an infinite medium calculation with a source $1/2S_f$. Use the two-group representation: fast group—$D = 2.0$ cm, $\Sigma_a = 0.006$ cm^{-1}, and $\Sigma^{1\to2} = 0.018$ cm^{-1}; thermal group—$D = 0.40$ cm and $\Sigma_a = 0.120$ cm^{-1}.

13.15. Repeat the derivation of Section 13.3 for multigroup diffusion theory.

13.16. Discuss how the result of Eq. (13.55) could be employed to calculate the response of a localized detector to a point neutron source some distance away if the adjoint function is known in the vicinity of the source.

13.17. Carry through the steps in deriving the variational synthesis approximation of Eq. (13.81).

13.18. Demonstrate that stationarity of the variational functional of Eq. (13.83) requires that the diffusion equation be satisfied and that the flux and current be continuous at the interface $y = y_i$. (*Hint*: Consider arbitrary and independent variations of the adjoint flux and current within the volume and on the interface.)

13.19. Derive the transport Eq. (13.100) and the associated boundary conditions of Eq. (13.101) from the stationarity of the functional of Eq. (13.99).

13.20. Consider a uniform slab reactor of thickness $2a$ with zero flux conditions at each boundary, which may be represented as a slab with a zero flux condition at $x = 0$ and a symmetry $\hat{n} \cdot \nabla \phi = 0$ condition at $x = a$. Use boundary perturbation theory to derive an estimate for the change in eigenvalue, λ_1, that would result from replacing the symmetry condition at $x = a$ with the condition $\hat{n} \cdot \nabla \phi + b_1 \phi = 0$.

13.21. In a critical uniform slab reactor in one-group theory, 10% of the neutrons leak from the reactor and the other 90% are absorbed. Use perturbation theory to calculate the reactivity worth of a 5% increase in absorption cross section over the right half of the reactor. Discuss the error in this estimate due to the failure to take into account the change in flux distribution caused by the increase in absorption cross section. Would the perturbation theory estimate be expected to underpredict or overpredict the reactivity worth because of this error? Discuss how the effect of this flux change on the reactivity worth could be taken into account without actually calculating the flux change.

14
Homogenization

Nuclear reactor cores are composed of a large number of fuel assemblies, each containing a large number of discrete fuel elements of differing composition and consisting of separate fuel and cladding regions, coolant, structural elements, burnable poisons, water channels, control rods, and so on—tens to hundreds of thousands of discrete, heterogeneous regions. On the other hand, most of the methods for calculating criticality and global flux distributions that are in use (in particular diffusion theory) are predicated on the existence of large (with respect to a mean free path) homogeneous regions. The methods employed to replace a heterogeneous lattice of materials of differing properties with an equivalent homogeneous mixture of these materials to which the previously discussed methods for the calculation of ultrafine group spectra, calculation of the diffusion of neutrons during the slowing-down process, and so on, is referred to as *homogenization theory*. Homogenization of a heterogeneous assembly usually proceeds in two steps: a lattice transport calculation to obtain the detailed heterogeneous flux distribution within a unit cell or fuel assembly, followed by the use of this detailed flux distribution to calculate average homogeneous cross sections for the unit cell or assembly.

The general procedure that is followed in nuclear reactor analysis is to perform very detailed energy and spatial calculations on a local basis to obtain cross sections averaged over energy and spatial detail which can be used in few group global core calculations. For example, for a thermal reactor, a pin-cell transport calculation of a cell consisting of the fuel, clad, coolant, and structural in a local region may be carried out in 20 to 100 fine groups to obtain homogenized 6 to 20 intermediate-group cross sections averaged over the pin-cell geometry and the 20 to 100 fine group spectrum. Several such pin-cell calculations may be needed for a fuel assembly and the adjacent water gaps and control rods. Intermediate-group assembly transport calculations are then performed for models that represent all fuel pins, control rods, water channels, can walls, and so on, associated with a given fuel assembly. It is important that the intermediate-group assembly transport calculation uses enough groups to represent the spectral interactions among fuel pins of different composition, control rods, water channels, and so on, at the intermediate-group level. Several such intermediate-group assembly calculations may be needed for the reactor core, and a large number of such calculations may be needed to represent different operating temperatures, depletion steps, void fractions, and so

Nuclear Reactor Physics. Weston M. Stacey
Copyright © 2007 WILEY-VCH Verlag GmbH & Co. KGaA, Weinheim
ISBN: 978-3-527-40679-1

on. The results of the intermediate-group assembly transport calculations are next averaged over the assembly spatial detail and the intermediate-group spectra to obtain two to six few-group homogenized assembly cross sections which can be used in few-group global core calculations of criticality and flux distribution.

We have discussed the procedure of group collapsing to obtain few-group cross sections from fine- or intermediate-group spectra in previous chapters. Here we are interested in the spatial-averaging procedures used to obtain homogenized cross sections appropriately averaged over spatial detail, and in the procedures used to construct effective diffusion theory cross sections for regions such as control rods in which the basic assumptions of diffusion theory are not satisfied.

14.1
Equivalent Homogenized Cross Sections

The general problem of homogenization can be illustrated by considering a symmetric, repeating array of fuel and moderator elements of volumes V_F and V_M. The average absorption cross section for the fuel–moderator unit cell is

$$\langle \Sigma_a \rangle_{\text{cell}} = \frac{\Sigma_a^F \phi_F V_F + \Sigma_a^M \phi_M V_M}{\phi_F V_F + \phi_M V_M} = \frac{\Sigma_a^F + \Sigma_a^M (V_M/V_F)\xi}{1 + (V_M/V_F)\xi} \tag{14.1}$$

where

$$\xi \equiv \frac{\phi_M}{\phi_F} \tag{14.2}$$

is referred to as the *flux disadvantage factor*, and ϕ_F and ϕ_M are the average neutron fluxes in the fuel and moderator, respectively. The homogenized cell average cross section of Eq. (14.1) is equivalent in the sense that if it is multiplied by the exact average cell flux

$$\phi_{\text{cell}} = \frac{\phi_F V_F + \phi_M V_M}{V_F + V_M} \equiv \frac{\phi_F V_F + \phi_M V_M}{V_{\text{cell}}} \tag{14.3}$$

and the cell volume, the result will be the exact absorption rate in the cell:

$$\langle \Sigma_a \rangle_{\text{cell}} \phi_{\text{cell}} V_{\text{cell}} \equiv \Sigma_a^F \phi_F V_F + \Sigma_a^M \phi_M V_M \tag{14.4}$$

This type of definition can obviously be extended to a multiregion heterogeneous assembly by defining

$$\langle \Sigma_a \rangle_{\text{cell}} = \frac{\Sigma_a^1 + \sum_{i=2}^{I} \Sigma_a^i (V_i/V_1)\xi_i}{1 + \sum_{i=2}^{I} (V_i/V_1)\xi_i}, \qquad \xi_i \equiv \frac{\phi_i}{\phi_1}$$

$$\phi_{\text{cell}} \equiv \frac{\sum_{i=1}^{I} \phi_i V_i}{\sum_{i=1}^{I} V_i} \equiv \frac{\sum_{i=1}^{I} \phi_i V_i}{V_{\text{cell}}} \tag{14.5}$$

The same type of definition defines equivalent cell average fission and scattering cross sections. The appropriate definition of the cell average diffusion coefficient is less straightforward. An equivalent cell average diffusion coefficient must represent the net leakage from the cell, but that depends on the calculational method which will be employed for that purpose. We will return to this subject when we consider the detailed homogenization procedures.

Thus the problem of cell homogenization reduces to the problem of determining the flux disadvantage factors, ξ_i, which will enable the homogenized model to predict the correct intracell reaction rates, and of determining the equivalent diffusion coefficients (or other leakage representation) which will enable the homogenized model to predict correct intercell leakage. Note that it is only necessary to know the relative value of the neutron flux in the different regions of the problem, not the absolute values, in order to calculate homogenized cross sections, which enables calculation of homogenized cross sections for local regions in a reactor before the absolute value of the flux is determined from a global calculation (which utilizes the homogenized cross sections). Calculation of the flux disadvantage factors from diffusion theory was discussed in Chapter 4. We turn now to methods that can be used when diffusion theory does not provide an adequate treatment of the heterogeneous problem, which is the usual case in a nuclear reactor.

14.2
ABH Collision Probability Method

The ABH collision probability method (named after its originators—Ref. 15) found widespread use for calculation of thermal disadvantage factors before the availability of assembly transport codes (discussed in Section 14.4) and provides physical insight into the unit cell transport problem. A unit cell of fuel (F) and moderator (M) with zero net current on the cell boundary is assumed. It is further assumed that the neutron slowing-down source is uniform in the moderator and zero in the fuel. We define

$P_{FM} \equiv$ average probability that a neutron born uniformly and isotropically

in region F will eventually be absorbed in region M

$P_F \equiv$ average probability that a neutron born uniformly and isotropically

in region F escapes from the fuel before being absorbed

$\beta_M \equiv$ conditional probability that a neutron, having escaped from F

into M, will then be absorbed in M

A similar probability P_{MF} can be defined for region M. As discussed in Chapter 11, there exists a reciprocity relation

$$V_F \Sigma_a^F P_{FM} = V_M \Sigma_a^M P_{MF} \tag{14.6}$$

Since neutrons are only slowing down to thermal in the moderator, the reciprocity relation can be used to write the thermal utilization factor in terms of P_{FM}:

$$f \equiv P_{FM} = \frac{V_F}{V_M} \frac{\Sigma_a^F}{\Sigma_a^M} P_{FM} = \frac{\Sigma_a^F V_F \phi_F}{\Sigma_a^F V_F \phi_F + \Sigma_a^M V_M \phi_M} \tag{14.7}$$

and to write the thermal disadvantage factor

$$\xi \equiv \frac{\phi_M}{\phi_F} = \frac{\Sigma_a^F}{\Sigma_a^M} \frac{V_F}{V_M} \left(\frac{1}{f} - 1 \right) = \frac{1}{P_{FM}} - \frac{\Sigma_a^F}{\Sigma_a^M} \frac{V_F}{V_M} \tag{14.8}$$

The probability that a neutron born uniformly and isotropically in the fuel escapes into the moderator without making a collision was discussed in Chapter 11 and is given approximately by

$$P_{F0} \simeq \frac{1}{1 + 4(V_F \Sigma_t^F / S_F)} \tag{14.9}$$

where S_F is the surface area of the fuel. If the neutron does not escape but has a scattering event [probability $(1 - P_{F0})\Sigma_s^F / \Sigma_t^F$], it also has a probability P_{F0} of escaping without a second collision. Continuing this line of argument, the total probability that the neutron escapes from the fuel into the moderator may be written

$$P_F = P_{F0}\left[1 + (1 - P_{F0})\frac{\Sigma_s^F}{\Sigma_t^F} + (1 - P_{F0})^2 \left(\frac{\Sigma_s^F}{\Sigma_t^F} \right)^2 + \cdots \right]$$

$$= \frac{P_{F0}}{1 - (1 - P_{F0})(\Sigma_s^F / \Sigma_t^F)} = \frac{1}{1 + \Sigma_a^F (1 - P_{F0}) / \Sigma_t^F P_{F0}} \tag{14.10}$$

A somewhat more accurate expression, which takes into account the nonuniform distribution of the first collisions for a cylindrical fuel rod of radius a, is

$$P_F = \left\{ 1 + \frac{\Sigma_a^F}{\Sigma_t^P}\left(\frac{1 - P_{F0}}{P_{F0}} + a \right)\left[1 + \alpha \frac{\Sigma_s^F}{\Sigma_t^F} + \beta \left(\frac{\Sigma_s^F}{\Sigma_t^F} \right)^2 \right] + a\Sigma_a^F \right\}^{-1} \tag{14.11}$$

where the parameters α and β are as given in Fig. 14.1.

It is apparent from the definitions that

$$P_{FM} = P_F \beta_M \tag{14.12}$$

Equation (14.7) can be rearranged to

$$\frac{1}{f} - 1 = \frac{\Sigma_a^M}{\Sigma_a^F} \frac{V_M}{V_F} \frac{1}{P_F} + \frac{1 - f - \beta_M}{f}$$

$$= \frac{\Sigma_a^M}{\Sigma_a^F} \frac{V_M}{V_F} \frac{1}{P_F} + \frac{1 - P_{MF}}{P_{MF}} - \frac{\beta_M}{P_{MF}} \tag{14.13}$$

Using the reciprocity relation of Eq. (14.6) and, for the purpose of estimating β_M only, approximating $P_F \approx P_{F0} \approx S_F / 4 V_F \Sigma_a^F$, yields an approximation for the conditional probability:

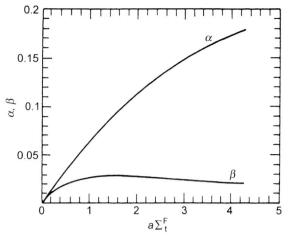

Fig. 14.1 Parameters α and β for use in calculation of ABH cylindrical escape probability. (From Ref. 11; used with permission of Wiley.)

$$\beta_M \simeq \frac{4\Sigma_a^M V_M}{S_M} P_{MF} \tag{14.14}$$

We approximate $P_{MF} \approx P_M$ ($\equiv$ probability that a neutron born in the moderator escapes from the moderator before being absorbed). We calculate P_M by solving the diffusion equation in the moderator

$$-D_M \nabla^2 \phi_M(\mathbf{r}) + \Sigma_a^M \phi_M(\mathbf{r}) = q_M \tag{14.15}$$

where q_M is the uniform slowing-down density in the moderator. The boundary conditions for Eq. (14.15) are symmetry at the cell boundary and a transport boundary condition at the fuel–moderator interface:

Cell boundary: $\qquad\qquad \mathbf{n}_s \cdot \nabla \phi_M = 0$

Fuel–moderator interface: $\quad \mathbf{n}_s \cdot \left(\dfrac{1}{\phi_M} \nabla \phi_M \right)\Big|_a = \dfrac{1}{d}$ $\qquad$ (14.16)

where $\mathbf{n}_s$ is the unit vector normal to the surface and d is a transport parameter related to the transport mean free path in the moderator and is given in Fig. 14.2 for a cylindrical unit cell. Assuming that all neutrons diffusing from the moderator into the fuel are absorbed, P_M is just the total neutron flow into the fuel from the moderator divided by the total neutron source in the moderator:

$$P_M = \frac{J_{out}^M S_F}{q_M V_M} = \frac{S_F}{q_M V_M} \mathbf{n}_s \cdot D_M \nabla \phi_M|_a = \left[\frac{a d V_M}{2 V_F L_M^2} + E\left(\frac{a}{L_M}, \frac{b}{L_M} \right) \right]^{-1} \tag{14.17}$$

where a is the thickness or radius of the fuel region, b is the thickness of the moderator region associated with a fuel element, $L_M^2 = D_M / \Sigma_a^M$, and $E(a/L_M, b/L_M)$ is the lattice function given in Table 3.6.

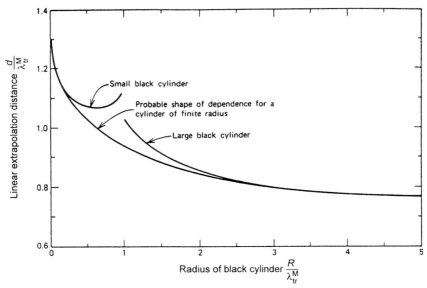

Fig. 14.2 Transport boundary condition for cylinders. (From Ref. 11; used with permission of Wiley.)

With these approximations for β_M and P_{MF} given by Eqs. (14.14) and (14.17), respectively, and the expression for P_F given by Eq. (14.11), Eq. (14.13) can be evaluated:

$$\frac{1}{f} - 1 = \frac{\Sigma_a^M V_F}{\Sigma_a^F V_F}\left\{1 + \frac{\Sigma_a^F}{\Sigma_t^F}\left(\frac{1 - P_{F0}}{P_{F0}} - a\Sigma_t^F\right)\left[1 + \alpha\frac{\Sigma_s^F}{\Sigma_t^F} + \beta\left(\frac{\Sigma_s^F}{\Sigma_t^F}\right)^2\right]\right\}$$
$$+ \left(\frac{ad}{2L_M^2} - a\Sigma_a^M\right)\frac{V_M}{V_F} + E\left(\frac{a}{L_M}, \frac{b}{L_M}\right) - 1 \qquad (14.18)$$

the disadvantage factor can be calculated from

$$\xi = \frac{\Sigma_a^F}{\Sigma_a^M}\frac{V_F}{V_M}\left(\frac{1}{f} - 1\right) \qquad (14.19)$$

and the homogenized cross section can be calculated from Eq. (14.1).

Although we have developed the ABH method in the context of thermal neutrons, the same general procedure can be applied to homogenize cross sections for any group of a multigroup scheme.

14.3
Blackness Theory

Blackness theory refers to a class of methods for matching an approximate (e.g., diffusion theory) solution in one region to a very accurate solution of the transport equation in an adjacent region in order to obtain an effective diffusion theory cross section that will preserve the transport theory accuracy in the calculation of reaction rate. Such a procedure is required in order to treat control rods, lumped burnable poisons, and so on, within the context of multigroup diffusion theory.

Consider a purely absorbing slab occupying the region $x_i \leq x \leq x_{i+1}$. The one-speed transport equation within the absorbing slab is

$$\mu \frac{\partial}{\partial x} \psi(x, \mu) + \Sigma_a \psi(x, \mu) = 0 \tag{14.20}$$

This equation may be solved for the exiting neutron fluxes $\psi^+(x_{i+1}, \mu)$ to the right and $\psi^-(x_i, \mu)$ to the left in terms of the entering fluxes from the left $\psi^+(x_i, \mu)$ and from the right $\psi^-(x_i, \mu)$, where the $+/-$ denotes $\mu > 0/\mu < 0$:

$$\psi^+(x_{i+1}, \mu) = \psi^+(x_i, \mu) \exp\left(-\Sigma_a \frac{\Delta}{\mu}\right), \quad 1 > \mu > 0$$

$$\psi^-(x_i, \mu) = \psi^-(x_{i+1}, \mu) \exp\left(\Sigma_a \frac{\Delta}{\mu}\right), \quad -1 < \mu < 0 \tag{14.21}$$

where $\Delta = x_{i+1} - x_i$. The incident fluxes into the purely absorbing region are assumed to have the P_l form that is consistent with a diffusion theory solution in the adjacent fuel-moderating region:

$$\psi^+(x_i, \mu) = \phi(x_i) + 3\mu J(x_i), \quad 1 > \mu > 0$$

$$\psi^-(x_{i+1}, \mu) = \phi(x_{i+1}) + 3\mu J(x_{i+1}), \quad -1 < \mu < 0 \tag{14.22}$$

The currents at the surfaces of the absorbing region can be written

$$J(x_i) \equiv \int_{-1}^{1} \frac{d\mu}{2} \mu \psi(x_i, \mu)$$

$$= \int_{-1}^{0} \frac{d\mu}{2} \mu \psi^-(x_i, \mu) + \int_{0}^{1} \frac{d\mu}{2} \mu \left[\phi(x_i) + 3\mu J(x_i)\right]$$

$$\tag{14.23}$$

$$J(x_{i+1}) \equiv \int_{-1}^{1} \frac{d\mu}{2} \mu \psi(x_{i+1}, \mu)$$

$$= \int_{-1}^{0} \frac{d\mu}{2} \mu \left[\phi(x_{i+1}) + 3\mu J(x_{i+1})\right] + \int_{0}^{1} \frac{d\mu}{2} \mu \psi(x_{i+1}, \mu)$$

Using Eqs. (14.21) to evaluate the exiting fluxes, these equations become

$$\frac{1}{2} J(x_i) = \frac{1}{4} \phi(x_i) - \frac{1}{2} E_3(\Sigma_a \Delta) \phi(x_{i+1}) + \frac{3}{2} E_4(\Sigma_a \Delta) J(x_{i+1})$$

$$\tag{14.24}$$

$$-\frac{1}{2} J(x_{i+1}) = \frac{1}{4} \phi(x_{i+1}) - \frac{1}{2} E_3(\Sigma_a \Delta) \phi(x_i) - \frac{3}{2} E_4(\Sigma_a \Delta) J(x_i)$$

where E_3 and E_4 are the exponential integral functions,

$$E_{n+2}(\xi) \equiv \int_{0}^{1} \mu^n \exp\left(-\frac{\Sigma_a \Delta}{\mu}\right) d\mu \tag{14.25}$$

Equations (14.24) can be rearranged to define the blackness parameters

$$\alpha(\Sigma_a \Delta) \equiv \frac{J(x_i) - J(x_{i+1})}{\phi(x_i) + \phi(x_{i+1})} = \frac{1 - 2E_3(\Sigma_a \Delta)}{2[1 + 3E_4(\Sigma_a \Delta)]}$$

$$\beta(\Sigma_a \Delta) \equiv \frac{J(x_i) + J(x_{i+1})}{\phi(x_i) - \phi(x_{i+1})} = \frac{1 + 2E_3(\Sigma_a \Delta)}{2[1 - 3E_4(\Sigma_a \Delta)]}$$

(14.26)

The parameter α is the ratio of the average inward current to the average flux at the surface of the absorbing slab. This quantity is used as a boundary condition for the diffusion theory calculation in the adjacent region,

$$\frac{J_M}{\phi_M} = -\frac{D^M \nabla \phi_M}{\phi_M} = \alpha$$

(14.27)

(e.g., the transport parameter d of the ABH method is $d/\lambda_{\text{tr}} = 1/3\alpha$). This transport boundary condition was used in Chapter 3 to derive an effective diffusion theory cross section for the control rod:

$$\Sigma_a^c = \frac{\Sigma_a^M}{a[(\Sigma_a^M/\alpha) + (1/L_M)\coth(a/L_M)] - 1}$$

(14.28)

where a is the half-thickness of the fuel–moderator region denoted by M. Since this development was for a purely absorbing slab, the results are valid at any energy, provided that the cross sections for that energy are used.

For a purely absorbing slab with a spatially dependent absorption cross section, the results above are valid if the following replacement is made:

$$\Sigma_a \Delta \rightarrow \int_{x_i}^{x_{i+1}} \Sigma_a(x)\,dx$$

(14.29)

14.4
Fuel Assembly Transport Calculations

Pin Cells

A fuel assembly consists of a large number of fuel pins of differing fuel loading, enrichment, burnup, and so on, each of which is clad and surrounded by moderator and perhaps other elements, such as structure and burnable poisons, as depicted in Fig. 14.3. At this most detailed level of heterogeneity, the assembly can be considered to be made up of a large number of units cells, or *pin cells*, consisting of a fuel pin, cladding, surrounding moderator, and perhaps structure and burnable poison. The first step in homogenizing the fuel assembly is to homogenize each of the pin cells, by calculating the multigroup flux distribution across the fuel, clad, moderator, and so on, and using it to calculate volume-averaged cross sections for the pin cell.

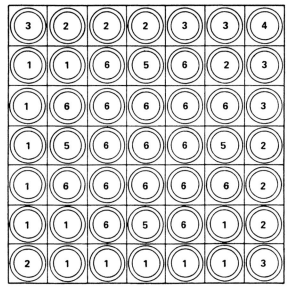

Fig. 14.3 Representative fuel assembly. (From Ref. 16; used with permission of American Nuclear Society.)

If the pin cell can be considered to be one of a large number of identical pin cells, reflective symmetry boundary conditions can be used. However, this assumption becomes questionable in the vicinity of gaps, control pins, burnable poisons, or fuel pins of very different composition (e.g., MOX pins near UO_2 pins). The influence of the surrounding environment can be introduced into the pin-cell calculation by specifying the partial inward current J^- (and a zero reflection, or black boundary conditions) or the net current $J = J^+ - J^-$ (and a perfectly reflecting boundary condition) on the cell boundary.

Wigner–Seitz Approximation

If the cell associated with each pin is defined symmetrically and such that the cells fill the volume of the assembly, the pin-cell boundary will have a noncylindrical shape depending on the lattice geometry, generally square or hexagonal. Since the pin geometry is cylindrical, it is convenient to approximate the actual pin-cell geometry by an equivalent cylindrical cell that preserves moderator volume. The approximate Wigner–Seitz cell has a radius R that depends on the pin-to-pin distance p as $R = p/\pi^{1/2}$ for a square pitch fuel lattice and $R = p(3^{1/2}/2\pi)^{1/2}$ for a hexagonal pitch lattice.

The change in geometry can lead to an anomalously high flux in the moderator of a cell with reflective boundary conditions because a neutron introduced into the cell traveling in the direction of a chord that does not pass through the innermost n shells before intersecting the reflecting cell boundary will never pass through these innermost n shells since spectral reflection from the cylindrical wall will result in

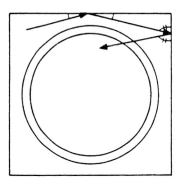
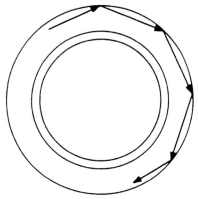

Fig. 14.4 Reflection misrepresentation in Wigner–Seitz approximation. (From Ref. 16; used with permission of American Nuclear Society.)

motion along a similar chord. On the other hand, as shown in Fig. 14.4, correct reflection from a square or hexagonal boundary will cause motion into the innermost shells. This problem can be corrected by "white reflection" in a cosine distribution with respect to the inward normal.

Collision Probability Pin-Cell Model

The collision probability methodology of Section 9.3 can be extended to handle the albedo (partial reflection) and incident current conditions that enable the environment to influence the pin-cell calculations. With reference to Fig. 14.5, consider a cylindrical pin-cell consisting of i annular regions. Using the notation of Section 9.3, define the probability, γ_{0i}, that an uniformly distributed isotropic flux of neutrons at the external surface (S_B) of the pin-cell will suffer a first collision in region i before exiting across surface S_B:

$$\gamma_{0i} \equiv \frac{\Sigma_{ti} \int_{V_i} d\mathbf{r}_i \int_{S_B} d\mathbf{r}_B \int_{\Omega \supset V_i} d\mathbf{\Omega}(\mathbf{n} \cdot \mathbf{\Omega})(1/4\pi)e^{-\alpha(\mathbf{r}_i, \mathbf{r}_B)}}{\int_{S_B} \int_{\mathbf{\Omega} \cdot \mathbf{n} < 0} d\mathbf{\Omega}(\mathbf{n} \cdot \mathbf{\Omega})(1/4\pi)} \tag{14.30}$$

where $\Omega \supset V_i$ indicates those values of Ω that intersect the volume V_i, $\mathbf{n}$ is the outward unit vector to the surface S_B, $\alpha(\mathbf{r}_i, \mathbf{r}_B)$ is the optical distance (e.g., distance measured in mean free paths) along the chord from $\mathbf{r}_B$ to $\mathbf{r}_i$, and $\mathbf{n} \cdot \mathbf{\Omega}/4\pi$ is the rate at which neutrons in an isotropic flux of unit strength will cross the surface at S_B into the pin-cell. This probability is related to the first-flight escape probability, P_{0i}, that a neutron introduced in volume V_i will exit the pin-cell across surface S_B without a collision:

$$P_{0i} \equiv \frac{\int_{V_i} d\mathbf{r}_i \int_{S_B} d\mathbf{r}_B \int_{\Omega \subset V_i} d\mathbf{\Omega}(\mathbf{n} \cdot \mathbf{\Omega})(1/4\pi)e^{-\alpha(\mathbf{r}_B, \mathbf{r}_i)}}{\int_{4\pi} d\mathbf{\Omega} \int_{V_i} d\mathbf{r}_i(1/4\pi)} \tag{14.31}$$

where $1/4\pi$ is the isotropic angular flux corresponding to unit scalar flux in V_i and $\Omega \subset V_i$ indicates those values of Ω for which a neutron could have reached $\mathbf{r}_B$ on a

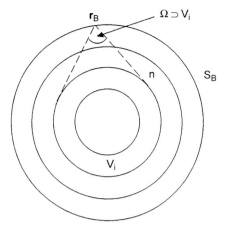

Fig. 14.5 Cylindrical pin-cell model.

first flight from within volume V_i. Except for the Σ_{ti}, the numerators are identical, reflecting the fact that the probabilities for neutrons traveling from the surface into volume V_i without collision and traveling in the opposite direction from within V_i to the surface are identical. This allows Eq. (14.30) to be written

$$\gamma_{0i} = \frac{\Sigma_{ti} V_i P_{0i}}{S_B \int_{-1}^{0} d\mu \int_{0}^{2\pi} d\phi\,(\mathbf{n} \cdot \mathbf{\Omega}/4\pi)} = \frac{\Sigma_{ti} V_i P_{0i}}{\frac{1}{4} S_B} \tag{14.32}$$

In terms of the probability $P^{ij}/\Sigma_{ti} V_i$ that neutrons introduced uniformly and isotropically within volume V_i have their first collision in volume V_j, this may be written

$$\gamma_{0i} = \frac{\Sigma_{ti} V_i}{\frac{1}{4} S_B} \left(1 - \sum_{j=1}^{I} \frac{P^{ij}}{\Sigma_{ti} V_i} \right) = \frac{4}{S_B} \left(\Sigma_{ti} V_i - \sum_{j=1}^{I} P^{ij} \right) \tag{14.33}$$

A similar line of argument leads to the result that the probability, R_i, that a uniformly distributed isotropic flux of neutrons crossing the external surface (S_B) of the pin-cell will be removed (absorbed or scattered to another group) by a collision in region i before exiting across surface S_B is related to the total escape probability, P_i, that a neutron introduced in volume V_i will escape (perhaps after multiple collisions) out of the pin-cell across surface S_B:

$$R_i = \frac{\Sigma_{ri} V_i P_i}{\frac{1}{4} S_B} \tag{14.34}$$

We now wish to construct source and current flux response functions in terms of which the flux in any one of the annular regions of the pin-cell can be constructed:

$$\phi_i = \sum_{k=1}^{I} Q_k X^{ki}(\beta) + j_{\text{ex}}^{-} Y_i(\beta) \tag{14.35}$$

where Q_k is the neutron source density in annular region k, $X^{ki}(\beta)$ is the neutron flux produced in region i by a unit neutron source density in region k, taking into account possible multiple reflections at the cell boundary with albedo β, and $Y^i(\beta)$ is the neutron flux produced in annular region i by unit neutron inward current across the cell boundary. The quantities $X^{ki}(0)$ and $Y_i(0)$ refer to the response functions above when the albedo of the region surrounding the pin-cell is zero (i.e., when there is no reflection of neutrons exiting the pin-cell back across surface S_B). The response functions $X^{ki}(\beta)$ and $Y^i(\beta)$ can be calculated in terms of $X^{ki}(0)$ and $Y_i(0)$ and the albedo, β.

For a neutron incident into the pin-cell across the boundary S_B, the cell has an effective albedo $(1 - R)$, where

$$R = \sum_{j=1}^{I} R_i = \sum_{i=1}^{I} \Sigma_{ri} V_i Y_i(0) \tag{14.36}$$

is the total removal (Σ_r is the cross section for absorption plus scatter to another group) probability for a neutron incident on the cell from outside. For a cohort of incident neutrons, a fraction R is removed and a fraction $(1 - R)$ is returned to the boundary S_B. Of the $(1 - R)$ returned to the boundary, a fraction β (the albedo of the surrounding assembly for neutrons exiting the pin-cell) is reflected back into the pin-cell. Of the fraction $(1 - R)\beta$ that enter the cell for a second time, a fraction R is removed and a fraction $(1 - R)$ return to the surface S_B a second time, and so on.

Thus an inward partial current of neutrons incident across S_B is effectively amplified by the factor $1 + (1 - R)\beta + [(1 - R)\beta]^2 + \cdots = 1/[1 - (1 - R)\beta]$. If $Y^i(0)$ is the neutron flux produced in annular region i by unit neutron inward current across the cell boundary, without taking into account reflection of exiting neutrons back into the pin-cell, the neutron flux due to a unit inward, current taking reflection into account, is

$$Y_i(\beta) = \frac{Y_i(0)}{1 - \beta(1 - R)} \tag{14.37}$$

The flux $X^{ki}(\beta)$ in volume V_i due to a unit neutron source density in volume V_k is made up of two components: the flux $X^{ki}(0)$ due to source neutrons from volume V_k which have not been reflected from the boundary S_B, and the flux due to the number of source neutrons $P_k V_k$ from volume V_k which do reach the boundary and are reflected with albedo β. These reflected neutrons can be treated as an incoming flux, and the flux produced by it in volume V_i is found by multiplying by $Y_i(\beta)$. The resulting expression is

$$X^{ki}(\beta) = X^{ki}(0) + \beta P_k V_k \frac{Y_i(0)}{1 - \beta(1 - R)} \tag{14.38}$$

The collision probability equations (9.54) were derived under the implicit assumption of no reflection from the external boundary (i.e., $\beta = 0$) and no incident current. Thus these equations are suitable for calculating the basic response functions $X^{ki}(0)$ and $Y_i(0)$ when a first collision source term to account for incident partial current density j_{ex}^- is included:

$$\Sigma_{ti} V_i \phi_i = \sum_{j=1}^{I} P^{ji} \left[\frac{(\Sigma_{sj} + \nu\Sigma_{fj})\phi_j}{\Sigma_{tj}} + \frac{Q_j}{\Sigma_{tj}} \right] + \gamma_{i0} j_{ex}^-$$

$$\equiv \sum_{j=1}^{I} P^{ji} \left(c_j \phi_j + \frac{Q_j}{\Sigma_{tj}} \right) + \gamma_{i0} j_{ex}^- \tag{14.39}$$

The collision probabilities P^{ji} for a cylindrical cell are given by Eqs. (9.63) to (9.65). In some applications it may be more convenient to treat the fission neutron source as a fixed source and include it in the Q_j term.

The quantities $X^{ki}(0)$ satisfy this equation with a unit source density in volume V_k only and no incident current density:

$$\Sigma_{ti} V_i X^{ki}(0) = \sum_{j=1}^{I} P^{ji} c_j X^{kj}(0) + \frac{P^{ki}}{\Sigma_k}, \quad i = 1, \ldots, I, \ k = 1, \ldots, I \tag{14.40}$$

This constitutes a set of I^2 equations to be solved for the $X^{ki}(0)$. The quantities $Y_i(0)$ satisfy Eq. (14.39) with no volumetric source but with a unit external current density:

$$\Sigma_i V_i Y_i(0) = \sum_{j=1}^{I} P^{ji} c_j Y_j(0) + \gamma_{i0}, \quad i = 1, \ldots, I \tag{14.41}$$

a set of I equations to be solved for the $Y_i(0)$.

In summary, the pin-cell calculation consists of: (1) solve Eqs. (14.40) and (14.41) for the isolated pin-cell flux response functions $X^{ki}(0)$ and $Y_i(0)$; (2) construct the flux response functions $X^{ki}(\beta)$ and $Y_i(\beta)$ which take into account reflection from the surrounding medium by the albedo β from Eqs. (14.37) and (14.38); (3) calculate the flux in each annular region of the pin-cell using Eq. (14.35); and (4) construct homogenized cross sections for the cell using Eq. (14.5).

Interface Current Formulation

The outward partial current density from the pin-cell across surface S_B consists of two components: (1) the source neutrons which are introduced within the pin-cell and which are crossing S_B for the first time ($\sum_{i=1}^{I} P_i V_i Q_i$), and (2) the incident neutrons (j_{ex}^-) which traverse the pin-cell without being removed with probability $(1-R)$—and both components are reflected with probability β and constitute an inward current that may traverse the cell without removal, and so on. The total outward partial current density due to neutron sources within the pin-cell and neutrons incident on the pin-cell from the surrounding medium is

$$j_{out}^+(\beta) = \frac{\sum_{i=1}^{I} V_i P_i Q_i + (1-R) j_{ex}^-}{1 - \beta(1-R)} \tag{14.42}$$

The inward partial current density across surface S_B also has two components: (1) the source neutrons that escape from the pin-cell to reach S_B for the first time ($\sum_{i=1}^{I} P_i V_i Q_i$) and are reflected with probability β, and (2) the incident neutrons (j_{ex}^{-}), both of which may traverse the pin-cell without removal with probability $(1-R)$ to reach surface S_B and be reflected with probability β, and so on. The total incident partial current density is

$$J_{\text{in}}^{-}(\beta) = \frac{\beta \sum_{i=1}^{I} V_i P_i Q_i + j_{\text{ex}}^{-}}{1 - \beta(1-R)} \tag{14.43}$$

The net current density (in the outward direction) across the surface of the pin-cell is

$$j(\beta) \equiv j_{\text{out}}^{+}(\beta) - j_{\text{in}}^{-}(\beta) = \frac{(1-\beta)\sum_{i=1}^{I} V_i P_i Q_i - R j_{\text{ex}}^{-}}{1 - \beta(1-R)} \tag{14.44}$$

Multigroup Pin-Cell Collision Probabilities Model

The pin-cell model above extends immediately to multigroup by making the replacements $\Sigma_{ti} \rightarrow \Sigma_{ti}^{g}$, $\Sigma_{ri} \rightarrow \Sigma_{ti}^{g} - \Sigma_{si}^{g \rightarrow g}$ (i.e., group removal cross section), $\gamma_{0i} \rightarrow \gamma_{0i}^{g}$, $P^{ji} \rightarrow P_{g}^{ji}$ and $R_i \rightarrow R_i^{g}$, and extending certain equations to multigroup. Equations (14.40) become

$$\Sigma_{ti}^{g} V_i X_g^{ki}(0) = \sum_{j=1}^{I} P_g^{ji} \frac{\sum_{g'=1}^{G} (\Sigma_{sj}^{g' \rightarrow g} + \chi^g \nu \Sigma_f^{g'}) X_{g'}^{ji}(0) + \delta_{jk}}{\Sigma_{tj}^{g}},$$

$$g = 1, \ldots, G; \ i, k = 1, \ldots, I \tag{14.45}$$

which can be written in matrix notation as

$$\mathbf{V}_i \Sigma_{ti} X^{ki}(0) = \sum_{j=1}^{I} \left[\mathbf{P}_{SF}^{ji} X^{ji}(0) + \mathbf{P}^{ji} \right], \quad i, j = 1, \ldots, I \tag{14.46}$$

and Eqs. (14.41) become

$$\Sigma_{ti}^{g} V_i Y_i^{g}(0) = \sum_{j=1}^{I} P_g^{ji} \frac{\sum_{g'=1}^{G} (\Sigma_{sj}^{g' \rightarrow g} + \chi^g \nu \Sigma_{fj}^{g'}) Y_j^{g'}(0)}{\Sigma_{tj}^{g}} + \gamma_{0i}^{g},$$

$$g = 1, \ldots, G; \ i, j = 1, \ldots, I \tag{14.47}$$

which can be written in matrix notation as

$$\mathbf{V}_i \Sigma_{ti} Y_i(0) = \sum_{j=1}^{I} \mathbf{P}_{SF}^{ji} \mathbf{Y}_j(0) + \gamma_{0i}, \quad i = 1, \ldots, I \tag{14.48}$$

Equations (14.37) and (14.38), with the appropriate group cross probabilities, can be used to correct the basic flux response functions $X_g^{ki}(0)$ and $Y_i^{g}(0)$ to account for reflection from the surface S_B, and the multigroup fluxes in each region of the pin-cell can be calculated from the multigroup version of Eq. (14.35):

$$\phi_i^g = \sum_{k=1}^{I} Q_k^g X_g^{ki}(\beta) + j_{ex}^{-g} Y_i^g(\beta), \quad i = 1, \ldots, I \qquad (14.49)$$

Resonance Cross Sections

Homogenized resonance cross sections are calculated at the pin-cell level using the methods discussed in Chapter 11.

Full Assembly Transport Calculation

Once the finest level of heterogeneity has been homogenized with a series of pin-cell calculations, the assembly is made up of a large number of homogeneous regions (e.g., the square pin-cells of Fig. 14.3), surrounded by structure, water gaps, control rods, other dissimilar assemblies, and so on (i.e., the assembly is still a heterogeneous medium embedded in a larger-scale heterogeneous medium, the reactor core). The next step in the homogenization process is to perform a multigroup transport calculation on the pin-cell-homogenized assembly for the purpose of obtaining average group fluxes for each homogenized pin-cell that can be used to calculate homogenized cross sections that will allow the entire assembly to be represented as a homogenized region.

Any of the transport methods discussed in Chapter 9 (collision probabilities, discrete ordinates, Monte Carlo) or even diffusion theory in some cases can be used for the full assembly transport calculation. Such calculations are normally performed using reflective conditions on the assembly boundary, or more correctly on the boundary defined by the centerline of the water gap or other medium separating adjacent assemblies, thus implicitly assuming an infinite array of identical assemblies. The fact that different assemblies have different homogenized properties is taken into account in the global core calculation based on a homogenized assembly model which follows assembly homogenization. However, the fact that the adjacent assembly is dissimilar or that there is a control rod nearby or that there is significant leakage out of or into an assembly affects the assembly calculation and hence the homogenized properties of the assembly. Stratagems such as extending the boundaries for an assembly calculation into adjacent assemblies or over a larger planar region have evolved for dealing with this problem.

14.5
Homogenization Theory

When used in the calculation for which they were intended, homogenized cross sections, should yield a result that is equivalent, in some sense, to the result that would have been obtained if the calculation could be performed with all the spatial detail without the need for homogenization. It is useful, in this regard, to develop homogenization procedures that would preserve the essential integral properties of a global heterogeneous transport calculation, the result of which is assumed known for the purpose of development of homogenization procedures, and then to evaluate the homogenized cross sections using an approximation to the global heterogeneous transport solution.

Homogenization Considerations

The neutron flux distribution and effective multiplication constant, k, can be described exactly by multigroup transport theory, which we write in the general form

$$\nabla \cdot \mathbf{J}_g(\mathbf{r}) + \Sigma_t^g(\mathbf{r})\phi_g(\mathbf{r})$$

$$= \frac{\chi^g}{k} \sum_{g=1}^{G} \nu\Sigma_f^{g'}(\mathbf{r})\phi_{g'}(\mathbf{r}) + \sum_{g'=1}^{G} \Sigma^{g' \to g}(\mathbf{r})\phi_{g'}, \quad g = 1, \dots, G \qquad (14.50)$$

Imagine that we know the solution to Eq. (14.50) and wish to use it to define homogenized cross sections which when used in the solution of the homogenized transport equation

$$\nabla \cdot \hat{\mathbf{J}}_g(\mathbf{r}) + \hat{\Sigma}_t^g(\mathbf{r})\hat{\phi}_g(\mathbf{r})$$

$$= \frac{\chi^g}{\hat{k}} \sum_{g'=1}^{G} \nu\hat{\Sigma}_f^{g'}(\mathbf{r})\hat{\phi}_{g'}(\mathbf{r}) + \sum_{g'=1}^{G} \hat{\Sigma}^{g' \to g}\hat{\phi}_{g'}, \quad g = 1, \dots, G \qquad (14.51)$$

yield the same result for certain important quantities as would be obtained if the detailed cross sections and the exact solution of Eq. (14.50) were used in their evaluation (i.e., preserves certain properties of the exact solution). The most important quantities to be preserved are the multiplication constant, k, the group reaction rates averaged over the homogenization region, and the group currents averaged over the surface of the homogenization region. Preservation of the last two quantities requires that

$$\int_{V_i} \hat{\Sigma}_x^g(\mathbf{r})\hat{\phi}_g(\mathbf{r}) \, d\mathbf{r} = \int_{V_i} \Sigma_x^g(\mathbf{r})\phi_g(\mathbf{r}) \, d\mathbf{r} \qquad (14.52)$$

$$\int_{S_i^k} \hat{\mathbf{J}}_g(\mathbf{r}) \cdot d\mathbf{S} = \int_{S_i^k} \mathbf{J}_g(\mathbf{r}) \cdot d\mathbf{S} \qquad (14.53)$$

where V_i is the volume of the homogenization region i and S_i^k is the kth surface of the homogenization region i. Satisfaction of Eqs. (14.52) and (14.53) would also ensure preservation of k.

If the homogenized cross sections are uniform over the homogenization region, an exact definition is

$$\hat{\Sigma}_{x_i}^g \equiv \frac{\int_{V_i} \Sigma_x^g(\mathbf{r})\phi_g(\mathbf{r}) \, d\mathbf{r}}{\int_{V_i} \hat{\phi}_g(\mathbf{r}) \, d\mathbf{r}} \qquad (14.54)$$

and when diffusion theory is to be used in the homogenized calculation,

$$\hat{D}_{ik}^g \equiv \frac{-\int_{S_i^k} \mathbf{J}_g(\mathbf{r}) \cdot d\mathbf{S}}{\int_{S_i^k} \nabla \hat{\phi}_g(\mathbf{r}) \cdot d\mathbf{S}} \tag{14.55}$$

The practical difficulty in using Eqs. (14.54) and (14.55), of course, is that the exact solution of the global transport equation is not known (and never will be, or we would not be bothering with homogenization) and the homogenized solution of the global diffusion equation is not known prior to solving Eq. (14.51), which requires the homogenized group constants as input. Another conceptual problem is that the integrals in Eq. (14.55) will generally be different for each surface, k, so that it is not possible to define a constant value of the homogenized diffusion coefficient which preserves the surface-averaged currents over all the surfaces.

Conventional Homogenization Theory

The conventional pin-cell or assembly homogenization procedure approximates the solution to the global core transport equation, $\phi^g(\mathbf{r})$ and $\mathbf{J}^g(\mathbf{r})$, with the solutions, $\phi_A^g(\mathbf{r})$ and $\mathbf{J}_A^g(\mathbf{r})$, to a pin-cell or assembly transport calculation, usually with symmetry boundary conditions, $\mathbf{n} \cdot \mathbf{J}_A^g(\mathbf{r}) = 0$. The numerator of Eq. (14.54) is then evaluated using $\phi_A^g(\mathbf{r})$ instead of the (unavailable) exact global transport solution $\phi^g(\mathbf{r})$. This assembly transport solution, $\phi_A^g(\mathbf{r})$, is also used to evaluate the flux integral in the denominator of Eq. (14.54). A possible choice of the homogenized diffusion coefficient is

$$\hat{D}_{iA}^g \equiv \frac{\int_{V_i} D^g(\mathbf{r})\phi_A^g(\mathbf{r})\,d\mathbf{r}}{\int_{V_i} \phi_A^g(\mathbf{r})\,d\mathbf{r}} \tag{14.56}$$

Rather large errors have been found in calculations that employed these conventional homogenization methods when compared with exact solutions for benchmark problems. The major source of error is in the treatment of the homogenized diffusion coefficients and the imposition of continuity of flux and current continuity boundary conditions at interfaces between homogenization regions. The source of the problem is that the homogenized diffusion equation, with continuity of current and flux imposed at interfaces, lacks sufficient degrees of freedom to preserve both surface currents and reaction rates.

14.6
Equivalence Homogenization Theory

It is possible to require that both the volume-integrated reaction rates and the surface-integrated currents from the heterogeneous problem be preserved in the homogenized problem [i.e., that Eqs. (14.52) and (14.53) be satisfied] if the conti-

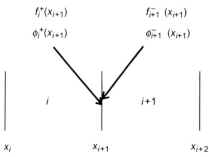

$f_i^+(x_{i+1})$ $\quad\quad\quad\quad$ $f_{i+1}^-(x_{i+1})$

$\phi_i^+(x_{i+1})$ $\quad\quad\quad\quad$ $\phi_{i+1}^-(x_{i+1})$

i $\quad\quad\quad\quad$ $i+1$

x_i $\quad\quad\quad\quad$ x_{i+1} $\quad\quad\quad\quad$ x_{i+2}

Fig. 14.6 Equivalence theory notation.

nuity of flux condition is relaxed. Instead of continuity of flux, the flux interface condition

$$\hat{\phi}_i^+(x_{i+1})f_i^{\,+}(x_{i+1}) = \hat{\phi}_{i+1}^-(x_{i+1})f_{i+1}^{\,-}(x_{i+1}) \tag{14.57}$$

is imposed at the interface at x_{i+1} between homogenization regions i and $i + 1$, where $\phi_i^+(x_{i+1})$ and $\phi_{i+1}^-(x_{i+1})$ are the homogenized fluxes in homogenization regions $x_i \le x \le x_{i+1}$ and $x_{i+1} \le x \le x_{i+2}$, respectively, both evaluated at the interface x_{i+1} between the two, as indicated in Fig. 14.6. Similarly, $f_{i+1}^{\,-}(x_{i+1})$ refers to the flux discontinuity factor at the lower (minus) interface x_{i+1} of the region $x_{i+1} \le x \le x_{i+2}$, and $f_i^{\,+}(x_{i+1})$ refers to the flux discontinuity factor at the upper (plus) interface x_{i+1} of the region $x_i \le x \le x_{i+1}$. The flux discontinuity factors on each side of the interface at x_{i+1} are defined by the ratios of the heterogeneous to homogeneous fluxes at this interface:

$$f_i^{\,+}(x_{i+1}) = \frac{\phi_i^+(x_{i+1})}{\hat{\phi}_i^+(x_{i+1})}, \quad\quad f_{i+1}^{\,-}(x_{i+1}) = \frac{\phi_{i+1}^-(x_{i+1})}{\hat{\phi}_{i+1}^-(x_{i+1})} \tag{14.58}$$

Equations (14.57) and (14.58) express the requirement that the heterogeneous flux is continuous at the interface and relate the homogeneous to heterogeneous fluxes at the interface. The discontinuity factors introduce additional degrees of freedom into the homogenization procedure, which permits the satisfaction of Eqs. (14.52) and (14.53).

Let us now consider the implementation of equivalence theory. For the moment, we continue to assume the existence of an exact heterogeneous solution for the entire core. The evaluation of homogenized cross sections from Eq. (14.54) is straightforward. We examine implementation of the requirement of Eq. (14.53) for the homogenized multigroup diffusion equation in two dimensions:

$$-\nabla \cdot \hat{D}_{ij}^g \nabla \hat{\phi}_{ij}^g(x, y) + \hat{\Sigma}_{tij}^g \hat{\phi}_{ij}^g(x, y)$$

$$= \sum_{g'=1}^{G} \hat{\Sigma}_{ij}^{g' \to g} \hat{\phi}_{ij}^{g'}(x, y) + \frac{\chi^g}{k} \sum_{g'=1}^{G} \nu \hat{\Sigma}_{fij}^{g'} \hat{\phi}_{ij}^{g'}(x, y), \quad g = 1, \ldots, G \tag{14.59}$$

where the homogenized cross sections for homogenization region (i, j) have been calculated from Eq. (14.54) and both homogenized cross sections and diffusion coefficients are constant within region (i, j). Integrating this equation over the y-dimension of the homogenization region (i, j), which is defined by $x_i \leq x \leq x_{t+1}$ and $y_j \leq y \leq y_{j+1}$, yields

$$
-\hat{D}_{ij}^g \frac{d^2}{dx^2} \int_{y_j}^{y_{j+1}} dy\, \hat{\phi}_{ij}^g(x, y) - \hat{D}_{ij}^g \int_{y_j}^{y_{j+1}} dy \frac{d^2}{dy^2} \hat{\phi}_{ij}^g(x, y)
$$

$$
+ \hat{\Sigma}_t^g \int_{y_j}^{y_{j+1}} dy\, \hat{\phi}_{ij}^g(x, y)
$$

$$
= \sum_{g'=1}^{G} \hat{\Sigma}_{ij}^{g' \to g} \int_{y_j}^{y_{j+1}} dy\, \hat{\phi}_{ij}^{g'}(x, y) + \frac{\chi^g}{k} \sum_{g'=1}^{G} v\hat{\Sigma}_{fij}^{g'} \int_{y_j}^{y_{j+1}} dy\, \hat{\phi}_{ij}^{g'}(x, y),
$$

$$
g = 1, \ldots, G \tag{14.60}
$$

Since the heterogeneous solution is assumed to be known, the heterogeneous y-direction leakage (L_{ijy}^g) is known, in principle, and may be used to evaluate the y-direction leakage term in Eq. (14.60); that is,

$$
\hat{L}_{ijy}^g(x) \equiv -\hat{D}_{ij}^g \int_{y_j}^{y_{j+1}} dy \frac{d^2}{dy^2} \hat{\phi}_{ij}^g(x, y) = L_{ijy}^g(x)
$$

$$
\equiv \int_{y_i}^{y_{j+1}} dy \frac{d}{dy} J^g(x, y) = J^g(x, y_{j+1}) - J^g(x, y_j) \tag{14.61}
$$

Furthermore, the known values of the heterogeneous currents (J_g) at x_{i+1} and x_i can be used as boundary conditions for the solution of Eq. (14.60) in the homogenization region (i, j):

$$
-\hat{D}_{ij}^g \frac{d}{dx} \int_{y_j}^{y_{j+1}} dy\, \hat{\phi}_{ij}^g(x_{i+1}, y) = \int_{y_j}^{y_{j+1}} J_g(x_{i+1}, y)\, dy
$$

$$
\tag{14.62}
$$

$$
-\hat{D}_{ij}^g \frac{d}{dx} \int_{y_j}^{y_{j+1}} dy\, \hat{\phi}_{ij}^g(x_i, y) = \int_{y_j}^{y_{j+1}} J_g(x_i, y)\, dy
$$

With the (assumed) known values of the heterogeneous fluxes at the interfaces and the calculated values of the homogeneous flux integrals, the discontinuity factors for region (i, j) at the surfaces at x_{i+1} and at x_i can be calculated as the ratio of heterogeneous-to-homogeneous flux integrals:

$$
f_{ig}^+ = \frac{\Phi_i^g(x_{i+1})}{\hat{\Phi}_i^g(x_{i+1})}, \qquad f_{ig}^- = \frac{\Phi_i^g(x_i)}{\hat{\Phi}_i^g(x_i)} \tag{14.63}
$$

where

$$
\hat{\Phi}_i^g(x) \equiv \int_{y_j}^{y_{j+1}} dy\, \hat{\phi}_{ij}^g(x, y), \qquad \Phi_i^g(x) = \int_{y_j}^{y_{j+1}} dy\, \phi(x, y) \tag{14.64}
$$

The global heterogeneous solution will not be known, of course, so the practical implementation of the prescriptions above requires their approximation using a local heterogeneous solution for an assembly or set of assemblies, usually performed with a zero current boundary condition. It is important that the same approximate heterogeneous solution be used to evaluate the leakage term of Eq. (14.61) in Eq. (14.60), to evaluate the boundary conditions of Eq. (14.62) for Eq. (14.60), and to evaluate the numerators of the flux discontinuity factors. A similar procedure yields the flux discontinuity factors f_j^+ and f_{j+1}^- for region (i, j) at the surfaces at $y = y_j$ and $y = y_{j+1}$. The four different flux discontinuity factors for region (i, j) will in general be different.

Note that this procedure can be implemented for any arbitrary definition of the homogenized diffusion coefficient. The choice of diffusion coefficient will, of course, affect the solution for the homogeneous flux in the calculation above, hence affect the value of the computed flux discontinuity factor. A common choice for the homogenized diffusion coefficient is the simple heterogeneous flux-weighted value:

$$\hat{D}_{ij}^g = \frac{\int_{x_i}^{x_i+1} dx \int_{y_j}^{y_{j+1}} dy\, D^g(x, y)\phi^g(x, y)}{\int_{x_i}^{x_i+1} dx \int_{y_j}^{y_{j+1}} dy\, \phi^g(x, y)} \tag{14.65}$$

The calculation of flux discontinuity factors can be implemented by using assembly calculations of both the heterogeneous and homogeneous fluxes and currents. The volume integral of flux over the assembly can be normalized to be the same in both calculations. If the homogeneous assembly calculation is carried out with zero current symmetry boundary conditions, the homogeneous flux distribution is uniform within the homogenization region. Under these approximations, the flux discontinuity factor can be calculated entirely from the results of the heterogeneous assembly calculation as the ratio of the surface integral of the heterogeneous assembly flux to the volume integral of the heterogeneous flux, as may be seen by considering

$$\frac{\int_{y_j}^{y_{j+1}} dy\, \phi_A^g(x_{i+1}, y)}{\int_{x_i}^{x_i+1} dx \int_{y_j}^{y_{j+1}} dy\, \phi_A^g(x, y)} = \frac{\Phi_{i+1}^g(x_{i+1})}{\int_{x_i}^{x_i+1} dx \int_{y_j}^{y_{j+1}} dy\, \hat{\phi}_{ij}^g(x, y)}$$

$$= \frac{\Phi_{i+1}^g(x_{i+1})}{\Delta x_i\, \hat{\Phi}_{ij}^g} \equiv \frac{f_{ig}^+}{\Delta x} \tag{14.66}$$

where $\phi_A^g(x, y)$ is the heterogeneous flux from the assembly calculation, the common normalization of the heterogeneous and homogeneous fluxes has been used in the second step, and the uniformity of the homogeneous assembly flux with symmetry boundary conditions has been used in the third step. The discontinuity factors calculated from Eq. (14.66), referred to as *assembly discontinuity factors*, will be accurate for assemblies in which the net current almost vanishes over the boundaries, but will be inaccurate for conditions in which there is significant leakage across assembly interfaces; this is an area of active research.

This formulation of equivalence theory is appropriate for any nodal method that uses surface-averaged fluxes [e.g., the quantities defined by Eq. (14.64)] in evaluating node-to-node coupling. The expression for the nodal interface current on the interface at x_{i+1} between nodes (i, j) and $(i + 1, j)$ is

$$J^+_{g(i,j)} = \frac{2D^g_{ij}D^g_{i+1j}}{\Delta_{xi}\Delta_{xi+1}} \frac{f^+_{gi}\Phi^g_i - f^-_{gi+1}\Phi^g_{i+1}}{f^-_{gi+1}D^g_{ij}/\Delta_{xi} + f^+_{gi}D^g_{i+1j}/\Delta_{xi+1}}$$ (14.67)

Similar expressions obtains for the other nodal interfaces.

14.7
Multiscale Expansion Homogenization Theory

A more formal development of homogenization theory builds on the spatial structure typical of a nuclear reactor, a repeating array of highly heterogeneous fuel assemblies within an almost periodic (symmetric) configuration with assembly-averaged properties that vary slowly from assembly to assembly. This suggests the introduction of two spatial scales—the fine scale of the intra-assembly heterogeneity (r_f) and the coarse scale of the global inter-assembly variation (r_c)—which are treated as independent spatial variables. The multiscale homogenization theory will be illustrated with one-group diffusion theory, the governing equation for which is written with r_f and r_c as formally independent spatial variables:

$$-\left(\frac{\partial}{\partial \mathbf{r}_c} + \frac{\partial}{\partial \mathbf{r}_f}\right) \cdot D(\mathbf{r}_c, \mathbf{r}_f)\left(\frac{\partial}{\partial \mathbf{r}_c} + \frac{\partial}{\partial \mathbf{r}_f}\right)\Phi(\mathbf{r}_c, \mathbf{r}_f)$$

$$+ \Sigma_a(\mathbf{r}_c, \mathbf{r}_f)\Phi(\mathbf{r}_c, \mathbf{r}_f) - \frac{1}{k}\nu\Sigma_f(\mathbf{r}_c, \mathbf{r}_f)\Phi(\mathbf{r}_c, \mathbf{r}_f) = 0$$ (14.68)

Normalized to a core average diffusion length, L, the spatial gradients are of different order: $O(Ld/d\mathbf{r}_c) \sim O(L\mathbf{r}_f/\mathbf{r}_c d/d\mathbf{r}_f) \sim \varepsilon O(Ld/d\mathbf{r}_f)$, where $\varepsilon \equiv \mathbf{r}_f/\mathbf{r}_c$ is a small parameter on the order of the ratio of the scale lengths of the intra-assembly heterogeneity to the assembly dimensions. Making flux and eigenvalue expansions in powers of the small parameter ε,

$$\Phi(\mathbf{r}_c, \mathbf{r}_f) = \sum_{n=0} \varepsilon^n \Phi_n(\mathbf{r}_c, \mathbf{r}_f), \qquad \frac{1}{k} = \sum_{n=0} \frac{\varepsilon^n}{k_n}$$ (14.69)

and substituting in Eq. (14.68) yields to leading order $O(\varepsilon^0)$:

$$L_{r_f}\Phi(\mathbf{r}_c, \mathbf{r}_f) \equiv -\frac{\partial}{\partial \mathbf{r}_f} \cdot D(\mathbf{r}_c, \mathbf{r}_f)\frac{\partial \Phi_0(\mathbf{r}_c, \mathbf{r}_f)}{\partial \mathbf{r}_f} + \Sigma_a(\mathbf{r}_c, \mathbf{r}_f)\Phi_0(\mathbf{r}_c, \mathbf{r}_f)$$

$$- \frac{1}{k_0}\nu\Sigma_f(\mathbf{r}_c, \mathbf{r}_f)\Phi_0(\mathbf{r}_c, \mathbf{r}_f) = 0$$ (14.70)

Equation (14.70) plus the periodic (symmetry) boundary conditions on an assembly defines the detailed heterogeneous intra-assembly flux for an assembly k; there will be K such heterogeneous assembly problems, corresponding to the K different fuel assembly types in the reactor core. The dependence on $\mathbf{r}_c$ indicated in

Eq. (14.70) is a dependence on the assembly for which the calculation is made; all intra-assembly spatial dependence is represented by the $\mathbf{r}_f$ dependence. Since no spatial gradients with respect to $\mathbf{r}_c$ occur in Eq. (14.70), the general solution is

$$\Phi_0(\mathbf{r}_c, \mathbf{r}_f) = A_0(\mathbf{r}_c)\phi_0(\mathbf{r}_c, \mathbf{r}_f) \tag{14.71}$$

where $A_0(\mathbf{r}_c)$ is an arbitrary function of the global spatial scale parameter which will be determined from a higher-order equation.

The first-order $O(\varepsilon^1)$ equation is

$$L_{r_f}\Phi_1(\mathbf{r}_c, \mathbf{r}_f) = \frac{\partial}{\partial \mathbf{r}_c} \cdot D(\mathbf{r}_c, \mathbf{r}_f)\frac{\partial}{\partial \mathbf{r}_f}\phi_0(\mathbf{r}_c, \mathbf{r}_f)A_0(\mathbf{r}_c)$$

$$+ \frac{\partial}{\partial \mathbf{r}_f} \cdot D(\mathbf{r}_c, \mathbf{r}_f)\frac{\partial \phi_0(\mathbf{r}_c, \mathbf{r}_f)}{\partial \mathbf{r}_c}A_0(\mathbf{r}_c)$$

$$+ \frac{1}{k_1}\nu\Sigma_f(\mathbf{r}_c, \mathbf{r}_f)\phi_0(\mathbf{r}_c, \mathbf{r}_f)A_0(\mathbf{r}_c) \tag{14.72}$$

which is an inhomogeneous equation of the same form as the homogeneous Eq. (14.70). By the Fredholm alternative theorem, Eq. (14.72) has a solution only if the right side is orthogonal to the solutions of the equation that is adjoint to Eq. (14.70). Since this equation is self-adjoint for one-group diffusion theory (it is not for multigroup diffusion theory or transport theory) with periodic boundary conditions, a solvability condition for Eq. (14.72) is

$$\frac{1}{k_1} = \langle\phi_0, \nu\Sigma_f\phi_0\rangle A_0 + \left\langle\phi_0, \left(\frac{\partial}{\partial \mathbf{r}_f}D + D\frac{\partial}{\partial \mathbf{r}_f}\right)\phi_0\right\rangle \cdot \frac{\partial A_0}{\partial \mathbf{r}_c}$$

$$+ \left\langle\phi_0, \frac{\partial \phi_0}{\partial \mathbf{r}_f} \cdot \frac{\partial D}{\partial \mathbf{r}_c} + D\frac{\partial}{\partial \mathbf{r}_f} \cdot \frac{\partial \phi_0}{\partial \mathbf{r}_c}\right\rangle A_0 \tag{14.73}$$

where $\langle\cdot\rangle$ indicates a spatial integral over $\mathbf{r}_f$ within node k. Equation (14.73) provides a calculation for k_1. The solution of Eq. (14.72) consists of a solution to the homogeneous equation, which is ϕ_0, with an arbitrary multiplier $A_1(\mathbf{r}_c)$, and particular solutions corresponding to the terms on the right side:

$$\Phi_1(\mathbf{r}_c, \mathbf{r}_f) = A_1(\mathbf{r}_c)\phi_0(\mathbf{r}_c, \mathbf{r}_f) + \sum_\xi g_\xi(\mathbf{r}_c, \mathbf{r}_f)\mathbf{n}_\xi \cdot \frac{\partial A_0(\mathbf{r}_0)}{\partial \mathbf{r}_c}$$

$$+ q(\mathbf{r}_c, \mathbf{r}_f)A_0(\mathbf{r}_c) \tag{14.74}$$

where the particular solutions satisfy

$$L_{rf}g_\xi(\mathbf{r}_c, \mathbf{r}_f) = \mathbf{n}_\xi \cdot \left[\frac{\partial}{\partial \mathbf{r}_f}\nabla\phi_0(\mathbf{r}_c, \mathbf{r}_f) + D(\mathbf{r}_c, \mathbf{r}_f)\frac{\partial}{\partial \mathbf{r}_f}\phi_0(\mathbf{r}_c, \mathbf{r}_f)\right] \tag{14.75}$$

$$L_{rf}q(\mathbf{r}_c, \mathbf{r}_f) = \frac{1}{k_1}\nu\Sigma_f(\mathbf{r}_c, \mathbf{r}_f)\phi_0(\mathbf{r}_c, \mathbf{r}_f)$$

with periodic assembly boundary conditions. There is an equation of the form of the first of Eqs. (14.75) for each coordinate direction.

The second-order $O(\varepsilon^2)$ equation is

$$L_{rf}\phi_2 = \frac{\partial}{\partial \mathbf{r}_c} \cdot D \frac{\partial}{\partial \mathbf{r}_c} \Phi_0 + \left(\frac{\partial}{\partial \mathbf{r}_f} \cdot D \frac{\partial}{\partial \mathbf{r}_c} + \frac{\partial}{\partial \mathbf{r}_c} \cdot D \frac{\partial}{\partial \mathbf{r}_f} \right) \Phi_1$$
$$+ \frac{1}{k_1} \nu \Sigma_f \Phi_1 + \frac{1}{k_2} \nu \Sigma_f \Phi_0 \qquad (14.76)$$

which has a solvability condition

$$\left\langle \phi_0, \left(\frac{1}{k_1} \nu \Sigma_f \Phi_1 + \frac{1}{k_2} \nu \Sigma_f \Phi_0 \right) \right\rangle + \left\langle \phi_0, \frac{\partial}{\partial \mathbf{r}_c} \cdot D \frac{\partial}{\partial \mathbf{r}_c} \Phi_0 \right\rangle$$
$$+ \left\langle \phi_0, \left(\frac{\partial}{\partial \mathbf{r}_f} \cdot D \frac{\partial}{\partial \mathbf{r}_c} + \frac{\partial}{\partial \mathbf{r}_c} \cdot D \frac{\partial}{\partial \mathbf{r}_f} \right) \Phi_1 \right\rangle = 0 \qquad (14.77)$$

that provides a solution for k_2. Integrating Eq. (14.76) over the $\mathbf{r}_f$ intra-assembly heterogeneous spatial scale yields the global diffusion equation with parameters averaged over the fuel assembly:

$$\frac{\partial}{\partial \mathbf{r}_c} \cdot \langle \mathbf{D} \rangle \frac{\partial}{\partial \mathbf{r}_c} A_0(\mathbf{r}_c) + \frac{1}{\varepsilon^2} \left(\frac{1}{k} \langle \nu \Sigma_f \rangle - \langle \Sigma_a \rangle \right) A_0(\mathbf{r}_c)$$
$$+ \langle \vec{\Gamma} \rangle \cdot \frac{\partial}{\partial \mathbf{r}_c} A_0(\mathbf{r}_c) + \langle S \rangle A_0(\mathbf{r}_c) = 0 \qquad (14.78)$$

where, defining the normalization $N \equiv \langle \phi_0, \phi_0 \rangle$, the appropriate assembly-averaged homogenized nu-fission and absorption cross section are flux-adjoint weighted with the detailed intra-assembly solutions

$$\langle \nu \Sigma_f \rangle = \frac{\langle \phi_0, \nu \Sigma_f \phi_0 \rangle}{N}$$
$$\langle \Sigma_a \rangle = \frac{\langle \phi_0, \Sigma_a \phi_0 \rangle}{N} \qquad (14.79)$$

the elements of the diffusion tensor for a two-dimensional problem are

$$\langle D_{11} \rangle = \left[\langle \phi_0, D\phi_0 \rangle + \left\langle \phi_0, D \frac{\partial}{\partial r_{f1}} g_2 + \frac{\partial}{\partial r_{f1}} D g_2 \right\rangle \right] / N$$
$$\langle D_{12} \rangle = \left\langle \phi_0, \left(\frac{\partial}{\partial r_{f1}} D + D \frac{\partial}{\partial r_{f1}} \right) g_1 \right\rangle / N$$
$$\langle D_{21} \rangle = \left\langle \phi_0, \left(\frac{\partial}{\partial r_{f2}} D + D \frac{\partial}{\partial r_{f2}} \right) g_2 \right\rangle / N \qquad (14.80)$$
$$\langle D_{22} \rangle = \left[\langle \phi_0, D\phi_0 \rangle + \left\langle \phi_0, \left(D \frac{\partial g_1}{\partial r_{f2}} + \frac{\partial}{\partial r_{f2}} D g_1 \right) \right\rangle \right] / N$$
$$\langle D_{33} \rangle = \frac{\langle \phi_0, D\phi_0 \rangle}{N}$$
$$\langle D_{13} \rangle = \langle D_{23} \rangle = \langle D_{31} \rangle = \langle D_{32} \rangle = 0$$

there is a source that acts like an effective fission or absorption cross section,

$$
\langle S \rangle = \left\langle \phi_0, \left(\frac{\partial}{\partial \mathbf{r}_c} \cdot D \frac{\partial}{\partial \mathbf{r}_f} + \frac{\partial}{\partial \mathbf{r}_f} \cdot D \frac{\partial}{\partial \mathbf{r}_c} \right) q \right\rangle + \frac{1}{k} \langle \phi_0, \nu \Sigma_f q \rangle
$$
$$
- \frac{1}{k} \langle \phi_0, \nu \Sigma_f \phi_0 \rangle - \left\langle \frac{\partial}{\partial \mathbf{r}_c} \phi_0, D \frac{\partial}{\partial \mathbf{r}_c} \phi_0 \right\rangle \tag{14.81}
$$

and there is a convection term (defined in Ref. 1). The source and convection terms arise because of the assembly-to-assembly variation of cross sections and diffusion coefficient. These terms, which vanish for a reactor with exactly periodic conditions associated with each assembly, account for the effect of inter-assembly leakage between adjacent assemblies, which is not accounted for in the calculation of ϕ_0.

Thus the solution of Eqs. (14.70) and (14.75), with periodic boundary conditions, for the detailed intra-assembly flux distribution ϕ_0 and supplementary intranodal functions g_ξ and q can be used to calculate flux-adjoint-weighted homogenized assembly parameters for a consistently formulated global diffusion equation (14.78). This type of multiscale procedure can also be employed to develop a global diffusion equation based on assembly homogenization with transport lattice calculations replacing Eq. (14.70).

14.8
Flux Detail Reconstruction

The homogenization procedure results in homogenized cross sections that can be used for an entire fuel assembly or collections of fuel assemblies (e.g., modules) in a full core calculation. The resulting flux distribution from the full core calculation reflects the global flux distribution, but not the local detailed flux distribution. The detailed assembly or module flux calculations that were used in the homogenization process must be superimposed on the global flux distribution, and the detailed pin-cell flux distributions must be further superimposed on the assembly or module flux distributions. It is important that the assumptions used in reconstructing the detailed flux distribution be consistent, if not identical, with the assumptions made in the homogenization process.

References

1 H. ZHANG, RIZWAN-UDDIN, and J. J. DORNING, "Systematic Homogenization and Self-Consistent Flux and Pin Power Reconstruction for Nodal Diffusion Methods, Part I: Diffusion Theory Based Theory," *Nucl. Sci. Eng.* 121, 226 (**1995**); "Transport-Equation-Based Systematic Homogenization Theory for Nodal Diffusion Methods with Self-Consistent Flux and Pin Power Reconstruction," *J. Transport Theory Stat. Phys.* 26, 433 (**1997**); "A Multiple-Scales Systematic Theory for the Simultaneous Homogenization of Lattice Cells and Fuel Assemblies," *J. Transport Theory Stat. Phys.* 26, 765 (**1997**).

2 A. HEBERT et al., "A Consistent Technique for the Global Homogenization of a Pressurized Water Reactor

Assembly," *Nucl. Sci. Eng. 109*, 360 (**1991**); "Development of a Third Generation SPH Method for the Homogenization of a PWR Assembly," *Proc. Conf. Mathematical Methods and Supercomputing in Nuclear Applications*, Karlsruhe, Germany (**1993**), p. 558; "A Consistent Technique for the Pin-by-Pin Homogenization of a Pressurized Water Assembly," *Nucl. Sci. Eng. 113*, 227 (**1993**).

3 K. S. SMITH, "Assembly Homogenization Techniques for Light Water Reactor Analysis," *Prog. Nucl. Energy 14*, 303 (**1986**).

4 A. JONSSON, "Control Rods and Burnable Absorber Calculations," in Y. Ronen, ed., *CRC Handbook of Nuclear Reactor Calculations III*, CRC Press, Boca Raton, FL (**1986**).

5 R. J. J. STAMM'LER and M. J. ABBATE, *Methods of Steady State Reactor Physics in Nuclear Design*, Academic Press, London (**1983**), Chap. VII.

6 A. KAVENOKY, "The SPH Homogenization Method," *Proc. Specialist's Mtg. Homogenization Methods in Reactor Physics*, Lugano, Switzerland, 1978, IAEA-TECDOC-231, International Atomic Energy Agency, Vienna (**1980**).

7 V. C. DENIZ, "The Theory of Neutron Leakage in Reactor Calculations," in Y. Ronen, ed., *CRC Handbook of Nuclear Reactor Calculations II*, CRC Press, Boca Raton, FL (**1986**), p. 409.

8 K. KOEBKE, "A New Approach to Homogenization and Group Condensation," *Proc. Specialist's Mtg. Homogenization Methods in Reactor Physics*, Lugano, Switzerland, 1978, IAEA-TECDOC-231, International Atomic Energy Agency, Vienna (**1980**).

9 R. T. CHIANG and J. DORNING, "A Homogenization Theory for Lattices with Burnup and Non-uniform Loadings," *Proc. Top. Mtg. Advances in Reactor Physics and Core Thermal-Hydraulics*, American Nuclear Society, La Grange Park, IL (**1980**), p. 240.

10 E. W. LARSEN, "Neutron Transport and Diffusion in Inhomogeneous Media, I," *J. Math. Phys. 16*, 1421 (**1975**); "Neutron Transport and Diffusion in Inhomogeneous Media, II," *Nucl. Sci. Eng. 60*, 357 (**1976**); "Neutron Drift in Heterogeneous Media," *Nucl. Sci. Eng. 65*, 290 (**1978**).

11 J. J. DUDERSTADT and L. J. HAMILTON, *Nuclear Reactor Analysis*, Wiley, New York (**1976**), Chap. 10.

12 A. F. HENRY, *Nuclear-Reactor Analysis*, MIT Press, Cambridge, MA (**1975**), Chap. 10.

13 J. R. ASKEW, F. J. FAYERS, and F. B. KEMSHELL, "A General Description of the Lattice Code WIMS," *J. Br. Nucl. Energy Soc. 5*, 564 (**1966**).

14 C. W. MAYNARD, "Blackness Theory for Slabs," in A. Radkowsky, ed., *Naval Reactors Physics Handbook*, U.S. Atomic Energy Commission, Washington, DC (**1964**), pp. 409–448.

15 A. AMOUYAL, P. BENOIST, and J. HOROWITZ, "New Method of Determining the Thermal Utilization Factor in a Unit Cell," *J. Nucl. Energy 6*, 79 (**1957**).

16 E. E. LEWIS and W. F. MILLER, *Computational Methods of Neutron Transport*, American Nuclear Society, La Grange Park, IL (**1993**).

Problems

14.1. Carry through the detailed derivation of the ABH method.

14.2. Consider a two-region slab geometry model of a unit cell consisting of a fuel plate of thickness $a = 1$ cm with a moderator region of thickness $b = 2$ cm on each side, with zero current cell boundary conditions and a uniform slowing-down source in the moderator. The fuel is UO_2, with thermal cross sections $\Sigma_a = 0.169 \text{ cm}^{-1}$,

$\Sigma_s = 0.372$ cm^{-1}, and $1 - \mu_0 = 0.9887$. The moderator is H$_2$O, with thermal cross sections $\Sigma_a = 0.022$ cm^{-1}, $\Sigma_s = 3.45$ cm^{-1}, and $1 - \mu_0 = 0.676$. Use the ABH method to calculate the thermal disadvantage factor, thermal utilization and homogenized scattering, and absorption cross sections for the cell.

14.3. Carry through the detailed derivation of blackness theory.

14.4. A reactor assembly consists of repeating arrays of three fuel–moderator unit cells of the type described in Problem 14.2, then a 0.1-cm-thick boron plate with thermal cross sections $\Sigma_a = 25$ cm^{-1}, $\Sigma_s = 0.346$ cm^{-1}, and $(1 - \mu_0) = 0.9394$, and then another three fuel–moderator unit cells. Use blackness theory to calculate an effective diffusion theory cross section to represent the boron slabs in the fuel–moderator plus boron plate array.

14.5. A reactor fuel assembly consists of five of the fuel–moderator boron arrays described in Problem 14.4. Use one-group diffusion theory to calculate the assembly detailed heterogeneous flux distribution. Calculate the homogenized assembly absorption and scattering cross sections and diffusion coefficient and the assembly flux discontinuity factors using equivalence theory.

14.6. Construct the Wigner–Seitz cell model for a fuel pin 1 cm in diameter within a 2-cm square of moderator.

14.7. Set up and solve the collision probability equations for Problem 14.6, in one-group theory. Use the fuel and moderator parameters given in Problem 14.2.

14.8. Calculate the homogenized cross sections for the pin-cell model of Problem 14.7, using conventional homogenization theory.

14.9. Consider a lattice made up of a repeating array of 1-cm-thick fuel plates separated by 2 cm of H$_2$O, as described in Problem 14.2, but with different fuel enrichments in different plates. Taking a fuel plate and 1 cm of H$_2$O on each side as an assembly, use diffusion theory to solve for the assembly heterogeneous flux, with zero current assembly boundary conditions. Calculate the homogenized assembly cross sections and diffusion coefficient and the assembly flux discontinuity factor, using equivalence homogenized theory.

14.10.* Write a one-dimensional S_4 code in slab geometry and repeat Problem 14.9 using an S_4 assembly heterogeneous flux.

---

* Problem 14.10 is a longer problem suitable for a take-home project.

15
Nodal and Synthesis Methods

Even after the local fuel pin, clad, coolant, and so on, heterogeneity is replaced by a homogenized representation, a reactor core remains a highly heterogeneous medium because of the intra-assembly and assembly-to-assembly variation in fuel composition, burnable poisons, control rods, water channels, structure and so on. The mesh spacing in a conventional few-group finite-difference model of such a core is constrained by two requirements: (1) it must be sufficiently fine to represent the remaining spatial heterogeneity adequately, and (2) it must be no larger than the shortest (thermal) group diffusion length in order to avoid numerical inaccuracy. A few-group finite-difference model that could adequately describe such a core might well have 10^5 to 10^6 unknowns (the fluxes in each group at each mesh point). The direct solution of such a problem, even in diffusion theory, remains a formidable computation that was unthinkable until very recently. For calculations such as fuel burnup or transient analysis, in which many full-core spatial solutions are needed, direct few-group finite-difference solutions remain impractical.

A large number of approximation methods have been developed to enable a more computationally tractable solution for the effective multiplication constant and neutron flux distribution in reactor cores. Following historical precedent, these methods can generally be classified as nodal, coarse-mesh, or synthesis methods, although the distinction among the categories may be largely a matter of perspective and sequencing of calculational steps.

Nodal methods characterize the global neutron flux distribution in terms of a small number of parameters in each of several large regions, or nodes, into which the reactor core is subdivided for this purpose. Such methods generally require detailed heterogeneous intranodal flux distributions to construct homogenized parameters for each of the many nodes into which a reactor core may be divided and to calculate coupling parameters that link the average flux solutions in adjacent nodes. The global average nodal fluxes must then be combined with the intranodal heterogeneous flux solutions if a heterogeneous flux distribution is required.

Coarse-mesh methods extend the numerical accuracy of conventional finite-difference methods by using higher-order approximations for the flux variation among mesh points. Like nodal methods, coarse-mesh methods generally require detailed regional heterogeneous flux distributions in order to construct homog-

Nuclear Reactor Physics. Weston M. Stacey
Copyright © 2007 WILEY-VCH Verlag GmbH & Co. KGaA, Weinheim
ISBN: 978-3-527-40679-1

enized parameters and to combine with the coarse-mesh solution to construct a detailed heterogeneous flux solution.

   Synthesis methods generally combine detailed heterogeneous two-dimensional planar flux distributions by means of a one-dimensional axial calculation to obtain a global heterogeneous flux solution. Such methods do not require a previous homogenization within large regions of the core as do nodal and coarse mesh methods, but in effect perform a homogenization in constructing the parameters to be used in the axial synthesis calculation, thus ensuring a certain consistency between the homogenization and the approximate model calculation.

15.1
General Nodal Formalism

Writing the multigroup neutron balance equations in the form

$$\nabla \cdot J_g(r) + \Sigma_t^g(r)\phi_g(r)$$

$$= \sum_{g'=1}^{G} \Sigma^{g' \to g}(r)\phi_{g'}(r) + \frac{\chi^g}{k} \sum_{g'=1}^{G} \nu\Sigma_f^{g'}(r)\phi_{g'}(r), \quad g = 1, \ldots, G \quad (15.1)$$

and integrating over the volume of node n (Fig. 15.1) yields an integral balance on node n:

$$\sum_{n'} L_{nn'}^g + \Sigma_{tn}^g \bar{\phi}_g^n V_n = \sum_{g'=1}^{G} \Sigma_n^{g' \to g} \bar{\phi}_{g'}^n V_n + \frac{\chi^g}{k} \sum_{g'=1}^{G} \nu\Sigma_{fn}^{g'} \bar{\phi}_{g'}^n V_n,$$

$$g = 1, \ldots, G, \ n = 1, \ldots, N \quad (15.2)$$

where the nodal average total, scattering, and fission cross sections are defined by expressions of the form

$$\Sigma_{tn}^g \equiv \frac{\int_{V_n} dr \, \Sigma_t^g(r)\phi_g(r)}{\int_{V_n} dr \, \phi_g(r)} \quad (15.3)$$

the average nodal flux is

$$\bar{\phi}_g^n \equiv \frac{\int_{V_n} \phi_g(r) \, dr}{V_n} \quad (15.4)$$

and the leakage between node n and adjacent node n' is defined by a surface integral over the common interface:

$$L_{nn'}^g \equiv \int_{r_s \in S_{nn'}} dr_s \, \mathbf{n} \cdot \mathbf{J}_g(r_s) \quad (15.5)$$

To be more specific, in discussion of the leakage term, we consider a parallelepiped node of dimensions Δx, Δy, and Δz, as shown in Fig. 15.1. The surface integrals

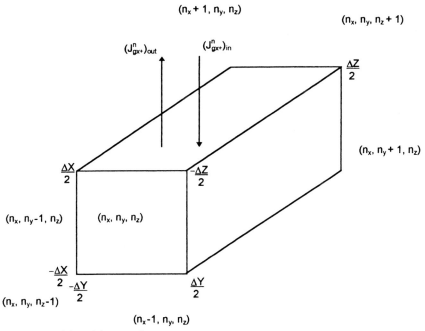

Fig. 15.1 Nodal model nomenclature.

of the net x-direction current at the node boundaries at $x = +\Delta x/2$ and at $-\Delta x/2$ are defined as

$$J^n_{gx\pm} \equiv \frac{\int_{-\Delta y/2}^{\Delta y/2} dy \int_{-\Delta z/2}^{\Delta z/2} dz\, \mathbf{n}_x \cdot \mathbf{J}_g(\pm\Delta x/2, y, z)}{\Delta y \Delta z} \tag{15.6}$$

with similar definitions for the surface integrals of net y- and z-direction currents at $\pm\Delta y/2$ and $\pm\Delta z/2$, respectively. Surface integrals of the outward and inward x-directed partial currents at $\pm x/2$ are defined in terms of the partial currents directed to the right ($\mathbf{J}_g^+$) and to the left ($\mathbf{J}_g^-$), respectively:

$$\left(\bar{J}^n_{gx\pm}\right)_{\text{out}} \equiv \frac{\int_{-\Delta y/2}^{\Delta y/2} dy \int_{-\Delta z/2}^{\Delta z/2} dz\, \mathbf{n}_x \cdot \mathbf{J}_g^{\pm}(\pm(\Delta x/2), y, z)}{\Delta y \Delta z}$$

$$\left(\bar{J}^n_{gx\pm}\right)_{\text{in}} \equiv -\frac{\int_{-\Delta y/2}^{\Delta y/2} dy \int_{-\Delta z/2}^{\Delta z/2} dz\, \mathbf{n}_x \cdot \mathbf{J}_g^{\mp}(\pm(\Delta x/2), y, z)}{\Delta y \Delta z} \tag{15.7}$$

with similar definitions for surface integrals of partial y- and z-direction currents at $\pm\Delta y/2$ and $\pm\Delta z/2$, respectively. The surface integrals of the net current are related to the surface integrals of the partial currents as the net current is related to the partial currents:

$$\bar{J}^n_{gx\pm} = \pm\left[\left(\bar{J}^n_{gx\pm}\right)_{\text{out}} - \left(\bar{J}^n_{gx\pm}\right)_{\text{in}}\right] \tag{15.8}$$

Using these definitions of surface integrals of the net current over the six faces bounding the node, the balance equations (15.2) can be written in the more explicit form

$$\frac{1}{\Delta x}\left(\bar{J}^n_{gx+} - \bar{J}^n_{gx-}\right) + \frac{1}{\Delta y}\left(\bar{J}^n_{gy+} - \bar{J}^n_{gy-}\right) + \frac{1}{\Delta z}\left(\bar{J}^n_{gz+} - \bar{J}^n_{gz-}\right) + \Sigma^g_{tn}\bar{\phi}^n_g$$

$$= \sum_{g'=1}^{G} \Sigma^{g' \to g}_{n}\bar{\phi}^n_{g'} + \frac{\chi^g}{k}\sum_{g'=1}^{G} \nu\Sigma^{g'}_{fn}\bar{\phi}^n_{g'}, \quad g = 1, \ldots, G \tag{15.9}$$

The various nodal formulations are distinguished primarily by the methods used to evaluate the surface currents in Eq. (15.9).

In diffusion theory approximation, the x-directed partial currents and the flux are related by

$$J^{\pm}_{gx}(x) = \frac{1}{4}\phi_g(x) \mp \frac{1}{2}D^g\frac{d\phi_g(x)}{dx} \tag{15.10}$$

with similar relations for the y- and z-directed partial currents. Thus the surface integrals of the flux at $\pm\Delta x/2$

$$\bar{\phi}^n_{gx\pm} \equiv \frac{\int_{-\Delta y/2}^{\Delta y/2} dy \int_{-\Delta z/2}^{\Delta z/2} dz\, \phi(\pm(\Delta x/2), y, z)}{\Delta y \Delta z} \tag{15.11}$$

are related to the corresponding surface integrals of the partial currents:

$$\bar{\phi}^n_{gx\pm} = 2\left[\left(J^n_{gx\pm}\right)_{\text{out}} + \left(J^n_{gx\pm}\right)_{\text{in}}\right] \tag{15.12}$$

with similar relations for the y- and z-directed partial currents at $\pm\Delta y/2$ and $\pm\Delta z/2$, respectively. All surface integrals with the node index n are evaluated in the limit as the surface is approached from within the nth node.

As mentioned, the various nodal formulations are distinguished primarily by the methods used to evaluate the surface currents in Eq. (15.9). Two rather distinct classes of nodal methods have evolved. The first class, often referred to as *conventional* or *simulation models*, makes use of detailed calculations or reactor operating experience to evaluate the surface current integrals in terms of differences in node-averaged fluxes for adjacent nodes, with empirically adjusted coupling coefficients. The second class, sometimes referred to as *consistently formulated models*, makes use of the concept of transverse integration and of higher-order (than ordinary finite-difference) approximations to evaluate the surface integrals of the current and the internodal coupling terms in order to derive nodal equations that can be expected to converge to the exact solution in the limit of small mesh spacing.

15.2
Conventional Nodal Methods

The first class of nodal models to be considered is based on relatively simple mathematical models with parameters that can be adjusted to match the results of more detailed calculation or measurement. Such methods are widely used in three-dimensional simulators, which play a key role in guiding and interpreting the operation of research and power reactors. The basis of such methods is the representation of the neutron flux or neutron fission rate within each of the many homogenized fuel assemblies by a single nodal average flux or fission rate that is coupled to the average flux or fission rate in adjacent nodes by the internodal diffusion of fast neutrons, which is represented by coupling coefficients. The reflector is usually represented by an albedo. Such methods are frequently based on $1\frac{1}{2}$-group theory. The coupling coefficients and the reflector albedo are normally adjusted to provide agreement with more detailed calculations or measurements.

The earlier versions of this class of nodal methods imposed a continuity of net current condition at interfaces:

$$\bar{J}^n_{gx}\left(\frac{\Delta x_n}{2}\right) = -D^g_n \left.\frac{d\bar{\phi}^n_{gx}}{dx}\right|(\Delta x_n/2) = -D^g_{n,\text{eff}}\left(\alpha^{n,n+1}_{gx}\bar{\phi}^{n+1}_g - \alpha^{n,n}_{gx}\bar{\phi}^n_g\right) \quad (15.13)$$

where $\bar{\phi}^n_g$ is the node-averaged flux, and chose the effective diffusion coefficients and coupling parameters α to match interface net currents or nodal average fluxes from detailed planar finite-difference calculations. The sometimes unphysical nature of the solution or the strong sensitivity to the properties of both adjacent fuel assemblies of the coupling coefficients obtained by such net current-matching procedures led to the development of coupling coefficients based on matching partial currents at node interfaces:

$$\bar{J}^{n+}_{gx}\left(\frac{\Delta x_n}{2}\right) \equiv \int_{(-\Delta y/2)}^{(\Delta y/2)} dy \int_{(-\Delta z/2)}^{(\Delta z/2)} dz\, \mathbf{n}_x \cdot \mathbf{J}^+_g\left(\frac{\Delta x}{2}, y, z\right) = \alpha^{n,n+1}_g \bar{\phi}^n_g \Delta x_n$$

$$\bar{J}^{n-}_{gx}\left(\frac{\Delta x_n}{2}\right) \equiv -\int_{(-\Delta y/2)}^{(\Delta y/2)} dy \int_{(-\Delta z/2)}^{(\Delta z/2)} dz\, \mathbf{n}_x \cdot \mathbf{J}^-_g\left(\frac{\Delta x}{2}, y, z\right) \qquad (15.14)$$

$$= \alpha^{n+1,n}_g \bar{\phi}^{n+1}_g \Delta x_{n+1}$$

The *gross coupling method* uses detailed finite-difference diffusion theory fluxes from a heterogeneous planar (x, y) model to calculate interface partial currents:

$$\mathbf{n}_x \cdot \mathbf{J}^{\pm}_g\left(\frac{\Delta x_n}{2}, y\right) = \frac{1}{4}\phi_g\left(\frac{\Delta x_n}{2}, y\right) \mp \frac{1}{2}D_g\left(\frac{\Delta x_n}{2}, y\right)\frac{\partial\phi_g((\Delta x_n/2), y)}{\partial x}$$

$$(15.15)$$

which are used to evaluate the coupling coefficients, α. For $\phi_g(\Delta x_n/2, y)$, $\bar{\phi}^n_g$ and $\bar{\phi}^{n+1}_g$ obtained from detailed planar calculations, Eqs. (15.14) and (15.6)—with the integral over Δz suppressed—are used to evaluate α^n_g and α^{n+1}_g. The nodal equations (15.9) in two-dimensional geometry may be written as

$$-\alpha^{n_x+1,n}_g \frac{\Delta x_{n_x+1}}{\Delta x_n}\bar{\phi}^{n_x+1}_g - \alpha^{n_y+1,n}_g \frac{\Delta y_{n_y+1}}{\Delta y_n}\bar{\phi}^{n_y+1}_g - \alpha^{n_x-1,n}_g \frac{\Delta x_{n_x-1}}{\Delta x_n}\bar{\phi}^{n_x-1}_g$$

$$-\alpha^{n_y-1,n}_g \frac{\Delta y_{n_y-1}}{\Delta y_n}\bar{\phi}^{n_y-1}_g$$

$$+ \left(\alpha_g^{n,n_x+1} + \alpha_g^{n,n_y+1} + \alpha_g^{n,n_x-1} + \alpha_g^{n,n_y-1} + \Sigma_{tn}^g\right)\bar{\phi}_g^n$$

$$-\sum_{g'=1}^{G} \Sigma_n^{g' \to g} \bar{\phi}_{g'}^n - \frac{\chi^g}{k} \sum_{g'=1}^{G} \nu\Sigma_f^{g'} \bar{\phi}_{g'}^n = 0 \qquad (15.16)$$

where the node n is designated by sub- and superscripts n_x and n_y so that the adjacent node in the x- and y-directions may be indicated by $n_x \pm 1$ and $n_y \pm 1$, respectively [e.g., n, $n_x + 1$ refers to the coupling between node n (n_x, n_y) and the adjacent node $(n_x + 1, n_y)$ at $x = +\Delta x_n/2$] (see Fig. 15.1).

Most of the conventional nodal models do not make use of detailed planar calculations to evaluate the internodal coupling coefficients. Instead, the coupling coefficients are reinterpreted in a manner that enables intranodal collision probability methods to be used in their evaluation. The one-group version of Eq. (15.16) may be rewritten as

$$-W^{n_x+1,n}S^{n_x+1} - W^{n_y+1,n}S^{n_y+1} - W^{n_x-1,n}S^{n_x-1} - W^{n_y-1,n}S^{n_y-1}$$

$$+ \left[W^{n,n_x+1} + W^{n,n_y+1} + W^{n,n_x-1} + W^{n,n_y-1} + \frac{k}{k_\infty^n} - 1\right]S^n = 0 \quad (15.17)$$

as a balance among the fission neutron production rates in the various nodes, where

$$S^n \equiv \nu\Sigma_{fn}\phi^n \Delta x_n \Delta y_n, \qquad k_\infty^n \equiv \frac{\nu\Sigma_{fn}}{\Sigma_{an}} \qquad (15.18)$$

and the coupling terms are of the form

$$W^{n,n_x+1} = \frac{\bar{J}_x^{n+}((\Delta x_n/2))}{\nu\Sigma_{fn}\bar{\phi}^n}$$

$$W^{n_x+1,n} = \frac{\bar{J}_x^{n-}((\Delta x_n/2))\Delta x_n}{\nu\Sigma_{fn}\bar{\phi}^{n+1}\Delta x_{n+1}} \qquad (15.19)$$

The new coupling coefficients W^{n,n_x+1} may be interpreted as the probability that a fission neutron born in node (n_x, n_y) escapes into node (n_{x+1}, n_y), and so on, quantities which readily lend themselves to calculation using collision probabilities or other methods. For example, the well-known FLARE code uses

$$W^{n,n_x+1} = (1-g)\frac{\sqrt{M_n^2}}{2\Delta x_n} + g\frac{M_n^2}{(\Delta x_n)^2} \qquad (15.20)$$

where M_n^2 is the migration area in node (n_x, n_y) and g is an adjustable parameter. The two terms correspond to the one-group transport and diffusion kernels for leakage from a slab of thickness Δx_n. A reformulation of the FLARE equations in $1\frac{1}{2}$-group theory leads to

$$W^{n,n_x+1} = \frac{M_n^2}{\Delta x_n^2} \cdot \frac{1}{k_\infty^n} \cdot \frac{2}{1 + M_{n+1}/M_n} \qquad (15.21)$$

Neutron conservation for an internal node requires that

$$S^n = \frac{k^n_\infty}{k} \sum_{m=1}^{6} W^{m,n} S^m \tag{15.22}$$

where the sum is over the six adjacent nodes. $W^{m,n}$ represents the probability that a neutron created from fission in node m will be absorbed in node n, since it has been assumed that a neutron escaping into an adjacent node is absorbed therein. (This assumption can be removed.) For nodes on the surface of the core, an albedo β_{nr} is used for each surface r which faces a reflector, so that the balance equation is

$$S^n = \frac{k^n_\infty}{k} \left[\sum_{m \neq r}^{6} W^{m,n} S^m + (1 - \beta_{nr}) W^{n,r} S^n \right], \quad n = 1, \dots, N \tag{15.23}$$

Equations (15.22) and (15.23) are solved iteratively, with the eigenvalue guess updated on each iteration by using the most recently calculated S^n in the neutron balance to evaluate

$$k = \frac{\sum_n S^n [1 - (1 - \beta_{nr}) W^{n,r}]}{\sum_n S^n / k^n_\infty} \tag{15.24}$$

Nodal methods of the type described in this section generally require parameter adjustment to obtain agreement with more detailed calculations or measurements of power distribution, effective multiplication constant, and so on. Computations based on these nodal methods run very fast and have found widespread use in three-dimensional reactor simulators.

15.3
Transverse Integrated Nodal Diffusion Theory Methods

A second class of nodal methods are those that have been formulated on the basis of integrating the three-dimensional diffusion equation over two transverse directions to obtain a one-dimensional diffusion equation, with transverse leakage terms, which can be solved within a node by approximating the dependence on the remaining spatial variable, usually with a polynomial. These methods are consistently formulated in that they reduce in the limit of small node sizes to the conventional finite-difference method for the homogenized reactor model.

Transverse Integrated Equations

Integration of the three-dimensional multigroup diffusion equations over the two transverse directions to obtain a one-dimensional equation in node n yields

$$\frac{d}{dx} \bar{J}^n_{gx}(x) + \frac{1}{\Delta y} L^g_{ny}(x) + \frac{1}{\Delta z} L^g_{nz}(x) + \Sigma^g_{tn} \bar{\phi}^n_{gx}(x)$$

$$= \sum_{g'=1}^{G} \Sigma_n^{g' \to g} \bar{\phi}_{g'x}^n(x) + \frac{\chi^g}{k} \sum_{g'=1}^{G} \nu \Sigma_{fn}^{g'} \bar{\phi}_{g'x}^n(x), \quad g = 1, \ldots, G \qquad (15.25)$$

The x-dependent flux and current averaged over the transverse directions are

$$\bar{\phi}_{gx}^n(x) \equiv \frac{\int_{-\Delta y/2}^{\Delta y/2} dy \int_{-\Delta z/2}^{\Delta z/2} dz\, \phi_g^n(x, y, z)}{\Delta y \Delta z} \qquad (15.26)$$

$$\bar{J}_{gx}^n(x) \equiv \frac{\int_{-\Delta y/2}^{\Delta y/2} dy \int_{-\Delta z/2}^{\Delta z/2} dz\, J_g^n(x, y, z)}{\Delta y \Delta z} \qquad (15.27)$$

and leakage terms transverse to the x-direction are

$$L_{ny}^g(x) = \frac{1}{\Delta z} \int_{-\Delta z/2}^{\Delta z/2} dz\, \mathbf{n}_y \cdot \left[\mathbf{J}_g\left(x, \frac{\Delta y}{2}, z\right) - \mathbf{J}_g\left(x, -\frac{\Delta y}{2}, z\right) \right]$$

$$= \frac{-1}{\Delta z} \int_{-\Delta z/2}^{\Delta z/2} dz \left[D_n^g \frac{\partial \phi_g(x, (\Delta y/2), z)}{\partial y} - D_n^g \frac{\partial \phi_g(x, -(\Delta y/2), z)}{\partial y} \right] \qquad (15.28)$$

$$L_{nz}^g(x) = \frac{1}{\Delta y} \int_{-\Delta y/2}^{\Delta y/2} dy\, \mathbf{n}_z \cdot \left[\mathbf{J}_g\left(x, y, \frac{\Delta z}{2}\right) - \mathbf{J}_g\left(x, y, -\frac{\Delta z}{2}\right) \right]$$

$$= \frac{-1}{\Delta y} \int_{-\Delta y/2}^{\Delta y/2} dy \left[D_n^g \frac{\partial \phi_g(x, y, (\Delta z/2))}{\partial z} - D_n^g \frac{\partial \phi_g(x, y, -(\Delta z/2))}{\partial z} \right] \qquad (15.29)$$

Making the diffusion theory approximation

$$\bar{J}_{gx}^n(x) = -D_n^g \frac{d}{dx} \bar{\phi}_{gx}^n(x) \qquad (15.30)$$

the multigroup diffusion theory x-direction transverse integrated equation for node n is

$$-\frac{d}{dx} D_n^g \frac{d}{dx} \bar{\phi}_{gx}^n(x) + \frac{1}{\Delta y} L_{ny}^g(x) + \frac{1}{\Delta z} L_{nz}^g(x) + \Sigma_{tn}^g \bar{\phi}_{gx}^n$$

$$= \sum_{g'=1}^{G} \Sigma_n^{g' \to g} \bar{\phi}_{g'x}^n(x) + \frac{\chi^g}{k} \sum_{g'=1}^{G} \nu \Sigma_{fn}^{g'} \bar{\phi}_{g'x}^n(x), \quad g = 1, \ldots, G \qquad (15.31)$$

The node-averaged values of the group flux and transverse leakage terms are

$$\bar{\phi}_g^n \equiv \frac{1}{\Delta x} \int_{-\Delta x/2}^{\Delta x/2} dx\, \bar{\phi}_{gx}^n(x)$$

$$= \frac{1}{\Delta x \Delta y \Delta z} \int_{-\Delta x/2}^{\Delta x/2} dx \int_{-\Delta y/2}^{\Delta y/2} dy \int_{-\Delta z/2}^{\Delta z/2} dz\, \phi_g(x, y, z) \qquad (15.32)$$

$$\bar{L}_{ny}^{g} \equiv \frac{1}{\Delta x} \int_{-\Delta x/2}^{\Delta x/2} dx \, L_{ny}^{g}(x) = J_{gy+}^{n} - J_{gy-}^{n}$$

$$\bar{L}_{nz}^{g} \equiv \frac{1}{\Delta x} \int_{-\Delta x/2}^{\Delta x/2} dx \, L_{nz}^{g}(x) = J_{gz+}^{n} - J_{gz-}^{n} \tag{15.33}$$

Integrating Eq. (15.25) over x and using Eqs. (15.32) and (15.33) yields the nodal balance equation (15.19). One-dimensional transverse integrated equations in the y- and z-directions are derived in a similar manner.

Polynomial Expansion Methods

The coarse mesh methods can obtain a higher-order accuracy than conventional finite-difference methods by expanding the x-dependence of the flux:

$$\bar{\phi}_{gx}^{n}(x) \simeq \bar{\phi}_{g}^{n} f_{0}(x) + \sum_{i=1}^{I} a_{gxi}^{n} f_{i}(x), \quad -\frac{\Delta x}{2} \leq x \leq \frac{\Delta x}{2} \tag{15.34}$$

where the polynomials

$$f_{0}(x) = 1, \qquad f_{1}(x) = \frac{x}{\Delta x} \equiv \xi$$

$$f_{2}(x) = 3\xi^{2} - \frac{1}{4}, \qquad f_{3}(x) = \xi\left(\xi - \frac{1}{2}\right)\left(\xi + \frac{1}{2}\right) \tag{15.35}$$

$$f_{4}(x) = \left(\xi^{2} - \frac{1}{20}\right)\left(\xi - \frac{1}{2}\right)\left(\xi + \frac{1}{2}\right), \qquad \cdots$$

are normalized so that the volume average of the polynomial representation of the flux is the volume average of the flux defined by Eq. (15.32):

$$\frac{1}{\Delta x} \int_{-\Delta x/2}^{\Delta x/2} dx \, f_{n}(x) = \begin{cases} 1, & n = 0 \\ 0, & n > 0 \end{cases} \tag{15.36}$$

and the surface average of the flux is equal to the surface-averaged flux defined by Eq. (15.11) at $x = \pm \Delta x/2$:

$$\bar{\phi}_{gx}^{n}\left(\pm \frac{\Delta x}{2}\right) = \phi_{gx\pm}^{n} \tag{15.37}$$

These requirements are satisfied by polynomial expansion coefficients,

$$a_{gx1}^{n} = \phi_{gx+}^{n} - \phi_{gx-}^{n}$$

$$a_{gx2}^{n} = \phi_{gx+}^{n} - \phi_{gx-}^{n} - 2\bar{\phi}_{g}^{n} \tag{15.38}$$

and the requirement that

$$f_{i}\left(\pm \frac{\Delta x}{2}\right) = 0, \quad n > 2 \tag{15.39}$$

on the polynomials.

In terms of these polynomials, the outgoing x-direction surface-averaged currents at $x = \pm\Delta x/2$ are

$$\left(J_{gx+}^n\right)_{\text{out}} = J_{gx+}^n + \left(J_{gx+}^n\right)_{\text{in}} = -D_n^g \frac{d}{dx} \bar{\phi}_{gx}^n \left(\frac{\Delta x}{2}\right) + \left(J_{gx+}^n\right)_{\text{in}}$$

$$= -\frac{D_n^g}{\Delta x}\left(a_{gx1}^n + 3a_{gx2}^n + \frac{1}{2}a_{gx3}^n + \frac{1}{5}a_{gx4}^n\right) + \left(J_{gx+}^n\right)_{\text{in}} \tag{15.40}$$

$$\left(J_{gx-}^n\right)_{\text{out}} = -J_{gx-}^n + \left(J_{gx-}^n\right)_{\text{in}} = D_n^g \frac{d}{dx} \bar{\phi}_{gx}^n \left(-\frac{\Delta x}{2}\right) + \left(J_{gx-}^n\right)_{\text{in}}$$

$$= \frac{D_n^g}{\Delta x}\left(a_{gx1}^n - 3a_{gx2}^n + \frac{1}{2}a_{gx3}^n - \frac{1}{5}a_{gx4}^n\right) + \left(J_{gx-}^n\right)_{\text{in}} \tag{15.41}$$

with similar expressions for the y- and z-direction surface-averaged currents at $\pm\Delta y/2$ and $\pm\Delta z/2$, respectively.

If the polynomial expansion of the x-direction flux in Eq. (15.34) is terminated at $I = 2$, and similarly for the y- and z-direction expansions, the transverse-integrated nodal equations are well posed in terms of node-averaged fluxes and incoming and outgoing partial currents over node boundaries (i.e., the number of equations and the number of unknowns agree). Equations (15.38) and (15.12) can be used to express Eqs. (15.40) and (15.41) in terms of node-averaged flux and partial currents at $x = \pm\Delta x/2$:

$$\left(J_{gx+}^n\right)_{\text{out}}\left(1 + \frac{8D_n^g}{\Delta x}\right) + \left(J_{gx-}^n\right)_{\text{out}}\frac{4D_n^g}{\Delta x} - \frac{6D_n^g}{\Delta x}\phi_g^n$$

$$= \left(J_{gx+}^n\right)_{\text{in}}\left(1 - \frac{8D_n^g}{\Delta x}\right) + \left(J_{gx-}^n\right)_{\text{in}}\left(-\frac{4D_n^g}{\Delta x}\right) \tag{15.42}$$

$$\left(J_{gx-}^n\right)_{\text{out}}\left(1 + \frac{8D_n^g}{\Delta x}\right) + \left(J_{gx+}^n\right)_{\text{out}}\frac{4D_n^g}{\Delta x} - \frac{6D_n^g}{\Delta x}\phi_g^n$$

$$= \left(J_{gx-}^n\right)_{\text{in}}\left(1 - \frac{8D_n^g}{\Delta x}\right) + \left(J_{gx+}^n\right)_{\text{in}}\left(-\frac{4D_n^g}{\Delta x}\right) \tag{15.43}$$

with similar expressions for the y- and z-direction surface-averaged currents at $\pm\Delta y/2$ and $\pm\Delta z/2$, respectively. Equation (15.8) can be used to replace the currents with partial currents in the nodal balance equation (15.9) to obtain

$$\frac{1}{\Delta x}\left\{\left[\left(J_{gx+}^n\right)_{\text{out}} + \left(J_{gx-}^n\right)_{\text{out}}\right] - \left[\left(J_{gx+}^n\right)_{\text{in}} + \left(J_{gx-}^n\right)_{\text{in}}\right]\right\}$$

$$+ \frac{1}{\Delta y}\left\{\left[\left(J_{gy+}^n\right)_{\text{out}} + \left(J_{gy-}^n\right)_{\text{out}}\right] - \left[\left(J_{gy+}^n\right)_{\text{in}} + \left(J_{gy-}^n\right)_{\text{in}}\right]\right\}$$

$$+ \frac{1}{\Delta z}\left\{\left[\left(J_{gz+}^n\right)_{\text{out}} + \left(J_{gz-}^n\right)_{\text{out}}\right] - \left[\left(J_{gz+}^n\right)_{\text{in}} + \left(J_{gz-}^n\right)_{\text{in}}\right]\right\}$$

$$+ \Sigma_{tn}^g \bar{\phi}_g^n = \sum_{g'=1}^{G} \Sigma_n^{g' \to g} \bar{\phi}_{g'}^n + \frac{\chi^g}{k} \sum_{g'=1}^{G} \nu \Sigma_{fn}^{g'} \bar{\phi}_{g'}^n, \quad g = 1, \dots, G \qquad (15.44)$$

Note that this equation could be derived directly by integrating Eq. (15.1) over the node.

The incoming x-direction partial currents to node n may be related to the outgoing partial currents from the adjacent node $n + 1$ at $\Delta x/2$. Using the flux discontinuity condition discussed in Chapter 14, the surface-averaged fluxes are related by

$$f_{gx+}^n \phi_{gx+}^n = f_{gx-}^{n+1} \phi_{gx-}^{n+1}$$
$$f_{gx+}^n \left[\left(J_{gx+}^n \right)_{\text{out}} + \left(J_{gx+}^n \right)_{\text{in}} \right] = f_{gx-}^{n+1} \left[\left(J_{gx-}^{n+1} \right)_{\text{out}} + \left(J_{gx-}^{n+1} \right)_{\text{in}} \right] \qquad (15.45)$$

where Eq. (15.12) has been used to write the second form of the equation. For unity flux discontinuity factors, Eq. (15.45) becomes the continuity of flux condition. The surface-averaged current continuity condition

$$J_{gx+}^n = J_{gx-}^{n+1}$$
$$\left(J_{gx+}^n \right)_{\text{out}} - \left(J_{gx+}^n \right)_{\text{in}} = \left(J_{gx-}^{n+1} \right)_{\text{in}} - \left(J_{gx-}^{n+1} \right)_{\text{out}} \qquad (15.46)$$

may be combined with the flux discontinuity condition to obtain

$$\left(J_{gx+}^n \right)_{\text{in}} = \frac{2 \left(J_{gx-}^{n+1} \right)_{\text{out}}}{1 + f_{gx+}^n / f_{gx-}^{n+1}} + \frac{1 - f_{gx+}^n / f_{gx-}^{n+1}}{1 + f_{gx+}^n / f_{gx-}^{n+1}} \left(J_{gx+}^n \right)_{\text{out}} \qquad (15.47)$$

Imposition of similar conditions at the interface with adjacent node $n - 1$ at $-\Delta x/2$ yields

$$\left(J_{gx-}^n \right)_{\text{in}} = \frac{2 \left(J_{gx+}^{n-1} \right)_{\text{out}}}{1 + f_{gx-}^n / f_{gx+}^{n-1}} + \frac{1 - f_{gx-}^n / f_{gx+}^{n-1}}{1 + f_{gx-}^n / f_{gx+}^{n-1}} \left(J_{gx-}^n \right)_{\text{out}} \qquad (15.48)$$

Similar expressions are obtained relating the incoming y- and z-direction surface-averaged partial currents at $\pm \Delta y/2$ and $\pm \Delta z/2$, respectively, to the outgoing partial currents from the adjacent nodes in the y- and z-directions.

The equations above can be derived directly from an expansion of the form

$$\phi_g^n(x, y, z) = \bar{\phi}_g + \sum_{i=1}^{2} \alpha_{gxi}^n f_i(x) + \sum_{j=1}^{2} \alpha_{gyj}^n f_j(y) + \sum_{k=1}^{2} \alpha_{gzk}^n f_k(z) \qquad (15.49)$$

without recourse to the transverse integration stratagem. In fact, Eqs. (15.44) follow directly from Eqs. (15.8) and (15.9), and the interface conditions of Eqs. (15.45) and (15.46) arise from other considerations. However, this transverse integration stratagem is essential for extending the formalism to higher order.

For polynomial expansions with $I > 2$ in Eq. (15.34), the transverse integrated equations are no longer well posed in the sense of having the same number of equations and unknowns. However, weighted residuals methods can be used to develop higher-order approximations, but this requires the further approximation of higher-order leakage moments. Multiplying Eq. (15.25) by the spatial function $w_i(x)$ and integrating yields

$$
\left\langle w_i(x), \frac{d}{dx} \bar{J}_{gx}^n(x) \right\rangle + \Sigma_{tn}^n \phi_{gxi}^n
$$

$$
= \sum_{g'=1}^{G} \Sigma_n^{g' \to g} \phi_{g'xi}^n + \frac{\chi^g}{k} \sum_{g'=1}^{G} \nu \Sigma_{fg'}^n \phi_{g'xi}^n - \frac{1}{\Delta y} L_{nyxi}^g - \frac{1}{\Delta z} L_{nzxi}^g \quad (15.50)
$$

where the ith spatial moment of the flux is

$$
\phi_{gxi}^n \equiv \left\langle w_i(x), \bar{\phi}_{gx}^n(x) \right\rangle \equiv \frac{1}{\Delta x} \int_{-(\Delta x/2)}^{(\Delta x/2)} dx\, w_i(x) \bar{\phi}_{gx}^n(x) \quad (15.51)
$$

and the ith spatial moment of the transverse leakage is

$$
L_{nyxi}^g \equiv \left\langle w_i(x), L_{ny}^g(x) \right\rangle \equiv \frac{1}{\Delta x} \int_{-(\Delta x/2)}^{(\Delta x/2)} dx\, w_i(x) L_{ny}^g(x) \quad (15.52)
$$

with a similar term for the z-direction transverse leakage.

The nodal balance equation results from choosing $w_0 = 1$ in Eq. (15.50). Numerical comparison with detailed finite-difference solutions indicates that the choices $w_1(x) = f_1(x)$ and $w_2(x) = f_2(x)$ yield good results. Using these two functions and integrating the first term in Eq. (15.50) by parts yields the two equations that must be solved for the higher-order flux moments:

$$
\frac{1}{2\Delta x} T_{nx}^g + \frac{D_n^g}{(\Delta x)^2} \alpha_{gx1}^n + \Sigma_{tn}^g \phi_{gx1}^n
$$

$$
= \sum_{g'=1}^{G} \Sigma_n^{g' \to g} \phi_{g'x1}^n + \frac{\chi^g}{k} \sum_{g'=1}^{G} \nu \Sigma_{fg'}^n \phi_{g'x1}^n - \frac{1}{\Delta y} L_{nyx1}^g - \frac{1}{\Delta z} L_{nzx1}^g \quad (15.53)
$$

$$
\frac{1}{2\Delta x} L_{nx}^g + \frac{3 D_n^g}{(\Delta x)^2} \alpha_{gx2}^n + \Sigma_{tn}^g \phi_{gx2}^n
$$

$$
= \sum_{g'=1}^{G} \Sigma_n^{g' \to g} \phi_{g'x2}^n + \frac{\chi^g}{k} \sum_{g'=1}^{G} \nu \Sigma_{fg'}^n \phi_{g'x2}^n - \frac{1}{\Delta y} L_{nyx2}^g - \frac{1}{\Delta z} L_{nzx2}^g \quad (15.54)
$$

where

$$
T_{nx}^g \equiv J_{gx+}^n + J_{gx-}^n, \qquad L_{nx}^g \equiv J_{gx+}^n - J_{gx-}^n \quad (15.55)
$$

Using $w_1(x) = f_1(x)$ and $w_2(x) = f_2(x)$ and Eq. (15.49) in Eq. (15.51) then yields the higher-order expansion coefficients

$$
\alpha_{gx3}^n = -120 \phi_{gx1}^n + 10 \alpha_{gx1}^n, \qquad \alpha_{gx4}^n = -700 \phi_{gx2}^n + 35 \alpha_{gx2}^n \quad (15.56)
$$

Solution of Eqs. (15.54) requires further approximation for the x-dependence of the x-direction transverse leakage (and similarly for the y- and z-direction transverse leakage terms). A number of approximations have been used, but the most successful has been the quadratic approximation

$$L_{ny}^g(x) = \bar{L}_{ny}^g + C_{gy1}^n f_1(x) + C_{gy2}^n f_2(x) \tag{15.57}$$

which is assumed, for the purpose of evaluating moments of the transverse leakage, to extend over node n and the two nodes adjacent to node n in the x-direction. Use of Eq. (15.57) in Eq. (15.52) then makes it possible to evaluate the transverse leakage moments in terms of the surface-averaged leakages (thus surface-averaged partial currents) in the adjacent nodes.

Combining results in the three coordinate directions leads to an interface current balance in each group of the form

$$J_g^{n,\text{out}} = P_g^n\left[Q_g^n - L_g^n\right] + R_g^n J_g^{n,\text{in}} \tag{15.58}$$

The column vectors $J_g^{n,\text{out}}$ and $J_g^{n,\text{in}}$ contain the six outgoing and incoming, respectively, surface-averaged partial currents for the nth node. The column vector Q_g^n contains the node-averaged scatter-in and fission sources to group g, and L_g^n contains the higher-order spatial moments of the transverse leakage computed using the quadratic fit or some other approximation. The matrices P_g^n and R_g^n contain nodal coupling coefficients. A variety of iterative schemes have been devised for solving Eq. (15.58), within an outer power iteration solution procedure. Generally, the three-dimensional geometry is subdivided into a number of axial planes, and the nodes within each plane are solved (swept) a few times using the most recent values for group fluxes in nodes in the adjacent planes. The number of planar sweeps required per group generally increases with the planar average diffusion length within the group.

The nodal procedure outlined above uses constant homogenized cross sections over the node. In applications where the actual cross sections vary significantly over the node, the use of constant cross sections introduces an error in calculating effects such as space-dependent internodal burnup. An extension to include low-order polynomial dependence of the cross sections over the node has been shown to lead to improved accuracy in such cases.

Analytical Methods

There are variants of the transverse integrated method in which an analytical solution is used in some part of the derivation of the transverse integrated nodal equations. In a variant known as the *analytical nodal method* the one-dimensional transverse integrated equation is integrated analytically to relate the nodal leakage in that dimension to the nodal average fluxes in the node and in the adjacent nodes in that dimension. In another variant known as the *nodal Green's function method*, the one-dimensional transverse integrated equation is formally solved by the method of Green's functions, resulting in expressions that can be used together with the polynomial expansion to evaluate coefficients. These are discussed more fully in Ref. 2.

Heterogeneous Flux Reconstruction

The results of the nodal calculation are global node-averaged fluxes, $\bar{\phi}_g^n$ flux distributions consisting of the polynomial flux distributions $\phi_g^n(x, y, z)$ within each node or assembly n [e.g., as constructed from Eqs. (15.34) for each direction] and nodal interface currents. These global fluxes and flux distributions are normalized to the reactor power level. To obtain a more detailed heterogeneous intra-assembly flux distribution, it is necessary to superimpose on these nodal average or smoothly varying polynomial flux distributions a detailed intranodal flux shape, $A_g^n(x, y)$, usually taken from a planar assembly transport calculation:

$$\Phi_g^n(x, y, z) = \phi_g^n(x, y, z) A_g^n(x, y) \tag{15.59}$$

The simplest such procedures use an assembly calculation with symmetry boundary conditions to determine $A_g^n(x, y)$ and Eq. (15.59). Improved accuracy has been obtained by using the first of Eqs. (15.59) to construct a gross intranodal flux distribution that approximates the gross intranodal flux shape from the global calculation. Use of the same intranodal flux shape for the nodal homogenization and flux reconstruction is necessary for consistency, but this is difficult to achieve in practice without an iteration among the homogenization, nodal solution, and flux reconstruction steps.

15.4
Transverse Integrated Nodal Integral Transport Theory Models

Transverse Integrated Integral Transport Equations

The concepts and procedures introduced in Section 15.3 can be extended to develop nodal methods based on integral transport theory. To limit the notational complexity, we discuss the development of integral transport nodal methods in two-dimensional rectangular geometry, although we note that three-dimensional models are in use for nuclear reactor analysis. Assuming that a detailed heterogeneous assembly transport calculation has been performed to produce homogenized multi-group constants that are uniform over the domain of node n $(-\Delta x/2 \leq x \leq \Delta x/2, -\Delta y/2 \leq y \leq \Delta y/2)$, the transport equation for the multi-group neutron flux within node n in two-dimensional Cartesian geometry may be written

$$\mu \frac{\partial}{\partial x} \psi_g^n(x, y, \mu, \phi) + \sqrt{1 - \mu^2} \cos\phi \frac{\partial}{\partial y} \psi_g^n(x, y, \mu, \phi) + \Sigma_{tn}^g \psi_g^n(x, y, \mu, \phi)$$

$$= \frac{1}{4\pi} S_g^n(x, y), \quad g = 1, \ldots, G \tag{15.60}$$

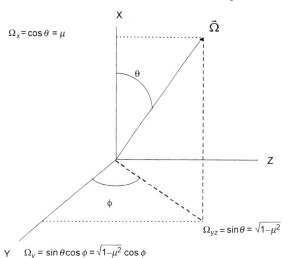

Fig. 15.2 Coordinate system for two-dimensional nodal transport model.

where, for notational convenience, the group in-scatter and fission terms have been written as a source term:

$$S_g^n(x, y) = \frac{\chi^g}{k} \sum_{g'=1}^{G} \nu\Sigma_f^{g'} \int_{-1}^{1} d\mu \int_{0}^{2\pi} d\phi \, \psi_{g'}^n(x, y, \mu, \phi)$$

$$+ \sum_{g'=1}^{G} \nu\Sigma_n^{g' \to g} \int_{-1}^{1} d\mu \int_{0}^{2\pi} d\phi \, \psi_{g'}^n(x, y, \mu, \phi) \qquad (15.61)$$

isotropic scattering has been assumed, and the coordinate system is defined such that

$$\Omega_x \equiv \mathbf{\Omega} \cdot \mathbf{n}_x = \mu, \qquad \Omega_y \equiv \mathbf{\Omega} \cdot \mathbf{n}_y = \sqrt{1 - \mu^2} \cos\phi \qquad (15.62)$$

The coordinate system and spatial domain of node n are depicted in Figs. 15.2 and 15.3.

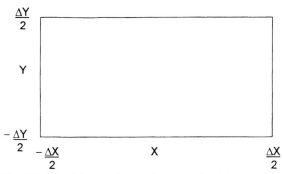

Fig. 15.3 Spatial domain for two-dimensional nodal model.

Integrating Eq. (15.60) over $-\Delta y/2 \leq y \leq \Delta y/2$ yields the one-dimensional x-direction transverse integrated transport equation for node n:

$$\mu \frac{\partial}{\partial x} \psi_{gx}^n(x, \mu, \phi) + \Sigma_{tn}^g \psi_{gx}^n(x, \mu, \phi) + L_{ny}^g(x, \mu, \phi)$$

$$= \frac{1}{4\pi} \int_{-\Delta y/2}^{\Delta y/2} dy\, S_g^n(x, y) \equiv \frac{1}{4\pi} S_g^n(x) \qquad (15.63)$$

where the x-direction angular flux is

$$\psi_{gx}^n(x, \mu, \phi) \equiv \frac{1}{\Delta y} \int_{-\Delta y/2}^{\Delta y/2} dy\, \psi_g^n(x, y, \mu, \phi) \qquad (15.64)$$

and the transverse leakage term defining the average net neutron loss rate across the node boundaries at $y = -\Delta y/2$ and $y = \Delta y/2$ is

$$L_{ny}^g(x, \mu, \phi)$$

$$\equiv \frac{1}{\Delta y} \sqrt{1 - \mu^2} \cos\phi \left[\psi_g^n\left(x, +\frac{\Delta y}{2}, \mu, \phi\right) - \psi_g^n\left(x, -\frac{\Delta y}{2}, \mu, \phi\right) \right] \quad (15.65)$$

Equation (15.63) can be integrated if the scattering, fission, and leakage are treated as a known source:

$$\psi_{gx}^n(x, \mu > 0, \phi) = \int_{-(\Delta x/2)}^{x} dx'\, e^{-\Sigma_{tn}^g(x-x')/\mu} \frac{1}{\mu} \left[\frac{1}{4\pi} S_g^n(x') - L_{ny}^g(x', \mu, \phi) \right]$$

$$+ \psi_{gx-}^{n,\text{in}}(\mu, \phi) e^{-\Sigma_{tn}^g(x+(\Delta x/2))/\mu}, \qquad \mu > 0 \qquad (15.66a)$$

$$\psi_{gx}^n(x, \mu < 0, \phi) = -\int_{x}^{(\Delta x/2)} dx'\, e^{-\Sigma_{tn}^g(x-x')/\mu}$$

$$\times \frac{1}{\mu} \left[\frac{1}{4\pi} S_g^n(x') - L_{ny}^g(x', \mu, \phi) \right]$$

$$+ \psi_{gx+}^{n,\text{in}}(\mu, \phi) e^{-\Sigma_{tn}^g(x-(\Delta x/2))/\mu}, \qquad \mu < 0 \qquad (15.66b)$$

where the inward-directed average angular fluxes at $x = \Delta x/2$ and $x = -\Delta x/2$ are

$$\psi_{gx\pm}^{n,\text{in}}(\mu, \phi) \equiv \psi_{gx}^n\left(\pm\frac{\Delta x}{2}, \mu \lessgtr 0, \phi\right) \qquad (15.67)$$

and the outward-directed average angular fluxes at $x = \Delta x/2$ and $x = -\Delta x/2$ are

$$\psi_{gx\pm}^{n,\text{out}}(\mu, \phi) \equiv \psi_{gx}^n\left(\pm\frac{\Delta x}{2}, \mu \gtrless 0, \phi\right) \qquad (15.68)$$

The average scalar flux in the x-direction problem is

$$\phi_{gx}^n(x) = \int_0^{2\pi} d\phi \left[\int_0^1 d\mu\, \psi_{gx}^n(x, \mu > 0, \phi) + \int_{-1}^0 d\mu\, \psi_{gx}^n(x, \mu < 0, \phi) \right]$$

$$= \frac{1}{2} \int_{-\Delta x/2}^{\Delta x/2} dx' \, E_1\left(\Sigma_{tn}^g |x - x'|\right) \left[S_g^n(x') - L_{ny}^{g,\text{iso}}(x')\right]$$

$$- \int_{-\Delta x/2}^{\Delta x/2} dx' \int_0^1 \frac{d\mu}{\mu} e^{-\Sigma_{tn}^g |x-x'|/\mu} \int_0^{2\pi} L_{ny}^{g,\text{anis}}(x', |\mu|, \phi) \, d\phi$$

$$+ \int_0^1 d\mu \, e^{-\Sigma_{tn}^g (x+(\Delta x/2))/\mu} \int_0^{2\pi} \psi_{gx-}^{n,\text{in}}(\mu, \phi) \, d\phi$$

$$+ \int_{-1}^0 d\mu \, e^{-\Sigma_{tn}^g (x-(\Delta x/2))/\mu} \int_0^{2\pi} \psi_{gx+}^{n,\text{in}}(\mu, \phi) \, d\phi \tag{15.69}$$

where the exponential integral function is

$$E_n(\xi) \equiv \int_0^1 d\mu \, \mu^{n-2} \exp\left(-\frac{\xi}{\mu}\right) \tag{15.70}$$

and the transverse leakage has been split into an isotropic and an anisotropic component:

$$L_{ny}^g(x', \mu, \phi) = \frac{1}{4\pi} L_{ny}^{g,\text{iso}}(x') + L_{ny}^{g,\text{anis}}(x', \mu, \phi) \tag{15.71}$$

Polynomial Expansion of Scalar Flux

Following the same general procedure used to develop the diffusion theory nodal model, the scalar flux for the x-direction problem is expanded:

$$\phi_{gx}^n(x) = \sum_{i=1}^I a_i \phi_{gxi}^n f_i(x), \quad I \le 2 \tag{15.72}$$

The expansion coefficients are normalized such that

$$\frac{1}{a_i} \equiv \frac{1}{\Delta x} \int_{-\Delta x/2}^{\Delta x/2} dx \, [f_i(x)]^2 \tag{15.73}$$

and the polynomials

$$f_0 = 1, \quad f_1(x) = \frac{x}{\Delta x}, \quad f_2(x) = 3\left(\frac{x}{\Delta x}\right)^2 - \frac{1}{4} \tag{15.74}$$

are used. The moments of the scalar flux are

$$\phi_{gxi}^n \equiv \int_{-\Delta x/2}^{\Delta x/2} dx \, \phi_{gx}^n(x) f_i(x) \tag{15.75}$$

so that ϕ_{gx0}^n is the node-averaged scalar flux.

Isotropic Component of Transverse Leakage

The surface average of the isotropic component of the transverse leakage is

$$
\bar{L}_{ny}^{g,\text{iso}} \equiv \frac{1}{\Delta x} \int_{-\Delta x/2}^{\Delta x/2} dx\, L_{ny}^{g,\text{iso}}(x)
$$

$$
= \left[\left(J_{gy+}^n \right)_{\text{out}} - \left(J_{gy+}^n \right)_{\text{in}} \right] - \left[\left(J_{gy-}^n \right)_{\text{in}} - \left(J_{gy-}^n \right)_{\text{out}} \right] \tag{15.76}
$$

where the surface average of the outward and inward partial currents at $+\Delta y/2$ and $-\Delta y/2$ are

$$
\left(J_{gy+}^n \right)_{\text{out}} \equiv \int_0^{2\pi} d\phi \int_0^1 d\mu\, \mu \psi_{gy+}^{n,\text{out}}(\mu,\phi)
$$

$$
\left(J_{gy-}^n \right)_{\text{out}} \equiv \int_0^{2\pi} d\phi \int_{-1}^0 d\mu\, \mu \psi_{gy-}^{n,\text{out}}(\mu,\phi)
\tag{15.77}
$$

and

$$
\left(J_{gy+}^n \right)_{\text{in}} \equiv \int_0^{2\pi} d\phi \int_{-1}^0 d\mu\, \mu \psi_{gy+}^{n,\text{in}}(\mu,\phi)
$$

$$
\left(J_{gy-}^n \right)_{\text{in}} \equiv \int_0^{2\pi} d\phi \int_0^1 d\mu\, \mu \psi_{gy-}^{n,\text{in}}(\mu,\phi)
\tag{15.78}
$$

respectively, with the directional neutron fluxes at $+\Delta y/2$ and $-\Delta y/2$, $\psi_{gy\pm}^{n,\text{in}}$ and $\psi_{gy\pm}^{n,\text{out}}$ defined by equations similar to Eqs. (15.67) and (15.68).

Double-P_n Expansion of Surface Fluxes

The angular dependence of the neutron flux on the surfaces of the node is approximated by a double-P_1 approximation, which allows independent linearly anisotropic distributions for the incident and exiting fluxes on a surface. In terms of the half-space polynomials, which are related to the Legendre polynomials by $p_n^+(\xi) = P_n(2\xi - 1)$ for $1 \geq \xi \geq 0$ and $p_n^-(\xi) = P_n(2\xi + 1)$ for $0 \geq \xi \geq -1$, the surface-averaged inward neutron fluxes at $\pm \Delta x/2$ are expanded:

$$
\psi_{gx\pm}^{n,\text{in}}(\mu,\phi) \equiv \psi_{gx}^n\left(\pm \frac{\Delta x}{2}, \mu \lessgtr 0, \phi \right)
$$

$$
\approx \frac{1}{2\pi} \left[\frac{1}{2} a_0^\pm p_0^\mp + \frac{3}{2} a_{1x}^\pm p_1^\mp(\Omega_x) + \frac{3}{2} a_{1y}^\pm p_1^\mp(\Omega_y) \right]
$$

$$
\approx \frac{1}{2\pi} \left(\frac{1}{2} C_0^\pm + \frac{3}{2} C_{1x}^\pm \mu + \frac{3}{2} C_{1y}^\pm \sqrt{1-\mu^2}\cos\phi \right)
$$

$$
= \frac{1}{2\pi} \left(4\bar{\phi}_{gx\pm}^{n,\text{in}} \pm 6 J_{gx\pm}^{n,\text{in}} \right) + \frac{1}{2\pi} \left(12 \bar{J}_{gx\pm}^{n,\text{in}} \pm 6 \psi_{gx\pm}^{n,\text{in}} \right) \mu
$$

$$
+ \frac{1}{2\pi} \left(3 \bar{J}_{gx\pm}^{n,\text{in}} \right) \sqrt{1-\mu^2}\cos\phi \tag{15.79}
$$

The angular moments of the surface-averaged inward fluxes that appear in Eq. (15.79) are

$$\bar{\psi}_{gx-}^{n,\text{in}} \equiv \int_0^{2\pi} d\phi \int_0^1 d\mu \, \psi_{gx-}^{n,\text{in}}(\mu, \phi)$$

$$\bar{\psi}_{gx+}^{n,\text{in}} \equiv \int_0^{2\pi} d\phi \int_{-1}^0 d\mu \, \psi_{gx-}^{n,\text{in}}(\mu, \phi)$$

$$\bar{J}_{gx-}^{n,\text{in}} \equiv \int_0^{2\pi} d\phi \int_0^1 d\mu \, \mu \psi_{gx-}^{n,\text{in}}(\mu, \phi)$$

$$\bar{J}_{gx+}^{n,\text{in}} \equiv \int_0^{2\pi} d\phi \int_{-1}^0 d\mu \, \mu \psi_{gx+}^{n,\text{in}}(\mu, \phi) \qquad (15.80)$$

$$\bar{J}_{gy-}^{n,\text{in}} \equiv \int_0^{2\pi} d\phi \int_0^1 d\mu \, \sqrt{1-\mu^2} \cos\phi \psi_{gy-}^{n,\text{in}}(\mu, \phi)$$

$$\bar{J}_{gy+}^{n,\text{in}} \equiv \int_0^{2\pi} d\phi \int_{-1}^0 d\mu \, \sqrt{1-\mu^2} \cos\phi \psi_{gy+}^{n,\text{in}}(\mu, \phi)$$

Using Eq. (15.79) to evaluate the integrals involving the incident fluxes in Eq. (15.69) yields

$$\phi_{gx}^n(x) = \int_{-\Delta x/2}^{\Delta x/2} dx' \, E_1\left(\Sigma_{tn}^g |x - x'|\right) \frac{1}{2}\left[S_g^n(x') - L_{ny}^{g,\text{iso}}(x')\right]$$

$$- \int_{-\Delta x/2}^{\Delta x/2} dx' \int_0^1 \frac{d\mu}{\mu} e^{-\Sigma_{tn}^g |x'-x'|/\mu} \int_0^{2\pi} d\phi \, L_{ny}^{g,\text{anis}}(x', |\mu|, \phi)$$

$$+ \bar{\psi}_{gx-}^{n,\text{in}}\left[4E_2\left(\Sigma_{tn}^g\left(x + \frac{\Delta x}{2}\right)\right) - 6E_3\left(\Sigma_{tn}^g\left(x + \frac{\Delta x}{2}\right)\right)\right]$$

$$+ \bar{J}_{gx-}^{n,\text{in}}\left[12E_3\left(\Sigma_{tn}^g\left(x + \frac{\Delta x}{2}\right)\right) - 6E_2\left(\Sigma_{tn}^g\left(x + \frac{\Delta x}{2}\right)\right)\right]$$

$$+ \bar{\psi}_{gx+}^{n,\text{in}}\left[4E_2\left(\Sigma_{tn}^g\left(\frac{\Delta x}{2} - x\right)\right) - 6E_3\left(\Sigma_{tn}^g\left(\frac{\Delta x}{2} - x\right)\right)\right]$$

$$+ \bar{J}_{gx+}^{n,\text{in}}\left[6E_2\left(\Sigma_{tn}^g\left(\frac{\Delta x}{2} - x\right)\right) - 12E_3\left(\Sigma_{tn}^g\left(\frac{\Delta x}{2} - x\right)\right)\right] \quad (15.81)$$

Angular Moments of Outgoing Surface Fluxes

The angular moments of the surface averaged outgoing flux and current at $\Delta x/2$ can be constructed from Eq. (15.66a), using Eq. (15.79) to expand the angular dependence of the incoming flux at $-\Delta x/2$:

$$\bar{\psi}_{gx+}^{n,\text{out}} \equiv \int_0^{2\pi} d\phi \int_0^1 d\mu \, \psi_{gx}^n\left(\frac{\Delta x}{2}, \mu > 0, \phi\right)$$

$$= \int_{-\Delta x/2}^{\Delta x/2} dx' \, E_1\left[\Sigma_{tn}^g\left(\frac{\Delta x}{2} - x'\right)\right] \frac{1}{2}\left[S_g^n(x') - L_{ny}^{g,\text{iso}}(x')\right]$$

$$- \int_{-\Delta x/2}^{\Delta x/2} dx' \int_0^1 \frac{d\mu}{\mu} e^{-\Sigma_{tn}^g((\Delta x/2)-x')/\mu} \int_0^{2\pi} L_{ny}^{g,anis}(x',\mu,\phi)\, d\phi$$

$$+ \bar{\psi}_{gx-}^{n,in}\left[4E_2\left(\Sigma_{tn}^g\Delta x\right) - 6E_3\left(\Sigma_{tn}^g\Delta x\right)\right]$$

$$+ \bar{J}_{gx-}^{n,in}\left[12E_3\left(\Sigma_{tn}^g\Delta x\right) - 6E_2\left(\Sigma_{tn}^g\Delta x\right)\right] \tag{15.82}$$

$$\bar{J}_{gx+}^{n,out} \equiv \int_0^{2\pi} d\phi \int_0^1 d\mu\, \mu \psi_{gx}^n\left(\frac{\Delta x}{2},\mu>0,\phi\right)$$

$$= \int_{-\Delta x/2}^{\Delta x/2} dx'\, E_2\left[\Sigma_{tn}^g\left(\frac{\Delta x}{2}-x'\right)\right]\frac{1}{2}\left[S_g^n(x') - L_{ny}^{g,iso}(x')\right]$$

$$- \int_{-\Delta x/2}^{\Delta x/2} dx' \int_0^1 d\mu\, e^{-\Sigma_{tn}^g((\Delta x/2)-x')/\mu} \int_0^{2\pi} L_{ny}^{g,anis}(x',\mu,\phi)\, d\phi$$

$$+ \bar{\psi}_{gx-}^{n,in}\left[4E_3\left(\Sigma_{tn}^g\Delta x\right) - 6E_4\left(\Sigma_{tn}^g\Delta x\right)\right]$$

$$+ \bar{J}_{gx-}^{n,in}\left[12E_4\left(\Sigma_{tn}^g\Delta x\right) - 6E_3\left(\Sigma_{tn}^g\Delta x\right)\right] \tag{15.83}$$

The angular moments of the surface-averaged outgoing flux and current at $-\Delta x/2$ can be constructed from Eq. (15.66b), using Eq. (15.79) to expand the angular dependence of the incoming flux at $+\Delta x/2$:

$$\bar{\psi}_{gx-}^{n,out} \equiv \int_0^{2\pi} d\phi \int_{-1}^0 d\mu\, \psi_{gx}^n\left(-\frac{\Delta x}{2},\mu<0,\phi\right)$$

$$= \int_{-\Delta x/2}^{\Delta x/2} dx'\, E_1\left[\Sigma_{tn}^g\left(\frac{\Delta x}{2}+x'\right)\right]\frac{1}{2}\left[S_g^n(x') - L_{ny}^{g,iso}(x')\right]$$

$$- \int_{-\Delta x/2}^{\Delta x/2} dx' \int_{-1}^0 \frac{d\mu}{\mu} e^{\Sigma_{tn}^g(\Delta x/2+x')/\mu} \int_0^{2\pi} d\phi\, L_{ny}^{g,anis}(x',\mu,\phi)$$

$$+ \bar{\psi}_{gx+}^{n,in}\left[4E_2\left(\Sigma_{tn}^g\Delta x\right) + 6E_3\left(\Sigma_{tn}^g\Delta x\right)\right]$$

$$+ \bar{J}_{gx+}^{n,in}\left[12E_3\left(\Sigma_{tn}^g\Delta x\right) + 6E_2\left(\Sigma_{tn}^g\Delta x\right)\right] \tag{15.84}$$

$$\bar{J}_{gx-}^{n,out} \equiv \int_0^{2\pi} d\phi \int_{-1}^0 d\mu\, \mu \psi_{gx}^n\left(-\frac{\Delta x}{2},\mu<0,\phi\right)$$

$$= \int_{-\Delta x/2}^{\Delta x/2} dx'\, E_2\left[\Sigma_{tn}^g\left(\frac{\Delta x}{2}+x'\right)\right]\frac{1}{2}\left[S_g^n(x') - L_{ny}^{g,iso}(x')\right]$$

$$+ \int_{-\Delta x/2}^{\Delta x/2} dx' \int_{-1}^0 d\mu\, e^{\Sigma_{tn}^g((\Delta x/2)+x')/\mu} \int_0^{2\pi} d\phi\, L_{ny}^{g,anis}(x',\mu,\phi)$$

$$+ \bar{\psi}_{gx+}^{n,in}\left[4E_3\left(\Sigma_{tn}^g\Delta\right) + 6E_4\left(\Sigma_{tn}^g\Delta x\right)\right]$$

$$+ \bar{J}_{gx+}^{n,in}\left[12E_4\left(\Sigma_{tn}^g\Delta x\right) + 6E_3\left(\Sigma_{tn}^g\Delta x\right)\right] \tag{15.85}$$

Nodal Transport Equations

These equations can be written, in terms of matrices and column vectors, in a form analogous to the diffusion theory relation of Eq. (15.58):

$$\boldsymbol{\psi}_g^{n,\text{out}} = \tilde{\boldsymbol{P}}_g^n [\boldsymbol{Q}_g^n - \boldsymbol{L}_g^n] + \tilde{\boldsymbol{R}}_g^n \boldsymbol{\psi}_g^{n,\text{in}} \qquad (15.86)$$

The column vectors $\boldsymbol{Q}_g^n$ and $\boldsymbol{L}_g^n$ are defined as for diffusion theory and represent the fission plus in-scatter source and the transverse leakage, respectively. The column vector $\boldsymbol{\psi}_g^{n,\text{out}}$ contains outgoing surface-averaged partial currents [Eqs. (15.83) and (15.85)] and half-angle integrated fluxes [Eqs. (15.82) and (15.84)]; and the column vector $\boldsymbol{\psi}_g^{n,\text{in}}$ contains incoming surface-averaged partial currents and half-angle integrated fluxes [Eqs. (15.80)] for each of the six (in three dimensions) nodal surfaces. The matrices $\boldsymbol{P}_g^n$ and $\boldsymbol{R}_g^n$ contain the nodal coupling coefficients.

The transverse integrated formulation allows for direct transmission of neutrons entering node n over the x-surface at $-\Delta x/2$ across the node to exit over the x-surface at $+\Delta x/2$ [e.g., the $\psi_{gx+}^{n,\text{in}}$ and $J_{gx+}^{n,\text{in}}$ terms in Eqs. (15.82) and (15.83)], but does not allow for the direct transmission of neutrons entering node n over an x-surface across the node to exit over a y- or z-surface.

15.5
Transverse Integrated Nodal Discrete Ordinates Method

A nodal transport equation can also be formulated in terms of the discrete ordinates approximation. The development is similar to that of Sections 15.3 and 15.4 and we will only briefly examine how the nodal coupling equations are formulated in terms of discrete ordinates. In two-dimensional Cartesian geometry, the multigroup discrete ordinates equations with isotropic scattering within a node of constant homogenized cross section may be written

$$\mu^m \frac{\partial \psi_g^m}{\partial x}(x,y) + \mu^m \frac{\partial \psi_g^m}{\partial y}(x,y) + \Sigma_t^g \psi_g^m(x,y) = \frac{\Sigma_s^g}{2\pi} \phi_g(x,y) + \frac{Q^g(x,y)}{2\pi}$$

$$(15.87)$$

where $\psi_g^m(x,y) = \psi_g(x,y,\boldsymbol{\Omega}_m)$ is the group flux in the ordinate direction $\boldsymbol{\Omega}_m$, $\mu^m = \mathbf{n}_x \cdot \boldsymbol{\Omega}_m$, and $\eta^m = \mathbf{n}_y \cdot \boldsymbol{\Omega}_m$. Letting the node extend over $-\Delta x/2 < x < \Delta x/2$, $-\Delta y/2 < y < \Delta y/2$ and integrating Eq. (15.87) over $-\Delta y/2 < y < \Delta y/2$ leads to the transverse-integrated one-dimensional discrete ordinates equation

$$\mu^m \frac{\partial \psi_{gx}^m(x)}{\partial x} + \Sigma_t^g \psi_{gx}^m(x) = \frac{\Sigma_s^g \phi_g(x)}{2\pi} + \frac{Q_x^g(x)}{2\pi} - \frac{L_{mg}^g(x)}{\Delta y} \equiv S_g^m(x) \quad (15.88)$$

where the y-direction transverse leakage is

$$L_{my}^g(x) \equiv \int_{-(\Delta y/2)}^{(\Delta y/2)} dy\, \eta^m \frac{\partial \psi_g^m(x,y)}{\partial y} = \eta^m \big[\psi_{gy+}^m(x) - \psi_{gy-}^m(x) \big] \qquad (15.89)$$

and $\psi_{gy\pm}^m(x) = \psi_{gy}^m(x, y = \pm\Delta y/2)$.

We now make a polynomial expansion of $S_g^m(x)$ within the node in the polynomials

$$f_0(x) = 1, \qquad f_1(x) = x, \qquad f_2(x) = x^2 - \frac{1}{12}, \qquad f_3(x) = x^3 - \frac{3}{20}x,$$

$$f_4(x) = x^4 - \frac{3}{14}x^2 + \frac{3}{560}, \qquad f_5(x) = x^5 - \frac{5}{18}x^3 + \frac{5}{336}x, \tag{15.90}$$

$$f_6(x) = x^6 - \frac{15}{44}x^4 + \frac{5}{176}x^2 - \frac{5}{14{,}784}$$

integrate Eq. (15.88) over $-\Delta x/2 < x < \Delta x/2$ in the direction of neutron flow (i.e., from $-\Delta x/2$ to $+\Delta x/2$ for $\mu^m > 0$ and in the opposite direction for $\mu^m < 0$), and make use of the orthogonality property

$$\int_{-\Delta x/2}^{\Delta x/2} f_i(x) f_j(x)\, dx = \left[\int_{-\Delta x/2}^{\Delta x/2} dx\, f_i^2(x) \right] \delta_{ij} \tag{15.91}$$

to obtain a relation among the outward-directed fluxes at one boundary, the inward-directed fluxes at the other boundary, and the group sources within the node

$$\psi_{gx\pm}^{m,\text{out}} = \frac{1}{\mu^m} \sum_{i=0}^{6} S_{gi}^m \int_{-\Delta x/2}^{\Delta x/2} f_i(\pm x) e^{(\Sigma_t^g (x - \Delta x/2))/\mu^m}\, dx$$

$$+ \psi_{gx\mp}^{m,\text{in}} e^{-(\Sigma_t^g \Delta x/|\mu^m|)} \tag{15.92}$$

where the S_{gi}^m are the coefficients of the polynomial expansion of $S_g^m(x)$.

Equation (15.88) can be integrated over $-\Delta x/2 < x' < x$ for $\mu^m > 0$ and in the opposite direction for $\mu^m < 0$ to obtain an expression for the flux $\psi_{gx}^m(x)$ similar to that given by Eq. (15.92) but with the upper limit of the integral replaced by x. This expression can expanded in polynomials, multiplied by f_i, and integrated over $-\Delta x/2 < x < \Delta x/2$ to obtain an expression for the ith node-averaged flux expansion coefficient in terms of the inward flux at $\pm\Delta x/2$ and the group sources within the node:

$$\psi_{gi}^m = \frac{1}{\mu^m D_i} \sum_{j=0}^{6} S_{gj}^m \int_{-\Delta x/2}^{\Delta x/2} dx\, f_i(\pm x) \int_{-\Delta x/2}^{x} dx'\, f_j(\pm x') e^{(\Sigma_t(x'-x)/|\mu^m|)}$$

$$+ \frac{1}{D_i} \psi_{gx\mp}^{m,\text{in}} \int_{-\Delta x/2}^{\Delta x/2} f_i(\pm x) e^{\Sigma_t^g (x - \Delta x/2)/|\mu^m|}, \qquad i = 0, \ldots, 6 \tag{15.93}$$

The fluxes at the interface between nodes $n-1$ and n are coupled by

$$\left(\psi_{gx+}^{m,\text{out}} \right)_{n-1} = \left(\psi_{gx-}^{m,\text{in}} \right)_n, \qquad \mu^m > 0$$

$$\left(\psi_{gx+}^{m,\text{in}} \right)_{n-1} = \left(\psi_{gx-}^{m,\text{out}} \right)_n, \qquad \mu^m < 0 \tag{15.94}$$

which enables the development of equations for solving the x-direction transverse integrated equations. A similar procedure is then applied to develop and solve the y-direction transverse integrated equations.

15.6
Finite-Element Coarse Mesh Methods

The finite-element methodology provides a systematic procedure for developing coarse mesh equations with higher-order accuracy than the conventional finite-difference equations. In the finite-element method, the spatial (or other) dependence of the neutron flux and current are represented by a supposition of *trial functions* which are nonzero only within a limited range of the spatial variables. These trial functions are continuous within volumes V_i, but may be discontinuous across the interfaces between adjacent volumes. The finite-element approximation will be developed from a variational principle that admits discontinuous trial functions, but it could also be derived from a weighted residuals development.

The development of finite-element approximations will be discussed for one-group P_1 and diffusion theory. The results can formally be extended to multigroup theory by replacing the total and fission cross sections with diagonal cross-section matrices, replacing the scattering cross section with the multigroup scattering matrix, and replacing fluxes and currents with column vectors of multigroup fluxes and currents.

Variational Functional for the P_1 Equations

The volume of the reactor core may be subdivided into volumes V_i within which the trial functions for the neutron flux and current are continuous. These regions are bounded by interfaces S_k across which the trial functions may be discontinuous. A variational functional for the one-group P_1 equations is

$$
\begin{aligned}
F_1 & \left\{ \mathbf{J}^*, \phi^*, \mathbf{J}, \phi \right\} \\
&= \sum_i \int_{V_i} \phi^* \left[\left(\Sigma_t - \Sigma_s - \frac{\nu \Sigma_f}{k} \right) \phi + \nabla \cdot \mathbf{J} \right] dr \\
&\quad + \sum_i \int_{V_i} \mathbf{J}^* \cdot \left[D^{-1} \mathbf{J} + \nabla \phi \right] dr + \sum_k \int_{S_k} \mathbf{n}_k \cdot \phi_k^* [\mathbf{J}_{k+} - J_{k-}] ds \\
&\quad + \sum_k \int_{S_k} \mathbf{J}^* \cdot \mathbf{n}_k [\phi_{k+} - \phi_{k-}] ds \qquad\qquad (15.95)
\end{aligned}
$$

where the first two terms are sums over the volumes within which the admissible trial functions are continuous and the last two terms are surface integrals over the interfaces between these volumes. The subscripts $k+$ and $k-$ refer to limiting values as the surface k is approached from the positive and negative sides, respectively.

The stationarity of this variational functional with respect to independent and arbitrary variations of the adjoint flux (ϕ^*) and current ($\mathbf{J}^*$) within the different volumes V_i requires that

$$\frac{\delta F_1}{\delta \phi_i} \delta \phi_i^* = \int_{V_i} \delta \phi^* \left[\left(\Sigma_t - \Sigma_S - \frac{\nu \Sigma_f}{k} \right) \phi + \nabla \cdot \mathbf{J} \right] dr = 0$$

$$\implies \nabla \cdot \mathbf{J} + \left(\Sigma_t - \Sigma_S - \frac{\nu \Sigma_f}{k} \right) \phi = 0, \quad r \in V_i \tag{15.96}$$

$$\frac{\delta F_1}{\delta \mathbf{J}_i^*} \delta \mathbf{J}_i^* = \int \delta \mathbf{J}^* \cdot [D^- \mathbf{J} + \nabla \phi] \, dr = 0$$

$$\implies \mathbf{J} = -D \nabla \phi, \quad r \in V_i \tag{15.97}$$

(i.e., that the P_1 equations are satisfied within the different volumes V_i).

The stationarity of this variational functional with respect to independent and arbitrary variations of the adjoint flux (ϕ_k^*) and current ($\mathbf{J}_k^*$) on the interfaces between volumes V_i requires that

$$\frac{\delta F_1}{\delta \phi_k^*} \delta \phi_k^* = \int_{S_k} \delta \phi_k^* \mathbf{n}_k \cdot [\mathbf{J}_{k+} - \mathbf{J}_{k-}] = 0$$

$$\implies \mathbf{n}_k \cdot \mathbf{J}_{k+} = \mathbf{n}_k \cdot \mathbf{J}_{k-} \tag{15.98}$$

$$\frac{\delta F_1}{\delta \mathbf{J}_k^*} \delta \mathbf{J}_k^* = \int_{S_k} \delta \mathbf{J}_k^* \mathbf{n}_k [\phi_{k+} - \phi_{k-}] = 0$$

$$\implies \phi_{k+} = \phi_{k-} \tag{15.99}$$

(i.e., that the normal component of the current and the flux are continuous across each interface).

Thus, the requirements that the variational functional of Eq. (15.95) is stationary with respect to arbitrary and independent variations of the adjoint flux and current in each volume V_i and on each interface S_k is equivalent to the requirements that the P_1 equations are satisfied within each volume V_i and that the normal component of the current and the flux are continuous across the interfaces bounding these volumes. This equivalence will now be exploited to develop finite-element approximations for the neutron flux distribution.

One-Dimensional Finite-Difference Approximation

Although it is not a finite-element approximation per se, it is instructive to derive variationally the conventional finite-difference approximation for a slab extending from $0 < x < a$ with zero flux boundary conditions. The slab is partitioned into N mesh intervals and the flux and current are expanded in piecewise constant functions

$$\phi(x) = \sum_{n=1}^{N-1} \phi_n H_n(x), \qquad \phi^*(x) = \sum_{n=1}^{N-1} \phi_n^* H_n(x) \tag{15.100}$$

$$J(x) = \sum_{n=0}^{N-1} J_n K_n(x), \qquad J^*(x) = \sum_{n=0}^{N-1} J_n^* K_n(x) \tag{15.101}$$

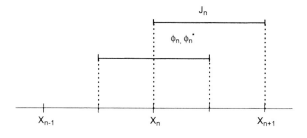

Fig. 15.4 Trial functions for finite-difference approximation.

where the H_n and K_n are Heaviside functions:

$$H_n(x) = \begin{cases} 1, & x_n - \frac{1}{2}h_{n-1} < x < x_n + \frac{1}{2}h_n \\ 0, & \text{otherwise} \end{cases}$$

$$K_n(x) = \begin{cases} 1, & x_n < x < x_{n+1} \\ 0, & \text{otherwise} \end{cases}$$

(15.102)

the domain of which is illustrated in Fig. 15.4. The volumes V_i over which the flux and adjoint flux trial functions are continuous are the mesh intervals $x_n - h_{n-1}/2 < x < x_n + h_n/2$, and the surfaces bounding these regions are at $x_n - h_{n-1}/2$ and $x_n + h_n/2$. The volumes V_i over which the current and adjoint current trial functions are continuous are the mesh intervals $x_n < x < x_{n+1}$, and the surfaces bounding these regions are at x_n and x_{n+1}.

For the piecewise constant adjoint flux trial functions of Eq. (15.100), the variations on the surfaces are not independent of the variations in the volumes; that is, instead of having separate Eqs. (15.96) and (15.98), these two equations must be combined, leading in this case to

$$\frac{\delta F_1}{\delta \phi^*} \delta \phi^* = \sum_{n=1}^{N-1} \delta \phi_n^* \left\{ \int_{x_n - \frac{1}{2}h_{n-1}}^{x_n + \frac{1}{2}h_n} dx \left[\left(\Sigma_t - \Sigma_S - \frac{\nu \Sigma_f}{k} \right) \phi + \frac{dJ}{dx} \right] \right.$$

$$\left. + (J_n - J_{n-1}) \right\}$$

$$= \sum_{n=1}^{N-1} \delta \phi_n^* \left\{ \left[\frac{1}{2}h_{n-1} \left(\Sigma_{tn-1} - \Sigma_{sn-1} - \frac{\nu \Sigma_{fn-1}}{k} \right) \phi \right. \right.$$

$$\left. \left. + \frac{1}{2}h_n \left(\Sigma_{tn} - \Sigma_{sn} - \frac{\nu \Sigma_{fn}}{k} \right) \right] \phi_n + (J_n - J_{n-1}) \right\} = 0$$

(15.103)

where the materials properties denoted by the subscript n have been taken to be uniform in the interval $x_n \leq x \leq x_{n+1}$. Similarly, the variations of the adjoint current trial functions in the volumes and on the surfaces are not independent, requiring that Eqs. (15.97) and (15.99) be combined to yield

$$
\frac{\delta F_1}{\delta J^*}\delta J^* = \sum_{n=0}^{N-1}\delta J_n^*\left[\int_{x_n}^{x_{n+1}}dx\left(D^{-1}J + \frac{d\phi}{dx}\right) + (\phi_{n+1} - \phi_n)\right]
$$

$$
= \sum_{n=0}^{N-1}\delta J_n^*\left[h_n D_n^{-1}J_n + (\phi_{n+1} - \phi_n)\right] = 0 \tag{15.104}
$$

Requiring that the variational functional be stationary with respect to arbitrary and independent variations $\delta\phi_n^*$ and δJ_n^* in each mesh interval yields

$$
\left[\frac{1}{2}h_{n-1}\left(\Sigma_{tn-1} + \Sigma_{sn-1} - \frac{\nu\Sigma_{fn-1}}{k}\right) + \frac{1}{2}h_n\left(\Sigma_{tn} - \Sigma_{sn} - \frac{\nu\Sigma_{fn}}{k}\right)\right]\phi_n
$$

$$
+ (J_n - J_{n-1}) = 0 \tag{15.105}
$$

$$
J_n = -D_n\frac{\phi_{n+1} - \phi_n}{h_n}
$$

which may be combined to obtain the standard form of the finite-difference diffusion equation.

$$
\left[\frac{1}{2}h_{n-1}\left(\Sigma_{tn-1} - \Sigma_{sn-1} - \frac{\nu\Sigma_{fn-1}}{k}\right) + \frac{1}{2}h_n\left(\Sigma_{tn} - \Sigma_{sn} - \frac{\nu\Sigma_{fn}}{k}\right)\right.
$$

$$
\left. + \frac{D_{n-1}}{h_{n-1}} + \frac{D_n}{h_n}\right]\phi_n - \frac{D_{n-1}}{h_{n-1}}\phi_{n-1} - \frac{D_{n+1}}{h_n}\phi_{n+1} = 0 \tag{15.106}
$$

The volumes V_i over which the flux and adjoint flux trial functions are continuous are the mesh intervals $x_n - h_{n-1}/2 < x < x_n + h_n/2$, and the surfaces bounding these regions are at $x_n - h_{n-1}/2$ and $x_n + h_n/2$.

Diffusion Theory Variational Functional

We shall restrict our attention to trial functions that are continuous over the volume of the reactor, which means that the last term in the variational functional of Eq. (15.95) is identically zero. We further restrict ourselves to current and adjoint current trial functions which satisfy Fick's law, so that the second term in Eq. (15.95) is identically zero and the current in the first and third terms may be replaced by $-D\nabla\phi$ to obtain the diffusion theory variational functional

$$
F_d\{\phi^*, \phi\} = \sum_i \int_{V_i} \phi^*\left[\left(\Sigma_t - \Sigma_s - \frac{\nu\Sigma_f}{k}\right)\phi - \nabla\cdot D\nabla\phi\right]dr
$$

$$
+ \sum_k \int_{S_k} \phi^*\mathbf{n}_s\left[(-D\nabla\phi)_{k+} - (-D\nabla\phi)_{k-}\right]dS
$$

$$
= \sum_i \int_{V_i}\left[\phi^*\left(\Sigma_t - \Sigma_s - \frac{\nu\Sigma_f}{k}\right)\phi + \nabla\phi^*\cdot D\nabla\phi\right]dr \tag{15.107}
$$

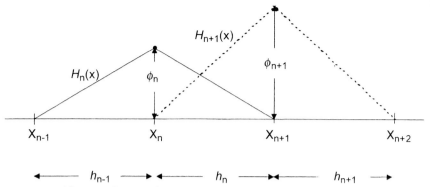

Fig. 15.5 Trial functions for linear finite-element approximation.

The second form of this functional resulted from integrating the divergence in the first term by parts over the various volumes to obtain terms that cancel identically with the interior surface terms in the second term and vanish on the outer boundary because of the physical boundary condition. Note that this second form of the functional admits trial functions which do not identically satisfy continuity of $-D\nabla\phi \cdot \mathbf{n}_s$ across interior surfaces.

Linear Finite-Element Diffusion Approximation in One Dimension

We consider the same problem as above, a slab reactor with zero flux boundary conditions in one-group diffusion theory. The neutron flux is expanded

$$\phi(x) = \sum_{n=1}^{N-1} \phi_n H_n(x), \qquad \phi^*(x) = \sum_{n=1}^{N-1} \phi_n^* H_n(x) \tag{15.108}$$

in tent function trial functions

$$H_n(x) = \begin{cases} \dfrac{x_{n+1} - x}{h_n}, & x_n < x < x_{n+1} \\[2mm] \dfrac{x - x_{n-1}}{h_{n-1}}, & x_{n-1} < x < x_n \\[2mm] 0, & \text{otherwise} \end{cases} \tag{15.109}$$

depicted in Fig. 15.5. The volumes V_i over which the trial functions ϕ and ϕ^* of Eq. (15.108) and the vectors $D\nabla\phi$ and $D\nabla\phi^*$ are continuous are just the mesh intervals $x_{n-1} < x < x_n$, and the surfaces are the x_n. Requiring that the variational functional of Eq. (15.107) be stationary with respect to arbitrary and independent variations in all the adjoint trial functions yields

$$\frac{\delta F_d}{\delta \phi_n^*}\delta\phi_n^* = \delta\phi_n^* \left[\int_{x_{n-1}}^{x_n} \left\{ \frac{x - x_n}{h_{n-1}}\left(\Sigma_{tn-1} - \Sigma_{sn-1} - \frac{\nu \Sigma_{fn-1}}{k} \right) \right. \right.$$

$$\left. \left. \times \sum_{n'=n-1}^{n+1} \phi_{n'} H_{n'}(x) + \frac{D_{n-1}}{h_{n-1}} \sum_{n'=n-1}^{n+1} \frac{d}{dx}[\phi_{n'} H_{n'}(x)] \right\} dx \right.$$

$$+ \int_{x_n}^{x_{n+1}} \left\{ \frac{x_{n+1} - x}{h_n} \left(\Sigma_{tn} - \Sigma_{sn} - \frac{\nu \Sigma_{fn}}{k} \right) \right.$$

$$\left. \times \sum_{n'=n-1}^{n+1} \phi_{n'} H_{n'}(x) - \frac{D_n}{h_n} \sum_{n'=n-1}^{n+1} \frac{d}{dx} [\phi_{n'} H_{n'}(x)] \right\} dx \Bigg]$$

$$= 0 \qquad\qquad\qquad\qquad\qquad\qquad\qquad\qquad\qquad (15.110)$$

Carrying out the integration results in a three-point coarse mesh equation for each mesh point:

$$\left[h_{n-1} \left(\Sigma_{tn-1} - \Sigma_{sn-1} - \frac{\nu \Sigma_{fn-1}}{k} \right) \right] \left(\frac{1}{3} \phi_n + \frac{1}{6} \phi_{n-1} \right)$$

$$+ \left[h_n \left(\Sigma_{tn} - \Sigma_{sn} - \frac{\nu \Sigma_{fn}}{k} \right) \right] \left(\frac{1}{3} \phi_n + \frac{1}{6} \phi_{n+1} \right)$$

$$+ \frac{D_{n-1}}{h_{n-1}} (\phi_n - \phi_{n-1}) - \frac{D_n}{h_n} (\phi_{n+1} - \phi_n) = 0 \qquad (15.111)$$

which are similar to the finite-difference equations (15.106), but with more coupling among mesh points.

Numerical studies reveal that Eqs. (15.111) can achieve the same accuracy as Eqs. (15.106) with much larger mesh spacing, h_n. This result is physically intuitive because the piecewise linear representation of the flux allowed by the trial functions of Eqs. (15.109) is more realistic than the step function representation allowed by the trial functions of Eqs. (15.100) and (15.101), as illustrated in Fig. 15.6. It stands to reason that higher-order polynomial trial functions should provide an even better representation of the flux and hence be more accurate.

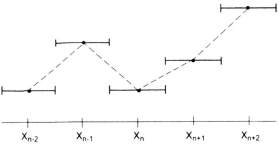

Fig. 15.6 Finite-difference (solid lines) and linear finite-element (dashed lines) representation of flux solution.

Higher-Order Cubic Hermite Coarse-Mesh Diffusion Approximation

The cubic Hermite interpolating polynomials

$$H_n^0(x) = \begin{cases} 3\left(\dfrac{x-x_{n-1}}{h_{n-1}}\right)^2 - 2\left(\dfrac{x-x_{n-1}}{h_{n-1}}\right)^3, & x_{n-1} \le x \le x_n \\ 3\left(\dfrac{x_{n+1}-x}{h_n}\right)^2 - 2\left(\dfrac{x_{n+1}-x}{h_n}\right)^3, & x_n \le x \le x_{n+1} \\ 0, & \text{otherwise} \end{cases}$$

$$H_n^-(x) = \begin{cases} \left[-\left(\dfrac{x-x_{n-1}}{h_{n-1}}\right)^2 + \left(\dfrac{x-x_{n-1}}{h_{n-1}}\right)^3\right]h_{n-1}, & x_{n-1} \le x \le x_n \\ 0, & \text{otherwise} \end{cases}$$

$$H_n^+(x) = \begin{cases} \left[\left(\dfrac{x_{n+1}-x}{h_n}\right)^2 - \left(\dfrac{x_{n+1}-x}{h_n}\right)^3\right]h_n, & x_n \le x \le x_{n+1} \\ 0, & \text{otherwise} \end{cases}$$

$$(15.112)$$

are frequently used for the development of coarse-mesh finite-element approximations. These polynomials have the properties

$$H_n^0(x_n) = \frac{dH_n^-}{dx}(x_n) = \frac{dH_n^+}{dx}(x_n) = 1$$
$$H_n^-(x_n) = H_n^+(x_n) = 0$$

$$(15.113)$$

These polynomials are used to construct trial functions:

$$\phi(x) = \sum_{n=1}^{N-1}\left[\phi_n^0 H_n^0(x) + \phi_n^-(x)H_n^-(x) + \phi_n^+ H_n^+(x)\right]$$

$$\phi^*(x) = \sum_{n=1}^{N-1}\left[\phi_n^{*0} H_n^0(x) + \phi_n^{*-} H_n^-(x) + \phi_n^{*+} H_n^+(x)\right]$$

$$(15.114)$$

The second property of Eq. (15.113) ensures that this trial function is continuous at the x_n. Thus the variational functional of Eq. (15.107) admits these trial functions.

Requiring stationarity of the variational functional with respect to arbitrary and independent variations of all the adjoint trial functions in all interior mesh intervals; that is, requiring $(\delta F_d/\delta\phi_n^{*0})\delta\phi_n^{*0} = 0$, $(\delta F_d/\delta\phi_n^{*-})\delta\phi_n^{*-} = 0$, and $(\delta F_d/\delta\phi_n^{*+})\delta\phi_n^{*+} = 0$ for $n = 1,\ldots,N-1$ yields three equations for each mesh point:

$$-\frac{D_n}{h_n}\left(\phi_{n+1}^0 - \phi_n^0\right) + \frac{D_{n-1}}{h_{n-1}}\left(\phi_n^0 - \phi_{n-1}^0\right)$$

$$+ h_n\left(\Sigma_{tn} - \Sigma_{sn-1} - \frac{\nu\Sigma_{fn}}{k}\right)\left(\frac{7}{20}\phi_n^0 + \frac{3}{20}\phi_{n+1}^0 - \frac{1}{30}h_n\phi_{n+1}^- + \frac{1}{20}h_n\phi_n^+\right)$$

$$+ h_{n-1}\left(\Sigma_{tn-1} - \Sigma_{sn-1} - \frac{\nu\Sigma_{fn-1}}{k}\right)$$

$$\times\left(\frac{7}{20}\phi_n^0 + \frac{3}{20}\phi_{n-1}^0 + \frac{1}{30}h_{n-1}\phi_{n-1}^+ - \frac{1}{20}h_{n-1}\phi_n^-\right) = 0$$

$$h_n\left(\Sigma_{tn} - \Sigma_{sn} - \frac{\nu\Sigma_{fn}}{k}\right)\left[-\frac{3}{140}\left(\phi_{n+1}^0 - \phi_n^0\right) + \frac{h_n}{420}\left(\phi_{n+1}^- + \phi_n^+\right)\right]$$

$$+ \frac{D_n}{h_n}\left[-\frac{1}{5}\left(\phi_{n+1}^0 - \phi_n^0\right) + \frac{h_n}{10}\left(\phi_{n+1}^- + \phi_n^+\right)\right] = 0$$

$$h_n\left(\Sigma_{tn} - \Sigma_{sn} - \frac{\nu\Sigma_{fn}}{k}\right)\left[-\frac{1}{2}\left(\phi_{n+1}^0 - \phi_n^0\right) + \frac{h_n}{60}\left(\phi_{n+1}^- - \phi_n^+\right)\right]$$

$$+ \frac{D_n}{6}\left(\phi_{n+1}^- + \phi_n^+\right) = 0$$

(15.115)

These equations are for the interior mesh intervals. The zero flux boundary condition requires ϕ_0^0 and ϕ_N^0 to vanish (or a symmetry boundary condition would require, for example, $\phi_0^0 = \phi_1^0$). However, additional constraints must be imposed to evaluate $\phi_0^\pm$ and $\phi_N^\pm$. Requiring stationarity of the variational functional at the external boundaries [i.e., $(\delta F_d/\delta\phi_N^{*\pm})\delta\phi_N^{*\pm} = 0$ and $(\delta F_d/\delta\phi_0^{*\pm})\delta\phi_0^{*\pm} = 0$] provides the additional equations that are necessary to specify the problem completely.

The use of cubic Hermite polynomials is found to increase the accuracy of the finite-element approximation relative to use of the linear polynomial of Eq. (15.109), but, of course, to increase the computing time because three equations per mesh point are involved instead of only one. The accuracy of which we are speaking is the error with respect to an exact solution of the homogenized problem, not with respect to an exact solution of the true heterogeneous problem. Although it seems plausible that a more accurate solution to the global homogeneous problem, when combined with a local heterogeneous solution, will yield a more accurate solution of the actual global heterogeneous problem, this is not obvious.

Multidimensional Finite-Element Coarse-Mesh Methods

In two dimensions, the volume of a core can be partitioned into region volumes V_i, which we refer to as *elements*. These elements can have a variety of shapes: triangles, quadrilaterals, tetrahedral, and so on. A finite-element approximation for the solution is represented by a linear combination of shape functions associated with each element, normally polynomials in the local coordinates within the element. A shape function has the value unity at its associated coarse mesh point and goes to zero at the surface of the volumes V_i associated with that element. For example, a quadratic polynomial

$$\phi(x, y) = a_1 + a_z y + a_3 y + a_4 x^2 + a_5 xy + a_6 y^2 \tag{15.116}$$

might be used to represent the flux within a triangular element. Usually, the polynomial is redefined so that the coefficients have the values of the flux at various support points throughout the element. A quadratic approximation clearly requires six support points, a linear approximation would require three support points, and so on. The value of the flux at each support point is an unknown in the resulting equations.

15.7
Variational Discrete Ordinates Nodal Method

The nodal and coarse mesh calculations described in Sections 15.2 to 15.4 proceed in three distinct steps: (1) the performance of local assembly two-dimensional transport calculation and the preparation of homogenized cross sections for each node, (2) the global solution of the nodal equations for the average flux in each node, and (3) reconstruction of the detailed heterogeneous intranode fluxes. We found in considering the coarse mesh methods that the higher-order polynomials which better represented the overall flux distribution within the coarse mesh region led to more accurate solutions *of the homogenized problem*, which is solved with nodal or coarse mesh equations in step 2.

It is possible to combine the three steps—homogenization, flux solution, detailed flux reconstruction—into a single, self-consistent procedure that uses the detailed heterogeneous assembly transport flux directly, instead of a polynomial approximation, to represent the flux distribution within the node or coarse mesh region. Since a relatively high order transport solution is needed for the heterogeneous assembly calculation, but a relatively low order transport calculation will usually suffice for the global nodal calculation, we illustrate the development of a methodology that can make use of high-order discrete ordinates heterogeneous two-dimensional assembly calculations as *trial functions* to develop a low-order discrete ordinates nodal calculational model.

Variational Principle

A variational principle for the neutron transport equation is

$$F\left[\psi(r, \mathbf{\Omega}), \psi^*(r, \mathbf{\Omega})\right]$$

$$= \sum_{\lambda=1}^{\Lambda}\left\{\iiint_{V_\lambda} dV \iint_{4\pi} d\mathbf{\Omega}\, \psi^*(r, \mathbf{\Omega})\left[-\mathbf{\Omega}\cdot\nabla\psi(r, \mathbf{\Omega}) - \Sigma_t(r)\psi(r, \mathbf{\Omega})\right.\right.$$

$$\left.\left. + \iint_{4\pi} d\mathbf{\Omega}'\, \Sigma_S\left(r, \mathbf{\Omega}' \to \mathbf{\Omega}\right)\psi\left(r, \mathbf{\Omega}'\right) + \frac{1}{4\pi}\iint_{4\pi} d\mathbf{\Omega}'\, \nu\Sigma_f(r)\psi\left(r, \mathbf{\Omega}'\right)\right]\right.$$

$$+ \sum_{\nu(\lambda)} \iint_{\sigma_{\lambda, \nu(\lambda)}} dS \iint_{4\pi} d\boldsymbol{\Omega} \, H(-\boldsymbol{\Omega} \cdot \mathbf{n})(\boldsymbol{\Omega} \cdot \mathbf{n}) \psi^*(r_{\lambda, \nu(\lambda)}, \boldsymbol{\Omega})$$

$$\times \left[\psi(r_{\lambda, \nu(\lambda)}, \boldsymbol{\Omega}) - \psi(r_{\nu(\lambda), \lambda}, \boldsymbol{\Omega}) \right]$$

$$+ \iint_{S_\lambda} dS \iint_{4\pi} d\boldsymbol{\Omega} \, H(-\boldsymbol{\Omega} \cdot \mathbf{n})(\boldsymbol{\Omega} \cdot \mathbf{n}) \psi^*(r_{\mathrm{ex}}, \boldsymbol{\Omega}) \psi(r_{\mathrm{ex}}, \boldsymbol{\Omega}) \Big\} \qquad (15.117)$$

The functional F is a sum over Λ volumetric reactor regions (or nodes). The first term of the sum is an integration over the nodal volume V_λ and the entire solid angle (4π). The second term of the sum is a sum over all the interior surfaces $\nu(\lambda)$ of node λ; this term is included to allow trial functions that are discontinuous across any surface. The notation $r_{\lambda, \nu(\lambda)}$ refers to the limit of all the points on the surface $\nu(\lambda)$ as approached from within node λ; similarly, $r_{\nu(\lambda), \lambda}$ refers to those same points as approached from the node adjacent to node λ [the node on the other side of surface $\nu(\lambda)$]. Each of the terms in this sum is an integral over the surface $\sigma_{\lambda, \nu(\lambda)}$ (formed by the points $r_{\lambda, \nu(\lambda)}$ and the solid angle 4π). The final term in the sum is an integral over the exterior surface S_λ of node λ (formed by the points r_{ex}). This term is included to allow trial functions that do not satisfy vacuum boundary conditions. In the functional F, $\mathbf{n}$ refers to the outward unit normal vector from node λ across an interior or exterior surface, and H is the Heaviside step function. In addition, $\Sigma_t(r)$, $\Sigma_S(r, \boldsymbol{\Omega}' \to \boldsymbol{\Omega})$, and $\nu\Sigma_f(r)$ are the usual cross sections for removal, scattering from angle $\boldsymbol{\Omega}'$ to $\boldsymbol{\Omega}$, and neutron production from fission, respectively. Although the functional F has been presented in a one-energy-group (or energy-independent) form, the extension of the results below to the multigroup case is straightforward.

The condition $(\delta F/\delta \psi^*)\delta \psi^* = 0$ requires that the stationary value of the trial function $\psi(r, \boldsymbol{\Omega})$ be identified as the forward angular flux satisfying the Boltzmann transport equation,

$$\boldsymbol{\Omega} \cdot \nabla \psi(r, \boldsymbol{\Omega}) + \Sigma_t(r)\psi(r, \boldsymbol{\Omega})$$

$$= \iint_{4\pi} d\boldsymbol{\Omega}' \, \Sigma_S(r, \boldsymbol{\Omega}' \to \boldsymbol{\Omega})\psi(r, \boldsymbol{\Omega}') + \frac{1}{4\pi} \iint_{4\pi} d\boldsymbol{\Omega}' \, \nu\Sigma_f(r)\psi(r, \boldsymbol{\Omega}')$$

$$(15.118)$$

as well as the interface continuity and vacuum boundary conditions

$$\psi(r_{\lambda, \nu(\lambda)}, \boldsymbol{\Omega}) - \psi(r_{\nu(\lambda), \lambda}, \boldsymbol{\Omega}) = 0 \qquad (15.119)$$

and

$$\psi(r_{\mathrm{ex}}, \boldsymbol{\Omega}) = 0, \qquad \boldsymbol{\Omega} \cdot n < 0 \qquad (15.120)$$

respectively.

The condition $(\delta F/\delta \psi)\delta \psi = 0$ requires that the stationary value of the trial function $\psi^*(r, \boldsymbol{\Omega})$ be identified as the adjoint angular flux satisfying the adjoint trans-

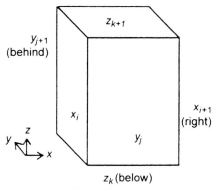

Fig. 15.7 Bounding surface notation.

port equation

$$-\boldsymbol{\Omega} \cdot \nabla \psi^*(r, \boldsymbol{\Omega}) + \Sigma_t(r)\psi^*(r, \boldsymbol{\Omega})$$

$$= \iint_{4\pi} d\boldsymbol{\Omega}' \, \Sigma_S(r, \boldsymbol{\Omega} \to \boldsymbol{\Omega}')\psi^*(r, \boldsymbol{\Omega}') + \frac{1}{4\pi} \iint_{4\pi} d\boldsymbol{\Omega}' \, \nu\Sigma_f(r)\psi^*(r, \boldsymbol{\Omega}') \tag{15.121}$$

as well as the interface continuity and vacuum boundary conditions

$$\psi^*(r_{\lambda, \nu(\lambda)}, \boldsymbol{\Omega}) - \psi^*(r_{\nu(\lambda), \lambda}, \boldsymbol{\Omega}) = 0 \tag{15.122}$$

and

$$\psi^*(r_{\text{ex}}, \boldsymbol{\Omega}) = 0, \qquad \boldsymbol{\Omega} \cdot n > 0 \tag{15.123}$$

respectively.

To apply the functional F to develop a nodal method, the reactor volume is first partitioned into $I \times J \times K$ regions, where I, J, and K are the number of partitions along the x, y, and z coordinates, respectively, and $I \times J \times K$ is equal to the Λ of the overall sum in F. Node ijk is bounded by the surfaces x_i, x_{i+1}, y_j, y_{j+1}, z_k, and z_{k+1} as illustrated in Fig. 15.7. The nodal or volumetric domain function, designated $\Delta_{ijk}(x, y, z)$, is defined as

$$\Delta_{ijk}(x, y, z) = \begin{cases} 1, & x_i < x < x_{i+1}, y_j < y < y_{j+1}, z_k < z < z_{k+1} \\ 0, & \text{otherwise} \end{cases} \tag{15.124}$$

The $I \times J$ regions in the radial (x–y) plane are called *channels*. The angular flux $\psi(x, y, z, \boldsymbol{\Omega})$ is represented in each channel ij as the product of a one-dimensional axial function $g_{ijk}(z, \boldsymbol{\Omega})$ and a precomputed two-dimensional planar function $f_{ijk}(x, y, \boldsymbol{\Omega})$ that is used over the axial domain k.

The angular dependence of the axial functions $g_{ijk}(z, \boldsymbol{\Omega})$ is discretized into eight functions $g_{ijk}^n(z)$, one for each octant of the unit sphere, and the octant domain function is designated $\Delta^n(\boldsymbol{\Omega})$, defined as

$$\Delta^n(\boldsymbol{\Omega}) = \begin{cases} 1, & \boldsymbol{\Omega} \text{ within octant } n \\ 0, & \text{otherwise} \end{cases} \tag{15.125}$$

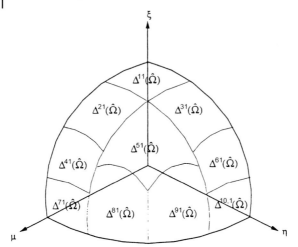

Fig. 15.8 Angular geometry notation for domain functions $\Delta^{mn}(\mathbf{\Omega})$ in an S_8 quadrature.

The angular geometry is illustrated in Fig. 15.8 with a hypothetical arrangement of the $10\Delta^{mn}(\mathbf{\Omega})$ in octant 1 ($\mu > 0$, $\eta > 0$, and $\xi > 0$) for a scheme with $M = 80$ (corresponding to an S_8 quadrature set). Note that the boundaries between adjacent domains are arbitrary.

The surface area of that part of the unit sphere corresponding to region m within octant n is designated w^{mn}, defined as

$$w^{mn} \equiv \frac{1}{4\pi} \iint_{4\pi} d\mathbf{\Omega}\, \Delta^n(\mathbf{\Omega}) \Delta^{mn}(\mathbf{\Omega}) = \frac{1}{4\pi} \iint_{\pi/2} d\mathbf{\Omega}\, \Delta^{mn}(\mathbf{\Omega}) \tag{15.126}$$

where Δ^{mn} is unity within region m in octant n and otherwise zero. Thus the w^{mn} have the same interpretation as standard discrete-ordinates weights, and $\sum_{n=1}^8 \sum_{m=1}^{M/8} w^{mn} = 1$. Note that the notation mn is shortened for "m within n" and that this notation scheme requires that the $M/8$ subregions of octant n be ordered symmetrically with respect to those of each of the other octants, thus effectively specifying the use of a standard level-symmetric quadrature set. This requirement may be eliminated with a suitable, though possibly more confusing change of notation and is in no way limiting.

As indicated in Fig. 15.9, the angle $\mathbf{\Omega}$ is decomposed into its three direction cosines as follows ($\mathbf{i}$, $\mathbf{j}$, and $\mathbf{k}$ are the unit vectors along the x, y, and z axes, respectively):

$$\mu = \mathbf{\Omega} \cdot \mathbf{i}, \qquad \eta = \mathbf{\Omega} \cdot \mathbf{j}, \qquad \xi = \mathbf{\Omega} \cdot \mathbf{k} \tag{15.127}$$

In addition, the azimuthal angle ω is defined (see Fig. 15.9) to be the angle between the z-axis and the projection of $\mathbf{\Omega}$ onto the y–z plane; thus

$$\eta = \sqrt{1 - \mu^2} \sin \omega$$
$$\xi = \sqrt{1 - \mu^2} \cos \omega \tag{15.128}$$

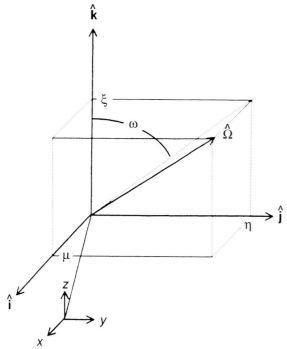

Fig. 15.9 Definition of angles.

Consistent averages of the $\boldsymbol{\Omega}$ direction cosines may be defined as follows:

$$\mu^{mn} \equiv \frac{1}{4\pi\, w^{mn}} \iint_{\substack{\text{octant} \\ n}} d\boldsymbol{\Omega}\, \Delta^{mn}(\boldsymbol{\Omega})\mu$$

$$\eta^{mn} \equiv \frac{1}{4\pi\, w^{mn}} \iint_{\substack{\text{octant} \\ n}} d\boldsymbol{\Omega}\, \Delta^{mn}(\boldsymbol{\Omega})\eta \qquad (15.129)$$

$$\xi^{mn} \equiv \frac{1}{4\pi\, w^{mn}} \iint_{\substack{\text{octant} \\ n}} d\boldsymbol{\Omega}\, \Delta^{mn}(\boldsymbol{\Omega})\xi$$

Using the domain functions defined in Eqs. (15.124) and (15.125), the trial function used for the forward angular flux in the functional F is

$$\psi(x, y, z, \boldsymbol{\Omega}) \approx g(z, \boldsymbol{\Omega}) f(x, y, \boldsymbol{\Omega})$$

$$= \sum_{i,j,k=1}^{I \times J \times K} \Delta_{ijk}(x, y, z)[g_{ijk}(z, \boldsymbol{\Omega}) f_{ijk}(x, y, \boldsymbol{\Omega})]$$

$$= \sum_{i,j,k=1}^{I \times J \times K} \Delta_{ijk}(x, y, z) \sum_{n=1}^{8} \Delta^{n}(\boldsymbol{\Omega}) g_{ijk}^{n}(z) \sum_{m=1}^{M/8} \Delta^{mn}(\boldsymbol{\Omega}) f_{ijk}^{mn}(x, y)$$

$$(15.130)$$

The adjoint angular flux $\psi^{*}(x, y, z, \boldsymbol{\Omega})$ is expanded analogously.

Using the trial function of Eq. (15.130) and the analogous expansion for the adjoint flux in F and requiring stationarity of the functional with respect to each of the adjoint axial functions $g_{ijk}^{*n}(z)$ yields reduced equations for the forward axial functions $g_{ijk}^{n}(z)$, with homogenized parameters defined in terms of the precomputed nodal basis functions $f_{ijk}^{mn}(x,y)$ and $f_{ijk}^{*mn}(x,y)$. The equation for each of the eight axial functions $g_{ijk}^{n}(z)$ is

$$\bar{\xi}_{ijk}^{n}\frac{dg_{ijk}^{n}(z)}{dz} + \bar{B}_{ijk}^{n}g_{ijk}^{n}(z) + \bar{\Sigma}_{ijk}(z)g_{ijk}^{n}(z)$$

$$= \sum_{n'=1}^{8} \nu\bar{\Sigma}_{f,ijk}^{nn'}(z)g_{ijk}^{n'}(z) + \sum_{n'=1}^{8} \bar{\Sigma}_{s,ijk}^{nn'}(z)g_{ijk}^{n'}(z)$$

$$- (1-\delta_{i1})H(\mu^{n})\big[\bar{\mu}_{ijk}^{n}(x_i)g_{ijk}^{n}(z) - \bar{\mu}_{(i,i-1)jk}^{n}(x_i)g_{(i-1)jk}^{n}(z)\big]$$

$$+ (1-\delta_{i1})H(-\mu^{n})\big[\bar{\mu}_{ijk}^{n}(x_{i+1})g_{ijk}^{n}(z) - \bar{\mu}_{(i,i+1)jk}^{n}(x_{i+1})g_{(i+1)jk}^{n}(z)\big]$$

$$- \delta_{i1}H(\mu^{n})\bar{\mu}_{1jk}^{n}(x_i)g_{1jk}^{n}(z) + \delta_{iI}H(-\mu^{n})\bar{\mu}_{Ijk}^{n}(x_{I+1})g_{Ijk}^{n}(z)$$

$$- (1-\delta_{j1})H(\eta^{n})\big[\bar{\eta}_{ijk}^{n}(y_j)g_{ijk}^{n}(z) - \bar{\eta}_{i(j,j+1)k}^{n}(y_j)g_{i(j+1)k}^{n}(z)\big]$$

$$+ (1-\delta_{jJ})H(-\eta^{n})\big[\bar{\eta}_{ijk}^{n}(y_{j+1})g_{ijk}^{n}(z) - \bar{\eta}_{i(j,j+1)k}^{n}(y_{j+1})g_{i(j+1)k}^{n}(z)\big]$$

$$- \delta_{j1}H(\eta^{n})\bar{\eta}_{i1k}^{n}(y_i)g_{i1k}^{n}(z) + \delta_{jJ}H(-\eta^{n})\bar{\eta}_{iJk}^{n}(y_{J+1})g_{iJk}^{n}(z) \qquad (15.131)$$

In Eq. (15.131), the homogenized total, fission, and scattering cross sections in channel ij and axial region k for octant n are defined consistently as

$$\bar{\Sigma}_{ijk}^{n}(z) = \frac{\int_{x_i}^{x_{i+1}} dx \int_{y_j}^{y_{j+1}} dy\, \Sigma(x,y,z) \sum_{m=1}^{M/8} w^{mn} f_{ijk}^{*mn}(x,y)f_{ijk}^{mn}(x,y)}{\int_{x_i}^{x_{i+1}} dx \int_{y_j}^{y_{j+1}} dy \sum_{m=1}^{M/8} w^{mn} f_{ijk}^{*mn}(x,y)f_{ijk}^{mn}(x,y)}$$

$$(15.132)$$

$$\bar{\Sigma}_{f,ijk}^{nn'}(z)$$

$$= \frac{\int_{x_i}^{x_{i+1}} dx \int_{y_j}^{y_{j+1}} dy\, \Sigma_f(x,y,z)\big[\sum_{m=1}^{M/8} w^{mn} f_{ijk}^{*mn}(x,y)\big]\big[\sum_{m'=1}^{M/8} w^{m'n'} f_{ijk}^{m'n'}(x,y)\big]}{\int_{x_i}^{x_{i+1}} dx \int_{y_j}^{y_{j+1}} dy \sum_{m=1}^{M/8} w^{mn} f_{ijk}^{*mn}(x,y)f_{ijk}^{mn}(x,y)}$$

$$(15.133)$$

and

$$\bar{\Sigma}_{s,ijk}^{mn'}(z) = \left(\int_{x_i}^{x_{i+1}} dx \int_{y_j}^{y_{j+1}} dy \sum_{l=0}^{L}(2l+1)\Sigma_s^l(x,y,z)\right.$$

$$\times \left\{\left[\sum_{m=1}^{M/8} w^{mn} P_l(\mu^{mn})f_{ijk}^{*mn}(x,y)\right]\right.$$

$$\times \left[\sum_{m'=1}^{M/8} w^{m'n'} P_l(\mu^{m'n'}) f_{ijk}^{m'n'}(x, y) \right]$$

$$+ 2 \sum_{k=1}^{l} \frac{(l-k)!}{(l+k)!} \left[\sum_{m=1}^{M/8} w^{mn} P_l^k(\mu^{mn}) \cos(k\omega^{mn}) f_{ijk}^{*mn}(x, y) \right]$$

$$\times \left[\sum_{m'=1}^{M/8} w^{m'n'} P_l^k(\mu^{m'n'}) \cos(k\omega^{m'n'}) f_{ijk}^{m'n'}(x, y) \right]$$

$$+ 2 \sum_{k=1}^{l} \frac{(l-k)!}{(l+k)!} \left[\sum_{m=1}^{M/8} w^{mn} P_l^k(\mu^{mn}) \sin(k\omega^{mn}) f_{ijk}^{*mn}(x, y) \right]$$

$$\times \left[\sum_{m'=1}^{M/8} w^{m'n'} P_l^k(\mu^{m'n'}) \sin(k\omega^{m'n'}) f_{ijk}^{m'n'}(x, y) \right] \right] \right\} \Bigg/$$

$$\int_{x_i}^{x_{i+1}} dx \int_{y_j}^{y_{j+1}} dy \sum_{m=1}^{M/8} w^{mn} f_{ijk}^{*mn}(x, y) f_{ijk}^{mn}(x, y) \tag{15.134}$$

respectively. Note that in deriving Eq. (15.134), the scattering cross section $\Sigma_s(r, \boldsymbol{\Omega}' \to \boldsymbol{\Omega})$ has been expanded in a Legendre polynomial of order L, and the addition theorem for spherical harmonics has been applied in the usual way. The isotropic portion of the homogenized scattering cross section has the same form as that of the fission cross section above.

The transverse leakage from node ijk (i.e., the leakage in the x- and y-directions) is defined consistently as

$$\bar{B}_{ijk}^n = \left[\int_{x_i}^{x_{i+1}} dx \int_{y_j}^{y_{j+1}} dy \sum_{m=1}^{M/8} w^{mn} \mu^{mn} f_{ijk}^{*mn}(x, y) \frac{\partial f_{ijk}^{mn}(x, y)}{\partial x} \right.$$

$$\left. + \int_{x_i}^{x_{i+1}} dx \int_{y_j}^{y_{j+1}} dy \sum_{m=1}^{M/8} w^{mn} \eta^{mn} f_{ijk}^{*mn}(x, y) \frac{\partial f_{ijk}^{mn}(x, y)}{\partial y} \right] \Bigg/$$

$$\int_{x_i}^{x_{i+1}} dx \int_{y_j}^{y_{j+1}} dy \sum_{m=1}^{M/8} w^{mn} f_{ijk}^{*mn}(x, y) f_{ijk}^{mn}(x, y) \tag{15.135}$$

The homogenized discrete ordinate for octant n is defined consistently as

$$\bar{\xi}_{ijk}^n = \frac{\int_{x_i}^{x_{i+1}} dx \int_{y_j}^{y_{j+1}} dy \sum_{m=1}^{M/8} w^{mn} \xi^{mn} f_{ijk}^{*mn}(x, y) f_{ijk}^{mn}(x, y)}{\int_{x_i}^{x_{i+1}} dx \int_{y_j}^{y_{j+1}} dy \sum_{m=1}^{M/8} w^{mn} f_{ijk}^{*mn}(x, y) f_{ijk}^{mn}(x, y)} \tag{15.136}$$

The four parameters required (for each octant) for coupling node ijk to its x-direction neighbors, nodes $(i-1)jk$ and $(i+1)jk$, are

$$\bar{\mu}_{ijk}^n(x_i) = \frac{\int_{y_j}^{y_{j+1}} dy \sum_{m=1}^{M/8} w^{mn} \mu^{mn} f_{ijk}^{*mn}(x_i, y) f_{ijk}^{mn}(x_i, y)}{\int_{x_i}^{x_{i+1}} dx \int_{y_j}^{y_{j+1}} dy \sum_{m=1}^{M/8} w^{mn} f_{ijk}^{*mn}(x, y) f_{ijk}^{mn}(x, y)}$$

$$\bar{\mu}_{(i,i-1)jk}^n(x_i) = \frac{\int_{y_j}^{y_{j+1}} dy \sum_{m=1}^{M/8} w^{mn} \mu^{mn} f_{ijk}^{*mn}(x_i, y) f_{(i-1)jk}^{mn}(x_i, y)}{\int_{x_i}^{x_{i+1}} dx \int_{y_j}^{y_{j+1}} dy \sum_{m=1}^{M/8} w^{mn} f_{ijk}^{*mn}(x, y) f_{ijk}^{mn}(x, y)}$$

$$\tag{15.137}$$

$$\bar{\mu}_{ijk}^n(x_{i+1}) = \frac{\int_{y_j}^{y_{j+1}} dy \sum_{m=1}^{M/8} w^{mn} \mu^{mn} f_{ijk}^{*mn}(x_{i+1}, y) f_{ijk}^{mn}(x_{i+1}, y)}{\int_{x_i}^{x_{i+1}} dx \int_{y_j}^{y_{j+1}} dy \sum_{m=1}^{M/8} w^{mn} f_{ijk}^{*mn}(x, y) f_{ijk}^{mn}(x, y)}$$

$$\bar{\mu}_{(i,i-1)jk}^n(x_{i+1}) = \frac{\int_{y_j}^{y_{j+1}} dy \sum_{m=1}^{M/8} w^{mn} \mu^{mn} f_{ijk}^{*mn}(x_{i+1}, y) f_{(i+1)jk}^{mn}(x_{i+1}, y)}{\int_{x_i}^{x_{i+1}} dx \int_{y_j}^{y_{j+1}} dy \sum_{m=1}^{M/8} w^{mn} f_{ijk}^{*mn}(x, y) f_{ijk}^{mn}(x, y)}$$

The four parameters required (for each octant) for coupling node ijk to its y-direction neighbors, nodes $i(j-1)k$ and $i(j+1)k$, are

$$\bar{\eta}_{ijk}^n(y_j) = \frac{\int_{x_i}^{x_{i+1}} dy \sum_{m=1}^{M/8} w^{mn} \eta^{mn} f_{ijk}^{*mn}(x, y_j) f_{ijk}^{mn}(x_i, y_j)}{\int_{x_i}^{x_{i+1}} dx \int_{y_j}^{y_{j+1}} dy \sum_{m=1}^{M/8} w^{mn} f_{ijk}^{*mn}(x, y) f_{ijk}^{mn}(x, y)}$$

$$\bar{\eta}_{i(j,j-1)k}^n(y_i) = \frac{\int_{x_i}^{x_{i+1}} dx \sum_{m=1}^{M/8} w^{mn} \eta^{mn} f_{ijk}^{*mn}(x, y_j) f_{i(j-1)k}^{mn}(x, y_j)}{\int_{x_i}^{x_{i+1}} dx \int_{y_j}^{y_{j+1}} dy \sum_{m=1}^{M/8} w^{mn} f_{ijk}^{*mn}(x, y) f_{ijk}^{mn}(x, y)}$$

$$\tag{15.138}$$

$$\bar{\eta}_{ijk}^n(y_{j+1}) = \frac{\int_{x_i}^{x_{i+1}} dx \sum_{m=1}^{M/8} w^{mn} \eta^{mn} f_{ijk}^{*mn}(x, y_{j+1}) f_{ijk}^{mn}(x, y_{j+1})}{\int_{x_i}^{x_{i+1}} dx \int_{y_j}^{y_{j+1}} dy \sum_{m=1}^{M/8} w^{mn} f_{ijk}^{*mn}(x, y) f_{ijk}^{mn}(x, y)}$$

$$\bar{\eta}_{i(j,j+1)k}^n(y_{j+1}) = \frac{\int_{x_i}^{x_{i+1}} dx \sum_{m=1}^{M/8} w^{mn} \eta^{mn} f_{ijk}^{*mn}(x, y_{j+1}) f_{i(j+1)k}^{mn}(x, y_{j+1})}{\int_{x_i}^{x_{i+1}} dx \int_{y_j}^{y_{j+1}} dy \sum_{m=1}^{M/8} w^{mn} f_{ijk}^{*mn}(x, y) f_{ijk}^{mn}(x, y)}$$

Note that the arguments of the Heaviside step functions H in Eqs. (15.131) are the direction cosines μ^{mn} and $\eta^{\mu\nu}$ of Eqs. (15.129), but with the single superscript n because only the octant needs to be identified.

The interface conditions that couple node ijk to its axially adjacent neighbors, nodes $ij(k-1)$ and $ij(k+1)$, are

$$\bar{\xi}_{ijk}^n g_{ijk}^n(z_k) = \bar{\xi}_{ij(k,k-1)}^n g_{ij(k-1)}^n(z_k), \qquad \bar{\xi}_{ijk}^n > 0$$

$$\bar{\xi}_{ijk}^n g_{ijk}^n(z_{k+1}) = \bar{\xi}_{ij(k,k+1)}^n g_{ij(k+1)}^n(z_{k+1}), \quad \bar{\xi}_{ijk}^n < 0$$

$$\tag{15.139}$$

where the coupling parameters are defined as

$$\bar{\xi}_{ij(k,k-1)}^n = \frac{\int_{x_i}^{x_{i+1}} dx \int_{y_j}^{y_{j+1}} dy \sum_{m=1}^{M/8} w^{mn} \xi^{mn} f_{ijk}^{*mn}(x, y) f_{ij(k-1)}^{mn}(x, y)}{\int_{x_i}^{x_{i+1}} dx \int_{y_j}^{y_{j+1}} dy \sum_{m=1}^{M/8} w^{mn} f_{ijk}^{*mn}(x, y) f_{ijk}^{mn}(x, y)}$$

$$\tag{15.140}$$

$$\bar{\xi}_{ij(k,k+1)}^n = \frac{\int_{x_i}^{x_{i+1}} dx \int_{y_j}^{y_{j+1}} dy \sum_{m=1}^{M/8} w^{mn} \xi^{mn} f_{ijk}^{*mn}(x, y) f_{ij(k+1)}^{mn}(x, y)}{\int_{x_i}^{x_{i+1}} dx \int_{y_j}^{y_{j+1}} dy \sum_{m=1}^{M/8} w^{mn} f_{ijk}^{*mn}(x, y) f_{ijk}^{mn}(x, y)}$$

Finally, the boundary conditions on Eqs. (15.131) are

$$
\begin{aligned}
g_{ij1}^n(z_1) &= 0, \qquad \bar{\xi}_{ij1}^n > 0 \\
g_{ijK}^n(z_{K+1}) &= 0, \quad \bar{\xi}_{ijK}^n < 0
\end{aligned}
\tag{15.141}
$$

Using the angular flux trial function of Eq. (15.130) in the usual definition of the isotropic flux specifies the heterogeneous flux reconstruction equation:

$$
\begin{aligned}
\phi(x, y, z) &\equiv \iint_{4\pi} d\mathbf{\Omega} \, \psi(x, y, x, \mathbf{\Omega}) \\
&\approx \iint_{4\pi} d\mathbf{\Omega} \sum_{n=1}^{8} \Delta^n(\mathbf{\Omega}) g_{ijk}^n(z) \sum_{m=1}^{M/8} \Delta^{mn}(\mathbf{\Omega}) f_{ijk}^{mn}(x, y) \\
&= \sum_{n=1}^{8} g_{ijk}^n(z) \sum_{m=1}^{M/8} f_{ijk}^{mn}(x, y) \iint_{4\pi} d\mathbf{\Omega} \, \Delta^n(\mathbf{\Omega}) \Delta^{mn}(\mathbf{\Omega}) \\
&= 4\pi \sum_{m=1}^{8} g_{ijk}^n(z) \sum_{m=1}^{M/8} w^{mn} f_{ijk}^{mn}(x, y)
\end{aligned}
\tag{15.142}
$$

Application of the Method

The steps required for application of the variational nodal discrete-ordinates method are the same as those required for standard nodal methods. First, a set of fine-mesh high-order two-dimensional calculations is performed for small heterogeneous local regions in the $x-y$ plane, such as assemblies or extended assemblies, for the purposes of homogenizing the nodes. (It is also possible to use full two-dimensional planar calculations to provide nodal trial functions.) In standard nodal methods, even for full three-dimensional global calculations, only two-dimensional local calculations are performed, but the manner of axial coupling required for the nodes is rarely specified. In the variational nodal discrete-ordinates method, the local calculations are performed using the discrete-ordinates (S_N) method. The fine-mesh S_N calculations yield the angular fluxes $f_{ijk}^{mn}(x, y)$, and $M = N(N + 2)$ for a three-dimensional problem. For this method it is also necessary to calculate the adjoint angular S_N fluxes $f_{ijk}^{*mn}(x, y)$ in each node. There is a different homogenized cross section defined for each of the eight S_2 directions; however, because of the axial symmetry obtained by the use of two-dimensional basis functions, only four are required. The basis functions $f_{ijk}^{mn}(x, y)$ and $f_{ijk}^{*mn}(x, y)$ are used with standard S_N ordinates and weights to compute homogenized parameters in accordance with the definitions of Eqs. (15.132) to (15.138) and (15.140).

The second step in standard nodal methodology is a global diffusion-theory calculation, which involves (in general, for the transverse integrated methods) three one-dimensional equations for the transverse integrated x-, y-, and z-direction fluxes. Usually, the problem is reduced to that of finding coefficients of fourth-order polynomials. In the variational nodal method, the global equations are one-dimensional (z-direction) S_2 first-order differential equations, which are equivalent in accuracy to the diffusion equations. The spatial discretization to use on the z-axis is not specified; thus any method may be used, including coarse-mesh, finite-difference, high-order polynomial expansion, or other standard method.

The final step in the nodal calculation for the standard and variational nodal methods is the reconstruction of heterogeneous fluxes or reaction rates from the homogeneous (global) calculation results. In the variational nodal method, the flux reconstruction is completely specified by Eq. (15.142).

15.8
Variational Principle for Multigroup Diffusion Theory

A complete mathematical description of the neutron distribution, within the context of multigroup diffusion theory, is provided by a coupled set of partial differential equations for the direct and adjoint flux (and current) with associated boundary, initial, final, and continuity conditions. An equivalent variational formulation must not only have the original equations as Euler equations, but must also embody the associated boundary, initial, final, and continuity conditions, either directly or indirectly through limitations on the admissible class of trial functions.

The following variational principle embodies all these conditions:

$$
\begin{aligned}
J = &\left[\int_{t_0}^{t_f} dt \int_V dr \left\{ \boldsymbol{\phi}^{*\mathrm{T}} \left[\boldsymbol{\Sigma} - (1-\beta)\boldsymbol{\chi}\, \boldsymbol{F}^{\mathrm{T}} \right] \boldsymbol{\phi} + \boldsymbol{\phi}^{*\mathrm{T}} \nabla \cdot \boldsymbol{j} + \boldsymbol{\phi}^{*\mathrm{T}} \tau \dot{\boldsymbol{\phi}} \right. \right. \\
&\qquad - \boldsymbol{\phi}^{*\mathrm{T}} \sum_{m=1}^{M} \lambda_m \chi_m C_m - \boldsymbol{j}^{*\mathrm{T}} \cdot 3\boldsymbol{\Sigma}_{\mathrm{tr}} \boldsymbol{j} - \boldsymbol{j}^{*\mathrm{T}} \cdot \nabla \boldsymbol{\phi} \\
&\qquad - \sum_{m=1}^{M} C_m^* \beta_m \boldsymbol{F}^{\mathrm{T}} \boldsymbol{\phi} + \sum_{m=1}^{M} C_m^* \lambda_m C_m \\
&\qquad \left. \left. + \sum_{m=1}^{M} C_m^* \dot{C}_m - \boldsymbol{\phi}^{*\mathrm{T}} \boldsymbol{S} + \boldsymbol{S}^{*\mathrm{T}} \boldsymbol{\phi} \right\} \right]_1 \\
&+ \left[\int_{t_0}^{t_f} dt \int_{S_{\mathrm{in}}} ds\, \mathbf{n} \cdot \left\{ (\gamma \boldsymbol{\phi}_+^{*\mathrm{T}} + (1-\gamma)\boldsymbol{\phi}_-^{*\mathrm{T}})(\boldsymbol{j}_+ - \boldsymbol{j}_-) \right. \right. \\
&\qquad\qquad \left. \left. - (\eta \boldsymbol{j}_+^{*\mathrm{T}} + (1-\eta)\, \boldsymbol{j}_-^{*\mathrm{T}})(\boldsymbol{\phi}_+ - \boldsymbol{\phi}_-) \right\} \right]_2 \\
&+ \left[\int_V dr \left\{ (a \boldsymbol{\phi}^{*\mathrm{T}}(+) + (1-a)\boldsymbol{\phi}^{*\mathrm{T}}(-)) \tau \left(\boldsymbol{\phi}(+) - \boldsymbol{\phi}(-) \right) \right. \right. \\
&\qquad\qquad \left. \left. + \sum_{m=1}^{M} \left(b C_m^*(+) + (1-b) C_m^*(-) \right) \left(C_m(+) - C_m(-) \right) \right\} \right]
\end{aligned}
$$

$$+ \left[\int_V d\mathbf{r} \left\{ \boldsymbol{\phi}^{*\mathrm{T}}(t_0) \boldsymbol{\tau} (\boldsymbol{\phi}(t_0) - \mathbf{g}_0) - \mathbf{g}_f^{*\mathrm{T}} \boldsymbol{\tau} \boldsymbol{\phi}(t_f) \right. \right.$$

$$\left. \left. + \sum_{m=1}^{M} \left[C_m^*(t_0) \big(C_m(t_0) - h_{m_0} \big) - h_{m_f}^* C_m(t_f) \right] \right\} \right]_4$$

$$+ \left[\int_{t_0}^{t_f} dt \int_{S_0} ds \left\{ (\mathbf{j}_{s_0}^{*\mathrm{T}} \cdot \mathbf{n})(\boldsymbol{\phi}_{s_0} + \ell \mathbf{j}_{s_0} \cdot \mathbf{n}) \right\} \right]_5$$

$$\equiv J_1 + J_2 + J_3 + J_4 + J_5 \qquad (15.143)$$

where

$\boldsymbol{\phi}^*, \boldsymbol{\phi}$ = $G \times 1$ column matrices of group adjoint and direct flux, respectively

$\mathbf{j}^*, \mathbf{j}$ = $G \times 1$ column matrices of group adjoint and direct current, respectively—vector quantities

$\mathbf{S}^*, \mathbf{S}$ = $G \times 1$ column matrices of group adjoint and direct source, respectively

C_m^*, C_m = scalar adjoint and direct delayed neutron precursor densities, respectively

$\boldsymbol{\Sigma}$ = $G \times G$ matrix of group removal and scattering cross sections

$\boldsymbol{\Sigma}_{\mathrm{tr}}$ = $G \times G$ diagonal matrix of group transport cross section

$\boldsymbol{\tau}$ = $G \times G$ diagonal matrix of inverse group neutron speeds

$\mathbf{F}$ = $G \times 1$ column matrix of group nu-fission cross sections

$\boldsymbol{\chi}, \boldsymbol{\chi}_m$ = $G \times 1$ column matrices of prompt- and delayed-fission neutron spectra, respectively

λ_m, β_m = delayed neutron precursor decay rate and precursor yield per fission, respectively

The term in the first set of brackets is an integral over the time of interest, $t_0 \leq t \leq t_f$, and the volume of the reactor. The Euler equations for this term are the direct and adjoint flux, current, and precursor equations, which result from the requirement that the first variations of J_1 with respect to each of the argument functions $(\boldsymbol{\phi}^*, \boldsymbol{\phi}, \mathbf{j}^*, \mathbf{j}, C_m^*, C_m)$ vanishes. In taking the first variation of J_1, integration by parts is required, which introduces certain additional terms. The requirement that these additional terms vanish, and hence that stationarity of J_1 implies satisfaction of the Euler equations, imposes restrictions on the admissible class of trial functions. The purpose of the additional terms, $J_2 - J_5$, is to remove these restrictions.

If direct and adjoint flux and current trial functions that are discontinuous across an internal interface, S_{in}, are admitted, a term of the general form of J_2 must be added to J_1 in order that stationarity of the functional $J_{12} \equiv J_1 + J_2$ implies satisfaction of the Euler equations and flux and current continuity conditions. The subscripts indicate limiting values on the $+$ and $-$ sides, with respect to the unit normal vector $\mathbf{n}$, of the surface S_{in}. γ and η are arbitrary constants.

Terms of the general form of J_2 have given rise to an overdetermination of interface conditions in synthesis applications. Consider, for example, the variation of J_{12} with respect to $\boldsymbol{\phi}^*$ (by a variation with respect to the column vector we intend separate and independent variations with respect to each element of the column matrix):

$$
\frac{\delta J_{12}}{\delta \boldsymbol{\phi}^{*\mathrm{T}}} \delta \boldsymbol{\phi}^{*\mathrm{T}} = 0 = \left[\int_{t_0}^{t_f} dt \int_V d\mathbf{r}\, \delta \boldsymbol{\phi}^{*\mathrm{T}} \left\{ \left[\boldsymbol{\Sigma} - (1-\beta)\boldsymbol{\chi} \boldsymbol{F}^{\mathrm{T}} \right] \boldsymbol{\phi} + \nabla \cdot \boldsymbol{j} + \tau \dot{\boldsymbol{\phi}} \right.\right.
$$

$$
\left.\left. - \sum_{m=1}^{M} \lambda_m \boldsymbol{\chi}_m C_m - \boldsymbol{S} \right\} \right]_1
$$

$$
+ \left[\int_{t_0}^{t_f} dt \int_{S_{\text{in}}} ds\, \mathbf{n} \cdot \left\{ \left[\gamma \delta \boldsymbol{\phi}_+^{*\mathrm{T}} + (1-\gamma) \delta \boldsymbol{\phi}_-^{*\mathrm{T}} \right] \right.\right.
$$

$$
\left.\left. \times (\boldsymbol{j}_+ - \boldsymbol{j}_-) \right\} \right]_2 \tag{15.144}
$$

For completely arbitrary $\delta \boldsymbol{\phi}^{*\mathrm{T}}$, the first term vanishes only if the expression within the first set of braces is identically zero, which is just the condition that the neutron balance equation is satisfied. Vanishing of the second term for arbitrary $\delta \boldsymbol{\phi}_+^{*\mathrm{T}}$ and $\delta \boldsymbol{\phi}_-^{*\mathrm{T}}$ appears to lead to two current continuity conditions. However, continuity of adjoint flux requires that $\delta \boldsymbol{\phi}_+^{*\mathrm{T}} = \delta \boldsymbol{\phi}_-^{*\mathrm{T}}$, and in fact there is only one current continuity condition. The difficulty in synthesis applications results from the failure to impose the condition $\delta \boldsymbol{\phi}_+^{*\mathrm{T}} = \delta \boldsymbol{\phi}_-^{*\mathrm{T}}$ on trial functions that are partially specified.

If direct and adjoint flux and precursor trial functions which are discontinuous in time at t_{in} are admitted, a term of the general form of J_3 must be added to J_1 in order that stationarity of the functional $J_{13} = J_1 + J_3$ implies the satisfaction of the Euler equations and flux and precursor time continuity conditions. The $+$ and $-$ arguments refer to times just after and just before, respectively, t_{in}. a and b are arbitrary constants. An overdetermination problem, analogous to that discussed for J_2, has also arisen in synthesis applications of J_3.

If direct flux and precursor trial functions that do not satisfy the known initial conditions g_0 and h_{m_0}, and adjoint flux and precursor trial functions that do not satisfy known final conditions g_f^* and $h_{m_f}^*$, are admitted, J_4 must be added to J_1 in order that stationarity of the resulting variational principle implies satisfaction of the Euler equations and the appropriate initial and final conditions. Similarly, stationarity of $J_{15} = J_1 + J_5$ implies satisfaction of the Euler equations and the external boundary conditions $\boldsymbol{\phi}_{s_0} + \ell \boldsymbol{j}_{s_0} \cdot \mathbf{n} = 0$, $\boldsymbol{\phi}_{s_0}^* + \ell \boldsymbol{j}_{s_0}^* \cdot \mathbf{n} = 0$, even if the flux and current trial functions do not satisfy these boundary conditions identically.

A general second-order variational principle for multigroup diffusion theory can also be written which admits the same extended class of trial functions as J of Eq. (15.143), and which leads to the same apparent interface overdetermination problem in synthesis applications. Using Fick's law to relate flux and current, and integrating by parts in Eq. (15.143), leads to

$$
\begin{aligned}
F = & \left[\int_{t_0}^{t_f} dt \int_V d\mathbf{r} \left\{ \boldsymbol{\phi}^{*T} [\boldsymbol{\Sigma} - (1-\beta)\boldsymbol{\chi} F^T] \boldsymbol{\phi} + \nabla \boldsymbol{\phi}^{*T} \cdot \boldsymbol{D} \nabla \boldsymbol{\phi} + \boldsymbol{\phi}^{*T} \boldsymbol{\tau} \boldsymbol{\phi} \right. \right. \\
& - \boldsymbol{\phi}^{*T} \sum_{m=1}^{M} \lambda_m \boldsymbol{\chi}_m C_m - \sum_{m=1}^{M} C_m^* \beta_m F^T \boldsymbol{\phi} \\
& \left. \left. + \sum_{m=1}^{M} C_m^* \lambda_m C_m + \sum_{m=1}^{M} C_m^* \dot{C} - \boldsymbol{\phi}^{*T} \boldsymbol{S} + \boldsymbol{S}^{*T} \boldsymbol{\phi} \right\} \right]_1 \\
& + \left[\int_{t_0}^{t_f} dt \int_{S_{\mathrm{in}}} ds\, \mathbf{n} \cdot \left\{ (\boldsymbol{\phi}_+^{*T} - \boldsymbol{\phi}_-^{*T})((1-\gamma)\boldsymbol{D}_+ \nabla \boldsymbol{\phi}_+ + \gamma \boldsymbol{D}_- \nabla \boldsymbol{\phi}_-) \right. \right. \\
& \left. \left. + (\eta \nabla \boldsymbol{\phi}_+^{*T} \boldsymbol{D}_+ + (1-\eta) \nabla \boldsymbol{\phi}_-^{*T} \boldsymbol{D}_-)(\boldsymbol{\phi}_+ - \boldsymbol{\phi}_-) \right\} \right]_2 \\
& + \left[\int_V d\mathbf{r} \left\{ (a\boldsymbol{\phi}^{*T}(+) + (1-a)\boldsymbol{\phi}^{*T}(-)) \boldsymbol{\tau} (\boldsymbol{\phi}(+) - \boldsymbol{\phi}(-)) \right. \right. \\
& \left. \left. + \sum_{m=1}^{M} (b C_m^*(+) + (1-b) C_m^*(-))(C_m(+) - C_m(-)) \right\} \right]_3 \\
& + \left[\int_V d\mathbf{r} \left\{ \boldsymbol{\phi}^{*T}(t_0) \boldsymbol{\tau} (\boldsymbol{\phi}(t_0) - \mathbf{g}_0) - \mathbf{g}_f^{*T} \boldsymbol{\tau} \boldsymbol{\phi}(t_f) \right. \right. \\
& \left. \left. + \sum_{m=1}^{M} \left[C_m^*(t_0)(C_m(t_0) - h_{m_0}) - h_{m_f}^* C_m(t_f) \right] \right\} \right]_4 \\
& + \left[-\int_{t_0}^{t_f} dt \int_{S_0} ds \left\{ \mathbf{n} \cdot \nabla \boldsymbol{\phi}_{s_0}^{*T} \boldsymbol{D} (\boldsymbol{\phi}_{s_0} - \ell \boldsymbol{D} \nabla \boldsymbol{\phi}_{s_0} \cdot \mathbf{n}) - \boldsymbol{\phi}_{s_0}^{*T} \boldsymbol{D} \nabla \boldsymbol{\phi}_{s_0} \cdot \mathbf{n} \right\} \right]_5 \\
\equiv & \ F_1 + F_2 + F_3 + F_4 + F_5 \tag{15.145}
\end{aligned}
$$

The diffusion coefficient matrix, $\boldsymbol{D} = \frac{1}{3} \boldsymbol{\Sigma}_{\mathrm{tr}}^{-1}$, has been introduced in Eq. (15.145).

15.9
Single-Channel Spatial Synthesis

The basic idea of single-channel synthesis is illustrated by the example of a uniform reactor with a rod (or bank of rods) partially inserted, as illustrated in Fig. 15.10. A few diffusion lengths above and below the rod tip the flux solution is essentially a one-dimensional radial flux shape ϕ_{rod} and ϕ_{unrod}, respectively. In the vicinity of the rod tip, it is plausible that some mixture of the two flux shapes will describe the

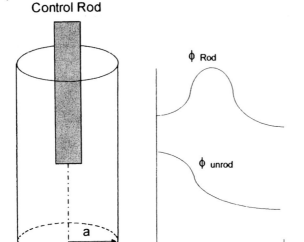

$$\phi\,(r,z) = a_{rod}\,(z)\phi_{rod}(r) + a_{unrod}\,(z)\phi_{unrod}(r)$$

Fig. 15.10 Single-channel synthesis example.

actual radial flux distribution. The synthesis approximation is developed by using trial functions of the form

$$\boldsymbol{\phi}(x, y, z, t) = \sum_{n=1}^{N} \boldsymbol{\psi}_n(x, y)\boldsymbol{\rho}_n(z, t) \tag{15.146}$$

$$\boldsymbol{j}(x, y, z, t) = \sum_{n=1}^{N} \boldsymbol{J}_{nx}(x, y)\boldsymbol{b}_n(z, t)\mathbf{i} + \boldsymbol{J}_{ny}(x, y)\boldsymbol{g}_n(z, t)\mathbf{j}$$
$$+ \boldsymbol{J}_{nz}(x, y)\boldsymbol{d}_n(z, t)\mathbf{k} \tag{15.147}$$

with similar expansions for the adjoint flux and current. $\boldsymbol{\psi}_n$ and the $\boldsymbol{J}_n$ are $G \times G$ diagonal matrices with elements given by the known group expansion functions $\psi_n^g(x, y)$ and $J_n^g(x, y)$, while $\boldsymbol{\rho}_n, \boldsymbol{b}_n, \boldsymbol{g}_n$, and $\boldsymbol{d}_n$ are $G \times 1$ column matrices with elements given by the corresponding unknown group expansion coefficients. (Direct and adjoint expansion functions must each be linearly independent, but similar functions may be used for direct and adjoint expansion functions.) Precursor trial functions of the form

$$C_m(x, y, z, t) = \frac{\beta_m}{\lambda_m} \boldsymbol{F}^{\mathrm{T}}(x, y, z, t) \sum_{n=1}^{N} \boldsymbol{\psi}_n(x, y)\boldsymbol{\pi}\, C_{m,n}(z, t) \tag{15.148}$$

$$C_m^*(x, y, z, t) = \boldsymbol{\chi}_m^{\mathrm{T}} \sum_{n=1}^{N} \boldsymbol{\psi}_n^*(x, y)\boldsymbol{\pi}\, C_{m,n}^*(z, t) \tag{15.149}$$

are used, where $\boldsymbol{\pi}$ is a $G \times 1$ column matrix with unit elements (i.e., a sum vector).

When the variational principle J of Eq. (15.143) is required to be stationary with respect to arbitrary variations in the trial functions, which are limited to variations in the expansion coefficients because the expansion functions are fixed, equations that must be satisfied by the expansion coefficients are obtained.

If the trial functions above are used throughout the reactor and at all times, then J_2 and J_3 are identically zero. In this case, equations valid for $0 < z < L$ and $t > t_0$ are obtained from J_1 and that part of J_5 contributed by the vertical (side) external surface.

$$\frac{\delta J}{\delta \boldsymbol{\rho}_{n'}^{*\mathrm{T}}} = 0, \quad n' = 1, \dots, N$$

$$\frac{\delta J}{\delta \boldsymbol{b}_{n'}^{*\mathrm{T}}} = 0, \quad n' = 1, \dots, N$$

$$\frac{\delta J}{\delta \boldsymbol{g}_{n'}^{*\mathrm{T}}} = 0, \quad n' = 1, \dots, N \qquad (15.150)$$

$$\frac{\delta J}{\delta \boldsymbol{d}_{n'}^{*\mathrm{T}}} = 0, \quad n' = 1, \dots, N$$

$$\frac{\delta J}{\delta C_{m,n'}^{*}} = 0, \quad m = 1, \dots, M; \; n' = 1, \dots, N$$

Equations (15.150) can be combined to eliminate $\boldsymbol{b}_n$, $\boldsymbol{g}_n$, and $\boldsymbol{d}_n$, leaving NG scalar equations, which can be written in matrix form as

$$\left(\boldsymbol{M} + \boldsymbol{R} - \frac{\partial}{\partial z} \boldsymbol{A} \frac{\partial}{\partial z}\right)\boldsymbol{\rho} + \boldsymbol{T}\dot{\boldsymbol{\rho}} = \sum_{m=1}^{M} \beta_m \boldsymbol{F}_m \boldsymbol{C}_m + \boldsymbol{S} \qquad (15.151)$$

where $\boldsymbol{A}$, $\boldsymbol{M}$, and $\boldsymbol{R}$ are $NG \times NG$ matrices, and $\boldsymbol{F}_m$ is an $NG \times N$ matrix. $\boldsymbol{R}$ and $\boldsymbol{A}$ are radial and axial leakage matrices resulting from the elimination of $\boldsymbol{b}_n$, $\boldsymbol{g}_n$, and $\boldsymbol{d}_n$. $\boldsymbol{\rho}$ and $\boldsymbol{S}$ are $NG \times 1$ column matrices and $\boldsymbol{C}_m$ is an $N \times 1$ column matrix. Thus G three-dimensional, time-dependent second-order PDEs (the multigroup diffusion equations) are replaced by NG one-dimensional time-dependent second-order PDEs [Eq. (15.151)]. The M three-dimensional, first-order ODEs (precursor equations) are replaced by NM one-dimensional first-order ODEs [last of Eqs. (15.150)].

Boundary conditions at the top ($z = L$) and bottom ($z = 0$) of the model are obtained by requiring stationarity of J with respect to arbitrary variations $\delta \boldsymbol{d}_n^{*\mathrm{T}}$ on the top and bottom surfaces:

$$\frac{\delta J}{\delta \boldsymbol{d}_{n'}^{*\mathrm{T}}(z = 0, L)} = 0, \quad n' = 1, \dots, N \qquad (15.152)$$

Initial conditions are derived by requiring stationarity of J with respect to arbitrary variations $\delta \boldsymbol{\rho}_N^{*\mathrm{T}}$ and $\delta C_{m,n}^{*}$ at $t = t_0$:

$$\frac{\delta J}{\delta \boldsymbol{\rho}_{n'}^{*\mathrm{T}}(t_0)} = 0, \quad n' = 1, \dots, N \qquad (15.153)$$

$$\frac{\delta J}{\delta C_{m,n'}^{*\mathrm{T}}(t_0)} = 0, \quad m = 1, \ldots, M; \ n' = 1, \ldots, N \tag{15.154}$$

A formally identical result could be obtained by deriving the synthesis equations from the second-order variational principle F, the only difference arising in the definition of the elements of the leakage matrices R and A in Eq. (15.151). Under certain restrictive conditions the two formulations are exactly identical.

Two-dimensional static flux solutions for $x-y$ slices through the reactor at various axial locations and/or for various conditions are normally chosen as expansion functions. For some problems different sets of expansion functions are appropriate for different axial regions, and it is convenient to use a discontinuous trial function formulation. In this case a term of the form of J_2 would be included in the variational principle for each $x-y$ planar surface at which the set of expansion functions changed. Equation (15.151) would again obtain within each axial zone, with the coefficients defined in terms of the expansion functions appropriate to that zone. Interface conditions result from the requirement that the variational principle be stationary with respect to arbitrary variations $\delta j^{*\mathrm{T}} \cdot \mathbf{n}$ and $\delta \phi^{*\mathrm{T}}$ on the interface, which results in

$$\int_{t_0}^{t_f} dt \int_{S_{\mathrm{in}}} dx\,dy\,\mathbf{n}\cdot\mathbf{k}\left\{\sum_{n'=1}^{zn}\left(\eta\delta d_{n'+}^{*\mathrm{T}} J_{n'z+}^{*\mathrm{T}} + (1-\eta)\delta d_{n-}^{*\mathrm{T}} J_{nz-}^{*\mathrm{T}}\right)\right.$$
$$\left. \times \sum_{n=1}^{N}(\psi_{n+}\rho_{n+} - \psi_{n-}\rho_{n-})\right\} = 0 \tag{15.155}$$

$$\int_{t_0}^{t_f} dt \int_{S_{\mathrm{in}}} dx\,dy\,\mathbf{n}\cdot\mathbf{k}\left\{\sum_{n'=1}^{N}\left(\gamma\delta\rho_{n'+}^{*\mathrm{T}} \psi_{n'+}^{*\mathrm{T}} + (1-\gamma)\delta\rho_{n'-}^{*\mathrm{T}} \psi_{n'-}^{*\mathrm{T}}\right)\right.$$
$$\left. \times \sum_{n=1}^{N}(J_{nz+}d_{n+} - J_{nz-}d_{n-})\right\} = 0 \tag{15.156}$$

If every $\delta\rho_{n'+}^{*\mathrm{T}}$ and $\delta\rho_{n'-}^{*\mathrm{T}}$ is assumed independent, $2N$ equations relating the Nd_n are obtained, and similarly the assumption that every $\delta d_{n'+}^{*\mathrm{T}}$ and $\delta d_{n'-}^{*\mathrm{T}}$ is independent leads to $2N$ equations relating the $N\rho_n$. Hence the system is overdetermined by a factor of 2. Several stratagems, have evolved for avoiding this difficulty.

By requiring that the flux (direct and adjoint) and current (direct and adjoint) trial functions not be discontinuous at the same interface, the overdetermination problem disappears. In this case $J_{nz+}^{*\mathrm{T}} \equiv J_{nz-}^{*\mathrm{T}}$ and $\delta d_{n+}^{*\mathrm{T}} \equiv \delta d_{n-}^{*\mathrm{T}}$, and so on. This technique of staggering the interfaces at which flux and current expansion functions are changed, which has been widely employed, has the disadvantage that S_{in} frequently corresponds to a physical interface in the reactor and it is desirable to change current and flux expansion functions at the same point. This may be accomplished, for all practical purposes, by allowing the two interfaces at which the current and flux trial functions are discontinuous to approach each other arbitrarily closely. A second strategy is to select γ and $\eta = 0, 1$. This is essentially what is done when a Lagrange multiplier principle is used in deriving the synthesis equations and the Lagrange multipliers are expanded in terms of the flux or current expansion functions on either the $+$ or $-$ side of the interface. Such interface conditions are not symmetric with respect to the arbitrary choice of the $+$ and $-$ sides of the interface. A third strategy consists of requiring that $\delta\rho_{n+}^{*\mathrm{T}} = \delta\rho_{n-}^{*\mathrm{T}}$ and $\delta d_{n+}^{*\mathrm{T}} = \delta d_{n-}^{*\mathrm{T}}$. With $\gamma = \eta = \frac{1}{2}$, the interface conditions are independent of the arbitrary choice of the $+$ and $-$ sides of the interface. Thus all of these stratagems for removing the overdetermination of interface conditions have certain unsatisfactory features.

As mentioned in Section 15.8, overdetermined interface conditions result because of failure to impose the restrictions $\delta\phi_+^* = \delta\phi_-^*$ and $\delta j_+^* = \delta j_-^*$, which must be satisfied if the adjoint flux and current are continuous, upon the trial functions. Although these restrictions cannot be imposed exactly, because the trial functions are partly specified, they can be imposed in an approximate manner to obtain relations among the variations $\delta\rho_{n+}^{*\mathrm{T}}$ and $\delta\rho_{n-}^{*\mathrm{T}}$ among $\delta d_{n+}^{*\mathrm{T}}$ and $\delta d_{n+}^{*\mathrm{T}}$. The requirement $\delta\phi_+^* = \delta\phi_-^*$ becomes

$$\sum_{n=1}^{N} \psi_{n+}^* \delta\rho_{n+}^* = \sum_{n=1}^{N} \psi_{n-}^* \delta\rho_{n-}^*$$

which cannot, in general, be satisfied exactly. However, this relation can be satisfied approximately. Multiplying by an arbitrary diagonal $G \times G$ matrix $\omega_{n'}(x, y)$ and integrating over the surface S_{in} yields one condition relating the $N\delta\rho_{n-}^*$ to the $N\delta\rho_{n+}^*$:

$$\sum_{n=1}^{N} \left(\int_{S_{\mathrm{in}}} dx\, dy\, \omega_{n'} \psi_{n+}^* \right) \delta\rho_{n+}^* = \sum_{n=1}^{N} \left(\int_{S_{\mathrm{in}}} dx\, dy\, \omega_{n'} \psi_{n-}^* \right) \delta p_{n-}^*$$

If this is repeated for N different matrix functions $\omega_{n'}$ the resulting set of equations may be written

$$A_+ \delta\rho_+^* = A_- \delta\rho_-^*$$

where A_+ is an $NG \times NG$ matrix and $\delta\rho_+$ is an $NG \times 1$ column matrix. This equation may be solved:

$$\delta\rho_+^* = (A_+)^{-1} A_- \delta\rho_-^* \equiv Q\delta\rho_-^*$$

The $NG \times NG$ matrix Q may be partitioned into N^2 diagonal $G \times G$ matrices $Q_{n',n}$ in terms of which the equation above may be written

$$\delta\rho_{n+}^* = \sum_{n'=1}^{N} Q_{nn'} \delta\rho_{n'-}^*, \quad n = 1, \ldots, N \tag{15.157}$$

$$\delta d_{n+}^* = \sum_{n'=1}^{N} P_{nn'} \delta d_{n'-}^*, \quad n = 1, \ldots, N \tag{15.158}$$

where $P_{nn'}$ is one of the N^2 diagonal $G \times G$ matrices analogous to the $Q_{nn'}$.

Making use of Eqs. (15.157) and (15.158), Eqs. (15.155) and (15.156) each yield N interface conditions. These interface conditions have several advantages relative to those described previously. The theoretical derivation is consistent and the problem of overdetermination never arises. Flux and current trial function discontinuities are allowed at the same interface. Unfortunately, these interface conditions are not, in general, symmetric with respect to the arbitrary choice of the $+$ and $-$ directions.

When the physical configuration of the reactor changes significantly during a transient, it may be plausible to use different sets of expansion functions over different intervals of time. For this situation a term of the form J_3 would be included in the variational principle for each time interface at which the set of expansion functions are changed. Requiring stationarity of the variational principle again results in Eqs. (15.150) and (15.151) within each time interval, with the coefficients defined in terms of the expansion functions appropriate to the time interval. Similarly, the boundary conditions of Eq. (15.152) and the spatial interface conditions of Eqs. (15.155) and (15.156), with the coefficients defined in terms of the expansion functions appropriate to the time interval, and the initial conditions of Eqs. (15.153) and (15.154) are obtained again. In addition, temporal interface conditions arise from the J_3 term, the variation of which is

$$\sum_{n'=1}^{N} \left\{ \int_{V} d\mathbf{r} \left[a\delta \boldsymbol{\rho}_{n'}^{*\mathrm{T}}(+)\boldsymbol{\psi}_{n'}^{*\mathrm{T}}(+) + (1-a)\delta \boldsymbol{\rho}_{n'}^{*\mathrm{T}}(-)\boldsymbol{\psi}_{n}^{*\mathrm{T}}(-) \right] \boldsymbol{\tau} \right.$$

$$\times \sum_{n=1}^{N} \left[\boldsymbol{\psi}_n(+)\boldsymbol{\rho}_n(+) - \boldsymbol{\psi}_n(-)\boldsymbol{\rho}_n(-) \right]$$

$$+ \boldsymbol{\pi}^{\mathrm{T}} \sum_{m=1}^{M} \int_{V} d\mathbf{r} \left[b\delta C_{m,n'}^{*}(+)\boldsymbol{\psi}_{n'}^{*\mathrm{T}}(+) + (1-b)\delta C_{m,n'}^{*\mathrm{T}}(-)\boldsymbol{\psi}_{n'}^{*\mathrm{T}}(-) \right]$$

$$\left. \times \boldsymbol{\chi}_m \frac{\beta_m}{\lambda_m} \boldsymbol{F}^{\mathrm{T}} \sum_{n=1}^{N} \left[\boldsymbol{\psi}_n(+)\boldsymbol{\pi} C_{m,n}(+) - \boldsymbol{\psi}_n(-)\boldsymbol{\pi} C_{m,n}(-) \right] \right\} = 0 \quad (15.159)$$

The same type of overdetermination, for the same reason [failure to impose the restrictions $\delta\boldsymbol{\phi}^*(+) = \delta\boldsymbol{\phi}^*(-)$ and $\delta C_m^*(+) = \delta C_m^*(-)$], has arisen in connection with the temporal interface conditions. If each variation $\delta\boldsymbol{\rho}_{n'}^{*}(+)$ and $\delta\boldsymbol{\rho}_{n'}^{*}(-)$ is (incorrectly) assumed to be independent, the $2N$ conditions are obtained relating the $N\rho_n(+)$ to the $N\rho_n(-)$, and similarly for the $\delta C_{m,n}^{*}(+) = \delta C_{m,n}^{*}(-)$. The stratagems that have been used to remove this apparent overdetermination parallel those employed for the spatial interface problem. Staggered discontinuities, in which the direct and adjoint flux (and precursor densities) are discontinuous at alternative times have been suggested, and the inconvenience of changing expansion functions at different times has been essentially eliminated by allowing alternative times to approach each other arbitrarily closely. As in the case of the spatial interface, the overdetermination never arises if the restrictions $\delta\boldsymbol{\phi}^*(+) = \delta\boldsymbol{\phi}^*(-)$ and $\delta C_m^*(+) = \delta C_m^*(-)$ are imposed in an approximate manner.

Adjoint synthesis equations may be derived by an analogous development, in this case by requiring stationarity of the functional with respect to the direct expansion coefficients. Similar results are obtained except that final conditions, rather than initial conditions, are obtained. It is necessary to impose approximately the conditions $\delta\boldsymbol{\phi}_{+} = \delta\boldsymbol{\phi}_{-}$, $\delta\boldsymbol{j}_{+} = \delta\boldsymbol{j}_{-}$ to obtain restrictions on the variations in the flux and current expansion coefficients at spatial interfaces, and to impose approximately the conditions $\delta\phi(+) = \delta\phi(-)$ and $\delta C_m(+) = \delta C_m(-)$ at temporal interfaces, to avoid an overdetermination difficulty.

15.10
Multichannel Spatial Synthesis

In Section 15.9 the idea of using different trial functions (i.e., different sets of expansion functions) in different axial regions was discussed. It is also possible to use different trial functions in different planar regions (channels), a procedure referred to as *multichannel synthesis*. This introduces two attractive possibilities. With expansion functions obtained from two-dimensional $(x-y)$ calculations based on a model encompassing the entire cross-sectional area of the reactor, the multichannel feature provides the additional flexibility of allowing different expansion coefficients to be used in different channels. Thus a greater range of planar flux shapes can be synthesized from a given set of expansion functions than is possible with the single-channel synthesis of Section 15.9. A second possibility, which has not been exploited, is the use of expansion functions in each channel obtained from two-dimensional $(x-y)$ calculations based on a model encompassing only the cross-sectional area of the channel.

The basic idea of multichannel synthesis can be illustrated by the simple example shown in Fig. 15.10. Let the radial dimension of the reactor model be divided into two channels, $0 \leq r \leq a/2$ and $a/2 \leq r \leq a$. Then the flux would be constructed by separately mixing ϕ_{rod} and ϕ_{unrod} in each channel:

$$\phi(r,z) = a_{\text{rod}}^1(z)\phi_{\text{rod}}(r) + a_{\text{unrod}}^1(z)\phi_{\text{unrod}}(r), \quad 0 \leq r \leq \frac{a}{2}$$

$$\phi(r,z) = a_{\text{rod}}^2(z)\phi_{\text{rod}}(r) + a_{\text{unrod}}^2(z)\phi_{\text{unrod}}(r), \quad \frac{a}{2} \leq r \leq a$$

The multichannel synthesis equations are derived by using separate trial functions for each channel, denoted by a superscript c, of the form

$$\phi^c(x,y,z,t) = \sum_{n=1}^{N} \psi_n^c(x,y)\rho_n^c(z,t) \tag{15.160}$$

$$\mathbf{j}^c(x,y,z,t) = \sum_{n=1}^{N}\left\{\left[J_{nx}^c(x,y)b_n^c(z,t) + \psi_n^c(x,y)B_n^c(z,t)\right]\mathbf{i}\right.$$

$$+ \left[J_{ny}^c(x,y) g_n^c(z,t) + \psi_n^c(x,y) G_n^c(z,t) \right] \mathbf{j}$$

$$+ J_{nz}^c(x,y) d_n^c(z,t) \mathbf{k} \Big\} \tag{15.161}$$

with similar expansions for the adjoint flux and current. The x and y components of the current are each expanded in two separate types of terms in anticipation of the frequent procedure of using $J_{nx} = -D(\partial \psi_n, \partial x)$, and so on. The second term, proportional to ψ_n, is included both for added flexibility and to ensure the existence of coupling between channels across an interface located where $\partial \psi_n / \partial x$ may be zero.

For the sake of illustration, the channel structure will be taken as concentric annuli, so that the interface terms J_2 are included for each vertical cylindrical surface, S_{in}, that separates channels. Because the derivation of initial conditions and interface conditions for axial and temporal trial function discontinuities, and the inclusion of external boundary terms, are identical to the derivation given in the preceding section, they will be omitted. Thus the multichannel synthesis equations are obtained from consideration of the stationarity properties of $J_{12} = J_1 + J_2$, with J_2 consisting of the terms discussed above. These equations may be written in matrix form as

$$\mathbf{M}^c \boldsymbol{\rho}^c + \mathbf{T}^c \dot{\boldsymbol{\rho}}^c + \mathbf{\Lambda}_x^c \mathbf{b}^c + \mathbf{\Lambda}_{x+}^c \mathbf{b}^{c+1} - \mathbf{\Lambda}_{x-}^c \mathbf{b}^{c-1} + \mathbf{\Gamma}_x^c \mathbf{B}^c + \mathbf{\Gamma}_{x+}^c \mathbf{B}^{c+1}$$

$$- \mathbf{\Gamma}_{x-}^c \mathbf{B}^{c-1} + \mathbf{\Lambda}_y^c \mathbf{g}^c + \mathbf{\Lambda}_{y+}^c \mathbf{g}^{c+1} - \mathbf{\Lambda}_{y-}^c \mathbf{g}^{c-1} + \mathbf{\Gamma}_y^c \mathbf{G} + \mathbf{\Gamma}_{y+}^c \mathbf{G}^{c+1}$$

$$- \mathbf{\Gamma}_{y-}^c \mathbf{G}^{c-1} + \mathbf{A}^c \frac{\partial \mathbf{d}^c}{\partial z} - \sum_{m=1}^{M} \beta_m \mathbf{F}_m^c \mathbf{C}_m^c - \mathbf{S}^c = 0 \tag{15.162}$$

$$\frac{\delta J_{12}}{\delta (\mathbf{b}_{n'}^{c*})^{\mathrm{T}}} = 0 \implies \boldsymbol{\ell}_x^c \mathbf{b}^c + \mathbf{L}_x^c \mathbf{B}^c + \mathbf{K}_x^c \boldsymbol{\rho}^c + \mathbf{K}_{x+}^c \boldsymbol{\rho}^{c+1} - \mathbf{K}_{x-}^c \boldsymbol{\rho}^{c-1} = 0 \tag{15.163}$$

$$\frac{\delta J_{12}}{\delta (\mathbf{g}_{n'}^{c*})^{\mathrm{T}}} = 0 \implies \boldsymbol{\ell}_y^c \mathbf{g}^c + \mathbf{L}_y^c \mathbf{G}^c + \mathbf{K}_y^c \boldsymbol{\rho}^c + \mathbf{K}_{y+}^c \boldsymbol{\rho}^{c+1} - \mathbf{K}_{y-}^c \boldsymbol{\rho}^{c-1} = 0 \tag{15.164}$$

$$\frac{\delta J_{12}}{\delta (\mathbf{B}_{n'}^{c*})^{\mathrm{T}}} = 0 \implies \mathbf{h}_x^c \mathbf{b}^c + \mathbf{H}_x^c \mathbf{B}^c + \mathbf{W}_x^c \boldsymbol{\rho}^c + \mathbf{W}_{x+}^c \boldsymbol{\rho}^{c+1} - \mathbf{W}_{x-}^c \boldsymbol{\rho}^{c-1} = 0 \tag{15.165}$$

$$\frac{\delta J_{12}}{\delta (\mathbf{G}_{n'}^{c*})^{\mathrm{T}}} = 0 \implies \mathbf{h}_y^c \mathbf{g}^c + \mathbf{H}_y^c \mathbf{G}^c + \mathbf{W}_y^c \boldsymbol{\rho}^c + \mathbf{W}_{y+}^c \boldsymbol{\rho}^{c+1} - \mathbf{W}_{y-}^c \boldsymbol{\rho}^{c-1} = 0 \tag{15.166}$$

$$\frac{\delta J_{12}}{\delta (d_{n'}^{c*})^{\mathrm{T}}} = 0 \quad \Longrightarrow \quad U^c d^c + V^c \frac{\partial \rho^c}{\partial z} = 0 \tag{15.167}$$

The matrices and column matrices in Eqs. (15.162) to (15.167) are of order $NG \times G$ and $NG \times 1$, respectively, except for F_m^c and C_m^c, which are $NG \times N$ and $N \times 1$, respectively.

Equations (15.162) to (15.167) may be combined to eliminate the current expansion coefficients, resulting in the set of one-dimensional (z) time-dependent matrix differential equations

$$M^c \rho^c + T^c \dot{\rho}^c + R^c \rho^c + R_+^c \rho^{c+1} + R_{++}^c \rho^{c+2} + R_-^c \rho^{c-1} + R_{--}^c \rho^{c-2}$$

$$- A^c \frac{\partial}{\partial z}\left[(U^c)^{-1} V^c \frac{\partial \rho^c}{\partial z} \right] - \sum_{m=1}^{M} \beta_m F_m^c C_m^c - S^c = 0$$

$$c = 1, \dots, \text{number of channels} \tag{15.168}$$

The matrices R_+ and R_{++} in Eqs. (15.168) result from elimination of the current-combining coefficients and serve to couple channel c radially to channels $c + 1$ and $c + 2$, and similarly, R_- and R_{--} couple channel c to channels $c - 1$ and $c - 2$. This general feature of nearest-neighbor and next-nearest-neighbor coupling is characteristic of the multichannel formulation, independent of the particular choice of channel structure.

Construction of the radial coupling matrices involves the evaluation of surface integrals containing normal derivative terms and considerable matrix inversion and matrix multiplication. The results are sensitive to the accuracy and consistency with which the surface integrals are evaluated, a fact that has hindered exploitation of the multichannel formalism. Moreover, the transport cross section is embedded in the matrices that lead to R_+, and so on, and a change in this quantity requires that the matrix inversions and multiplications involved in the construction of R_+, and so on, be repeated. These factors tend to mitigate the advantages of extra flexibility and increased accuracy inherent in the multichannel formulation. See Refs. 10 and 13 for a more detailed description.

15.11
Spectral Synthesis

In previous sections the emphasis has been on synthesizing the spatial dependence of the neutron flux, and the approximations that were discussed have, in fact, found their greatest application in problems where it was important, but uneconomical, to represent the detailed spatial variation of the flux. Another class of problems exists wherein it is important, but uneconomical, to represent the spectral variation of the flux in great detail. For such problems an attempt to synthesize the detailed spectrum from a few spectral functions is appealing. The general basis of the method is a trial function expansion in each spatial region, or channel c, of the form

$$\boldsymbol{\phi}^c(x, y, z, t) = \sum_{n=1}^{N} \boldsymbol{\psi}_n^c \rho_n^c(x, y, z, t) \tag{15.169}$$

$$\boldsymbol{j}^c(x, y, z, t) = \sum_{n=1}^{N} \boldsymbol{J}_n^c \boldsymbol{b}_n^c(x, y, z, t) \tag{15.170}$$

with similar expansions for the adjoint flux and current. Here $\boldsymbol{\psi}_n$ and $\boldsymbol{J}_n$ are known $G \times 1$ column matrices, and a single expansion coefficient ρ_n applies to all the G group components of the corresponding expansion function ψ_n. Because the objective of the method is to approximate the spectral dependence, it is not necessary to make an expansion of the precursor trial function.

Requiring stationarity of the variational principle with respect to the adjoint expansion coefficients yields the spectral synthesis equations within each channel c:

$$\frac{\delta J_1}{\delta(\rho_{n'}^{c*})} = 0, \quad n' = 1, \ldots, N \tag{15.171}$$

$$\frac{\delta J_1}{\delta(b_{n'}^{c*})} = 0, \quad n' = 1, \ldots, N \tag{15.172}$$

These two sets of N equations may be written as two matrix equations, and combined to eliminate the current combining coefficients,

$$\boldsymbol{M}^c \boldsymbol{\rho}^c - \boldsymbol{\Lambda}^c \nabla \cdot (\boldsymbol{\ell}^c)^{-1} \boldsymbol{K}^c \nabla \boldsymbol{\rho}^c + \boldsymbol{T}^c \dot{\boldsymbol{\rho}}^c - \sum_{m=1}^{M} \beta_m \boldsymbol{F}_m^c C_m - \boldsymbol{S}^c = 0 \tag{15.173}$$

$\boldsymbol{M}^c$, $\boldsymbol{\Lambda}^c$, $\boldsymbol{\ell}^c$, $\boldsymbol{K}^c$, and $\boldsymbol{T}^c$ are $N \times N$ matrices, while $\boldsymbol{\rho}^c$, $\boldsymbol{F}_m^c$ and $\boldsymbol{S}^c$ are $N \times 1$ column matrices.

For each spatial interface between channels a term of the form of J_2 must be added to the variational principle. The variation of such a term leads to

$$\sum_{n'=1}^{N} \sum_{n=1}^{N} n \cdot \left\{ \left[\gamma \delta \boldsymbol{\rho}_{n'+}^* \boldsymbol{\psi}_{n'+}^{*\mathrm{T}} + (1-\gamma) \delta \boldsymbol{\rho}_{n'-}^* \boldsymbol{\psi}_{n'-}^{*\mathrm{T}} \right] (\boldsymbol{J}_{n+} \boldsymbol{b}_{n+} - \boldsymbol{J}_{n-} \boldsymbol{b}_{n-}) \right.$$

$$\left. + \left[\eta \delta \boldsymbol{b}_{n'+}^* \boldsymbol{J}_{n'+}^{*\mathrm{T}} + (1-\eta) \delta \boldsymbol{b}_{n'-}^{*\mathrm{T}} \boldsymbol{J}_{n'-}^{*\mathrm{T}} \right] (\boldsymbol{\psi}_{n+} \rho_{n+} - \boldsymbol{\psi}_{n-} \rho_{n-}) \right\} = 0 \tag{15.174}$$

As was the case with the spatial synthesis, it is necessary to impose some form of restriction among the allowable variations $\delta \rho_{n+}^*$ and $\delta \rho_{n-}^*$, and $\delta \mathbf{b}_{n+}^*$ and $\delta \mathbf{b}_{n-}^*$, or to resort to some stratagem such as staggering the interfaces at which flux and current may be discontinuous or to set $\gamma, \eta = 1, 0$; otherwise, the interface conditions appear to be overdetermined. A restriction among the variations arises naturally from the requirement that variations $\delta \boldsymbol{\phi}_+^*$ and $\delta \boldsymbol{\phi}_-^*$ be equal to ensure continuity, and similarly, that $\delta j^*_+ = \delta j^*_-$. This requirement must be imposed in an approximate manner (unless $N = G$, in which case there is no advantage whatsoever to using spectral synthesis), and leads to

$$\delta\rho^*_{n+} = \sum_{n'=1}^{N} Q_{nn'}\delta\rho^*_{n'-}, \qquad \delta\mathbf{b}^*_{n+}\cdot\mathbf{n} = \sum_{n'=1}^{N} P_{nn'}\delta\mathbf{b}^*_{n-}\cdot\mathbf{n}$$

Using these relations to eliminate $\delta\rho^*_{n+}$ and $\delta\mathbf{b}^*_{n+}$, Eq. (15.174) can then be required to be satisfied for arbitrary and independent variations $\delta\rho^*_{n-}$ and $\delta\mathbf{b}^*_{n-}$, which yields the proper number of interface conditions.

The other strategies that have been suggested may be considered as special cases with $\gamma, \eta = 0, 1$ and/or $P_{nn'} = Q_{nn'} = \delta_{nn'}$. Thus, in general, the spatial interface conditions may be written in the form

$$\sum_{n=1}^{N}\mathbf{n}\cdot\left\{\left[A^{\mathrm{T}}_{n'}J_{n+}\right]\mathbf{b}_{n+} - \left[A^{\mathrm{T}}_{n'}J_{n-}\right]\mathbf{b}_{n-}\right\} = 0, \quad n' = 1,\ldots,N \qquad (15.175)$$

$$\sum_{n=1}^{N}\left\{\left[B^{\mathrm{T}}_{n'}\psi_{n+}\right]\rho_{n+} - \left[B^{\mathrm{T}}_{n'}\psi_{n-}\right]\rho_{n-}\right\} = 0, \qquad n' = 1,\ldots,N \qquad (15.176)$$

Note that unless

$$\left[A^{\mathrm{T}}_{n'}J_{n+}\right] = \left[A^{\mathrm{T}}_{n'}J_{n-}\right], \quad n' = 1,\ldots,N \qquad (15.177)$$

$$\left[B^{\mathrm{T}}_{n'}\psi_{n+}\right] = \left[B^{\mathrm{T}}_{n'}\psi_{n-}\right], \quad n' = 1,\ldots,N \qquad (15.178)$$

Eqs. (15.175) and (15.176) do not reduce to continuity requirements of the form

$$\mathbf{n}\cdot(\mathbf{b}_{n+} - \mathbf{b}_{n-}) = 0$$

$$\rho_{n+} - \rho_n = 0$$

The conventional few-group approximation is a special case of the spectral synthesis approximation in which an expansion function ψ_n or J_n has nonzero elements only for those groups that are to be collapsed into few-group n. Thus the result above indicates that continuity of few-group flux and normal current is generally not the proper interface condition, obtaining only under the special circumstances whereby Eqs. (15.177) and (15.178) are satisfied.

If different sets of spectral expansion functions are used in different time intervals, it is necessary to include terms of the form J_3, the stationarity of which yield temporal continuity conditions on the flux expansion coefficients at the time when the expansion functions are changed. To avoid an apparent overdetermination of continuity conditions, it is necessary either to resort to some stratagem such as requiring that the adjoint and direct flux expansion functions change at different times or setting $a = 0, 1$, or to impose in an approximate fashion the continuity condition $\delta\phi^*(+) = \delta\phi^*(-)$ to relate $\delta\rho^*_n(+)$ and $\delta\rho^*_n(-)$.

The continuity conditions resulting from this or other derivations can be written in the form

$$\sum_{n=1}^{N}\left\{\left[\boldsymbol{D}_{n'}^{\mathrm{T}}\boldsymbol{\psi}_n(+)\right]\rho_n(+) - \left[\boldsymbol{D}_{n'}^{\mathrm{T}}\boldsymbol{\psi}_n(-)\right]\rho_n(-)\right\}, \quad n' = 1, \ldots, N \qquad (15.179)$$

where $\boldsymbol{D}_n$ is an $N \times 1$ column matrix. Thus, in general, $\rho_n(+) = \rho_n(-)$ is not the continuity condition. Consequently, recalling that a few-group approximation is a special case of the spectral synthesis approximation, continuity of few-group fluxes at times when the expansion functions (within-group fine-group fluxes) changes is generally not the proper continuity condition and obtains only when

$$\boldsymbol{D}_{n'}^{\mathrm{T}}\boldsymbol{\psi}_n(+) = \boldsymbol{D}_{n'}^{\mathrm{T}}\boldsymbol{\psi}_n(-), \quad n' = 1, \ldots, N$$

The synthesis approximations lack, in general, the positivity properties associated with the multigroup diffusion equations. A consequence of this is that there is no a priori assurance that the fundamental eigenvalue (the one associated with an everywhere nonnegative flux solution) is larger in absolute value than any of the harmonic eigenvalues. Most numerical iteration schemes used in the solution of the synthesis equations converge to the eigenvalue, and corresponding eigenfunction, with the largest magnitude, but it is possible that a calculation will not converge to the fundamental solution.

References

1 J. A. FAVORITE and W. M. STACEY, "A Variational Synthesis Nodal Discrete-Ordinates Method," *Nucl. Sci. Eng.* 132, 181 (**1999**).

2 T. M. SUTTON and B. N. AVILES, "Diffusion Theory Methods for Spatial Kinetics Calculations," *Prog. Nucl. Energy* 30, 119 (**1996**).

3 R. T. ACKROYD et al., "Foundations of Finite Element Applications to Neutron Transport," *Prog. Nucl. Energy* 29, 43 (**1995**); "Some Recent Developments in Finite Element Methods for Neutron Transport," *Adv. Nucl. Sci. Technol.* 19, 381 (**1987**).

4 R. D. LAWRENCE, "Progress in Nodal Methods for the Solution of the Neutron Diffusion and Transport Equations," *Prog. Nucl. Energy* 17, 271 (**1986**); "Three-Dimensional Nodal Diffusion and Transport Methods for the Analysis of Fast-Reactor Critical

Experiments," *Prog. Nucl. Energy 18,* 101 (**1986**).

5 J. J. STAMM'LER and M. J. ABBATE, *Methods of Steady-State Reactor Physics in Nuclear Design,* Academic Press, London (**1983**), Chap. XI.

6 N. K. GUPTA, "Nodal Methods for Three-Dimensional Simulators," *Prog. Nucl. Energy* 7, 127 (**1981**).

7 J. J. DORNING, "Modern Coarse-Mesh Methods: A Development of the 70's," *Proc. Conf. Computational Methods in Nuclear Engineering,* Williamsburg, VA, American Nuclear Society, La Grange Park, IL (**1979**), p. 3-1.

8 M. R. WAGNER, "Current Trends in Multidimensional Static Reactor Calculations," *Proc. Conf. Computational Methods in Nuclear Engineering,* Charleston, SC, CONF-750413, American Nuclear Society, La Grange Park, IL (**1975**), p. I-1.

9 A. F. Henry, *Nuclear-Reactor Analysis*, MIT Press, Cambridge, MA (1975), Chap. 11; "Refinements in Accuracy of Coarse-Mesh Finite-Difference Solutions of the Group-Diffusion Equations," *Proc. Semin. Numerical Reactor Calculations*, International Atomic Energy Agency, Vienna (**1972**), p. 447.

10 W. M. Stacey, "Flux Synthesis Methods in Reactor Physics," *Reactor Technol. 15*, 210 (1972); "Variational Flux Synthesis Methods for Multigroup Neutron Diffusion Theory," *Nucl. Sci. Eng. 47*, 449 (**1972**); "Variational Flux Synthesis Approximations," *Proc. IAEA Semin. Numerical Reactor Calculations*, International Atomic Energy Agency, Vienna (1972), p. 561; *Variational Methods in Nuclear Reactor Physics*, Academic Press, New York (**1974**), Chap. 4.

11 R. Froehlich, "A Theoretical Foundation for Coarse Mesh Variational Techniques," *Proc. Int. Conf. Research on Reactor Utilization and Reactor Computation*, Mexico, D. F., CNM-R-2 (**1967**), p. 219.

12 S. Kaplan, "Synthesis Methods in Reactor Analysis," *Adv. Nucl. Sci. Technol. 3* (**1966**); "Some New Methods of Flux Synthesis," *Nucl. Sci. Eng. 13*, 22 (**1962**).

13 E. L. Wachspress et al., "Multichannel Flux Synthesis," *Nucl. Sci. Eng. 12*, 381 (**1962**); "Variational Synthesis with Discontinuous Trial Functions," *Proc. Conf. Applications of Computational Methods to Reactor Problems*, USAEC report ANL-7050, Argonne National Laboratory, Argonne, IL (**1965**), p. 191; "Variational Multichannel Synthesis with Discontinuous Trial Functions," USAEC report KAPL-3095, Knolls Atomic Power Laboratory, Schenectady, NY (**1965**).

Problems

15.1 Derive the nodal fission rate balance equations of Eq. (15.17) from the nodal flux balance equations of Eq. (15.16).

15.2 Use the rational approximation for the escape probability to calculate the coupling terms $W^{n,n+1}$ for cubic nodes.

15.3 Consider a slab reactor consisting of two core regions each 50 cm thick described by the parameters given for core 1 and core 2 in Table 15.1, with zero flux conditions on both external boundaries. Solve for the exact solution in two-group diffusion theory.

15.4 Construct a two-node conventional nodal model for the slab reactor of Problem 15.1. Solve for the multiplication constant and compare with the exact result of Problem 15.3.

15.5 Derive the transverse integrated nodal diffusion equations given by Eq. (15.31) and similar equations in the y- and z-directions.

15.6 Construct a two-node transverse integrated model for the slab reactor of Problem 15.3. Solve for the multiplication constant and compare with the exact result.

15.7 Derive the elements of the matrices P_g^n and R_g^n in the interface current balance of Eq. (15.58) for nodal diffusion theory.

Table 15.1

Group	Core 1		Core 2	
Constant	Group 1	Group 2	Group 1	Group 2
χ	1.0	0.0	1.0	0.0
$\nu\Sigma_f$ (cm^{-1})	0.0085	0.1851	0.006	0.150
Σ_a (cm^{-1})	0.0121	0.121	0.010	0.100
$\Sigma_s^{1\rightarrow2}$ (cm^{-1})	0.0241	–	0.016	–
D (cm)	1.267	0.354	1.280	0.400

15.8 Derive the nodal balance Eqs. (15.44) directly by integrating the transport equation (15.1) for each group over the node.

15.9 Derive the elements of the matrices $\boldsymbol{P}_g^n$ and $\boldsymbol{R}_g^n$ in the interface current balance of Eq. (15.86) for nodal GP_1 transport theory.

15.10 Construct a two-coarse-mesh finite-element model for the slab reactor of Problem 15.3. Solve for the multiplication constant and compare with the exact result.

15.11 Prove that the two forms of the variational functional F_d of Eq. (15.107) are equivalent in that the stationarity of both forms with respect to arbitrary and independent variations requires that the diffusion equation is satisfied within the volumes V_i and that the diffusion theory current is continuous across the surfaces separating adjacent volumes.

15.12 Derive a finite-element coarse-mesh approximation, based on a quadratic polynomial expansion, for the one-dimensional one-group diffusion equation.

15.13 Carry through the derivation to prove that stationarity of the variational functional of Eq. (15.143) with respect to arbitrary and independent variations in ϕ^*, $\boldsymbol{j}^*$, and C_m^* requires that the stationary functions ϕ, $\boldsymbol{j}$, and C_m satisfy the time-dependent transport equation, Fick's law relation, and precursor balance equation.

15.14 Derive the time-dependent equations for ϕ^*, $\boldsymbol{j}^*$, and C_m^* by requiring stationarity of the variational function of Eq. (15.143) with respect arbitrary and independent variations in ϕ, $\boldsymbol{j}$, and C_m.

15.15 Construct a single-channel synthesis model for the slab reactor of Problem 15.3, but in one-group diffusion theory. Obtain the one-group constants by using the two-group constants of Problem 15.3 in an infinite-medium spectrum calculation for ϕ_1 and ϕ_2, which can be used to construct effective one-group cross sections. Using the trial function $\phi(x) = a\cos(\pi x/100)$ for the flux and adjoint flux, calculate

the multiplication constant and compare with the exact
result of Problem 15.3.

15.15 Repeat Problem 15.15 using a two-channel synthesis model.

16
Space–Time Neutron Kinetics

The discussion of reactor dynamics in Chapter 5 was based on the implicit assumption that the spatial neutron distribution remained fixed and only the total neutron population changed in time. However, when a critical reactor is perturbed locally, the spatial neutron flux distribution, as well as the total neutron population, will change, and the change in the spatial flux distribution will affect the change in the total neutron population. A very local perturbation (e.g., the withdrawal of a control rod) will obviously affect the neutron flux in the immediate vicinity of the perturbation. However, a local or regional perturbation can also affect the global neutron flux distribution (i.e., produce a flux tilt), which will, in turn, alter the reactivity and affect the global neutron population. Moreover, for a transient below prompt critical, the largest part of the neutron source is due to the decay of delayed neutron precursors, which tends to hold back a flux tilt until the delayed neutron precursor distribution also tilts. The point kinetics equations discussed in Chapter 5 can be extended to treat flux tilts and delayed neutron holdback effects by recomputing the point kinetic parameters during the course of a transient. The various methods that have been discussed for calculating the spatial distribution of the neutron flux can also be extended to calculate the space- and time-varying neutron flux distribution by adding neutron density time derivative and delayed neutron precursor source terms and appending a set of equations to calculate local delayed neutron precursor densities. The methods of stability analysis and control can also be extended to include spatial dependence, as illustrated by an analysis of xenon spatial oscillations.

16.1
Flux Tilts and Delayed Neutron Holdback

Physical insight into the flux tilting and delayed neutron holdback phenomena can be obtained by considering a step local perturbation in the material composition of an initially critical reactor. In multigroup diffusion theory, the initial critical state of the reactor is described by

Nuclear Reactor Physics. Weston M. Stacey
Copyright © 2007 WILEY-VCH Verlag GmbH & Co. KGaA, Weinheim
ISBN: 978-3-527-40679-1

$$-\nabla \cdot D^g(\mathbf{r}, t)\nabla\phi_g(\mathbf{r}, t) + \Sigma_t^g(\mathbf{r}, t)\phi_g(\mathbf{r}, t) - \sum_{g'=1}^{G}\Sigma^{g' \to g}(\mathbf{r}, t)\phi_{g'}(\mathbf{r}, t)$$

$$= \chi^g \sum_{g'=1}^{G} \nu\Sigma_f^{g'}(\mathbf{r}, t), \quad g = 1, \ldots, G \tag{16.1}$$

which will be written in operator notation as

$$A_0\phi_0 = M_0\phi_0 \tag{16.2}$$

where the zero subscript is used to indicate the initial critical state.

Now we consider a spatially nonuniform change in materials properties which is represented by the changes ΔA in the destruction operator and ΔM in the fission operator so that $A_0 \to A = A_0 + \Delta A$ and $M_0 \to M = M_0 + \Delta M$. For changes producing reactivities well below prompt critical, the prompt jump approximation may be used to describe the neutron kinetics. Making the further approximation of a single delayed neutron precursor group, the neutron kinetics is described by

$$0 = (-A + \{1 - \beta\}M)\phi + \lambda C \tag{16.3}$$

$$\dot{C} = \beta M\phi - \lambda C \tag{16.4}$$

Expanding about the initial critical distributions

$$\phi(r, t) = \phi_0(r) + \Delta\phi(r, t) \tag{16.5}$$

$$C(r, t) = C_0(r) + \Delta C(r, t) \equiv \frac{\beta M_0\phi_0}{\lambda} + \Delta C \tag{16.6}$$

linearizing (i.e., ignoring quadratic terms $\Delta M \Delta\phi$, etc.), Laplace transforming, and combining the two equations results in an equation for the time dependence of the neutron flux $\Delta\phi$ in the frequency domain:

$$0 = \left[-A_0 + \left(1 - \frac{s\beta}{s + \lambda}\right)M_0\right]\Delta\tilde{\phi}(\mathbf{r}, s) + \frac{1}{s}\left[-\Delta A + \left(1 - \frac{s\beta}{s + \lambda}\right)\Delta M\right]\phi_0 \tag{16.7}$$

Modal Eigenfunction Expansion

We now expand the time-dependent flux,

$$\Delta\phi(r, t) = \sum_{n=0} a_n(t)\psi_n(r) \tag{16.8}$$

where the ψ_n are the spatial eigenfunctions of the initial critical reactor and satisfy

$$A_0\psi_n = \frac{1}{k_n}M_0\psi_n \tag{16.9}$$

[e.g., in a uniform slab reactor of width a, $\psi_n = \sin(n\pi x/a)$]. The corresponding adjoint eigenfunctions of the initial critical reactor are defined by

$$A_0^*\psi_n^* = \frac{1}{k_n}M_0^*\psi_n^* \tag{16.10}$$

From the definition of the adjoint operator discussed in Chapter 13, the orthogonality property

$$\langle \psi_m^*, M_0 \psi_n \rangle = \delta_{mn} \qquad (16.11)$$

the relationship

$$\langle \psi_m^*, A_0 \psi_n \rangle = \frac{1}{k_n} \langle \psi_m^*, M_0, \psi_n \rangle \qquad (16.12)$$

can be established, where $\langle XX \rangle$ indicates integration over space and summation over groups.

Using the eigenfunction expansion of Eq. (16.8) in Eq. (16.7), multiplying the resulting equation by ψ_m^*, integrating over space and summing over groups, and using Eqs. (16.11) and (16.12) yields

$$\tilde{a}_m(s) = \frac{(s+\lambda)\langle \psi_m^*, (-\Delta A + (1 - \frac{s\beta}{s+\lambda})\Delta M)\phi_0 \rangle}{s[(s+\lambda)(\frac{1-k_m}{1-(1-\beta)k_m})](\frac{1-(1-\beta)k_m}{k_m})\langle \psi_m^*, M_0\psi_m \rangle} \qquad (16.13)$$

which may be inverse Laplace transformed to obtain

$$a_m(t) = \frac{\rho_m k_m}{1-k_m}\left\{1 - \frac{\beta k_m}{1-(1-\beta)k_m}\exp\left[\frac{-\lambda(1-k_m)t}{1-(1-\beta)k_m}\right]\right\}$$
$$- \frac{\beta k_m \langle \psi_m^*, \Delta M \phi_0 \rangle}{[1-(1-\beta)k_m]\langle \psi_m^*, M_0\psi_m \rangle}\exp\left[\frac{-\lambda(1-k_m)t}{1-(1-\beta)k_m}\right] \qquad (16.14)$$

where

$$\rho_m \equiv \frac{\langle \psi_{m'}^*(-\Delta A + \Delta M)\phi_0 \rangle}{\langle \psi_m^*, M_0\psi_m \rangle} \qquad (16.15)$$

is the mth-mode reactivity.

Flux Tilts

If $\rho_m \neq 0$, a nonuniform perturbation in materials properties in a critical reactor will introduce higher harmonic eigenfunctions into the flux distribution, which becomes after the transient terms in Eq. (16.14) have died out

$$\phi(\mathbf{r}, \infty) = \left[1 + a_0(\infty)\right]\phi_0(r) + \sum_{n=1} \frac{\rho_n k_n}{1-k_n}\psi_n(r) \qquad (16.16)$$

For a uniform slab reactor in $1\frac{1}{2}$-group diffusion theory, the results of Chapter 3 can be used to write the nth-mode eigenvalue:

$$k_n = \frac{k_\infty}{1 + M^2 B_n^2} = \frac{k_\infty}{1 + M^2[(n+1)\pi/a]^2} = \frac{1 + M^2(\pi/a)^2}{1 + M^2[(n+1)\pi/a]^2} \qquad (16.17)$$

where M^2 is now the migration area and we have taken advantage of the fact that $k_0 = 1$ to write the last form of the equation.

The amplitude of the first harmonic eigenfunction, which would be the main component of a flux tilt, depends on the magnitude of the first harmonic reactivity, ρ_1, and on the first harmonic eigenvalue separation, $1 - k_1$ (note that $k_0 = 1$). Using Eq. (16.17), the $1\frac{1}{2}$-group diffusion theory estimate for the first harmonic eigenvalue separation of a uniform slab reactor is

$$1 - k_1 = \frac{3(M\pi/a)^2}{1 + (2M\pi/a)^2} \approx 3\left(\frac{M\pi}{a}\right)^2 \tag{16.18}$$

Thus reactors that are very large in units of migration length ($a/M \gg 1$) will have a small first harmonic eigenvalue separation and will be very "tilty."

Delayed Neutron Holdback

As indicated by Eq. (16.14), a tilt will not occur instantaneously upon the introduction of a nonuniform step change in materials properties into a critical reactor, but will gradually build in over a time $t \approx 2$ to $3\tau_{\text{tilt}}$, where

$$\tau_{\text{tilt}} = \frac{1 - (1 - \beta)k_1}{\lambda(1 - k_1)} > \lambda^{-1} \tag{16.19}$$

Physically, the prompt neutrons respond essentially instantaneously (on the neutron lifetime scale) to the change in materials properties, but the delayed neutron source only gradually changes from the initial fundamental mode distribution into the asymptotic distribution.

16.2
Spatially Dependent Point Kinetics

The multigroup diffusion theory approximation for the space and time dependence of the neutron flux within a nuclear reactor is described by the set of G equations

$$\frac{1}{v^g}\frac{\partial \phi^g(\mathbf{r}, t)}{\partial t} = \nabla \cdot D^g(\mathbf{r}, t)\nabla \phi^g(\mathbf{r}, t) - \Sigma_t^g(\mathbf{r}, t)\phi^g(\mathbf{r}, t)$$

$$+ \sum_{g'=1}^{G} \Sigma^{g' \to g}(\mathbf{r}, t)\phi^{g'}(\mathbf{r}, t)$$

$$+ \lambda_0(1 - \beta)\chi_p^g \sum_{g'=1}^{G} \nu\Sigma_f^{g'}(\mathbf{r}, t)\phi^{g'}(\mathbf{r}, t)$$

$$+ \sum_{m=1}^{M} \lambda_m \chi_m^g C_m(\mathbf{r}, t), \quad g = 1, \ldots, G \tag{16.20a}$$

which for notational convenience we shall write in operator notation as

$$\frac{1}{v}\frac{\partial \phi(\mathbf{r}, t)}{\partial t} = -A(\mathbf{r}, t)\phi(\mathbf{r}, t) + \lambda_0(1 - \beta)F_p(\mathbf{r}, t)\phi(\mathbf{r}, t)$$

$$+ \sum_{m=1}^{M} \lambda_m C_m(\mathbf{r}, t) \tag{16.20b}$$

The space and time dependence of the M groups of delayed neutron precursors are described by

$$\frac{\partial C_m(\mathbf{r}, t)}{\partial t} = \beta_m \sum_{g=1}^{G} \nu \Sigma_f^g(\mathbf{r}, t) \phi^g(\mathbf{r}, t) - \lambda_m C_m(\mathbf{r}, t) \qquad (16.21a)$$

which in operator notation becomes

$$\frac{\partial C_m(\mathbf{r}, t)}{\partial t} = \lambda_0 \beta_m F(\mathbf{r}, t) \phi(\mathbf{r}, t) - \lambda_m C_m(\mathbf{r}, t), \quad m = 1, \ldots, M \qquad (16.21b)$$

where

$A, F =$ loss and production operator, respectively
$\phi(\mathbf{r}, t) =$ neutron flux
$C_m(\mathbf{r}, t) =$ precursor density of type m
$\nu =$ neutron speed
$\chi_m, \lambda_m, \beta_m =$ fission spectrum, decay constant, and delayed neutron fraction, respectively, for precursor type m
$F_p \equiv \chi_p F =$ fission source for prompt neutrons (χ_p is the fission spectrum for prompt neutrons; $F_m \equiv \chi_m F$ will be the fission source for delayed neutrons from precursor group m in subsequent equations)
$\lambda_0 =$ eigenvalue adjusted to render the system critical at time $t = 0$.

In the multigroup form of Eqs. (16.20a) and (16.21a), $\phi(\mathbf{r}, t)$ represents a column vector of group fluxes, and A and F are matrices.

For the initial, static configuration, these equations reduce to

$$(A_0 - \lambda_0 F_0)\phi_0 = 0 \qquad (16.22)$$

For the perturbed static configuration (i.e., after the delayed neutrons reach equilibrium), these equations reduce to

$$(A_e - \lambda_e F_e)\phi_e = 0 \qquad (16.23)$$

and the quantity $-\Delta\lambda = \lambda_0 - \lambda_e \equiv (k_e - k_0)/k_e k_0$ is called the *static reactivity worth* of the perturbation ($k = \lambda^{-1}$). [Note that because Eq. (16.23) is an eigenvalue problem, the word *static* here refers only to the flux distribution, not the amplitude.] The static reactivity worth of the perturbation is

$$\rho_e \equiv -\Delta\lambda = \frac{\langle \phi_0^*, (\lambda_0 \Delta F - \Delta A)\phi_e \rangle}{\langle \phi_0^*, F\phi_e \rangle} \qquad (16.24)$$

where the static flux adjoint function ϕ_0^* satisfies

$$\left(A_0^* - \lambda_0 F_0^*\right)\phi_0^* = 0 \qquad (16.25)$$

(The inner product notation $\langle , \rangle$ indicates an integration over volume and a sum over energy groups.)

Derivation of Point Kinetics Equations

The exact space–time equations are reduced to the point reactor kinetics model by writing the flux as a product of a shape function and an amplitude function; that is,

$$\phi(\mathbf{r}, t) = \psi(\mathbf{r}, t) n(t) \tag{16.26}$$

The point kinetics equations are derived by weighting Eqs. (16.20) and (16.21) with the static adjoint flux and integrating over volume and summing over energy:

$$\dot{n}(t) = \frac{\rho(t) - \bar{\beta}(t)}{\Lambda(t)} n(t) + \sum_{m=1}^{M} \lambda_m P_m(t) \tag{16.27}$$

$$\dot{P}_m(t) = \frac{\beta_m \gamma_m(t)}{\Lambda(t)} n(t) - \lambda_m P_m(t), \quad m = 1, \ldots, M \tag{16.28}$$

where the dynamic reactivity, prompt neutron generation time, and delayed neutron effectiveness are defined as

$$\rho(t) = \frac{\langle \phi_0^*, (\lambda_0 \Delta F - \Delta A) \psi(\mathbf{r}, t) \rangle}{\langle \phi_0^*, F \psi(\mathbf{r}, t) \rangle}$$

$$= \int d\mathbf{r} \sum_{g=1}^{G} \phi_0^{g^*}(\mathbf{r}) \chi_p^g \left[\left\{ \lambda_0 \sum_{g'=1}^{G} \Delta \left(\nu \Sigma_f^{g'}(\mathbf{r}, t) \right) \psi^{g'}(\mathbf{r}, t) \right\} \right.$$

$$- \left\{ \nabla \cdot \Delta D^g(\mathbf{r}, t) \nabla \psi^g(\mathbf{r}, t) + \Delta \Sigma_t^g(\mathbf{r}, t) \psi^g(\mathbf{r}, t) \right.$$

$$\left. \left. - \sum_{g'=1}^{G} \nabla \Sigma^{g' \to g}(\mathbf{r}, t) \psi^{g'}(\mathbf{r}, t) \right\} \right] \Bigg/$$

$$\int d\mathbf{r} \sum_{g=1}^{G} \phi_0^{g^*}(\mathbf{r}) \chi_p^g \sum_{g'=1}^{G} \nu \Sigma_f^{g'}(\mathbf{r}, t) \psi^{g'}(\mathbf{r}, t) \tag{16.29}$$

$$\Lambda^{-1}(t) = \frac{\langle \phi_0^*, F \psi(\mathbf{r}, t) \rangle}{\langle \phi_0^*, \mathbf{v}^{-1} \psi(\mathbf{r}, t) \rangle}$$

$$= \frac{\int d\mathbf{r} \sum_{g=1}^{G} \phi_0^{g^*}(\mathbf{r}) \chi_p^g \sum_{g'=1}^{G} \nu \Sigma_f^{g'}(\mathbf{r}, t) \psi^{g'}(\mathbf{r}, t)}{\int d\mathbf{r} \sum_{g=1}^{G} \phi_0^{g^*}(\mathbf{r})(1/v^g) \psi^g(\mathbf{r}, t)} \tag{16.30}$$

and

$$\gamma_m(t) = \frac{\langle \phi_0^*, F_m \psi(\mathbf{r}, t) \rangle}{\langle \phi_0^*, F \psi(\mathbf{r}, t) \rangle} = \frac{\int d\mathbf{r} \sum_{g=1}^{G} \phi_0^{g^*}(\mathbf{r}) \chi_m^g \sum_{g'=1}^{G} \nu \Sigma_f^{g'}(\mathbf{r}, t) \psi^{g'}(\mathbf{r}, t)}{\int d\mathbf{r} \sum_{g=1}^{G} \phi_0^{g^*}(\mathbf{r}) \chi_p^g \sum_{g'=1}^{G} \nu \Sigma_f^{g'}(\mathbf{r}, t) \psi^{g'}(\mathbf{r}, t)} \tag{16.31}$$

respectively ($\bar{\beta} = \gamma_1\beta_1 + \cdots + \gamma_M\beta_M$, $\Delta A = A - A_0$, and $\Delta F = F - F_0$). In principle, the point kinetics equations can be used to calculate the exact space–time neutron flux, if the correct spatial flux shape is used at all times to evaluate the parameters defined by Eqs. (16.29) to (16.31). Note that these parameters do not depend on the amplitude of the flux, only the flux distribution.

In a large LWR core, the flux is slow to reach equilibrium in its perturbed static distribution, due to the holdback effect of the delayed neutrons. Thus, for the first few seconds after a perturbation, the time-dependent flux shape $\psi(\mathbf{r}, t)$ differs from the static perturbed flux shape ϕ_e, and the dynamic reactivity of Eq. (16.29) differs from the static reactivity of Eq. (16.24).

In the standard implementation of the point kinetics method, the parameters are estimated using the initial static flux distribution ϕ_0. This approximation corresponds to first-order perturbation theory, and for the reactivity, it is denoted

$$\rho_0 = \frac{\langle \phi_0^*, (\lambda_0 \Delta F - \Delta A)\phi_0 \rangle}{\langle \phi_0^*, F\phi_0 \rangle} \tag{16.32}$$

This expression can be shown (Chapter 13) to be a first-order approximation of the static reactivity [i.e., ρ_0 is an estimate of the difference ($-\Delta\lambda = \lambda_0 - \lambda_e$) of reciprocal eigenvalues for the initial and perturbed core static configurations that is accurate to first order in the flux perturbation $\Delta\phi = \phi_e - \phi_0$ (i.e., error $\sim \Delta\phi$)].

Adiabatic and Quasistatic Methods

If parameters of Eqs. (16.29) to (16.31) calculated with the initial spatial flux shape are used throughout the transient calculation, the result is the standard point kinetics approximation of Chapter 5. If the parameters are recomputed at selected times during the transient, using a static neutron flux solution corresponding to the instantaneous conditions of the reactor, the result is an improvement to the standard point kinetics known as the *adiabatic method*.

In the quasistatic (QS) method, the point-kinetics equations are used for the flux amplitude, but the flux shape is recomputed (at time steps $t = t_n$) using

$$\left[A - \lambda_0(1 - \beta)F_p + \frac{1}{v}\left(\frac{\dot{n}}{n} + \frac{1}{\Delta t_n}\right) \right]_{t_n} S_n$$

$$= \frac{1}{v\Delta t_n}S_{n-1} + \frac{1}{n(t_n)}\sum_{m=1}^{M}\chi_m\lambda_m C_m(\mathbf{r}, t_n) \tag{16.33}$$

where $\Delta t_n = t_n - t_{n-1}$ is the shape time step. (The precursor density is computed directly from the flux history.) When the flux shape from the nth such recalculation S_n is used directly to estimate the reactivity using the inner-product definition [Eq. (16.29)], the result is

$$\rho_n(t) \equiv \frac{\langle \phi_0^*, (\lambda_0 \Delta F - \Delta A)S_n \rangle}{\langle \phi_0^*, FS_n \rangle} \tag{16.34}$$

a potentially accurate estimate of the dynamic reactivity, depending, of course, on the accuracy of the flux calculation. It is more accurate to use a flux shape interpolated from the most recent known shape S_{n-1} and the best guess for the next shape S_n^ℓ, where ℓ represents the most recent (the ℓth) calculation of S_n using Eq. (16.33). (The S_n^ℓ are considered converged when the last one satisfies a normalization constraint.) Regardless of the approximate shape that is used, $\rho_n(t)$ is a first-order estimate of the static reactivity corresponding to the reactor conditions at time t, and it will be referred to as such.

Variational Principle for Static Reactivity

A variational estimate, accurate to second order [error $\sim (\Delta\phi)^2$], for the static reactivity worth of a perturbation to an altered system (i.e., a system other than the one for which ϕ_0 and ϕ_0^* were calculated) is

$$\rho_{v,e} = \frac{\langle \phi_0^*, (\lambda\,\Delta F - \Delta A)S \rangle}{\langle \phi_0^*, F'S \rangle} \times \left[1 - \langle \phi_0^*, (A - \lambda F)\Gamma \rangle - \langle \Gamma^*, (A' - \lambda'F')S \rangle \right]$$

(16.35)

where the generalized adjoint function Γ^* is calculated using

$$\left(A_0^* - \lambda_0 F_0^* \right)\Gamma^* = \frac{(\Delta A^* - \lambda_0 \Delta F^*)\phi_0^*}{\langle \phi_0^*, (\Delta A - \lambda_0 \Delta F)\phi_0 \rangle} - \frac{F_0^* \phi_0^*}{\langle \phi_0^*, F_0\phi_0 \rangle}$$

(16.36)

and the function Γ is calculated using

$$(A_0 - \lambda_0 F_0)\Gamma = \frac{(\Delta A - \lambda_0 \Delta F)\phi_0}{\langle \phi_0^*, (\Delta A - \lambda_0 \Delta F)\phi_0 \rangle} - \frac{F_0 \phi_0}{\langle \phi_0^*, F_0\phi_0 \rangle}$$

(16.37)

In Eq. (16.35), the unprimed operators and eigenvalue refer to the altered system at time t_n and the primed operators and eigenvalue refer to the altered (by previous changes from the initial) system plus a perturbation (i.e., $\Delta A = A' - A$ and $\Delta F = F' - F$) at time $t > t_n$. The variational functional $\rho_{v,e}$ provides an estimate of the static reactivity worth of the perturbation, $-\Delta\lambda = \lambda - \lambda'$, in the altered system. The functional $\rho_{v,e}$ is stationary about the altered and static perturbed altered adjoint and direct eigenvalue equations, respectively [as well as being stationary about the equations for Γ^* and Γ, for which Eqs. (16.36) and (16.37) are approximations].

When $\rho_{v,e}$ is used to estimate the reactivity for the point-kinetics method without updating the flux shapes, the initial configuration described by Eqs. (16.22) and (16.25) is considered the altered system and ϕ_0 is used for S. In this case, $\rho_{v,e}$ provides a second-order estimate of the static reactivity of Eq. (16.24), rather than the dynamic reactivity of Eq. (16.29). In so doing, it ignores the delayed neutron holdback effect, an omission that leads to errors in reactivity estimates and consequent errors in power calculations.

When $\rho_{v,e}$ is used to estimate the reactivity for the QS method, the configuration at the time t_n of the most recent shape calculation is considered the altered system, and the S_n is used for S. In this case, $\rho_{v,e}$ provides an estimate accurate to second order of the static reactivity worth of perturbations made since time t_n. This estimate ignores the delayed neutron holdback effect. The total reactivity worth of all perturbations (and alterations) is found by adding this perturbed reactivity worth in the altered system to the best available estimate of the dynamic reactivity worth of the alteration, which is $\rho_n[S_n(r, t_n)]$. Because it is necessary to use the flux shape corresponding to the altered system, it is not appropriate to use the variational static reactivity estimate with interpolated flux shapes.

Variational Principle for Dynamic Reactivity

To account for the delayed neutron holdback effect on the reactivity, a variational principle should be stationary about the solutions of the time-dependent diffusion and precursor equations, rather than stationary about the solution of the perturbed static diffusion equation. To this end, the following functional was constructed:

$$
\begin{aligned}
\rho_v &\left[\psi, \psi^*, \xi_m, \xi_m^*, \Gamma, \Gamma^*\right] \\
&= \frac{\langle \psi^*, (\lambda_0 \Delta F - \Delta A)\psi \rangle}{\langle \psi^*, F\psi \rangle} \\
&\quad \times \left\{ 1 - \langle \psi^*, (A_0 - \lambda_0 F_0)\Gamma \rangle - \left\langle \Gamma^*, \left[A - \lambda_0(1-\beta)F_p + \frac{1}{v}\frac{\partial}{\partial t} \right]\psi \right\rangle \right. \\
&\quad\quad + \left\langle \Gamma^*, \sum_{m=1}^{M} \lambda_m \chi_m \xi_m \right\rangle - \left\langle \sum_{m=1}^{M} \xi_m^*, \left(\lambda_m + \frac{\partial}{\partial t} \right)\xi_m \right\rangle \\
&\quad\quad \left. + \left\langle \sum_{m=1}^{M} \xi_m^*, \lambda_0 \beta_m F\psi \right\rangle \right\}
\end{aligned}
\tag{16.38}
$$

The usual procedure is to require that the functional be stationary with respect to arbitrary and independent variations of the trial functions over all the independent variables. However, in order to retain the time dependence of the dynamic reactivity, the integrals in ρ_v (indicated by $\langle, \rangle$), are only over space and energy, not time. Thus, the stationarity conditions for the functional are established by requiring that it be stationary with respect to arbitrary and independent variations of only the space and energy dependencies of the functions Γ^*, ξ_m^*, Γ, ψ^*, ψ, and ξ_m. The following equations result:

$$
A\psi_s - \lambda_0(1-\beta)F_p\psi_s - \sum_{m=1}^{M} \lambda_m \chi_m \xi_{m,s} + \frac{1}{v}\frac{\partial \psi_s}{\partial t} = 0
\tag{16.39}
$$

$$
\frac{\partial \xi_{m,s}}{\partial t} = \lambda_0 \beta_m F\psi_s - \lambda_m \xi_{m,s}
\tag{16.40}
$$

$$
A_0^* \psi_s^* - \lambda_0 F_0^* \psi_s^* = 0
\tag{16.41}
$$

$$
(A_0 - \lambda_0 F_0)\Gamma_s = \frac{(\Delta A - \lambda_0 \Delta F)\psi_s}{\langle \psi_s^*, (\Delta A - \lambda_0 \Delta F)\psi_s \rangle} - \frac{F\psi_s}{\langle \psi_s^*, F\psi_s \rangle}
\tag{16.42}
$$

$$A^* \Gamma_s^* - \lambda_0 (1 - \beta) F_p^* \Gamma_s^* - \sum_{m=1}^{M} \lambda_0 \beta_m F \xi_{m,s}^*$$
$$= \frac{(\Delta A^* - \lambda_0 \Delta F^*) \psi_s^*}{\langle \psi_s^*, (\Delta A - \lambda_0 \Delta F) \psi_s \rangle} - \frac{F^* \psi_s^*}{\langle \psi_s^*, F \psi_s \rangle} \tag{16.43}$$

and

$$\lambda_m \xi_{m,s}^* - \lambda \chi_m^{\mathrm{T}} \Gamma_s^* = 0 \tag{16.44}$$

respectively. Comparing Eqs. (16.39) and (16.20), Eqs. (16.40) and (16.21), and Eqs. (16.41) and (16.25), it is clear that ψ_s and $\xi_{m,s}$ can be identified as the solutions $\phi(\mathbf{r}, t)$ and $C_m(\mathbf{r}, t)$ of the exact time-dependent diffusion and precursor equations and that ψ_s^* can be identified as the unperturbed static adjoint flux ϕ_0^*.

The stationary value of ρ_v is

$$\rho_{v,s} = \frac{\langle \phi_0^*, (\lambda_0 \Delta F - \Delta A) \phi(\mathbf{r}, t) \rangle}{\langle \phi_0^*, F \phi(\mathbf{r}, t) \rangle} \tag{16.45}$$

the exact, dynamic reactivity worth of a perturbation. To adapt the functional ρ_v for use with the QS method, we introduce as a trial function

$$\psi(\mathbf{r}, t) \approx S(\mathbf{r}, t) n(t) \tag{16.46}$$

and note that the best available approximation for the time derivative of the precursor density is

$$\frac{\partial C_m(\mathbf{r}, t)}{\partial t} \approx \beta_m F S(\mathbf{r}, t) n(t) - \lambda_m C_m(\mathbf{r}, t) \tag{16.47}$$

Under these conditions [and noting that $\psi^* = \phi_0^*$ is available from Eq. (16.25)], the functional becomes

$$\rho_v = \frac{\langle \phi_0^*, (\lambda_0 \Delta F - \Delta A) S \rangle}{\langle \phi_0^*, F S \rangle} \left\{ 1 - \left\langle G^*, [A - \lambda_0 (1 - \beta) F_p] S \right\rangle - \left\langle G^*, \frac{1}{\mathrm{v}} \frac{\partial S}{\partial t} \right\rangle \right.$$
$$\left. - \frac{\dot{n}}{n} \left\langle G^*, \frac{1}{\mathrm{v}} S \right\rangle + \frac{1}{n} \left\langle G^*, \sum_{m=1}^{M} \lambda_m \chi_m C_m \right\rangle \right\} \tag{16.48}$$

where the quantity $\Gamma^*(\mathbf{r}, t) n(t)$ has been replaced by a trial function $G^*(\mathbf{r}, t)$. Note that ΔA and ΔF here refer to the total perturbation, not the perturbation since the most recent shape calculation, and that $A = A_0 + \Delta A$, $F = F_0 + \Delta F$.

Using Eqs. (16.44) and (16.46) in Eq. (16.43) results in the following equation for $G^*(\mathbf{r}, t)$:

$$\left(A^* - \lambda_0 F^* \right) G^*(\mathbf{r}, t) = \frac{(\Delta A^* - \lambda_0 \Delta F^*) \phi_0^*}{\langle \phi_0^*, (\Delta A - \lambda_0 \Delta F) S(\mathbf{r}, t) \rangle} - \frac{F^* \phi_0^*}{\langle \phi_0^*, F S(\mathbf{r}, t) \rangle} \tag{16.49}$$

It is computationally economical to compute the generalized adjoint function G^* only once for a particular core configuration. In this case, the initial static configuration is used, resulting in the following approximation:

$$\left(A_0^* - \lambda_0 F_0^*\right)G^* = \frac{(\Delta A^* - \lambda_0 \Delta F^*)\phi_0^*}{\langle \phi_0^*, (\Delta A - \lambda_0 \Delta F)S_0 \rangle} - \frac{F_0^* \phi_0^*}{\langle \phi_0^*, F_0 S_0 \rangle} \qquad (16.50)$$

(any magnitude perturbation ΔA and/or ΔF can be used since these operators appear in both the numerator and denominator of the same term). Thus $G^*(\mathbf{r})$ differs only in amplitude from $\Gamma^*(\mathbf{r})$ of Eq. (16.36).

The form of the functional represented by Eq. (16.48) is well suited for use with the QS method. In the QS method, the point-kinetics equations are used for the flux amplitude $n(t)$, the precursor concentration densities $C_m(\mathbf{r}, t)$ are updated at each time step and are therefore available for use in the variational estimate, and the flux shape $S(\mathbf{r}, t)$ is recomputed periodically using Eq. (16.33). The variational dynamic reactivity estimate can be used with or without flux shape interpolations.

It should be noted that the G^* of Eq. (16.50) satisfies the orthogonality condition

$$\langle G^*, F_0 S_0 \rangle = 0 \qquad (16.51)$$

As a consequence, when the initial flux shape S_0 is used in ρ_v and if the precursor density functions $C_m(\mathbf{r}, t)$ have the same shape as S_0, the variational estimate for dynamic reactivity reduces to the variational estimate for static reactivity, $\rho_{v,e}$ of Eq. (16.35) [in which the second term in the square brackets disappears because of Eq. (16.25)]. The effect of this reduction is that until the flux shape is recomputed or until some other approximation is made to replace S_0, the new variational functional still ignores the delayed neutron holdback effect.

Numerical tests on a large LWR model indicate that the flux shape computational effort required with the QS method can be reduced by a factor of 3 to 4 by using the variational estimate of dynamic reactivity. In addition, use of a variational reactivity estimate rather than the standard first-order estimate of static reactivity can improve the accuracy of the QS method enough that the time-consuming flux shape interpolation/recomputation procedure may not be necessary.

16.3
Time Integration of the Spatial Neutron Flux Distribution

The various methods that have been discussed for calculating the spatial neutron flux distribution (finite-difference, nodal, finite-element, synthesis, etc.) can be extended to calculate the space–time neutron flux distribution by adding a neutron density time derivative, distinguishing between prompt and delayed neutron sources in the neutron balance equation and appending equations to calculate the delayed neutron precursor densities [e.g., Eqs. (16.20) and (16.21)]. Writing the group fluxes and precursor densities at every spatial point (e.g., mesh point, node) as a column vector $\boldsymbol{\psi}$, and writing the terms of the multigroup neutron and delayed neutron precursor balance equations at each spatial point as a matrix $\boldsymbol{H}$, the space–time neutron kinetics equations can be written as a coupled set of ordinary differential equations

$$H\psi = \dot{\psi} \tag{16.52}$$

Explicit Integration: Forward-Difference Method

The simplest approximate solution to Eq. (16.52) is obtained by a simple *forward-difference algorithm*,

$$\psi(p+1) = \psi(p) + \Delta t H(p)\psi(p) \tag{16.53}$$

where the argument p denotes the value at time t_p, and $\Delta t = t_{p+1} - t_p$. In terms of the multigroup diffusion equations, this algorithm is

$$\phi^g(p+1) = \phi^g(p) + \Delta t v^g \left\{ \nabla \cdot D^g(p)\nabla\phi^g(p) - \left[\Sigma_a^g(p) + \Sigma_s^g(p)\right]\phi^g(p) \right.$$

$$+ \sum_{g'=1}^{G} \Sigma_s^{g'\to g}(p)\phi^{g'}(p)$$

$$+ (1-\beta)\chi_p^g \sum_{g'=1}^{G} v\Sigma_f^{g'}(p)\phi^{g'}(p)$$

$$\left. + \sum_{m=1}^{M} \lambda_m \chi_m^g C_m(p) \right\}, \quad g = 1,\dots,G \tag{16.54}$$

and for the precursors,

$$C_m(p+1) = C_m(p) + \Delta t\left[\beta_m \sum_{g=1}^{G} v\Sigma_f^g(p)\phi^g(p) - \lambda_m C_m(p)\right],$$

$$m = 1,\dots,M \tag{16.55}$$

where the spatial dependence is implicit.

This algorithm suffers from a problem of numerical stability, which requires the use of such small time steps that the advantage offered by the simplicity of the algorithm is usually more than offset by the large number of time steps required. The nature of this problem is seen by considering an expansion of $\psi(p)$ in the eigenfunctions of the operator H:

$$\psi(p) = \sum_n a_n \Omega_n \tag{16.56}$$

where

$$H\Omega_n = \omega_n \Omega_n \tag{16.57}$$

Substituting Eq. (16.56) into Eq. (16.53) yields

$$\boldsymbol{\psi}(p+1) = \sum_n a_n (1 + \omega_n \Delta t) \boldsymbol{\Omega}_n \tag{16.58}$$

The condition for numerical stability is that the fundamental mode $\boldsymbol{\Omega}_1$ grow more rapidly than the harmonics $\boldsymbol{\Omega}_n, n \geq 2$. This requires that

$$|1 + \omega_1 \Delta t| > |1 + \omega_n \Delta t|, \quad n \geq 2 \tag{16.59}$$

To ensure this, $|\omega_n \Delta t|$ must be much less than unity. The eigenvalue problem of Eq. (16.57) is a generalization to several groups and many spatial points of the in-hour equation of Section 5.3. The magnitude of the fundamental eigenvalue is on the order of the precursor decay constant, except for highly supercritical transients, in which case small time steps must be used in any case. Numerical studies have shown that the smallest eigenvalues can be on the order of $-(v^g \Sigma_a^g)$, which can be about -10^4 for thermal neutrons and about -10^7 for fast neutrons. Thus $\Delta t < 10^{-7}$ may be required for stability. When the time derivative terms for the epithermal groups are assumed to vanish (a useful approximation since $1/v^G \gg 1/v^g$, $g \neq G$), $\Delta t < 10^{-4}$ may be required.

Implicit Integration: Backward-Difference Method

The numerical stability problem associated with the preceding method can be all but eliminated by the *backward-difference algorithm*:

$$\boldsymbol{\psi}(p+1) = [\boldsymbol{I} - \Delta t \boldsymbol{H}(p+1)]^{-1} \boldsymbol{\psi}(p) \tag{16.60}$$

In terms of the precursor and multigroup diffusion equations, this algorithm is

$$C_m(p+1) = \frac{C_m(p)}{1 + \lambda_m \Delta t} + \frac{\beta_m}{1 + \lambda_m \Delta t} \sum_{g=1}^{G} v \Sigma_f^g(p+1) \phi^g(p+1),$$
$$m = 1, \ldots, M \tag{16.61}$$

$$\nabla \cdot D^g(p+1) \nabla \phi^g(p+1) + \left[\Sigma_a^g(p+1) + \Sigma_s^g(p+1) \right] \phi^g(p+1)$$

$$+ \sum_{g'=1}^{G} \Sigma_s^{g' \rightarrow g}(p+1) \phi^{g'}(p+1) + (1 - \beta) \chi_p^g \sum_{g'=1}^{G} v \Sigma_f^{g'}(p+1) \phi^{g'}(p+1)$$

$$+ \sum_{m=1}^{M} \frac{\lambda_m \chi_m^g \beta_m}{1 + \lambda_m \Delta t} \sum_{g'=1}^{G} v \Sigma_f^{g'}(p+1) \phi^{g'}(p+1) - \frac{1}{v^g \Delta t} \phi^g(p+1)$$

$$= -\frac{1}{v^g \Delta t} \phi^g(p) - \sum_{m=1}^{M} \frac{\lambda_m \chi_m^g C_m(p)}{1 + \lambda_m \Delta t}, \quad g = 1, \ldots, G \tag{16.62}$$

An expansion of the type of Eq. (16.56) substituted into Eq. (16.60) yields

$$\boldsymbol{\psi}(p+1) = \sum_n a_n (1 - \Delta t \omega_n)^{-1} \boldsymbol{\Omega}_n \tag{16.63}$$

and the condition for stability is

$$\left| (1 - \Delta t \omega_1)^{-1} \right| > \left| (1 - \Delta t \omega_n)^{-1} \right|, \quad n \geq 2 \tag{16.64}$$

The method is unconditionally stable if $0 > \text{Re}\{\omega_1\} > \text{Re}\{\omega_n\}, n \geq 2$. For $\text{Re}\{\omega_1\} > 0$, the stability requirement is determined by the requirement that $\boldsymbol{\psi}(p+1)$ be a positive vector, which necessitates that

$$\Delta t < \frac{1}{\omega_1} \tag{16.65}$$

This requirement is restrictive only for large ω_1 that correspond to fast transients where small time steps would be necessary in any case.

The difficulty with the backward-difference method arises from the necessity of inverting a matrix at each time step. The actual matrix that must be inverted is the coefficient matrix for the left side of Eq. (16.62); the delayed neutrons can be determined directly. Thus, although much larger time steps can be taken with the implicit method than with the explicit method, the computation time needed for the matrix inversions may more than offset this advantage. The size time step used in the backward-difference method is usually limited by the effect of truncation error (of order Δt^2) upon the accuracy of the solution rather than by numerical stability.

Implicit Integration: θ Method

For a constant $\boldsymbol{H}$ in the interval $t_p \leq t \leq t_{p+1}$, Eq. (16.52) has the formal solution

$$\boldsymbol{\psi}(p+1) = \exp(\Delta t \boldsymbol{H})\boldsymbol{\psi}(p) = \left(\boldsymbol{I} + \Delta t \boldsymbol{H} + \frac{\Delta t^2}{2!} \boldsymbol{H}^2 + \cdots \right) \boldsymbol{\psi}(p) \tag{16.66}$$

The algorithms of Eqs. (16.53) and (16.60) may be considered as approximations to Eq. (16.66). An improved algorithm results from the prescription

$$\boldsymbol{\psi}(p+1) - \boldsymbol{\psi}(p) = \Delta t [\boldsymbol{M}\boldsymbol{\psi}(p+1) + (\boldsymbol{H} - \boldsymbol{M})\boldsymbol{\psi}(p)] \tag{16.67}$$

with matrix elements of $\boldsymbol{M}$ and $\boldsymbol{H}$ related by

$$m_{ij} = \theta_{ij} h_{ij} \tag{16.68}$$

where the m_{ij}, thus the θ_{ij}, are chosen so that $\boldsymbol{\psi}(p+1)$ calculated from Eq. (16.67) agrees with $\boldsymbol{\psi}(p+1)$ calculated from Eq. (16.66). This requires that

$$\boldsymbol{M} = \frac{1}{\Delta t} \boldsymbol{I} - \boldsymbol{H}[\exp(\Delta t \boldsymbol{H}) - \boldsymbol{I}]^{-1} \tag{16.69}$$

Assuming that H has distinct eigenvalues, it may be diagonalized by the transformation

$$(J^+)^T H J = \Gamma \tag{16.70}$$

where J and J^+ are the modal matrices corresponding to H and H^T (i.e., the columns of J and J^+ are the eigenvectors of H and H^T, respectively), and Γ is a diagonal matrix composed of the eigenvalues of H. Thus

$$(J^+)^T M J = \frac{1}{\Delta t} I - \Gamma [\exp(\Delta t \Gamma) - I]^{-1} = L \tag{16.71}$$

with L diagonal. From this it follows that

$$M = J L (J^+)^T \tag{16.72}$$

and the factors θ_{ij} can be determined from

$$\theta_{ij} = \frac{m_{ij}}{h_{ij}}$$

after the m_{ij} are found from Eq. (16.72).

Because solving for the θ_{ij} rigorously would entail a great deal of effort, several approximations are made in employing this method to arrive at an algorithm for solutions of the multigroup kinetics equations. The delayed neutrons are treated as sources, and thus are neglected in the determination of the θ_{ij}. An average space-independent value of θ_{ij} is calculated based on a flux square weighting procedure. The delayed neutron precursors have a separate θ_{ij}. Denoting the θ_{ij} associated with groups g and g' as $\theta_{gg'}$ and θ_{ij} associated with the delayed neutrons as θ_d, the following algorithm results:

$$C_m(p+1) = \frac{1 - (1 - \theta_d)\lambda_m \Delta t}{1 + \theta_d \lambda_m \Delta t} C_m(p) + \frac{\Delta t \beta_m}{1 + \theta_d \lambda_m \Delta t}$$

$$\times \left[\sum_{g=1}^{G} \nu \Sigma_f^g(p+1) \phi^g(p+1)\theta_{1g} \right.$$

$$\left. + \sum_{g=1}^{G} \nu \Sigma_f^g(p) \phi^g(p)(1 - \theta_{1g}) \right], \quad m = 1, \ldots, M \tag{16.73}$$

$$\theta_{gg}\left\{ \nabla \cdot D^g(p+1)\nabla \phi^g(p+1) - \left[\Sigma_a^g(p+1) + \Sigma_s^g(p+1) \right]\phi^g(p+1) \right\}$$

$$+ \sum_{g'=1}^{G} \theta_{gg'} \Sigma_s^{g' \to g}(p+1)\phi^{g'}(p+1)$$

$$+ \chi_p^g(1 - \beta) \sum_{g'=1}^{G} \theta_{gg'} \nu \Sigma_f^{g'}(p+1)\phi^{g'}(p+1) - \frac{1}{\Delta t \nu^g}\phi^g(p+1)$$

$$+ \sum_{m=1}^{M} \frac{\chi_m^g \lambda_m \Delta t \beta_m \theta_d}{1 + \theta_d \lambda_m \Delta t} \sum_{g'=1}^{G} \nu \Sigma_f^{g'}(p+1)\phi^{g'}(p+1)\theta_{1g'}$$

$$= -(1 - \theta_{gg})\{\nabla \cdot D^g(p)\nabla\phi^g(p) - [\Sigma_a^g(p) + \Sigma_s^g(p)]\phi^g(p)\}$$

$$- \sum_{g'=1}^{G} (1 - \theta_{gg'})\Sigma_s^{g' \to g}(p)\phi^{g'}(p)$$

$$- (1 - \beta)\chi_p^g \sum_{g'=1}^{G} (1 - \theta_{gg'})\nu \Sigma_f^{g'}(p)\phi^{g'}(p)$$

$$- \frac{1}{\Delta t \nu^g}\phi^g(p) - \sum_{m=1}^{M} \frac{\chi_m^g \lambda_m C_m(p)}{1 + \theta_d \lambda_m \Delta t}$$

$$- \sum_{m=1}^{M} \frac{\chi_m^g \theta_d \lambda_m \Delta t \beta_m}{1 + \theta_d \lambda_m \Delta t} \sum_{g'=1}^{G} \nu \Sigma_f^{g'}(p)\phi^{g'}(p)(1 - \theta_{1g'})$$

$$g = 1, \ldots, G \tag{16.74}$$

In the limit $\theta_{gg'}$, $\theta_d \to 1$ Eqs. (16.73) and (16.74) reduce to the backward-difference algorithms of Eqs. (16.61) and (16.62), while Eqs. (16.73) and (16.74) reduce to the forward-difference algorithms of Eqs. (16.54) and (16.55) in the limit $\theta_{gg'}$, $\theta_d \to 0$. As mentioned, a number of approximations are made in arriving at Eqs. (16.73) and (16.74), so the mathematical properties associated with Eqs. (16.67) to (16.72) are not rigorously retained by Eqs. (16.73) and (16.74).

Insight into the stability properties of the θ-method can be gained by considering the situation for a constant matrix $\boldsymbol{H}$ and a constant time step Δt. Expanding the exact solutions of Eq. (16.52) in the eigenfunctions $\boldsymbol{\Omega}_n$, of $\boldsymbol{H}$ given by Eq. (16.57),

$$\boldsymbol{\psi}(t_p) = \sum_{n=1}^{N} a_n \boldsymbol{\Omega}_n e^{\omega_n t_p} = \sum_{n=1}^{N} a_n \boldsymbol{\Omega}_n e^{\omega_n(p\Delta t)} \tag{16.75}$$

where the expansion coefficients a_n are determined from the initial conditions and where $\omega_1 > \omega_2 > \cdots > \omega_N$. For the same eigenfunctions to satisfy Eq. (16.67), which becomes

$$\gamma_n \boldsymbol{\Omega}_n = (\boldsymbol{I} - \Delta t \boldsymbol{M})^{-1}(\boldsymbol{I} + \Delta t \boldsymbol{H} - \Delta t \boldsymbol{M})\boldsymbol{\Omega}_n \tag{16.76}$$

the eigenvalues must be related by

$$\gamma_n = \frac{1 + (1 - \theta)\omega_n \Delta t}{1 - \theta \omega_n \Delta t} \tag{16.77}$$

The general solution for the θ-approximation of Eq. (16.67) may be written

$$\psi(t_p) = \sum_{n=1}^{N} a_n \gamma_n^p \Omega_n \tag{16.78}$$

where $t_p = p \Delta t$. Comparison with the exact solution of Eq. (16.75) indicates that $\exp(\omega_n t) = \exp(\omega_n p \Delta t)$ has been replaced by γ_n^p in the approximate solution. For a stable θ approximation, $\gamma_n > -1$; otherwise, γ_n^p will oscillate and diverge as time increases. Thus, Eq. (16.77) and the eigenvalues ω_n can be used to determine a maximum stable step size Δt.

Numerical experience indicates that the algorithm of Eqs. (16.73) and (16.74) is (1) numerically stable for time steps two orders of magnitude greater than are required for stability of Eqs. (16.54) and (16.55), and (2) somewhat more accurate than the algorithm of Eqs. (16.61) and (16.62) for the same time steps. The algorithm of Eqs. (16.74) requires inversion of the same type of matrix as does the backward-difference algorithm of Eqs. (16.62), and, in addition, requires computation of $\theta_{gg'}$ and θ_d, although the latter computation is negligible with respect to the time required for the matrix inversion. In practice, the θ's are predetermined based on experience or intuition.

Implicit Integration: Time-Integrated Method

The delayed neutron precursor equations may, in principle, be integrated directly between t_p and t_{p+1}:

$$C_m(p+1) = \exp(-\lambda_m \Delta t) C_m(p)$$

$$+ \beta_m \int_{t_p}^{t_{p+1}} dt \, \exp[-\lambda_m(t_{p+1}-t)] \sum_{g=1}^{G} \nu \Sigma_f^g(t) \phi^g(t) \tag{16.79}$$

If the assumption is made that the group-fission rate at each point varies linearly in time in the interval $t_p \leq t \leq t_{p+1}$, Eq. (16.79) yields an implicit integration algorithm for the precursors,

$$C_m(p+1) = \exp(-\lambda_m \Delta t) C_m(p)$$

$$+ \frac{\beta_m}{\lambda_m} \left\{ \left[\frac{1 - \exp(-\lambda_m \Delta t)}{\lambda_m \Delta t} - \exp(-\lambda_m \Delta t) \right] \sum_{g=1}^{G} \nu \Sigma_f^g(p) \phi^g(p) \right.$$

$$\left. - \left[\frac{1 - \exp(-\lambda_m \Delta t)}{\lambda_m \Delta t} - 1 \right] \sum_{g=1}^{G} \nu \Sigma_f^g(p+1) \phi^g(p+1) \right\}$$

$$\tag{16.80}$$

Integration of the multigroup diffusion equation over the interval $t_p \leq t \leq t_{p+1}$, with the assumption that all reaction rates vary linearly in that interval, results in an implicit integration algorithm for the neutron flux,

$$\nabla \cdot D^g(p+1)\nabla\phi^g(p+1) - \left[\Sigma_a^g(p+1) + \Sigma_s^g(p+1)\right]\phi^g(p+1)$$

$$+ \sum_{g'=1}^{G} \Sigma_s^{g' \to g}(p+1)\phi^{g'}(p+1)$$

$$+ \left\{ \chi_p^g - \sum_{m=1}^{M} \beta_m\left(\chi_p^g - \chi_m^g\right) + \sum_{m=1}^{M} \frac{2}{\Delta t}\frac{\chi_m^g \beta_m}{\lambda_m}\left[\frac{1 - \exp(-\lambda_m \Delta t)}{\lambda_m \Delta t} - 1\right]\right\}$$

$$\times \sum_{g'=1}^{G} \nu\Sigma_f^{g'}(p+1)\phi^{g'}(p+1) - \frac{2}{v^g \Delta t}\phi^g(p+1)$$

$$= -\frac{2}{\Delta t}\sum_{m=1}^{M} \chi_m^g[1 - \exp(-\lambda_m \Delta t)]C_m(p) - \frac{2}{v^g \Delta t}\phi^g(p)$$

$$- \left\{\chi_p^g - \sum_{m=1}^{M} \beta_m\left(\chi_p^g - \chi_m^g\right)\right.$$

$$\left. - \sum_{m=1}^{M} \frac{2}{\Delta t}\frac{\chi_m^g \beta_m}{\lambda_m}\left[\frac{1 - \exp(-\lambda_m \Delta t)}{\lambda_m \Delta t} - \exp(-\lambda_m \Delta t)\right]\right\}$$

$$\times \sum_{g'=1}^{G} \nu\Sigma_f^{g'}(p)\phi^{g'}(p) - \nabla \cdot D^g(p)\nabla\phi^g(p) + \left[\Sigma_a^g(p) + \Sigma_s^g(p)\right]\phi^g(p)$$

$$- \sum_{g'=1}^{G} \Sigma_s^{g' \to g}(p)\phi^{g'}(p) \tag{16.81}$$

In arriving at Eq. (16.81), integration of the precursors was treated as in Eq. (16.80) (i.e., the group-fission rate was assumed to vary linearly).

Equations (16.80) and (16.81) define the time-integrated algorithm, which, like Eqs. (16.73) and (16.74), represents an attempt to reduce the truncation error associated with the simple implicit integration formulas of Eqs. (16.61) and (16.62) without materially increasing the computational time required to obtain a solution.

All three implicit integration algorithms require inversion (at each time step) of roughly the same matrix. Numerical experience indicates that the θ-method and the time-integrated method yield essentially identical results, and that both methods are somewhat more accurate than the backward-difference method.

Implicit Integration: GAKIN Method

The mathematical properties of this method derive directly from the properties of the spatial finite-difference approximation. This approximation is

$$\dot{\theta} = K\theta \tag{16.82}$$

where

$$
\theta = \begin{bmatrix} \boldsymbol{\psi}^1 \\ \vdots \\ \boldsymbol{\psi}^G \\ \boldsymbol{d}_1 \\ \vdots \\ \boldsymbol{d}_M \end{bmatrix}
\tag{16.83}
$$

with $\boldsymbol{\psi}^g$ and $\boldsymbol{d}_m$ representing $N \times 1$ column vectors of group fluxes and m-type precursor densities, respectively, at each of N spatial mesh points. The matrix $\boldsymbol{K}$ can be written in terms of $N \times N$ submatrices K_{ij}:

$$
\boldsymbol{K} = \begin{bmatrix} \boldsymbol{K}_{11} & \boldsymbol{K}_{12} & \boldsymbol{K}_{13} & \cdots & \boldsymbol{K}_{1,G+M} \\ \boldsymbol{K}_{21} & \boldsymbol{K}_{22} & \boldsymbol{K}_{23} & \cdots & \boldsymbol{K}_{2,G+M} \\ \vdots & & & & \vdots \\ \boldsymbol{K}_{G+M,1} & \boldsymbol{K}_{G+M,2} & & \cdots & \boldsymbol{K}_{G+M,G+M} \end{bmatrix}
\tag{16.84}
$$

The $N \times N$ matrices $\boldsymbol{K}_{ij}$ are split,

$$
\boldsymbol{K}_{ij} = \boldsymbol{\Gamma}_{ij} + v^i \boldsymbol{D}^i, \quad 1 \leq i \leq G
\tag{16.85}
$$

where $\boldsymbol{D}^i$ represents the coupling among mesh points due to the diffusion term.

By splitting $\boldsymbol{K}$ into a matrix $\boldsymbol{L}$, which contains all the submatrices below the diagonal block; a matrix $\boldsymbol{U}$, which contains all the submatrices above the diagonal block; and into the block diagonal matrices $\boldsymbol{\Gamma}$ and $\boldsymbol{D}$,

$$
\boldsymbol{L} = \begin{bmatrix} 0 & 0 & 0 & \cdots & 0 \\ \boldsymbol{K}_{21} & 0 & 0 & \cdots & 0 \\ \vdots & & & & \vdots \\ \boldsymbol{K}_{G+M,1} & \boldsymbol{K}_{G+M,2} & & \cdots & 0 \end{bmatrix}
\tag{16.86}
$$

$$
\boldsymbol{U} = \begin{bmatrix} 0 & \boldsymbol{K}_{12} & \boldsymbol{K}_{13} & \cdots & \boldsymbol{K}_{1,G+M} \\ 0 & 0 & \boldsymbol{K}_{23} & \cdots & \boldsymbol{K}_{2,G+M} \\ \vdots & & & & \\ 0 & & \cdots & & 0 \end{bmatrix}
\tag{16.87}
$$

$$
\boldsymbol{\Gamma} = \begin{bmatrix} \boldsymbol{\Gamma}_{11} & 0 & 0 & \cdots & & 0 \\ 0 & \boldsymbol{\Gamma}_{22} & 0 & \cdots & & 0 \\ 0 & \cdots & \boldsymbol{\Gamma}_{GG} & \cdots & & 0 \\ 0 & & \cdots & \boldsymbol{K}_{G+1,G+1} & & 0 \\ \vdots & & & & & \vdots \\ 0 & \cdots & & & & \boldsymbol{K}_{G+M,G+M} \end{bmatrix}
\tag{16.88}
$$

$$D = \begin{bmatrix} v^1 D^1 & & 0 & 0 & \cdots & 0 \\ 0 & v^2 D^2 & \cdots & 0 & \cdots & 0 \\ \vdots & & & & & \vdots \\ 0 & & v^G D^G & \cdots & \cdots & 0 \\ 0 & & & \cdots & 0 & \cdots & 0 \\ \vdots & & & & & \vdots \\ 0 & & & & \cdots & 0 \end{bmatrix} \qquad (16.89)$$

Equation (16.82) may be written

$$\dot{\theta} - \boldsymbol{\Gamma}\theta = (L + U)\theta + D\theta \qquad (16.90)$$

This equation may formally be integrated over the interval $t_p \le t \le t_{p+1}$:

$$\theta(t_{p+1}) = \exp(\Delta t \, \boldsymbol{\Gamma})\theta(t_p) + \int_0^{\Delta t} dt' \exp[(\Delta t - t')\boldsymbol{\Gamma}](L + U)\theta(t_p + t')$$

$$+ \int_0^{\Delta t} dt' \exp[(\Delta t - t')\boldsymbol{\Gamma}]D\theta(t_p + t') \qquad (16.91)$$

In the first integral of Eq. (16.91), the approximation

$$\theta(t_p + t') = \exp(\omega t')\theta(t_p) \qquad (16.92)$$

is made, and the second integral is performed with the approximation

$$\theta(t_p + t') = \exp[-\omega(\Delta t - t')]\theta(t_{p+1}) \qquad (16.93)$$

In general, ω is a diagonal matrix. Using Eqs. (16.92) and (16.93) in Eq. (16.91) results in

$$[I - (\omega - \boldsymbol{\Gamma})^{-1}(I - \exp[(\boldsymbol{\Gamma} - \omega)\Delta t])D]\theta(t_{p+1})$$

$$= [\exp(\boldsymbol{\Gamma}\Delta t) + (\omega - I)^{-1}(\exp(\omega\Delta t) - \exp(\boldsymbol{\Gamma}\Delta t))(L + U)]\theta(t_p) \qquad (16.94)$$

which may be written

$$\theta(t_{p+1}) = A\theta(t_p) \qquad (16.95)$$

If all the diagonal elements of ω are equal to ω_1, which is the eigenvalue of

$$K\theta_n = \omega_n\theta_n \qquad (16.96)$$

with largest real part, then from Eq. (16.90),

$$(L + U)\theta_1 = (\omega_1 I - \boldsymbol{\Gamma} - D)\theta_1 \qquad (16.97)$$

From the definition of A (with $\omega = \omega_1 I$) it can be shown that

$$A\theta_1 = \exp(\Delta t \omega_1)\theta_1 \qquad (16.98)$$

It can be shown that ω_1 is real and simple, and that $\boldsymbol{\theta}_1$ is positive. For all real values of ω, hence for $\omega = \omega_1$, $\boldsymbol{A}$ can be shown to be nonnegative, irreducible, and primitive. From the Perron–Frobenius theorem it follows that $\boldsymbol{A}$ has a simple, real, largest eigenvalue ρ_1 and a corresponding positive eigenvector. The eigenvalue $\rho_1 = \exp(\Delta t \omega_1)$ is seen from Eq. (16.98) to have a positive eigenvector that is the fundamental-mode solution of the kinetics equations (16.96). If it can be shown that ρ_1 is the largest eigenvalue of $\boldsymbol{A}$, Eq. (16.98) indicates that the asymptotic solution of the integration algorithm of Eq. (16.95) is the asymptotic solution of Eq. (16.82) for a step change in properties, which shows that the method is unconditionally numerically stable.

The transpose matrix $\boldsymbol{A}^{\mathrm{T}}$ has the same properties and eigenvalue spectrum as $\boldsymbol{A}$:

$$\boldsymbol{A}^{\mathrm{T}} \boldsymbol{q}_n = \rho_n \boldsymbol{q}_n \tag{16.99}$$

By the Perron–Frobenius theorem, $\boldsymbol{A}^{\mathrm{T}}$ has a real, simple eigenvalue, ρ_k, which is larger than the real part of the other eigenvalues, and the corresponding eigenvector is positive. Premultiplying Eq. (16.93) for $n = k$ by $\boldsymbol{\theta}_1^{\mathrm{T}}$, premultiplying Eq. (16.98) by $\boldsymbol{q}_1^{\mathrm{T}}$, and subtracting yields

$$0 = [\exp(\Delta t \omega_1) - \rho_1] \boldsymbol{\theta}_1^{\mathrm{T}} \boldsymbol{q}_1 \tag{16.100}$$

Because $\boldsymbol{\theta}_1$ and $\boldsymbol{q}_1$ are positive, Eq. (16.100) is satisfied only if

$$\exp(\Delta t \omega_1) = \rho_1$$

is the real eigenvalue. Thus the method is numerically unconditionally stable.

Inversion of the matrix on the left of Eq. (16.94) to obtain $\boldsymbol{A}$ can be accomplished by the inversion of $GN \times N$ matrices. In practice, an approximation to ω_1 is obtained by an expression of the form

$$\omega_1 = \frac{1}{\Delta t} \ln \frac{\theta_i(t_p)}{\theta_i(t_{p-1})} \tag{16.101}$$

where i indicates some component or components of the $\boldsymbol{\theta}$ vector, and different values of ω_1 are used in different parts of the reactor (i.e., $\omega \neq \omega_1 \boldsymbol{I}$).

Alternating Direction Implicit Method

The implicit integration methods of previous sections all reduced to an algorithm for the neutron flux which required the inversion of a matrix at each time step. When the finite-difference spatial approximation is employed, this matrix is $NG \times NG$, where N is the number of mesh points and G is the number of energy groups. In one-dimensional problems, the matrix to be inverted becomes block tridiagonal with $G \times G$ blocks, and inversion can be accomplished by the backward-elimination/forward-substitution method and requires the inversion of N $G \times G$ matrices. In the GAKIN method, this matrix inversion can be accomplished by inverting G $N \times N$ matrices.

However, for multidimensional problems, the matrix inversion associated with the implicit methods poses a formidable and time-consuming task. Alternative formulations of the θ and GAKIN methods have been proposed to reduce the time required for this matrix inversion. Another technique, designed to eliminate this same problem, is the *alternating direction implicit* (ADI) *method*. The basis of the ADI method is to make the algorithm implicit for one space dimension at a time and to alternate the space dimension for which the algorithm is implicit. The ideas involved are illustrated by a two-dimensional problem. The equation for the group g neutron flux can be written in the notation of Section 16.2 as

$$\dot{\psi}^g = \left(v^g D_x^g + \frac{1}{2} \Gamma_{gg} \right) \psi^g + \left(v^g D_y^g + \frac{1}{2} \Gamma_{gg} \right) \psi^g$$

$$+ \sum_{g'=1}^{G} K_{gg'} \psi^{g'} + \sum_{m=1}^{M} K_{g,G+m} d_m \qquad (16.102)$$

where the $N \times N$ diffusion matrix D^g, which represents

$$\frac{\partial}{\partial x} D^g \frac{\partial}{\partial x} + \frac{\partial}{\partial y} D^g \frac{\partial}{\partial y}$$

has been separated into D_x^g, which represents

$$\frac{\partial}{\partial x} D^g \frac{\partial}{\partial x}$$

and D_y^g, which represents

$$\frac{\partial}{\partial y} D^g \frac{\partial}{\partial y}$$

For the time step t_p to t_{p+1}, an integration algorithm which is implicit in the x-direction and explicit in the y-direction is chosen. First define

$$H_x^g \equiv v^g D_x^g + \tfrac{1}{2} \Gamma_{gg}$$
$$H_y^g \equiv v^g D_y^g + \tfrac{1}{2} \Gamma_{gg} \qquad (16.103)$$

then the algorithm is written

$$\psi^g(p+1) - \psi^g(p)$$

$$= \Delta t \left[H_x^g(p+1)\psi^g(p+1) + H_y^g(p)\psi^g(p) + \sum_{g'-1}^{G} K_{gg'}(p)\psi^{g'}(p) \right.$$

$$\left. + \sum_{m-1}^{M} K_{g,G+m}(p)d_m(p) \right], \quad g = 1, \ldots, G$$

or

$$\left[I - \Delta t \, H_x^g(p+1) \right] \psi^g(p+1)$$

$$= \left[I + \Delta t \, H_y^g(p) \right] \psi^g(p)$$

$$+ \Delta t \left[\sum_{g'=1}^{G} K_{gg'}(p) \psi^{g'}(p) + \sum_{m=1}^{M} K_{g,G+m}(p) d_m(p) \right], \quad g = 1, \ldots, G$$

$$(16.104)$$

For the time step t_{p+1} to t_{p+2}, an algorithm that is implicit in the y-direction and in the removal, scattering, fission, and precursor terms is chosen:

$$\psi^g(p+2) - \psi^g(p+1)$$

$$= \Delta t \left[H_x^g(p+1) \psi^g(p+1) + H_y^g(p+2) \psi^g(p+2) \right.$$

$$+ \sum_{g'=1}^{G} K_{gg'}(p+2) \psi^{g'}(p+2)$$

$$\left. + \sum_{m=1}^{M} K_{g,G+m}(p+2) d_m (p+2) \right], \quad d = 1, \ldots, G \qquad (16.105)$$

Use, for the sake of definiteness, the implicit integration formulas of Eq. (16.61) for the precursors,

$$d_m(p+2) = \frac{1}{1 + \lambda_m \Delta t} d_m(p+1) + \frac{\beta_m}{1 + \lambda_m \Delta t} \sum_{g=1}^{G} F^g(p+2) \psi^g(p+2)$$

$$(16.106)$$

where F^G is an $N \times N$ diagonal matrix representing $\nu \Sigma_f^g$ associated with each point. Using Eq. (16.106), Eq. (16.105) becomes

$$\left[I - \Delta t \, H_y^g(p+2) \right] \psi^g(p+2)$$

$$- \Delta t \left[\sum_{g'=1}^{G} K_{gg'}(p+2) \psi^{g'}(p+2) + \sum_{m=1}^{M} K_{g,G+m}(p+2) \frac{\beta_m}{1 + \lambda_m \Delta t} \right.$$

$$\left. \times \sum_{g'=1}^{G} F^{g'}(p+2) \psi^{g'}(p+2) \right]$$

$$= \left[I + \Delta t \, H_x^g(p+1) \right] \psi^g(p+1)$$

$$+ \Delta t \sum_{m=1}^{M} \frac{1}{1 + \lambda_m \Delta t} K_{g,G+m}(p+2) d_m(p+1), \quad g = 1, \ldots, G$$

$$(16.107)$$

The solution proceeds by alternating between the algorithms of Eqs. (16.104) and (16.107). If there are $N^{1/2}$ mesh points in both the x- and y-directions, the matrices that must be inverted in order to solve Eqs. (16.104) and (16.107) can be partitioned so that, rather than inverting an $NG \times NG$ matrix, $N^{1/2}$ $N^{1/2}G \times N^{1/2}G$ matrices are inverted. This happens because the matrix to be inverted in Eqs. (16.104) couples mesh points only in the x-direction, and the matrix to be inverted in Eq. (16.107) couples mesh points only in the y-direction. In the case of Eq. (16.104), each of the $N^{1/2}G \times N^{1/2}G$ matrices can be further partitioned into G $N^{1/2} \times N^{1/2}$ matrices, because the neutron source terms due to fission, scattering, and precursor decay are treated explicitly in this step. More general algorithms treat these source terms implicitly in both steps.

Stiffness Confinement Method

The set of neutron and delayed neutron precursor equations are referred to as *stiff* because of the great difference in the time constants that govern the prompt neutron and precursor responses. The accuracy and stability of numerical integration methods are usually determined by the shortest time constant, the prompt neutron lifetime, which has little effect on the precursor solution. The *stiffness confinement method* seeks to confine the difficulty to the neutron equations by decoupling the precursor equations through the definition of *dynamic frequencies*:

$$\omega_\phi^g(\mathbf{r}, t) \equiv \frac{1}{\phi_g} \frac{\partial \phi_g}{\partial t}, \qquad \omega_c^m \equiv \frac{1}{C_m} \frac{\partial C_m}{\partial t} \tag{16.108}$$

These definitions can be used to replace the time derivatives in the multigroup diffusion and precursor equations, which allow the latter to be formally solved and used to evaluate the precursor densities in the multigroup diffusion equations, resulting in

$$\nabla \cdot D^g(\mathbf{r}, t)\phi_g(\mathbf{r}, t) - \left[\Sigma_t^g + \frac{\omega_\phi^g(\mathbf{r}, t)}{v_g} \right]\phi_g(\mathbf{r}, t) + \sum_{g'=1}^{G} \Sigma^{g' \to g}(\mathbf{r}, t)\phi_{g'}(\mathbf{r}, t)$$

$$+ \left\{ (1 - \beta)\chi_p^g + \sum_{m=1}^{M} \frac{\beta_m \lambda_m \chi_m^g}{\omega_c^m(\mathbf{r}, t) + \lambda_m} \right\} \sum_{g'=1}^{G} v\Sigma_f^{g'}(\mathbf{r}, t)\phi_{g'}(\mathbf{r}, t) = 0,$$

$$g = 1, \ldots, G \tag{16.109}$$

These equations are identical to the static multigroup diffusion equations, but with modified total and fission cross sections which include the dynamic frequencies. Thus, to advance the solution in time, an estimate is made of the dynamic frequencies, Eqs. (16.109) are solved for the group fluxes, the precursors are updated, an improved guess of the dynamic frequencies is calculated using the new flux and precursor values, and the iteration is repeated until convergence.

Symmetric Successive Overrelaxation Method

Successive over-relaxation is combined with an exponential transformation to decouple stiffness in the symmetric successive over-relaxation (SSOR) method. The matrix $\boldsymbol{H}$ is first decomposed into a lower $\boldsymbol{L}$, a diagonal $\boldsymbol{D}$, and an upper $\boldsymbol{U}$ matrix:

$$\boldsymbol{H} = \boldsymbol{L} + \boldsymbol{D} + \boldsymbol{U} \tag{16.110}$$

The solution is then advanced iteratively over the $(p + 1)$ time step by a forward sweep:

$$\boldsymbol{\psi}_{n+1/2}(p + 1) = \theta[\boldsymbol{I} - \Delta t_{p+1}(\boldsymbol{L} + \boldsymbol{D})]^{-1}[\Delta t_{p+1}\boldsymbol{U}\boldsymbol{\psi}_n(p + 1) + \boldsymbol{\psi}(p)]$$
$$+ (1 - \theta)\boldsymbol{\psi}_n(p + 1) \tag{16.111}$$

followed by a backward sweep:

$$\boldsymbol{\psi}_{n+1}(p + 1) = \theta[\boldsymbol{I} - \Delta t_{p+1}(\boldsymbol{D} + \boldsymbol{U})]^{-1}[\Delta t_{p+1}\boldsymbol{L}\boldsymbol{\psi}_{n+1/2}(p + 1) + \boldsymbol{\psi}(p)]$$
$$+ (1 - \theta)\boldsymbol{\psi}_{n+1/2}(p + 1) \tag{16.112}$$

where $1 \leq \theta \leq 2$ and n refers to the iteration number.

An exponential transformation of the multigroup fluxes and the precursor densities

$$\boldsymbol{\psi}(p + 1) = \exp(\Delta t_{p+1}\boldsymbol{\omega})\hat{\boldsymbol{\psi}}(p + 1) \tag{16.113}$$

may be made first, using dynamic frequencies calculated from local flux and precursor values for the present and previous times:

$$\omega_\phi^g \equiv \frac{1}{\Delta t_{p+1}} \ln \frac{\phi_g(p)}{\phi_g(p - 1)}, \qquad \omega_c^m(p + 1) \equiv \frac{1}{\Delta t_{p+1}} \ln \frac{C_m(p)}{C_m(p - 1)} \tag{16.114}$$

With the transformation of Eq. (16.113), Eq. (16.52) becomes

$$\frac{\partial}{\partial t}\hat{\boldsymbol{\psi}} = \exp(-\Delta t_{p+1}\boldsymbol{\omega})[\boldsymbol{H} - \boldsymbol{\omega}]\exp(\Delta t_{p+1}\boldsymbol{\omega})\hat{\boldsymbol{\psi}} \tag{16.115}$$

which is integrated using the over-relaxation procedure of Eqs. (16.111) and (16.112).

The dynamic frequencies are estimated at the beginning of the time step from Eqs. (16.114) to determine ω_0. A global frequency correction factor $\Delta\omega_n$ is computed on each iteration by considering Eqs. (16.111) and (16.112) to each advance the solution a half time step. The dynamic frequency is then corrected:

$$\boldsymbol{\Omega}_n = \boldsymbol{\Omega}_0 + \omega_n \boldsymbol{I} \tag{16.116}$$

where now $\boldsymbol{\Omega}$ is a matrix containing the local values of the frequencies ω.

Generalized Runge–Kutta Methods

Runge–Kutta methods have long been popular for integrating ordinary differential equations, but the requirement for small time steps to achieve sufficient accuracy has limited their application in solving space-discretized space–time neutron kinetics problems. However, generalizations of these methods to allow larger time steps and increased stability (Ref. 3) have recently been applied to these problems. The Runge–Kutta method is based on an explicit time differencing of Eq. (16.52) and a linear Taylor's series approximation:

$$\boldsymbol{\psi}(p+1) = \boldsymbol{\psi}(p) + \Delta t_{p+1} \boldsymbol{H}(p+1)\boldsymbol{\psi}(p+1)$$

$$\simeq \boldsymbol{\psi}(p) + \Delta t_{p+1}\left[\boldsymbol{H}(p)\boldsymbol{\psi}(p) + \Delta t_{p+1}\left\{\frac{\partial}{\partial\boldsymbol{\psi}}(\boldsymbol{H}\boldsymbol{\psi})\bigg|_p\right.\right.$$

$$\left.\left. \times \frac{1}{\Delta t_{p+1}}[\boldsymbol{\psi}(p+1) - \boldsymbol{\psi}(p)]\right\}\right] \qquad (16.117)$$

where the term $\partial(\boldsymbol{H}\boldsymbol{\psi})/\partial\boldsymbol{\psi}|_p$ is the partial derivative of the left side of Eq. (16.52) with respect to the appropriate multigroup neutron flux or delayed neutron precursor density evaluated at the beginning of the time step, $t = t_p$.

The generalized Runge–Kutta methods are based on the algorithm

$$\boldsymbol{y}(p+1) = \boldsymbol{y}(p) + \sum_{i=1}^{s} c_i \boldsymbol{K}_i(p+1) \qquad (16.118)$$

for advancing the solution from t_p to t_{p+1}, where s is the number of stages, c_i are fixed expansion coefficients, and the column vectors $\boldsymbol{K}(p+1)$ are found by solving a system of N (the number of energy groups times discrete spatial points plus the number of delayed neutron precursor groups times the number of discrete spatial points) linear equations for each of the s stages (i.e., for s different right sides for each time step):

$$\left[\boldsymbol{I} - \gamma\Delta t_{p+1}\frac{\partial}{\partial\boldsymbol{\psi}}(\boldsymbol{H}\boldsymbol{\psi})\bigg|_p\right]\boldsymbol{K}_i(p+1)$$

$$= \Delta t_{p+1}\boldsymbol{H}\boldsymbol{\psi}|_{p^*} + \Delta t_{p+1}\left[\frac{\partial}{\partial\boldsymbol{\psi}}(\boldsymbol{H}\boldsymbol{\psi})\bigg|_{p^*}\sum_{m=1}^{i-1}\gamma_{im}\boldsymbol{K}_m(p+1)\right], \quad i = 1, \ldots, I$$

$$(16.119)$$

where $\boldsymbol{H}\boldsymbol{\psi}|_{p^*}$ is the evaluation of the left side of Eq. (16.52) at the intermediate points t_{p^*} where the solution vector is given by

$$\boldsymbol{\psi}(p^*) = \boldsymbol{\psi}(p) + \sum_{m=1}^{i-1}\alpha_{im}\boldsymbol{K}_m(p+1) \qquad (16.120)$$

where γ, γ_m, and α_{im} are fixed constants. The scheme is well suited for a variable time step because it employs an embedded Runge–Kutta–Fehlberg estimate for $\psi(p + 1)$, which provides the capability to monitor truncation error without increasing computational time.

16.4
Stability

In a nuclear reactor operating at steady-state conditions, an equilibrium obtains among the interacting neutronic, thermodynamic, hydrodynamic, thermal, xenon, and so on, phenomena. The state of the reactor is defined in terms of the values of the state functions[1] associated with each of these phenomena (e.g., the neutron flux, the coolant enthalpy, the coolant pressure). If a reactor is perturbed from an equilibrium state, will the ensuing state (1) remain bounded within some specified domain of the state functions, (2) return to the equilibrium state after a sufficiently long time, or (3) diverge from the equilibrium state in that one or more of the state functions takes on a shape outside a specified domain of state functions? This is the question of stability.

In this section we extend the concepts of Section 5.9 to outline a theory appropriate for the stability analysis of spatially dependent reactor models. First, we consider the stability analysis of the coupled system of ordinary differential equations that results when the spatial dependence is discretized by a finite-difference, nodal, or other approximation. Then the extended Lyapunov theory for the stability analysis of the coupled partial differential equations which describe spatially continuous systems is discussed.

Classical Linear Stability Analysis

The finite-difference, time-synthesis, nodal, or point kinetics approximations, and the corresponding approximations to the other state function equations, may be written as a coupled set of ordinary differential equations relating the discrete state variables y_i:

$$\dot{y}_i(t) = f_i\big(y_i(t), \ldots, y_N(t)\big), \quad i = 1, \ldots, N \tag{16.121}$$

where, for instance, y_i may be the neutron flux at node i and y_{l+j} may be the coolant enthalpy at node j. The coupling among the equations arises because the cross sections in the neutronics equations depend on the local temperature, density, and xenon concentration, because the temperature, density, and xenon concentration depend on the local flux, and because neutron and heat diffusion and coolant transport introduces a coupling among the value of the state variables at different locations.

---

1) In a spatially dependent system such as a nuclear reactor, the state of the system is defined in terms of spatially dependent state functions. When the spatial dependence is discretized by one of the approximations discussed in previous sections, the state of the system is defined in terms of discrete state variables.

Equations (16.121) may be written as a vector equation,

$$\dot{\mathbf{y}}(t) = \mathbf{f}(\mathbf{y}(t)) \tag{16.122}$$

where the components of the column vectors $\mathbf{y}$ and $\mathbf{f}$ are the y_i and f_i, respectively. The equilibrium state $\mathbf{y}_e$ satisfies

$$\mathbf{f}(\mathbf{y}_e) = 0 \tag{16.123}$$

If the solution of Eq. (16.122) is expanded about $\mathbf{y}_e$,

$$\mathbf{y}(t) = \mathbf{y}_e + \hat{\mathbf{y}}(t) \tag{16.124}$$

and the part of the right-hand side of Eq. (16.122) that is linear in $\hat{\mathbf{y}}$ is separated out, Eq. (16.122) may be written

$$\dot{\hat{\mathbf{y}}}(t) = \mathbf{h}(\mathbf{y}_e)\hat{\mathbf{y}}(t) + \mathbf{g}(\mathbf{y}_e, \hat{\mathbf{y}}(t)) \tag{16.125}$$

The matrix $\mathbf{h}$ has constant elements, some of which may depend on the equilibrium state.

Classical linear stability analysis proceeds by ignoring the nonlinear term g in Eq. (16.125). It is readily shown that the condition for the stability of the linearized equations is that the real part of all eigenvalues of the matrix $\mathbf{h}$ are negative. To illustrate this, apply a permutation transformation that diagonalizes $\mathbf{h}$ to the linear approximation to Eq. (16.125):

$$\mathbf{P}^{\mathrm{T}}\dot{\hat{\mathbf{y}}}(t)\mathbf{P} = \mathbf{P}^{\mathrm{T}}\mathbf{h}\mathbf{P}\mathbf{P}^{\mathrm{T}}\hat{\mathbf{y}}(t)\mathbf{P} \tag{16.126}$$

since

$$\mathbf{P}^{\mathrm{T}}\mathbf{P} = \mathbf{P}\mathbf{P}^{\mathrm{T}} = \mathbf{I} \tag{16.127}$$

Define $\mathbf{X}(t) = \mathbf{P}^{\mathrm{T}}\hat{\mathbf{y}}(y)\mathbf{P}$. Then the transformed equations are decoupled:

$$\dot{X}_i(t) = \omega_i X_i(t), \quad i = 1, \ldots, N \tag{16.128}$$

where ω_i are the eigenvalues of $\mathbf{h}$. The solutions of these equations subject to $X_i(0) = X_{i0}$ are

$$X_i(t) = X_{i0}e^{\omega_i t}, \quad i = 1, \ldots, N \tag{16.129}$$

which may be written in vector notation as

$$\mathbf{X}(t) = \mathbf{\Gamma}(t)\mathbf{X}_{i0} \tag{16.130}$$

where $\mathbf{\Gamma}(t) = \mathrm{diag}(\exp(\omega_i t))$. Hence

$$\hat{\mathbf{y}}(t) = \mathbf{P}\mathbf{X}(t)\mathbf{P}^{\mathrm{T}} = \mathbf{P}\mathbf{\Gamma}(t)\mathbf{X}_{i0}\mathbf{P}^{\mathrm{T}} \tag{16.131}$$

If $\text{Re}\{\omega_i\} < 0$, $\lim_{t\to\infty} \hat{\mathbf{y}}(t) = 0$ (i.e., the state of the system returns to the equilibrium state). If $\text{Re}\{\omega_i\} > 0$, one or more of the components of $\hat{\mathbf{y}}$ approach ∞ as $t \to \infty$, and the system is unstable. Thus stability analysis of the linearized equations amounts to determining if the eigenvalues of the $\mathbf{h}$ matrix are in the left (stable) or right (unstable)-half complex plane. This determination may be accomplished most readily by Laplace transforming the linearized equation into the frequency domain and then applying one of the methods of linear control theory (e.g., Bode, Nyquist, root locus, Hurwitz) that have been developed explicitly for this purpose. This methodology was applied in the stability analyses of Chapter 5.

Lyapunov's Method

The method of Lyapunov attempts to draw certain conclusions about the stability of the solution of Eq. (16.125) without any knowledge of this solution. Essential to this method is the choice of a scalar function $V(\hat{\mathbf{y}})$ which is a measure of a metric distance of the state $\mathbf{y} = \mathbf{y}_e + \hat{\mathbf{y}}$ from the equilibrium state $\mathbf{y}_e$. Let $\hat{\mathbf{y}}\{t, \hat{\mathbf{y}}_0\}$ be the solution of Eq. (16.125) for the initial condition $\hat{\mathbf{y}}(t = 0) = \hat{\mathbf{y}}_0$. If it can be shown that $V(\hat{\mathbf{y}}(t, \hat{\mathbf{y}}_0))$ will be small when $V(\hat{\mathbf{y}}_0)$ is small, then $\mathbf{y}_e$ is a stable equilibrium state. If, in addition, it can be shown that $V(\hat{\mathbf{y}}(t, \hat{\mathbf{y}}_0))$ approaches zero for large times, $\mathbf{y}_e$ is an asymptotically stable equilibrium state.

Define a scalar function $V(\hat{\mathbf{y}})$ that depends on all the state variables $\hat{y}_i$ and which has the following properties in some region $\mathcal{R}$ about the equilibrium state $\mathbf{y}_e$:

1. $V(\hat{\mathbf{y}})$ is positive definite [i.e., $V(\hat{\mathbf{y}}) > 0$ if $\hat{\mathbf{y}} \neq 0$, $V(\hat{\mathbf{y}}) = 0$ if $\hat{\mathbf{y}} = 0$].
2. $\lim_{\hat{\mathbf{y}}\to 0} V(\hat{\mathbf{y}}) = 0$, $\lim_{\hat{\mathbf{y}}\to\infty} V(\hat{\mathbf{y}}) = \infty$.
3. $V(\hat{\mathbf{y}})$ is continuous in all its partial derivatives (i.e., $\partial V/\partial y_i$ exist and are continuous for $i = 1, \ldots, N$).
4. $\dot{V}(\hat{\mathbf{y}})$ evaluated along the solution of Eq. (16.125) is nonpositive; that is,

$$\dot{V}(\hat{\mathbf{y}}) = \sum_{i=1}^{M} \frac{\partial V}{\partial \hat{y}_i} \dot{\hat{y}}_i = \sum_{i=1}^{N} \frac{\partial V}{\partial \hat{y}_i} f_i \leq 0 \tag{16.132}$$

A scalar function $V(\hat{\mathbf{y}})$ satisfying properties 1 to 4 is a Lyapunov function.

Three theorems based on the Lyapunov function can be stated about the equilibrium solution of Eq. (16.125).

Theorem 16.1 (Stability Theorem). *If a Lyapunov function exists in some region $\mathcal{R}$ about $\mathbf{y}_e$, this equilibrium state is stable for all initial perturbations in $\mathcal{R}$ [i.e., for all initial perturbations $\hat{\mathbf{y}}_0$ in $\mathcal{R}$, the solution of Eq. (16.125), $\mathbf{y}(t, \hat{\mathbf{y}}_0)$, remains within the region $\mathcal{R}$ for all $t > 0$].*

Theorem 16.2 (Asymptotic Stability Theorem). *If a Lyapunov function exists in some region $\mathcal{R}$ about $\mathbf{y}_e$, and in addition $\dot{V}$ evaluated along the solution of Eq. (16.125) is negative definite ($\dot{V} < 0$ if $\hat{\mathbf{y}} \neq 0$, $\dot{V} = 0$ if $\hat{\mathbf{y}} = 0$) in $\mathcal{R}$, this equilibrium state is* *asymptotically stable for all initial perturbations in $\mathscr{R}$ [i.e., for all initial perturbations $\hat{\mathbf{y}}_0$ in $\mathscr{R}$, the solution of Eq. (16.125) is $\hat{\mathbf{y}}(t, \hat{\mathbf{y}}_0) = 0$ after a sufficiently long time].*

Theorem 16.3 (Instability Theorem). *If a scalar function $V(\hat{\mathbf{y}})$ which has properties 1 to 3 exists in a region $\mathscr{R}$, and $\dot{V}$ evaluated along the solution of Eq. (16.125) does not have a definite sign, the equilibrium state $\mathbf{y}_e$ is unstable for initial perturbations in $\mathscr{R}$ [i.e., for initial perturbations $\hat{\mathbf{y}}_0$ in $\mathscr{R}$, the solution of Eq. (16.125), $\hat{\mathbf{y}}(t, \hat{\mathbf{y}}_0)$, does not remain in $\mathscr{R}$ for all $t > 0$].*

Mathematical proofs of these theorems can be constructed. Rather than repeat these proofs, which may be found in the literature (e.g., Ref. 14), it is more informative to consider a topological argument. Properties 1 to 3 define a concave upward surface (the function V) in the phase space defined by the $\hat{y}_i$. This surface has a minimum within the region $\mathscr{R}$ at $\hat{y}_1 = \cdots = \hat{y}_N = 0$ by property 1, and increases monotonically in value as the $\hat{y}_i$ increase, by properties 2 and 3. Thus contours can be drawn in the hyperplane of the $\hat{y}_i$ representing the locus of points at which V has a given value. These contours are concentric about the equilibrium state $\hat{y}_i = 0, i = 1, \ldots, N$. Proceeding outward from this origin, the value of V associated with each contour is greater than the value associated with the previous contour. In other words, $V(\mathbf{y})$ is a bowl in the hyperspace of the y_i, with center at $\hat{y}_i = 0, i = 1, \ldots, N$.

The outward normal to those contours is

$$\sum_{i=1}^{N} \frac{\partial V}{\partial \hat{y}_i} \mathbf{i}$$

where $\mathbf{i}$ denotes the unit vector in the direction in phase space associated with the state variable $\hat{y}_i$. The direction in which the state of the system is moving in phase space is given by

$$\sum_{i=1}^{N} \dot{y}_i \mathbf{i} = \sum_{i=1}^{N} f_i \mathbf{i} \tag{16.133}$$

For stability, the direction in which the state of the system is moving must never be toward regions in which V is larger (i.e., never away from the equilibrium state):

$$\left(\sum_{i=1}^{N} \frac{\partial V}{\partial \hat{y}_i} \mathbf{i} \right) \cdot \left(\sum_{j=1}^{N} f_j \mathbf{j} \right) = \sum_{i=1}^{N} \frac{\partial V}{\partial \hat{y}_i} f_i \leq 0 \tag{16.134}$$

For asymptotic stability, the state of the system must always move toward regions in which V is smaller (i.e., always move toward the equilibrium state). Thus the inequality must always obtain in the foregoing relation. If the system can move away from the equilibrium state into regions of larger V, the $\leq$ is replaced by $>$ in the foregoing relation and the equilibrium state is unstable.

The Lyapunov method yields the same results obtained in the preceding section in the limit in which the nonlinear terms are small. The function

$$V(\hat{\boldsymbol{y}}) = \hat{\boldsymbol{y}}^{\mathrm{T}}\hat{\boldsymbol{y}} = \sum_{i=1}^{N}(\hat{y}_i)^2 \qquad (16.135)$$

satisfies properties 1 to 3. Making use of Eq. (16.125) yields

$$\dot{V}(\hat{\boldsymbol{y}}) = \sum_{i=1}^{N}\frac{\partial V}{\partial \hat{y}_i}\dot{\hat{y}}_i = 2\sum_{i=1}^{N}\hat{y}_i\dot{\hat{y}}_i = 2\hat{\boldsymbol{y}}^{\mathrm{T}}\dot{\hat{\boldsymbol{y}}} = 2\{\hat{\boldsymbol{y}}^{\mathrm{T}}\boldsymbol{h}\hat{\boldsymbol{y}} + \hat{\boldsymbol{y}}^{\mathrm{T}}\boldsymbol{g}\} \qquad (16.136)$$

If the region $\mathscr{R}$ is defined such that

$$\left|\hat{\boldsymbol{y}}^{\mathrm{T}}\boldsymbol{h}\hat{\boldsymbol{y}}\right| > \left|\hat{\boldsymbol{y}}^{\mathrm{T}}\boldsymbol{g}\right|$$

a sufficient condition for $\dot{V}$ to be negative definite in $\mathscr{R}$ is that $\hat{\boldsymbol{y}}^{\mathrm{T}}\boldsymbol{h}\hat{\boldsymbol{y}}$ is negative definite, a sufficient condition for which is that the eigenvalues of $\boldsymbol{h}$ have negative real parts. This is the same result obtained in the linear analysis of the preceding section. In this case, the Lyapunov method provides, in addition, the region $\mathscr{R}$ within which the linear analysis is valid.

In applying the Lyapunov method, construction of a suitable Lyapunov function is the main consideration. Because the Lyapunov function for a system of equations is not unique, the analysis yields sufficient, but not necessary, conditions for stability.

Lyapunov's Method for Distributed Parameter Systems

A more basic characterization of a reactor system is in terms of spatially distributed state functions, rather than discrete state variables. These state functions satisfy coupled partial differential equations, which may be written

$$\dot{y}_i(r,t) = f_i\big(y_1(r,t),\ldots,y_N(r,t),r\big), \quad i = 1,\ldots,N \qquad (16.137)$$

where y_i is a state function (e.g., neutron group flux) and f_i denotes a spatially dependent operation involving scalars and spatial derivatives on the state functions. These equations may be written

$$\dot{\boldsymbol{y}}(r,t) = \boldsymbol{f}\big(\boldsymbol{y}(r,t),r\big) \qquad (16.138)$$

where $\boldsymbol{y}$ is a column vector of the y_i and $\boldsymbol{f}$ is a column vector of the operations denoted by the f_i.

The extension of Lyapunov's methods to systems described by state functions involves the choice of a functional that provides a measure of the distance of the vector of state functions $\boldsymbol{y}$ from a specified equilibrium state, $\boldsymbol{y}_{\mathrm{eq}}$. The distance between two states $\boldsymbol{y}_a$ and $\boldsymbol{y}_b$, $d[\boldsymbol{y}_a,\boldsymbol{y}_b]$, is defined as the metric on the product state function space consisting of all possible functions of position that the component state functions can take on.

An equilibrium state $\mathbf{y}_{eq}(r)$ satisfying

$$f\left(\mathbf{y}_{eq}(r), r\right) = 0 \tag{16.139}$$

is stable if, for any number $\varepsilon > 0$, it is possible to find a number $\delta > 0$ such that when

$$d[\mathbf{y}_0(r), \mathbf{y}_{eq}(r)] < \delta$$

then

$$d[\mathbf{y}(r, t; \mathbf{y}_0), \mathbf{y}_{eq}(r)] < \varepsilon \quad \text{for } t \geq 0 \tag{16.140}$$

where $\mathbf{y}(r, t; \mathbf{y}_0)$ is the solution of Eq. (16.138) with the initial condition $\mathbf{y}(r, 0) = \mathbf{y}_0(r)$. If in the limit of large t, the distance $d[\mathbf{y}(r, t; \mathbf{y}_0), \mathbf{y}_{eq}]$ approaches zero, then $\mathbf{y}_{eq}$ is asymptotically stable.

Theorem 16.4 (Stability Theorem). *For an equilibrium state $\mathbf{y}_{eq}(r)$ to be stable, it is necessary and sufficient that in some neighborhood of $\mathbf{y}_{eq}(r)$ that includes the equilibrium state there exists a functional $V[\mathbf{y}]$ with the following properties:*

1. *V is positive definite with respect to $d[\mathbf{y}, \mathbf{y}_{eq}]$; that is, for any $C_1 > 0$, there exists a $C_2 > 0$ depending on C_1 such that when $d[\mathbf{y}, \mathbf{y}_{eq}] > C_1$, then $V[\mathbf{y}] > C_2$ for all $t \geq 0$, and $\lim_{d[\mathbf{y}, \mathbf{y}_{eq}] \to 0} V[\mathbf{y}] = 0$.*
2. *V is continuous with respect to $d[\mathbf{y}, \mathbf{y}_{eq}]$; that is, for any real $\varepsilon > 0$, there exists a real $\delta > 0$ such that $V[\mathbf{y}] < \varepsilon$ for all $\mathbf{y}$ in the state function space for $0 < t < \infty$, when $d[\mathbf{y}_0, \mathbf{y}_{eq}] < \delta$.*
3. *$V[\mathbf{y}]$ evaluated along any solution $\mathbf{y}$ of Eq. (16.138) is nonincreasing in time for all $t > 0$ provided that $d[\mathbf{y}_0, \mathbf{y}_{eq}] < \delta_0$, where δ_0 is a sufficiently small positive number.*

Theorem 16.5 (Asymptotic Stability Theorem). *If, in addition to these three conditions, $V[\mathbf{y}]$ evaluated along any solution to Eq. (16.138) approaches zero for large t, the equilibrium state is asymptotically stable.*

The same type of topological arguments made above in support of the theorems for the discrete representation of spatial dependence by coupled ODES are appropriate here, if the state space is generalized to a state function space. Construction of a suitable Lyapunov functional is the essential aspect of applying the theory of this section. Although the conditions cited in the theorems are necessary and sufficient for stability, the V-functional chosen may result in more restrictive stability criteria than would be obtained from another V-functional. Thus stability analyses employing Lyapunov functionals yield only sufficient conditions for stability.

Control

An intended change in the operating state of a nuclear reactor is produced by a control action (e.g., withdrawing a bank of control rods, increasing the coolant flow). The nature of the change in operating state depends on the control action, of course, and a great deal of practical experience exists on how to effect a desired change. However, in some cases the intuitive control action can exacerbate, rather than correct, a problem—the control-induced xenon spatial oscillations in the large production reactors being a good example. The methodology of control theory has found some application in nuclear reactor control, and a brief review is provided in this section.

Variational Methods of Control Theory

When discrete spatial approximations (e.g., nodal, finite-difference) are employed, the dynamics of a spatially dependent nuclear reactor model are described by a system of ordinary differential equations

$$\dot{y}_i(t) = f_i(y_1, \ldots, y_N, u_1, \ldots, u_R), \quad i = 1, \ldots, N \tag{16.141}$$

with the initial conditions

$$y_i(t = t_0) = y_{i0}, \quad i = 0, \ldots, N \tag{16.142}$$

The y_i are the state variables (e.g., nodal neutron flux, temperature) and the u_r are control variables (e.g., control rod cross section in a node). Equation (16.141) may be written more compactly by defining vector variables $\mathbf{y}$, $\mathbf{u}$, and $\mathbf{f}$:

$$\dot{\mathbf{y}}(t) = \mathbf{f}(\mathbf{y}(t), \mathbf{u}(t)) \tag{16.143}$$

Many problems in control may be formulated as a quest for the control vector $\mathbf{u}^*$ that causes the solution of Eq. (16.143), $\mathbf{y}^*$, to minimize a functional:[2]

$$J[\mathbf{y}] = \int_{t_0}^{t_f} dt\, F(\mathbf{y}(t), \dot{\mathbf{y}}(t)) \tag{16.144}$$

This control problem may be formulated within the framework of the classical calculus of variations by treating the control variables as equivalent to the state variables. The theory of the calculus of variations is restricted to variables that are continuous in time, which limits the admissible set of control variables.

---

2) Functionals of this form may arise when the objective of the control program is to correct a flux perturbation in such a manner as to minimize the deviation from the nominal flux distribution, at the same time minimizing the rate of change of local flux densities. Other typical control problems are those in which the objective is to attain a given final state in a minimum time; a functional with $F = 1$ and an additional term that provides a measure of the deviation from the specified final state is appropriate in this case.

The system equations are treated as constraints or subsidiary conditions, and are included in the functional with Lagrange multiplier variables:

$$J'[\mathbf{y},\mathbf{u},\boldsymbol{\lambda}] = \int_{t_0}^{t_f} dt \left[F(\mathbf{y}(t),\dot{\mathbf{y}}(t)) + \sum_{i=1}^{N}\lambda_i(t)\{\dot{y}_i(t) - f_i(\mathbf{y}(t),\mathbf{u}(t))\} \right]$$
(16.145)

Variations of the modified functional J' (with respect to each y_i and u_r) are required to vanish at the minimum:

$$\delta J' = 0 = \int_{t_0}^{t_f} dt \left\{ \sum_{i=1}^{N}\left[\frac{\partial F}{\partial y_i}\delta y_i + \frac{\partial F}{\partial \dot{y}_i}\delta\dot{y}_i + \lambda_i\delta\dot{y}_i - \sum_{j=1}^{N}\lambda_j\frac{\partial f_j}{\partial y_i}\delta y_i \right. \right.$$

$$\left. \left. - \sum_{r=1}^{R}\lambda_i\frac{\partial f_i}{\partial u_r}\delta u_r \right] \right\}$$
(16.146)

Integrating the $\delta\dot{y}_i$ terms by parts and using the initial conditions to set $\delta y_i(t_0) = 0$, this expression becomes

$$\delta J' = 0 = \int_{t_0}^{t_f} dt \sum_{i=1}^{N}\left\{ \left[\frac{\partial F}{\partial y_i} - \frac{\partial}{\partial t}\frac{\partial F}{\partial \dot{y}_i} - \dot{\lambda}_i - \sum_{j=1}^{N}\lambda_i\frac{\partial f_j}{\partial y_i} \right]\delta y_i - \sum_{r=1}^{R}\lambda_i\frac{\partial f_i}{\partial u_r}\delta u_r \right\}$$

$$+ \sum_{i=1}^{N}\left(\lambda_i + \frac{\partial F}{\partial \dot{y}_i} \right)\delta y_i \bigg|_{t=t_f}$$
(16.147)

In order that Eq. (16.147) be satisfied for arbitrary (but continuous) variations δy_i and δu_r, it is necessary that

$$\dot{\lambda}_i(t) = -\sum_{j=1}^{N}\lambda_j(t)\frac{\partial f_j}{\partial y_i}(\mathbf{y},\mathbf{u})$$

$$+ \left[\frac{\partial F}{\partial y_i}(\mathbf{y},\mathbf{f}) - \frac{\partial}{\partial t}\frac{\partial F}{\partial \dot{y}_i}(\mathbf{y},\mathbf{f}) \right], \quad i = 1,\ldots,N$$
(16.148)

$$\sum_{i=1}^{N}\lambda_i(t)\frac{\partial f_i}{\partial u_r}(\mathbf{y},\mathbf{u}) = 0, \quad r = 1,\ldots,R$$
(16.149)

and that λ_i satisfy the final conditions

$$\lambda_i(t_f) + \frac{\partial F}{\partial \dot{y}_i}\bigg|_{t_f} = 0, \quad i = 1,\ldots,N$$
(16.150)

Equations (16.141), (16.148), and (16.149) must be solved simultaneously, subject to the initial conditions of Eqs. (16.142) and the final conditions of Eq. (16.150), for the optimal controls $u_r^*(t)$ and the optimal solutions $y_i^*(t)$.

In many problems, additional constraints are placed on the allowable values that may be taken on by the state variables and control variables. Constraints of the form

$$\phi_m(\mathbf{y}(t), \mathbf{u}(t)) = 0, \quad m = 1, \ldots, M < N$$

or

$$\phi_m\big(\mathbf{y}(t), \dot{\mathbf{y}}(t), \mathbf{u}(t), \dot{\mathbf{u}}(t)\big) = 0, \quad m = 1, \ldots, M < N$$

may be added to the functional of Eq. (16.144) with Lagrange multiplier variables and treated in the same fashion as before. Equations for additional Lagrange multiplier variables and the additional constraint equations are included with Eqs. (16.141), (16.148), and (16.149) in this case.

When integral constraints of the form

$$\int_{t_0}^{t_f} dt\, \phi_m\big(\mathbf{y}(t), \dot{\mathbf{y}}(t), \mathbf{u}(t), \dot{\mathbf{u}}(t)\big) = 0, \quad m = 1, \ldots, M < N$$

are present, the functional of Eq. (16.144) is modified with Lagrange multiplier constants ω_m,

$$J \to \hat{J} = \int_{t_0}^{t_f} dt\left(F + \sum_{m=1}^{M} \omega_m \phi_m\right) = \int_{t_0}^{t_f} dt\, \hat{F} \tag{16.151}$$

and the derivation proceeds as before with $F \to \hat{F}$. In addition to Eqs. (16.141), (16.148), and (16.149), the constraint equations and expressions for the ω_m are obtained.

Inequality constraints (e.g., maximum control rod shim rates) are encountered frequently. Although these can sometimes be reduced to equivalent equality constraints of one of the three types discussed, they generally constitute a class of problems that are difficult to treat within the framework of the calculus of variations. Another class of such problems is those for which the optimal control is discontinuous.

Dynamic Programming

An alternative treatment of the variational problem that circumvents the requirement for continuous control variables is provided by dynamic programming. Consider the problem of determining the control vector $\mathbf{u}^*(t)$ that causes the solution $\mathbf{y}^*(t)$ of Eq. (16.143) to minimize the functional of Eq. (16.144), subject to constraints on the allowable values of the control variable that may be represented by

$$\mathbf{u}(t) \in \mathbf{\Omega}(t) \tag{16.152}$$

To develop the dynamic programming formalism, consider the functional of Eq. (16.144) evaluated between a variable lower limit $(t, \mathbf{y}(t))$ and a fixed upper limit $(t_f, \mathbf{y}(t_f))$. Define the minimum value of this functional as S, a function of the lower, variable limit $(t, \mathbf{y}(t))$:

$$S(t, \mathbf{y}(t)) = \min_{u \in \Omega} \int_t^{t_f} dt'\, F\big(\mathbf{y}(t'), \dot{\mathbf{y}}(\mathbf{y}(t'), \mathbf{u}(t'))\big) \qquad (16.153)$$

In writing Eq. (16.153), $\dot{\mathbf{y}}$ is written as an explicit function of $\mathbf{y}$ and $\mathbf{u}$ to indicate that Eq. (16.143) must be satisfied in evaluating the integrand.

By definition of S, for $\Delta t > 0$,

$$S(t, \mathbf{y}(t)) \le S(t + \Delta t, \mathbf{y}(t + \Delta t)) + \int_t^{t+\Delta t} dt'\, F\big(\mathbf{y}(t'), \dot{\mathbf{y}}(\mathbf{y}(t'), \mathbf{u}(t'))\big) \qquad (16.154)$$

where $\mathbf{y}(t + \Delta t)$ and $\mathbf{y}(t)$ are related by Eq. (16.143); that is,

$$\mathbf{y}(t + \Delta t) = \mathbf{y}(t) + \Delta t\, \mathbf{f}\big(\mathbf{y}(t), \mathbf{u}(t)\big) + O(\Delta t^2) \qquad (16.155)$$

For the optimal choice of $\mathbf{u}(t') = \mathbf{u}^*$ in the interval $t \le t' \le t + \Delta t$, the equality obtains in Eq. (16.154). Approximating the integral in Eq. (16.154) by taking the integrand constant at its value at t, this equation becomes [3]

$$S(t, \mathbf{y}(t)) = \min_{\mathbf{u}(t) \in \boldsymbol{\Omega}(t)} \Big[S(t + \Delta t, \mathbf{y}(t + \Delta t)) + \Delta t\, F\big(\mathbf{y}(t), \mathbf{f}(\mathbf{y}(t), \mathbf{u}(t))\big) \Big] \qquad (16.156)$$

Equation (16.156) can be solved by retrograde calculation, starting with the final condition

$$S(t_f, \mathbf{y}(t_f)) = \min_{u \in \Omega} \int_{t_f}^{t_f} dt'\, F(\mathbf{y}, \dot{\mathbf{y}}) = 0 \qquad (16.157)$$

In each step of the retrograde solution, the optimal manner to proceed from each possible state $y(t)$ to time t_f is computed. Thus, when the initial time is reached, the optimal control at each discrete time and the corresponding sequence of states constituting the optimal trajectory are known.

Pontryagin's Maximum Principle

When a Taylor's series expansion of the first term on the right of Eq. (16.156) is made, this equation becomes

$$0 = \min_{\mathbf{u}(t) \in \boldsymbol{\Omega}(t)} \Bigg[\frac{\partial S}{\partial t}(t, \mathbf{y}(t)) + \sum_{i=1}^N \frac{\partial S}{\partial y_i}(t, \mathbf{y}(t)) f_i\big(\mathbf{y}(t), \mathbf{u}(t)\big)$$

$$+ F\big(\mathbf{y}(t), \mathbf{f}(\mathbf{y}(t), \mathbf{u}(t))\big) \Bigg] \qquad (16.158)$$

---

3) In Eq. (16.156), the minimization is with respect to the values of the control vector at time t. These values are assumed constant over the interval t to $t + \Delta t$. On the other hand, the minimization in Eq. (16.153) is with respect to the values taken on by the control vector at all times t', $t \le t' \le t_f$.

Define the variables

$$\psi_i(t) \equiv -\frac{\partial S}{\partial y_i}(t, \mathbf{y}(t)), \quad i = 1, \ldots, N \tag{16.159}$$

$$\psi_{N+1}(t) \equiv -\frac{\partial S}{\partial t}(t, \mathbf{y}(t)) \tag{16.160}$$

With these definitions, Eq. (16.158) becomes

$$0 = \min_{u(t) \in \Omega(t)} \left[-\psi_{N+1}(t) - \sum_{i=1}^{N} \psi_i(t) f_i(\mathbf{y}(t), \mathbf{u}(t)) + F(\mathbf{y}(t), \mathbf{f}(\mathbf{y}(t), \mathbf{u}(t))) \right]$$

which may be written

$$0 = \max_{u(t) \in \Omega(t)} \left[\psi_{N+1}(t) + \sum_{i=1}^{N} \psi_i(t) f_i(\mathbf{y}(t), \mathbf{u}(t)) - F(\mathbf{y}(t), \mathbf{f}(\mathbf{y}(t), \mathbf{u}(t))) \right]$$

$$\tag{16.161}$$

This is the maximum principle of Pontryagin.

When the vector $\mathbf{u}(t)$ takes on its optimal value, derivatives of the quantity within the square brackets with respect to t and y_i must vanish, which requires that

$$\frac{\partial \psi_{N+1}}{\partial y_j} + \sum_{i=1}^{N} \frac{\partial \psi_i}{\partial y_j} f_i = -\sum_{i=1}^{N} \psi_i \frac{\partial f_i}{\partial y_j} + \left(\frac{\partial F}{\partial y_j} + \sum_{i=1}^{N} \frac{\partial F}{\partial \dot{y}_i} \frac{\partial f_i}{\partial y_j} \right), \quad j = 1, \ldots, N$$

$$\frac{\partial \psi_{N+1}}{\partial t} + \sum_{i=1}^{N} \frac{\partial \psi_i}{\partial t} f_i = -\sum_{i=1}^{N} \psi_i \frac{\partial f_i}{\partial t}$$

Using the identities

$$\frac{d\psi_j}{dt} = -\sum_{i=1}^{N} \frac{\partial^2 S}{\partial y_i \partial y_j} f_i - \frac{\partial^2 S}{\partial t \partial y_j} = \sum_{i=1}^{N} \frac{\partial \psi_i}{\partial y_j} f_i + \frac{\partial y_{N+1}}{\partial y_j}, \quad j = 1, \ldots, N$$

$$\frac{d\psi_{N+1}}{dt} = -\sum_{i=1}^{N} \frac{\partial^2 S}{\partial y_i \partial t} f_i - \frac{\partial^2 S}{\partial t^2} = \sum_{i=1}^{N} \frac{\partial \psi_i}{\partial t} f_i + \frac{\partial \psi_{N+1}}{\partial t}$$

these equations become

$$\frac{d\psi_j}{dt} = -\sum_{i=1}^{N} \psi_i \frac{\partial f_i}{\partial y_j} + \left(\frac{\partial F}{\partial y_j} + \sum_{i=1}^{N} \frac{\partial F}{\partial \dot{y}_i} \frac{\partial f_i}{\partial y_j} \right), \quad j = 1, \ldots, N \tag{16.162}$$

$$\frac{d\psi_{N+1}}{dt} = -\sum_{i=1}^{N} \psi_i \frac{\partial f_i}{\partial t} \tag{16.163}$$

Appropriate final conditions for the ψ_i and ψ_{N+1} can be shown to be

$$\psi_1(t_f) = \cdots = \psi_N(t_f) = \psi_{N+1}(t_f) = 0 \tag{16.164}$$

Thus Eqs. (16.141), (16.161), (16.162), and (16.163) are solved simultaneously, subject to the initial and final conditions of Eqs. (16.142) and (16.164), respectively.

The computational procedure for solving either the calculus of variations or maximum principle equations is generally iterative. At $t = t_0$, the y_i are known from the initial conditions. When the maximum principle formulation is used, initial values of ψ_i are guessed, and the initial value of the control variables are determined from Eq. (16.161). Then the y_i and ψ_i are calculated at $t_0 + \Delta t$ from Eqs. (16.141) and (16.162) and (16.163) and the control is found from Eq. (16.161), and so on. This procedure is repeated in small time increments until the final time t_j. Then $\psi_i(t_f)$ and $\psi_{N+1}(t_f)$ are compared with the final conditions:

$$\psi_1(t_f) = \cdots = \psi_N(t_f) = \psi_{N+1}(t_f) = 0$$

and the initial values of ψ_i and ψ_{N+1} are changed and the entire process is repeated. This is continued until a set of initial values $\psi_i(t_0)$ and $\psi_{N+1}(t_0)$ are found that yield the correct final values.

Variational Methods for Spatially Dependent Control Problems

The basic description of the transient neutron flux and temperature distributions within a nuclear reactor is in terms of partial differential equations. It is not clear that the optimal control computed by first reducing these equations to ordinary differential equations by discretizing the spatial variable and then using the methods above is the same as would be obtained if the optimal control were determined directly from the partial differential equation description of the reactor dynamics. The variational formalism can be extended to the partial differential equation description of the reactor dynamics.

The state of the system is specified in terms of state functions $y_i(r, t)$ rather than discrete state variables as previously. The function space Γ_i, consisting of all possible functions of position that the state function y_i can take on, is a component function space, and the product space $\Gamma = \Gamma_1 \otimes \Gamma_2 \otimes \cdots \otimes \Gamma_N$ of all such component function spaces is the state function space on which the vector state function $\mathbf{y} = (y_1, \ldots, y_N)$ is defined. Similarly, the vector control function $\mathbf{u} = (u_1, \ldots, u_R)$ is defined on the product space of the component function spaces defined by all possible functions of position that the control functions u_r can take on. The distance between two states $\mathbf{y}_a$ and $\mathbf{y}_b$ is defined as the metric on Γ.

Equations for nuclear reactor dynamics can be written in the form

$$\dot{y}_i(r, t) = L_i(r)y_i(r, t) + f_i(\mathbf{y}, \mathbf{u})$$

$$y_i(r, t_0) = y_{i0}(r) \tag{16.165}$$

$$y_i(R, t) = 0, \quad i = 1, \ldots, N$$

where y_i denotes a state function, L_i contains a spatial differential operator acting on y_i and f_i is a spatially dependent function of $\mathbf{y}$ and $\mathbf{u}$. The outer boundary of the reactor is denoted by R. These equations may be written in matrix form:

$$\dot{\mathbf{y}}(r, t) = \mathbf{L}(r)\mathbf{y}(r, t) + \mathbf{f}(\mathbf{y}, \mathbf{u})$$
$$\mathbf{y}(r, 0) = \mathbf{y}_0(r), \qquad \mathbf{y}(R, t) = 0 \tag{16.166}$$

Many control problems may be formulated as the quest for the control vector function $\mathbf{u}$ for which the solution of Eq. (16.165) minimizes a functional

$$J[\mathbf{y}, \mathbf{u}] \int_{t_0}^{t_f} dt \int_V dr \, F\big(\mathbf{y}(r, t), \dot{\mathbf{y}}(r, t)\big) \tag{16.167}$$

The standard calculus-of-variations formulation of this problem begins by adding Eq. (16.166) to the integrand of Eq. (16.167) with a Lagrange multiplier vector function $\lambda(r, t) = (\lambda_i, \ldots, \lambda_N)$:

$$J'[\mathbf{y}, \mathbf{u}, \lambda] = \int_{t_0}^{t_f} dt \int_V dr \big[F(\mathbf{y}, \dot{\mathbf{y}}) + \lambda^{\mathrm{T}} (\dot{\mathbf{y}} - \mathbf{L}\mathbf{y} - \mathbf{f}(\mathbf{y}, \mathbf{u})) \big] \tag{16.168}$$

The control functions u_r are treated in the same fashion as the state functions y_i. Next, the variation of J' is required to vanish:

$$\delta J' = \int_{t_0}^{t_f} dt \int_V dr \sum_{i=1}^N \left(\frac{\partial F}{\partial y_i} \delta y_i + \frac{\partial F}{\partial \dot{y}_i} \delta \dot{y}_i + \lambda_i \delta \dot{y}_i - \lambda_i L_i \delta y_i \right.$$
$$\left. - \sum_{j=1}^N \lambda_i \frac{\partial f_i}{\partial y_j} \delta y_j - \sum_{r=1}^R \lambda_i \frac{\partial f_i}{\partial u_r} \delta u_r \right) = 0$$

Integration by parts of the terms involving $\delta \dot{y}_i$ and $L_i, \delta y_i$,[4] and use of the initial conditions $\delta y_i(r, t_0) = 0$ leads to

$$\delta J' = \int_{t_0}^{t_f} dt \int_V dt \sum_{i=1}^N \left[\delta y_i \left(\frac{\partial F}{\partial y_i} - \frac{\partial}{\partial t} \frac{\partial F}{\partial \dot{y}_i} - \dot{\lambda}_i - L_i^+ \lambda_i - \sum_{j=1}^N \lambda_j \frac{\partial f_j}{\partial y_i} \right) \right.$$
$$\left. - \sum_{r=1}^R \lambda_i \frac{\partial f_i}{\partial u_r} \delta u_r \right] + \int_V dr \sum_{i=1}^N \frac{\partial F}{\partial \dot{y}_i} \delta y_i \bigg|_{t=t_f}$$
$$+ \int_V dr \sum_{i=1}^N \lambda_i \delta y_i \bigg|_{t=t_f} + \int_{t_0}^{t_f} dt \sum_{i=1}^N P_i(\lambda_i) = 0 \tag{16.169}$$

In arriving at Eq. (16.169), the adjoint operator L_i^+ and the bilinear concomitant P_i are defined by the relation

$$\int_V dr \, \lambda_i L_i \delta y_i = \int_V dr \, \delta y_i L_i^+ \lambda_i + P_i(\lambda_i) \tag{16.170}$$

---

4) Commutability of the variational operator δ and the operators $\partial/\partial t$ and L_i imply an assumption of continuous variations δy_i, as does the existence of the integrals involving these terms.

$\delta J'$ must vanish for arbitrary variations of y_i and u_r, which requires that the Lagrange multiplier functions satisfy the partial differential equations

$$\dot{\lambda}_i(r, t) = -L_i^+(r)\lambda_i(r, t) - \sum_{j=1}^{N} \lambda_j(r, t)\frac{\partial f_j(\mathbf{y}, \mathbf{u})}{\partial y_i}$$

$$+ \left(\frac{\partial F(\mathbf{y}, \dot{\mathbf{y}})}{\partial y_i} - \frac{\partial}{\partial t}\frac{\partial F(\mathbf{y}, \dot{\mathbf{y}})}{\partial \dot{y}_i}\right), \quad i = 1, \ldots, N \qquad (16.171)$$

the final conditions

$$\left(\lambda_i + \frac{\partial F}{\partial \dot{y}_i}\right)_{t=t_f} = 0, \quad i = 1, \ldots, N \qquad (16.172)$$

and the boundary conditions

$$P_i\big(\lambda_i(R, t)\big) = 0, \quad i = 1, \ldots, N \qquad (16.173)$$

In addition,

$$\sum_{i=1}^{N} \lambda_i(r, t)\frac{\partial f_i(\mathbf{y}, \mathbf{u})}{\partial u_r} = 0, \quad r = 1, \ldots, R \qquad (16.174)$$

must be satisfied.

In this formulation, the u_r, as well as the y_i, are treated as continuous functions. This imposes artificial restrictions on the u_r. In some problems the control is discontinuous.

Dynamic Programming for Spatially Continuous Systems

Proceeding as above, the dynamic programming formalism is developed by considering the minimum value of the functional of Eq. (16.167) evaluated between a fixed upper limit and a variable lower limit as a function of the lower limit:

$$S\big(t, \mathbf{y}(r, t)\big) = \min_{\mathbf{u} \in \Omega} \int_t^{t_f} dt' \int_V dr\, F\big(\mathbf{y}(r, t'), \dot{\mathbf{y}}(\mathbf{y}(r, t'), \mathbf{u}(r, t'))\big) \qquad (16.175)$$

In writing Eq. (16.175), the dependence of the integrand upon Eq. (16.166) is shown implicitly, and any constraints on the control vector function are implied by $\mathbf{u} \in \Omega$. By definition,

$$S\big(t, \mathbf{y}(r, t)\big) \leq S\big(t + \Delta t, \mathbf{y}(r, t + \Delta t)\big)$$

$$+ \int_t^{t+\Delta t} dt' \int_V dr\, F\big(\mathbf{y}(r, t'), \dot{\mathbf{y}}(\mathbf{y}(r, t'), \mathbf{u}(r, t'))\big)$$

For the optimal control, the equality obtains. Approximating the integral over time, this becomes[5]

$$S(t, \mathbf{y}(r, t)) = \min_{\mathbf{u}(t) \in \Omega(t)} \left[S(t + \Delta t, \mathbf{y}(r, t + \Delta t)) \right.$$

$$\left. + \Delta t \int_V dr \, F(\mathbf{y}(r, t), \dot{\mathbf{y}}(\mathbf{y}(r, t), \mathbf{u}(r, t))) \right] \qquad (16.176)$$

Equation (16.176) is the dynamic programming algorithm for the partial differential equation description of reactor dynamics. It is solved retrogressively, with the final condition

$$S(t_f, \mathbf{y}(r, t)) = 0 \qquad (16.177)$$

which is apparent from the defining Eq. (16.175).

Pontryagin's Maximum Principle for a Spatially Continuous System

Using a Taylor's series expansion

$$S(t + \Delta t, \mathbf{y}(r, t + \Delta t))$$

$$= S(t, \mathbf{y}(r, t)) + \Delta t \frac{\partial S}{\partial t}(t, \mathbf{y}(r, t))$$

$$+ \Delta t \int_V dr \sum_{i=1}^N \frac{\partial S}{\partial y_i}(t, \mathbf{y}(r, t)) \left[L_i(r) y_i(r, t) + f_i(\mathbf{y}(r, t), \mathbf{u}(r, t)) \right]$$

Eq. (16.176) becomes

$$0 = \min_{\mathbf{u}(t) \in \Omega(t)} \left\{ \frac{\partial S}{\partial t}(t, \mathbf{y}(r, t)) + \sum_{i=1}^N \int_V dr \frac{\partial S}{\partial y_i}(t, \mathbf{y}(r, t)) \right.$$

$$\times \left[L_i(r) y_i(r, t) + f_i(\mathbf{y}(r, t), \mathbf{u}(r, t)) \right]$$

$$\left. + \int_V dr \, F(\mathbf{y}(r, t), \dot{\mathbf{y}}(\mathbf{y}(r, t), \mathbf{u}(r, t))) \right\} \qquad (16.178)$$

Define the functions

$$\psi_i(r, t) = -\frac{\partial S}{\partial y_i}(t, \mathbf{y}(r, t)), \quad i = 1, \ldots, N \qquad (16.179)$$

$$\psi_{N+1}(r, t) = -\frac{\partial S}{\partial t}(t, \mathbf{y}(r, t)) \qquad (16.180)$$

---

5) The minimization in Eq. (16.175) is with respect to the control vector function over the time interval $t \le t' \le t_f$, whereas the minimization in Eq. (16.176) is with respect to the control vector function evaluated at time t.

Then Eq. (16.178) becomes

$$
\begin{aligned}
0 = \max_{\boldsymbol{u}(t) \in \boldsymbol{\Omega}(t)} \Bigg\{ & \psi_{N+1}(r, t) \\
& + \sum_{i=1}^{N} \int_V dr \, \psi_i(r, t) \big[L_i(r) y_i(r, t) + f_i(\boldsymbol{y}(r, t), \boldsymbol{u}(r, t),) \big] \\
& - \int_V dr \, F\big(\boldsymbol{y}(r, t), \dot{\boldsymbol{y}}(\boldsymbol{y}(r, t), \boldsymbol{u}(r, t))\big) \Bigg\}
\end{aligned}
$$

(16.181)

This is the extension of Pontryagin's maximum principle to the partial differential equation description of the reactor dynamics.

When the optimal $\boldsymbol{u}^*(t)$ is chosen, variational derivatives of the quantity within the square brackets must vanish. This leads to the boundary conditions

$$
P_i\big(\psi_i(R, t)\big) = 0
$$

(16.182)

where P_i is the bilinear concomitant defined in Eq. (16.170), and to the equations

$$
\frac{d\psi_j}{dt} = -L_j^+ \psi_j - \sum_{i=1}^{N} \psi_i \frac{\partial f_i}{\partial y_j}
$$

$$
+ \left[\frac{\partial F}{\partial y_j} + \sum_{i=1}^{N} \frac{\partial F}{\partial \dot{y}_i} \frac{\partial}{\partial y_j}(L_i y_i + f_i) \right], \quad j = 1, \dots, N
$$

(16.183)

$$
\frac{d\psi_{N+1}}{dt} = -\sum_{i=1}^{N} \psi_i \frac{\partial}{\partial t}(L_i y_i + f_i)
$$

(16.184)

Identities similar to those just before Eq. (16.162) have been used in arriving at these equations. Appropriate final conditions for the ψ_i and ψ_{N+1} are

$$
\psi_1(t_f) = \cdots = \psi_N(t_f) = \psi_{N+1}(t_f) = 0
$$

(16.185)

The optimal control functions must be found by solving Eqs. (16.166) and (16.182) to (16.184). The initial conditions associated with the y_i and the final conditions associated with the ψ_i and ψ_{N+1} produce a system of equations that must, in general, be solved iteratively. This formulation allows discontinuous control functions and can incorporate constraints on the control functions readily, which are its principle advantages with respect to the calculus-of-variations-formulation.

16.5
Xenon Spatial Oscillations

Xenon-135, with a thermal absorption cross section of 2.6×10^6 barns and a half-life against β-decay of 9.2 h, is produced by the fission product decay chain

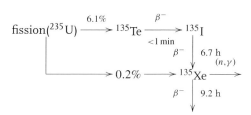

The instantaneous production rate of ^{135}Xe depends on the ^{135}I concentration and hence on the local neutron flux history over the past 50 h or so. On the other hand, the destruction rate of ^{135}Xe depends on the instantaneous flux through the neutron absorption process and on the flux history through the ^{135}Xe decay process. When the flux is suddenly reduced in a reactor that has been operating at a thermal flux level $>10^{13}$ n/cm$^2 \cdot$ s, the xenon destruction rate decreases dramatically while the xenon production rate is initially unchanged, thus increasing the xenon concentration. The xenon concentration passes through a maximum and decreases to a new equilibrium value as the iodine concentration decays away to a new equilibrium value (see Section 6.2).

When a flux tilt is introduced into a reactor, the xenon concentration will initially increase in the region in which the flux is reduced, and initially decrease in the region of increased flux, for similar reasons. This shift in the xenon distribution is such as to increase (decrease) the multiplication properties of the region in which the flux has increased (decreased), thus enhancing the flux tilt. After a few hours the increased xenon production due to the increasing iodine concentration in the high-flux region causes the high-flux region to have reduced multiplicative properties, and the multiplicative properties of the low-flux region increase due to the decreased xenon production associated with a decreasing iodine concentration. This decreases, and may reverse, the flux tilt. In this manner it is possible, under certain conditions, for the delayed xenon production effects to induce growing oscillations in the spatial flux distribution. Such oscillations were common in the large production reactors at Hanford and Savannah River, and measures are required to control them in most thermal power reactors.

Because of the time scale of the iodine and xenon dynamics, prompt and delayed neutron dynamics may be neglected (i.e., changes in the neutron flux are assumed to occur instantaneously, and the delayed neutron precursors are assumed to be always in equilibrium). Moreover, ^{135}I can be assumed to be formed directly from fission. The appropriate equations are

$$\nabla \cdot D^g(r,t)\nabla\phi^g(r,t) - \left[\Sigma_a^g(r,t) + \Sigma_s^g(r,t) + \sigma_x^G X(r,t)\delta_{g,G}\right]\phi^g(r,t)$$

$$+ \sum_{g'=1}^{G} \Sigma_s^{g' \to g}(r,t)\phi^{g'}(r,t) + \chi_p^g \sum_{g'=1}^{G} \nu\Sigma_f^{g'}(r,t)\phi^g(r,t) = 0, \quad g = 1,\ldots,G$$

$$(16.186)$$

$$\gamma_i \sum_{g=1}^{G} \Sigma_f^g(r,t)\phi^g(r,t) - \lambda_i I(r,t) = \dot{I}(r,t) \tag{16.187}$$

$$\gamma_x \sum_{g=1}^{G} \Sigma_f^g(r,t)\phi^g(r,t) + \lambda_i I(r,t) - \lambda_x X(r,t) - \sigma_x^G X(r,t)\phi^G(r,t) = \dot{X}(r,t)$$

$$(16.188)$$

In writing these equations it is assumed that the xenon absorption cross section is zero except in the thermal group ($g = G$). The absorption cross section, Σ_a^g, does not include xenon. The quantity σ_x^G is the microscopic absorption cross section of xenon for thermal neutrons, γ and λ denote yields and decay constants, and I and X are the iodine and xenon concentrations. Changes in the macroscopic cross sections and diffusion coefficients are due to control rod motion or temperature feedback.

Linear Stability Analysis

One of the features of Eqs. (16.186) to (16.188) that makes their solution by analytical methods difficult is the nonlinearity introduced by the xenon absorption term (implicit nonlinearities are also introduced by the dependence of the cross sections on the flux via the temperature feedback). Linearizing Eqs. (16.186) to (16.188) reduces their complexity but also reduces their applicability to a small region about the equilibrium point. The linearized equations are used principally for investigations of stability; that is, if a small flux tilt is introduced, will this flux tilt oscillate spatially with an amplitude that diminishes or grows in time?

The linearized equations are obtained by expanding about the equilibrium point, denoted by a zero subscript:

$$\phi^g(r,t) = \phi_0^g(r) + \delta\phi^g(r,t)$$

$$I(r,t) = I_0(r) + \delta I(r,t)$$

$$X(r,t) = X_0(r) + \delta X(r,t)$$

making use of the fact that the equilibrium solutions satisfy the time-independent version of Eqs. (16.186) to (16.188), and neglecting terms that are nonlinear in $\delta\phi^g$ and δX:

$$\nabla \cdot D^g(r)\nabla\delta\phi^g(r,t) - \left[\Sigma_a^g(r) + \Sigma_s^g(r) + \sigma_x^G(r)X_0(r)\delta_{g,G}\right]\delta\phi^g(r,t)$$

$$+ \sum_{g'=1}^{G} \Sigma_s^{g' \to g}(r)\delta\phi^{g'}(r,t) - \sigma_x^G(r)\phi_0^G(r)\delta X(r,t)\delta_{g,G}$$

$$+ \chi_p^g \sum_{g'=1}^{G} \nu \Sigma_f^{g'}(r) \delta\phi^{g'}(r,t) = 0, \quad g = 1, \ldots, G \tag{16.189}$$

$$\gamma_i \sum_{g=1}^{G} \Sigma_f^g(r)\delta\phi^g(r,t) - \lambda_i \delta I(r,t) = \delta\dot{I}(r,t) \tag{16.190}$$

$$\gamma_x \sum_{g=1}^{G} \Sigma_f^g(r)\delta\phi^g(r,t) + \lambda_i \delta I(r,t) - \lambda_x \delta X(r,t)$$

$$- \sigma_x^G(r) X_0(r)\delta\phi^G(r,t) - \sigma_x^G(r)\phi_0^G(r)\delta X(r,t) = \delta\dot{X}(r,t) \tag{16.191}$$

The effect of temperature feedback has been neglected momentarily in writing Eqs. (16.189) to (16.191), in that the time dependence of the cross sections has been suppressed. Feedback effects will be reintroduced later.

Upon Laplace transforming the time dependence, Eqs. (16.189) to (16.191) become

$$\nabla \cdot D^g(r)\nabla\delta\phi^g(r,p) - \left[\Sigma_a^g(r) + \Sigma_s^g(r) + \sigma_x^G(r) X_0(r)\delta_{g,G}\right]\delta\phi^g(r,p)$$

$$+ \sum_{g'=1}^{G} \Sigma_s^{g' \to g}(r)\delta\phi^{g'}(r,p) - \sigma_x^G(r)\phi_0^G(r)\delta X(r,p)\delta_{g,G}$$

$$+ \chi_p^g \sum_{g'=1}^{G} \nu \Sigma_f^{g'}(r)\delta\phi^{g'}(r,p) = 0, \quad g = 1, \ldots, G \tag{16.192}$$

$$\gamma_i \sum_{g=1}^{G} \Sigma_f^g(r)\delta\phi^g(r,p) - (p + \lambda_i)\delta I(r,p) = -\delta I(r,t=0) \tag{16.193}$$

$$\gamma_x \sum_{g=1}^{G} \Sigma_f^g(r)\delta\phi^g(r,p) + \lambda_i \delta I(r,p) - \left(p + \lambda_x + \sigma_x^G(r)\phi_0^G(r)\right)\delta X(r,p)$$

$$- \sigma_x^G(r) X_0(r)\delta\phi^G(r,p) = -\delta X(r,t=0) \tag{16.194}$$

Equations (16.192) to (16.194) may be written

$$H\delta y = \delta y_0 \tag{16.195}$$

where

$$\delta y(r,p) \equiv \begin{bmatrix} \delta\phi^1(r,p) \\ \vdots \\ \delta\phi^G(r,p) \\ \delta I(r,p) \\ \delta X(r,p) \end{bmatrix}, \quad \delta y_0 \equiv \begin{bmatrix} 0 \\ \vdots \\ 0 \\ -\delta I(r,t=0) \\ -\delta X(r,t=0) \end{bmatrix}$$

and H is composed of the coefficient terms on the left side of Eqs. (16.192) to (16.194).

The solution of Eq. (16.195) is formally

$$\delta y(r, p) = H^{-1}(r, p)\delta y_0(r, t = 0) \tag{16.196}$$

Thus the solutions of Eqs. (16.192) to (16.194) are related to the initiating perturbations by a transfer function matrix, H^{-1}. The condition that the solutions diminish[6] in time is equivalent to the condition that the poles of the transfer function (thus the roots of H) lie in the left-half complex plane. The roots of H are the eigenvalues, p, of Eqs. (16.192) to (16.194), with a homogeneous right-hand side. These homogeneous equations are known as the p-mode equations. The p-mode equations generally have complex eigenfunctions and eigenvalues and must be calculated numerically except for the simplest geometries. Numerical determination of the p-eigenvalues requires special codes and has been successful only for slab geometries. For practical reactor models, it is necessary to resort to approximate methods to evaluate the p-eigenvalues. Two methods that have been employed successfully are the μ- and λ-mode approximations.

μ-Mode Approximation

The μ-mode approximation is motivated by recognition that the only manner in which Eq. (16.192) differs from a standard static diffusion theory problem is through the additional term $-\sigma_x^G \phi_0^G \delta X$ in the thermal group balance equation. Using the homogeneous versions of Eqs. (16.193) and (16.194), this term may be written

$$\sigma_x^G(r)\phi_0^G(r)\delta\phi X(r, p) = N(r, p)\Sigma_f^G(r)\delta\phi^G(r, p) \tag{16.197}$$

where

$$N(r, p) = \frac{[1 + \eta(r)][\gamma_x p + \lambda_i(\gamma_x + \gamma_i)]\delta f(r, p) - \eta(r)(\gamma_x + \gamma_i)(p + \lambda_i)f_0(r)}{[1 + (p/\lambda_x) + \eta(r)][p + \lambda_i)(1 + 1/\eta(r)]} \tag{16.198}$$

with

$$\eta(r) \equiv \frac{\sigma_x^G(r)\phi_0^G(r)}{\lambda_x} \tag{16.199}$$

$$\delta f(r, p) \equiv \sum_{g=1}^{G} \frac{\Sigma_f^g(r)}{\Sigma_f^G(r)} \frac{\delta\phi^g(r, p)}{\delta\phi^G(r, p)} \tag{16.200}$$

$$f_0(r) \equiv \sum_{g=1}^{G} \frac{\Sigma_f^g(r)}{\Sigma_f^G(r)} \frac{\phi_0^g(r)}{\phi_0^G(r)} \tag{16.201}$$

6) The solutions of Eqs. (16.192) to (16.194) have an oscillatory time dependence if the roots of H have an imaginary component. The requirement that these roots lie in the left-half complex plane ensures that these solutions oscillate with a diminishing amplitude.

In applications, the quantity $\delta f(r, p)$ is usually assumed equal to $f_0(r)$.

Using these definitions, the p-mode equations [homogeneous versions of Eq. (16.192) to (16.194)] may be written in the equivalent form

$$\nabla \cdot D^g(r)\nabla \delta \phi^g(r, p) - \left[\Sigma_a^g(r) + \Sigma_s^g(r) + \sigma_x^G(r)X_0(r)\delta_{g,G}\right]\delta \phi^g(r, p)$$

$$+ \sum_{g'=1}^{G} \Sigma_s^{g' \rightarrow g}(r)\delta \phi^{g'}(r, p) + \chi_p^g \sum_{g'=1}^{G} \nu\Sigma_f^{g'}(r)\delta \phi^{g'}(r, p)$$

$$= N(r, p)\Sigma_f^G(r)\delta \phi^{g'}(r, p)\delta_{g,G}, \quad g = 1, \ldots, G \quad (16.202)$$

If $N(r, p)$ is real, the term $N\Sigma_f^G$ in Eq. (16.202) is formally like a distributed poison, and Eq. (16.202) can be solved with standard multigroup diffusion theory codes. In general, $N(r, p)$ is complex because the p-eigenvalues are complex. The essential assumption of the μ-mode approximation is that $N(r, p)$ is real.

There are two types of μ-mode approximations and they differ in the treatment of the spatial dependence of $N(r, p)$. In the first approximation the spatial dependence is retained explicitly and $N(r, p)\Sigma_f^G(r)$ is treated as a distributed poison, in which case Eqs. (16.202) become the standard multigroup criticality equations. A value of p is guessed, $N(r, p)$ is evaluated, and Eqs. (16.202) are solved for the eigenvalue k ($1/k$ multiplies the fission term in the eigenvalue problem). This procedure is repeated until the calculated eigenvalue agrees with the known critical eigenvalue; the corresponding value of p is an approximation to the p-eigenvalue with the largest real part.

An alternative μ-mode approximation (and the one that gives rise to the name μ-mode) results when $N(r, p)$ is assumed to be spatially independent:

$$N(r, p) = \mu(p) \quad (16.203)$$

In this case, Eqs. (16.202) define an eigenvalue problem for the μ-eigenvalues, which can be solved, with a slight modification to the coding, by conventional multigroup diffusion theory codes. To obtain an estimate of the p-eigenvalue from the calculated μ-eigenvalue requires definitions of effective values of $\bar{\eta}$ and $\bar{f}_0$ which account for the spatial dependence of these quantities. In practice, an effective $\bar{\eta}$ is usually defined as

$$\bar{\eta} \equiv \frac{\int dr\, \phi_0^{G*}(r)\Sigma_f^G(r)\eta(r)\phi_0^G(r)}{\int dr\, \phi_0^{G*}(r)\Sigma_f^G(r)\phi_0^G(r)} \quad (16.204)$$

an expression that can be motivated by perturbation theory. The asterisk denotes adjoint. Temperature feedback effects are included in the calculation of μ-eigenvalues by perturbation theory.

λ-Mode Approximation

The λ-mode approximation begins with Eqs. (16.192) to (16.194) and expands the spatial dependence in the eigenfunctions of the neutron balance operator at the equilibrium point (i.e., λ-modes):

$$\nabla \cdot D^g(r) \nabla \psi_n^g(r) - \left[\Sigma_a^g(r) + \Sigma_s^g(r) + \sigma_x^G(r) X_0(r) \delta_{g,G}\right] \psi_n^g(r)$$

$$+ \sum_{g'=1}^{G} \Sigma_s^{g' \to g}(r) \psi_n^{g'}(r) + \frac{1}{k_n} \chi_p^g \sum_{g'=1}^{G} \nu \Sigma_f^{g'}(r) \psi_n^{g'}(r) = 0, \quad g = 1, \ldots, G$$

(16.205)

normalized such that

$$\int dr \left[\sum_{g'=1}^{G} \chi_p^{g'} \psi_m^{g'*}(r)\right]\left[\sum_{g=1}^{G} \nu \Sigma_f^g(r) \psi_n^g(r)\right] = \delta_{m,n}$$

(16.206)

where ψ_m^{g*} satisfy equations adjoint to Eq. (16.205) with appropriate adjoint boundary conditions. It is convenient to treat thermal feedback explicitly in this approximation by including a power feedback term

$$+ \alpha \delta f(r) \Sigma_f^G(r) \delta \phi^G(r, p) \phi_0^G(r)$$

on the left side of Eq. (16.192) for group G.

When the iodine is eliminated between Eqs. (16.193) and (16.194), and the flux and xenon are expanded in λ-modes,

$$\delta \phi^g(r, p) = \sum_{n=1}^{N} A_n(p) \psi_n^g(r), \quad g = 1, \ldots, G$$

(16.207)

$$\delta X(r, p) = \sum_{n=1}^{N} B_n(p) \Sigma_f^G(r) \psi_n^G(r)$$

(16.208)

the biorthogonality relation of Eq. (16.206) may be used to reduce Eqs. (16.192) to (16.194) to a set of $2N$ algebraic equations in the unknowns A_n and B_n, with inhomogeneous terms involving spatial integrals containing $\delta X(r, t = 0)$ and $\delta I(r, t = 0)$. These equations may be written as a transfer function relation between the inhomogeneous terms $\boldsymbol{R}$ and the column vector $\boldsymbol{A}(p)$ containing the A_n and B_n:

$$\boldsymbol{A}(p) = \hat{\boldsymbol{H}}(p) \cdot \boldsymbol{R}$$

(16.209)

Again, the condition for stability is that the poles of $\hat{\boldsymbol{H}}$ lie in the left-half complex p-plane. When $N = 1$ in the expansion of Eqs. (16.207) and (16.208), Eq. (16.209) may be reduced to the scalar relation

$$A_1(p) = \hat{H}'(p) \cdot R$$

(16.210)

where

$$\hat{H}'(p) = [(p - p_1)(p - p_2)]^{-1}$$

(16.211)

and

$$p_1 = -p_r + i\left(c - p_r^2\right)^{1/2}$$
$$p_2 = -p_r - i\left(c - p_r^2\right)^{1/2}$$

(16.212)

with

$$p_r = \frac{\lambda_x}{2}\left\{\left(1 + \frac{\lambda_i}{\lambda_x} + \eta\right) - \frac{\eta}{\Omega}\left[\frac{(\gamma_i + \gamma_x)\eta}{1 + \beta} - \gamma_x\right]\right\}$$
$$c = \lambda_i \lambda_x \left[(1 + \eta) + \frac{\eta(\gamma_i + \gamma_x)}{\Omega}\left(1 - \frac{\eta}{1 + \beta}\right)\right]$$

The parameters η, Ω, and β, which characterize the reactor in this formulation, are defined as

$$\eta \equiv \frac{1}{\lambda_x}\frac{\int dr\, \psi_1^{G*}(r)\sigma_x^G(r)\Sigma_f^G(r)\phi_0^G(r)\psi_1^G(r)}{\int dr\, \psi_1^{G*}(r)\Sigma_f^G(r)\psi_1^G(r)}$$

(16.213)

$$\Omega \equiv \frac{1/k_1 - 1/k_0}{\delta f \int dr\, \psi_1^{G*}(r)\Sigma_f^G(r)\psi_1^G(r)} - \frac{\int dr\, \psi_1^{G*}(r)\alpha(r)\Sigma_f^G(r)\phi_0^G(r)\psi_1^G(r)}{\int dr\, \psi_1^{G*}(r)\Sigma_f^G(r)\psi_1^G(r)}$$

(16.214)

$$\beta \equiv \frac{\delta f(\gamma_i + \gamma_x)\int dr\, \psi_1^{G*}(r)\Sigma_f^G(r)\phi_0^G(r)\psi_1^G(r)}{\int dr\, \psi_1^{G*}(r)\lambda_x X_0(r)\psi_1^G(r)} - 1$$

(16.215)

The quantity δf was defined previously as the ratio of the total fission rate to the thermal group fission rate and an effective spatially independent value has been assumed. The fundamental and first harmonic λ-eigenvalues are denoted by k_0 and k_1, respectively.

The requirement that the poles of $\hat{H}'(p)$ lie in the left-half complex p-plane (i.e., that $p_r > 0$) defines a relationship among η, Ω and β. In practice, $\beta \simeq \eta$ has been found to be a good approximation, so that the stability requirement defines a curve in the η–Ω phase plane, as shown in Fig. 16.1.

The effect of physical parameters upon xenon spatial stability can be traced through Eqs. (16.213) and (16.214) and Fig. 16.1. The quantity Ω is primarily determined by the eigenvalue separation $1/k_1 - 1/k_0$. A reactor becomes less stable when the eigenvalue separation decreases, which occurs when the dimensions are increased, when the migration length is decreased, or when the power distribution is flattened. A negative power coefficient ($\alpha < 0$) increases Ω, thus making a reactor more stable. The quantity η is proportional to the thermal flux level, ϕ_0^G. An increase in thermal flux level is generally destabilizing (increasing η), but may be stabilizing if $\alpha < 0$ (increasing Ω); that is, for $\alpha < 0$, an increase in thermal flux moves the point characterizing a given reactor in Fig. 16.1 to the right and up. It is interesting that an increase in thermal flux level can, under some circumstances, be stabilizing, although this is not generally the case.

If $\Omega > \gamma_i$, Fig. 16.1 predicts stability independent of the value of η. Physically, Ω is a measure of the reactivity required to excite the first harmonic λ-mode in

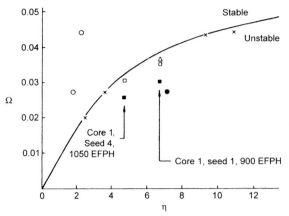

Fig. 16.1 λ-mode linear xenon stability criterion. PWR results: open square, calculated with feedback; solid square, calculated, no feedback; open triangle, inferred from experiment. Calculated transients: open circle, decaying oscillation; cross, neutral oscillation; solid circle, growing oscillation. (From Ref. 9(c); used with permission of Academic Press.)

the presence of power feedback, and γ_i is a measure of the maximum reactivity that can be introduced by iodine decay into xenon. The parameters η and Ω can be evaluated using standard multigroup diffusion theory codes. A fundamental λ-mode flux and first harmonic λ-mode flux and adjoint calculations are required. The integrals in Eqs. (16.213) and (16.214) may be performed with any code that computes perturbation theory–type integrals. Computation of first harmonic flux and adjoint requires either that the problem is symmetric so that zero flux boundary conditions may be located on node lines or that the Wielandt iteration scheme be employed. Several comparisons with experiment and numerical simulation are indicated in Fig. 16.1. The location of the symbol indicates the prediction of the stability criterion, and the type symbol indicates the experimental or numerical result.

At 900 effective full power hours (EFPH), Core 1 Seed 1 of the Shippingport reactor experienced planar xenon oscillations with a doubling time of 30 h. Using this doubling time and the calculated value for η, an experimental Ω may be inferred that agrees with the calculated Ω to within 3%. Core 1 Seed 4 of the Shippingport reactor was observed to be quite unstable at 893 EFPH, and to be slightly unstable at 1397 EFPH. These observations are consistent with the predictions of the stability criterion at 1050 EFPH.

The finite-difference approximations to Eqs. (16.186) to (16.188) were solved numerically for a variety of two-dimensional three-group reactor models. These same reactor models were evaluated for stability with the λ-mode stability criterion. The results depicted in Fig. 16.1 indicate that the predictions of the stability criterion were generally reliable.

In the analysis of this section the total power was assumed to be held constant and the effects of nonlinearities and control rod motion on the stability were neglected. Although the effects of xenon dynamics upon the total power in an uncontrolled reactor can be evaluated, most reactors can be controlled to yield a constant power output. The treatment of nonlinearities and control rod motion is discussed next.

Nonlinear Stability Criterion

The extended methods of Lyapunov, which were discussed in Section 16.4, are applied to derive a stability criterion which includes the nonlinear terms that were neglected in the preceding section. Employing a one-group neutronics model and retaining the prompt neutron dynamics and expanding the flux, iodine, and xenon about their equilibrium states, the equations governing the reactor dynamics may be written in matrix form as

$$\dot{\boldsymbol{y}}(r,t) = \boldsymbol{L}(r)\boldsymbol{y}(r,t) + \boldsymbol{g}(r,t) \tag{16.216}$$

where

$$\boldsymbol{y}(r,t) \equiv \begin{bmatrix} \delta\phi(r,t) \\ \delta X(r,t) \\ \delta I(r,t) \end{bmatrix}, \qquad \boldsymbol{g}(r,t) \equiv - \begin{bmatrix} v\sigma_x \delta X \delta\phi + \alpha v \Sigma_f (\delta\phi)^2 \\ \sigma_x \delta X \delta\phi \\ 0 \end{bmatrix}$$

$$\boldsymbol{L}(r) \equiv \begin{bmatrix} (v\nabla \cdot D\nabla - v\Sigma_a + v\nu\Sigma_f - v\sigma_x X_0 - \alpha v \Sigma_f \phi_0) & -v\sigma_x \phi_0 & 0 \\ \gamma_x \Sigma_f - \sigma_x X_0 & -(\lambda_x + \sigma_x \phi_0) & \lambda_i \\ \gamma_i \Sigma_f & 0 & -\lambda_i \end{bmatrix} \tag{16.217}$$

where v is the neutron speed, α is the power feedback coefficient, and the other notation is as defined previously.

A Lyapunov functional may be chosen as

$$V[\boldsymbol{y}] = \frac{1}{2} \int_R dr \, \boldsymbol{y}^{\mathrm{T}}(r,t)\boldsymbol{y}(r,t) \tag{16.218}$$

The condition for stability (asymptotic stability) in the sense of Lyapunov is that $\dot{V}$ evaluated along the system trajectory defined by Eq. (16.216) is negative semidefinite (definite).

$$\dot{V} = \frac{1}{2} \int_R dr \, [\dot{\boldsymbol{y}}^{\mathrm{T}}\boldsymbol{y} + \boldsymbol{y}^{\mathrm{T}}\dot{\boldsymbol{y}}]$$

$$= \frac{1}{2} \int_R dr \, \boldsymbol{y}^{\mathrm{T}}(\boldsymbol{L}^* + \boldsymbol{L})\boldsymbol{y} + \int_R dr \, \boldsymbol{g}^{\mathrm{T}}\boldsymbol{y}$$

$$\leq -\mu \int_R dr \, \boldsymbol{y}^{\mathrm{T}}\boldsymbol{y} + \left(\int_R dr \, \boldsymbol{g}^{\mathrm{T}}\boldsymbol{g}\right)^{1/2} \left(\int_R dr \, \boldsymbol{y}^{\mathrm{T}}\boldsymbol{y}\right)^{1/2} \tag{16.219}$$

where μ is the smallest eigenvalue of

$$\frac{1}{2}(\boldsymbol{L}^* + \boldsymbol{L})\boldsymbol{\varphi}_n = -\mu_n \boldsymbol{\varphi}_n \tag{16.220}$$

Thus the condition for stability is

$$\mu \geq \frac{(\int_R dr\, \mathbf{g}^T \mathbf{g})^{1/2}}{(\int_R dr\, \mathbf{y}^T \mathbf{y})^{1/2}} \tag{16.221}$$

For a given reactor model and equilibrium state, characterized by μ, relation (16.221) defines the domain of perturbations for which a stable response will be obtained. For asymptotic stability, the inequality must obtain in relation (16.221).

The linear eigenvalue problem, Eq. (16.220), which must be solved for μ, involves the matrix L of Eq. (16.217) and its Hermitean adjoint L^*. The matrix operator $\frac{1}{2}(L^* + L)$ is self-adjoint with a spectrum of real eigenvalues and a complete set of orthogonal eigenfunctions.

The foregoing choice of Lyapunov functional is not unique. As a consequence, this type of analysis provides sufficient, but not necessary, conditions for stability.

Control of Xenon Spatial Power Oscillations

Inclusion of the control system in a stability analysis is difficult primarily because of the difficulty encountered in analytically representing the motion of discrete control rods required to maintain criticality. Control rod motion has a profound effect on the transient response to a perturbation in the equilibrium state in many cases, however, and neglect of this effect may invalidate the stability analysis completely.

Variational Control Theory of Xenon Spatial Oscillations

When the spatial dependence is represented by the nodal approximation, a general optimality functional may be written (for a M-node model)

$$J[\phi_1, \ldots, \phi_M, u_1, \ldots, u_M] = \sum_{m=1}^{M} \int_{t_0}^{t_f} dt\left\{[\phi_m(t) - N_m(t)]^2 + K u_m^2(t)\right\} \tag{16.222}$$

where ϕ_m and N_m represent the actual and the desired, respectively, time-dependent fluxes in node m, u_m is the control in node m, and K is a constant that can be varied to influence the relative importance of the two types of terms in the optimality functional. The purpose of the control program is to find the $u_m(t)$ that minimizes the optimality functional, subject to the constraints that the reactor remain critical,

$$0 = \sum_{m' \neq m}^{M} I_{mm'}[\phi_{m'}(t) - \phi_m(t)] + [\nu \Sigma_{fm} - \Sigma_{am} - \sigma_x X_m(t) - u_m(t)]\phi_m(t)$$

$$= f_{1m} \tag{16.223}$$

and the iodine and xenon dynamics equations are satisfied,

$$\dot{I}_m(t) = \gamma_i \Sigma_{fm} \phi_m(t) - \lambda_i I_m(t) = f_{2m}, \qquad I_m(0) = I_{m0} \tag{16.224}$$

$$\dot{X}_m(t) = \gamma_x \Sigma_{fm}\phi_m(t) + \lambda_m I_m(t) - [\lambda_x + \sigma_x\phi_m(t)]X_m(t) = f_{3m}$$

$$X_m(0) = X_{m0}, \quad m = 1, \dots, M \tag{16.225}$$

The m subscript denotes node m and $I_{mm'}$ is the internodal coupling coefficient of the type discussed in Sections 15.2 and 15.3.

Equations (16.148) become

$$0 = 2[\phi_m(t) - N_m(t)] - [\nu\Sigma_{fm} - \Sigma_{am} - \sigma_x X_m(t) - u_m(t)]\omega_{1m}(t)$$

$$+ \sum_{m'\neq m}^{M} I_{mm'}[\omega_{1m}(t) - \omega_{1m'}(t)] - \gamma_i\Sigma_{fm}(t)\omega_{2m}(t)$$

$$- [\gamma_x\Sigma_{fm} - \sigma_x X_m(t)]\omega_{3m}(t) \tag{16.226}$$

$$\dot{\omega}_{2m}(t) = \lambda_i[\omega_{2m}(t) - \omega_{3m}(t)] \tag{16.227}$$

$$\dot{\omega}_{3m}(t) = \sigma_x\phi_m(t)\omega_{1m}(t) + [\lambda_x + \sigma_x\phi_m(t)]\omega_{3m}(t), \quad m = 1, \dots, M \tag{16.228}$$

(The symbol ω has been used to denote the Lagrange multipliers, since λ is conventionally used to represent the decay constants.) The final conditions corresponding to Eqs. (16.150) are

$$\omega_{2m}(t_f) = 0, \quad \omega_{3m}(t_f) = 0, \quad m = 1, \dots, M \tag{16.229}$$

Equations (16.149) are modified somewhat in this case because the optimality functional depends on the control. The more general relation is

$$\frac{\partial F}{\partial u_r} + \sum_{i=1}^{N} \lambda_i(t)\frac{\partial f_i}{\partial u_r}(\mathbf{y}, \mathbf{u}) = 0, \quad r = 1, \dots, R \tag{16.230}$$

which becomes

$$2Ku_m(t) + \omega_{1m}(t)\phi_m(t) = 0, \quad m = 1, \dots, M \tag{16.231}$$

Equations (16.231) can be used to eliminate the u_m from Eqs. (16.223) and (16.226). The modified equations, plus Eqs. (16.224), (16.225), (16.227), and (16.228), constitute a set of $6M$ equations which, together with the initial and final conditions specified above, can be solved for the optimal flux, iodine, xenon, and Lagrange multiplier trajectories. The optimal control can then be determined from Eqs. (16.231).

If no approximation is made for the spatial dependence, an equivalent optimality functional is

$$J[\phi, u] = \int_V dr \int_{t_0}^{t_f} dt\left[(\phi(r,t) - N(r,t))^2 + Ku^2(r,t)\right] \tag{16.232}$$

and the constraints are

$$0 = \nabla \cdot D(r)\nabla\phi(r,t) + [\nu\Sigma_f(r) - \Sigma_a(r) - \sigma_x X(r,t) - u(r,t)]\phi(r,t) = f_1 \tag{16.233}$$

$$\dot{I}(r,t) = \gamma_i \Sigma_f(r)\phi(r,t) - \lambda_i I(r,t) = f_2, \qquad I(r,0) = I_0(r) \tag{16.234}$$

$$\dot{X}(r,t) = \gamma_x \Sigma_f(r)\phi(r,t) + \lambda_i I(r,t) - [\lambda_x + \sigma_x \phi(r,t)]X(r,t) = f_3$$

$$X(r,0) = X_0(r) \tag{16.235}$$

Equations (16.171) become

$$0 = 2[\phi(r,t) - N(r,t)] - \nabla \cdot D(r)\nabla \omega_1(r,t)$$

$$- [\nu\Sigma_f(r) - \Sigma_a(r) - \sigma_x X(r,t) - u(r,t)]\omega_1(r,t)$$

$$- \gamma_i \Sigma_f(r)\omega_2(r,t) - [\gamma_x \Sigma_f(r) - \sigma_x X(r,t)]\omega_3(r,t) \tag{16.236}$$

$$\dot{\omega}_2(r,t) = \lambda_i[\omega_2(r,t) - \omega_3(r,t)] \tag{16.237}$$

$$\dot{\omega}_3(r,t) = \sigma_x \phi(r,t)\omega_1(r,t) + [\lambda_x + \sigma_x \phi(r,t)]\omega_3(r,t) \tag{16.238}$$

The final conditions of Eqs. (16.172) are

$$\dot{\omega}_2(r,t_f) = \omega_3(r,t_f) = 0 \tag{16.239}$$

and the boundary condition of Eq. (16.173) is

$$\omega_1(R,t) = 0 \tag{16.240}$$

Because the optimality functional contains the control functions, Eqs. (16.174) must be modified to

$$\frac{\partial F}{\partial u_r} + \sum_{i=1}^{N} \lambda_i(r,t)\frac{\partial f_i}{\partial u_r}(y,u) = 0 \tag{16.241}$$

which becomes

$$2Ku(r,t) + \omega_1(r,t)\phi(r,t) = 0 \tag{16.242}$$

16.6
Stochastic Kinetics

The evolution of the state of a nuclear reactor is essentially a stochastic process and should, in general, be described mathematically by a set of stochastic kinetics equations. For most problems in reactor physics it suffices to describe the mean value of the state variables in a deterministic manner and to ignore the stochastic aspects. However, the stochastic features of the state variables are important in the analysis of reactor startups in the presence of a weak source and underlie some experimental techniques, such as the measurement of the dispersion of the number of neutrons born in fission, the Rossi-α measurement, and the measurement and interpretation of reactor noise. The purpose of this section is to present a computationally tractable formalism for the calculation of stochastic phenomena in a space- and energy-dependent time-varying zero-power reactor model.

Forward Stochastic Model

The spatial domain of a reactor may be partitioned into I space cells, and the energy range of interest may be partitioned into G energy cells. Subject to this partitioning, the state of the reactor is defined by the set of numbers

$$N \equiv \{n_{ig} c_{im}\}, \quad i = 1, \ldots, I; \ g = 1, \ldots, G; \ m = 1, \ldots, M$$

where n_{ig} is the number of neutrons in space cell i and energy cell g, and c_{im} is the number of m-type delayed neutron precursors in space cell i.

Define the transition probability $P(N't'|Nt)$ that a reactor that was in state N' at time t' will be in state N at time t. The probability generating function for this transition probability is defined by the relation

$$G(N't'|Ut) \equiv \sum_N P(N't'|Nt) \prod_{igm} u_{ig}^{n_{ig}} v_{im}^{c_{im}} \tag{16.243}$$

The summation over N implies a summation over all values of n_{ig} and c_{im} for all i, g, and m. The quantities u_{ig} and v_{im} play the role of transform variables.

The transition probability will be written

$$P(N't'|Nt) = \left[\prod_{ig} P_{ig}(N't'|n_{ig}t) \right] \left[\prod_{im} \hat{P}_{im}(N't'|C_{im}t) \right] \tag{16.244}$$

for mnemonic reasons. This formalism does not denote product probabilities and is used only to facilitate the distinction between states that differ only by the number of neutrons in one space-energy cell or the number of m-type precursors in one space cell.

Some properties of the probability generating function that will be needed in the subsequent analysis are:

$$G(N't'|Ut)_{U=1} = \sum_N P(N't'|Nt) \equiv 1 \tag{16.245}$$

$$\frac{\partial G}{\partial u_{ig}}(N't'|Ut)|_{U=1} = \sum_N n_{ig} P(N't'|Nt) \equiv \bar{n}_{ig}(t) \tag{16.246}$$

$$\frac{\partial G}{\partial v_{im}}(N't'|Ut)|_{U=1} = \sum_N c_{im} P(N't'|Nt) \equiv \bar{c}_{im}(t) \tag{16.247}$$

$$W_{ig,i'g'}(t) \equiv \frac{\partial^2 G(N't'|Ut)}{\partial u_{ig} \partial u_{i'g'}}\bigg|_{U=1} = \begin{cases} \overline{n_{ig}(t)(n_{ig}(t)-1)}, & ig = i'g' \\ \overline{n_{ig}(t)n_{i'g'}(t)}, & ig \neq i'g' \end{cases} \tag{16.248}$$

$$Y_{im,i'g'}(t) \equiv \frac{\partial^2 G(N't'|Ut)}{\partial v_{im} \partial u_{i'g'}}\bigg|_{U=1} = \overline{n_{i'g'}(t)c_{im}(t)} \tag{16.249}$$

$$Z_{im,i'm'}(t) \equiv \frac{\partial^2 G(N't'|Ut)}{\partial v_{im} \partial v_{i'm'}}\bigg|_{U=1} = \begin{cases} \overline{c_{im}(t)(c_{im}(t)-1)}, & im = i'm' \\ \overline{c_{im}(t)c_{i'm'}(t)}, & im \neq i'm' \end{cases} \tag{16.250}$$

The notation $U = 1$ indicates that the expression is evaluated for all u_{ig} and v_{im} equal to unity. The overbar denotes an expectation value, as defined explicitly in Eqs. (16.246) and (16.247). In the foregoing equations and in the subsequent development, the dependence of the expectation values at time t on the state of the reactor at time t' is implicit. By considering the events that could alter the state of the reactor during the time interval $t \to t + \Delta t$, balance equations for the transition probability and the probability generating function may be derived. In the limit $\Delta t \to 0$, the probability of more than one event occurring during Δt becomes negligible, and the balance equations can be constructed by summing over all single event probabilities.

- *Source neutron emission:*

$$\left.\frac{\partial P}{\partial t}\right|_s = \sum_{ig} S_{ig}[P_{ig}(n_{ig}-1) - P_{ig}(n_{ig})]\left(\prod' P_{i'g'}\right)\left(\prod \hat{P}_{i'm'}\right)$$

$$\left.\frac{\partial G}{\partial t}\right|_s = \sum_{ig} S_{ig}[u_{ig}-1]G$$

- *Capture event* (includes capture by detectors):

$$\left.\frac{\partial P}{\partial t}\right|_c = \sum_{ig} \Lambda_{cig}[(n_{ig}+1)P_{ig}(n_{ig}+1) - n_{ig}P_{ig}(n_{ig})]\left(\prod' P_{i'g'}\right)\left(\prod \hat{P}_{i'm'}\right)$$

$$\left.\frac{\partial G}{\partial t}\right|_c = \sum_{ig} \Lambda_{cig}[1 - u_{ig}]\frac{\partial G}{\partial u_{ig}}$$

- *Transport event:*

$$\left.\frac{\partial P}{\partial t}\right|_T = \sum_{ig}\sum_{i'} l_{ii'}^g [(n_{ig}+1)P_{ig}(n_{ig}+1)P_{i'g}(n_{i'g}-1)$$
$$- n_{ig}P_{ig}(n_{ig})P_{i'g}(n_{i'g})]\left(\prod' P_{i''g''}\right)\left(\prod \hat{P}_{i''m}\right)$$

$$\left.\frac{\partial G}{\partial t}\right|_T = \sum_{ig}\sum_{i'} l_{ii'}^g [u_{i'g} - u_{ig}]\frac{\partial G}{\partial u_{ig}}$$

- *Scattering event:*

$$\left.\frac{\partial P}{\partial t}\right|_s = \sum_{ig} \Lambda_{sig}\left[(n_{ig}+1)P_{ig}(n_{ig}+1)\sum_{g'} K_i^{gg'} P_{ig}(n_{ig'}-1)\left(\prod' P_{i''g''}\right)\right.$$
$$\left. - n_{ig}P_{ig}(n_{ig})\left(\prod' P_{i'g'}\right)\right]\left(\prod \hat{P}_{i'm'}\right)$$

$$\left.\frac{\partial G}{\partial t}\right|_s = \sum_{ig} \Lambda_{sig}\left[\sum_{g'} K_i^{gg'} u_{ig'} - u_{ig}\right]\frac{\partial G}{\partial u_{ig}}$$

- *Delayed neutron emission:*

$$\frac{\partial P}{\partial t}\bigg|_d = \sum_{im} \lambda_m \bigg[(c_{im}+1)\hat{P}_{im}(c_{im}+1) \sum_g \chi_m^g P_{ig}(n_{ig}-1)\bigg(\prod{}' P_{i'g'}\bigg)$$

$$- c_{im}\hat{P}_{im}(c_{im})\bigg(\prod{}' P_{i'g'}\bigg)\bigg]\bigg(\prod \hat{P}_{i'm'}\bigg)$$

$$\frac{\partial G}{\partial t}\bigg|_d = \sum_{im} \lambda_m \bigg[\sum_g \chi_m^g u_{ig} - v_{im}\bigg]\frac{\partial G}{\partial v_{im}}$$

- *Fission event:*

$$\frac{\partial P}{\partial t}\bigg|_f = \sum_{ig} \Lambda_{fig}\bigg[(n_{ig}+1)P_{ig}(n_{ig}+1)\sum_{vp} p_g(v_p)$$

$$\times \bigg\{(1-\beta' v_p)\sum_{g'} \chi_p^{g'} P_{ig}(n_{ig'}-v_p)\bigg(\prod{}' P_{i''g''}\bigg)\bigg(\prod \hat{P}_{i'm'}\bigg)$$

$$+ \sum_m \beta_m' v_p \sum_{g'} \chi_p^{g'} P_{ig'}(n_{ig'}-v_p) \cdot \hat{P}_{im}(c_{im}-1)$$

$$\times \bigg(\prod{}' P_{i''g''}\bigg)\bigg(\prod{}' \hat{P}_{i'm'}\bigg)\bigg\}$$

$$- n_{ig}P_{ig}(n_{ig})\bigg(\prod{}' P_{i'g'}\bigg)\bigg(\prod \hat{P}_{i'm'}\bigg)\bigg]$$

$$\frac{\partial G}{\partial t}\bigg|_f = \sum_{ig} \Lambda_{fig}\bigg[\sum_{g'} f_g \chi_p^{g'} - \bigg\{\beta' - \sum_m \beta_m' v_{im}\bigg\}\sum_{g'} u_{ig'}\frac{\partial f_g}{\partial u_{ig'}}\chi_p^{g'} - u_{ig}\bigg]$$

$$\times \frac{\partial G}{\partial u_{ig}}$$

The quantity Λ_{-ig} represents a reaction frequency per neutron, in space cell i and energy cell g, and the subscripts c, s, and f refer to capture, scattering, and fission, respectively.[7] $K^{gg'}$ is the probability that a scattering event which occurred in energy cell g transfers a neutron to energy cell g', while χ_p^g and χ_m^g are the probabilities that a neutron produced by fission and m-type precursor decay, respectively, has energy within energy cell g. The decay constant for precursor type m is λ_m, and β_m' is the average ratio of the number of m-type precursors to the number of prompt neutrons produced in a fission ($\beta' = \Sigma_m \beta_m'$). S_{ig} is the neutron source rate in space cell i and energy cell g. The quantity $l_{ii'}^g$ represents the frequency per neutron at which neutrons in space cell i and energy cell g will diffuse into space cell i' (without a change in energy). The prime on the product operator, $\prod$, indicates that the product is taken over all i, g, and m except those explicitly shown in the same

---

7) For example, $\Lambda_{fig} = v^g \Sigma_f^g$, v^g = neutron speed; Σ_f^g = fission cross section.

term. The quantity f_g is the probability generating function for $p_g(\nu_p)$, which is the probability distribution function for the number of prompt neutrons emitted in a fission that was caused by a neutron in energy cell g:

$$f_g(u_{ig'}) = \sum_{n_{ig'}} u_{ig'}^{n_{ig'}} p_g(\nu_p) \tag{16.251}$$

A single fissionable species is assumed for simplicity.

Appropriate balance equations for the transition probability, P, and its probability generating function, G, may be constructed from these terms:

$$\frac{\partial P}{\partial t}(N't'|Nt) = \left.\frac{\partial P}{\partial t}\right|_S + \left.\frac{\partial P}{\partial t}\right|_c + \left.\frac{\partial P}{\partial t}\right|_T + \left.\frac{\partial P}{\partial t}\right|_s + \left.\frac{\partial P}{\partial t}\right|_d + \left.\frac{\partial P}{\partial t}\right|_f \tag{16.252}$$

$$\frac{\partial G}{\partial t}(N't'|Ut) = \left.\frac{\partial G}{\partial t}\right|_S + \left.\frac{\partial G}{\partial t}\right|_c \left.\frac{\partial G}{\partial t}\right|_T + \left.\frac{\partial G}{\partial t}\right|_s + \left.\frac{\partial G}{\partial t}\right|_d + \left.\frac{\partial G}{\partial t}\right|_f \tag{16.253}$$

Means, Variances, and Covariances

By differentiating Eq. (16.253) with respect to u_{ig} and v_{im} and evaluating the resulting expressions for $U = 1$, equations for the mean value of the neutron and precursor distribution, respectively, are obtained [see Eqs. (16.246) and (16.247)]:

$$\frac{\partial \bar{n}_{ig}(t)}{\partial t} = S_{ig}(t) - [\Lambda_{cig}(t) + \Lambda_{sig}(t) + \Lambda_{fig}(t)]\bar{n}_{ig}(t)$$

$$+ \sum_{g'=1}^{G} \Lambda_{sig}(t) K_i^{g'g} \bar{n}_{ig'}(t) + \chi_p^g \sum_{g'=1}^{G} \bar{\nu}_p^{g'} \Lambda_{fig'}(t) \bar{n}_{ig'}(t)$$

$$+ \sum_{m=1}^{M} \chi_m^g \lambda_m \bar{c}_{im}(t) + \sum_{i'=1}^{I} l_{i'i}^g(t)[\bar{n}_{i'g}(t) - \bar{n}_{ig}(t)] \tag{16.254}$$

$$\frac{\partial \bar{c}_{im}(t)}{\partial t} = -\lambda_m \bar{c}_{im}(t) + \beta_m' \sum_{g=1}^{G} \bar{\nu}_p^{g'} \Lambda_{fig}(t) \bar{n}_{ig}(t)$$

$$g = 1, \ldots, G; \; i = 1, \ldots, I; \; m = 1, \ldots, M \tag{16.255}$$

Making use of the identities $\bar{\nu}_p^g \equiv (1 - \beta)\bar{\nu}^g$, where $\bar{\nu}^g$ is the average number of neutrons (prompt and delayed) per fission induced by a neutron in energy cell g, and $\beta' = \beta/(1 - \beta)$, it is apparent that these are the conventional space- and energy-dependent neutron and precursor kinetics equations in the finite-difference multigroup approximation.

By taking second partial derivatives of Eq. (16.253) with respect to u_{ig} and v_{im}, and evaluating the result for $U = 1$, equations for the quantities defined by Eqs. (16.248) to (16.250) are derived:

$$\frac{\partial W_{ig,i'g'}}{\partial t} = S_{ig}\bar{n}_{i'g'} + S_{i'g'}\bar{n}_{ig'} - (\Lambda_{cig} + \Lambda_{ci'g'})W_{ig,i'g'}$$

$$+ \sum_j l_{ji}^g (W_{jg,i'g'} - W_{ig,i'g'}) + \sum_j l_{ji'}^{g'} (W_{jg',ig} - W_{jg,ig})$$

$$+ \sum_{g''} (\Lambda_{sig''} K_i^{g''g} W_{i'g,ig} + \Lambda_{si'g''} K_{i'}^{g''g'} W_{i'g'',ig})$$

$$- (\Lambda_{sig} + \Lambda_{si'g'})W_{ig,i'g'} + \sum_m \lambda_m \left(\chi_m^g Y_{im,i'g'} + \chi_m^{g'} Y_{i'm,ig}\right)$$

$$+ \chi_p^g \sum_g \bar{v}_p^{g''} \Lambda_{fig''} W_{i'g',ig''}$$

$$+ \chi_p^{g'} \sum_{g''} \bar{v}_p^{g''} \Lambda_{fi'g''} W_{ig,i'g''} (\Lambda_{fig} + \Lambda_{fi'g'})W_{ig,i'g'}$$

$$+ \chi_p^g \sum_{g''} \Lambda_{fig''} \overline{v_p^{g''}(v_p^{g''} - 1)} \bar{n}_{ig'} \delta_{ig,i'g'}$$

$$i, i' = 1, \ldots, I; \; g, g' = 1, \ldots, G \qquad (16.256)$$

$$\frac{\partial Y_{im,i'g'}}{\partial t} = S_{i'g'}\bar{c}_{im} - \Lambda_{ci'g'} Y_{im,i'g'} + \sum_j l_{ji'}^{g'} (Y_{im,jg'} - Y_{im,i'g'})$$

$$+ \sum_{g''} \Lambda_{si'g''} K_{i'}^{g''g'} Y_{im,i'g'} - \Lambda_{si'g'} Y_{im,i'g'} - \lambda_m Y_{im,i'g'}$$

$$+ \sum_{m'} \lambda_{m'} \chi_{m'}^{g'} Z_{im,i'm'} + \beta_m' \sum_{g''} \bar{v}_p^{g''} \Lambda_{fi'g''} W_{i'g',ig''}$$

$$+ \chi_p^{g'} \sum_{g''} \bar{v}_p^{g''} \Lambda_{fi'g''} Y_{im,i'g''} - \Lambda_{fi'g'} Y_{im,i'g'}$$

$$+ \beta_m \sum_{g''} \overline{(v_p^{g''})^2} \Lambda_{fi'g''} \bar{n}_{i'g''} \delta_{i,i'}$$

$$i, i' = 1, \ldots, I; \; m = 1, \ldots, M; \; g' = 1, \ldots, G \qquad (16.257)$$

$$\frac{\partial Z_{im,i'm'}}{\partial t} = -\lambda_m Z_{im,i'm'} - \lambda_{m'} Z_{im,i'm'} + \beta_m' \sum_g \bar{v}_p^g \Lambda_{fi'g} Y_{im,i'g}$$

$$+ \beta_{m'}' \sum_g \bar{v}_p^g \Lambda_{fig} Y_{i'm',ig}, \quad i, i' = 1, \ldots, I; \; m, m' = 1, \ldots, M$$

$$(16.258)$$

Equations (16.256) to (16.258) are coupled.

From Eqs. (16.248) to (16.250) it is apparent that the solutions of Eqs. (16.256) to (16.258) are related to the variances and covariances of the neutron and precursor distributions; for example,

$$\sigma_{ig}^2 \equiv \overline{(n_{ig} - \bar{n}_{ig})^2} = W_{ig,ig} - \bar{n}_{ig}(\bar{n}_{ig} - 1) \tag{16.259}$$

$$\sigma_{im}^2 \equiv \overline{(c_{im} - \bar{c}_{im})^2} = Z_{im,im} - \bar{c}_{im}(\bar{c}_{im} - 1) \tag{16.260}$$

$$\sigma_{igm}^2 \equiv \overline{(n_{ig} - \bar{n}_{ig})(c_{im} - \bar{c}_{im})} = Y_{im,ig} - \bar{n}_{ig}\bar{c}_{im} \tag{16.261}$$

Correlation Functions

Define the correlation functions

$$\overline{n_{ig}(t)n_{i'g'}(t')} \equiv \sum_N \sum_{N'} n_{ig}n_{i'g'} \, P\big(N't'|Nt\big) \tag{16.262}$$

$$\overline{c_{im}(t)n_{i'g'}(t')} \equiv \sum_N \sum_{N'} c_{im}n_{i'g'} \, P\big(N't'|Nt\big) \tag{16.263}$$

$$\overline{n_{ig}(t)c_{i'm'}(t')} \equiv \sum_N \sum_{N'} n_{ig}c_{i'm'} \, P\big(N't'|Nt\big) \tag{16.264}$$

$$\overline{c_{im}(t)c_{i'm'}(t')} \equiv \sum_N \sum_{N'} c_{im}c_{i'm'} \, P\big(N't'|Nt\big) \tag{16.265}$$

By differentiating Eqs. (16.262) to (16.265) with respect to t, and using Eqs. (16.245) to (16.248), (16.254), and (16.255), equations satisfied by the correlation functions may be obtained.

$$\frac{\partial}{\partial t}\overline{n_{ig}(t)n_{i'g'}(t')} = S_{ig}(t)\bar{n}_{i'g'}(t) - [\Lambda_{cig}(t) + T_{ig}(t) + \Lambda_{sig}(t)$$

$$+ \Lambda_{fig}(t)]\overline{n_{ig}(t)n_{i'g'}(t)} + \sum_{m=1}^{M} \lambda_m \chi_m^g \overline{c_{im}(t)n_{i'm'}(t')}$$

$$+ \sum_{g''=1}^{G}\left[\Lambda_{sig''}(t)K_i^{g''g}(t) + \chi_p^g \bar{\nu}_p^{g''}\Lambda_{fig''}(t)\right]$$

$$\times \overline{n_{ig''}(t)n_{i'g'}(t')} \tag{16.266}$$

$$\frac{\partial}{\partial t}\overline{c_{im}(t)n_{i'g'}(t')} = \beta_m' \sum_{g=1}^{G} \bar{\nu}_p^g \Lambda_{fig}(t)\overline{n_{ig}(t)n_{i'g'}(t')} - \lambda_m\overline{c_{im}(t)n_{i'g'}(t')}$$

$$\tag{16.267}$$

$$\frac{\partial}{\partial t}\overline{n_{ig}(t)c_{i'm'}(t')} = S_{ig}(t)\bar{c}_{i'm'}(t') + \sum_{g''=1}^{G}\left[\Lambda_{sig''}(t)K_i^{g''g}(t) + \chi_p^g \bar{\nu}_p^{g''}\Lambda_{fig''}(t)\right]$$

$$\times \overline{n_{ig''}(t)c_{i'm'}(t)} - [\Lambda_{cig}(t) + T_{ig}(t) + \Lambda_{sig}(t)$$

$$+ \Lambda_{fig}(t)]\overline{n_{ig}(t)c_{i'm'}(t')} + \sum_{m=1}^{M} \lambda_m \chi_m^g \overline{c_{im}(t)c_{i'm'}(t')}$$

$$\tag{16.268}$$

$$\frac{\partial}{\partial t}\overline{c_{im}(t)c_{i'm'}(t')} = \beta'_m \sum_{g=1}^{G} \bar{v}_p^g \Lambda_{fig}(t)\overline{n_{ig}(t)c_{i'm'}(t')} - \lambda_m \overline{c_{im}(t)c_{i'm'}(t')}$$

$$i, i' = 1, \ldots, I; \; m, m' = 1, \ldots, M \tag{16.269}$$

Equations (16.266) and (16.267) are coupled, as are Eqs. (16.268) and (16.269). The operator T_{ig} is defined by the operation

$$T_{ig}\overline{n_{ig}n_{i'g'}} = \sum_{i''=1}^{I} l_{i''i}^g \overline{(n_{i''g}n_{i'g'} - n_{ig}n_{i'g'})}$$

Physical Interpretation, Applications, and Initial and Boundary Conditions

If all members of a large ensemble of identical reactors are known to be in an identical state, N', at time t', if all reactors are operated identically subsequent to time t', and if the state, N, of each reactor could be determined at a later time t, the number of reactors in the ensemble that would be found to have a given state, N, would approach the distribution $P(N't'|Nt)$. Thus $\bar{n}_{ig}(t)$ is the ensemble average for the number of neutrons in space cell i and energy cell g at time t, and similarly, $\bar{c}_{im}(t)$ is the ensemble average for the number of m-type precursors in space cell i at time t.

Alternatively, consider a single reactor that is brought to a known state N' at a reference time t', and subsequently operated in a given manner until a reference time t. If this procedure is repeated a large number of times, the distribution of the number of times the reactor is in a given state N at reference time t approaches $P(N't'|Nt)$. Consequently, $\bar{n}_{ig}(t)$ and $\bar{c}_{im}(t)$ are mean values of the neutron and precursor populations, and $\sigma_{ig}^2(t)$ and $\sigma_{im}^2(t)$ are the mean-squared deviations in these populations. These deviations are an indication of the uncertainty associated with the usual assumption that the actual population is equal to the mean value, the latter being predicted by conventional kinetics equations. Such considerations are important in analyzing weak-source startup problems.

A set of initial conditions for Eqs. (16.254) to (16.258) may be obtained from the identity

$$P\left(N^0 t_0 | N t_0\right) = \delta_{NN^0} \tag{16.270}$$

where the zero superscript indicates the known state at t_0. From Eqs. (16.243) and (16.245) to (16.250), the following initial conditions may be deduced:

$$G\left(N^0 t_0 | U t_0\right) = \prod_{igm} u_{ig}^{n_{ig}^0} u_{im}^{c_{im}^0} \tag{16.271}$$

$$\bar{n}_{ig}(t_0) = \frac{\partial G}{\partial u_{ig}}\left(N^0 t_0 | U t_0\right)\bigg|_{U=1} = n_{ig}^0 \tag{16.272}$$

$$\bar{c}_{im}(t_0) = \frac{\partial G}{\partial v_{im}}\left(N^0 t_0 | U t_0\right)\bigg|_{U=1} = c_{im}^0 \tag{16.273}$$

$$W_{ig,i'g'}(t_0) = \begin{cases} n_{ig}^0 \left(n_{ig}^0 - 1 \right), & i'g' = ig \\ n_{ig}^0 n_{i'g'}^0, & i'g' \neq ig \end{cases} \tag{16.274}$$

$$Y_{im,i'g'}(t_0) = n_{i'g'}^0 c_{im}^0 \tag{16.275}$$

$$Z_{i'm',im}(t_0) = \begin{cases} c_{im}^0 \left(c_{im}^0 - 1 \right), & i'm' = im \\ c_{im}^0 c_{i'm'}^0, & i'm' \neq im \end{cases} \tag{16.276}$$

In practice, it is not possible to ascertain the "known" initial conditions. This difficulty may be circumvented by using homogeneous initial conditions and, in a subcritical system, taking the asymptotic solution of Eqs. (16.254) to (16.258) as the initial conditions for further calculations involving changes in operating conditions. Alternatively, the time-independent versions of Eqs. (16.254) to (16.258) may be solved to provide initial conditions.

External boundary conditions may be treated by assuming that the space cells on the exterior of the reactor are contiguous to a fictitious external space cell in which the mean value, variance, or covariance is zero, for the purpose of evaluating the net leakage operator. This is equivalent to the familiar extrapolated boundary condition of neutron-diffusion theory.

The interpretation of $P(N't'|Nt)$ just discussed leads to an interpretation of the correlation functions. For example, $\overline{n_{ig}(t)n_{i'g'}(t')}$ is the expectation (mean) value of the product of the number of neutrons in space cell i' and energy cell g' at t', and the number of neutrons in space cell i and energy cell g at t. When the reactor properties are time independent, the ensemble average may be replaced by an average over time in a single reactor (the ergodic theory).[8] In this case, $\overline{n_{ig}(t'+\tau)n_{i'g'}(t')}$ is amenable to experimental measurement if the energy and space cells are chosen to conform with the detector resolution. The corresponding theoretical quantity is obtained by solving the time-independent versions of Eqs. (16.266) using the same type of external boundary treatment discussed before, and employing corrections for the detection process and counting circuit statistics.

Numerical Studies

Equations (16.254) to (16.258) have been solved numerically for the special case of one energy cell, one delayed neutron precursor type, and one spatial dimension, to study the characteristics of the neutron and precursor distributions under a variety of static and transient conditions. The results of these studies may be characterized in terms of the mean value of the neutron ($\bar{n}_i$) and precursor ($\bar{c}_i$) distributions in region i and in terms of the relative variances in the neutron and precursor distributions in region i, which are defined by the relations

$$\mu_i \equiv \frac{\overline{(n_i - \bar{n}_i)^2}}{\bar{n}_i^2} = \frac{W_{i,i} - \bar{n}_i(\bar{n}_i - 1)}{\bar{n}_i^2} \tag{16.277}$$

8) For a subcritical reactor.

$$\varepsilon_i \equiv \frac{\overline{(c_i - \bar{c}_i)^2}}{\bar{c}_i^2} = \frac{Z_{i,i} - \bar{c}_i(\bar{c}_i - 1)}{\bar{c}_i^2} \tag{16.278}$$

The quantities μ_i and ε_i are measurements of the relative dispersion in the neutron and precursor statistical distributions in region i.

Certain general trends emerge from the numerical studies that have been performed:

1. When the reactor is subcritical, the asymptotic values of μ_i and ε_i vary from region to region, and within a given region $\varepsilon_i < \mu_i$.

2. When the reactor is subcritical, the asymptotic values of μ_i and ε_i depend on the source level and distribution and the degree of subcriticality. In general, increasing the source level or the multiplication factor reduces μ_i and ε_i.

3. When the reactor is supercritical, μ_i and ε_i attain asymptotic values that are identical in all regions, and $\mu_i = \varepsilon_i$.

4. When the reactor is brought from a subcritical to a supercritical configuration, μ_i generally decreases and ε_i generally increases.

5. The asymptotic value of μ_i and ε_i in a supercritical reactor is sensitive to the manner in which the reactor is brought supercritical.
   a. For the withdrawal of a single rod (or group of rods) between fixed limits, the more rapid the withdrawal the larger the asymptotic value of μ_i and ε_i.
   b. When a number of rods are to be withdrawn, each rod at the same rate, withdrawing the rods on one side of the reactor and then withdrawing the rods on the other side of the reactor results in a larger asymptotic value for μ_i and ε_i than if all the rods are withdrawn simultaneously.
   c. Withdrawing a rod (group of rods) from position a to position c, then reinserting it (them) to position b $(a > b > c)$ results in a larger asymptotic value of μ_i and ε_i than if the rod (group of rods) was withdrawn at the same rate from position a to position b.

6. The time at which μ_i and ε_i obtain an asymptotic value may differ from region to region, particularly if flux tilting is significant.

7. When the reactor is brought from a subcritical to a supercritical configuration, the asymptotic value of μ_i and ε_i depends on the source level and the initial subcritical multiplication factor.

8. The more supercritical the configuration obtained before μ_i and ε_i attain their asymptotic value, the larger this asymptotic value is.

9. For a supercritical reactor, μ_i and ε_i generally attain their asymptotic value when $\bar{n}_i$ is of the order of 10^5 n/cm^3.

In a subcritical reactor, the neutron fluctuations are governed by fluctuations in the neutron sources, which are the instantaneous natural and neutron-induced fission rates and delayed neutron precursor decay rates, as well as by the fluctuations of the fission, capture, and diffusion processes. The precursor fluctuations are governed by an integral of the fission fluctuations over several mean lifetimes for the precursors ($\tau_{\text{mean}} = \lambda^{-1}$). This integral dependence of the precursor fluctuations on the fluctuations in the fission process tends to smooth out the fluctuations in the former relative to fluctuations in the latter:

$$C_m(r,t) = \int_{t-n\tau}^{t} dt' \, e^{-\lambda(t-t')} \beta \Sigma_f(t,t') n(r,t') v, \quad n \sim 10$$

In a supercritical reactor, the precursor fluctuations still depend on an integral of the fission fluctuations over the last few mean precursor lifetimes. However, the major contribution to the integral now comes from times close to the upper limit of the integral. Thus the precursor fluctuations tend to depend on the instantaneous fission fluctuations. In a supercritical reactor the major source of prompt neutrons very quickly becomes the neutron-induced fission rate. Thus the neutron and precursor fluctuations are governed by fluctuations in the instantaneous fission rates, and it is plausible that these fluctuations are statistically identical.[9]

In a subcritical reactor in which the relative fission and the capture and diffusion probabilities vary from region to region, it is reasonable to expect the fluctuations in the neutron population to exhibit different statistical characteristics from region to region. Similarly, when the relative absorption and scattering probabilities and the fission spectrum differ for the various energy groups in a subcritical reactor, the fluctuations in the neutron populations in the different energy groups plausibly exhibit different statistical characteristics. It is interesting that in a supercritical reactor the fluctuations in the neutron population exhibit asymptotically the same statistical characteristics at all spatial positions and in all energy groups.

From the numerical results, the behavior of the stochastic distribution of the neutron and precursor populations within a reactor can be deduced. In subcritical reactors the stochastic neutron distribution is spatially and energy dependent, and the stochastic precursor distribution is spatially dependent. In general, in a subcritical reactor, the stochastic neutron distribution is more disperse than the stochastic precursor distribution at the same spatial location.

In a supercritical reactor, the asymptotic stochastic neutron distribution is space and energy independent and is identical to the asymptotic stochastic precursor distribution. As a reactor is brought from a subcritical to a supercritical configuration, the stochastic neutron distribution generally becomes less disperse, whereas the stochastic precursor distribution becomes more disperse. The dispersion of the asymptotic distribution in a supercritical reactor depends on the manner in which

9) Have the same relative variances.

the reactor attains its final configuration as well as on the multiplicative properties of the initial and final configurations and the source level. The dispersion of the asymptotic distribution is more sensitive to changes that are made to the reactor configuration when the mean neutron and precursor densities are small than to later changes made in the presence of larger mean neutron and precursor densities.

Startup Analysis

The essential problem of the analysis of a reactor startup is determination of the probability that the actual neutron population is within a prescribed band about the mean neutron population predicted by the deterministic kinetics equations. As a specific example, consider a startup excursion that is terminated by a power level trip actuating the scram mechanism. The scram is initiated at a finite time after the trip point is reached, during which time interval the neutron density continues to increase. If the startup procedure consists of shimming out control rods, the principal concern is that the actual neutron population is less than the mean population, in which case the neutron density at which the trip point is reached occurs later, with the reactor being more supercritical and thus on a shorter period than is predicted by the deterministic kinetics equations. Consequently, the power excursion is more severe than would be predicted deterministically.

Startup analyses may be separated into two phases, stochastic and deterministic. The first phase is analyzed with stochastic kinetics, and the results are used as initial conditions, with associated probabilities, for the second phase, which is analyzed with deterministic kinetics. Feedback effects generally may be ignored during the stochastic phase. A reasonable time to switch from the stochastic to the deterministic phase is the time at which the neutron and precursor distributions obtain their asymptotic shape. This time may probably be approximated by the time at which μ_i and ε_i of Eqs. (16.277) and (16.278) attain their asymptotic value. If the neutron and precursor distributions [i.e., $P(N't'|Nt_s)$] were known at the switchover time t_s, the probability that the actual neutron and precursor densities are less than some specified values could be calculated.

The asymptotic neutron and precursor distributions in a reactor with large multiplication and no feedback can be approximated by the gamma distribution, which is completely characterized by the mean and variance of the distribution (i.e., by $\bar{n}_i$ and μ_i and $\bar{c}_i$, and ε_i). Use of the gamma distribution is suggested theoretically by the fact that the stationary probability distribution of a variate in a stationary multiplicative process approaches a gamma distribution as the multiplication increases without limit, and is justified empirically by the fact that its use in conjunction with a point reactor kinetics model leads to results that are in reasonable agreement with the GODIVA weak-source transient data.

The gamma distribution is

$$F(x)\,dx = \frac{r^r}{\Gamma(r)} x^{r-1} e^{-rx}\,dx \tag{16.279}$$

where Γ is the gamma function, x the ratio of the actual value of the variate to the mean value of the variate, and r the ratio of the mean value of the variate to the square root of the variance. For example,

$$x = \frac{n_i}{\bar{n}_i}, \qquad r = \mu_i^{-1/2} \tag{16.280}$$

for the monoenergetic model.

From Eq. (16.279), the probability that $x < \Delta$ can be computed.

$$\text{Prob}\{x < \Delta\} = \int_0^{\Delta} F(x)\,dx = 1 - \frac{\Gamma_{\text{in}}(r, \Delta r)}{\Gamma(r)} \tag{16.281}$$

where Γ_{in} is the incomplete gamma function. This can be written entirely in terms of tabulated functions by using certain identities,

$$\text{Prob}\{x < \Delta\} = \frac{(\Delta r)^r e^{-\Delta r} M(1, r+1, \Delta r)}{r \Gamma(r)} \tag{16.282}$$

where M is the confluent hypergeometric function.

Based on the results of the stochastic phase, initial conditions for the deterministic phase can be assigned from

$$n_i^0 = \Delta \bar{n}_i(t_s), \qquad c_i^0 = \Delta \bar{c}_i(t_s) \tag{16.283}$$

where $\bar{n}_i$ and $\bar{c}_i$ are the mean values of the neutron and precursor densities at the switchover time, t_s. For a given value of Δ, Eq. (16.282) yields the probability that $n_i(t_s) < \Delta \bar{n}_i(t_s), c_i(t_s) < \Delta \bar{c}_i(t_s)$.

References

1 J. A. FAVORITE and W. M. STACEY, "Variational Estimates of Point Kinetics Parameters," *Nucl. Sci. Eng. 121*, 353 (**1995**); "Variational Estimates for Use with the Improved Quasistatic Method for Reactor Dynamics," *Nucl. Sci. Eng. 126*, 282 (**1997**).

2 T. M. SUTTON and B. N. AVILES, "Diffusion Theory Methods for Spatial Kinetics Calculations," *Prog. Nucl. Energy 30*, 119 (**1996**).

3 P. KAPS and P. RENTROP, "Generalized Runge–Kutta Methods of Order Four with Step-Size Control for Stiff Ordinary Differential Equations," *Numer. Math. 33*, 55 (**1979**); W. H. PRESS, S. A. TEUKOLSKY, W. T. VETTERLING, and B. P. FLANNERY, *Numerical Recipes in Fortran: The Art of Scientific Computing*,

2nd Ed., Cambridge University Press, Cambridge (**1992**).

4 W. WERNER, "Solution Methods for the Space–Time Dependent Neutron Diffusion Equation," *Adv. Nucl. Sci. Technol. 10*, 313 (**1977**).

5 H. L. DODDS, "Accuracy of the Quasistatic Method for Two-Dimensional Thermal Reactor Transients with Feedback," *Nucl. Sci. Eng. 59*, 271 (**1976**).

6 A. F. HENRY, *Nuclear Reactor Analysis*, MIT Press, Cambridge, MA (**1975**), Chap. 7.

7 D. R. FERGUSON, "Multidimensional Reactor Dynamics: An Overview," *Proc. Conf. Computation Methods in Nuclear Engineering*, CONF-750413, VI, 49 (**1975**).

8 D. C. WADE and R. A. RYDIN, "An Experimentally Measurable Relationship between Asymptotic Flux Tilts and Eigenvalue Separation," in D. L. Hetrick, ed., *Dynamics of Nuclear Systems*, University of Arizona Press, Tuscon, AZ **(1972)** p. 335.

9 W. M. STACEY, "Space- and Energy-Dependent Neutronics in Reactor Transient Analysis," *Reactor Technol. 14*, 169 **(1971)**; "Xenon-Induced Spatial Power Oscillations," *Reactor Technol. 13*, 252 **(1970)**; *Space–Time Nuclear Reactor Kinetics*, Academic Press, New York **(1969)**.

10 K. O. OTT and D. A. MENELEY, "Accuracy of the Quasistatic Treatment of Spatial Reactor Kinetics," *Nucl. Sci. Eng. 36*, 402 **(1969)**; D. A. MENELEY et al., "A Kinetics Model for Fast Reactor Analysis in Two Dimensions," in D. L. HETRICK, ed., *Dynamics of Nuclear Systems*, University of Arizona Press, Tuscon, AZ **(1972)**.

11 J. LEWINS and A. L. BABB, "Optimum Nuclear Reactor Control Theory," *Adv. Nucl. Sci. Technol. 4*, 252 **(1968)**.

12 A. A. FEL'DBAUM, *Optimal Control Systems*, Academic Press, New York **(1965)**.

13 L. S. PONTRYAGIN, V. G. BOLTYANSKII, R. V. GAMKNELIDZE, and E. F. MISHCHENKO, *The Mathematical Theory of Optimum Processes*, Wiley-Interscience, New York **(1962)**.

14 J. LASALLE and S. LEFSCHETZ, *Stability by Lyapunov's Direct Method*, Academic Press, New York **(1961)**.

15 R. BELLMAN, *Dynamic Programming*, Princeton University Press, Princeton, NJ **(1957)**.

16 J. N. GRACE, ed., "Reactor Kinetics" in *Naval Reactors Physics Handbook*, A. Radkowsky, ed., USAEC, Washington **(1964)**.

Problems

16.1 Estimate the relative "tiltiness" of graphite- and H_2O-moderated thermal reactors by estimating $1 - k_1$ as a function of slab reactor thickness over the range $1 \leq a \leq 5$ m. Calculate the associated time constant for the tilt to take place due to delayed neutron holdback.

16.2 Derive the orthogonality property given by Eq. (16.11) and the relationship of Eq. (16.12).

16.3 Calculate and plot the delayed neutron holdback time constant τ_{tilt} as a function of the ratio of reactor thickness to migration area for a uniform slab reactor.

16.4 Derive the point-kinetics equations from the multigroup diffusion equations. Discuss the physical significance of the point-kinetics parameters.

16.5 Consider a uniform bare slab reactor in one group diffusion theory ($D = 1.2$ cm, $\Sigma_a = 0.12$ cm^{-1}, $\nu\Sigma_f = 0.14$ cm^{-1}) that is perturbed over the left one-half of the slab by a 1% increase in absorption cross section. Calculate the critical slab thickness and the unperturbed flux distribution. Calculate the generalized adjoint function of Eq. (16.36). Calculate the first-order perturbation theory estimate, the variational estimate, and the exact value of the reactivity worth of the perturbation.

16.6 Numerically integrate the point-kinetics equations for the transient ensuing from the perturbation in Problem 16.5, using the three different reactivity estimates. Use the prompt jump approximation and one group of delayed neutrons ($\lambda = 0.08$ s^{-1}, $\beta = 0.0075$).

16.7 Derive a two-node kinetics model for the slab reactor of Problem 16.5. Numerically integrate the kinetics equations for the transient ensuing from the perturbation. Use the time-integrated method for the integration of one group of delayed neutron precursors and a prompt-jump approximation.

16.8 Repeat Problem 16.7, but retaining the time derivative in the neutron equations and approximating it by the θ-method. Solve the problem with $\theta = 0, 0.5$, and 1.

16.9 Assume that the absorption and fission cross sections in each node of Problem 16.7 have power temperature feedback coefficients and that the temperature in each node is determined by a balance between fission heating and conductive cooling. Analyze the linear stability of the two-node model as a function of the feedback coefficient values.

16.10. It is wished to linearly increase the power in node 1 of the reactor of Problems 16.5 and 16.7 by 25% and in node 2 by 50% over 10 s, by withdrawing separate control rods in nodes 1 and 2, and then to maintain constant power. Determine the time history of the change in control rod cross section in each node which will best approximate this desired power trajectory. Use the prompt-jump approximation and assume one group of delayed neutrons.

16.11. Construct a Lyapunov functional for the point kinetics equations with one delayed neutron precursor group. What can you say about the stability of these equations?

16.12. Consider a reactor described by the point kinetics equation with one group of delayed neutron precursors, a conductive heat removal equation, and a temperature coefficient of reactivity α_T. Analyze the linear stability of this reactor model.

16.13. Construct a Lyapunov function for the reactor model of Problem 16.12 and analyze the stability.

16.14. Carry through the derivation of the λ-mode linear stability criterion for xenon spatial oscillations discussed in Section 16.6.

16.15. Analyze the stability with respect to xenon spatial oscillations of the reactor of Problem 16.5 as a function of equilibrium flux level and power feedback coefficient. Use the λ-mode stability criterion.

16.16. Write a two-node dynamics code for one neutron energy group and one delayed neutron precursor group to solve for the time dependences of the means and variances in the neutron and precursor populations in a low-source startup problem. Use the properties $(D = 1.5 \text{ cm}, \Sigma_f = 0.008 \text{ cm}^{-1}, \Sigma_{ac} = 0.0125 \text{ cm}^{-1})$ and $(D = 0.1 \text{ cm}, \Sigma_f = 0.008 \text{ cm}^{-1}, \Sigma_c = 0.005 \text{ cm}^{-1})$ for two adjacent slab regions of thickness 150 cm each, the delayed neutron parameters $\beta = 0.0075$, $\lambda = 0.088 \text{ s}^{-1}$, and the prompt neutron parameters $\bar{v}_p = 2.41$, $\overline{v_p(v_p - 1)} = 3.84$. Calculate the startup of the reactor with a source of $S = 5 \times 10^2 \text{ s}^{-1}$ in the first regions.

16.17. Calculate the probability that the actual value of the neutron flux is less than 110% of the mean value as a function of μ_i, the mean-squared variance in the density to the square of the mean value of the density.

Appendix A
Physical Constants and Nuclear Data

I. Miscellaneous Physical Constants

Avogadro's number, N_A	6.022045×10^{23} mol^{-1}
Boltzmann constant, k	1.380662×10^{-23} J/K
	0.861735×10^{-4} eV/K
Electron rest mass, m_e	9.109534×10^{-31} kg
	0.5110034 MeV
Elementary charge, e	$1.6021892 \times 10^{-19}$ C
Gas constant, R	8.31441 J mol^{-1}/K
Neutron rest mass, m_n	$1.6749544 \times 10^{-27}$ kg
	939.5731 MeV
Planck's constant, h	6.626176×10^{-34} J/Hz
Proton rest mass, m_p	$1.6726485 \times 10^{-27}$ kg
	938.2796 MeV
Speed of light, c	2.99792458×10^{8} m/s

II. Some Useful Conversion Factors

1 eV	$1.6021892 \times 10^{-19}$ J
1 MeV	10^{6} eV
1 amu	$1.6605655 \times 10^{-27}$ kg
	931.5016 MeV
1 W	1 J/s
1 day	86,400 s
1 mean year	365.25 days
	8766 h
	3.156×10^{7} s
1 Ci	3.7000×10^{10} disintegrations/s
1 K	8.617065×10^{-5} eV

Nuclear Reactor Physics. Weston M. Stacey
Copyright © 2007 WILEY-VCH Verlag GmbH & Co. KGaA, Weinheim
ISBN: 978-3-527-40679-1

III. 2200-m/s Cross Sections for Naturally Occurring Elements [From Reactor Physics Constants, ANL-5800 (1963)]

Atomic No.	Element or Compound	Atomic or Mol. Wt.	Density (g/cm³)	Nuclei per Unit Volume (×10⁻²⁴)	$1 - \bar{\mu}_0$	ξ	Microscopic Cross Section (barns) σ_a	σ_s	σ_t	Macroscopic Cross Section (cm⁻¹) Σ_a	Σ_s	Σ_t
1	H	1.008	8.9†	5.3†	0.3386	1.000	0.33	38	38	1.7†	0.002	0.002
	H_2O	18.016	1	0.0335‡	0.676	0.948	0.66	103	103	0.022	3.45	3.45
	D_2O	20.030	1.10	0.0331‡	0.884	0.570	0.001	13.6	13.6	3.3†	0.449	0.449
2	He	4.003	17.8†	2.6†	0.8334	0.425	0.007	0.8	0.807	0.02†	2.1†	2.1†
3	Li	6.940	0.534	0.0463	0.9047	0.268	71	1.4	72.4	3.29	0.065	3.35
4	Be	9.013	1.85	0.1236	0.9259	0.209	0.010	7.0	7.01	124‡	0.865	0.865
	BeO	25.02	3.025	0.0728‡	0.939	0.173	0.010	6.8	6.8	73†	0.501	0.501
5	B	10.82	2.45	0.1364	0.9394	0.171	755	4	759	103	0.346	104
6	C	12.011	1.60	0.0803	0.9444	0.158	0.004	4.8	4.80	32†	0.385	0.385
7	N	14.008	0.0013	5.3†	0.9524	0.136	1.88	10	11.9	9.9†	50†	60†
8	O	16.000	0.0014	5.3†	0.9583	0.120	20†	4.2	4.2	0.000	21†	21†
9	F	19.00	0.0017	5.3†	0.9649	0.102	0.001	3.9	3.90	0.01†	20†	20†
10	Ne	20.183	0.0009	2.6†	0.9667	0.0968	<2.8	2.4	5.2	7.3†	6.2†	13.5†
11	Na	22.991	0.971	0.0254	0.9710	0.0845	0.525	4	4.53	0.013	0.102	0.115
12	Mg	24.32	1.74	0.0431	0.9722	0.0811	0.069	3.6	3.67	0.003	0.155	0.158
13	Al	26.98	2.699	0.0602	0.9754	0.0723	0.241	1.4	1.64	0.015	0.084	0.099
14	Si	28.09	2.42	0.0522	0.9762	0.0698	0.16	1.7	1.86	0.008	0.089	0.097
15	P	30.975	1.82	0.0354	0.9785	0.0632	0.20	5	5.20	0.007	0.177	0.184
16	S	32.066	2.07	0.0389	0.9792	0.0612	0.52	1.1	1.62	0.020	0.043	0.063
17	Cl	35.457	0.0032	5.3†	0.9810	0.0561	33.8	16	49.8	0.002	80†	0.003
18	A	39.944	0.0018	2.6†	0.9833	0.0492	0.66	1.5	2.16	1.7†	3.9	5.6†
19	K	39.100	0.87	0.0134	0.9829	0.0504	2.07	1.5	3.57	0.028	0.020	0.048
20	Ca	40.08	1.55	0.0233	0.9833	0.0492	0.44	3.0	3.44	0.010	0.070	0.080

21	Sc	44.96	2.5	0.0335	0.9852	0.0438	24	24	48	0.804	0.804	1.61
22	Ti	47.90	4.5	0.0566	0.9861	0.0411	5.8	4	9.8	0.328	0.226	0.555
23	V	50.95	5.96	0.0704	0.9869	0.0387	5	5	10.0	0.352	0.352	0.704
24	Cr	52.01	7.1	0.0822	0.9872	0.0385	3.1	3	6.1	0.255	0.247	0.501
25	Mn	54.94	7.2	0.0789	0.9878	0.0359	13.2	2.3	15.5	1.04	0.181	1.22
26	Fe	55.85	7.86	0.0848	0.9881	0.0353	2.62	11	13.6	0.222	0.933	1.15
27	Co	58.94	8.9	0.0910	0.9887	0.0335	38	7	45	3.46	0.637	4.10
28	Ni	58.71	8.90	0.0913	0.9887	0.0335	4.6	17.5	22.1	0.420	1.60	2.02
29	Cu	63.54	8.94	0.0848	0.9896	0.0309	3.85	7.2	11.05	0.0326	0.611	0.937
30	Zn	65.38	7.14	0.0658	0.9897	0.0304	1.10	3.6	4.70	0.072	0.237	0.309
31	Ga	69.72	5.91	0.0511	0.9925	0.0283	2.80	4	6.80	0.143	0.204	0.347
32	Ge	72.60	5.36	0.0445	0.9909	0.0271	2.45	3	5.45	0.109	0.134	0.243
33	As	74.91	5.73	0.0461	0.9911	0.0264	4.3	6	10.3	0.198	0.277	0.475
34	Se	78.96	4.8	0.0366	0.9916	0.0251	12.3	11	23.3	0.450	0.403	0.853
35	Br	79.916	3.12	0.0235	0.9917	0.0247	6.7	6	12.7	0.157	0.141	0.298
36	Kr	83.80	0.0037	2.6†	0.9921	0.0236	31	7.2	38.2	81†	19†	99†
37	Rb	85.48	1.53	0.0108	0.9922	0.0233	0.73	12	12.7	0.008	0.130	0.138
38	Sr	87.63	2.54	0.0175	0.9925	0.0226	1.21	10	11.2	0.021	0.175	0.195
39	Yt	88.92	5.51	0.0373	0.9925	0.0223	1.313	4.3	4.3	0.049	0.112	0.160
40	Zr	91.22	6.4	0.0423	0.9927	0.0218	0.185	8	8.2	0.008	0.338	0.347
41	Nb	92.91	8.4	0.0545	0.9928	0.0214	1.16	5	6.16	0.063	0.273	0.336
42	Mo	95.95	10.2	0.0640	0.9931	0.0207	2.70	7	9.70	0.173	0.448	0.621
43	Tc	98.0	—	—	0.9932	0.0203	22	—	—	—	—	—
44	Ru	101.1	12.2	0.0727	0.9934	0.0197	2.56	6	8.56	0.186	0.436	0.622
45	Rh	102.91	12.5	0.0732	0.9935	0.0193	149	5	154	10.9	0.366	11.3
46	Pd	106.4	12.16	0.0689	0.9937	0.0187	8	3.6	11.6	0.551	0.248	0.799
47	Ag	107.88	10.5	0.0586	0.9938	0.0184	63	6	69	3.69	0.352	4.04
48	Cd	112.41	8.65	0.0464	0.9940	0.0178	2450	7	2457	114	0.325	114
49	In	114.82	7.28	0.0382	0.9942	0.0173	191	2.2	193	7.30	0.084	7.37
50	Sn	118.70	6.5	0.0330	0.9944	0.0167	0.625	4	4.6	0.021	0.132	0.152

(Continued)

III. (Continued)

Atomic No.	Element or Compound	Atomic or Mol. Wt.	Density (g/cm³)	Nuclei per Unit Volume (×10⁻²⁴)	$1 - \bar{\mu}_0$	ξ	Microscopic Cross Section (barns)			Macroscopic Cross Section (cm⁻¹)		
							σ_a	σ_s	σ_t	Σ_a	Σ_s	Σ_t
51	Sb	121.76	6.69	0.0331	0.9945	0.0163	5.7	4.3	10.0	0.189	0.142	0.331
52	Te	127.61	6.24	0.0295	0.9948	0.0155	4.7	5	9.7	0.139	0.148	0.286
53	I	126.91	4.93	0.0234	0.9948	0.0157	7.0	3.6	10.6	0.164	0.084	0.248
54	Xe	131.30	0.0059	2.7†	0.9949	0.0152	35	4.3	39.3	95†	12†	0.001
55	Cs	132.91	1.873	0.0085	0.9950	0.0150	28	20	48	0.238	0.170	0.408
56	Ba	137.36	3.5	0.0154	0.9951	0.0145	1.2	8	9.2	0.018	0.123	0.142
57	La	138.92	6.19	0.0268	0.9952	0.0143	8.9	15	24	0.239	0.403	0.642
58	Ce	140.13	6.78	0.0292	0.9952	0.0142	0.73	9	9.7	0.021	0.263	0.283
59	Pr	140.92	6.78	0.0290	0.9953	0.0141	11.3	4	15.3	0.328	0.116	0.444
60	Nd	144.27	6.95	0.0290	0.9954	0.0138	46	16	62	1.33	0.464	1.79
61	Pm	145.0	–	–	0.9954	0.0137	60	–	–	–	–	–
62	Sm	150.35	7.7	0.0309	0.9956	0.0133	5600	5	5605	173	0.155	173
	Sm₂O₃	348.70	7.43	0.0128†‡	0.974	0.076	16,500	22.6	16,500	211	0.289	211
63	Eu	152.0	5.22	0.0207	0.9956	0.0131	4300	8	4308	89.0	0.166	89.2
	Eu₂O₃	352.00	7.42	0.0127†‡	0.978	0.063	8740	30.2	8770	111	0.383	111
64	Gd	167.26	7.95	0.0305	0.9958	0.0127	46,000	–	–	1403	–	–
65	Tb	158.93	8.33	0.0316	0.9958	0.0125	46	–	–	1.45	–	–
66	Dy	162.51	8.56	0.0317	0.9959	0.0122	950	100	1050	30.1	3.17	33.3
	Dy₂O₃	372.92	7.81	0.0126†‡	0.993	0.019	2200	214	2414	27.7	2.7	30.4
67	Ho	164.94	8.76	0.0320	0.9960	0.0121	65	–	–	2.08	–	–
68	Er	167.27	9.16	0.0330	0.9960	0.0119	173	15	188	5.71	0.495	6.20
69	Tm	168.94	9.35	0.0333	0.9961	0.0118	127	7	134	4.23	0.233	4.46
70	Yb	173.04	7.01	0.0244	0.9961	0.0115	37	12	49	0.903	0.293	1.20

71	Lu	174.99	9.74	0.0335	0.9962	0.0114	112	–	–	3.75	–	–
72	Hf	178.5	13.3	0.0449	0.9963	0.0112	105	8	113	4.71	0.0359	5.07
73	Ta	180.95	16.6	0.0553	0.9963	0.0110	21	5	26	1.16	0.277	1.44
74	W	183.86	19.3	0.0632	0.9964	0.0108	19.2	5	24.2	1.21	0.316	1.53
75	Re	186.22	20.53	0.0664	0.9964	0.0107	86	14	100	5.71	0.930	6.64
76	OS	190.2	22.48	0.0712	0.9965	0.0105	15.3	11	26.3	1.09	0.783	1.87
77	Ir	192.2	22.42	0.0703	0.9965	0.0104	440	–	–	30.9	–	–
78	Pt	195.09	21.37	0.0660	0.9966	0.0102	8.8	10	18.8	0.581	0.660	1.24
79	Au	197.0	19.32	0.0591	0.9966	0.0101	98.8	9.3	107.3	5.79	0.550	6.34
80	Hg	200.61	13.55	0.0407	0.9967	0.0099	380	20	400	15.5	0.814	16.3
81	Ti	204.39	11.85	0.0349	0.9967	0.0098	3.4	14	17.4	0.119	0.489	0.607
82	Pb	207.21	11.35	0.0330	0.9968	0.0096	0.170	11	11.2	0.006	0.363	0.369
83	Bi	209.0	9.747	0.0281	0.9968	0.0095	0.034	9	9	0.001	0.253	0.256
84	Po	210.0	9.24	0.0265	0.9968	0.0095	–	–	–	–	–	–
85	At	211.0	–	–	0.9968	0.0094	–	–	–	–	–	–
86	Rn	222.0	0.0097	2.6†	0.9970	0.0090	0.7	–	–	–	–	–
87	Fr	223.0	–	–	0.9980	0.0089	–	–	–	–	–	–
88	Ra	226.05	5	0.0133	0.9971	0.0088	20	–	–	0.266	–	–
89	Ac	227.0	–	–	0.9971	0.0088	510	–	–	–	–	–
90	Th	232.05	11.3	0.0293	0.9971	0.0086	7.56	12.6	20.2	0.222	0.369	0.592
91	Pa	231.0	15.4	0.0402	0.9971	0.0086	200	–	–	8.04	–	–
92	U	238.07	18.9	0.04783	0.9972	0.0084	7.68	8.3	16.0	0.367	0.397	0.765
	UO$_2$	270.07	10	0.0223‡	0.9887	0.036	7.6	16.7	24.3	0.169	0.372	0.542
93	Np	237.0	–	–	0.9972	0.0084	170	–	–	–	–	–
94	Pu	239.0	19.74	0.0498	0.9972	0.0083	1026	9.6	1036	51.1	0.478	51.6
95	Am	242.0	–	–	0.9973	0.0082	8.000	–	–	–	–	–

† Value has been multiplied by 10^5.
‡ Molecules/cm^3.

IV. 2200-m/s Cross Sections of Special Interest

^{10}B:	$\sigma_a = 3837\mathrm{b}$	
^{11}B:	$\sigma_a = 0.005$	
^{135}Xe:	$\sigma_a = 2.7 \times 10^6$	
^{233}U:	$\sigma_\gamma = 49$	$\sigma_f = 524$
^{235}U:	$\sigma_\gamma = 101$	$\sigma_f = 577$
^{238}U:	$\sigma_\gamma = 2.73$	
^{239}Pu:	$\sigma_\gamma = 274$	$\sigma_f = 741$
^{240}Pu:	$\sigma_\gamma = 286$	$\sigma_f = 0.03$
^{241}Pu:	$\sigma_\gamma = 425$	$\sigma_f = 950$
^{242}Pu:	$\sigma_\gamma = 30$	$\sigma_f < 0.2$

This appendix is adapted by permission of John Wiley & Sons from James J. Duderstadt and Louis J. Hamilton, *Nuclear Reactor Analysis*, copyright © 1976 by John Wiley & Sons, Inc.

Appendix B
Some Useful Mathematical Formulas

(1) *Solution of First-Order Linear Differential Equations*:

$$\frac{df}{dx} + a(x)f(x) = g(x) \tag{B.1}$$

$$f(x) = e^{-A(x)}\left[\int^{x} dx'\, e^{A(x')} g(x') + C\right], \quad A(x) = \int^{x} dx'\, a(x') \tag{B.2}$$

(2) *Differentiation of a Definite Integral*:

$$\frac{d}{dx}\int_{b(x)}^{a(x)} dx'\, F(x,x') = F(x,a)\frac{da}{dx} - F(x,b)\frac{db}{dx} + \int_{b(x)}^{a(x)} dx'\, \frac{\partial F(x,x')}{\partial x} \tag{B.3}$$

(3) *Representation of Laplacian ∇^2 in Various Coordinate Systems*:
  (a) *Cartesian*:

$$\nabla^2 = \frac{\partial^2}{\partial x^2} + \frac{\partial^2}{\partial y^2} + \frac{\partial^2}{\partial z^2} \tag{B.4}$$

  (b) *Cylindrical*:

$$\nabla^2 = \frac{1}{r}\frac{\partial}{\partial r} r \frac{\partial}{\partial r} + \frac{1}{r^2}\frac{\partial^2}{\partial \theta^2} + \frac{\partial^2}{\partial z^2} \tag{B.5}$$

Nuclear Reactor Physics. Weston M. Stacey
Copyright © 2007 WILEY-VCH Verlag GmbH & Co. KGaA, Weinheim
ISBN: 978-3-527-40679-1

(c) *Spherical*:

$$\nabla^2 = \frac{1}{r^2}\frac{\partial}{\partial r}r^2\frac{\partial}{\partial r} + \frac{1}{r^2\sin\theta}\frac{\partial}{\partial\theta}\left(\sin\theta\frac{\partial}{\partial\theta}\right)$$
$$+ \frac{1}{r^2\sin^2\theta}\frac{\partial^2}{\partial\phi^2} \tag{B.6}$$

(4) *Gauss' Divergence Theorem*:

$$\int_V d^3r\,\nabla\cdot\mathbf{A} = \int_S dS\,\hat{\mathbf{e}}_s\cdot\mathbf{A} \tag{B.7}$$

where $\hat{\mathbf{e}}_s$ is the unit vector normal to the surface element dS.

(5) *Green's Theorem*:

$$\int d^3r\,\nabla\phi\cdot\nabla\psi = \int dS\,\phi\hat{\mathbf{e}}_s\cdot\nabla\psi - \int d^3r\,\phi\nabla^2\psi \tag{B.8}$$

$$\int d^3r\left(\phi\nabla^2\psi - \psi\nabla^2\phi\right) = \int dS\,\hat{\mathbf{e}}_s\cdot(\phi\nabla\psi - \psi\nabla\phi) \tag{B.9}$$

(6) *Taylor Series Expansion*:

$$f(x) = f(x_0) + (x - x_0)f'(x_0) + \frac{(x - x_0)^2}{2!}f''(x_0) + \cdots \tag{B.10}$$

(7) *Fourier Series Expansion*:

$$f(x) = \sum_{n=1}^{\infty} a_n\sin\frac{n\pi x}{l} + \frac{1}{2}b_0 + \sum_{n=1}^{\infty} b_n\cos\frac{n\pi x}{l} \tag{B.11}$$

where

$$a_n \equiv \frac{1}{l}\int_{-l}^{l} dx'\,f(x')\sin\frac{n\pi x'}{l}, \quad b_n \equiv \frac{1}{l}\int_{-l}^{l} dx'\,f(x')\cos\frac{n\pi x'}{l} \tag{B.12}$$

Appendix C
Step Functions, Delta Functions, and Other Functions

C.1
Introduction

Consider the discontinuous function $\Theta(x)$ defined by the properties

$$\Theta(x) = \begin{cases} 0, & x < 0 \\ 1, & x \geq 0 \end{cases} \tag{C.1}$$

$\Theta(x)$ is the unit "step function" introduced by Heaviside in his development of operational calculus (now known as integral transform analysis). One can perform numerous operations on $\Theta(x)$. In particular in can be integrated to yield the ramp function

$$\eta(x) = \int_{-\infty}^{x} dx' \, \Theta(x') = \begin{cases} 0, & x < 0 \\ x, & x \geq 0 \end{cases} \tag{C.2}$$

Let's try something a bit more unusual by taking the derivative of $\Theta(x)$. Clearly this is ridiculous, because this derivative, call it $\delta(x)$, is undefined at $x = 0$ because $\Theta(x)$ is discontinuous at this point:

$$\delta(x) = \Theta'(x) = \lim_{\varepsilon \to 0} \left[\frac{\Theta(x + \varepsilon) - \Theta(x)}{\varepsilon} \right] = \begin{cases} 0, & x \neq 0 \\ \infty, & x = 0 \end{cases} \tag{C.3}$$

Nevertheless Dirac, Heaviside, and others have made very good use of this strange "function." To be more specific, the Dirac δ-function, $\delta(x)$, has the properties

$$\delta(x - x_0) = \begin{cases} 0, & x \neq x_0 \\ \infty, & x = x_0, \end{cases} \qquad \int_{-\infty}^{\infty} dx \, \delta(x - x_0) = 1 \tag{C.4}$$

Nuclear Reactor Physics. Weston M. Stacey
Copyright © 2007 WILEY-VCH Verlag GmbH & Co. KGaA, Weinheim
ISBN: 978-3-527-40679-1

In a sense, it resembles a generalization of the Kronecker δ-function

$$\delta_{mn} = \begin{cases} 0, & m \neq n \\ 1, & m = n \end{cases}$$

The most useful property of the Dirac δ-function occurs when it is integrated along with a well-behaved function, say $f(x)$:

$$\int dx \; f(x)\delta(x - x_0) = f(x_0) \tag{C.5}$$

This property not only is very interesting, but extremely useful in mathematical physics. Unfortunately the proof of this property—and, indeed, all of the *theory* of such generalized functions—requires a rather potent dose of mathematics. [Such generalized functions are really not functions at all, but rather a class of linear functionals called "distributions" defined on some set of suitable test functions (which are "infinitely differentiable with compact support").]

Fortunately one does not need all of this high-powered mathematics in order to *use* δ-functions. Only a knowledge of their properties is necessary.

C.2
Properties of the Dirac δ-Function

A. Alternative Representations

$$\delta(x - x_0) = \frac{1}{\pi} \lim_{\lambda \to \infty} \frac{\sin \lambda(x - x_0)}{(x - x_0)} \tag{C.6}$$

$$\delta(x - x_0) = \frac{1}{\pi} \lim_{\varepsilon \to 0^+} \frac{\varepsilon}{(x - x_0)^2 + \varepsilon^2} \tag{C.7}$$

B. Properties

$$\delta(x) = \delta(-x) \tag{C.8}$$

$$\delta(ax) = \frac{1}{|a|}\delta(x), \quad a \neq 0 \tag{C.9}$$

$$\delta[g(x)] = \sum_n \frac{1}{|g'(x_n)|}\delta(x - x_n) \quad [g(x_n) = 0, \; g'(x_n) \neq 0] \tag{C.10}$$

$$x\delta(x) = 0 \tag{C.11}$$

$$f(x)\delta(x - a) = f(a)\delta(x - a) \tag{C.12}$$

$$\int \delta(x - y)\delta(y - a)\,dy = \delta(x - a) \tag{C.13}$$

$$\delta(x) = \frac{1}{2\pi} \int_{-\infty}^{\infty} dk\, e^{ikx} \tag{C.14}$$

Actually these properties only make sense when inserted in an integral. For example, property (C.8) really should be interpreted as

$$\int dx\, f(x)\delta(x) = \int dx\, f(x)\delta(-x) = f(0) \tag{C.15}$$

C. Derivatives

One can differentiate a δ-function as many times as one wishes. The mth derivative is defined by

$$\int_{-\infty}^{\infty} \delta^{[m]}(x - a) f(x)\,dx = (-1)^m \frac{d^m f}{dx^m}\bigg|_{x=a} \tag{C.16}$$

One can show

$$\delta^{[m]}(x) = (-1)^m \delta^{[m]}(-x) \tag{C.17}$$

$$\int \delta^{[m]}(x - y)\delta^{[n]}(y - a)\,dy = \delta^{[m+n]}(x - a) \tag{C.18}$$

$$x^{m+1}\delta^{[m]}(x) = 0 \tag{C.19}$$

Perhaps of more direct use is the application of these properties to the first derivative

$$\int_{-\infty}^{\infty} \delta'(x) f(x)\,dx = -f'(0) \tag{C.20}$$

$$\delta'(x) = -\delta'(-x) \tag{C.21}$$

$$\int \delta'(x - y)\delta(y - a)\,dy = \delta'(x - a) \tag{C.22}$$

$$x\delta'(x) = -\delta(x) \tag{C.23}$$

One can generalize the concept of a δ-function to several dimensions. For example, we would define the three-dimensional δ-function by

$$\int d^3r'\, \delta(\mathbf{r}' - \mathbf{r}) f(\mathbf{r}') = f(\mathbf{r}) \tag{C.24}$$

Note that we could write this in Cartesian coordinates as

$$\delta(\mathbf{r} - \mathbf{r}') = \delta(x - x')\delta(y - y')\delta(z - z') \tag{C.25}$$

Such multidimensional δ-functions are of very considerable use in vector calculus.

More detailed discussions of the Dirac δ-function and its relatives are found in the following references.

References

1   J. W. DETTMAN, *Mathematical Methods in Physics and Engineering*, 2nd Edition, McGraw-Hill, New York **(1969)**.

2   M. J. LIGHTHILL, *Fourier Analysis and Generalized Functions*, Cambridge University Press, Cambridge **(1959)**.

3   A. MESSIAH, *Quantum Mechanics*, Vol. I, Wiley, New York **(1965)**, pp. 468–470.

Appendix D
Some Properties of Special Functions

(1) *Legendre Functions*:
   (a) *Defining equation*:

$$(1 - x^2)f'' - 2xf' + l(l+1)f = 0, \quad l = \text{integer} \tag{D.1}$$

   (b) *Representation*:

$$P_l(x) = \frac{1}{2^l l!} \frac{d^l}{dx^l} (x^2 - 1)^l \tag{D.2}$$

   (c) *Properties*:

$$P_0(x) = 1, \quad P_1(x) = x, \quad P_2(x) = \frac{1}{2}(3x^2 - 1)$$

$$P_3 = \frac{1}{2}(5x^3 - 3x), \quad \dots \tag{D.3}$$

$$\int_{-1}^{+1} P_l(x) P_{l'}(x)\, dx = \frac{2}{2l+1}\delta_{ll'} \tag{D.4}$$

   (d) *Recurrence relations*:

$$P'_{l+1}(x) - x P'_l(x) = (l+1) P_l(x) \tag{D.5}$$

$$(l+1)P_{l+1}(x) - (2l+1)x P_l(x) + l P_{l-1}(x) = 0 \tag{D.6}$$

(2) *Associated Legendre Polynomials*:
   (a) *Defining equation*:

$$(1 - x^2)f'' - 2xf' + \left[l(l+1) - \frac{m^2}{1 - x^2} \right] f = 0 \tag{D.7}$$

   (b) *Representation*:

$$P_l^m(x) = (1 - x^2)^{(m/2)} \frac{d^m}{dx^m} P_l(x) \tag{D.8}$$

Nuclear Reactor Physics. Weston M. Stacey
Copyright © 2007 WILEY-VCH Verlag GmbH & Co. KGaA, Weinheim
ISBN: 978-3-527-40679-1

(c) *Spherical harmonics:*

$$Y_{lm}(\Omega) = \left[\frac{(2l+1)(l-m)!}{4\pi(l+m)!} \right]^{(1/2)} P_l(\cos\theta)e^{im\phi}$$

(D.9)

(d) *Properties:*

$$\int_{4\pi} d\Omega\, Y_{lm}^*(\Omega)Y_{l'm'}(\Omega) = \delta_{ll'}\delta_{mm'}$$

(D.10)

$$P_l(\Omega \cdot \Omega') = \frac{4\pi}{(2l+1)} \sum_{m=-l}^{l} Y_{lm}^*(\Omega)Y_{lm}(\Omega')$$

(D.11)

(3) *Bessel Functions:*
   (a) *Defining equation:*

$$x^2 f'' + xf' + (x^2 - n^2)f = 0$$

(D.12)

   (b) *Solution:* $J_n(x)$, Bessel function of first kind
   $Y_n(x)$, Bessel function of second kind

   (c) *Representation:*

$$J_n(x) = \sum_{k=0}^{\infty} \frac{(-1)^k}{\Gamma(k+1)\Gamma(k+n+1)} \left(\frac{x}{2}\right)^{n+2k}$$

$$Y_n(x) = \frac{J_n(x)\cos(n\pi) - J_{-n}(x)}{\sin n\pi}$$

(D.13)

   (d) *Hankel functions:*

$$H_n^{(1)}(x) = J_n(x) + iY_n(x)$$

(D.14)

$$H_n^{(2)}(x) = J_n(x) - iY_n(x)$$

(D.15)

(4) *Modified Bessel Functions:*
   (a) *Defining equation:*

$$x^2 f'' + xf' - (x^2 + n^2)f = 0$$

(D.16)

   (b) *Solution:* $I_n(x)$, modified Bessel function of first kind
   $K_n(x)$, modified Bessel function of second kind

   (c) *Representation:*

$$I_n(x) = i^{-n} J_n(ix) = i^n J_n(-ix)$$

$$K_n(x) = \frac{\pi}{2}i^{n+1} H_n^{(1)}(ix) = \frac{\pi}{2}i^{-n-1} H_n^{(2)}(-ix)$$

(D.17)

(5) *Useful Expansions of Bessel Functions for small x:*

$$J_0(x) = 1 - \frac{x^2}{4} + \frac{x^4}{64} - \frac{x^6}{2304} + \cdots \tag{D.18}$$

$$J_1(x) = \frac{x}{2} - \frac{x^3}{16} + \frac{x^5}{384} - \cdots \tag{D.19}$$

$$Y_0(x) = \frac{2}{\pi}\left[\left(\gamma + \ln\frac{x}{2}\right)J_0(x) + \frac{x^2}{4} + \cdots\right], \quad \gamma \equiv 0.577216 \tag{D.20}$$

$$Y_1(x) = \frac{2}{\pi}\left[\left(\gamma + \ln\frac{x}{2}\right)J_1(x) - \frac{1}{x} - \frac{x}{4} + \cdots\right] \tag{D.21}$$

$$I_0(x) = 1 + \frac{x^2}{4} + \frac{x^4}{64} + \frac{x^6}{2304} + \cdots \tag{D.22}$$

$$I_1(x) = \frac{x}{2} + \frac{x^3}{16} + \frac{x^5}{384} + \cdots \tag{D.23}$$

$$K_0(x) = -\left(\gamma + \ln\frac{x}{2}\right)I_0(x) + \frac{x^2}{4} + \frac{3x^4}{128} + \cdots \tag{D.24}$$

$$K_1(x) = \left(\gamma + \ln\frac{x}{2}\right)I_1(x) + \frac{1}{x} - \frac{x}{4} - \frac{5x^3}{64} + \cdots \tag{D.25}$$

(a) *Asymptotic expansions for large x:*

$$I_0(x) = \frac{e^x}{\sqrt{2\pi x}}\left(1 + \frac{1}{8x} + \cdots\right) \tag{D.26}$$

$$I_1(x) = \frac{e^x}{\sqrt{2\pi x}}\left(1 - \frac{3}{8x} + \cdots\right) \tag{D.27}$$

$$K_0(x) = \sqrt{\frac{\pi}{2x}}e^{-x}\left(1 - \frac{1}{8x} + \cdots\right) \tag{D.28}$$

$$K_1(x) = \sqrt{\frac{\pi}{2x}}e^{-x}\left(1 + \frac{3}{8x} + \cdots\right) \tag{D.29}$$

(b) *Recurrence relations:*

$$x J_n' = n J_n - x J_{n+1} = -n J_n + x J_{n-1} \tag{D.30}$$

$$2n J_n = x J_{n-1} + x J_{n+1} \tag{D.31}$$

$$x I_n' = n I_n + x I_{n+1} = -n I_n + x I_{n-1} \tag{D.32}$$

$$x K_n' = n K_n - x K_{n+1} = -n K_n - x K_{n-1} \tag{D.33}$$

$$J_0' = -J_1, \quad Y_0' = -Y_1, \quad I_0' = I_1, \quad K_0' = -K_1 \tag{D.34}$$

(c) *Integrals:*

$$\int x^n J_{n-1}(x)\,dx = x^n J_n, \qquad \int x^n Y_{n-1}(x)\,dx = x^n Y_n \tag{D.35}$$

$$\int x^n I_{n-1}\,dx = x^n I_n, \qquad \int x^n K_{n-1}\,dx = -x^n K_n \tag{D.36}$$

(6) *Gamma Function:*
   (a) *Definition:*

$$\Gamma(z) = \int_0^\infty dt\, e^{-t} t^{z-1} \tag{D.37}$$

   (b) *Properties:*

$$\Gamma(z+1) = z\Gamma(z)$$
$$\Gamma(1/2) = \sqrt{\pi}, \quad \Gamma(0) = \infty, \quad \Gamma(1) = 1, \quad \dots, \quad \Gamma(n) = (n-1)! \tag{D.38}$$

(7) *Error Function:*
   (a) *Definition:*

$$\mathrm{erf}(x) = \frac{2}{\sqrt{\pi}} \int_0^x dt\, e^{-t^2} \tag{D.39}$$

   (b) *Complementary error function:*

$$\mathrm{erfc}(x) = 1 - \mathrm{erf}(x) = \frac{2}{\sqrt{\pi}} \int_0^\infty dt\, e^{-t^2} \tag{D.40}$$

(8) *Exponential Integrals:*
   (a) *Definition:*

$$E_n(x) = \int_1^\infty dt\, \frac{e^{-xt}}{t^n}, \quad E_1(x) = \int_1^\infty dt\, \frac{e^{-xt}}{t} = \int_x^\infty dt\, \frac{e^{-t}}{t} \tag{D.41}$$

   (b) *Properties:*

$$E_0(x) = \frac{e^{-x}}{x} \tag{D.42}$$

$$E_n'(x) = -E_{n-1}(x) \tag{D.43}$$

$$E_n(x) = \frac{1}{n-1}[e^{-x} - xE_{n-1}(x)], \quad n > 1 \tag{D.44}$$

$$E_1(x) = -\gamma - \ln x - \sum_{n=1}^\infty \frac{(-1)^n x^n}{nn!} \tag{D.45}$$

References

1 M. ABRAMOWITZ and I. STEGUN (Eds.), *Handbook of Mathematical Functions*, Dover, New York (**1965**).

2 H. MARGENAU and G. M. MURPHY, *The Mathematics of Physics and Chemistry*, 2nd Ed., Vol. I, Van Nostrand, Princeton, NJ (**1956**).

3 I. S. GRADSHTEYN and I. M. RYZHIK, *Table of Integrals, Series, and Products*, 4th Ed., Academic Press, New York (**1965**).

4 P. M. MORSE and H. FESHBACH, *Methods of Theoretical Physics*, Vols. I and II, McGraw-Hill, New York (**1953**).

Appendix E
Introduction to Matrices and Matrix Algebra

E.1
Some Definitions

One defines a *matrix of order* $(m \times n)$ to be a rectangular array of m rows and n columns

$$
A = \begin{bmatrix}
a_{11} & a_{12} & a_{13} & \cdots & a_{1n} \\
a_{21} & & \vdots & & \vdots \\
\vdots & & a_{ij} & & \vdots \\
a_{m1} & \cdots & \cdots & \cdots & a_{mn}
\end{bmatrix}
\tag{E.1}
$$

The *matrix elements* a_{ij} will be identified by subscripts denoting their row i and column j. If the matrix has the same number of rows as columns, it is said to be a *square matrix*; for example,

$$
A = \begin{pmatrix}
a_{11} & a_{12} & a_{13} \\
a_{21} & a_{22} & a_{23} \\
a_{31} & a_{32} & a_{33}
\end{pmatrix}
\tag{E.2}
$$

A *diagonal matrix* has nonzero elements only along its main diagonal:

$$
A = \begin{bmatrix}
a_{11} & 0 & 0 \\
0 & a_{22} & 0 \\
0 & 0 & a_{33}
\end{bmatrix}
\tag{E.3}
$$

A *tridiagonal matrix* would have nonzero elements only along its central three diagonals:

$$
A = \begin{bmatrix}
a_{11} & a_{12} & 0 & 0 & \cdots \\
a_{21} & a_{22} & a_{23} & 0 & \cdots \\
0 & a_{32} & a_{33} & a_{34} & \cdots \\
\vdots & \vdots & \vdots & &
\end{bmatrix}
\tag{E.4}
$$

Nuclear Reactor Physics. Weston M. Stacey
Copyright © 2007 WILEY-VCH Verlag GmbH & Co. KGaA, Weinheim
ISBN: 978-3-527-40679-1

The *unit matrix* is the diagonal matrix with elements $a_{ij} = 1$, $i = j$:

$$I = \begin{bmatrix} 1 & 0 & 0 & \cdots \\ 0 & 1 & 0 & \cdots \\ 0 & 0 & 1 & \cdots \\ \vdots & \vdots & \vdots & \end{bmatrix} \tag{E.5}$$

For two matrices to be equal, each of their matrix elements must be equal:

$$A = \begin{bmatrix} a_{11} & a_{12} & a_{13} & \cdots \\ a_{21} & a_{22} & & \cdots \\ \vdots & & & \end{bmatrix} = \begin{bmatrix} b_{11} & b_{12} & b_{13} & \cdots \\ b_{21} & b_{22} & & \cdots \\ \vdots & \vdots & & \end{bmatrix} = B \tag{E.6}$$

The *transpose* of a matrix is obtained by interchanging its rows and columns:

$$\left[A^{\mathrm{T}} \right]_{ij} = \left[A \right]_{ji} \tag{E.7}$$

or

$$A^{\mathrm{T}} = \begin{bmatrix} a_{11} & a_{12} & a_{13} & \cdots \\ a_{21} & \ddots & & \\ a_{31} & & & \\ \vdots & & & \end{bmatrix}^{\mathrm{T}} = \begin{bmatrix} a_{11} & a_{21} & a_{31} & \cdots \\ a_{12} & \ddots & & \\ a_{13} & & & \\ \vdots & & & \end{bmatrix} \tag{E.8}$$

The *determinant* of a matrix is formed by taking the determinant of the elements of the matrix:

$$\det A \equiv |A| = \begin{vmatrix} a_{11} & a_{12} & a_{13} & \cdots \\ a_{21} & \vdots & & \\ a_{31} & & & \\ \vdots & & & \end{vmatrix} \tag{E.9}$$

Of course, the determinant of a matrix is a scalar—that is, just a number.

One defines the *cofactor* of a square matrix for an element a_{ij} by deleting the ith row and jth column, calculating the determinant of the remaining array, and multiplying by $(-1)^{i+j}$:

$$(\text{cof } A)_{23} = \text{cof} \begin{bmatrix} a_{11} & a_{12} & a_{13} & a_{14} & \cdots \\ \cancel{a_{21}} & \cancel{a_{22}} & \cancel{a_{23}} & \cancel{a_{24}} & \cdots \\ a_{31} & a_{32} & a_{33} & a_{34} & \cdots \\ a_{41} & a_{42} & a_{43} & a_{44} & \cdots \\ \vdots & \vdots & \vdots & \vdots & \end{bmatrix}$$

$$= (-1)^{i+j} \begin{vmatrix} a_{11} & a_{12} & a_{14} & \cdots \\ a_{31} & a_{32} & a_{34} & \cdots \\ a_{41} & a_{42} & a_{44} & \cdots \\ \vdots & \vdots & \vdots & \end{vmatrix} \tag{E.10}$$

We can construct the *adjoint* or *Hermitean conjugate* of a matrix by complex-conjugating each of its elements and then transposing as

$$\boldsymbol{A}^{\dagger} = \left(\boldsymbol{A}^{*}\right)^{\mathrm{T}} \quad \text{or} \quad (a_{ij})^{\dagger} = \left(a_{ji}^{*}\right) \tag{E.11}$$

For example,

$$\boldsymbol{A}^{\dagger} = \begin{pmatrix} a_{11} & a_{12} \\ a_{21} & a_{22} \end{pmatrix}^{\dagger} = \begin{pmatrix} a_{11}^{*} & a_{12}^{*} \\ a_{21}^{*} & a_{22}^{*} \end{pmatrix}^{\mathrm{T}} = \begin{pmatrix} a_{11}^{*} & a_{21}^{*} \\ a_{12}^{*} & a_{22}^{*} \end{pmatrix}$$

If the determinant of a matrix vanishes, $\det(\boldsymbol{A}) = 0$, then the matrix $\boldsymbol{A}$ is said to be *singular*. If $\det(\boldsymbol{A}) \neq 0$, the matrix is said to be *nonsingular*.

E.2
Matrix Algebra

Two matrices of the same order may be *added* by adding their corresponding elements (the same holds for *subtraction*):

$$\boldsymbol{A} + \boldsymbol{B} = \begin{bmatrix} a_{11} & a_{12} & \cdots \\ a_{21} & & \\ \vdots & & \end{bmatrix} + \begin{bmatrix} b_{11} & b_{12} & \cdots \\ b_{21} & & \\ \vdots & & \end{bmatrix}$$

$$= \begin{bmatrix} a_{11} + b_{11} & a_{12} + b_{12} & \cdots \\ a_{21} + b_{21} & & \\ \vdots & & \end{bmatrix} \tag{E.12}$$

In order for *matrix multiplication* to be possible, the number of columns of the first matrix must equal the number of rows of the second matrix. One then calculates the matrix elements of $\boldsymbol{C} = \boldsymbol{A} \cdot \boldsymbol{B}$ as

$$c_{ij} = \sum_{k=1}^{n} a_{ik} b_{kj} \tag{E.13}$$

or more explicitly

$$\boldsymbol{A} \cdot \boldsymbol{B} = \begin{bmatrix} a_{11} & a_{12} & a_{13} & \cdots \\ a_{21} & a_{22} & a_{23} & \cdots \\ a_{31} & a_{32} & a_{33} & \cdots \\ \vdots & \vdots & \end{bmatrix} \begin{bmatrix} b_{11} & b_{12} & b_{13} & \cdots \\ b_{21} & b_{22} & b_{23} & \cdots \\ b_{31} & b_{32} & b_{33} & \cdots \\ \vdots & \vdots & \end{bmatrix}$$

$$= \begin{bmatrix} a_{11}b_{11} + a_{12}b_{21} + \cdots \\ a_{21}b_{12} + a_{22}b_{22} + \cdots \\ \vdots \end{bmatrix} \tag{E.14}$$

Notice that matrix multiplication is not commutative—that is, $\boldsymbol{A} \cdot \boldsymbol{B} \neq \boldsymbol{B} \cdot \boldsymbol{A}$ in general.

A very important matrix concept is the *inverse* of a square matrix, A^{-1}, which is defined by the relation

$$A^{-1} \cdot A = A \cdot A^{-1} = I \tag{E.15}$$

The inverse can be calculated as

$$A^{-1} = \frac{1}{|A|} (\operatorname{cof} A)^{\mathrm{T}} \tag{E.16}$$

For example, consider

$$A = \begin{pmatrix} 2 & 1 \\ -1 & 1 \end{pmatrix}$$

Then

$$|A| = 3$$

while

$$(\operatorname{cof} A)^{\mathrm{T}} = \begin{pmatrix} 1 & 1 \\ -1 & 2 \end{pmatrix}^{\mathrm{T}} = \begin{pmatrix} 1 & -1 \\ 1 & 2 \end{pmatrix}$$

Hence

$$A^{-1} = \frac{1}{3} \begin{pmatrix} 1 & -1 \\ 1 & 2 \end{pmatrix} = \begin{pmatrix} \frac{1}{3} & -\frac{1}{3} \\ \frac{1}{3} & \frac{2}{3} \end{pmatrix}$$

Notice that if a matrix is singular, that is, $\det(A) = 0$, then it has no inverse.

Appendix F
Introduction to Laplace Transforms

F.1
Motivation

Differential equations play a central role in the description of most scientific phenomena. Moreover, in many cases these phenomena can be approximately described by a particularly simple type of differential equation—namely, those with constant coefficients. In this Appendix we will try to develop one of the most powerful tools for solving such equations: the application of integral transforms, and more specifically, the use of Laplace transforms to solve differential equations.

The analogy between the use of transform methods to solve differential equations and the use of logarithms to simplify arithmetic operations is quite striking. Suppose we wish to multiply two complicated numbers a and b together. Then an easy way to do this is to use logarithms

$$a \longrightarrow \log a$$

$$a \times b \longrightarrow \boxed{\text{``Transform''}} \longrightarrow \log a + \log b \longrightarrow \boxed{\text{``Invert''}} \longrightarrow e^{(\log a + \log b)}$$

$$= a \times b$$

That is, by first taking logs we have simplified the original problem, reducing it to a simple sum.

This is essentially the idea behind integral transform techniques. Suppose we symbolically represent the transform operation on a function as

$$f(t) \longrightarrow \tilde{f}(s)$$

Then the idea is to transform the differential equation of interest

$$\frac{df}{dt} + \cdots \longrightarrow \boxed{\text{Transform}} \longrightarrow s\tilde{f}(s) + \cdots \longrightarrow \boxed{\text{Invert}} \longrightarrow f(t)$$

In this manner, the integral transform can be used to convert this differential equation into a simpler problem (frequently an algebraic equation) that can then be solved rather easily for the transformed solution. We then must somehow "invert" the transform to obtain the actual solution of interest.

Nuclear Reactor Physics. Weston M. Stacey
Copyright © 2007 WILEY-VCH Verlag GmbH & Co. KGaA, Weinheim
ISBN: 978-3-527-40679-1

Example. Consider the very simple ordinary differential equation (familiar from prompt neutron reactor kinetics)

$$\frac{dn}{dt} - \left(\frac{\rho}{\Lambda}\right)n(t) = 0, \qquad n(0) = n_0 \tag{F.1}$$

Now define the Laplace transform of $n(t)$ as

$$\tilde{n}(s) = \int_0^\infty dt\, e^{-st} n(t) \equiv \mathcal{L}\{n\} \tag{F.2}$$

To transform the ordinary differential equation (F.1), multiply by e^{-st} and integrate over t

$$\int_0^\infty dt\, e^{-st} \frac{dn}{dt} - \left(\frac{\rho}{\Lambda}\right)\int_0^\infty dt\, e^{-st} n(t) = 0$$

or using integration by parts

$$s\tilde{n}(s) - n(0) - (\rho/\Lambda)\tilde{n}(s) = 0$$

but this is now just an algebraic equation which can be easily solved for

$$\tilde{n}(s) = \frac{n_0}{s - (\rho/\Lambda)} \tag{F.3}$$

We must now "invert" $\tilde{n}(s)$ to find

$$n(t) = \mathcal{L}^{-1}\{\tilde{n}(s)\} \tag{F.4}$$

By noting that

$$\mathcal{L}\{e^{-at}\} = \int_0^\infty dt\, e^{-st} e^{-at} = \frac{1}{s+a} \Rightarrow \mathcal{L}^{-1}\left\{\frac{1}{s+a}\right\} = e^{-at}$$

we find

$$n(t) = n_0 \mathcal{L}^{-1}\left\{\frac{1}{s - (\rho/\Lambda)}\right\} = n_0 \exp[(\rho/\Lambda)t] \tag{F.5}$$

Example. Integral transforms can also be applied to the solution of partial differential equations. Consider, for example, the initial value problem for a nonmultiplying slab in one-speed diffusion theory

$$\frac{1}{v}\frac{\partial\phi}{\partial t} = D\frac{\partial^2\phi}{\partial x^2} - \Sigma_a\phi(x, t)$$

Initial condition: $\phi(x, 0) = \phi_0(x)$

Boundary condition: $\phi(0, t) = \phi(l, t) = 0$ \qquad (F.6)

Define the Laplace transform of $\phi(x,t)$ with respect to t by

$$\tilde{\phi}(x,s) = \int_0^\infty dt\, e^{-st} \phi(x,t) \tag{F.7}$$

Now multiplying (F.6) by e^{-st} and integrating over all times t, we find the transformed partial differential equation becomes

$$\frac{1}{v}\left[s\tilde{\phi}(x,s) - \phi(x,0)\right] = D\frac{d^2\tilde{\phi}}{dx^2} - \Sigma_a\tilde{\phi}(x,s)$$

Since the boundary conditions also depend on time, we must transform them to find:

$$\tilde{\phi}(0,s) = \tilde{\phi}(l,s) = 0$$

Hence if we regard s only as a parameter, the application of Laplace transforms has reduced our original *partial* differential equation (F.6) to an inhomogeneous *ordinary* differential equation in x

$$D\frac{d^2\tilde{\phi}}{dx^2} - \left(\Sigma_a + \frac{s}{v}\right)\tilde{\phi}(x,s) = \phi_0(x)$$

Boundary condition: $\tilde{\phi}(0,s) = \tilde{\phi}(l,s) = 0$ \hfill (F.8)

We can now solve this in any of the standard ways (e.g., eigenfunction expansions or Green's functions) to find $\tilde{\phi}(x,s)$, and then invert to find

$$\phi(x,t) = \mathcal{L}^{-1}\{\tilde{\phi}(x,s)\} \tag{F.9}$$

Hence as should be apparent from these simple examples, Laplace transforms can be used to greatly simplify the solution of differential equations by: (a) transforming the original differential equation, (b) solving the transformed equation (which is now presumably a simpler equation such as an algebraic equation or ordinary differential equation) for the transformed solution, and (c) finally inverting the transformed solution to obtain the desired solution of the original equation. It is usually a straightforward task to complete the first two steps. The final step, that of inversion, can frequently be accomplished in a "cookbook" fashion by merely looking up the inverse in a table of Laplace transforms that some other fellow has had to work out. The general theory of how to perform such inversions from scratch is important, however, since the inverses of many of the functions one encounters in practice are not tabulated. However since it is heavily steeped in the theory of functions of a complex variable, we will avoid a detailed discussion of Laplace transform inversion via contour integration here and simply refer the reader to one of several standard texts (see Refs. 1–3).

F.2
"Cookbook" Laplace transforms

We will now set up the recipes for solving differential equations with Laplace transforms. First we must determine just what types of equations we can consider:

(a) This can be any linear differential equation (ordinary or partial) in which the variable to be transformed runs from 0 to ∞. (Such as an initial value problem in time or a half-space problem in space.)

(b) We will further restrict ourselves to the study of differential equations with constant coefficients (i.e., the coefficients in the equation do not depend on the variable to which we are applying the transform). This restriction can sometimes be relaxed; however we will not consider the more general problem of differential equations with variable coefficients here.

We will define the Laplace transform of a function $f(t)$ by

$$\tilde{f}(s) = \int_0^\infty dt \, e^{-st} f(t) \tag{F.10}$$

There are of course some restrictions on the type of function $f(t)$ and the ranges of values of s for which this integral will be properly defined, but let's not worry about details at this stage of the game.

The general scheme for transforming the differential equation we are interested in solving is the same as before—namely, multiply by e^{-st} and integrate over all t, using liberal integration by parts. One then solves the resulting transformed equation and attempts to invert the solution.

To facilitate in the preparation of a table of Laplace transforms (a cookbook), one merely takes the transforms of as many different functions as possible. Several useful transforms of general functions are (see Refs. 4, 5):

- Derivatives:

$$\mathcal{L}\left\{\frac{df}{dt}\right\} = s\tilde{f}(s) - f(0) \tag{F.11}$$

Recall that we obtained this by integration by parts. Further integration by parts yields

$$\mathcal{L}\left\{\frac{d^n f}{dt^n}\right\} = s^n \tilde{f}(s) - s^{n-1} f(0) - s^{n-2} f'(0) - \cdots - f^{[n-1]}(0) \tag{F.12}$$

- Integration:

$$\mathcal{L}\left\{\int_0^t dt' \, f(t')\right\} = \frac{1}{s}\tilde{f}(s) \tag{F.13}$$

Proof:

$$\mathcal{L}\left\{\int_0^t dt'\, f(t')\right\} = \int_0^\infty dt\, e^{-st} \int_0^t dt'\, f(t')$$

$$= -\frac{e^{-st}}{s} \int_0^t dt'\, f(t')\Big|_0^\infty + \frac{1}{s} \int_0^\infty dt\, e^{-st}\, f(t)$$

$$= \frac{1}{s}\tilde{f}(s)$$

- Differentiation by s:

$$\mathcal{L}\{t f(t)\} = -\frac{d\tilde{f}}{ds} \tag{F.14}$$

Proof:

$$\frac{d\tilde{f}}{ds} = \int_0^\infty dt\, f(t)\frac{d}{ds}\left(e^{-st}\right) = -\int_0^\infty dt\, e^{-st}[t f(t)]$$

- Complex translation:

$$\mathcal{L}\{e^{at} f(t)\} = \tilde{f}(s - a) \tag{F.15}$$

Proof:

$$\int_0^\infty dt\, e^{at}\, e^{-st}\, f(t) = \int_0^\infty dt\, e^{-(s-a)t}\, f(t) = \tilde{f}(s - a)$$

- Real translation:

$$\mathcal{L}\{f(t - a)\Theta(t - a)\} = e^{-as}\tilde{f}(s) \tag{F.16}$$

where $\Theta(t)$ is the step function,

$$\Theta(t) = \begin{cases} 1, & t \geq 0 \\ 0, & t < 0 \end{cases}$$

Several examples of more specific transform pairs are presented in Table F.1.
  Several other very useful relations (see Refs. 4, 5) are:
- Convolution theorem:

$$\mathcal{L}\left\{\int_0^t d\tau\, f(t - \tau)g(\tau)\right\} = \tilde{f}(s) = \tilde{g}(s) \tag{F.17}$$

(This result is useful for relating the inverse of the product of
two transformed functions.)
- Initial value theorem:

$$\lim_{t \to 0} f(t) = \lim_{s \to \infty} s\tilde{f}(s) \tag{F.18}$$

Table F.1

$f(t)$	$\tilde{f}(s)$
1	$\frac{1}{s}$
e^{-at}	$\frac{1}{s+a}$
$\delta(t)$	1
$\delta(t-t_1)$	e^{-st_1}
t^n	$\frac{n!}{s^{n+1}}$
$\frac{t^{n-1}e^{-at}}{(n-1)!}$	$\frac{1}{(s+a)^n}$
$\frac{(e^{-bt}-e^{-at})}{a-b}$	$\frac{1}{(s+a)(s+b)}$
$\frac{(be^{-bt}-ae^{-at})}{(b-a)}$	$\frac{s}{(s+a)(s+b)}$
$\sin at$	$\frac{a}{(s^2+a^2)}$
$\cos at$	$\frac{s}{(s^2+a^2)}$
$t\sin at$	$\frac{2as}{(s^2+a^2)^2}$
$t\cos at$	$\frac{(s^2-a^2)}{(s^2+a^2)^2}$
$\Theta(t)$	$\frac{1}{s}$

- Final value theorem:

$$\lim_{t\to\infty} f(t) = \lim_{s\to 0} s\,\tilde{f}(s) \tag{F.19}$$

There are a number of reasonably complete tables of such transform pairs (see Refs. 4, 5). After obtaining the transformed solution, one can then turn to such tables in an effort to locate the desired inverse. However in many cases it will be necessary to proceed with a direct inversion calculation.

References

1 P. M. Morse and H. Feshbach, *Methods of Theoretical Physics*, Vol. 1, McGraw Hill, New York (**1953**), Chapter 4.

2 W. Kaplan, *Operational Methods for Linear Systems*, Addison-Wesley, Reading, MA (**1962**).

3 H. S. Carslaw and J. C. Jaegar, *Operational Methods in Applied Mathematics*, Dover, New York (**1948**).

4 P. A. McCollum and B. F. Brown, *Laplace Transform Tables and Theorems*, Holt, Rinehart, and Winston, New York (**1965**).

5 F. E. Nixon, *Handbook of Laplace Transforms*, Prentice-Hall, Englewood Cliffs, NJ (**1960**).

Index

ABH method, *see* Homogenization
Absorption, 26
Absorption probability, 87, 314
Actinides, *see* Transuranics
Adiabatic method, *see* Point kinetics
Acelerator transmutation of waste reactor, 241
Adjoint
  eigenvalue, 489
  function, 486, 491, 496, 498, 501, 508, 563, 567, 571, 580, 588, 603
  generalized adjoint function Γ, 606
  operator, 486
Albedo
  boundary condition, 52
  diffusion theory, 52
Asymptotic period, 149
  measurement, 156
Asymptotic shape, 58

Bare reactors, *see* Diffusion theory
Barn, 6
Bessel functions, 682
Beta decay, *see* Radioactive decay
Bethe–Tait model, 188
Bickley function, 318
Binding energy, 3
Blackness theory, *see* Homogenization
Breeding ratio, *see* Fuel composition
Boltzmann equation, 307
Boundary and interface conditions
  albedo, *see* Albedo
  diffusion theory, *see* Diffusion theory
  extrapolated, *see* Extrapolation distance boundary condition
  Mark, *see* Mark boundary conditions
  Marshak, *see* Marshak boundary conditions

transport theory, *see* Neutron transport theory
Breit–Wigner resonance scattering cross section, 14, 20, 430
Buckling
  geometric, 57, 59, 147
  material, 58
Burnable poison, 208, 250, 253

Cadmium ratio, 72
Capture, 13
Capture-to-fission ratio, α, 33
Center-of-mass system, 27
Central limit theorem, 374
Closing nuclear fuel cycle, 244
Collision probabilities method
  ABH method, 517
  collision probability, 88, 320
  collision probability annular geometry, 322
  collision probability slab geometry, 320
  collision probability two dimensions, 320
  pin-cell model, 524, 528
  reciprocity, 320
  thermalization in heterogeneous lattices, 474
  transmission probabilities, 320
Compound nucleus, 5
  formation, 13
Control rod
  cross sections, effective diffusion theory, 73
  follower, 257, 297
  scram, 250, 257, 285, 295, 297
  windowshade model, 76
Control theory
  dynamic programming, 633
  Pontryagin's maximum principle, 635
  variational, 631

Nuclear Reactor Physics. Weston M. Stacey
Copyright © 2007 WILEY-VCH Verlag GmbH & Co. KGaA, Weinheim
ISBN: 978-3-527-40679-1

Conversion/breeding ratios, 219. *See also* Fuel composition
Correlation methods, 179
Coupling coefficients, 85
Criticality
  critical, 37, 58
  delayed critical, 151
  prompt critical, 153
  subcritical, 37, 58, 151
  supercritical, 37, 58
  superprompt-critical, 152
Criticality condition
  bare homogeneous reactor, 58, 60
  Monte Carlo, 378
  power iteration, 80, 138
  reflected homogeneous reactor, 65
  reflected slab, 64
  rodded cylindrical reactor, 77
  two region, two-group reactor, 132
Criticality minimum volume, 61
Criticality safety, *see* Nuclear reactor analysis
Cross sections
  2200 m/s values, 110, 670
  absorption, 45, 670
  capture, 14, 24, 201, 670
  definition, 5
  elastic scattering, 20, 24, 670
  evaluated, *see* Evaluated nuclear data files
  fission, 5, 24, 201, 670
  for important nuclides, 26
  low-energy summary, 24
  macroscopic, 24
  spectrum-averaged, 24, 25, 63
  total, 24, 670
  transport, *see* Transport cross section
  units, 5
Cross spectral density, 180
Current
  net current, 45, 310
  partial current, 45, 310

Delayed critical, 151
Delayed neutrons
  decay constants, 143
  holdback, 599
  kernel, 155
  neutron kinetics effects, 39, 150
  precursor, 147
  yield, 143
Densities, elements and reactor materials, 670

Depletion model, 219
Detailed balance principle, 109, 458, 467
Diffusion coefficient, 45, 342
  directional, 397
  multigroup, 128, 398
Diffusion cooling, 480
Diffusion length, 48, 53, 56
Diffusion parameters, 56
Diffusion theory
  applicability, 47
  bare homogeneous reactor, 57
  boundary and interface conditions, 46, 94
  derivation, 43, 342
  directional, 397
  kernels, 50
  lethargy-dependent, 396
  multigroup theory, 127, 398, 599
  nonmultiplying media solutions, 48
  numerical solution, 77, 137
  one-dimensional geometry, 93, 347
  reflected reactors, 62, 134
  two-region reactors, 130
Dirac δ-function, 677
Discrete ordinates methods
  acceleration of convergence, 362
  cylindrical and spherical geometries, 359
  diamond difference scheme, 358, 360, 367
  equivalence with P_L equations, 95, 356
  level-symmetric quadrature, 365, 574
  nodal, 571
  ordinates and quadratures, multidimensional, 363
  ordinates and quadratures, P_L and D-P_L, 95, 355
  ordinates and quadratures, S_N, 365
  slab geometry, 94, 354
  spatial finite differencing and iteration, S_N method in 2D Cartesian geometry, 366
  spatial finite differencing and iteration, slab geometry, 357
  spatial mesh size limitations, 358
  sweeping mesh grid, 360, 367
Doppler broadening, 119. *See also* Resonance and Reactivity
Dynamic programming, 638

Eigenvalue separation, 602, 647
Elastic scattering
  average cosine of scattering angle, 389
  average logarithmic energy loss, 30, 389
  cross sections, 22, 670

energy–angle correlation, 28, 373
kernel, 29, 102, 386
kinematics, 27, 385
Legendre moments of transfer function, *see* Legendre moments of elastic scattering transfer function
moderating ratio, 30
potential, 20
relation between CM and lab scattering angles, 28
resonance, 20
transfer function, 386
Emergency core cooling, 285, 295
Energy release from fission, 12
Error function, 684
Escape probability, 87
Eta (number of neutrons per absorption in fuel), 34
Equivalence theory, *see* Homogenization
Escape probability, *see* Integral transport theory; Interface current methods; Resonance
Evaluated nuclear data files, 24, 115
Even-parity transport theory, *see* Neutron transport theory
Excitation energy for fission, 4
Exoergic reactions, 286
Exponential integral function, 684
Extrapolation distance boundary condition, 47, 342

Fermi age, *see* Neutron slowing down
Fertile isotopes, 198
Few group approximations, 115, 128
Fick's law, 45, 342, 353, 397
Finite difference equations
diamond difference relation, 358, 360, 367
diffusion equation, one-dimensional slab, 78
diffusion equation, two-dimensional Cartesian, 80
discrete ordinates, rectangle, 366
discrete ordinates, slab, 357
discrete ordinates, sphere, 360
limitations on mesh spacing, 358
Finite element methods
cubic Hermite approximation, 569
finite-difference approximation, 564
linear approximation, 567
First collision source, *see* Integral transport theory

Fissile isotopes, 5
Fission
cross sections, 6, 201, 670
energy release, 12
fast, 34
neutron chain fission reaction, *see* Neutron chain fission reaction
neutron yield, 8
probability per neutron absorbed, 236
process, 4
products, 8, 197
products significant in accidents, 284
spectrum, 11
spontaneous, 4, 201, 202
threshold, 4
yields, 10
Flux, scalar, 310
Flux tilts, 599, 641
Four-factor formula, 37
Flux disadvantage factor, *see* Thermal disadvantage factor
Fuel assemblies, 68, 251, 254, 257, 259, 261–263, 522
Fuel burnup
composition changes, 205
depletion model, 219
energy extraction, 235
fission products, *see* Fission products
in-core fuel management, 210
reactivity changes, *see* Reactivity
transmutation-decay chains, *see* Transmutation-decay chains
units, 205
Fuel composition
discharged UO_2, 205
equilibrium distribution in recycled fuel, 236
fertile-to-fissile conversion and breeding, 217
plutonium buildup, 206
power distribution, 209
reactor-grade uranium and plutonium, 233
recycled LWR fuel, 221
recycled plutonium physics differences, 207, 225
recycled uranium physics differences, 224
weapons-grade uranium and plutonium, 233
Fuel lumping, 37
Fuel recycling, *see* Fuel composition
Fuel reprocessing, 221

Gamma function, 684
Gauss' divergence theorem, 676
Gauss–Seidel, 82
Generalized perturbation theory, *see* Variational methods
Gaussian elimination, 79
Green's theorem, 676
Group collapsing, 117, 395, 403

Hazard index, 226. *See also* Radioactive waste
Heterogeneity, *see* Homogenization
High level waste repository
  decay heat, 243
  isotopes, 242
Homogenization
  ABH method, 517
  blackness theory, 520
  collision probabilities pin-cell model, 524
  conventional theory, 531
  cross sections, equivalent homogeneous, 69, 516
  diffusion theory, 67
  diffusion theory lattice functions F and E, 70
  equivalence theory, 531
  flux discontinuity factor, 532
  flux reconstruction, 538, 554
  flux (thermal) disadvantage factor, 69, 516, 518. *See also* Self-shielding
  interface current pin-cell method, 527
  multiscale expansion theory, 535
  pin-cell model, 522, 528
  resonance cross sections, 423
  spatial self-shielding, *see* Self-shielding
  transport boundary conditions, 520, 522
  Wigner–Seitz cell, 523

Importance function, 145, 375, 487. *See also* Adjoint function
Infinite multiplication constant, k_∞, 37, 113
Inhour equation, 149
Integral transport theory
  absorption probability, 314
  anisotropic plane source, 312
  distributed volumetric scattering and fission sources, 315
  escape probability, 314
  first-collision source, 315
  isotropic line source, 317
  isotropic plane source, 311
  isotropic point source, 311

probability of traveling a distance t from a line source, 318
  scattering and fission, inclusion of, 315
  transmission probability, 314. *See also* Transmission probability
Interface current methods
  boundary conditions, 329
  emergent currents, 326, 327, 331
  escape probabilities in slab geometry, 328
  escape probabilities in two-dimensional geometries, 333, 335
  escape probabilities rational approximations, 337
  pin-cell model, 527
  reflection probability in slab geometry, 328
  response matrix, 329
  transmission probabilities in slab geometry, 328
  transmission probabilities in two-dimensional geometries, 333
Iteration methods
  acceleration of convergence, 362, 369
  alternating direction implicit, 619
  forward elimination/backward substitution (Gauss elimination), 79
  power, for criticality problems, 79, 82, 363, 378, 413
  scattering, for discrete ordinates equations, 358, 413
  successive over-relaxation, 82, 623
  successive relaxation (Gauss–Seidel), 82, 137
  sweeping over mesh points for one-dimensional discrete ordinates, 360
  sweeping over mesh points for two-dimensional discrete ordinates, 367

$J(\xi, \beta)$ resonance function, 125

Lagrange multiplier, 632
Laguerre polynomials, 479
Laplace transforms, 691
Laplacian representation, 675
Legendre moments of elastic scattering transfer function
  anisotropic scattering in CM, 389
  definition, 387
  isotropic scattering in CM, 388, 389
Legendre polynomials
  associated Legendre functions, 339, 681
  definition and properties, 338, 681

half-angle Legendre polynomials, 347
Lethargy, 385
Loss of coolant accident, *see* Reactor safety
Loss of flow accident, *see* Reactor safety
Lyapunov's method for stability analysis, 627, 629, 649

Macroscopic cross section, 44
Mark boundary conditions, *see* Spherical harmonics
Marshak boundary condition, *see* Spherical harmonics
Mass defect, 3
Matrix algebra, 687
Maxwellian distribution, 109
Maxwellian energy distribution, 19
Mean chord length, 426
Mean free path, 425
Mesh spacing limit, 83
Migration length, 55
Minimum critical volume, 61
Mixed oxide fuel, 222, 233, 239
Moderator properties, 30
Moderating ratio, *see* Elastic scattering
Monte Carlo methods
  absorption weighting, 377
  analog simulation of neutron transport, 372
  correlated sampling, 378
  criticality problems, 378
  cumulative probability distribution functions, 371
  exponential transformation, 376
  flux and current estimates, 377
  forced collisions, 376
  importance sampling, 375
  probability distribution functions, 371
  Russian roulette, 377
  splitting, 377
  statistical estimation, 373
  variance reduction, 375
Multigroup theory, 127
  collision probabilities for thermalization, 476
  cross-section definition, 113, 403, 413
  cross-section preparation, 115
  diffusion theory, 398, 599
  few group constants, 117, 395
  few group solutions, infinite medium, 114
  mathematical properties, 113

one-and-one-half-group diffusion theory, 129
perturbation diffusion theory, 168, 483
pin-cell collision probabilities model, 528
resonance cross sections, *see* Resonance
two-group diffusion theory, 128, 130, 134
Multiplication constant, k_{eff}, 37, 60, 129, 137

Neutron balance schematic, 36
Neutron chain fission reaction
  criticality, 37
  delayed neutron effect on, 38
  effect of fuel lumping, 37
  effective multiplication constant, 37
  neutron balance in a thermal reactor, 34
  process, 33
  prompt neutron dynamics, 38
  resonance escape, 36. *See also* Resonance
  source multiplication, 39
  utilization, 34
Neutron diffraction, 24
Neutron emission, 19
Neutron energy distribution
  fission energy range analytical solution, 101
  multigroup calculation, 111
  resonances, 123
  slowing-down range analytical solutions, 102
  spectra in UO_2 and MOX fuel cells, 222
  spectra typical for LMFBR and LWR, 41
  thermal range analytical solutions, 108
Neutron lifetime, 61
Neutron scalar flux, 44
Neutron slowing down
  average cosine of scattering angle, 389
  average lethargy increase, 389
  B_1 theory, 394
  consistent P_1 approximation, 405
  continuous slowing down theory, 400
  diffusion theory, 127, 397
  discrete ordinates, 411
  elastic scattering kernel, 386
  Fermi age, 107
  hydrogen, 103
  isotropic CM scattering, 388
  Legendre moments, *see* Legendre moments of elastic scattering transfer function
  P_1 theory, 390
  P_l continuous slowing, 407, 410

slowing down density, *see* Neutron slowing down density
weak absorption, 106
without absorption, 104
Neutron slowing down density
age approximation, 404
anisotropic scattering, 407
definition, 105, 400
extended age approximation, 405
Grueling–Goertzel approximation, 406
hydrogen, 403
scattering resonances, 409
Selengut–Goertzel approximation, 405
weak absorption, 106
Neutron sources
accelerator-spallation, 273
tokamak D–T fusion, 273
Neutron temperature, 109
Neutron thermalization
collision probability methods for heterogeneous lattices, 474
differential scattering cross section, 453, 457
effective neutron temperature, 110
energy eigenfunctions of scattering operator, 477
free-hydrogen model, 455
Gaussian representation, 459
heavy gas model, 456, 466
incoherent approximation, 459
intermediate scattering function, 458
measurement of scattering functions, 460
moments expansion, 470
monatomic Maxwellian gas, 454
multigroup calculation, 473
numerical solution, 468
pair distribution function, 457
pulsed neutron, 477
Radkowsky model, 455
scattering function, 457
spatial eigenfunction expansion, 477
thermalization parameters for carbon, 472
Wigner–Wilkins model, 463
Neutron transport equation
Boltzmann, 90
integral, 88
Neutron transport theory
boundary conditions, 310
collision probabilities, *see* Collision probabilities methods
current, 87, 310

discrete ordinates, *see* Discrete ordinates methods
equation, 305
even-parity, 369, 505
integral, *see* Integral transport theory
interface current, *see* Interface current methods
Monte Carlo, *see* Monte Carlo methods
partial current, 310
scalar flux, 310
spherical harmonics, *see* Spherical harmonics methods
streaming operator in various geometries, 309
Neutron wavelength, 20, 430
Nodal methods, 83
conventional methods, 544
double-P_n expansion, 558
formalism, 542
gross coupling, 545
polynomial expansion, 549, 557
transverse integrated diffusion theory methods, 547
transverse integrated transport theory models, 554
transverse integrated discrete ordinates methods, 561
transverse leakage, 548, 556, 561
variational discrete ordinates methods, 571
Noise analysis, 181
Nonleakage probability, 34, 60, 166
Nu (number of neutrons per fission), 11
Nuclear reactor analysis
core operating data, 279
criticality and flux distribution, 276
criticality safety, 279
fuel cycle, 277
homogenized cross sections, 275. *See also* Homogenization
safety, *see* Reactor safety
transient, 278
Nuclear reactors
advanced, 269
advanced gas-cooled reactor AGR, 260
boiling water reactor BWR, 250, 299
characteristics of power reactors, 265
classification by coolant, 41
classification by neutron spectrum, 40
high-temperature gas-cooled reactor HTGR, 260
integral fast reactor IFR, 272, 300

light water breeder reactor LWBR, 265
liquid-metal fast breeder reactor LMFBR, 261
MAGNOX, 260
molten salt breeder reactor MSBR, 265
pebble bed reactor, 265
pressure tube graphite-moderated reactor RBMK, 258
pressure tube heavy water reactor CANDU, 255
pressurized water reactor PWR (AP-600, PIUS), 249, 299
representative parameters, 266
Nuclear reactors, advanced
advanced boiling water reactor (ABWR), 266
advanced liquid metal reactor (ALMR), 264
advanced pressure tube reactor, 268
advanced pressurized water reactor (APWR, EPR, AP-600, AP-1000, APR-1400), 267
gas-cooled fast reactor (GFR), 270
generation-IV reactors (GEN-IV), 269
integral fast reactor (IFR), 264, 300
lead-cooled fast reactor (LFR), 271
modular high-temperature gas-cooled reactor (GT-MHR), 268
molten salt reactor (MSR), 271
pebble bed modular reactor (PBMR), 268
sodium-cooled fast reactor (SFR), 272
sub-critical reactors, 273
super-critical water reactor (SCWR), 272
very high temperature reactor (VHTR), 272
Nuclear stability, 4
Nuclides, 3
Number of fission neutrons, η, 33

ODE solution, 675
Optical path length, 311
Orthogonality conditions
associated Legendre functions, 345
half-range Legendre polynomials, 347
Legendre polynomials, 338
reactor eigenfunctions (λ-modes), 600
spherical harmonics, 351

Perturbation theory
adjoint function, *see* Adjoint
boundary, 508

generalized, *see* Variational methods
multigroup diffusion theory, 168, 483
reactivity worth, 169, 486, 490
samarium reactivity worth, 212
xenon reactivity worth, 216
Photoneutrons, 146
Physical constants, 473
Plutonium
buildup, 206
composition in spent UO_2 fuel, 207
composition—reactor-grade, 233
composition—weapons-grade, 233
concentrations in recycled PWR fuel, 221
physics differences between weapons- and reactor-grade, 234
recycle physics effects, 225
Point kinetics
adiabatic method, 605
approximate solutions for fast excursions, 186
approximate solutions with feedback, 183
approximate solutions without feedback, 150
derivation of equations, 602
equations, 147
quasistatic method, 605
transfer functions, *see* Transfer functions
Poison
burnable, *see* Burnable poison
control rods, *see* Control rods
fission products, *see* Fuel burnup
samarium, *see* Samarium
soluble, *see* Soluble poison
xenon, *see* Xenon
Pontryagin's maximum principle, 639
Power autocorrelation function, 181
Power coefficients, 178
Power distribution
fuel burnup, 210
peaking, 73
thermal hydraulics, 280
xenon spatial oscillations, *see* Xenon spatial oscillations
Power iteration, *see* Iteration methods
Power peaking, *see* Power distribution
Prompt jump, 151, 152
Prompt jump approximation, 153, 184
Prompt neutron generation time, 147
Prompt neutron lifetime, 38
Pulsed neutron measurement, 157
PUREX separation technology, 239

PWR typical composition and cross sections, 63, 140

Pyrometallurgical separation technology, 239

Quasi-static method, *see* Point kinetics

Radiative capture, 13
Radioactive decay, 8, 19, 39, 143, 198, 209, 211, 213, 217, 226, 239, 283, 641
Radioactive waste
  cancer dose per Curie in spent fuel, 230
  hazard potential, 226
  radioactivity of LWR and LMFBR spent fuel, 227
  radiotoxic inventory decay of spent fuel, 240
  risk factor, 226
  toxicity factor, 230
Reactivity
  autocorrelation function, 180
  control rod worth, *see* Control rod
  definition, 147, 604
  feedback, 161
  fuel burnup penalty, 206
  measurement of, 149, 156
  penalty, 208
  perturbation estimate, *see* Perturbation theory
  samarium worth, 212
  spectral density, 180
  temperature defect, 167
  variational estimate, *see* Variational methods
  xenon worth, 215, 216
Reactivity coefficients
  delay time constants, 178
  Doppler, 161, 162, 169
  expansion, 164, 170
  fuel bowing, 171
  fuel motion, 170
  nonleakage, 166
  power, 178
  representative values, 166, 171
  sodium void, 169
  temperature, 162, 170
  thermal utilization, 165
Reactivity control
  BWRs, 250
  CANDUs, 257
  gas-cooled reactors, 260
  LMFBRs, 261
  PWRs, 249

RBMKs, 259
Reactor accidents
  anticipated transients without scram, 288
  Chernobyl, 297
  energy sources, 285
  loss of coolant, 287, 295
  loss of flow, 287
  loss of heat sink, 287, 294
  predicted frequency of fatality, 293
  reactivity insertion, 287, 297
  Three Mile Island, 294
Reactor noise, *see* Noise analysis
Reactor safety
  accidents, *see* Reactor accidents
  analysis, *see* Reactor safety analysis
  defense in depth, 285
  multiple barriers, 283
  passive, 299
  passive safety demonstration, 300
  radionuclides of concern, 283
  risks, 291
Reactor safety analysis
  event tree, 289
  fault tree, 289
  probabilistic risk assessment, 288
  radiological assessment, 291
Reactor startup analysis, 663
Reflected reactors, *see* Diffusion theory
Reflector savings, 64
Resonance
  Adler–Adler approximation, 443
  Breit–Wigner, multilevel formula, 442
  Breit–Wigner, single-level formula, 14, 430, 441
  cross sections, 6, 117, 417
  Dancoff correction, 428
  Doppler broadening, 119, 127
  equivalence relations, 422
  escape probability, 36, 122
  escape probability, closely packed lattices, 427
  escape probability, isolated fuel element, 425
  heterogeneous fuel–moderator cell, 415
  heterogeneous resonance escape probability, 423
  homogenized resonance cross section, 423
  infinite dilution resonance integral, 422
  integral, 122, 419
  intermediate resonance approximation, 424, 500

$J(\xi, \beta)$ function, 125
multiband theory, 433
multigroup cross sections, 122, 423, 430, 432
narrow resonance approximation, 123, 419
overlap of different species, 432
pole representation, 445
Porter–Thomas distribution, 429
practical width, 122
R-matrix representation, 439
rational approximation, 427
reciprocity, 418
Reich–Moore formalism, 443
Sauer rational approximation, 427
self-overlap effects, 431
self-shielding, 415, 433
statistical resonance parameters, 431
strength function, 430
unresolved resonances, 428
wide resonance approximation, 123, 420
Response matrix, 329
Rod drop measurement, 157
Rod oscillator measurement, 158, 179
Rossi-α measurement, 159

Samarium, 211
Sauer rational approximation, 427
Self-shielding
resonance, 103, 415, 422, 433
spatial, 65, 433
Separation of variables, 54, 58
Soluble poison, 208, 250
Source jerk measurement, 157
Space-dependent nuclear reactor kinetics
delayed flux tilts, 601
direct time-Integration, *see* Time integration methods
dynamic programming, 638
linear analysis, 642
Lyapunov's method for nonlinear stability analysis, 629, 649
modal eigenfunction expansion, 600
Pontryagin's maximum principle, 639
stochastic, *see* Stochastic kinetics
variational control theory, 636
xenon spatial oscillations, *see* Xenon spatial oscillations
Spent nuclear fuel, 238
Spherical harmonics methods
associated Legendre functions, *see* Legendre polynomials

boundary and interface conditions, P_L theory, 91, 340
boundary and interface conditions, D-P_L theory, 349
diffusion equations in one-dimensional geometry, 93, 347
diffusion theory, from P_1 theory, 342
diffusion theory, in multidimensional geometries, 353
double-P_L theory, 348
extrapolated boundary condition, 342
half angle Legendre polynomials, *see* Legendre polynomials
Legendre polynomials, *see* Legendre polynomials
Mark boundary conditions, 341
Marshak boundary conditions, 340, 343, 399
multidimensional geometry, 350
P_L equations in slab geometry, 91, 339
P_L equations in spherical and cylindrical geometries, 344
simplified P_L theory, 343
spherical harmonic functions, 350, 682
Stability
criteria, 175, 178
feedback delay, 178
instability conditions for two-temperature model, 176
linear analysis, 625
Lyapunov method, 627
threshold power level, 174
transfer function analysis, 171
xenon spatial oscillations, *see* Xenon spatial oscillations
Stochastic kinetics
correlation functions, 658
forward stochastic model, 653
means, variances, and covariances, 656
reactor startup analysis, 663
transition probability, 653
transition probability generating function, 653
Synthesis methods
formalism, 502
multichannel, 589
single-channel, 583
spectral, 591

Temperature defect, *see* Reactivity
Thermal disadvantage factor, 65, 520

Thermal-hydraulics
  interaction with reactor physics, 280
  reactor safety, 285
  reactor stability, 172
Thermal utilization, 34, 72, 165, 518
Thorium fuel cycle, 200
Time eigenvalues, 58
Time integration methods
  alternating direction implicit, 619
  explicit forward-difference, 610
  implicit backward-difference, 611
  implicit GAKIN, 616
  implicit θ, 612
  implicit time-integrated, 615
  Runge–Kutta, generalized, 624
  stiffness confinement, 622
  symmetric successive overrelaxation, 623
Transfer functions
  measurement, 179, 182
  phase angle, 158
  with feedback, 171, 182
  zero-power, 155, 159
Transmission probability, 87, 88, 305, 314.
    See also Integral transport theory and
    Interface current methods
Transmutation–decay chains
  cross sections and decay data, 201
  fission products, 203
  fuel, 199
Transmutation of spent nuclear fuel, 237
Transport boundary condition, 74, 519, 522
Transport cross section, 45, 93, 342, 395
Transuranics
  cancer dose per Curie in spent fuel, 230
  equilibrium distribution in continuously
    recycled fuel, 236
  probability of fission per neutron, 237
  risk factor in spent fuel, 233
  transmutation, 237
Transverse leakage, 577
TRISO fuel particles, 268, 270

Unit conversion, 669
Uranium
  composition natural, 233
  composition reactor-grade, 233
  composition weapons-grade, 234
  fuel cycle, 199
  physics effects of recycle, 224
  resource utilization, 235

Variational methods
  collision probability theory, 502
  construction of variational functionals, 500
  control theory, 631, 636
  diffusion theory, 583
  discontinuous trial function, 575
  discontinuous trial functions, 563, 565,
    569, 589
  dynamic reactivity, 607
  even-parity transport theory, 505
  flux correction factor, 492
  functional, 491
  functional admitting discontinuous trial
    functions, 504, 563, 566, 572, 580, 582
  heterogeneity reactivity, 502
  interface and boundary terms, 504
  intermediate resonance integral, 500
  multigroup diffusion theory, 580
  P_1 equations, 563, 580
  Rayleigh quotient, 499, 503
  reaction rate ratios, 495
  reaction rates, 497
  reactivity worth, 490, 492
  Ritz procedure, 506
  Roussopolos functional, 498
  Schwinger functional, 499, 501
  static reactivity, 490, 606
  stationarity, 498
  synthesis, 504. *See also* Synthesis
  transport equation, 571
  trial functions, 499, 501, 504, 505, 584, 592,
    608

Weapons grade plutonium and uranium, *see*
  Fuel composition
Wigner rational approximation, 427
Wigner–Seitz approximation, *see* Homoge-
  nization

Xenon, 213, 641
Xenon spatial oscillations
  λ-mode stability analysis, 645
  linear stability analysis, 642
  μ-mode stability analysis, 644
  nonlinear stability criterion, 649
  variational control, 650

ZEBRA composition, 497

Related Titles

Lilley, J. S.

Nuclear Physics

Principles and Applications

412 pages
2001
Hardcover
ISBN-13: 978-0-471-97935-7

Thomas, A. W., Weise, W.

The Structure of the Nucleon

303 pages with 94 figures and 9 tables
2001
Hardcover
ISBN-13: 978-3-527-40297-7

Turner, J. E.

Atoms, Radiation, and Radiation Protection

575 pages
1995
Hardcover
ISBN-13: 978-0-471-59581-6

Griffiths, D.

Introduction to Elementary Particles

399 pages with 100 figures
1987
Hardcover
ISBN-13: 978-0-471-60386-3